MICROELECTRONIC APPLICATIONS OF CHEMICAL MECHANICAL PLANARIZATION

BICENTENNIAL 1807 WILEY 2007 BICENTENNIAL

THE WILEY BICENTENNIAL–KNOWLEDGE FOR GENERATIONS

Each generation has its unique needs and aspirations. When Charles Wiley first opened his small printing shop in lower Manhattan in 1807, it was a generation of boundless potential searching for an identity. And we were there, helping to define a new American literary tradition. Over half a century later, in the midst of the Second Industrial Revolution, it was a generation focused on building the future. Once again, we were there, supplying the critical scientific, technical, and engineering knowledge that helped frame the world. Throughout the 20th Century, and into the new millennium, nations began to reach out beyond their own borders and a new international community was born. Wiley was there, expanding its operations around the world to enable a global exchange of ideas, opinions, and know-how.

For 200 years, Wiley has been an integral part of each generation's journey, enabling the flow of information and understanding necessary to meet their needs and fulfill their aspirations. Today, bold new technologies are changing the way we live and learn. Wiley will be there, providing you the must-have knowledge you need to imagine new worlds, new possibilities, and new opportunities.

Generations come and go, but you can always count on Wiley to provide you the knowledge you need, when and where you need it!

WILLIAM J. PESCE
PRESIDENT AND CHIEF EXECUTIVE OFFICER

PETER BOOTH WILEY
CHAIRMAN OF THE BOARD

MICROELECTRONIC APPLICATIONS OF CHEMICAL MECHANICAL PLANARIZATION

Edited by

YUZHUO LI

WILEY-INTERSCIENCE
A JOHN WILEY & SONS, INC., PUBLICATION

Published by John Wiley & Sons, Inc., Hoboken, New Jersey
Published simultaneously in Canada

For general information on our other products and services or for technical support, please contact our Customer Care Department within the United States at (800) 762-2974, outside the United States at (317) 572-3993 or fax (317) 572-4002.

Wiley also publishes its books in a variety of electronic formats. Some content that appears in print may not be available in electronic formats. For more information about wiley products, visit our web site at www.wiley.com.

Wiley Bicentennial Logo: Richard J. Pacifico

Library of Congress Cataloging-in-Publication Data:

Microelectronic applications of chemical mechanical planarization / edited by Yuzhuo Li.
p. cm.
"Wiley Interscience."
Includes bibliographical references.
ISBN 978-0-471-71919-9
1. Integrated circuits–Design and construction. 2. Chemical mechanical planarization. 3. Microelectronics–Materials. I. Li, Yuzhuo.
TK7874.M4675 2007
621.3815–dc22

2007015557

MIX
Paper from responsible sources
FSC® C013604

CONTENTS

FOREWORD

Chemical mechanical planarization, or CMP, has become one of the newest and most important fabrication technologies adopted by the semiconductor industry worldwide, despite a remarkably nontraditional and somewhat controversial developmental history. Begun as a mere research and development curiosity more than 20 years ago at IBM, the technique borrows heavily from the traditional mechanical wet polishing processes for silicon substrate wafers and optical glass lenses. Introduced for production at a time when dry fabrication processes were overwhelmingly favored, the completely wet CMP process was initially considered unconventional and incompatible with the rest of the manufacturing processes, to say the least. In addition, to an industry that is meticulously conscientious about particle contamination, a process that intentionally uses slurry saturated with particles seemingly adds insult to injury and thus qualifies as a true disruptive technology. This was well before the world became familiar with such a catchy yet descriptive term "disruptive technology" popularized through a series of articles and books by renowned Harvard Business School Professor Clayton Christiansen on innovations in commercial enterprises. Some examples of his works include *Disruptive Technologies: Catching the Wave*, coauthored by Joseph L. Bower, Harvard Business Review, January–February 1995 and *The Innovator's Dilemma: When New Technologies Cause Great Firms to Fail*, Harvard Business School Press, 1997. Like many other major disruptive technologies seen by society throughout history, CMP indeed has lived up to its reputation. It disrupted the conventional thought process but enabled an industry to overcome many technological challenges, significantly advanced the processing capability for ever diverse and complex semiconductor devices, and inspired innovations in several associated fields such as wafer cleaning, defect inspection, and complex chemical delivery.

Although the initial impetus for CMP was to enable lithographic patterning by reducing depth-of-focus variations and the ability to stack multiple BEOL (back end of the line) levels on those flat surfaces, the technology also enabled a number of other advancements that were not obvious to the original developers. These include both the ability to form multilevel Cu wiring via single and dual damascene processes and the capability of fabricating shallow trench isolation (STI) structures. In addition, CMP has proven to be remarkably adaptable beyond the traditional silicon-based IC (integrated circuit) world. This is evidenced by its increasing use today in the fabrication of MEMS devices, 3D chips, and in the integration of optoelectronic devices. In addition, the introduction of CMP technology has inspired an array of engineering solutions and true innovations in peripheral semiconductor infrastructural arenas. These include the development of novel post-CMP cleaning processes and solutions that make the dream dry-in/dry-out process possible. Furthermore, advances in defect, thickness, and polishing end-point metrologies have been vitally important in both CMP process optimization and day-to-day manufacturing line management. Finally, the broad range of novel chemical and particle types employed by today's CMP stations in a modern fab has spurred a host of engineering solutions to the multiple issues of complex chemical delivery, filtration, tool/facility cleaning, and waste disposal of spent process fluids.

But how could this technique go from a quirky novelty used by several U.S.-based semiconductor manufacturers to a set of processes adopted, used, and optimized throughout the world? In a single word: extendibility. Beginning in the early 1990s, it was found by a steadily increasing number of, first, industry and then academic researchers that the original CMP techniques could be readily applied to other fabrication problems of interest: for existing processes reduced to smaller dimensional ground rules as per Semiconductor Industry Roadmaps; for new processes used for different insulators, metals, and semiconductors; to new device cross-sectional architectures such as damascene for metals such as Cu or insulators defining STI devices; and further adaptations to wafer types, sizes, and applications extending beyond traditional integrated circuits.

In conjunction with the realization that CMP was becoming a *required* global semiconductor fabrication technology, during that time frame there was an increasingly sophisticated, active, and expanding infrastructure being developed. That infrastructure would eventually supply the polishing and metrology tools, process consumables, and cleaning equipment necessary to enable existing and new users of these techniques to concentrate solely on the customized process development and process integration activities that would lead to a further explosion in novel uses and applications of the technology.

As indicated earlier, similar to other disruptive technologies, CMP also endured and overcame skepticisms. Many questions were asked. Can CMP process deliver consistent wafer-to-wafer and run-to-run reproducibility? Would the abrasive particles introduce cross-contamination in the device fabrication manufacturing facility? Can the particle-induced defects such as scratch and delamination be minimized? Can CMP serve as a long-term, robust semiconductor manufacturing technology? During the last 20 years, the CMP community has answered these and many other questions with performance-driven research, development, and implementation. The fact that the introduction and scale-up of the technique, *in manufacturing*, for fabricating tungsten studs and planarization of interlayer dielectric surfaces took less than half a dozen years after the initial research and development activities had begun, stands as a remarkable testament both to the robustness of those initial processes and to the remarkable motivation and dedication of those process engineers, technicians, and scientists who believed in the fundamental promise of the new paradigm-shifting technology.

Despite the collective 20-year invention, development, and manufacturing experience now dedicated to this technology, there has been a noticeable lack of archival information available and dedicated for teaching about CMP. Currently, CMP technology is discussed in numerous forums throughout the world involving scientific conferences, workshops, user groups, trade shows, and technical articles in the scientific literature and in patent publications. Any of these can give a snapshot in time of the development and status of the technology for a user willing to mine those resources.

However, what is really needed is a high-quality textbook to summarize the current state-of-the-art CMP technology. There has also been a lack of archival material of this type available to and appropriate for both existing and new users alike. The current work, conceived, edited, partially written, and organized by Professor Yuzhuo Li of Clarkson University along with a distinguished list of coauthors, promises to add significantly to the current archival record dedicated to chemical mechanical planarization. I say this because of the broad range of useful topics covered in this text, in addition to the fact that in concentrating on and organizing around *chemical* aspects of this technology, the Editor has focused on a key feature of this process technology that has, to my knowledge, not been adequately dealt with in other works. In the opinion of this researcher, one of the reasons that CMP has been so successful to date has been the variety of chemistries that have been found to be usefully applicable in the technology.

Hopefully, the work discussed within will help to ensure that the next 10 years of planarization technology development will be as fascinating, interesting, and useful as the first 20 have been.

FRANK B. KAUFMAN, PhD.

Geneva, Illinois
frank.b.kaufman@gmail.com
December 2006

CONTRIBUTING AUTHORS

Leonard Borucki
3831 E. Ivy Street
Mesa, AZ 85205, USA

Craig Burkhard
Center for Advanced Materials Processing
Clarkson University
8 Clarkson Avenue
Potsdam, NY 13699, USA

Ahmed A. Busnaina
NSF Center for Microcontamination Control
Northeastern University
Boston, MA 02115, USA

Krishnayya Cheemalapati
Intel Corporation
Hillsboro, OR 97124, USA

Laertis Economikos
IBM Systems and Technology Group
Semiconductor Research & Development Center
Hopewell Junction, NY 12533, USA

Norm Gitis
Center for Tribology, Inc.
1715 Dell Ave.
Campbell, CA 95008, USA

Suresh Kumar Govindaswamy
Micron Technology, Inc.
Mail Stop 3-314
9600 Godwin Drive
Manassas, VA 20110, USA

Ronald J. Gutmann
Professor Emeritus
Rensselaer Polytechnic Institute
CII 6015
110 8th St
Troy, NY 12180, USA

Yi-Koan Hong
Division of Materials and Chemical Engineering
Hanyang University
Ansan 426-791, Korea

Jinshan (Jason) Huo
Fujimi Corporation
11200 SW Leveton Dr.
Tualatin, OR 97062, USA

Jason Keleher
Cabot Microelectronics
870 N. Commons Drive
Aurora, IL 60504, USA

Toshihiro Kobayashi
Mipox International Corporation
25821 Industrial Blvd., Suite 200
Hayward, CA 94545, USA

Yongqing Lan
Department of Chemistry
Clarkson University
8 Clarkson Avenue
Potsdam, NY 13699, USA

Paul Lefevre
Fujimi Corporation
12929 SW Wilmington Lane
Tigard, OR 97224, USA

Yuzhuo Li
Center for Advanced Materials Processing
Department of Chemistry
Clarkson University
8 Clarkson Avenue
Potsdam, NY 13699, USA

Jian-Qiang Lu
Center for Integrated Electronics
Rensselaer Polytechnic Institute
CII 6015
110 8th St
Troy, NY 12180, USA

J. Jay McMahon
Center for Integrated Electronics
Rensselaer Polytechnic Institute
CII 6015
110 8th St
Troy, NY 12180, USA

Mansour Moinpour
Intel Corporation
2200 Mission College Blvd
Mail Stop SC3-06
Santa Clara, CA 95054, USA

Raghu Mudhivarthi
Center for Tribology, Inc.
1715 Dell Ave.
Campbell, CA 95008, USA

Jin-Goo Park
Division of Materials and Chemical Engineering
Hanyang University
Ansan 426-791, Korea

Ed Paul
Department of Chemistry
Stockton College
Pomona, NJ 08240, USA

Ara Philipossian
Department of Chemical and Environmental Engineering
University of Arizona
PO Box 210011
Tuscan, AZ 85721, USA

John H. Rydzewski
Intel Corporation
Strategic Facilities Technology Development
RA1-220
2501 N.W. 229th Avenue
Hillsboro, OR 97124, USA

Udo Schwalke
Institute for Semiconductor Technology
Darmstadt University of Technology
Schlossgartenstr. 8
64289 Darmstadt, Germany

Rakesh K. Singh
Liquid Microcontamination Control
Entegris, Inc.
129 Concord Road, Bldg. 2
Billerica, MA 01821, USA

Yordan Stefanov
Institute for Semiconductor Technology
Darmstadt University of Technology
Schlossgartenstr. 8
64289 Darmstadt, Germany

Joseph M. Steigerwald
Intel Corporation
RA1-234
2501 N.W. 229th Street
Hillsboro, OR 97124, USA

Manabu Tsujimura
Ebara Corporation
4-2-1 Honfujisawa
Fujisawa-shi 251-8502, Japan

Jeff Visser
ATDF
2706 Montopolis Drive
Austin, Texas 78741, USA

Changxue Wang
Center for Advanced Materials Processing
Clarkson University
8 Clarkson Avenue
Potsdam, NY 13699, USA

Gerfried Zwicker
Fraunhofer Institut fuer Siliziumtechnologie ISIT
Fraunhoferstr. 1
D-25524 Itzehoe

1

WHY CMP?

Yuzhuo Li

1.1 INTRODUCTION

Technology wonders have permeated into every facet of our daily life: fast computers with dual core processors and terabit hard drives, cell phones with cameras and GPS functions, video games with vivid graphics and superior sound, personal entertainment gadgets that go where we go, and smart implants that dose medicine on demand—just to name a few. These technology marvels that enable us to do things faster, more efficiently, and sometimes effortlessly all benefit from the advancements of semiconductor manufacturing processes. None of the advanced microelectronic devices could be built today without the continuous progress in shrinking the minimum feature size and increasing the circuitry complexity at the wafer level. The manufacturability of the smallest features or structures on a wafer is predominately determined or limited by the capability of the photolithographic step. To image lines or features accurately and precisely across the wafer, a photolithographic tool must be able to focus at all points of interest. For technology node dealing with relatively large features (>0.5 μm), the photolithographic process with relatively high depth of focus can tolerate certain levels of topography on the surface. With the reduction in minimum feature size, the depth of focus is also sharply reduced. A minute surface topography or step height may lead to a loss in yield [1,2]. To overcome such a challenge, the microelectronic industry revitalized a set of polishing skills that have been serving mankind for generations and brought the craft to a state-of-the-art level to meet the challenges faced by the semiconductor industry. This rejuvenated process is now known as chemical–mechanical polishing or planarization (CMP). More

Microelectronic Applications of Chemical Mechanical Planarization, Edited by Yuzhuo Li

specifically, a CMP step was added in between each metallization and dielectric layer in wafer production to address the depth-of-focus issue in photolithography [3]. Soon the technique also enabled the implementation of copper as a better electric conductor, ending more than 40 years of monopoly of aluminum as an interconnect [4]. Since the publication of the first book dedicated to this topic in 1997 [5], the field has been flourished with innovations, discoveries, breakthroughs, and successful implementations. Part of this book will cover the new breakthroughs and discoveries with an emphasis on the chemistry behind the processes and the materials used in the applications. Furthermore, a focus will be placed on the correlation between the use of various consumables and their impact on the polishing outcome. The outlook of the technology will also be discussed in light of new applications and new solutions to persistent problems. This introductory chapter is organized according to the three major utilities of CMP—preparation of planar surfaces, formation of functional microstructures, and elimination of surface defects.

1.2 PREPARATION OF PLANAR SURFACE

1.2.1 Multilevel Metallization and the Need for Planarization

In a state-of-the-art integrated circuit, there are many active and passive elements including millions of transistors, capacitors, and resistors on a single chip [5]. In this ultra-large-scale integration (ULSI) era, the number of transistors per chip has already crossed the 40 million mark and is expected to increase to more than a billion over the next decade [6]. These discrete elements must be connected with conductive wiring to form a circuit. As chips become smaller and more complex, the demand for more efficient interconnect systems has also increased dramatically. One solution is to have multilevel wiring over the devices. A multilevel wiring scheme offers more direct routing and reduces the average length of connections among devices. This leads to a significant reduction in signal processing delays and improvement in chip performance (see Section 1.3.1 for details). Figure 1.1 shows a cross section of such a multilevel interconnect network in which metal lines are isolated by the dielectric and connected by vertical vias [5,7]. It is noted that the metal lines on the lower levels are much narrower in order to match the dimensions of the transistors and other microstructures. At top levels, the need for high-line density is reduced. Therefore, there are more rooms for wider lines. A wider line also helps to avoid the mismatch with the vertical vias. With the implementation of a multilevel metallization scheme, the packing density of the metal lines need not keep pace with the packing density at the gate level. Hence, interconnect dimensions need not shrink at the same pace as the gate-level dimensions [8]. This offers a potential for chip performance improvement without revamping the entire IC layout.

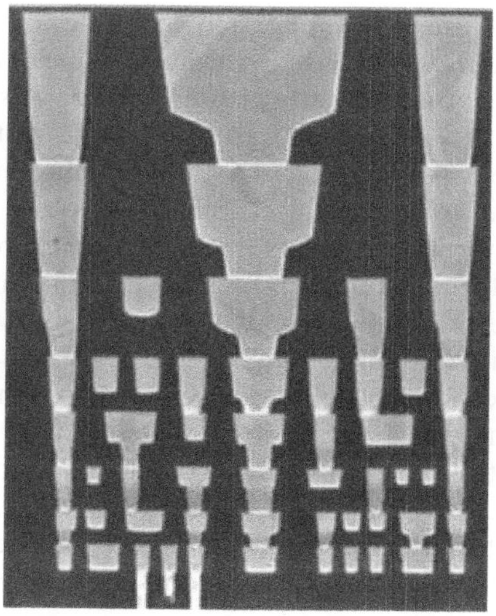

FIGURE 1.1 A cross section SEM image of a representative multilevel interconnect network (from Ref. 9).

The implementation of multilevel metallization presented immense opportunities for performance increase at the chip level. At the same time, the scheme also created enormous challenges in fabrication at the wafer level. The major source of such a challenge is the rugged topography buildup as the number of interconnect levels increases as shown in Fig. 1.2a [10]. The surface roughness has a direct negative impact on the accuracy and efficiency of pattern transfer onto photoresist with contact photolithography [11–19]. As the critical dimension of the device reduces, the depth of focus in photolithography also

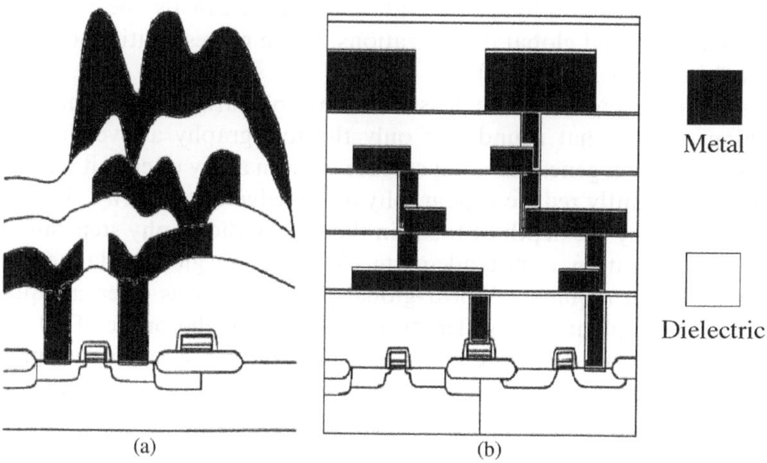

FIGURE 1.2 Devices fabricated without (left) and with (right) planarization (from Ref. 10).

diminishes. In other words, the topography or surface roughness will lead to a much wider distribution in focusing accuracy, which in turn translates to inaccurate patterning at significantly greater number of sites. For example, if the depth of focus for a particular feature size is in the order of 0.5 μm determined by an optical lithography tool, any step heights larger than 0.5 μm on the surface of pre- or intermetal dielectrics will cause improper patterning on the photoresist layer. Subsequently, the multilevel interconnect network will fail. The depth-of-focus limitation became insurmountable by any other techniques available at a fab when the critical dimensions dropped below 0.35 μm, which requires the surfaces to be planar within the same range. Driven by necessity, an effective planarization process was sought, envisioned, tested, and subsequently implemented. The process was CMP.

A comparison between a planarized and nonplanarized surface topography is shown in Fig. 1.2. By meeting the depth of focus requirement for the photolithographic step, CMP eliminated several yield-related issues such as missing contacts, undesired current leaks, and electromigrations [11–19].

1.2.2 Degrees of Planarization

The topography buildup on wafers is a combination of accumulated unevenness at feature, die, and wafer level. Other terms such as nanotopography, micro- or macrowaviness, and wharf have been used to describe such unevenness of a wafer at different length scales. To meet the requirement set by the depth of focus for subquarter micron technology, the roughness to be eliminated is in the regime of nanotopography and microwaviness. In other words, the step height of interest has an average wavelength of several microns to millimeters [1,5,20–23]. Similarly, depending on the net effectiveness on various types of topography, planarization processes can also be categorized as smoothing, local, and global planarizations. Some representative scenarios are illustrated in Fig. 1.3 [5,20–23].

As shown in Fig. 1.3, the least effective planarization is the so-called smoothing process that rounds off only the topography above the features. Local planarization generates a flat surface over an array of circuit features but does not significantly reduce topography at the edge of the array. To meet the requirement set by the depth of focus in the photolithography step, smoothing or local planarization is not adequate. A complete global planarization is desirable, but not required. A near-global planarization is often adequate. In other words, the planarization length is preferred in the order of 20–30 mm, which is the size of a typical die. As of today, there are no known processes that produce this effect over widely varying surface topographies and pattern layout densities other than CMP. CMP is the only technique that can produce planarization results that meet the requirements of lithography. The above discussion can be quantified by using a planarization length R (μm) and its corresponding angle θ (degrees) that are illustrated in Fig. 1.4.

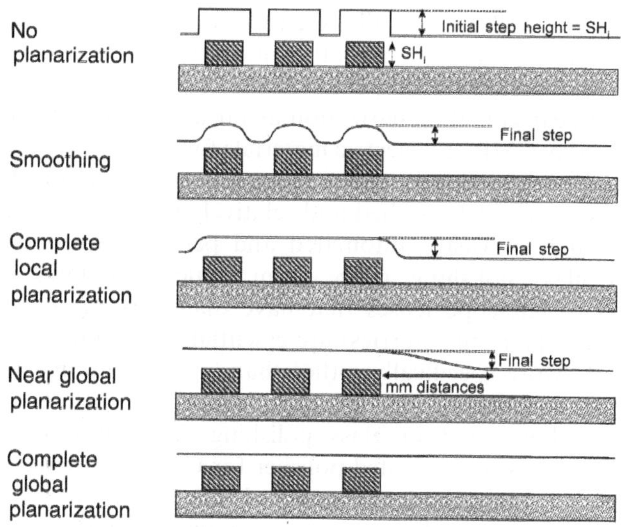

FIGURE 1.3 Levels of planarization that are relevant to semiconductor processing (from Ref. 5).

According to the definition given in Fig. 1.4, the following values of R and θ can be used to categorize degrees of planarization:

- Surface smoothing: $R = 0.1$–2.0 and $\theta > 30°$.
- Local planarization: $R = 2.0$–100 and $30° > \theta > 0.5°$.
- Global planarization: $R \gg 100$ and $\theta < 0.5°$.

1.2.3 Methods of Planarization

Several contending technologies are presently being used to achieve local and global planarizations that include spin on deposition (SOD), reflow of boron phosphorous silicate glass (BPSG), spin etch planarization (SEP), reactive ion etching and etch back (RIE EB), spin on deposition and etch back

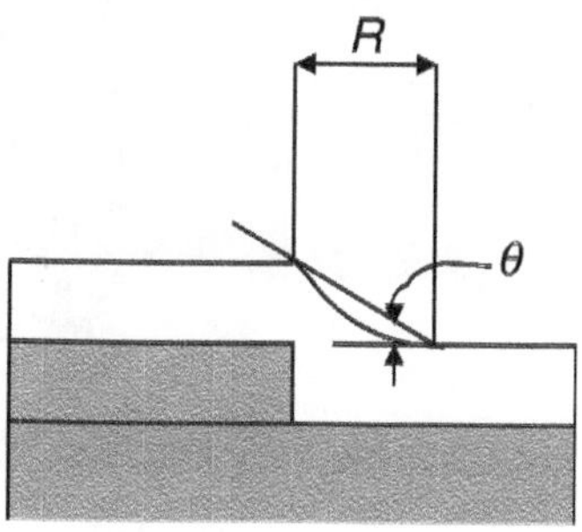

FIGURE 1.4 Planarization length R and slope θ (from Ref. 24).

(SOD + EB), and CMP. Among all these techniques, CMP is the only one that can offer excellent local and global planarities at the same time. More specifically, CMP can yield local planarization of features in the order of tens of microns and near-global planarization as far as tens of millimeters [5,20].

The modern-day CMP of dielectric materials for wafer processing has a root in glass polishing that has been practiced throughout civilization. The polishing mechanism has been widely studied and relatively well understood [5,20]. The process has also been vastly automated and perfected over the years. The substrates of glass polishing range from optical windows measured in submillimeters to telescope lenses that have diameters measured in meters. The consumables (pads and slurries) are essentially the same as those used in dielectric CMP. More specifically, other than some additional requirements, the silica- and ceria-based slurries used today for dielectric CMP bear resemblance to those used in glass polishing. Though more primitive in comparison to today's sophisticated polisher for CMP in a semiconductor fab, the glass polishing tool had the essential features even for the earliest applications. For example, Fig. 1.5 shows a picture illustrates the type of polisher used to polish the telescope lens in the Galileo era. In 1609, Galileo heard of the telescope while in Venice, and on his return, constructed one for himself. In 1610, Galileo published his telescopic discoveries in *The Starry Messenger* [25].

One who is well versed in CMP may choose to believe that the machine has the functions detailed below [26]. Can you identify them?

1. Variable speed platen.
2. Variable speed quill.

FIGURE 1.5 A highly ornamented Lens-grinding lathe on display at the Institute and Museum of the History of Science in Florence, Italy (from Ref. 26).

3. Vertical motion of the quill with variable downforce control.
4. Slurry dam and variable control of work piece slurry immersion.
5. Slurry drain.
6. Optional quill offset to provide eccentric polish head motion.

Prior to the implementation of CMP, various grinding and polishing techniques had been used in the semiconductor industry to planarize raw silicon wafers. In addition, to achieve a global flatness, the planarization process also removes the damage and defects caused by the sawing process to the single crystal. Because of the fact that silica and ceria do not chemically react with bare silicon, except the top oxidized silicon dioxide layer, the grinding and polishing process for this application is dominated by mechanical events.

1.2.4 Chemical and Mechanical Planarization of Dielectric Films

The most commonly implemented and extensively investigated CMP steps are the preparation of planar premetal dielectrics (PMD) and interlayer dielectrics (ILD) films on wafer. Together they are labeled as "oxide" CMP, as they both use the same materials that are based on silicon dioxide. Both processes share the integration concerns in deposition, planarity, and defectivity.

PMD CMP was designed to provide planarization between the front-end active devices and the back-end metallization. Several reasons for the planarization are (a) enabling contact lithography, (b) enabling contact etch uniformity, and (c) enabling contact tungsten CMP [5,20,21]. ILD CMP is meant to provide planarization between the increasing numbers of metal layers in the back end. The motivation is twofold: (a) enabling via lithography and (b) enabling via tungsten CMP. PMD and ILD CMP are "stop-in-film" processes [1,5,20,22–24,27,28]. There are no interfaces on which for CMP to stop. Therefore, the overall performance of the process is extremely dependent on consistent removal rate, within-die, within-wafer, and lot-to-lot uniformity.

In addition to the construction of a multilevel interconnect network, the semiconductor industry also improves the performance of IC chips by incorporating low-resistivity metal wiring such as copper and new dielectric materials with lower *k* constant (see Section 1.3.1 for details). The added benefit of using low-*k* dielectric materials includes a reduction in the crosstalk [29–31] and power dissipation [29–33]. The key challenge for the implementation of low-*k* materials is related to their intrinsic weak mechanical properties. Furthermore, in order to achieve a *k* value below 2.2, practically all materials are made with pores that exacerbate mechanical stability issue [29–33]. This is a particular concern for the CMP community as the operation invariably involves mechanical stress and shear force. In addition, practically all low-*k* dielectric materials are hydrophobic in nature. Upon exposure to moisture or wetness, the dielectric constant tends to increase. Therefore, unlike silicon-dioxide-based dielectric, the effective *k* constant may change after CMP. To

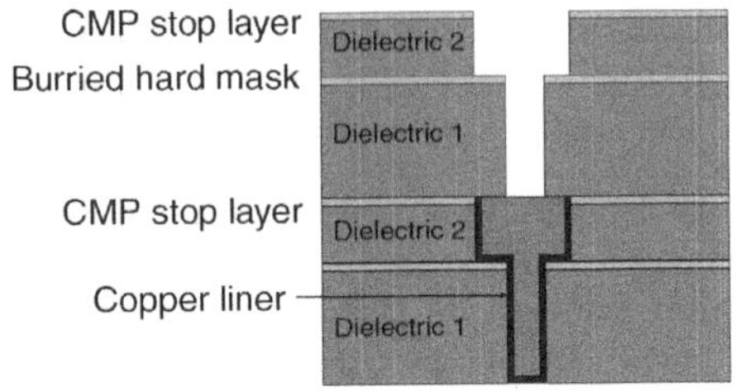

FIGURE 1.6 Incorporation of hard masks to protect the low-*k* dielectric materials (from Ref. 23).

overcome these challenges, an array of possible solutions has been explored and implemented. To mechanically protect the low-*k* dielectric material, a cap material sometimes is incorporated into the design of the device as shown in Fig. 1.6 [23]. Hard masks such as SiCN are incorporated to avoid the exposure of the low-*k* material to CMP consumables. This leads to the diversity of thin films that a CMP process will encounter. The hard masks also help to simplify lithography, etch, and clean.

1.2.5 Preparation of Planar Thin Films for Non-IC Applications Using CMP

Nearly every laptop or desktop computer in use today contains one or more hard disk drives. Every mainframe server and supercomputer is normally connected to hundreds of them. You can even find DVR, iPod, and camcorders that use hard disks instead of tape or flash memory. The computer hard drives store changing digital information on rigid magnetic memory disks. Figure 1.7 shows a stack of platens that have magnetic layers on them. Figure 1.8 shows a typical cross section of the rigid disk. In order to deposit the magnetic materials

FIGURE 1.7 A side view of a multiplaten computer hard drive (from Ref. 34).

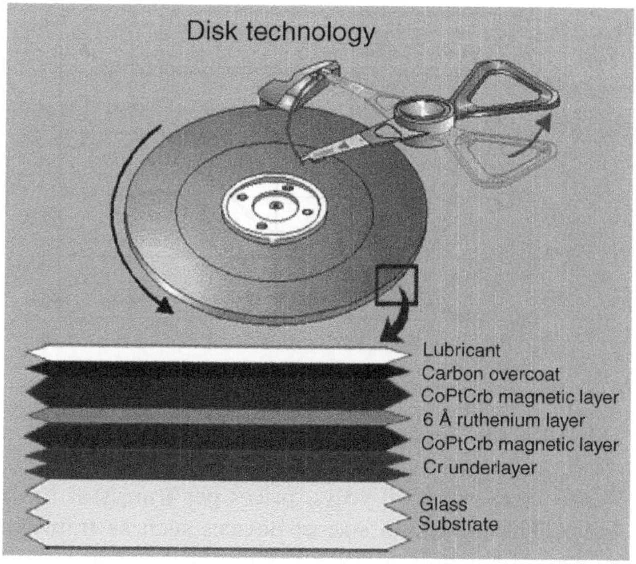

FIGURE 1.8 A cross section of a typical computer hard drive disk (from Ref. 35).

properly, the substrate must be perfectly flat and free of defects such as pits, scratch, and bumps. Any of these defects not only lower the effectiveness of the magnetic layer to store the information but also can cause the crash of read–write heads that are flying over the platen at a tremendous speed and impressive low altitude. The operation can be compared to a situation where a large aircraft is flying at the top speed, less than a meter above the ground. Any nanoasperity on the computer hard drive disk is equivalent to an insurmountable mountain for the aircraft to avoid. Therefore, a CMP process has been used to planarize the substrates for the computer hard drives. There are two major types of substrates used in today's computer hard drives. One is glass based: ceria (CeO_2) particles are the most commonly used abrasive for this application. The other is aluminum coated with NiP. The NiP layer is usually electrochemically plated and then subsequently planarized with alumina-based slurry followed by silica-based slurry to remove the defects and nanoasperities. The surface roughness after the CMP process is often required to be less than 1 Å.

1.3 FORMATION OF FUNCTIONAL MICROSTRUCTURES

1.3.1 *RC* Delay and New Interconnect Materials

Miniaturization of semiconductor devices has been a continuous trend in the microelectronics industry. The decrease in minimum feature length reduces the overall device size, increases the packing density, and thus reduces the cost of

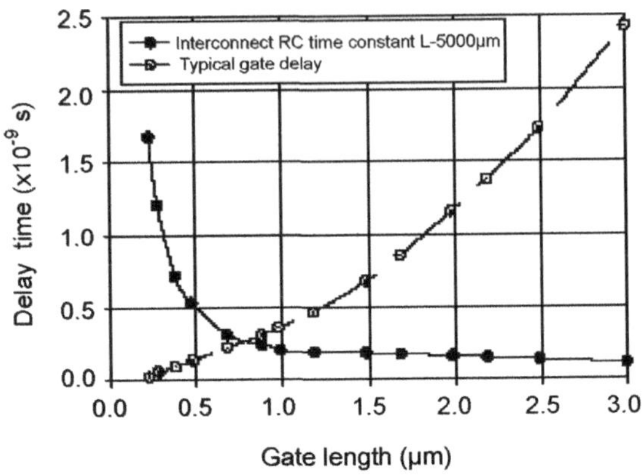

FIGURE 1.9 Delay time vs. gate length (from Ref. 41).

function [5,37,38]. In the past 50 years, prices per transistor have gone down 100 million times. The minimum size of devices such as transistors has been reduced by a factor of a billion [19,39]. However, as the feature size scales down to below 0.5 μm, the improvement of device performance such as speed is hindered by the delays in signal processing. In a typical device, there are two major sources of processing delays—intrinsic gate delay and interconnect delay [36]. The intrinsic gate delay is the time required to switch the transistor on or off [40]. Interconnects are the metal wires that connect different devices on a chip among themselves and the outside world [20]. The interconnect delay is the time spent for a signal to propagate from the source to its destination in a circuit. The total delay in signal processing is the sum of interconnect delay and the device delay. As shown in Fig. 1.9, the gate delays typically decrease as the gate length decreases. The interconnect delays on the contrary increase as the gate length decreases. As the device sizes reduce below the sub-micron level (below 0.5 μm), the total delay is dominated by the interconnect delay.

The two key components in interconnect delays include the inherent resistance (R) of the metal lines and the capacitance (C) of the dielectric material in between the lines. The so-called RC delay is defined as the time required for the voltage at one end of a metal line to reach 63 % of its final value when a step input is presented at the other end of the line [18]:

$$RC = \rho \varepsilon l^2 / td \tag{1.1}$$

where R is the resistance of the interconnect, C is the capacitance of the dielectric in between the lines, ρ is the resistivity of the interconnect, ε is the permittivity of the insulator, t is the thickness of the insulator, and d is the thickness of the metal line or interconnect.

There are two types of capacitances associated with interconnect—the line-to-ground capacitance and line-to-line capacitance as illustrated in Fig. 1.10. Although line-to-substrate capacitance decreases as the feature size decreases,

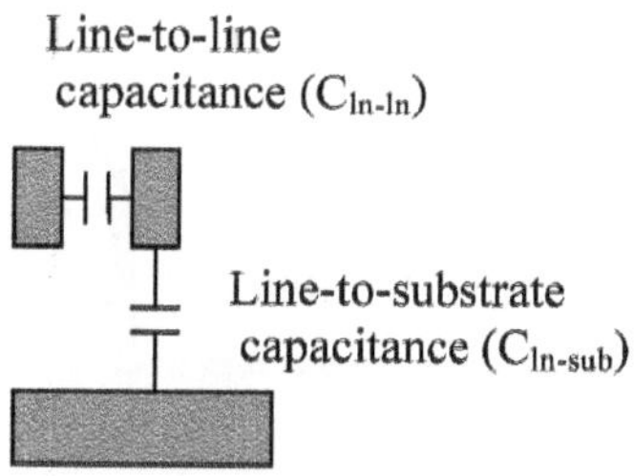

FIGURE 1.10 Two categories of capacitance (from Ref. 41).

the line-to-line capacitance (or the interconnect delay) increases with the reduction of the feature size. To reduce the total delay in signal processing along with the chip miniaturization, the industry took a parallel approach—replacing the traditional interconnect material (Al) with a better conductor (Cu) and substituting traditional silicon dioxide with low-*k* dielectric materials.

The first generation of the interconnect material is aluminum with a resistivity of $\rho = 2.66\ \mu\Omega\ \text{cm}$. One approach to reduce *RC* delay is to switch to an interconnect material with lower resistivity as indicated by Eq. (1.1). A wide range of metals was considered as a potential candidate in the early 1990s. Gold has excellent resistance to corrosion and electromigration but its conductivity is similar to that of aluminum. Silver has the lowest resistivity ($\rho = 1.59\ \mu\Omega\ \text{cm}$) but poor resistance to corrosion and electromigration. Hence, copper that has a resistivity of $1.67\ \mu\Omega\ \text{cm}$ and excellent resistance to electromigration was selected. Compared to aluminum, copper has one drawback. It cannot be deposited by RIE. Therefore, a copper interconnect is typically formed via a damascene process in which a pattern is first etched into the dielectric and overfilled with copper. The excess copper above the

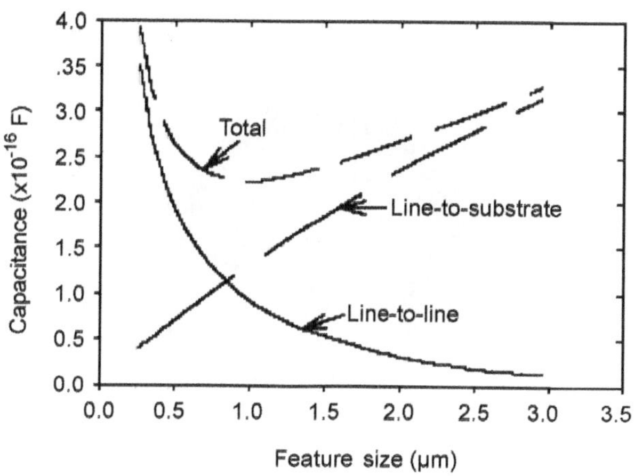

FIGURE 1.11 Capacitance vs. feature size (from Ref. 41).

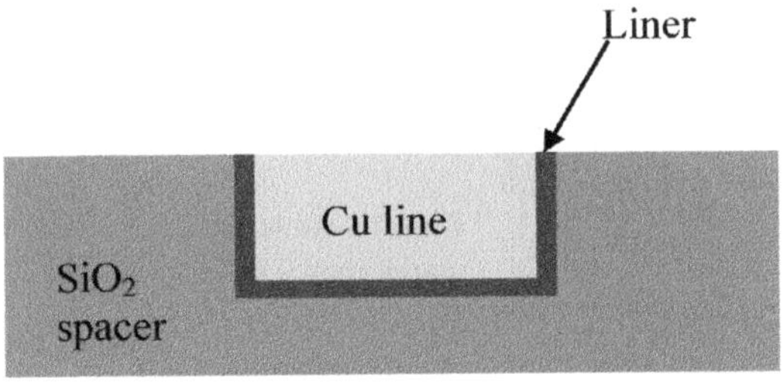

FIGURE 1.12 Typical layout of a trench showing Cu, dielectric, and barrier (Ta or TaN) (from Ref. 42).

trench is then removed. The copper remaining in the trench forms individual lines (Fig. 1.12). Copper has poor adhesion to dielectric materials such as silicon dioxide. Compared to aluminum, copper is also more liable to diffuse into SiO_2. To address the adhesion and diffusion issues, a barrier is placed between the copper and the dielectric [1,42]. There are several possible candidates for barrier materials, a combination of Ta and TaN has been the choice for many successful manufacturing processes.

1.3.2 Damascene and Dual Damascene [11]

Damascene "Damasquinado de Oro" or "Damasquino" is an art of decorating nonprecious metals with gold. It has roots in the Middle Ages and originates from the oriental-style artisan work done in Damascus, Syria. The craft, perfected by the Arabs and brought with them to Spain, has remained virtually unchanged over the centuries. Figure 1.13 shows a piece of jewelry made with a damascene process.

FIGURE 1.13 A typical piece of jewelry made with a damascene process (from Ref. 43).

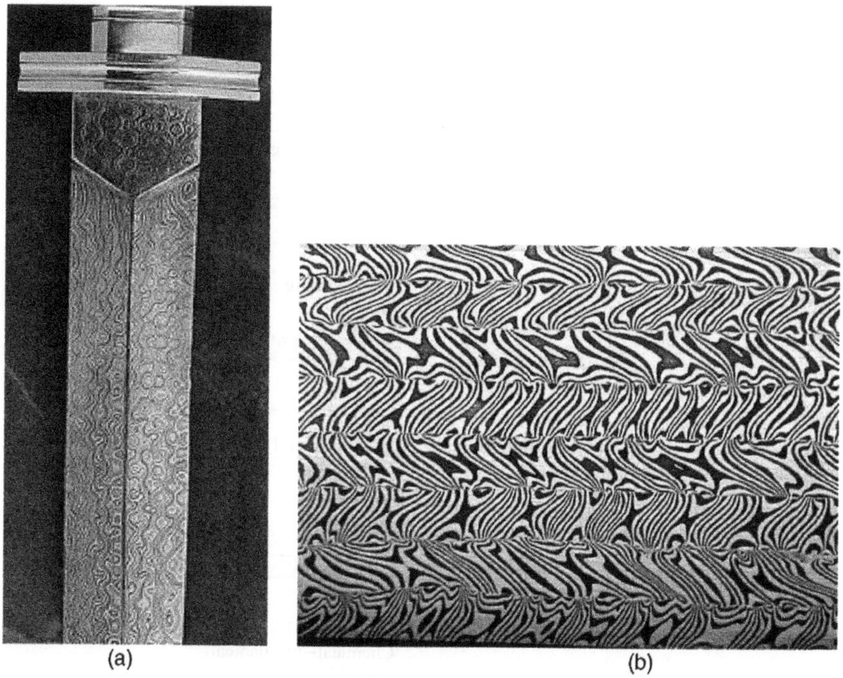

FIGURE 1.14 Sword made with a damascene process (a) and typical patterns on a damascus metal (b) (from Ref. 44,43).

The technique was apparently also used to make the legendary Damascus swords. The details for making Damascus steel remain a mystery even with the presence of numerous well-preserved samples. Recent research into the structure and composition of the steel reveals that the strength of the steel was a result of carbon nanotubes and carbide nanowires present in the structure of the forged metal. Damascus swords often had an obvious patterned texture on their surfaces (Fig. 1.14).

The semiconductor industry borrowed the word damascene to describe the patterned metal line formation process. Figure 1.15 illustrates a basic process for the formation of a copper line via a damascene process. The advantage of using copper is that it could be used as both an interconnect and a via; hence, the method of dual damascene comes into play. This method has come into use after the introduction of copper. In short, it can be said as opposite to that of RIE used for patterning aluminum. The oxide is etched to form patterns required for patterns of wires or vias. The barrier is then deposited followed by copper. The excess burden of copper is removed by using CMP, believed to be the only technique that gives global planarization. The process eliminates the etching of copper and maintains planar surfaces necessary for multilevel metallization. The process of dual damascene eliminates complexity by reducing the number of steps in the patterning process. It also reduces the

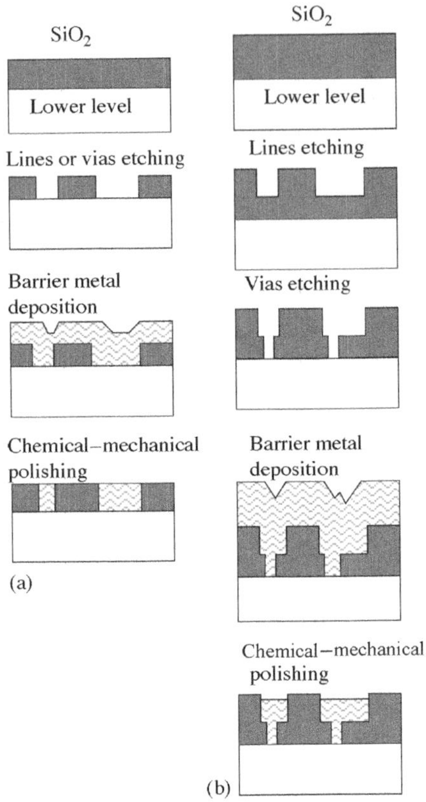

FIGURE 1.15 Damascene and dual damascene techniques employed (from Ref. 45).

risk of failure between metal and via. The schematics of both single and dual damascene are shown in Fig. 1.15.

The low-resistivity and high-electromigration properties have made copper the material of choice for the fabrication of interconnects in present-day IC

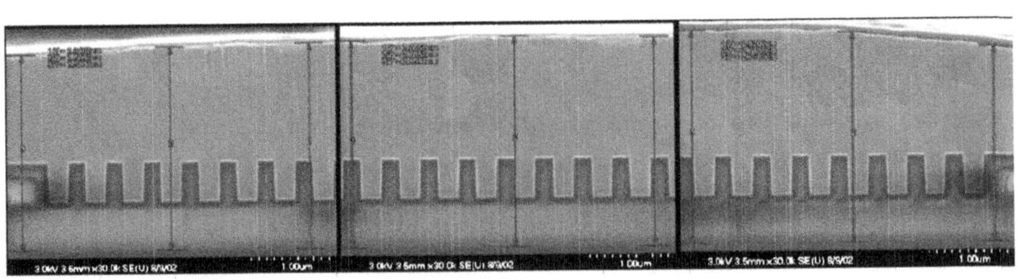

FIGURE 1.16 Cross section SEM image of copper wafer showing overburden Cu with underlying features. The features shown are 50% in metal:dielectric density and 2 μm in width (from Ref. 46).

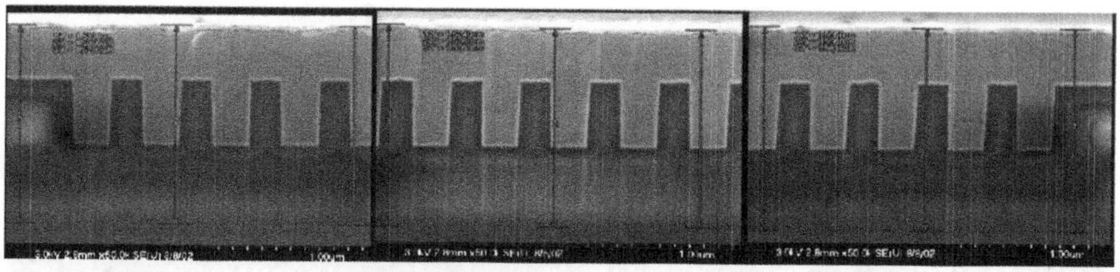

FIGURE 1.17 Cross section SEM image of copper wafer after the removal of the overburden with the achievement of planarization. The features shown are 50% in density and 2 μm in width (from Ref. 46).

chips. The inability of copper to form volatile compounds at lower pressures to assist RIE has left damascene as the only viable process to incorporate copper through CMP. Because of the copper migration issue, the interconnect lines are not directly in contact with the dielectric. A diffusion barrier is required to protect the integrity of the line. Therefore, after the removal of overburden copper, the barrier is also removed. A typical multistep Cu CMP process involves three steps: the overburden copper is initially planarized, which is followed by a Cu-clearing step. The third step involves the clearing of the barrier metal. Figure 1.16–1.19 clearly illustrate the three steps described [46]. Figures 1.20 and 1.21 show a closer view of typical features before and after the barrier CMP.

After the removal of the copper barrier layer (usually made of Ta and TaN), the feature needs to be perfectly flat between the three materials (dielectric, barrier, and copper line). A representative SEM image of such a result is shown in Fig. 1.22.

1.3.3 Tungsten CMP

The main application of tungsten CMP is to create the so-called tungsten plugs that provide the vertical links between in-line wiring. As shown in Fig. 1.1, the number of such plugs decreases as the size of such plugs increases at higher metallization level. Figure 1.22 shows a representative tungsten plug [48]. It is

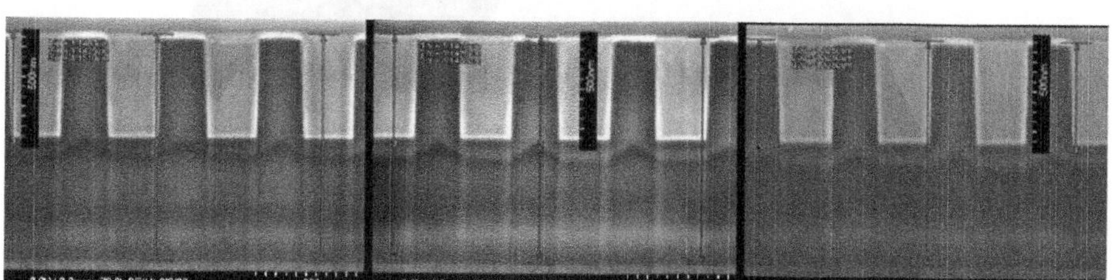

FIGURE 1.18 Cross section SEM image of copper wafer after copper clearing step. The barrier is still present at this stage. The features shown are 50% in density and 2 μm in width (from Ref. 46).

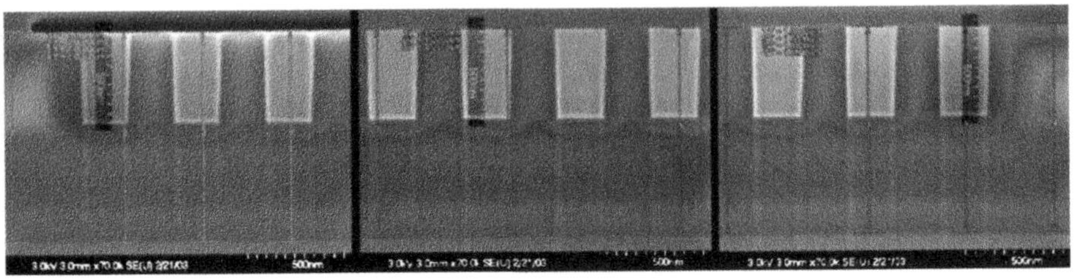

FIGURE 1.19 Cross section SEM image of copper wafer after the removal of barrier (from Ref. 46).

noted that it will take three damascene processes to create such a structure: first construction of a copper line, then a tungsten plug, and then another copper line [49]. Similar to copper CMP, tungsten plug also requires an adhesion and diffusion layer (Ti and TiN) [1,50]. Therefore, a W CMP process is actually a combination of tungsten, titanium, and titanium nitride removal, all in one step.

1.3.4 STI

Another important microstructure in IC manufacturing process is shallow trench isolation (STI) that allows the effective separation of active devices and increase of packing densities. Figure 1.23 shows a schematic of an STI structure before and after polishing [51]. It is important for the dishing of the oxide in the trench and the nitride loss to be as low as possible.

With the various types of CMP described above (dielectric and metal CMP), a multilevel interconnect network can be constructed. Impressive progress

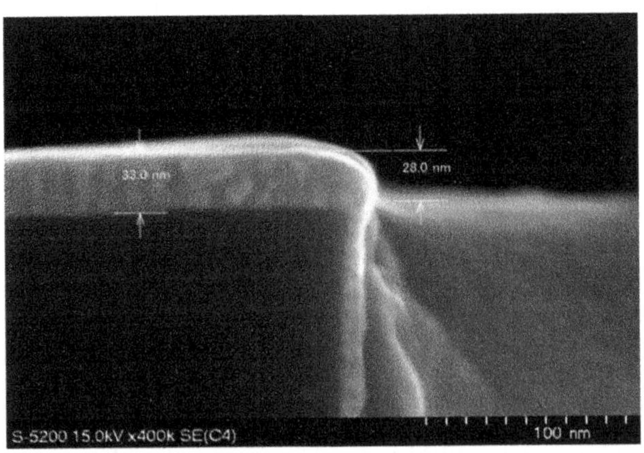

FIGURE 1.20 Cross section SEM image of a copper interconnect after the removal of overburden copper and before the removal of barrier layer (from Ref. 47).

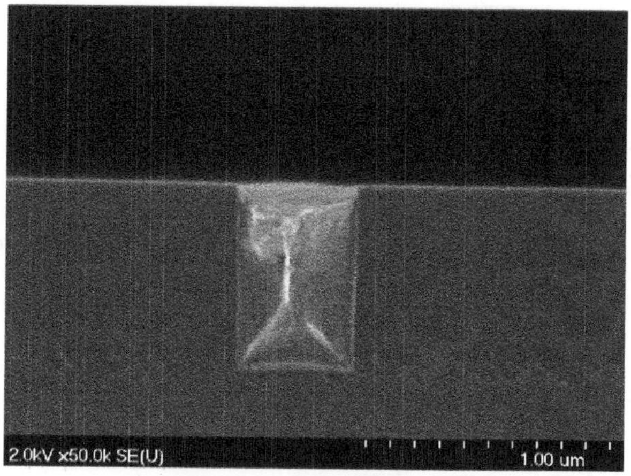

FIGURE 1.21 Cross section SEM image of a copper interconnect after the removal of overburden copper and barrier layer (from Ref. 47).

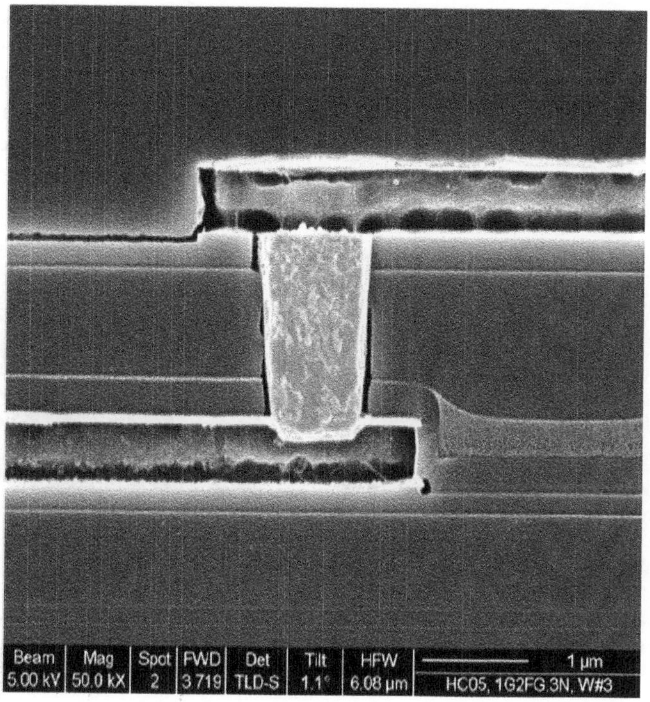

FIGURE 1.22 Cross section SEM image of a representative tungsten plug in between two copper lines (from Ref. 48).

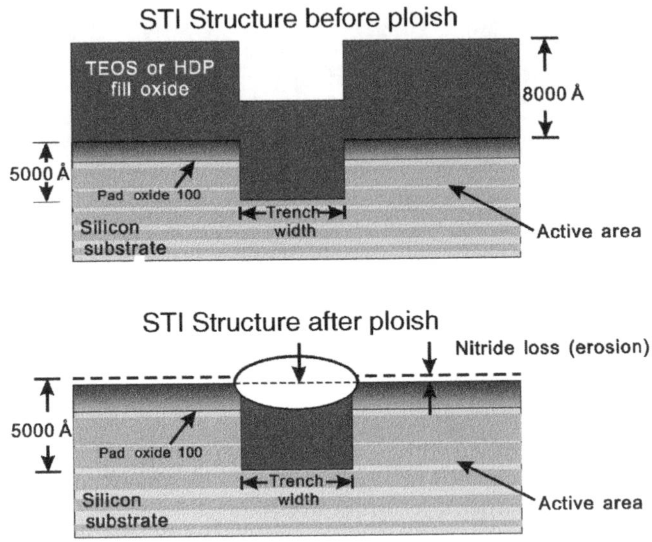

FIGURE 1.23 Schematic of an STI structure before and after polishing (from Ref. 51).

has been made over the past decade in constructing such a complex and dense network that provides the much needed boost to the IC performance. Fig. 1.24 shows the sharp contrast of the level of complexity in IC chip manicuring. Figure 1.24a shows the very first IC with four transistors on a single level of metal connection. Figure 1.24b shows, 37 years later, over 40 millions of transistors packed into a single IC with multilevel interconnects [6].

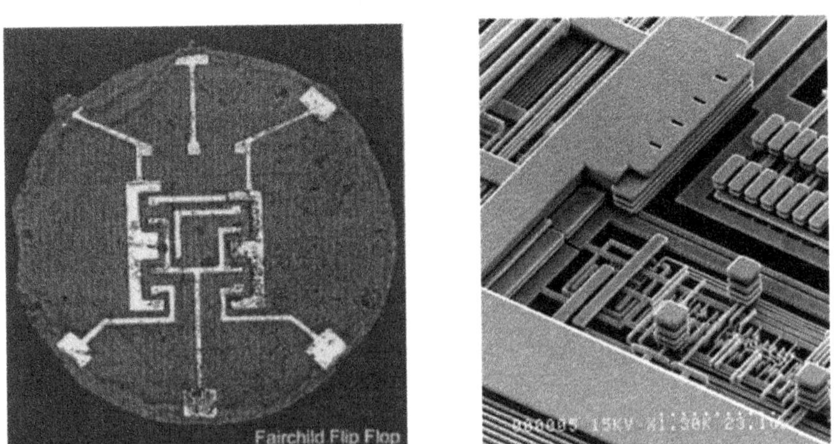

FIGURE 1.24 The first IC built on single layer of metal connect that links four transistors (a) and the IC with multilevel interconnect (b) (from Ref. 52).

1.4 CMP TO CORRECT DEFECTS

The application of CMP could also be extended to the reduction of surface defects in addition to the preparation of planar surfaces and fabrication of functional microstructures. As matter of fact, these types of applications have already been implemented in some cases as a part of the planarization process. For example, at the end of a copper or tungsten CMP process, a buffing step is inserted to remove residues, particles, and correct some minor defects such as shallow scratches. The buffing process is typically carried out on the last platen using DI water or a solution that is similar to those used in a post-CMP cleaning. In most cases, a buffing procedure is performed on a much softer pad [53,54]. Sometimes, owing to tool limitation or other concerns, the same pad or platen is used. Chen and co-workers [53] employed a buffing process on the same pad employed for polishing to reduce the residue silica abrasives. The silica abrasives were believed to be chemisorbed onto the copper oxide surface. Instead of DI water, a solution of HNO_3/BTA was used in this buffing process. The presence of nitric acid helped to etch a thin layer of copper oxide and loosen the particle adhesion to the surface. The presence of BTA as a passivating agent protects the copper surface from excessive etching or corrosion. The wafers were subsequently scrubbed to eliminate the residual particles. Cheemalapati et al. demonstrated the usefulness of an in situ buffing step to reduce the organic residue left by a copper CMP process. More specifically, at the near end of a copper clearing process, the copper slurry was substituted with a post-CMP cleaning solution for a short period of time. The extent of the organic residue was significantly reduced. This is particularly useful if the organic residue becomes difficult to clean after the wafer is exposed to air [55]. The elimination of preexisting scratches using a CMP step on copper blanket wafers was also shown by Hegde and Babu [56]. Different copper CMP slurries with and without the abrasives were studied for the effectiveness of removing the preexisting scratches. The ratio between the removal rate and the static etch rate was found to be the dominating factor in determining the depth of scratch that could possibly be removed. The application of such processes could possibly become useful for a three-step Cu CMP process that employs multiple slurries.

For some applications, the crystalline structure of a surface has a significant impact on the proper growth of the next layer of materials. The surface not only must be perfectly planar but also must be free from crystal lattice defects. For example, sapphire is a widely used material for blue emitting diode, laser diode devices, visible–infrared window, and random applications. Although there is a large mismatch in the lattice constants and thermal expansion coefficient between nitride and sapphire, sapphire is still known as the most commonly used substrate in the GaN device for its physical robustness and high-temperature stability. The performance of these devices is highly dependent on the quality of the substrate surface processing. Wang et al.

TABLE 1.1 Advantages of CMP.

Benefits	Remarks
Planarization	Achieves global planarization
Planarize various materials	Wide range of wafer surfaces can be planarized
Planarize multimaterial surfaces	Useful for planarizing multiple materials during the same polish step
Reduce severe topography	Reduces severe topography to allow fabrication with tighter design rules an additional interconnection levels
Alternative method of metal patterning	Provides an alternative means of patterning metal, eliminating the need to plasma etch, difficult to etch metals and alloys
Improved metal step coverage	Improves metal step coverage due to reduction in topography
Increased IC reliability	Contributes to increasing IC reliability, speed, yield (lower defect density) of sub-0.5 μm circuits
Reduce defects	CMP is a subtractive process and can remove surface defects
No hazardous gases	Does not use hazardous gas common in dry etch process

demonstrated that CMP followed by a chemical etching yields the best quality sapphire substrate surfaces [57–60].

1.5 ADVANTAGES AND DISADVANTAGES OF CMP

A list of advantages and disadvantages of CMP are shown in Tables 1.1 and 1.2, respectively [15]. By no means are the lists complete, but they offer some useful comparisons with other associated or competing technologies.

TABLE 1.2 Disadvantages of CMP.

Disadvantages of CMP	Remarks
New technology	CMP is a new technology for wafer planarization. There is relatively poor control over process variables with narrow process latitude
New defects	New types of defects from CMP can affect die yield. These defects become more critical for sub-0.25 μm feature sizes
Need for additional process development	CMP requires additional process development for process control and metrology. An example is the endpoint of CMP is difficult to control for desired thickness
Cost of ownership is high	CMP processes materials require high maintenance and frequent replacements of chemicals and parts

1.6 CONCLUSION

CMP emerged as an enabling technique for the semiconductor industry to overcome the depth-of-focus challenge for the implementation of a multilevel interconnect scheme. Soon, the technique was adapted to assist the formation of STI microstructures and vertical tungsten via. The introduction of copper as a new interconnect material helped launch CMP as an independent field with broad participation of scientists and engineers from a wide range of disciplines including chemistry, physics, materials science, and chemical and mechanical engineering. The number of patents, publications, and conferences dedicated to CMP processes has dramatically increased over the past 15 years. From an application point of view, CMP is able to not only prepare planar surfaces with impressive palanarization length but also enable the formation of microstructures such as copper lines, tungsten vias, and STI. The process can be so well controlled that the technique could also be implemented to remove surface defects from prior manufacturing steps. From the operations point of view, the industry has built an infrastructure consisting of polishers, metrology tools, slurry delivery, consumable management, and matching supply chains. The cost of tool ownership is declining. This will help the implementation of this process in the fab for both routine techniques and new applications.

QUESTIONS

1. Fundamentally, other than the three major types of applications of CMP described in this chapter, what other types of application also exist or can be developed?
2. Why is the planarization length desirable at die size? Will a planarization length at wafer diameter scale really be an advantage?
3. Other than the damascene process, is there any other way to form microstructures such as copper lines, tungsten vias, and STI?
4. In addition to Tables 1.1 and 1.2, what are the other potential advantages and disadvantages of CMP in relationship to competing technologies?

REFERENCES

1. Doering R, Nishi Y, editors. Handbook of Semiconductor Manufacturing. Marcel Dekker; 2000.
2. Chiu GL-T, Shaw JM, Guest editors. Optical lithography: introduction. IBM J Res Dev 1997; 41(1–2); p 3–6.
3. Dornfeld DA, Luo J. Integrated Modeling of Chemical Mechanical Planarization for Sub-Micron IC Fabrication. Springer; 2004. p 16.
4. Steigerwald JM, Murarka SP, Gutmann RJ, Duquette DJ. Chemical processes in the chemical mechanical polishing of copper. Mater Chem Phys 1995;41:217–228.

5. Steigerwald JM, Murarka SP, Gutmann RJ. Chemical Mechanical Planarization of Microelectronic Materials.New York: Wiley; 1996.
6. Nair R. Effect of increasing chip density on the evolution of computer architectures. IBM J Res Develop 2002;46(2–3); p 223–224.
7. Muraka SP. Metallization Theory and Practice for VLSI and ULSI. Massachusetts (MA): Butterworth-Heinemann; 1993.
8. Sheats JR, Smith BW, editors. Microlithography: Science and Technology. CRC Press; 1998. p 49.
9. Available at http://www.intel.com/technology/silicon/65nm_technology.htm.
10. Wolters P. Available at http://www. peter-wolters.com/cmp/cmpmultilevel.htm; 2003.
11. Landis H, Burke P, Cote W, Hill W, Hoffman C, Kaanta C, Koburger C, Lange W, Leach M, Luce S. Integration of chemical–mechanical polishing into CMOS integrated circuit manufacturing. Thin Solid Films 1992;220(1–2); p 1–7.
12. Treichel H, Eckstein E, Kern W. New dielectric materials and insulators for microelectronic applications. Ceram Int 1996;22(5):435–442.
13. Hu YZ, Yang G-R, Chow TP, Gutmann RJ. Chemical–mechanical polishing of PECVD silicon nitride. Thin Solid Films 1996;290–291:453–455.
14. Deleonibus S. Is there LOCOS after LOCOS? Solid State Electron 1997;41(7):1027–1039.
15. Zantye PB, Kumar A, Sikder AK. Chemical mechanical planarization for microelectronics applications. Mater Sci Eng R: Rep 2004;45(3–6):89–220.
16. Tay FEH, editor. Materials & Process Integration for MEMS. Kluwer Academic Publishers; 2002. p 160.
17. Schwartz GC, Srikrishnan KV, Gross A, editors. Handbook of Semiconductor Interconnection Technology. Marcel Dekker; 1997. p 287.
18. Holloway PH, McGuire GE. Handbook of Compound Semiconductors: Growth, Processing, Characterization, and Devices. Noyes Publications; 1996. p 415.
19. Madou MJ. Fundamentals of Microfabrication: The Science of Miniaturization. CRC Press; 2002. p 331.
20. Oliver MR, editor. Chemical–Mechanical Planarization of Semiconductor Materials. Springer; 2004.
21. Radojcic R, Pecht MG, Rao G. Guidebook for Managing Silicon Chip Reliability. CRC Press; 1999. p 82.
22. Kareh B-E. Fundamentals of Semiconductor Processing Technology. Kluwer Academic Publishers; 1995. p 568.
23. Franssila S. Introduction to Microfabrication. Wiley; 2004. p 169.
24. Freeman JL, Tracy CJ, Wilson SR, editors. Handbook of Multilevel Metallization for Integrated Circuits: Materials, Technology, and Applications. Noyes Publications; 1993. p 352.
25. Reddy F, Walz-Chojnacki G. Celestial Delights: The Best Astronomical Events Through 2010.Celestial Arts; 2002. p 133.
26. Personal communication with Allan Paterson of Strasbaugh; Jan 2004 and permission from Photo Franca Principe, IMSS – Florence Italy, August 2007.

27. Seshan K, editor. Handbook of Thin Film Deposition Techniques Principles, Methods, Equipment and Applications. 2nd ed. William Andrew Inc.; 2002. p 553.
28. Borst CL, Gill WN, Gutmann RJ. Chemical–Mechanical Polishing of Low Dielectric Constant Polymers and Organosilicate Glasses. Kluwer Academic Publishers; 2002. p 111.
29. Lee WW, Ho PS, Leu J. Low Dielectric Constant Materials for IC Applications. Springer; 2003.
30. Tung C-H, Sheng GTT, Lu C-Y. ULSI Semiconductor Technology Atlas. Wiley IEEE; 2003. p 217.
31. Jess J, Reis R, editors. Design of System on a Chip: Devices & Components. Springer; 2004. p 255.
32. Zschech E, Whelan C, Mikolajick T. Materials for Information Technology: Devices, Interconnects and Packaging. Springer; 2005. p 461.
33. Tummala RR. Fundamentals of Microsystems Packaging.McGraw-Hill Professional; 2001. p 75.
34. Available at http://www.pcguide.com/ref/hdd/.
35. Available at http://www.hitachigst.com/hdd/research/storage/adt/index.html.
36. McGuire GE. Semiconductor Materials and Processing Technology Hand Book. William Andrew Publishing; 1988.
37. Institute of Electrical and Electronics Engineers. Proceedings of the IEEE; 1913. p 633.
38. Kang S-M, Leblebici Y. CMOS Digital Integrated Circuits Analysis & Design: Analysis and Design. 3rd ed. McGraw-Hill Professional; 2002. p 115.
39. Ohring M. Reliability and Failure of Electronic Materials and Devices.Elsevier; 1998. p 6.
40. Sah C-T. Fundamentals of Solid-State Electronics: Study Guide. World Scientific; 1993. p 257.
41. Nguyen VH, Kranenburg HV, Woerlee PH. Copper for advanced interconnect. Proceedings of Third International Workshop on Materials Science; Hanoim;1999 November 2–4.
42. Wrschka P, Hernandez J, Oehrlein G. Chemical mechanical planarization of copper damascene structures. J Electrochem Soc 2000;147(2):706–712.
43. Zamorano M. Fabrica de Espadas y Armas Blanca "Damascene Technique in Metal Working," http://www.tf.uni-kiel.de/matwis/amat/def_en/kap_5/advanced/t5_1_1.html.
44. Available at http://www.realarmorofgod.com/damascus-sword-making.html.
45. Muraka SP, Verner IV, Gutmann RJ. Copper—Fundamental Mechanism for Microelectronic Applications. Wiley; 2000.
46. Oliver MR. Integration Issues with Cu CMP. CMP Users Group; 2003.
47. Li Y. CMP slurry developments.CMP for ULSI Multilevel Interconnection Short Course 2005; Fremont, CA; (2005 with courtesy from P. LeFevre).
48. Rhoades RL. Outsourced CMP Foundry Capabilities for Process-Level Development Through Full Production. CMP Users Group; 2004.
49. Schmitz JEJ. Chemical Vapor Deposition of Tungsten and Tungsten Silicides for VLSI/ULSI Applications. Noyes Publications; 1992. p 15.

50. Seo Y.-J, Lee W-S. Effect of oxidants for exact selectivity control of W- and Ti-CMP process. Microelectron Eng 2005;77:132–138.
51. Schlueter J. Trench warfare: CMP and shallow trench isolation. Semiconductor International; October 1999.
52. Available at http://www-03.ibm.com/chips/.
53. Chen P-L, Chen J-H, Tsai M-S, Dai B-T, Yeh C-F. Post-Cu CMP cleaning for colloidal silica abrasive removal. Microelectron Eng 2004;75:352–360.
54. Shen JJ, Costas WB, Cook LM. The effect of post chemical mechanical planarization buffing on defect density of tungsten and oxide wafers. J Electorchem Soc 1998;145(12):4240–4243.
55. Bartosh K, Peters D, Hughes M, Li Y, Cheemalapati K, Chowdhury R. Organic residue removal through novel surface preparation chemistries and processes for CMP and post CMP applications. Proceedings of 10th International VLSI/ULSI Multilevel Interconnection Conference; 2003. p 533–539.
56. Hegde S, Babu SV. Removal of shallow and deep scratches and pits from polished copper films. Electrochem Solid-State Lett 2003;6(10):G216–G219.
57. Ning XJ, Chien FR, Pirouz P, Yang JW, Asif Khan M. Growth defects in GaN films on sapphire: the probable origin of threading dislocations. J Mater Res 1996;11(3):580.
58. Lagerlof KPD, Grimes RW. The defect chemistry of sapphire (α-Al_2O_3). Acta Mater 1998;46(16):5689–5700.
59. Zhu H, Tessaroto LA, Sabia R, Greenhut VA, Smith M, Niesz DE. Chemical mechanical polishing (CMP) anisotropy in sapphire. Appl Surf Sci 2004;236(1–4): 120–130.
60. Wang Y, Zhang L, Zhou S, Xu J. Surface treatment effects of sapphire for high-quality III-nitride film growth. In Yao J-Q, Chen YJ, Lee S, editors. Semiconductor Lasers and Applications II. Proc SPIE 2005;5628:228–233.

2

CURRENT AND FUTURE CHALLENGES IN CMP MATERIALS

Mansour Moinpour

2.1 INTRODUCTION

Chemical–mechanical polishing (CMP) has emerged as the premier technique for achieving both local and global planarizations in silicon integrated circuit (Si IC) manufacturing. With the transition of Si IC fabrication industry to using sub-half-micron devices in the late 1990s, the CMP market size has grown rapidly, from about \$300–400 millions in 1997 to over \$2 billions in 2002 and is predicted to be over \$3 billions by 2008 [1]. Among the \$3 billions, roughly half belongs to equipment related to CMP, such as polishers and metrology tools. The other half is associated with materials such as slurries and pads (Fig. 2.1). Another recent market research study predicts the CMP slurry/pad market size to be around \$1.8 billions by 2009 presenting a CAGR of 16% and 17% for slurries and pads, respectively, driven primarily by the rise of copper CMP and transition to 300 mm wafer size (Fig. 2.2) [2]. Similar to other semiconductor sectors, the CMP community faces constant challenges in the identification, selection, characterization, and qualification of materials. They are vital to the success of implementing and sustaining the CMP processes in the ever-competitive global semiconductor manufacturing environment.

To put the topic in perspective, the market size of the materials related to semiconductor manufacturing is now over 10 billions USD [3]. Among these key materials, the sum of CMP slurry and pad is over 11%, which is approaching the combination of photoresist and stripping chemicals (Fig. 2.3). On the one hand, this is a strong indication that CMP technology has become a

Microelectronic Applications of Chemical Mechanical Planarization, Edited by Yuzhuo Li

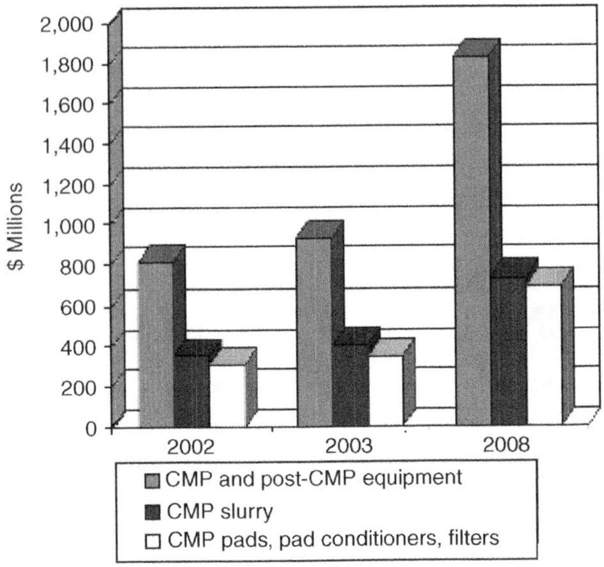

FIGURE 2.1 Worldwide markets for CMP and post-CMP equipment, CMP slurries, CMP pads, and other consumables, 2003–2008 ($ millions) (from Ref. 1).

key component in the semiconductor manufacturing process. On the other hand, there is a strong indication that the CMP community is carrying a significant share of the burden in keeping the overall semiconductor manufacturing process more cost effective. A key factor in this equation is

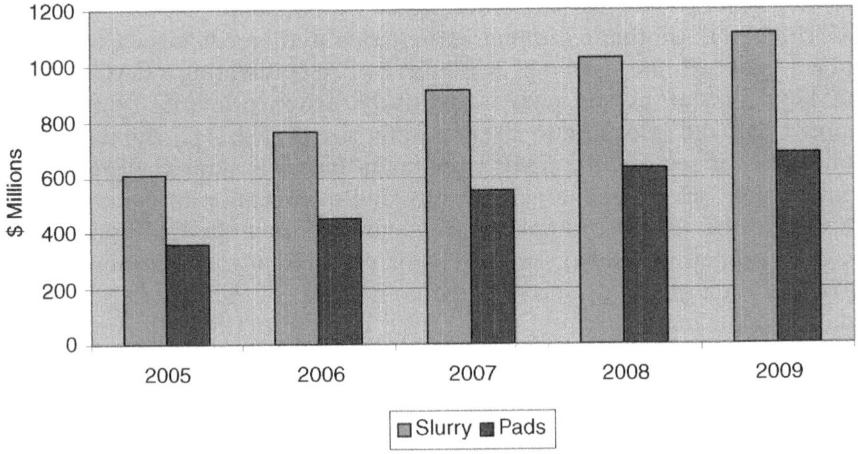

FIGURE 2.2 Worldwide markets for CMP slurries and pads, 2005–2009 ($ millions) (from Ref. 2).

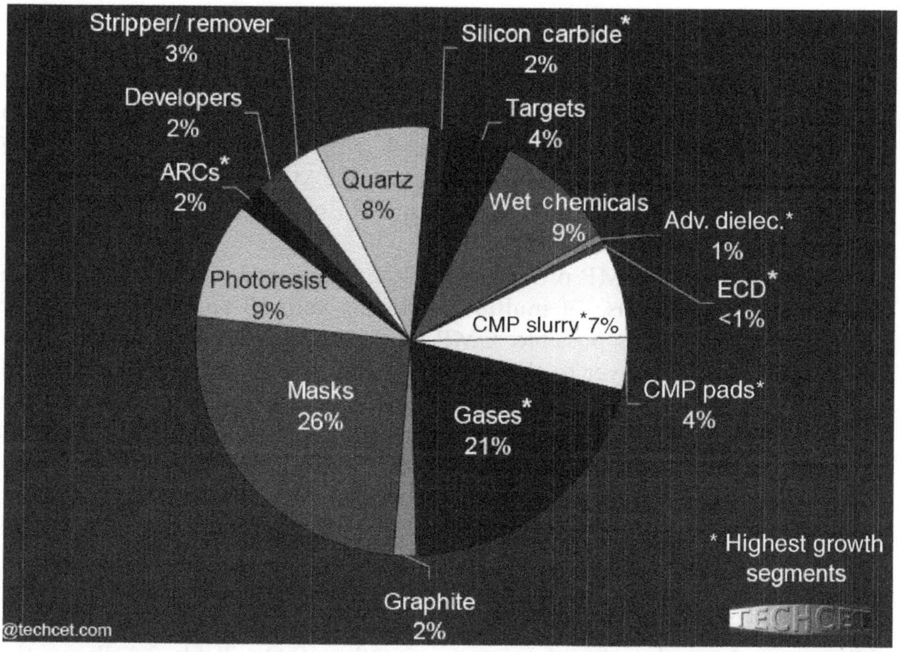

FIGURE 2.3 Key materials used in the IC manufacturing process (from Ref. 3).

the materials employed in the process today and that are to be used in the future. This chapter will provide an overview of the challenges associated with CMP-related materials throughout their development and implementation history. Some case studies will then be presented in which novel analytical techniques are used to characterize the CMP-related materials.

2.2 HISTORIC PERSPECTIVE AND FUTURE TRENDS

From a historical perspective, it was the oxide CMP as an introductory planarization technology that enabled the fabrication of logic and DRAM devices with feature sizes less than (or equal to) 0.8 μm [4]. Subsequently, CMP provides a technological advantage in front-end process modules such as shallow trench isolation [5] and polysilicon polish [6] as well as back-end-of-line (BEOL) processing, where CMP's ability to planarize, achieve high selectivity, and leave smooth surfaces provides a significant advantage over competing technologies. For logic devices with feature sizes <0.35 μm, the BEOL process consists of multiple interlayer dielectric (ILD) CMP steps integrated with subtractive Al etch and W plug technologies [7]. Advanced DRAM devices employ at least three layers of ILD polish [6]. W CMP became a technology enabler for <0.35-μm devices [8–10]. The legacy of W etch-back technology had large plug recesses and was susceptible to incoming W deposition defects. Tungsten (W) polish gave sub-100-nm plug recesses and actually removed the incoming W deposition defects. Subsequently, integrated circuits (IC) fabrication technology was converted to copper (Cu) dual damascene for BEOL processing in sub-0.18-μm technology nodes. ILD and W polishing steps have been replaced with Cu CMP [11–13], making Cu polishing the fastest growing segment of CMP industry. In addition, DRAM manufacturers are looking at using Al dual damascene for advanced sub-90-nm memory devices.

The importance of CMP process for BEOL processing is often described in terms of enabling the use of multiple vertically stacked layers of metal that keep the die as small as possible. CMP works to prevent the propagation of topography from one layer to the next layer and so on. The lack of undesired topography enabled the optical lithography tools with inherently decreasing depth of focus to accurately define multilayer interconnects. Figure 2.4 shows the historical trend for the number of metals as a function of logic technology generation. Multiple layers of metal interconnect reduce the average metal line length and thus decrease the signal delay time. The subsequent technological challenge for BEOL processing came from the successful introduction of low dielectric constant (low-k) ILD films in high-volume manufacturing for sub-0.1-μm technology nodes. These materials tend to have substantially different mechanical and surface properties than SiO_2. Table 2.1 shows the evolution of CMP process steps.

In a CMP process, a rotating polymer-based pad is pressed against the polished metal/oxide layer surface of a wafer while slurry or a combination of slurry and other chemicals is introduced into the polish platen. In a post-CMP cleaning process, residual abrasive particles and residues of polished layers are removed via flushing of the post-CMP solution and gentle mechanical

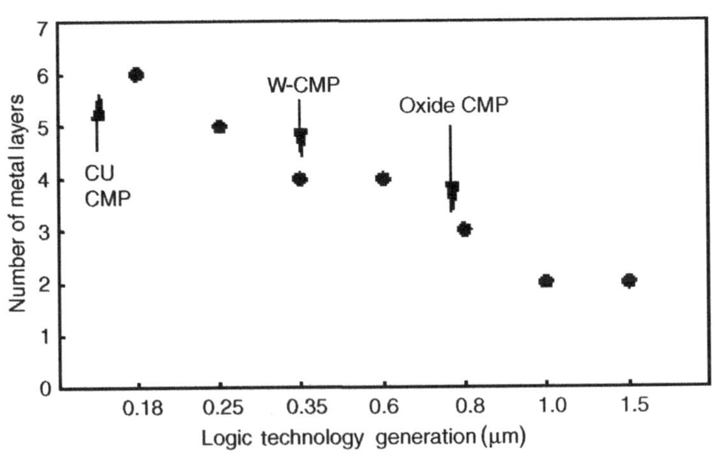

FIGURE 2.4 The impact of CMP on the number of metal layers as a function of logic technology generation.

TABLE 2.1 CMP Process and Equipment Evolution.

	Application	CMP Equipment	Post-CMP Clean
1st Generation 0.8–0.5 μm	Oxide (ILD)	Single platen/ single head One-step polish	Conventional wafer cleaning (wet stations) Wafer scrubbing/DI water
2nd Generation <0.5 μm	Above + ILD0 W CMP + STI	Multiplaten/ Multihead Two-step polish (Buff step) End-point detection On-board metrology	Wafer scrubbing/DI water, NH_4OH
3rd Generation <0.25 μm	Above + Cu, doped ILD (e.g., SiOF)	Integrated dry-in/dry-out Multiplaten/ multihead Nonrotary (e.g., orbital, linear CMP) Multistep polish End-point detection On-board metrology	Integrated dry-in/dry-out Wafer scrubbing/DIW, NH_4OH, HF New cleaning methods and new chemistries
4th Generation <0.1 μm	Above + low κ/ULK CMP and new applications (both FE and BE)	Integrated dry-in/dry-out Multiplaten/ Multihead Nonrotary (e.g., orbital, linear CMP) Multistep polish End-point detection On-board metrology	Integrated dry-in/dry-out Wafer scrubbing/DIW, NH_4OH, HF New cleaning methods and new chemistries

interaction between two pairs of brushes and silicon wafers being cleaned. Normally, CMP and post-CMP occur in the presence of a chemically active slurry and post-CMP solution/DIW. The removal of a polished layer and surface planarization is achieved due to the combination of various factors, such as consumable parameters, and CMP tool design, which jointly affect the polishing performance. During the introduction of each technology generation, a new set of materials entered the scene. The material's growth will continue to be strong for the next decade, supported by technology nodes of 90 nm and below. These materials include new slurries and pads, new post-CMP cleaning

TABLE 2.2 Interconnect Process Trends in 2006 [3].

Application	2006 DRAM 65 nm Node	2006 MPU 65-nm Node	2009 Projection for Both DRAM and MPU
Metal wire interconnect	PVD Al alloy subtractive etch %Cu CMP damascene	Cu dual damascene EPD or CVD, CMP patterned	Cu dual damascene EPD or CVD, CMP patterned
Plugs (Vias) local interconnect	W Plugs W DD interconnect CVD, CMP patterned	W plugs W DD interconnect CVD, CMP patterned	W plugs W DD interconnect CVD, CMP patterned
Barrier metals	Ti/TiN or Ta/TaN PVD CMP Patterned	Ta, TaN PVD with CMP Patterned	TaN, NB N, TiW, or Electroless CoWP, or?
Interconnect dielectrics	CVD dielectric $4.0 \geq k \geq 2.5$; most 2.8–3.0 CVD, CMP planarization RIE patterning	CVD dielectric $4.0 \geq k \geq 2.5$; most 2.8–3.0 CVD, CMP planarization RIE patterning	SiO_2 + C+? ($4.0 \geq k \geq 2.3$) CVD, CMP planarization RIE patterning
Device isolation	Shallow trench isolation CVD SiO_2 and CMP	Shallow trench isolation CVD SiO_2 and CMP	Shallow trench isolation CVD SiO_2 and CMP
Capacitors	PolySi or med-high *k* dielectric capacitors		PolySi or high *k* dielectric (Ru?) capacitors
Gate dielectric	SiO_2, Si_3N_4, SiO_xN_y		Higher *k* dielectric, HfO_x

chemistries, low-*k* and high-*k* dielectrics, barrier/liner materials, plating chemistries, spin-on polymers, as well as photoresists, strippers, and residue removers aimed at 90 nm, 65 nm, and 45 nm technology nodes and beyond. Table 2.2 lists some of the trends assembled by Holland [3].

In terms of dielectrics, there will be a continuous effort to employ materials with lower *k* constants in more and more applications. It has been seen that SiO_2, SiOF, and CVD OSG (organosilicate glass, also known as CDO—carbon doped oxide) dominate the 90-nm technology node. It is anticipated that the 65-nm technology node will utilize more CVD OSG. While the fabs continue to work to delay the use of porous low-*k* materials, the research and development of porous spin-on and CVD dielectric materials continue. Figure 2.5 shows that although spin-on low-*k* materials is gaining momentum, the field is still dominated by the CVD-based process [3].

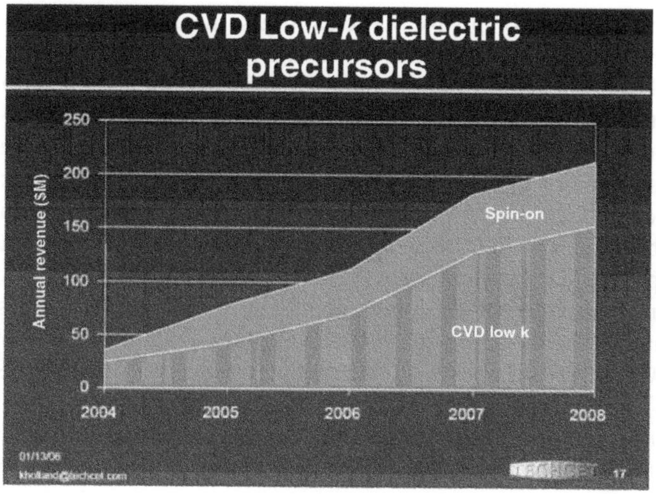

FIGURE 2.5 The rapid growth for both spin-on and CVD low-*k* dielectric materials (from Ref. 3).

For CMP consumables, according to one market report, as shown in Fig. 2.6, the slurry market is growing at a faster pace than that for pads, whereas a market study by Linx Consulting predicts almost equal growth rate for both pads and slurries (with pads actually having slightly faster growth rate, see Reference 2). By the end of this decade, the total market for these two interrelated technologies will be over $1.5 billions. It is important to point out that the materials related to the post-CMP cleaning process have seen even faster growth than those used in the CMP slurry and pad. This is a direct result of the higher demand for defect reduction at more advanced technology nodes.

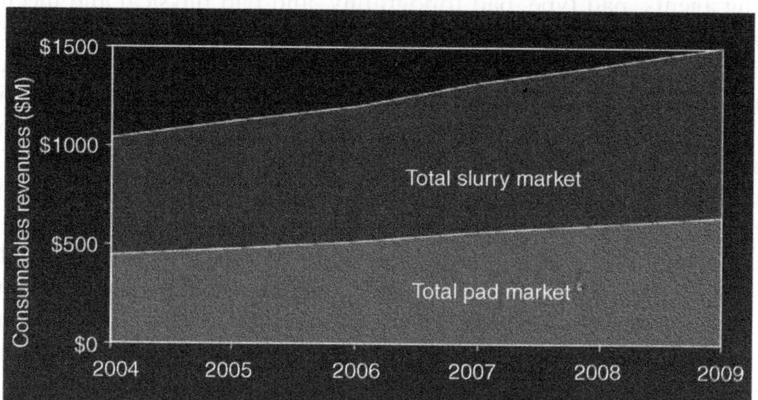

FIGURE 2.6 Growth of CMP slurry and pad markets (from Ref. 3).

In 2005, the total use of post-CMP cleaning solution was estimated at about 2.5 million liters that translates to roughly a $30 million market. It is anticipated that the post-CMP cleaning solution will soon become the third key component in CMP operation.

There are two other CMP materials (consumables) that are normally included as a part of CMP market. These are pad-conditioning disks (array of diamonds embedded and/or held in place in a metal matrix through the use of sintering, brazing, or CVD technologies) and brushes (primarily made out of PVA) for post-CMP cleaning steps. Pad conditioning helps in maintaining the desired surface texture of the pad during the CMP process; it also helps in slurry transport, hence achieving a uniform and constant film removal profile. The post-CMP brush assists in mechanical removal of particles and film residues off the wafer surface, and hence it is an integral part of the post-CMP cleaning process. Although both these materials are of paramount importance in the overall performance of CMP process, they are not discussed in detail here.

2.3 CMP MATERIAL CHARACTERIZATION

The current momentum in integrating CMP into existing and new processes, as well as the rate of introduction of better and faster integrated circuits into the marketplace, continues to exceed the fundamental understanding of the physical and chemical attributes of CMP materials. Optimization of CMP process depends on the optimization of the properties of CMP consumables. Thus, it is important to establish a basic understanding of CMP consumables in terms of physical, chemical, thermal, and mechanical properties and behaviors. CMP is a complex chemical and mechanical process that depends heavily on consumable parameters such as solution pH, abrasive type, particle charge and size, oxidizers, complexing agents, surfactants, corrosion inhibitors, buffering agents, pad type, pad topography, and pad physical and mechanical properties [13–16]. In addition, the CMP outcome is also significantly affected by the chemical interactions among slurry, pads, polished films, heat due to mechanical friction, and slurry flow distribution. Thus, the optimization of CMP process starts with the selection of CMP consumables based on their properties. Therefore, it is critical to establish a basic understanding of CMP consumables (pads, slurries, brushes, pad conditioners, etc.) in terms of their physical, chemical, thermomechanical, and rheological properties.

In this section, some case studies will be presented on the characterization of CMP pad and slurry [17–20] using such advanced analytical techniques as dynamic mechanical analysis (DMA), modulated differential scanning calorimetry (MDSC), thermal gravimetric analysis (TGA), thermal mechanical analysis (TMA), dynamic rheometry, dual emission laser induced fluorescence (DELIF), and the dynamic nuclear magnetic resonance (DNMR). More specifically, these techniques were used to characterize (a) the effect of heat treatment on thermal and mechanical properties of pads, (b) the impact of applied shear on slurry rheometry and particle size distribution, (c) the influence of slurry and DI water absorption on mechanical and thermal properties of pads, (d) the importance of surface adsorption of chemicals onto abrasive particles, (e) the consequence of pad grooving on mechanical properties and on slurry flow characteristics, and (f) the slurry film thickness, friction measurement, and real-time imaging of pad–wafer contact. The motivation in each case is to identify key material characteristics that can be utilized by consumable manufactures in process control and fine tuning the material performance. The suitability of these techniques for evaluating the dynamic behavior of consumables in CMP processes will be discussed. Furthermore, slurry stability with respect to defectivity is briefly discussed, and several characterization case studies are presented.

2.3.1 Thermal Effects

During CMP, a pad can be subjected to high temperature as a result of mechanical friction between the polymer-based pad and a silicon wafer in the solid–solid contact mode [21]. This heating effect is partially alleviated in the hydrodynamical contact mode due to the slurry flow. Pad heating caused by exothermic chemical reaction between slurry and polished metal has also been reported [22]. It has been found empirically that the slurry temperature increases by approximately 20–30 °C during CMP. It has also been demonstrated [23] that the temperature increase is also a function of the polishing time and wafer size (Fig. 2.7). It is important to realize that these data reflect the average temperature over the pad–wafer contact, whereas a purely elastic model that should be valid for the soft polymer-based pads [24] predicts that only about 1% of the pad surface is in contact with the wafer during CMP [25]. As such, the local pad temperature during CMP could be much higher, especially at the localized points of contacts between the wafer and the pad. Pad heating can substantially and irreversibly change the physical and mechanical properties of the pads and their chemical structure [26].

DMA, MDSC, TMA, and TGA tests have been conducted using the samples of a concentrically grooved polyurethane (PU) pad annealed at various temperatures. Annealing was done to simulate the pad heating during CMP due to the mechanical friction between a polymer-based pad and a silicon wafer, or exothermic chemical reactions between the slurry and polished metal layers. The effect of annealing at various temperatures and conditioning times was first studied using DMA. Relative decreases in storage modulus were measured at 25 and 50 °C (Fig. 2.8). As shown in Fig. 2.8, the glass transition temperature was assigned as the peak of G'' [27] and the macromolecular mobility was assigned as the height of the damping curve, tan δ [28]. In the temperature range from 30 to 50 °C, the storage modulus decreases by approximately 30%. Therefore, pad modulus and pad

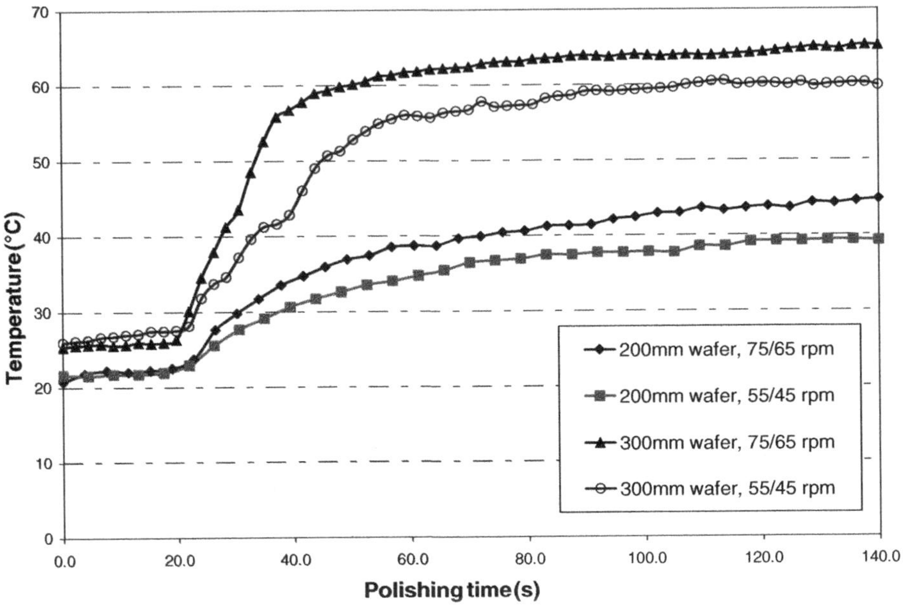

FIGURE 2.7 IC1000 pad surface temperature profiles during the polishing of 200-mm and 300-mm blanket oxide wafers using silica-based slurry under 6 psi downforce and 200 ml/min slurry flow rate and with two different table/carrier speeds (Strasbaugh n-Hance Polisher).

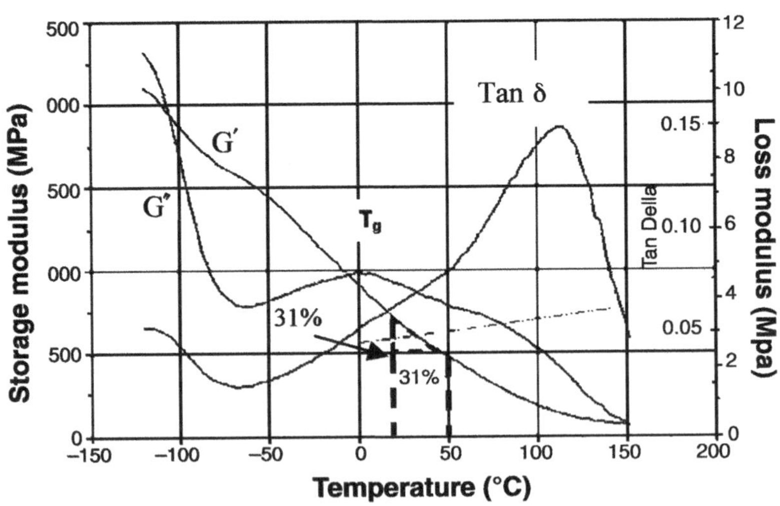

FIGURE 2.8 DMA scan for the pad conditioned at RT and tested at a frequency of 100 Hz. G' drops by 30 % in the temperature range of 30–50 °C.

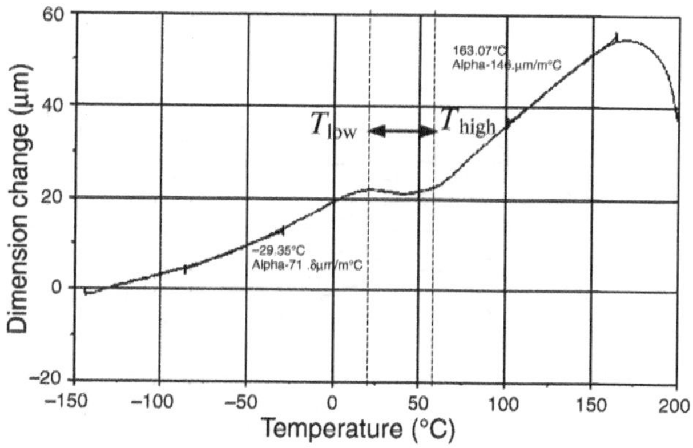

FIGURE 2.9 TMA scan for the pad conditioned at RT and tested using a penetration microprobe. Temperature dependence of CTE shows three different ranges:

1. below 25 °C, CTE = 70 µm/m*°C,
2. above 50 °C, CTE = 145 µm/m*°C,
3. between 25 and 50 °C, CTE = 0 µm/m*°C.

compressibility are likely to change during the CMP process causing instability of a CMP process.

In case of TMA tests, as shown in Fig. 2.9, coefficients of thermal expansion (CTE) were measured at temperatures below 25 °C, α_1, and above 50 °C, α_2. In order to avoid additional pressure exerted on the wafer due to thermal expansion of the pad during CMP, it is preferable to operate a pad in the temperature range within which the pad's CTE is equal to zero. Therefore, also shown in Fig. 2.9 were the low and high limit temperatures (T_{low} and T_{high}) and the temperature range ($T_{dif} = T_{high} - T_{low}$) within which CTE was equal to approximately 0 µm/m °C. The widest T_{dif} of 63 °C (from 11 to 74 °C) was observed for the pad thermally conditioned (annealed) at 110 °C The effect of various annealing times (1, 2, 4, 8, 24 h) on T_{dif} was also studied, and it was shown that the widest T_{dif} was observed for the pad conditioned for 8 h. The glass transition temperatures, measured at three different frequencies for both longitudinal and transverse specimens, also decreased as the conditioning temperature increased (Fig. 2.10). This is consistent with the observation done on T_{dif}.

2.3.2 Slurry Rheology Studies

CMP slurries—especially those used in Cu polish processes—are complex mixtures typically consisting of abrasives, oxidizers, corrosion inhibitors, buffers, and surfactants. Slurry properties are highly sensitive to the chemical composition, temperature, and shearing in the delivery line and/or during CMP process. Shearing can also lead to particle agglomeration [29] or, conversely, desegregation. Particle agglomeration may be responsible for the presence of microscratching in shallow trench isolation (STI) polishing [30].

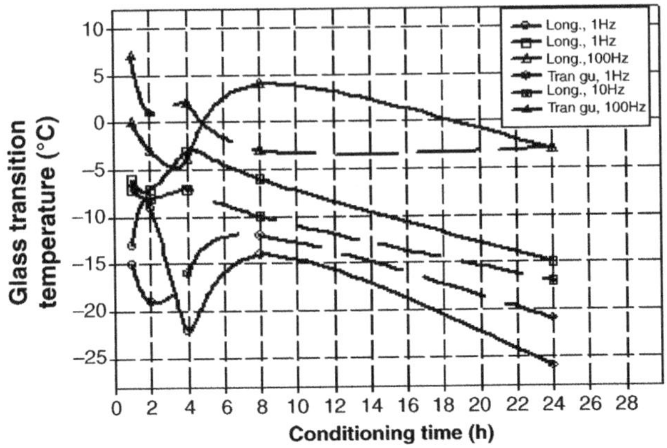

FIGURE 2.10 Effect of conditioning time on glass transition temperature, T_g, assigned as a temperature at the peak of the DMA loss modulus. T_g, measured at three different oscillating frequencies of 1, 10, and 100 Hz, is shown for the specimens with longitudinal and transverse groove orientations.

Detecting and characterizing subtle differences in slurry properties due to the above effects are an ongoing challenge.

Traditionally, the effect of slurry shearing is determined using viscometry tests. However, viscometry provides only one parameter, that is, viscosity, obtained at a fixed temperature and a fixed low shear rate in a steady-state mode. In order to characterize the effect of shearing on slurry properties, dynamic rheometry tests were conducted. In the steady-state measuring mode, dynamic rheometry measures viscosity in a wide range of shear rates and temperatures. In the frequency sweep mode, it determines dynamic shear storage and loss moduli, tangent of mechanical losses, complex viscosity, and phase angle. The time sweep mode allows to measure viscosity changes under certain shear over a period of time, while temperature/frequency sweep mode allows the prediction of the time- and temperature-dependent slurry properties using the time–temperature superposition principle. Therefore, dynamic rheometry appears to be a powerful technique for slurry characterization that provides an insight into both mechanical (such as dynamic shear moduli, viscosity, phase angle) and chemical (such as degree of the particle association) aspects of the slurry performance [31].

An example of the steady-state rheometric test at room temperature for various slurries of similar chemical compositions is shown in Fig. 2.11. Slurry viscosity decreases as shear rate increases; hence, the tested slurry is a non-Newtonian liquid and cannot be characterized by measuring viscosity at a single shear rate. Decrease in the slurry viscosity can also be ascribed to the deagglomeration of the slurry particles caused by shearing. A steady-state rheometric test distinguishes the slurries with only minor differences in their compositions, as shown in Fig. 2.11. One area of further opportunity in terms of

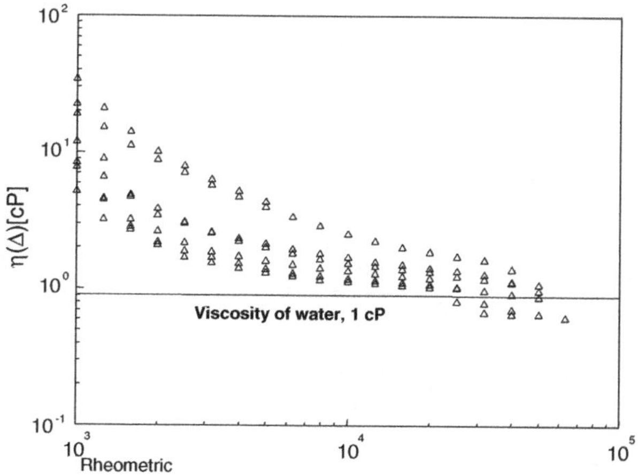

FIGURE 2.11 Rheometric steady-state test: dependence of slurry viscosity η on shear rate. The test was conducted for seven slurry samples with only subtle differences in slurry compositions and/or manufacturing processes.

studying rheological behavior of slurries is the measurement of high shear rate. Most of the reported data in the literature as well as what normally gets done by slurry manufacturers are standard viscosity measurements and/or steady-state rheometric tests under shear rates of 10^3–10^4 1/s. This range of shear rate (or even smaller values) could mimic gentle shearing during the slurry delivery to the polishers. Based on the measurement of slurry film thickness between the wafer and the pad—reported to be in the range of 10–40 microns [32,33]—and nominal velocities during polishing, shear rates of 10^5–10^7 1/s are not uncommon. Slurry shearing characterizations at such high shear rates are not regularly reported, so the key is to be able to test slurry at very high shear rates ($>10^5$ 1/s) to mimic the actual shearing conditions during the CMP process.

An example of a frequency sweep test of a slurry at two different temperatures, 5 and 30 °C, is shown in Fig. 2.12 [34]. As the slurry temperature decreases, the storage (G') and loss (G'') shear moduli and viscosity (η) increase. Transient changes in slurry shearing can cause intermittent and undesirable particle agglomeration leading to defects and microscratches. It is known that, for a silica-based slurry, particle agglomerates of above 1 μm are the major cause of the wafer defects. It was also shown that even particle agglomerates or "soft" oversized particles can adversely affect the polishing performance. Generally, particle agglomeration in slurry should be avoided since they interfere with slurry filtering and blending in slurry-delivery systems and cause large defects during polishing. A big challenge is to monitor and detect subtle variations in slurry properties including the rheological characteristics during slurry manufacturing and before introducing it to the polishing tool. It is shown, for example, that the viscosity and behavior of a slurry sheared in the slurry supply line can be different from the viscosity of "as received" slurry [34].

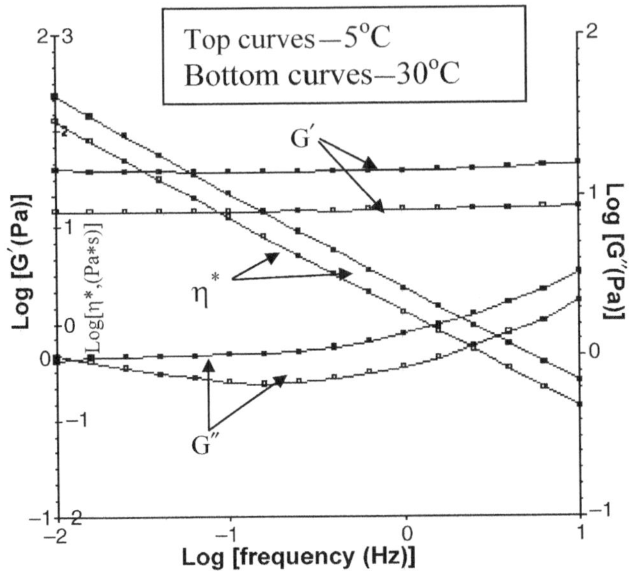

FIGURE 2.12 Rheometric frequency sweep test of slurries conducted at 5–30 °C. G', G'', and complex $\eta^* = \eta' + i\eta''$ reduce as temperature increases.

2.3.3 Slurry–Pad Interactions

The dependence of the performance of CMP process on the mechanical properties, shape, porosity, bending, and grooving of CMP pads has been widely reported [25,35–37]. However, the effect of interaction between the pad and slurry or rinsing water has not been extensively studied. Pad interaction with slurry and rinsing deionized water (DIW) during CMP can substantially and irreversibly change the physical and mechanical properties and the chemical structure of the porous pads [36]. The effect of pad and slurry interaction was studied using pads soaked in polishing slurry or DIW. In addition, pads were also soaked in buffered solutions at different pHs to simulate the different ranges of slurry acidity. The effect of soaking in various solutions on the thermal and mechanical properties of the pads was studied using DMA and MDSC [19]. Diffusion of aqueous media to the polyurethane pad was described using Fickian diffusion model [38]. The average weight gain of the pad specimens exposed to four different aqueous media followed the Fickian behavior [28], as shown in Fig. 2.13. Diffusion coefficients were then calculated [38,39]. The highest diffusivity D was found for slurry, followed by DIW, buffer solution at pH 4, that at and pH 11. The proximity of diffusivities for slurry and DIW can be explained by the large percentage of water in slurry composition. Higher diffusivity of slurry was probably due to the chemical interaction between the slurry components and the PU-based resin of the pad.

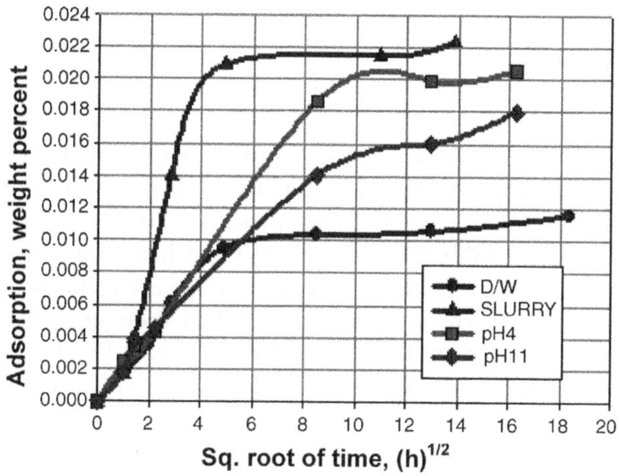

FIGURE 2.13 Weight gain of the pad specimens exposed to slurry, deionized water, pH 4 and pH 11 buffer solutions.

DMA tests were conducted with the specimens soaked in various environments for 0, 24, 72, 168, and 320 h. Pads exposed to slurry showed the lowest storage modulus G' (especially at the temperatures below 0 °C) and the highest chain mobility reflected by the highest peak of the damping curve. The effect of time for which pads were soaked in pH 4 buffer solution on the reduction of the pad's dynamic storage modulus (i.e., pad softening) is shown in Fig. 2.14. The pad softening due to the increase in the chain mobility can be

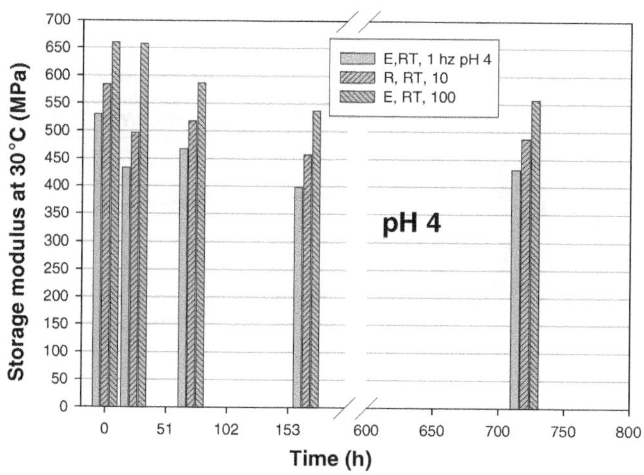

FIGURE 2.14 Effect of soaking time in pH 4 buffer solution on the reduction of the pad dynamic storage modulus (pad softening). DMA tests are conducted at 1, 10, and 100 Hz.

related to the resin's reversible plasticizing or to nonreversible chemical reactions due to the exposure to an aqueous media. The highest impact of the exposure to slurry on the pad softening can be explained by the highest rate of slurry diffusion to the pads.

The heat peaks of the nonreversing and reversing MDSC traces can be associated with nonreversible reactions such as a chemical reaction or cure and reversible reactions such as plasticizing processes in PU resin. Pad samples exposed to all tested media showed nonreversing heat peaks between 70 and 100 °C. Therefore, nonreversible chemical reactions are responsible for the pad softening. Endothermic irreversible heats reached their maximum value after approximately 180 h of exposure, as shown in Fig. 2.15. This suggests that chemical reactions that lead to pad softening are complete after approximately 180 h of exposure.

PU-based pads used in this study are what are normally known in the CMP industry as "hard" pads. Similar pad–slurry interactions are also observed with "soft" pads. Soft pads are normally made of porous multilayer polyurethane (PU) based polymeric material with an embossed surface. In the case of soft pads, the changes in the polymeric material could be more drastic. In addition, because of the multilayered structure of soft pads, the interpretation of DMA, TMA, and MDSC results are more challenging. It was shown that thermal and mechanical properties of soft pads are affected by soaking in slurry and water [40]. Absorption in the soft pad could not be described by the Fickian diffusion (Fig. 2.16), since it was dominated by the fast filling of the pad cavities with liquid (Fig. 2.17). Pad shrinkage due to heating and soaking in slurry and DIW was observed. Pad softening of

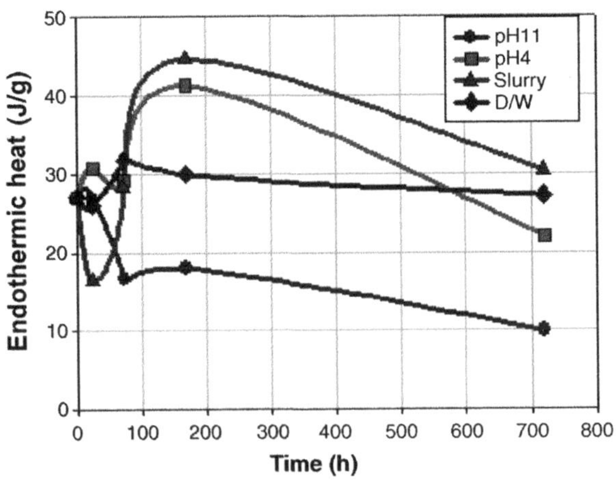

FIGURE 2.15 Time dependence of endothermic heat related to an irreversible chemical reaction. Endothermic heat of pads soaked in pH 4, pH 11 buffer solutions, slurries, and DIW was measured using MDSC.

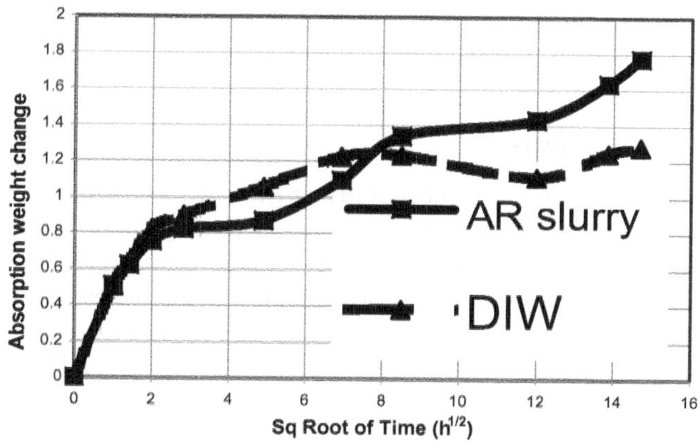

FIGURE 2.16 Absorption of DIW and slurry in soft pad.

approximately 50%, caused by pad heating, was observed within the typical operating temperature range of 30–70 °C. Simultaneous pad cross-linking and plasticizing due to soaking in DIW and slurry were assumed on the basis of the analysis of pad moduli, macromolecular mobility, and irreversible heat of exothermic reaction.

In general, not much has been published in the area of pad–slurry interactions. As the device geometries shrink, new materials are introduced and film stacks become more complex, so do the slurry formulations in various CMP steps in both front-end-of-the-line (FEOL) and back-end-of-the-line (BEOL) processes. Examples include, but are not limited to, new complex formulations for STI, multistack insulating layers (such as SiO_2, SiN, and SiCN films), and new barrier/liner materials. Consequently, there is even a higher need to characterize and understand the interaction of slurry chemistry with the polishing pads. Because of the complex and proprietary nature of slurries, using "model" chemical solutions is a better approach to systematically

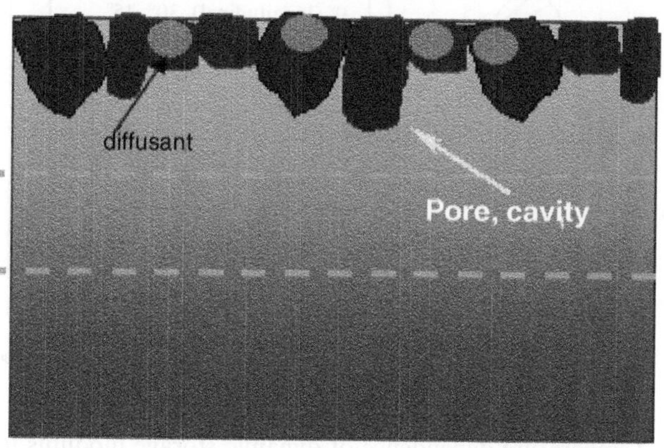

FIGURE 2.17 Schematic diagram of filling cavities in porous soft pads with water.

characterize changes in chemical, structural, and mechanical properties of pads as a result of to exposure to chemical environment [41].

2.3.4 Pad Groove Effects

Pad groove quality, depth, and design affect the pad performance and pad life [42] by influencing the slurry flow distribution during the CMP process [43]. The effect of the groove geometry on CMP performance has been evaluated using a model or full-scale polishing process [42]. We have evaluated the effect of the groove orientation on the pad's thermal and mechanical properties using dynamicl mechanical analysis. The rectangular specimens for DMA tests were cut from the circularly grooved polyurethane-based pads in such a way that grooves were oriented at various angles with respect to the long side of the specimen. The specimens were tested in a flexural deformation mode, at three various oscillating frequencies, within a temperature range of −120 to 180 °C. Pad parameters such as flexural dynamic storage and loss moduli (G' and G'') and pad-damping properties were monitored throughout the DMA test. The effect of groove orientation on storage modulus of the pad is shown in Fig. 2.18. Samples with longitudinal (0°) and 30° groove orientation, with respect to the long side of the rectangular specimen, showed the highest modulus at a temperature range from −120 to 75 °C. In this range, storage modulus decreased as orientation angle increased. This indicates that groove orientation impacts the mechanical properties of CMP pads. Different groove orientation results in different storage modulus and pad-damping properties. It is also reported that grooved pads exhibit different frictional heating effects as compared to flat pads [43], which could be another factor contributing to the transient nature of thermomechanical properties of pads during polishing.

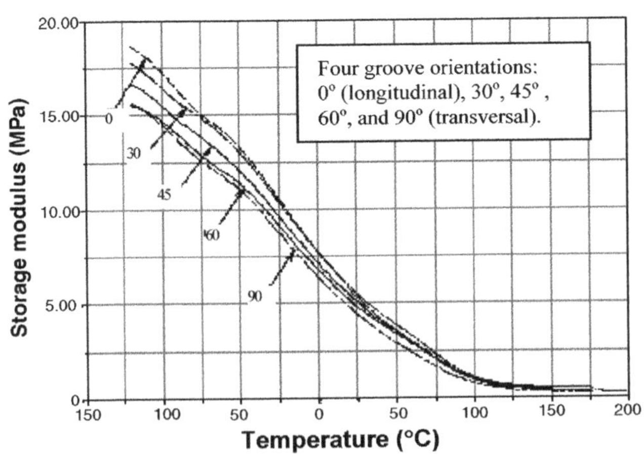

FIGURE 2.18 Temperature dependence of dynamic storage modulus for the samples with different groove orientation.

During actual CMP process, on a macroscale, these effects may cancel out each other. However, the transient, microscale effects on CMP pad properties and hence material removal are not yet clear and need to be further studied.

2.3.5 Pad–Wafer Contact and Slurry Transport: Dual Emission Laser Induced Fluorescence

The other aspect of rheological characterization of CMP process is a better understanding of slurry transport through pad asperities in the space between pad and wafer, pad–wafer contact mechanics, and the measurements of slurry coefficient of friction (COF). Extensive studies have been carried out in these areas [44–48]. DELIF technique was developed to study slurry transport between the polishing pad and the wafer. DELIF is an optical technique that allows the measurement of micronscale pad–wafer gap widths to be observed during the polishing process [49–51]. The technique works on the basis of the difference in the amount of light reflected by the dyed pad and/or slurry and correlating it to scalar parameters such as pH and film thickness. In case of slurry film thickness between pad and wafer, it is able to measure a relative difference with a precision of more than a micron. Because of measurement issues, however, the absolute distance is accurate only to within 5 μm. Friction measurements between the pad and the wafer have also been made during the polishing in addition to DELIF measurements [48]. The DELIF technique is used to instantaneously capture the slurry film thickness during CMP. The fluorescence ratio is correlated with a slurry film thickness by constructing a film of known thickness. One can then correlate slurry film thickness with friction measurements, both measured *in situ* using DELIF. One such study is shown in Fig. 2.19. The images collected during DELIF are averaged over the total number of pixels. Each DELIF image correlates with a specific point of the friction spectrum, and the average ratio can be plotted versus the instantaneous friction measurement. Preliminary results comparing film thickness and instantaneous friction measurements show no correlation at standard CMP-operating conditions. There is evidence of a correlation between friction

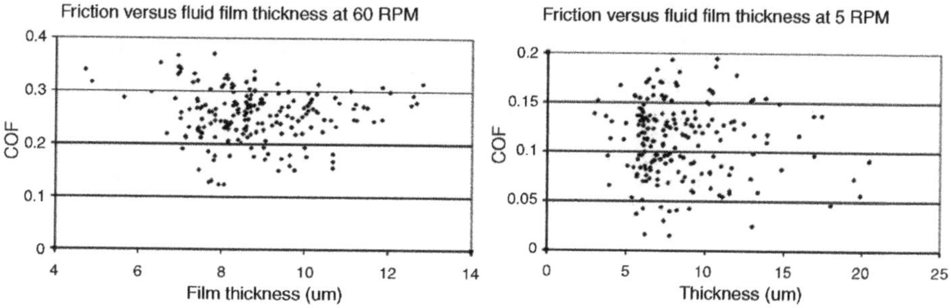

FIGURE 2.19 The coefficient of friction at (**a**) 60 RPM and (**b**) 5 RPM, measured using DELIF technique, show no correlation with instantaneous fluid film thickness.

and slurry film thickness at low pad and wafer rotation speed, but more experimentation is required to determine if that relationship is repeatable.

Modeling work indicates that there is a delicate balance between hydrodynamic lubrication, mixed solid–liquid contact, and direct solid–solid contact [52]. Modeling efforts have been employed to predict the film thickness and removal rate based on the contact regime. Higher removal rates can be obtained with more solid–solid contact, but with a higher incidence of wafer scratching. Changing experimental parameters to increase the film thickness will result in less scratching of the wafer and also a lower removal rate [53]. We have employed the DELIF technique to measure slurry film thickness between pad asperities and wafer surface [54]. Figure 2.20 is a typical DELIF image. Typical CMP pad has a textured surface consisting of many peaks and valleys. These asperities are protrusions that aid in the distribution of slurry and removal of waste. These asperities range in size, with 10–15 μm being a common height. They range in diameter too, many being around 40 μm in diameter. We believe that the asperity behavior will indicate when and to what degree the wafer–pad contact is occurring. The pad's profile is essentially a Gaussian distribution of points. There are outliers on either end, such as very tall peaks or very deep holes in the pad. The bulk of the asperities lie clustered around some mean pad height. The dark areas of the image reflect a lack of

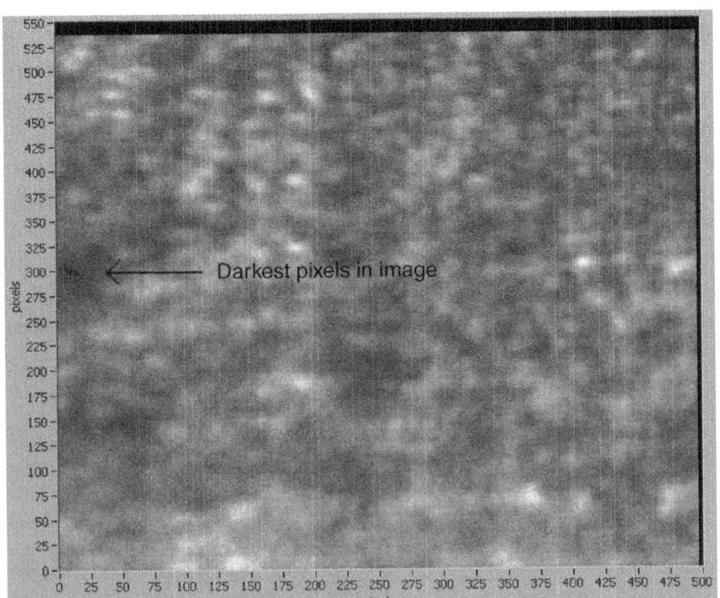

FIGURE 2.20 DELIF static image taken at 10 psi (from Ref. 54). The darkest pixels in the image represent the thinnest fluid layer, and therefore the contact region. The high pressure causes so much compression of the pad that the image gets slightly out of focus.

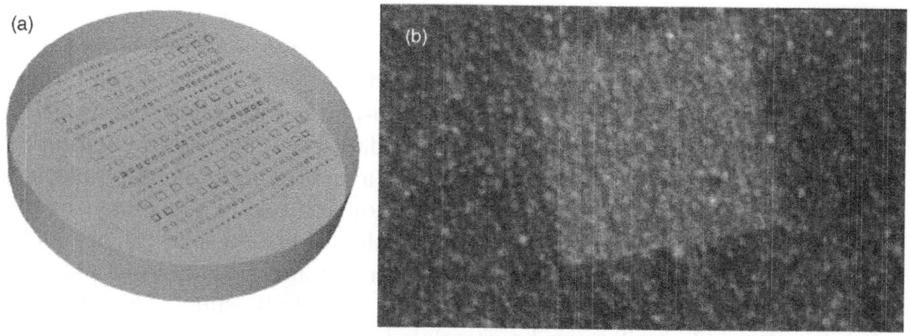

FIGURE 2.21 (**a**) Wafer-etch geometry. (**b**) An image of the slurry layer between a patterned wafer and a Fruedenburg FX9 polishing pad.

fluid, meaning those are the high peaks that reach nearly up to the wafer. The bright regions are deeper holes in the pad where a thick layer of fluid sits. In addition to studying friction beneath a flat surface, DELIF can be used to study CMP of a patterned surface; an example is shown in Fig. 2.21 [55]. Insight into the evolution of pad–wafer contact during CMP process allows better and more efficient design of pad and pad conditioner, improved slurry transport, and better understanding of interactions between pad, slurry, and wafer under various processing conditions.

2.3.6 Dynamic Nuclear Magnetic Resonance

With increasing complexity of slurry formulation and incorporation of more chemical additives (oxidizers, rate and corrosion inhibitors, surfactants, etc.) into slurry, there exists a greater need to understand how these additives interact with abrasive particles. For example, we need to examine if different chemical additives in a slurry adsorbs particles or not and if so, would there be a change in the effective concentration of these additives. Besides, there might be side reactions before or during the CMP process between these additives, impacting the slurry performance. For example, packed column technique has been used to study interaction of Cu and Ta surfaces with slurry chemistry using single abrasive systems as well as mixed abrasive systems [56,57]. The technique, combined with zeta potential and static etch rate measurement and wafer polishing experiments, has shed light on the correlation between slurry additives and polishing performance metrics such as film removal rate and planarization efficiency. Dynamic NMR has become an increasingly important and visible technique in characterizing colloidal dispersions [58–61]. For example, using dynamic NMR technique, rotational and translational motions of molecules can be easily measured as relaxation times or diffusion coefficients. In a colloidal dispersion, these molecules as chemical additives may be in equilibrium between free and adsorbed states. DNMR measurements can easily reveal any subtle change in such dynamic behavior such as enhanced adsorption or desorption. The two most often measured relaxation parameters are spin–lattice (T_1) and spin–spin relaxation (T_2) times. When dealing with a slurry system, a two-phase fast exchange model is commonly used in analyzing the relaxation data for a species in equilibrium between adsorbed and dissolved states. Such a species could be the molecules that make up the continuous phase or a minor additive to the slurry. In this model, there are only two distinctive molecular layers—a layer of molecules that are strongly adsorbed on the particle surface and a continuous phase of molecules that are essentially the same as their bulk state. The observed relaxation time (T_1 or T_2) can be correlated with the relaxation times of the surface adsorbed layer and bulk phase:

$$1/T_{\text{obs}} = X_s/T_s + X_b/T_b \quad (2.1)$$

where T_{obs} is the observed relaxation time, T_s is the relaxation time for molecules adsorbed on surface, T_b is the bulk phase relaxation time, and X_s and X_b are the molar fractions [62].

It has been demonstrated that the model works well with a wide range of heterogeneous environments including pores, channels, surfaces, and gels and a collection of hosting materials such as silica, alumina, titania, and clays [63–66]. As the exact molar fraction and relaxation time for the surface-adsorbed molecules are difficult to determine, the experimental results are often used as a relative comparison among samples with similar compositions. For example, relative surface area or particle size can be estimated for two slurry samples with a similar T_s value based on their observed relaxation times [67].

As the relaxation time for the surface-adsorbed molecules is mainly determined by the molecular interaction between the adsorbed molecules and the surface of interest, a comparison among samples with different surface properties may yield information on these at a molecular level. For example, when two silica slurry samples are investigated, a sample with a rich content of hydroxyl group should possess a smaller T_1 or T_2 corrected with particle size and surface area differences.

The polishing of tantalum film is an important process in CMP as tantalum is a popular choice as a barrier material in damascene interconnect scheme. It is generally accepted that the removal rate for tantalum is directly related to the availability of the surface hydroxyl groups on the abrasive particles. There is a good correlation between the total surface area of the silicas and the total hydroxyl content (determined by $LiAlH_4$ method) [68]. A comparison between Aerosil™ (Degussa Corp.) 50 and 130 shows that the material removal rate (MRR) of tantalum is directly related to the increase in surface area and total hydroxyl content. However, the MRR did not increase much when a silica with even higher surface area and greater hydroxyl content (Aerosil 200) was used. This phenomenon is consistent with the fact that the T_1 slope for Aerosil 200 is almost the same as that for Aerosil 130 (Fig. 2.22). As the T_1 slope is directly

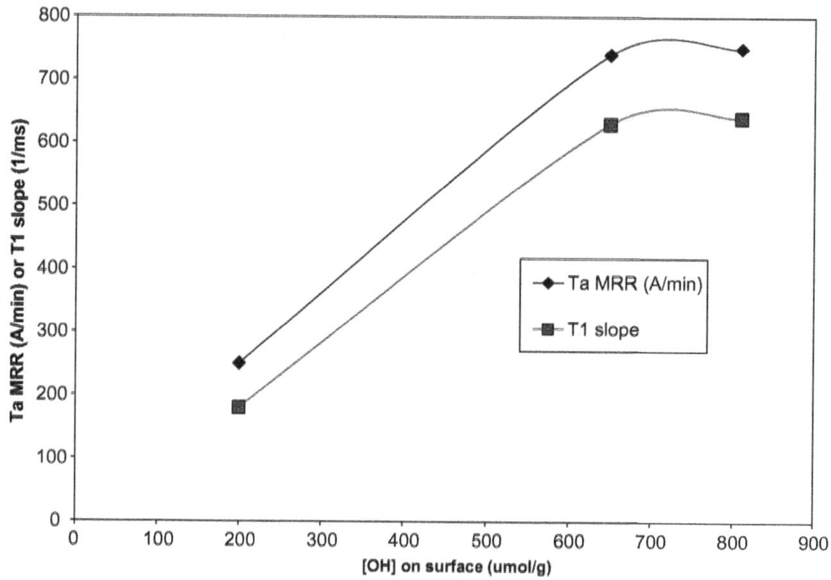

FIGURE 2.22 Correlation among hydroxyl content on surface, material removal rates for T_a, and T_1 relaxation slopes measured using DMNR technique.

correlated with the relative amount of bound water molecules that are in equilibrium with the bulk water, the hydroxyl groups that hold water molecules too tight to allow their exchange with the bulk phase will not be included in such a dynamic NMR measurement. It is very likely that those hydroxyl groups will not participate in the interaction with tantalum surface. Therefore, the dynamic NMR measurement of water T_1 could serve as a good indicator for the silica–tantalum interaction during CMP.

Pulsed field gradient NMR (PFG-NMR) is a powerful, nondestructive technique of measuring self-diffusion coefficients in a colloidal dispersion [69–71]. Molecules associated with an aggregate or a particle will diffuse more slowly than their free dissolving state. More specifically, when a water-soluble species is partially adsorbed onto an abrasive particle, the measured overall diffusion coefficient (D^{app}) of the species is decreased. If the diffusion coefficient of the free dissolving species (D^{free}) can be measured in the absence of abrasive particles, the partition coefficient or surface adsorption tendency can be calculated based on simple equations (Eqs. 2.2 and 2.3):

$$D_{\mathrm{A}}^{\mathrm{app}} = p.D_{\mathrm{A}}^{\mathrm{micelle}} + (1-p)D_{\mathrm{A}}^{\mathrm{free}} \tag{2.2}$$

or

$$p = (D_{\mathrm{A}}^{\mathrm{free}} - D_{\mathrm{A}}^{\mathrm{app}})/(D_{\mathrm{A}}^{\mathrm{free}} - D_{\mathrm{A}}^{\mathrm{micelle}}) \tag{2.3}$$

where p represents the fraction of molecules A that are associated with a particle such as a micelle ($0 \leq p \leq 1$). A CMP slurry often contains surfactant molecules at a concentration above its critical micelle concentration (CMC). The surfactant molecules can aggregate and form micelles with hydrophobic "pockets" that can encapsulate other relatively hydrophobic molecules in the slurry. For example, benzotriazole (BTA) is relatively hydrophobic and has a water solubility less than 2%. In a CMP slurry, it is likely that some of the BTA molecules will be partitioned into the micelles. Therefore, the effective concentration of this corrosion inhibitor may change as the concentration of surfactant changes. Measuring the actual partition coefficient in such complex slurry is a difficult task. Dynamic NMR technique provides a solution to this problem. As shown in Fig. 2.23, the partition coefficient of BTA is a function of surfactant concentration and the presence of abrasive particles and complexing ions such as copper. The practical implication is that the effective concentration of this hydrophobic corrosion inhibitor is not static. It could change during polishing as the effective concentration of surfactant may change. Furthermore, during copper CMP, with the introduction of copper ions, the effective concentration of BTA may further decrease due to the complexation with copper ions. It is important to point out that, in Fig. 2.23, the copper–BTA is combined with free BTA as they have similar diffusion

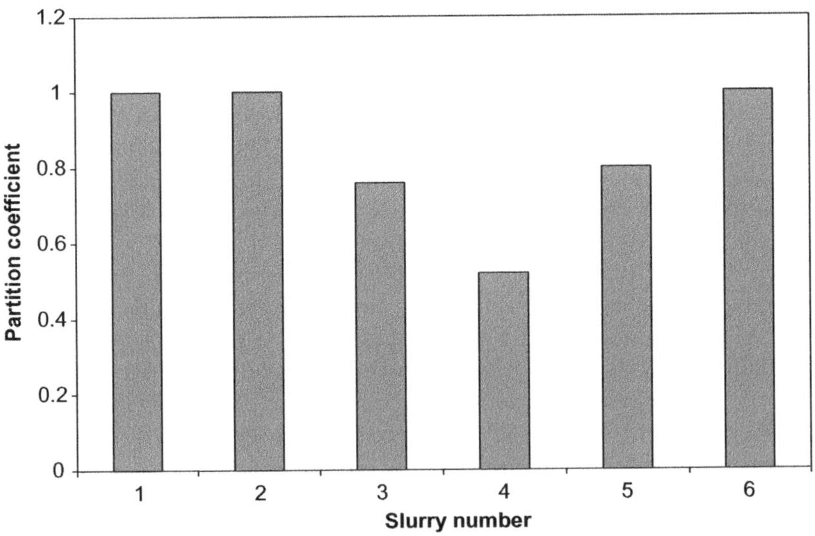

FIGURE 2.23 Dependence of partition coefficient on the concentration of BTA in an aqueous environment, where 1 = 100% free dissolving in water and 0 = 100% encapsulated. Sample 1 contains 1 mM of BTA and no surfactant. Sample 2 contains 1 mM of BTA and 5 mM (below cmc) of sodium dodecyl sulfate (SDS). Sample 3 contains 1 mM of BTA and 15 mM of SDS. Sample 4 contains 1 mM of BTA and 30 mM of SDS. Addition of 100 ppm of copper (II) nitrate to sample 4 results in sample 5. Addition of 300 ppm of copper ions to sample 5 results in sample 6 (from Ref. [72]).

coefficients. Therefore, in this case, a higher partition coefficient does not directly translate to a higher free BTA concentration [72]. HPLC analysis of a silica-based slurry containing BTA and H_2O_2, conducted in our laboratory, has also shown that effective BTA concentration in the slurry does not remain constant over time [73].

2.3.7 CMP Slurry Stability and Correlation with Defectivity

One of the major drawbacks of CMP is the tendency of abrasive particles in slurries to form aggregates, which have the potential to cause defects on wafer surfaces. Therefore, it is crucial to understand the mechanisms by which aggregates are formed so that appropriate metrology can be used to identify defect-causing slurries before they are used in the fab. Single particle optical sensing (SPOS) techniques are commonly used to obtain large particle counts (LPC) for slurries prior to their use in a fab. Other techniques that can be used to characterize slurries are static light scattering, dynamic light scattering, and zeta potential measurements. All of these techniques usually require that the slurry be diluted prior to measuring. However, diluting with water changes the ionic strength and the pH of solution, and both properties have been shown to affect aggregation and electric double layer characteristics of particles [68]. We have shown that both known defect-causing silica-based slurries and defect-free slurries demonstrate similar zeta potential, mean particle sizes, and LPC using standard water dilutions [74]. To quantify the effects of water dilution on zeta potential and mean particle size, an alternative diluting solution that simulates the ionic strength and pH of the original slurry was evaluated. While the effects on mean particle size are slight, the alternative diluting solution demonstrates an average increase of 30 mV in zeta potential as compared to the water-diluted slurries. In addition, it has been shown that relatively low concentrations of electrolytes can induce and propagate particle aggregation. The effects of doping silica-based slurries with aluminum added as a salt, an oxide, and a hydroxide were also quantified [74]. The results indicate that rapid aggregation takes place when silica-based slurries are doped with 50 ppm Al added as aluminum chloride, as verified by SPOS. Aluminum added as either oxide or hydroxide to slurries demonstrates no measurable particle aggregation using SPOS.

A critical physical property of CMP slurries that affords to handle the performance optimization is the size distribution of the slurry's abrasive particles. The region of the particle size distribution with diameters greater than 0.5 μm has been of particular interest. Analyses of the cumulative number of particles with polystyrene-equivalent, light scattering intensity diameters $\geq$0.56 μm, referred to as the LPC, are routinely performed via SPOS, and the LPC is often used as the primary particle size distribution metric in correlation with defect metrology. Although the LPC represents a convenient metric for relating the size distribution characteristics of abrasive particles with defect creation, this parameter provides no direct morphological analysis of defect-creating particles. Detected particles are binned into channels of specific size during a SPOS analysis on the basis of their silica or polystyrene sphere-equivalent, light scattering intensity. Consequently, particles of different *shape* cannot be resolved or characterized according to these heterogeneities in a conventional SPOS analysis. A new metrology for CMP slurries and the development of analytical instrumentation capable of yielding information about heterogeneities in particle shape are currently underway [75]. The field-flow fractionation (FFF) separation technique offers a means to resolve complex particle-size distributions into narrowed hydrodynamic size fractions prior to undergoing SPOS analysis. Two different types of particle diameters are measured by these two techniques. The FFF elution times are directly correlated with hydrodynamic diameters d_h through existing theoretical equations or by establishing calibration curves. Hence, the FFF elution profile is a direct reflection of d_h present in the slurry mixture. The SPOS detection system, on the contrary, provides the number distribution of light scattering, spherical-equivalent diameters $N(d_l)$. Simultaneous measurement of d_h and $N(d_l)$ across a slurry's elution profile creates the opportunity to determine the ratio of measured diameters, $r = d_h/d_l$, for every elution slice. The value of r is indicative of heterogeneity in particle shape in an elution slice. So far, the basic concept and feasibility has been demonstrated using model particle systems. Method development and application to real CMP abrasive systems will be carried out in near future.

Stability of CMP slurries has implications in LPCs and particle size distribution. LPCs are known to have a negative impact on scratch defects, and a change in the abrasive size can influence removal rates; therefore, slurry stability is important for both defectivity performance as well as for slurry shelf life. Although the stability of colloidal suspensions has been investigated for decades via the stability ratio, its application to CMP slurries has been limited. We explored simultaneous turbidity and low-angle light scattering to quantify the stability ratio under well-defined mixing conditions [76]. A commercially available particle-sizing instrument was used to follow blue and red laser light turbidity and intensity of low angle scattered light during early-stage homoaggregation of an electrostatically stabilized polystyrene latex. Rate constants for doublet formation were calculated and presented in terms of the stability ratio. Stability is systematically decreased by reducing the electrostatic barrier through traditional means of increasing the background electrolyte concentration. Stability ratios were found to show differences in absolute value using red and blue light turbidity and low angle scattered light methods. Better agreement at low salt concentrations was found by normalizing each data set with stability ratios measured at high salt concentration. Critical coagulation concentrations from turbidity and low-angle light scattering were found to be in reasonable agreement. The methodology is demonstrated for a commercially available ceria-based CMP slurry and extended to the measurement of ceria heterostability as a function of silica particle size and concentration to gain insight into ceria interaction with oxide and the propensity of such systems to form large agglomerates [76].

We have also conducted adhesion measurements between real CMP abrasive particles (not "model" particles) and various surfaces as well as particle hardness and elastic modulus measurements, using colloidal AFM and nanoindentation AFM, respectively, in an attempt to correlate CMP defectivity with mechanical and adhesion properties of CMP abrasives [77,78]. For example, in a carefully designed experiment, we have been able to demonstrate that softer particles, indeed, result in fewer scratches [78].

2.4 CONCLUSIONS

In this chapter, an overview of the challenges in the identification, selection, characterization, and qualification of materials for CMP-related semiconductor manufacturing has been provided. Some case studies using novel analytical techniques have also been presented. It is clear that, as the CMP process is complex and dynamic, analytical techniques that can provide only static properties of the materials are not adequate. They will not be able to distinguish consumables that have different dynamic properties that have greater impact on the CMP performance during polishing. Therefore, the techniques used in these case studies that can reveal dynamic aspects of the physical and mechanical properties are more relevant to the evaluation and qualification of CMP materials. With the increasing demand for new materials or materials with tighter specifications to reduce defects for future technology nodes, there will be an increase in the use of novel analytical techniques that can distinguish subtle dynamic properties.

ACKNOWLEDGMENT

The author wish to acknowledge Dr. Alex Tregub, Dr. Anthony Kim, Dr. Jam Soroshian, Daniel Apone, Darren DeNardis, Caprice Gray, Grace Ng, Yasa Sampurno, and James Vlahakis for their help in conducting the experiments and in the preparation of this manuscript. The author would also like to thank Mr. Yongqing Lan and Dr. Changxue Wang for their invaluable assistance in editing this chapter.

QUESTIONS

1. As measurements on the dynamic properties of slurries and pads are important, can you list all analytical techniques you know or are familiar with that can be implemented in a CMP system?
2. Referring to question 1, what would be the advantages and limitations of your techniques, considering the possible interference from mechanical movements and chemical reactions?

3. What are the major challenges of introducing or implementing new consumable materials into an existing CMP process?
4. What are the general guidelines for selecting consumable materials for a new CMP process?

REFERENCES

1. Chemical mechanical polishing equipment and materials: a technical and market analysis. http://www.bccresearch.com/avm/AVM047A.asp, December 2003.
2. CMP technologies and markets to the 45 nm node: a technical and market analysis, linx consulting.http://www.linx-consulting.com, 2005.
3. Holland K. Critical materials: enabling the future of integrated circuits. http://www.techcet.com/SMCTechcet2006.pdf, January 13, 2006.
4. Renteln P, Thomas M, Pierce J. VMIC Proceedings; 1991. p 57–60.
5. Davari B, Koburger CW, Schulz R, Warnock JD, Furukawa T, Jost M, Taur Y, Schwittek WG, DeBrosse JK, Kerbaugh ML, Mauer JL. IEDM Technical Digest; 1989. p 61–64.
6. Koh GH, Ha DW, Cho CH, Jeong HS, Jeong GT, Yang WS, Lee KH, Lee JG, Park BJ, Lee JK, Bae JS, Sim JH, Kim KN. CMP-MIC Proceedings; 1998. p15–19.
7. Barla K, Gounelle C, Lair C, Lafarges Y, Lasserre V, Lis S, Maddalon C, Verove C, Lous E, Morand Y, Passemard G, Pires F, Demolliens O. VMIC Proceedings; 1998. p 25–30.
8. Rutten M, Huynh C. SRC Topical Conference on CMP; July1995.
9. Kaufman F et al. Chemical–mechanical polishing for fabricating patterned W metal features as chip interconnects. J Electrochem Soc 1991;138 (11):3460– 3465.
10. Sivaram S, Bath H, Legget R, Maury A, Monnig K, Tolles R. Solid State Technology 1992. p 87–91.
11. Edelstein D, Heidenreich J, Goldblatt R, Cote W, Uzoh C, Lustig N, Roper P, McDevitt T, Motsiff W, Simon A, Dukovic J, Wachnik R, Rathore H, Schulz R, Su L, Luce S, Slattery J. Tech Digest IEEE Intern Electron Devices Mtg 1997; 773–776.
12. Goldblatt RD, Agarwala B, Anand MB, Barth EP, Biery GA, Chen ZG, Cohen S, Connolly JB, Cowley A, Dalton T, Das SK, Davis CR, Deutsch A, DeWan C, Edelstein DC, Emmi PA, Faltermeier CG, Fitzsimmons JA, Hedrick J, Heidenreich JE, Hu CK, Hummel JP, Jones P, Kaltalioglu E, Kastenmeier BE, Krishnan M, Landers WF, Liniger E, Liu J, Lustig NE, Malhotra S, Manger DK, McGahay V, Mih R, Nye HA, Purushothaman S, Rathore HA, Seo SC, Shaw TM, Simon AH, Spooner TA, Stetter M, Wachnik RA, Ryan JG. Proc. IEEE, 2000; p 261–263.
13. Tsai TC, Leung P, Lin KC, Naujok M, Clevenger LA, Chan HC. CMP–MIC Proceedings; 2002. p 19–24.
14. Philipossian A, Moinpour M, Oehler A. CMP-MIC Proceedings; 1996. p 13–19.

15. Ouma Dennis O, et al. SPIE Microelectronics Conference, Microelectronic Device Technology Session, Austin, TX, October 1997, Conference Proceedings; 1997. p 1–12.
16. Steigerwald JM, Murarka SP, Gutmann RJ. Chemical Mechanical Planarization of Microelectronics Materials; New York: John Wiley & Sons, Inc.,1997.
17. Tregub A, Moinpour M, Sorooshian J. Proceedings of CMPUG Annual Symposium; Oct 11,2001; Santa Clara, CA: 2001.
18. Tregub A, Moinpour M, Sorooshian J. Proceedings of 18th VMIC Nov 2001; Santa Clara, CA: 2001. p 275–280.
19. Tregub A, Moinpour M. AVS 3rd International Conference on Microelectronics and Interfaces; Feb 11–14; Santa Clara, CA: 2002. p 72–74.
20. Tregub A, Moinpour M, Sorooshian J. I5.4, 2002 Spring Meeting Proceedings, Symposium I, Chemical–Mechanical Planarization; Babu SV, Singh R, Hayasaka N, Oliver M, editors. MRS Proceedings Volume 732E, (published; presented at MRS 2002, May, San Francisco).
21. Luo J, Dornfeld DA. IEEE Transa Semicond Manufacturing 2001;14 (2):112–133.
22. Wang Y-L, Liu C, Feng M-S, Tseng W-T. The exothermic reaction and temperature measurement for tungsten CMP technology and its application on endpoint detection. Mater Chem Phys 1998; 52:17–22.
23. Li Y, Cheemalapati K, Wang C, Burkhard C, Jun W, Kodaka I, Atsushi H, Toshihiro K, Hozumi K. VMIC; Fremont, CA: Sept 26–28, 2006. p 444–450.
24. Shi FG, Zhao B. Appl Phys 1998; A67:249–252.
25. Lawing AS. Poster Presentation; Semicon West, San Francisco: July 2001.
26. Li W, Shin DW, Tomozawa M, Muraka SP. The effect of the polishing pad treatments on the chemical–mechanical polishing of SiO_2 films. Thin Solid Films 1995; 270:601–606.
27. Nielsen LE, Landel RF. Mechanical Properties of Polymers and Composites. 2nd ed. new York: Marcel Dekker, Inc.; 1991, p 141.
28. Tregub A, Inglehart L, Pham C, Friedrich R. 29th International SAMPE Technical Conference, Oct–Nov, 1997, Orlando, Florida: 1997. p 787–799.
29. Oehler AC, Flores-Snyder C. Clarkson University CAMP 4th International Symposium on CMP, Lake Placid, NY: Aug1999.
30. Flores-Snyder C, Oehler AC, Malik V. Clarkson University CAMP 3rd International Symposium on CMP; Lake Placid, NY: Aug1998.
31. Larson RG, Gubbins KE. Structure and Rheology of Complex Fluids. Oxford University Press; 1999. p 656.
32. Lu J, Rogers C, Manno VP, Philipossian A, Anjur S, Moinpour M. Measurements of slurry film thickness and wafer drag during CMP. Electrochem Soc 2004; 151 (4):G241–G247.
33. Lu J, Coppeta J, Rogers C, Manno VP, Racs L, Philipossian A, Moinpour M, Kaufman F. Mater Res Soc Symp Proc 2000; 613:E1.2.1–E.1.2.6.
34. Moinpour M, Tregub A, Oehler A, Cadien K. Advances in characterization of CMP consumables. MRS Bulletin 2002; 27(10):766–771.
35. Tseng W-T, Liu C-W, Dai B-T, Yeh C-F. Effects of mechanical characteristics on the chemical–mechanical polishing of dielectric thin films. Thin Solid Films 1996; 290–291:458–463.

36. Zhou Y-y, Davis EC. Mater Sci Eng 1999, B68:91–98.
37. Thakurta G, Borst CL, Schwendeman DW, Gutmann RJ, Gill WN. Pad porosity, compressibility and slurry delivery effects in chemical–mechanical planarization: modeling and experiments. Thin Solid Films 2000; 366:181–190.
38. Shen C, Springer GS. Moisture absorption and desorption of composite materials. J Comp Mater 1976; 10:2–20.
39. Tregub A, Renolayan A. CDCC'98; Sherbrooke,Quebec, Canada: Aug1998, 129–137.
40. Tregub A, Ng G, Moinpour M. 2003 MRS Spring Meeting Proceedings, Symposium I; Chemical–Mechanical Planarization.
41. Bozak A, Moinpour M. (unpublished work).
42. Huey S, Mear ST, Wang Y, Jin RR, Ceresi J, Freeman P, Johnson D, Vo T, Eppert S. IEEE/SEMI Advanced Semiconductor Manufacturing Conference; Boston, MA: Sept1999. p 54–58.
43. Coppeta J, Racz L, Philipossian A, Kaufman F, Rogers C. Third International Chemical Mechanical Polish Planarization for ULSI Multilevel Interconnection Conference; Santa Clara, CA: Feb1998.
44. Borucki L, Charms L, Philipossian A. Analysis of frictional heating of grooved and flat CMP polishing pads. Electrochem Soc 2004; 151(12):G809–813.
45. Philipossian A, Charms L. Proceedings of the CMP Symposium of the 204th Meeting of the Electrochemical Society; Philadelphia, PA: 2002.
46. Li Z, Borucki L, Philipossian A. Proceedings of the CMP Symposium of the 204th Meeting of the Electrochemical Society; Orlando, FL: 2003.
47. Sorooshian J, Hetherington D, Philipossian A. Effect of process temperature on coefficient of friction during CMP. Electrochem Solid State Lett 2004; 7(10):G222–G224.
48. Gray C, Apone D, Barns C, Moinpour M, Anjur S, Manno V, Rogers C. 2004 MRS Spring Meeting Proceedings; Chemical–Mechanical Planarization Symposium.
49. Coppeta JR. Investigating fluid behavior beneath a wafer during chemical mechanical polishing process. PhD Thesis; Tufts University; 1999.
50. Lu J, Coppeta J, Rogers C, Manno V, Racz L, Philipossian A, Moinpour M, Kaufman F. Mater Res Soc Symp 2000; 613:E1.2.1–E1.2.6.
51. Lu JC. Fluid film lubrication in chemical mechanical planarization. Master's Thesis; Tufts University; 2001.
52. Runnels SR, Eyman LM. Tribology analysis of chemical–mechanical polishing. Electrochem Soc 1994; 141(6):1698–1701.
53. Thakurta DG, Borst CL, Schwendeman DW, Gutmann RJ, Gill WN. Thin Solid Films 2000; 336:181–190.
54. Apone D, Gray C, Rogers C, Manno VP, Barns C, Moinpour M, Anjur S, Philipossian A. 2005 MRS Spring Meeting Proceedings; Chemical–Mechanical Planarization Symposium.
55. Gray C, Rogers C, Manno V, Vlahakis J, Barns C, Moinpour M, Anjur S, Philipossian A, Borucki L. CMP–MIC Proceedings; 2006.

56. Lu Z, Ryde NP, Babu SV, Matijević E. Particle adhesion studies relevant to chemical mechanical polishing. Langmuir 2005; 21:9866–9872.

57. Gorantla VRK, Goia D, Matijević E, Babu SV. Role of amine and carboxyl functional groups of complexing agents in slurries for chemical mechanical polishing of copper. J Electrochem Soc 2005; 152 (12):G912–G916.

58. Koenig JL. Spectroscopy of Polymers. Washington, DC: American Chemical Society; 1992.

59. Jonas J, Adamy ST, Grandinetti PJ, Masuda Y, Morris SJ, Campbell DM, Li Y. High pressure NMR study of transport and relaxation in complex liquids of 2-ethylhexyl cyclohexanecarboxylate and 2-ethylhexyl benzoate. J Phys Chem 1990; 94:1157.

60. Hakansson B, Soderman O, Balinov B. Nuclear magnetic resonance of emulsions. In:Hubbard AT, editor. Encyclopedia of Surface and Colloid Science. Vol. 3, New York: Marcel Dekker; 2002.

61. Grandjean J. Nuclear magnetic resonance spectroscopy of molecules and ions at clay surfaces. In: Hubbard AT, editor. Encyclopedia of Surface and Colloid Science. Vol.3, Hubbard, New York: Marcel Dekker; 2002.

62. Liu G, Li Y, Jonas J. Reorientational dynamics of molecular liquids in confined geometries. J Chem Phys 1989; 90:5881.

63. Liu G, Mackowiak M, Li Y, Jonas J. NMR deuteron relaxation study of low-dimensional effects on molecular liquids in restricted geometries. Chem Phys 1990; 149:65.

64. Liu G, Mackowiak M, Li Y, Jonas J. Rotational diffusion of liquid toluene in confined geometry. J Chem Phys 1991; 94:239.

65. Gay ID. Adsorbed species: spectroscopy and dynamics. In: Grant DM, Harris RK, editors. Encyclopedia of Nuclear Magnetic Resonance. New York: John Wiley & Sons, Inc.; 1996. p 733–738.

66. Kirkpatrick DJ. Geological applications. In: Grant DM, Harris RK, editors. Encyclopedia of Nuclear Magnetic Resonance. New York: John Wiley & Sons, Inc.; 1996. p 2194–2202.

67. Maciel GE. Silica surface: characterization. In: Grant DM, Harris RK, editors. Encyclopedia of Nuclear Magnetic Resonance. New York: John Wiley & Sons, Inc.; 1996. p 4370–4386.

68. Iler RK. The Colloid Chemistry of Silica and Silicates. Ithaca, New York: Cornell University; 1955.

69. Stilbs P. Fourier transform NMR pulsed-gradient spin—echo (FT-PGSE) self-diffusion measurements of solubilization equilibria in SDS solutions. J Colloid Interface Sci 1982; 87 (2):385–394.

70. Odeh F, America W, Dhane S, Li Y. Langmuir submitted(2005).

71. Heldt N et al. Characterization of a polymer-stabilized liposome system. Reactive Funct Polymers 2001; 48:181–191.

72. Li Y. State-of-the-art short course. Chemical Mechanical Planarization for ULSI Multilevel Interconnection: Fremont, CA; March 5,2007.

73. Choi H, Rawat A, Moinpour M. (unpublished results).

74. DeNardis D, Choi H, Kim A, Moinpour M, Oehler A. *2005 MRS Spring Meeting Proceedings*, Chemical–Mechanical Planarization Symposium.
75. Kim S, Williams R, Park I, Remsen EE, Moinpour M, Kim A. (to be presented at the 2007 MRS Spring Meeting, Chemical–Mechanical Planarization Symposium, San Francisco.
76. Sampurno Y, Philipossian A, Choi H, Moinpour M, Rawat A, Kim A. Clarkson University CAMP 11th International Symposium on CMP; Lake Placid NY: August 2006.
77. Burtovyy R, Liu Y, Zdyrko B, Tregub A, Moinpour M, Buehler M, Luzinov I. (to be published in the Journal of the Electrochemical Society).
78. Armini S, Moinpour M, Whelan CM, Maex K, Hernandez JL. (to be presented at 2007 CMP–MIC, Fremont CA March 2007).

3

PROCESSING TOOLS FOR MANUFACTURING

Manabu Tsujimura

3.1 CMP OPERATION AND CHARACTERISTICS

In today's wafer fabrication plant, chemical–mechanical polishing or planarization (CMP) is an integral part of the manufacturing flow (Fig. 3.1) [1]. Rotary polishing platforms were the initial polishing machines on which the semiconductor CMP processes were developed. A majority of the polishers used in industry and academic institutions today are rotary tools, even though other polishing platforms have been implemented. A typical schematic illustration of the rotary polishing platform used is shown in Fig. 3.2 [2]. The basic operating principle behind the rotary platform is that the wafer is held on a rotating carrier while being pressed face down against a rotating polishing pad, while a chemically and mechanically (abrasive) active slurry planarizes the wafer.

Typically, both the wafer carrier and platen are rotated in the same direction. A downforce is applied while the wafer carrier and platen are rotated on their own axes ω_c and ω_p, respectively. The polishing slurry is dispensed from a tube located at the center of the pad, and as the platen rotates the slurry is transported between the wafer and the pad [3–6]. In addition to the rotary platform, an orbital design has also been implemented (Fig. 3.3) [2]. The operating principle for the orbital design is similar to the rotary platform, except that the polishing head and table are in orbital motion to each other. In addition, the slurry is usually delivered through the pad.

Microelectronic Applications of Chemical Mechanical Planarization, Edited by Yuzhuo Li

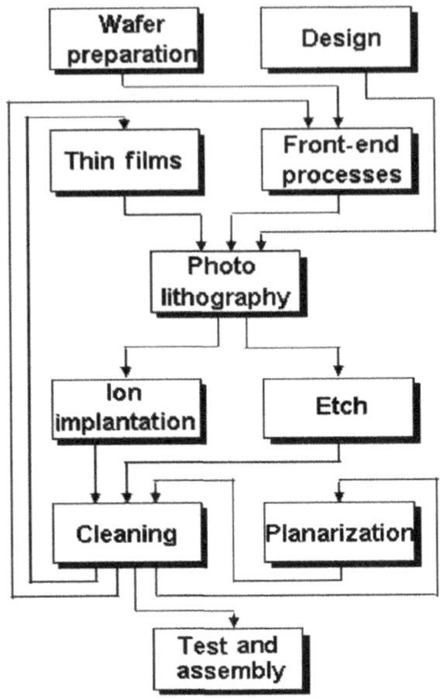

FIGURE 3.1 A typical wafer processing flow that shows that CMP is an integral part of the manufacturing process (from Ref. 1).

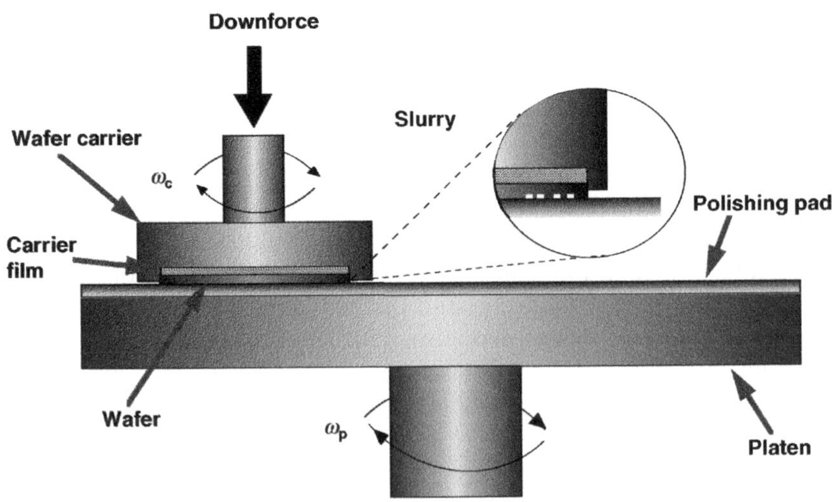

FIGURE 3.2 Schematic of a typical rotary chemical–mechanical planarization tool (from Ref. 2).

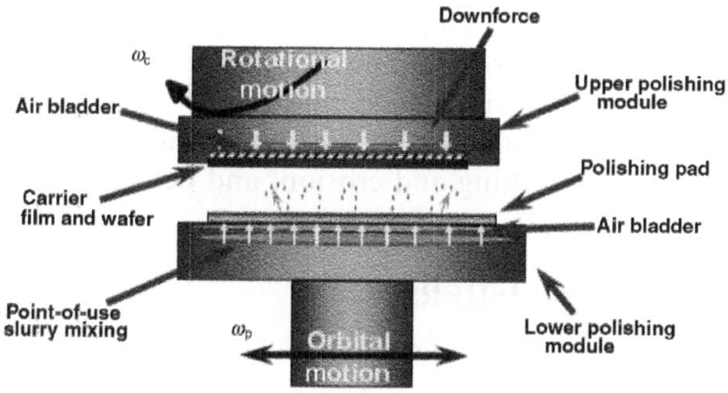

FIGURE 3.3 Schematic of a typical orbital chemical–mechanical planarization tool (from Ref. 2).

3.2 DESCRIPTION OF THE CMP PROCESS

Chemical–mechanical planarization occurs when the surface of the wafer to be polished is forced against a polishing pad. Aqueous slurry that contains abrasive particles is placed on a polishing pad. The wafer is moved relative to the slurry-covered pad and the rate at which material is removed is often described by the heuristic equation called Preston's law:

$$\mathrm{RR} = K_\mathrm{p} \times P \times V \tag{3.1}$$

where RR is the removal rate, K_p is a constant, called Preston's coefficient, P is the local pressure on the surface, and V is the relative velocity of the point on the surface of the wafer versus the pad [7–12]. The equation works well for systems where material was removed mainly by abrasion. In particular, Preston's equation provides a reasonably good fit to the silicon dioxide CMP data. Several modified Preston equations have been proposed for dielectric and metal CMP in the literature [13,14]. Each has its own advantages and limitations.

TABLE 3.1 Input Variables for CMP Processes.

Input Variables	
Machine parameters	Downpressure, linear velocity, and slurry flow
Particle characteristics	Size, size distribution, shape, mechanical properties, surface chemistry, dispersion stability, concentration, agglomeration, and oversize particle count
Slurry chemistry	Oxidizers, pH and pH drifts, pH stabilizers, complexing agents, dispersants, and corrosion inhibitors
Pad characteristics	Mechanical properties, topography, conditioning, and pad uniformity
Substrate characteristics	Wafer size, wafer stacks, feature size, feature density, and mechanical strength of each stack layer

CMP is a complex process and its outcome is influenced by many input variables. An incomplete list of these input variables is shown in Table 3.1.

The CMP output parameters include removal rate, planarization efficiency, surface finish, material removal rate selectivity, wafer-to-wafer uniformity, within-wafer uniformity, dishing and erosion, and defect levels [15].

3.3 OVERVIEW OF POLISHERS

3.3.1 CMP System [16]

Figure 3.4 shows subsystems and related materials in a CMP system. In addition to a polisher and post-CMP cleaning station, a CMP system encompasses slurry supply, waste treatment, monitors, slurry, and pad. The performance of a CMP system is measured by its output in (a) wafer uniformity, (b) polishing rate, (c) planarization efficiency within chip, and (d) defect count. From a mechanistic point of view, the tool uptime, throughput, and reliability of the system are also very important [16]. In order to satisfy these requirements, all subsystems described in Fig. 3.4 are considered as one total system of CMP and should be upgraded as a whole whenever needed.

A slurry supply subsystem must be able to deliver slurry to the point of use without agglomeration and keep its CMP performance consistent. If the slurry is agglomerated or its pH has changed, the CMP outcome could be significantly altered even if the polisher and other units are kept in perfect operating conditions. There are many technical challenges in slurry supply, such as filtration, reuse, recycle, and point-of-use blending versus central supply.

Post-CMP cleaning and drying have been recently highlighted as one of the most difficult challenges in terms of yield improvement [17]. In addition to some of the basic requirements such as removal of slurry residues and contaminants from the polished wafers, a comprehensive post-CMP cleaning process must also address issues such as watermark-less drying for

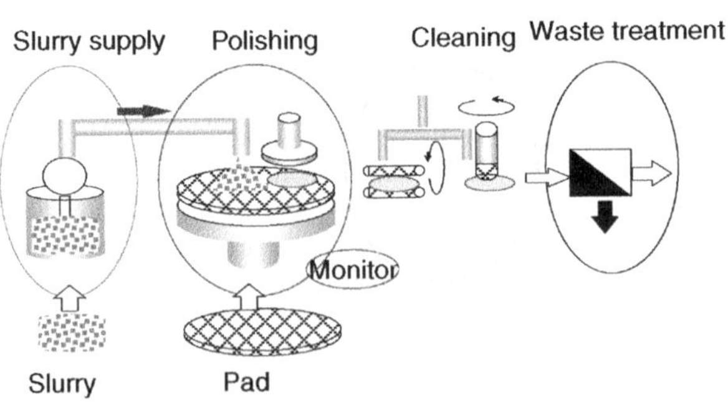

FIGURE 3.4 Schematic illustration of a typical CMP system that includes a slurry supply system, a polisher, a post-CMP cleaning unit, and a CMP waste treatment unit.

hydrophobic dielectric [18], corrosion on Cu metal, foreign materials (FM) that are larger than half critical dimension (CD), and wafer edge cleaning.

The function of a monitor [19] in a CMP polisher is to stop polishing at the target range and to observe polishing conditions. As within-wafer and wafer-to-wafer nonuniformity are almost unavoidable in polishing, overpolishing to compensate for such nonuniformity is often required. Therefore, a monitoring system has become an essential and integral part of a CMP polisher. Recently, *in situ* and *ex situ* monitoring systems have been implemented as a part of advance process control (APC) to minimize the process variation. In addition, as increasing attention has been paid to the nanotopography [20] of polished wafers, a range of *in situ* and *ex situ* measurement schemes on the nanotopography of incoming and polished wafers have also been investigated.

The treatment of CMP waste stream poses a serious challenge to the environment. Prior to the adoption of CMP processes, there were only a few processes in the fab that generated solid wastes. The introduction of CMP changed the chemical composition of the fab waste stream significantly. Currently, all solid CMP wastes are treated collectively in a central waste treatment system. It is conceivable that the future CMP polisher may contain an internal waste treatment module that can capture or eliminate some wastes before the spent slurry is discharged to the central waste treatment system.

3.3.2 Brief History of CMP Systems

Figure 3.5 shows a brief history of CMP systems. By and large, the CMP tools today have evolved from a typical polisher that was used for bare Si polishing, which was a stand-alone unit without postcleaning and drying.

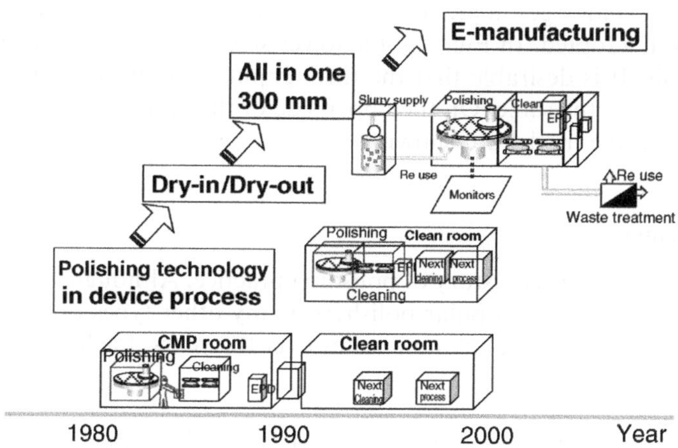

FIGURE 3.5 Schematic illustration of the evolution of CMP tools from stand-alone, dry-in/dry-out to those with advanced process control (APC).

In the early 1980s, CMP tools were typically installed in a special CMP room separated from the usual clean room, as they were regarded as the dirtiest tools in semiconductor manufacturing. When the adopted CMP process advanced significantly in the 1990s, the requirement for the wet CMP polisher to meet the same criteria as the other dry tools in the fab was raised. The dry-in/dry-out concept satisfied those requirements, in which a wafer comes in dry condition and goes out after drying through cleanings. At that time, the cleaning area and the polishing space were typically regarded as the cleanest and the dirtiest areas, respectively. The two processes were hence carried out away from each other. The technology to integrate these cleanest and dirtiest areas was one of the interesting challenges faced in the fab. The challenge was eventually overcome by adopting a hydrodynamic airflow design in the CMP area. After 2000, an all-in-one design including APC and e-manufacturing became mandatory. By 2012, 450-mm wafer tools are expected to be implemented according to the ITRS roadmap [21]. Although rotary-type polishers have often been used as a standard for processes involving wafers up to 300 mm, they may face challenges from several other types of CMP designs with 450-mm wafer application.

3.3.3 Diversity in CMP Tools

Figure 3.6 shows several types of CMP tools that have so far appeared in the market. The drawing at the center of Fig. 3.6 represents a bare-bone model, which has one wafer carrier (head) on one table. The left-hand side shows a model that has multiple wafer carriers on one table in order to increase throughput, that is, a multicarrier model. The right-hand side shows a model with several single-headed tables, that is, multitable model.

In addition to a higher throughput, a multicarrier model has a merit of lowering slurry usage and cost. Although nonuniformity among carriers is one of the biggest challenges in practice and the tools costlier, the multitable model has been well adopted. In Fig. 3.6, the lower side shows a built-in single-wafer cleaning unit. It is desirable that the throughput of both the polisher and the cleaning unit are the same. On the upper side of the figure, a small table model and a linear table model are shown.

3.3.4 Polisher

Figure 3.7 shows a brief history of polishing practice. Although the rotary table type is one of the most popular polishers, many other types of polishers have been introduced by several manufactures over the years. Each type has its own characteristic features [22].

- *Rotary type*: This is the most popular type. The wafer is set face down. There are variations, such as one head on one table, multiple heads on one table, and multiple tables with one head.

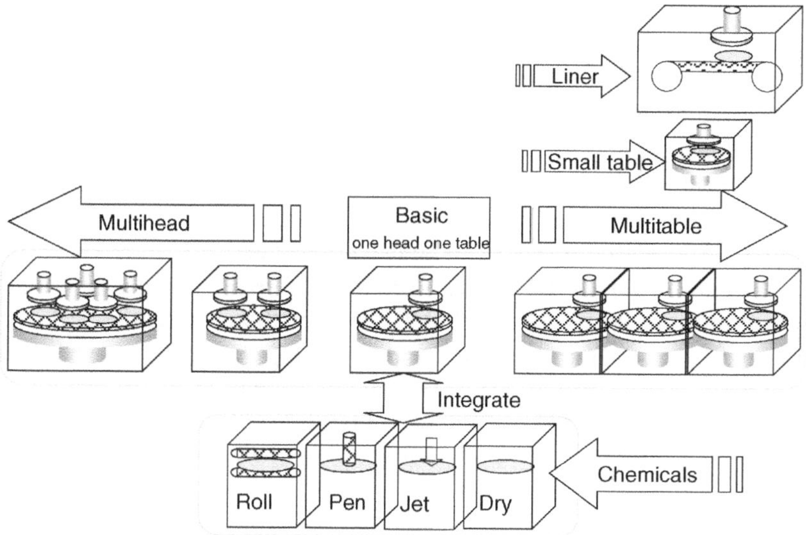

FIGURE 3.6 Schematic illustration of various CMP tool designs including multihead system (left), single head and platen system (center), multitable system (right), built-in single-wafer post-CMP cleaning system (bottom), and a small table and linear belt model (top).

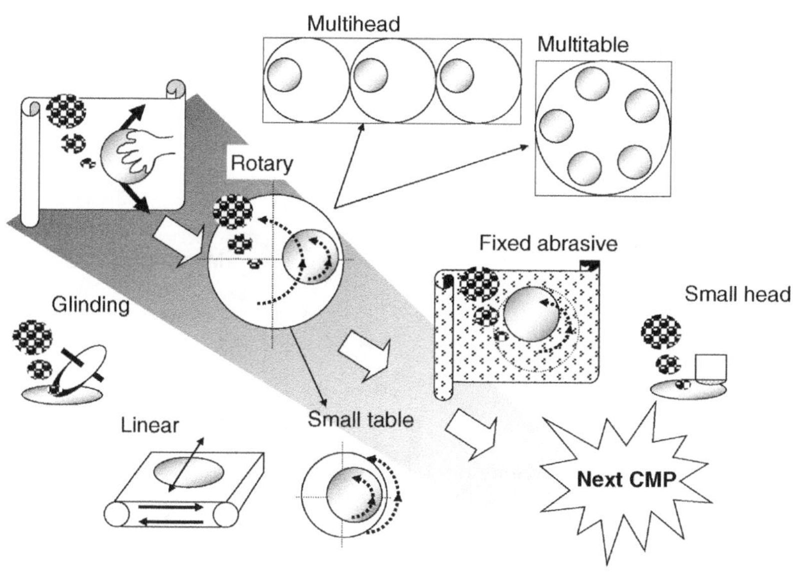

FIGURE 3.7 Schematic illustration of different types of polishers: rotary with multitable and multihead (1), small head (2), small table (3), linear motion type (4), fixed abrasive type (5), and grinding type (6).

- *Small table type*: Wafer center covers the center of a table. The wafer is set face down. This method conserves space. Slurry is generally introduced from the lower side of the table. The orbital type shown in Fig. 3.3 is one of these categories.
- *Small head type*: The wafer is set face up. The Nikon tool is one of these categories [23].
- *Linear type*: Table has linear motion. The wafer is set face down. The tool of Lam Research is one of these categories [24].
- *Fixed abrasive type*: Involves a fixed abrasive pad typically deployed in a web motion. Web tool of AMAT is one of these categories [25].
- *Grinding type*: Both grinding and polishing produces can be used to prepare a smooth surface. The main difference between grinding and polishing is the rigidity of the grinding abrasive and the elasticity of the polishing media. This results in a lower contact pressure for polishing.

3.3.5 Cleaning Module in a Dry-in/Dry-out System [26]

As described previously, a postcleaning tool was typically installed away from the polisher due to the notorious tendency of polishers for generating dust and the stringent requirement of a cleaning tool to minimize dust exposure. At that time, combining the polishing and cleaning functions into one framework was regarded as an impossible task. The huge difference in particulate contamination requirements for polishing and cleaning exceeds an order of 10^5. This may represent the widest requirement gap of any field. Before the dry-in/dry-out concept, budge-type RCA cleaning or single-wafer system as stand-alone was used. After the advent of the dry-in/dry-out concept, a variety of cleaning schemes such as bath, vertical, and function waters have been applied (Figs. 3.8 and 3.9).

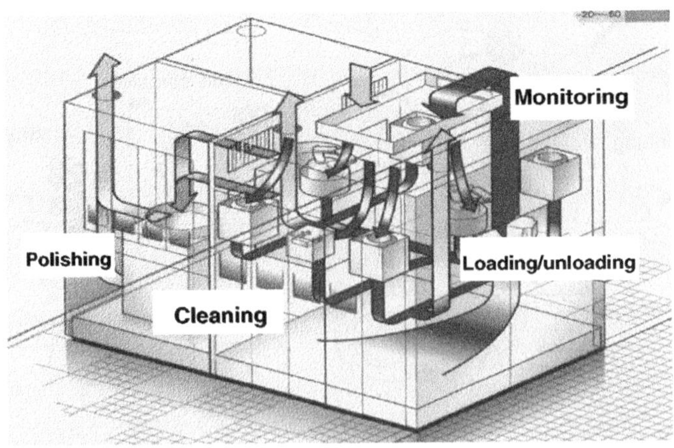

FIGURE 3.8 Schematic illustration of a polisher with the integrated cleaning. The concept of dry-in/dry-out is illustrated here.

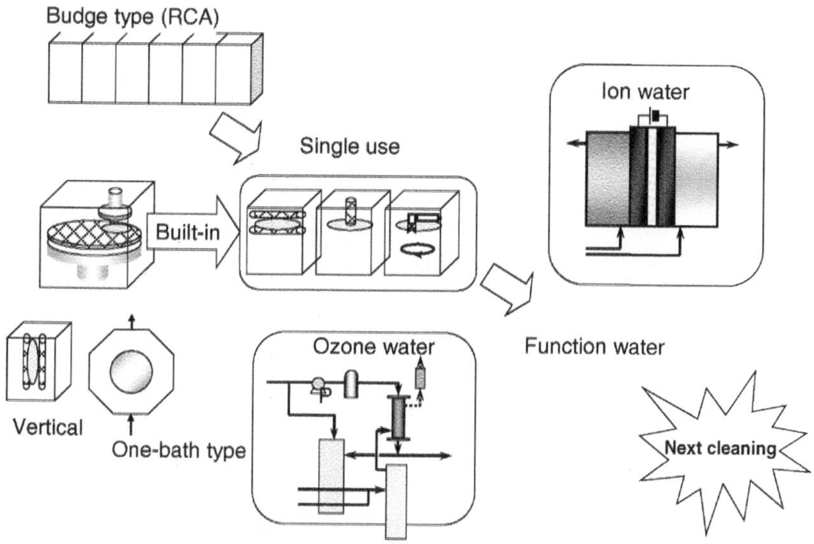

FIGURE 3.9 Schematic illustration of the several cleaning types available: RCA budge type (1), integrated cleaning with polisher type (2), vertical type (3), bath type (4), and functional water type (5).

3.4 CARRIERS AND DRESSERS

3.4.1 Functions of Carriers and Dressers

The principal polishing components are illustrated in Fig. 3.10. The non-uniformity of the polishing rate is governed by Preston's law, which states that the polishing rate is proportional to the pressure on the wafer and the relative velocity between the wafer and the pad. The wafer carrier has a function of controlling the pressure on the wafer or to make it uniform. In principle, the relative speed between the wafer and the pad shall remain constant if the carrier speed and the table speed are the same. In practice, the nonuniformity of the polishing rate would change depending on the pad surface condition. The main purpose of dressers is to keep the pad surface condition consistent. In other words, the carriers and the dressers are critical parts in obtaining and maintaining a good polishing performance. The details of the carriers and the dressers are discussed in the next two sections.

3.4.2 Carrier

As stated above, the primary function of the carrier is to hold the wafer and control the pressure between the wafer and the pad. Some representative

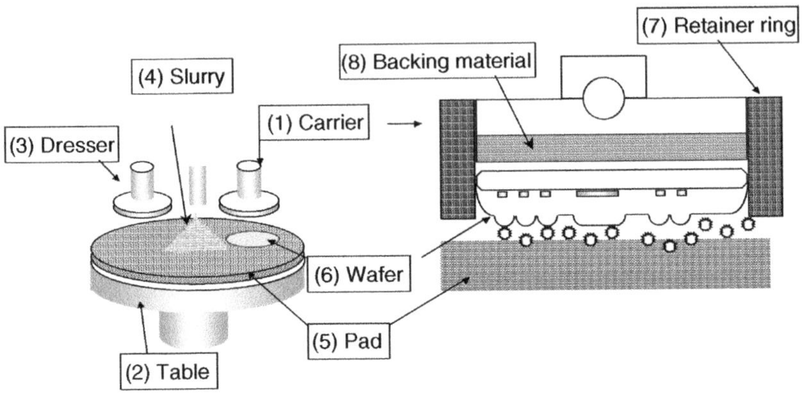

FIGURE 3.10 Schematic illustration of key working components in a rotary polisher.

carrier designs are shown in Fig. 3.11. It should be noted that there is a rich collection of patents and open literatures on various designs to hold the wafers. There are two common types of holders, one is a gimbal type (carrier A) and the other is a floating type (carriers B and C). The gimbal-type carrier may easily swing around its gimbals. In the case of the floating-type carrier, the wafer may be supported by a membrane floated by the fixed carrier.

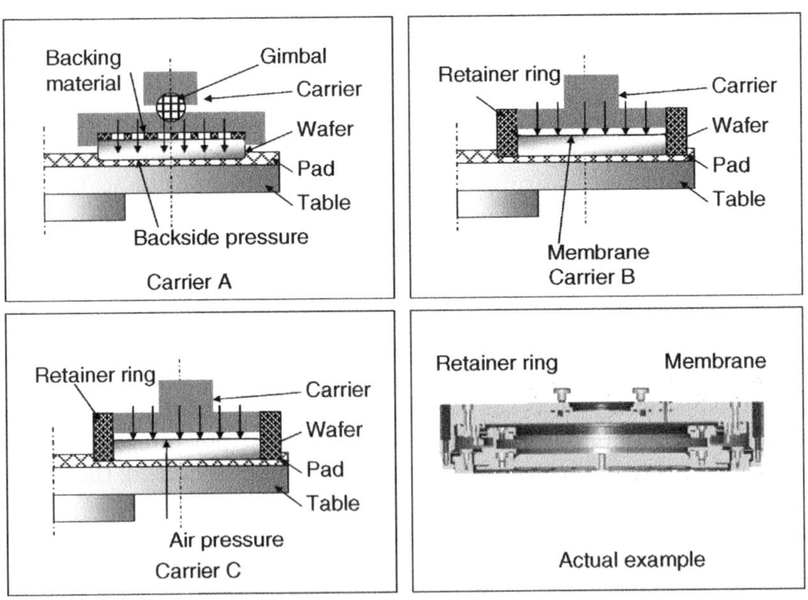

FIGURE 3.11 Schematic illustration of the cross section of different carrier models: gimbal type (A), floating type (B and C), and actual example of a carrier (D).

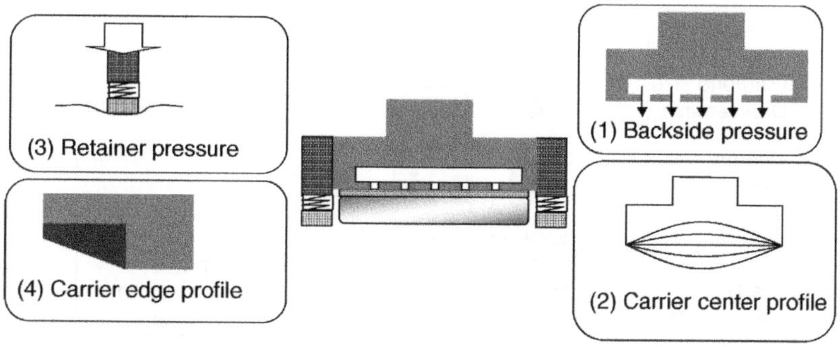

FIGURE 3.12 Schematic illustration of different ways to control polish pressure on a wafer.

The retainer ring in carriers B and C is one of the most important parts used to control the polishing profile, which is shown in Fig. 3.12.

1. *Carrier A*: The carrier is supported by gimbals and can be swung around its gimbals. The backside pressure is adopted in this carrier example to control the polishing pressure (rate).
2. *Carrier B* [27]: The wafer is supported by a membrane that is floated by a fixed carrier. Hydraulic pressure loaded on the membrane will control the polishing pressure (rate). The retainer ring is mounted in this carrier example, which can control the polishing pressure (rate) of the wafer edge portion.
3. *Carrier C* [28]: The wafer is supported by the fixed carrier. Hydraulic pressure is loaded directly on the wafer to control the polishing pressure (rate). The retainer ring is mounted in this carrier example, which can control the polishing pressure (rate) of the wafer edge portion.

An actual carrier example is also shown in Fig. 3.11.

Figure 3.12 shows the four methods that are used to control the polishing pressure, that is, polishing rate.

1. *Backside pressure*: Fluid pressure will be loaded on the wafer from the back side of the wafer through holes that were set on the back side of the carrier. The pressure on the wafer will be changed by selecting holes. In this way, the center portion of pressure (rate) will be controlled.
2. *Carrier center profile*: The center profile of the carrier back side will be controlled as a concave or convex shape. In this way, the center portion of pressure (rate) will also be controlled.
3. *Retainer pressure*: Reaction force on the wafer by the pad will be controlled by pressuring the retainer ring. In this way, the edge portion of pressure (rate) will be controlled.

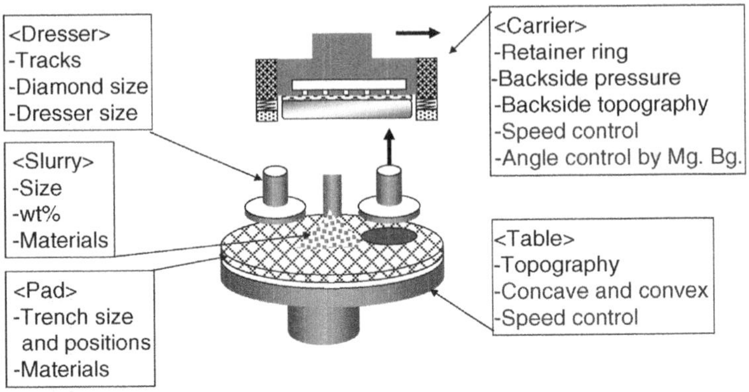

FIGURE 3.13 Several methods to control the polishing profile such as process conditions, carrier design, pad roughness, conditioner (dresser) design, and slurry.

4. *Carrier edge profile*: The edge profile of the carrier back side will be controlled as an inclined shape. In this way, the edge portion of pressure (rate) will also be controlled.

3.4.3 Profile Control by Carriers [29]

The incoming wafer nonuniformity may vary depending on many factors. Especially in case of Cu plating, the nonuniformity of the wafer edge is often poor. Therefore, a CMP process is required to control its polishing characteristics in accordance with the incoming wafer profile. In this section, several profile control methods are introduced as shown in Fig. 3.13.

Figure 3.14 shows the polishing profile control by carrier design. The polishing profile of the wafer center area can be modulated by backside pressure and modification at the carrier center. The polishing profile of the

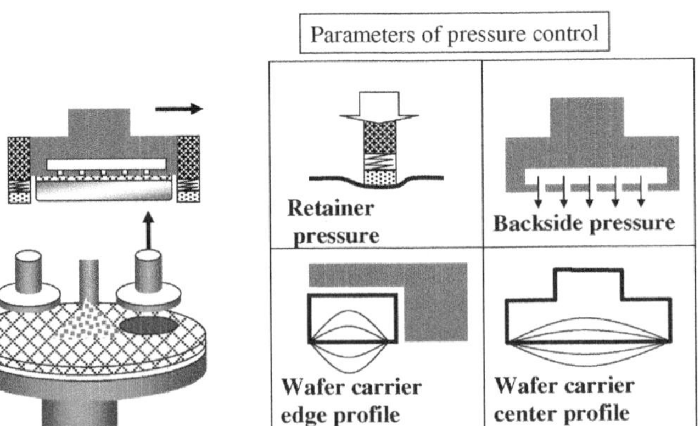

FIGURE 3.14 Schematic illustration of how carrier design can affect the polishing profile.

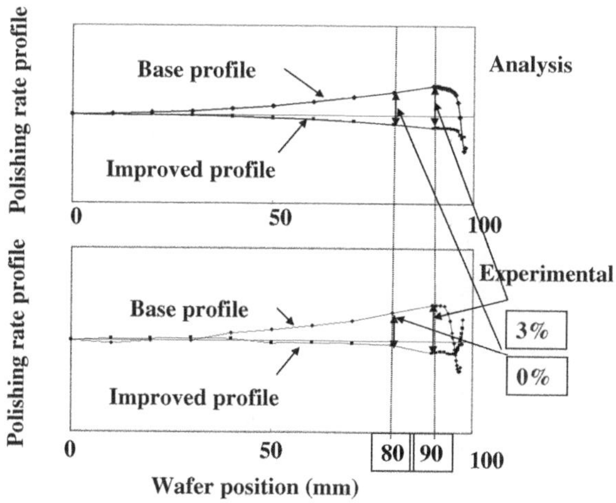

FIGURE 3.15 Finite element analysis result of the edge profile control achieved by changing the retainer ring design.

wafer edge can be varied by retainer pressure and modification at the carrier edge.

The result of edge profile control by the retainer ring is shown in Fig. 3.15. The analysis is performed using the latest retainer ring design under the following conditions:

Pad	IC1000/Suba 400
Backing film	NF200
Carrier force	500 g/cm^2
Retainer ring force	0–700 g/cm^2

The edge profile is predicted by the finite element analysis with a precision of 3 % on the edge as an example.

Figure 3.16 shows the polishing profile control by carrier and table speeds. Three cases are illustrated: (1) table speed is higher than carrier speed, (2) table speed is the same as carrier speed, and (3) table speed is lower than carrier speed. Each velocity vector is shown.

3.4.4 Dressers

The effect of dresser (pad conditioning) is shown in Fig. 3.17. There are two kinds of conditioning pads, that is, seasoning and dressing. A seasoning is typically conducted for about 5–10 min until an initial pad condition becomes consistent. A dressing is often performed for about 10 s just after polishing until the pad condition has returned to the previous baseline condition. In the case of polyurethane pads such as Rohm & Hass IC1000, a diamond dresser is often used to create cuts on the pad. In the case of nonwoven-type pads such as

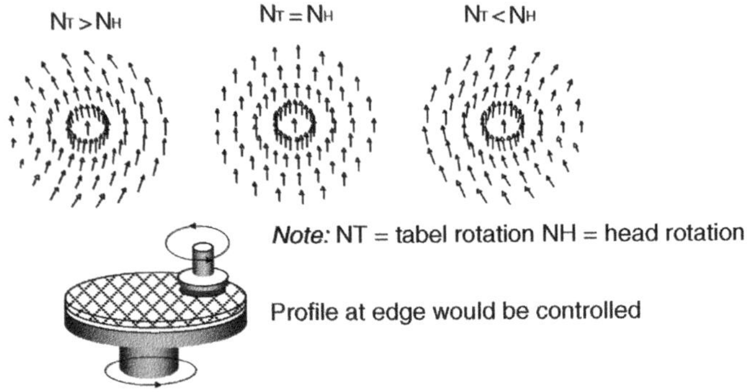

FIGURE 3.16 Schematic illustration of the impact of velocity vectors on the polishing profile for three different cases of velocity combinations of the table and the head.

Suba pads, a brush-type dresser can be adopted with which bristle is regenerated by rubbing between the pad and the dresser.

Figure 3.18 shows several types of diamond dressers. These are classified based on the following:

1. *Size of diamonds*: A conditioner can be classified by the size of diamond particles found on the surface. For example, #100 means 100-grit diamond particles were used on the conditioner.
2. *Geometric arrangement of diamonds*: The diamond particles are either randomly arranged by Ni plating or precisely decorated according to a particular matrix design.

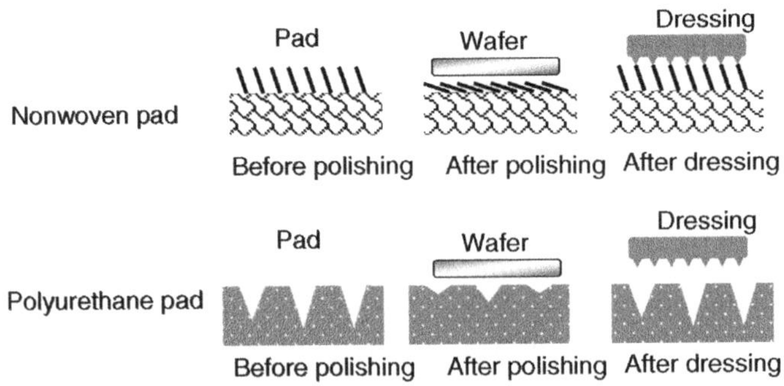

FIGURE 3.17 Schematic illustration of the pad asperity variation after polishing and conditioning on nonwoven and polyurethane pads. The polishing process decreases the number of asperities that are regenerated by the conditioning process.

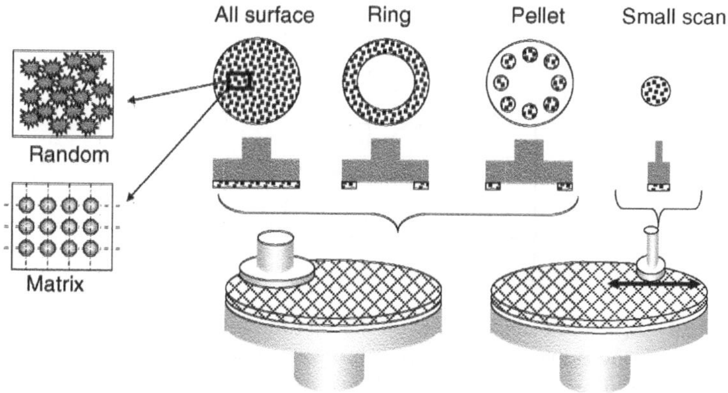

FIGURE 3.18 Schematic illustration of different varieties of dressers based on the arrangement, size, shape, and coverage.

3. Shape of the conditioner:
   (a) *All surface type*: Diamonds cover the entire surface around the conditioner. Within this type, the conditioner surface may be very small (few millimeters), which is called scan type as scanning is necessary in order to condition the surface covered by the wafers.
   (b) *Ring type*: Diamond is arranged only on edge portion.
   (c) *Pellet type*: Diamond is arranged in small pellets.

Figure 3.19 shows an actual example of a dresser setting. This dresser is rotated and swung on the pad.

Figure 3.20 shows the conditioning effect. A large-diameter conditioner has a relatively nonuniform track. Therefore, it requires swinging in order to make its tracks uniform. A ring-type conditioner has relatively better uniformity

FIGURE 3.19 Photograph showing the working principle of a conditioner. The conditioner rotates along its center while traversing to and from the center to the edge of the rotating pad.

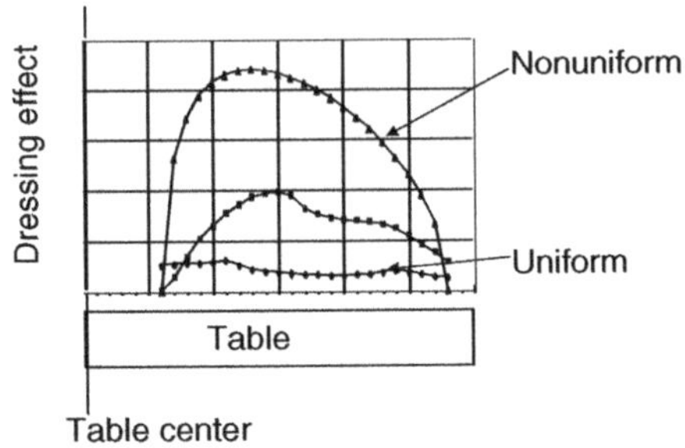

FIGURE 3.20 Schematic illustration of the conditioning effect across the pad area using different conditioner types. Large and ring-type diamond conditioners show nonuniform and uniform wear on the pad, respectively.

although the removal amount is smaller than a larger-diameter conditioner. A small-diameter type can achieve uniform removal by scanning.

There are several types of dressing tools other than diamond dressers. Jet cleaning [30] is one of the dressing methods in which high-pressure water and/or gas will be adopted in order to remove slurries or other residual contaminants.

3.5 *IN SITU* AND *EX SITU* METROLOGIES

3.5.1 Application

In spite of the low precision of most CMP hardware on the order of micrometers, CMP is required to perform tasks with nanometer accuracy. Therefore, monitoring technology that can control how and when to stop the polishing is essential for the proper operation of a polisher.

Figure 3.21 shows an application of such a monitoring component. In the process of fabricating an aluminum wire like memory device, CMP for interlayer dielectric (ILD) is adopted. The initial step height caused by the deposition process should be planarized. In this case, the residual oxide film thickness is measured by a monitoring unit. In other applications where Cu or W wires and W plugs are used, polishing should stop after metal is cleared. In this case, the thickness of residual metal is monitored. For shallow trench isolation (STI) process, CMP should stop just after the stopping layer such as silicon nitride. Thus, as the film materials and their configurations may be quite different in various CMP processes, different monitoring techniques or schemes may also have to be selected to optimize their specific process.

3.5.2 Representative Monitors

A CMP process may fall into one of the two categories: one is a blind process such as ILD CMP and the other is a recess process such as metal CMP. In the

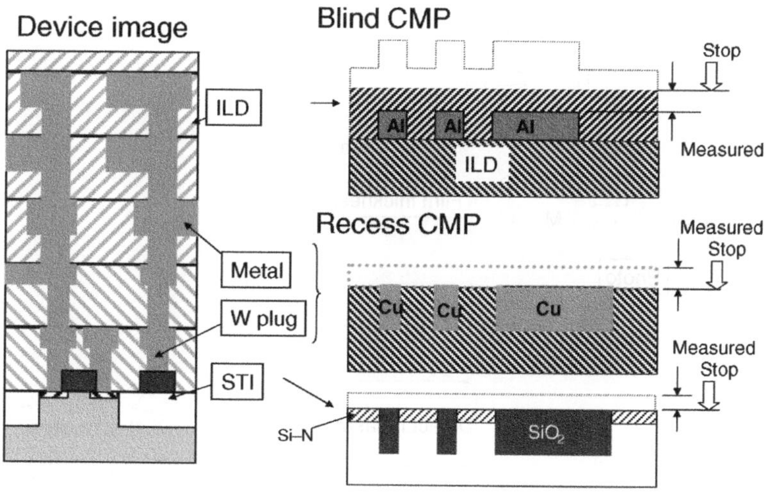

FIGURE 3.21 Schematic illustration of the requirement of monitors depending on the CMP process. The top figure shows ILD CMP, where the process needs to be stopped after achieving a target thickness. The figure at the middle shows Cu CMP that needs to be stopped after the metal gets cleared. The figure at the bottom shows STI CMP that needs to be stopped after the nitride layer is detected.

case of blind process, a film thickness monitor is typically used. In the case of recess process, it is possible to detect a different signal by a change in friction, vibration, and refection when the polishing reaches a different layer of materials (Fig. 3.22). Some representative monitoring methods are listed here:

1. *Friction change detected by motor current [31]*: This is one of the most popular recess process methods. In the example of W CMP, friction forces differ for W metals, barrier metals such as Ti and/or TiN, and substrates. These friction differences can be detected by the motor current of a wafer carrier or a table.
2. *Vibration-level change detected by vibration sensor [32]*: Carrier vibration changes when different materials are polished or when different topographies are polished. This method can be used for both blind and recess processes.
3. *Eddy current induced by magnetism [33]*: This is used only for metal polishing. When metal thickness changes, the eddy currents induced by magnetism also change. This system also measures metal thickness.
4. *Optical detection* [34]: The reflection difference generated by different materials is detected. This method is applicable to recess processes.
5. *Film thickness measurement*: Film thickness is measured after polishing, and the polishing data are then fed back to the next polishing. This is called an in-line monitoring. It is often used in ILD CMP.

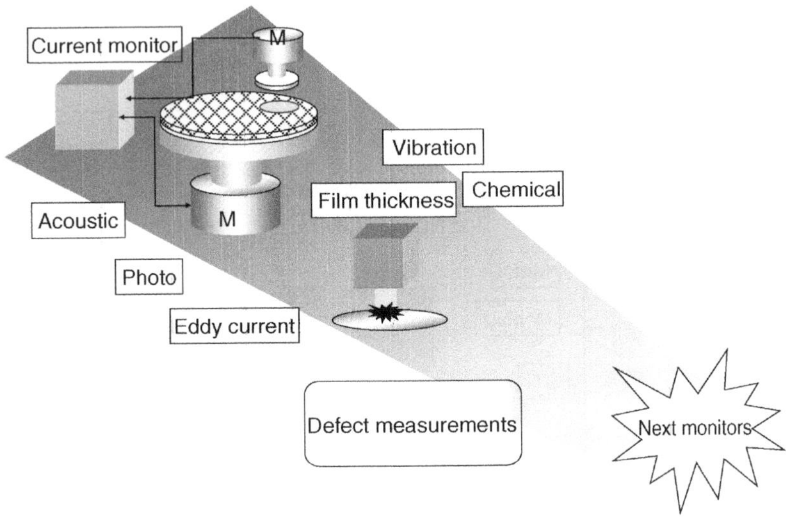

FIGURE 3.22 Schematic illustration of different types of monitors that could be implemented during a CMP process.

6. *Other methods*: Other approaches exist in addition to the methods listed above. In the case of STI, ammonia, which is expelled from the Si–N layer, is detected [35]. Acoustic sound differences can also be detected.

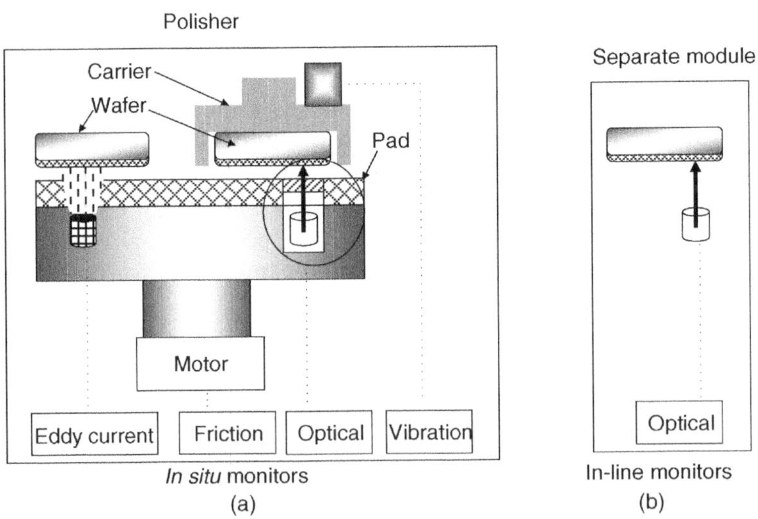

FIGURE 3.23 Schematic illustration of different types of monitors, including the position where these monitors are generally located in a tool system.

Figure 3.23 shows the examples of mounting of friction, eddy current, optical, and vibration types of monitors.

Figure 3.24 shows the examples of output data of an eddy current monitor, a friction monitor, an optical monitor, and a vibration monitor. In the eddy current monitor output, the first-layer clear point is shown in actual data and differential data. In the friction monitor output the, Cu clear point is also shown in actual data and differential data. In the optical monitor output, the residual oxide thickness output data using 100-nm waves and 60-nm waves are adopted. Shorter waves can give us the finer information of the film thickness. In the vibration monitor, the end point of STI is shown as an example.

3.5.3 Other Applications of the Monitors

There are other applications of the monitors described above. One application is in decreasing the raw process time (RPT) as shown in Fig. 3.25. Case A shows that the polisher, the cleaner, and the monitor are used stand-alone. Case B shows that the cleaner is integrated in CMP (dry-in/dry-out type). Case C shows that the monitor is also integrated in CMP. In case A, RPT is 120 min and the ratio of the actual polishing process is only 30% of RPT. In case B, RPT is shortened to 80 min and the ratio of the actual polishing process is up to 44 % of RPT. In case C, RPT is more shortened to 40 min and the ratio of the actual polishing process is moved up to 93% of RPT. In cases A and B, rework polishing would be needed if requested by a stand-alone monitor. If an *in situ* monitor is not used, the main polishing step often stops prematurely in order to avoid overpolish. However, this process usually leads to underpolishing. Therefore, the rework polishing is needed in cases A and B. In case C, RPT can be minimized as the rework is eliminated. The other merit of monitors is the APC feature. By using the information of monitors, the polishing process can be precisely controlled, and it can also be controlled by e-manufacturing.

3.5.4 Communication

Figure 3.26 shows a typical CMP control system. The functions of a control system include ensuring process and mechanical stability and conducting diagnosis for automation and maintenance purposes. In practice, selected parameters are detected on-site and relayed upstream using e-manufacturing systems.

For diagnostic purposes, relevant parameters are collected and logged into the control system in order to determine appropriate maintenance periods and its service timing. For example, noise and vibration can serve as indicators of the status of health for a polisher. Large pressure and temperature valuations may provide alarms for malfunctions or problems within fluid lines. Such techniques have been developed and implemented in other technical fields, such as power generation machinery. The adoption of these mature technologies significantly improves the uptime of CMP system.

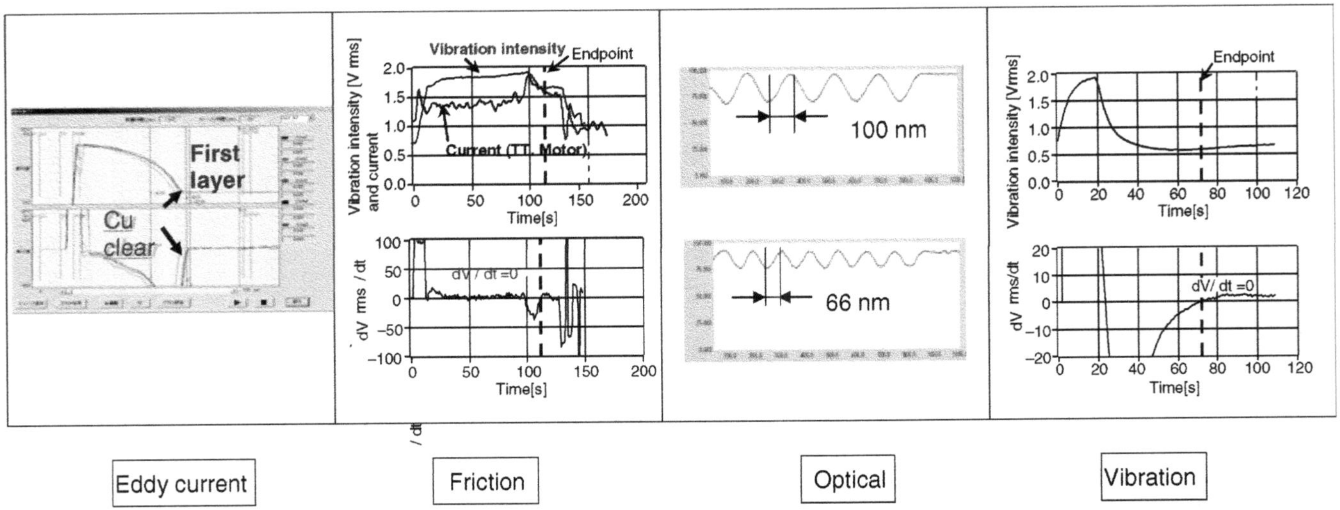

FIGURE 3.24 Schematic illustration of representative output characteristics of the various monitors.

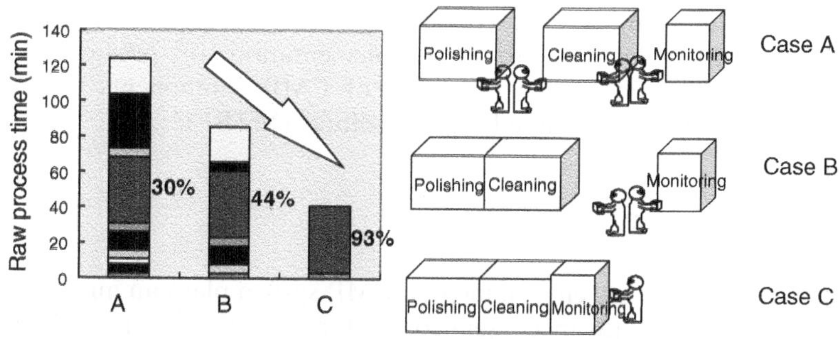

FIGURE 3.25 Schematic and graphical representations of the dependence of the raw process time and its dependence on polishing time, cleaning time, and monitoring.

Another important benefit of implementing an advanced control system is the reduction in maintenance requirements and decrease in unexpected failures. Despite the advanced state of automation technologies and/or diagnosis technologies, the most important automation parameter is the reliability of each component. In the case of 300-mm fab, minimal direct human intervention is required. An advantage of this mode of operation is the elimination of failure caused by accidental human errors. The disadvantage on the contrary is that there is no scope for any human intervention to correct any failures immediately or reduce the severity of the failure. Therefore, the overall

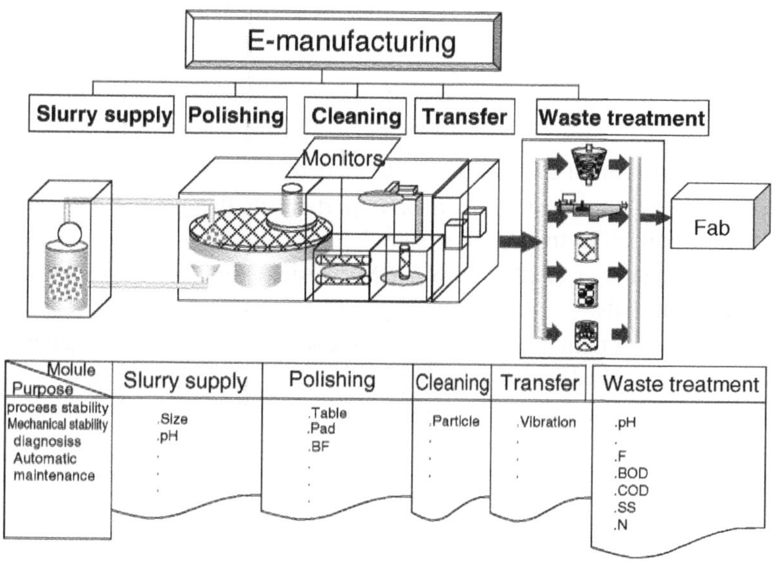

FIGURE 3.26 Schematic representation of e-manufacturing that includes online monitoring of slurry, cleaning solution, tool and process health, and combing the slurry waste disposal into the system. E-manufacturing reduces the cycle time and the human intervention.

reliability requirements for all components used in a CMP system have increased significantly due to the implementation of greater level of automation. For example, the reliability of CMP components used today must exceed 500 h of mean time between failures (MTBF).

3.6 CONCLUSIONS

In today's manufacturing environment, a CMP system plays an important role in the fabrication of advanced semiconductor devices that use sub-100-nm technology and beyond. The industry has gone through a period of diverse platforms and segmented market. The rotary type has emerged as a dominating polishing platform that may extend the applications of CMP the 450-mm wafer generation. A CMP system is now highly automated and integrated with sophisticated *in situ* monitoring components that improve the operating efficiency by cutting down the maintenance requirement as well as polishing yield by adjusting polishing parameters on the fly through a self-learning process. In addition, the seamless integration of post-CMP cleaning module has also remarkably improved the quality of the polished wafer in meeting the ever-increasing demand for defectivity reduction. The key challenges for a CMP system in the near future include the integration of porous low-*k* dielectric materials and accommodation of wider range of new materials used in the 45- and 32-nm technology nodes. In the foreseeable future, the CMP community must also address the issues related to the 450-mm wafer generation.

QUESTIONS

1. Describe the essential technologies and their roles in the CMP system.
2. What would be the most difficult or important challenges the 450-mm wafer generation will bring to a CMP system?
3. What still remains a challenge to a CMP system for the integration of low-*k* dielectric materials?
4. List all components in a CMP system that could contribute to the generation of defects on a polished wafer. How?
5. Design your own dream CMP system.

REFERENCES

1. Sorenson CT. Semiconductor manufacturing technology: semiconductor manufacturing process. MME Course Material. NSF/SRC Engineering Research Center for Environment Benign Semiconductor Manufacturing. University of Arizona. Tucson (AZ).
2. Jarrell S. Training Manual for Novellus CMP Polishers; 2005.

3. Kim S-D, Hwang I-S, Choi K-S. Hard-pad-based CMP of premetal dielectric planarization. J Electrochem Soc 2003;150(8):G450–G455.
4. Nguyen V, Van Kranenburg H, Worlee P. Dependency of dishing on polish time and slurry chemistry in Cu CMP. Microelectr Eng 2000;50:403–410.
5. Fayolle M, Romagna F. Copper CMO evaluation: planarization issues. Microelectr Eng 1997;37/38:135–141.
6. Kao Y-C, Yu C-C, Shen S-H. Robust operation of copper chemical mechanical polishing. Microelectr. Eng 2003;65:61–75.
7. Murarka SP, Steigerwald JM, Gutmann RJ. Chemical Mechanical Planarization of Microelectronic Materials. Wiley; 1997.
8. Oliver MR, editor. Chemical–Mechanical Planarization of Semiconductor Materials. Springer; 2004.
9. Radojcic R, Pecht MG, Rao G. Guidebook for Managing Silicon Chip Reliability. CRC Press; 1999.
10. Cook LM. Chemical processes in glass polishing. J Non-Crystalline Solids 1990;120 (1–3):152–171.
11. Steigerwald JM, Murarka SP, Gutmann RJ, Duquette DJ. Chemical processes in the chemical mechanical polishing of copper. Mater Chem Phys 1995;41(3):217–228.
12. Tseng W, Liu C, Dai B, Yeh C. Effects of mechanical characteristics on the chemical–mechanical polishing of dielectric thin films. Thin Solid Films 1996;290–291:458–463.
13. Luo Q, Ramarajan S, Babu SV. Modification of the Preston equation for the chemical–mechanical polishing of copper. Thin Solid Films 1998;335(1–2):160–167.
14. Tsu W, Wang Y. Re-examination of pressure and speed dependences of removal rate during chemical–mechanical polishing processes. Electrochem Soc Lett 1997;144(2):L15.
15. Zantye PB, Kumar A, Sikder AK. Chemical mechanical planarization for microelectronics applications. Mater Sci Eng 2004;45(3–6):89–220.
16. Tsujimura M. Present state of CMP from the viewpoint of a CMP manufacturer. SEMI Technology Symposium 94 Proceedings. 1994. p 251–257.
17. Tsujimura M. Defect free cleaning in the 65 nm TN beyond. Semicon Korea 2005 Proceedings; 2005.
18. Mertens P, Meuris M, Heyns M. Method and apparatus for removing a liquid from a surface of a rotating substrate.US patent 6,491,764. 2002 Dec 10.
19. Tsujimura M. Embedded process monitor and control in CMP tool. SEMI Technology Symposium; 2000. p 2-51-56.
20. Fukuda T, Shimizu Y, Yoshise M, Hashimoto M, Kumagai T. Proceedings of the Third International Symposium on Advanced Science and Technology of Silicon Materials. The Japan Society for the Promotion of Science; 2000. p 382.
21. ITRS (International Technology Roadmap for Semiconductors); 2005.
22. Doi T. Details of Semiconductor CMP Technology. Japan; 2000. p 61–66.
23. Hoshino S, Uda Y, Yamamoto E. Characteristics of copper CMP process for ultra-low pressure. Ninth International CMP Planarization for ULSI Multilevel Interconnection Conference; 2004 Feb 24–26.

24. Pant AK, Jairath R, Mishra K, Chadda S, Krusell WC. Integrated pad and belt for chemical mechanical polishing. US patent 6,328,642. 2001 Dec 11.

25. Ekonomikos L, Jamjin F-F, Simpson A, Ticknor A. STI planarization using fixed abrasive technology. Fut Fab Int 2002;12:217–220.

26. Okumura K, Aoki R, Yajima H, Ishikawa S, Tsujimura M. Method and apparatus for dry-in dry-out Polishing and washing of a semiconductor device.US patent 5,885,138. 1999 Mar 23.

27. Kajiwara J, Moloney GS, Wang HM, Hansen DA, Reyes A. System and method for pneumatic diaphragm CMP head having separate retaining ring and multi-region wafer pressure control. US patent 6,506,105. 2003 Jan 14.

28. Inaba T. Wafer polishing apparatus with retainer ring. US patent 6,033,292. 2000 Mar 7.

29. Tsujimura M, Ishii Y, Kimura N, Ota M. Polish profile control using magnetic control head. MRS 2004 Spring Meeting Unpublished.

30. Jeong HD, Lee SH. A new concept of conditioner for CMP. CMP-MIC Conference; 1988. p 209–215.

31. Kimura N, Sakata F, Takahashi T. Polishing endpoint detection method.US patent 5,639,388. 1997 June 27.

32. Kojima T. End point polishing apparatus and polishing method. US patent 5,876,265. 1999 Mar 2.

33. Li L, Barbee SG, Halperin A, Heinz TF. In-situ monitoring of the change in thickness of films. US patent 5,731,697. 1998 Mar 24.

34. Kobayashi Y, Nakai S, Tsuji H, Tsukuda W, Yamauchi H. Substrate polishing apparatus. US patent 6,758,723. 2004 July 6.

35. Li L, Barbee SG, Lee EJ, Martin FA, Wei C. Endpoint detection in chemical–mechanical polishing of cloisonné structures.US patent 6,291,351. 2001 Sep 18.

4

TRIBOMETROLOGY OF CMP PROCESS

Norm Gitis and Raghu Mudhivarthi

4.1 INTRODUCTION

Studies of mechanical and chemical–mechanical surface finishing treatments have long been a part of tribology, a science of friction and wear (material removal). During chemical–mechanical polishing or planarization (CMP), surfaces of both the object being treated (e.g., the semiconductor wafer) and the finishing tool (e.g., the polishing pad), together with the polishing slurry between them, constitute a rather classical tribological system. This is sometimes referred to as a three-body interface [1–3], because it includes two solids in relative motion and the slurry containing abrasive particles at the interface. Modern CMP machines may have an abrasive pad conditioner rubbing the pad during wafer polishing, which keeps one of the surfaces in a state of constant change. A tribological system is typically characterized by classical parameters of the coefficient of friction between the surfaces and by their wear rate.

The field of tribology has seen a number of theoretical models for various materials and interfaces [4]. Based on these models, the formulas for the calculation of both friction and wear have been derived. Unfortunately, these equations cannot be used directly by process and equipment design engineers unless the coefficient is known. Indeed, friction and wear properties of a tribological interface depend on macro-, micro-, and nanogeometry of the

Microelectronic Applications of Chemical Mechanical Planarization, Edited by Yuzhuo Li

rubbing surfaces, speed and acceleration of relative motion, stiffness and dampening characteristics of the mechanical system, physical and mechanical characteristics of ultrathin surface layers, organic films spontaneously formed on the surfaces, lubricating films formed by the polishing slurry, temperature, and humidity, and so on. The related fields of physics, mechanics, chemistry, and physical chemistry do not provide the exact knowledge of all these parameters. As a result, the only practical way of elucidating the crucial surface tribological parameters of the CMP interface is via *in situ* measurement during polishing.

4.2 TRIBOMETROLOGY OF CMP

Coefficient of friction, wear, acoustic emission, and surface roughness form the set of metrology data that are most crucial for a tribological phenomenon. To characterize surface friction behavior, measurement of its coefficient is preferred instead of simply measuring the force resisting the relative motion of the rubbing surfaces. Indeed, there are cases when changes in normal force cause substantial changes in the friction force, while the coefficient of friction remains constant. Alternatively, sometimes important changes in the coefficient of friction cannot be observed by monitoring the friction force due to periodic fluctuations of downforce.

The friction coefficient is defined as the ratio of the tangential friction force, resisting relative motion of the surfaces, to the normal load pressing the surfaces together. In the case of high adhesion of wet smooth surfaces, the normal load shall be considered as a sum of the externally applied normal force and the internal adhesion (stiction) between surfaces. For simplification, the normal load is usually considered equal to the externally applied normal force.

The coefficient of friction at the polishing interface reflects the nature of interaction of the wafer–abrasive pad materials. Since the material property of each constitutive material influences the coefficient of friction significantly, changes in the wafer surface can be analyzed by monitoring the coefficient of friction data either *in situ* or *ex situ*. The shear force at the interface and the friction coefficient depend on many aspects such as the pad's mechanical properties, the kinematic parameters of the polishing process, slurry viscosity, and chemical properties. For instance, under a constant normal force, a softer pad will experience a greater shear force at the leading edge of the wafer during polishing. The reason for this is twofold. First, on a macroscale, a softer pad will experience greater compression at the leading edge of the wafer in response to the applied normal load. This compression will subsequently result in the formation of a physical junction, which the wafer has to overcome continuously during its motion on the surface of the pad. Second, on a microscale, the pad asperities in the wafer–pad region will tend to collapse due to the relative softness of the pad. These two phenomena will combine to increase the net shear force between the wafer and the pad, and hence will result in the higher coefficient of friction for the softer pad as compared to the harder pad. The total wear of the interface, typically measured perpendicularly to the rubbing surfaces, consists of wafer and pad linear wear. The former is a useful process goal, whereas the latter outcome is a negative accompanying effect. At the wafer–pad interface, there is negligible pad linear wear (though there may be large pad glazing and deterioration) and supposedly substantial wafer material removal (up to the target level of wafer layer thickness). At the conditioner–pad interface, there is a significant pad linear wear with no conditioner wear (except for unavoidable dulling and deterioration of a conditioner surface, as well as loss of its abrasive particles).

The coefficient of friction used to generate Stribeck curves [5] offers an efficient means of monitoring a tribological process. Stribeck curves are generated using coefficient of friction data and the Sommerfeld number (see Fig. 4.1). These are very useful in determining the lubrication regime at the polishing interface.

The Sommerfeld number is defined as

$$S_f = \frac{\mu U}{p \delta_{\mathrm{eff}}}$$

where μ is the viscosity of the lubricant, U is the relative velocity, p is the applied pressure, and δ_{eff} is the effective lubricant film thickness. Applying the above formula to the CMP environment, the viscosity of the slurry can be easily found, since the pressure and velocity are known as they are the input process parameters. However, the fluid film thickness is the toughest to estimate. In a recent research [6] the fluid film thickness was estimated approximately using the pad surface roughness. To account for the deviations in the slurry film thickness on different grooved pads, a dimensionless factor has also been suggested.

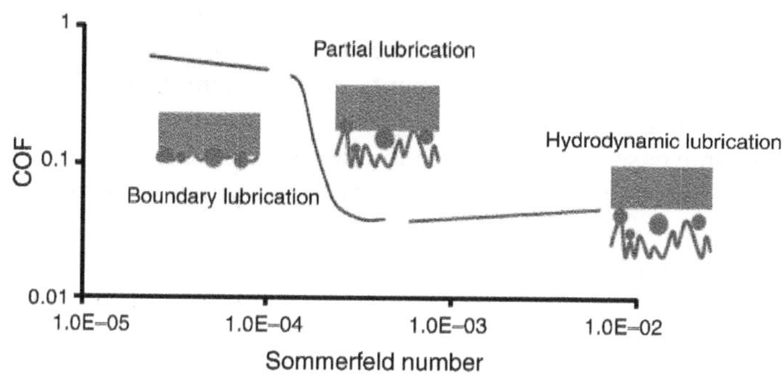

FIGURE 4.1 The generic Stribeck curve and the three modes of contact envisaged during CMP (from Ref. 6).

There are three main regimes of lubrication, namely (a) boundary lubrication, (b) mixed lubrication, and (c) hydrodynamic lubrication at a lubricated frictional interface, even though there are other minor regimes called hydrostatic and elastohydrodynamic lubrication regimes [2,3]. During CMP, solid–solid contact exists between the wafer and the pad during boundary lubrication, where the removal process is dominated by surface abrasion. In this regime, polishing results in severe surface damage due to aggressive abrasion by slurry particles and the polishing pad. Also, the thermal energy dissipated in this case must be very high, resulting in nonuniform and inconsistent material removal rate. In the mixed lubrication regime, there is a thin film of slurry that supports the applied pressure to some extent, and thus prevents aggressive abrasion. It is beneficial to ensure that the CMP process is conducted in this lubrication regime, as it would reduce the surface damage to a great extent. A hydrodynamic lubrication regime or a hydroplaning mode of polishing results from the applied pressure being totally supported by the slurry film present at the interface. This may result in a very low coefficient of friction but at the same time dramatically reduces the removal rate as there is practically no abrasion. Knowledge of the lubrication regime of polishing is thus highly beneficial to understand the polishing process in greater detail.

As mentioned earlier, the coefficient of friction reflects the nature of the surface features and the material that constitutes the interface. Coefficient of friction (COF) and contact acoustic emission (AE) (the AE signal detects acoustic energy dissipated at the interface due to mechanical interactions) can be constantly monitored and recorded during the polishing process in order to understand the interfacial tribology. Either *in situ* or postpolish analysis of the friction coefficient data allows to calculate the time taken to remove a particular thin film layer and thus the rate of material removal. However, these calculations can take place only after the entire layer is removed. But such monitoring of the coefficient of friction prevents overpolishing, thus avoiding defects such as dishing and erosion.

The removal of the last atomic layer of the exposed material to an underlying layer does not happen instantaneously. The time taken for this transition can be a measure of the nonuniformity of material removal: the longer the time of transition, the higher the polishing nonuniformity. The parallel acoustic emission measurements complement those of the friction coefficient (see Fig. 4.2) in detecting the end point of polishing [7]. For example, in some cases the difference in friction coefficients of the upper and lower wafer layers may be insignificant, but their acoustic response to CMP may be sufficient to detect the transition from one layer to the other. Also, in the determination of the end point of planarization, the high-frequency acoustics may sometimes be more informative than friction. In the following sections, a discussion of the applicability of tribometrology in consumable characterization, quality control, pad conditioning process optimization, and polishing process optimization is presented. Also, the effect of various factors on the friction at the interface will be briefly discussed.

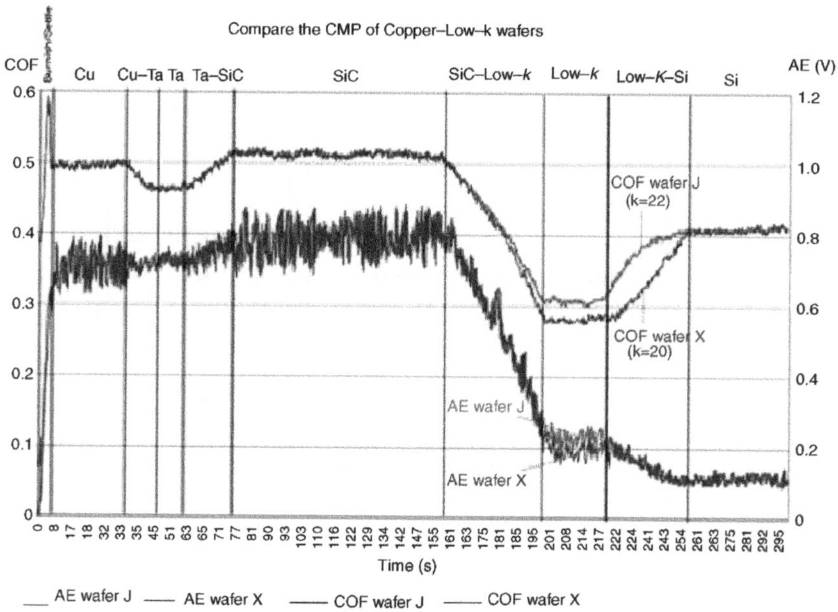

FIGURE 4.2 *In situ* CMP monitoring on copper–low-*k* wafers (from Ref. 7).

4.3 FACTORS INFLUENCING THE TRIBOLOGY DURING CMP

4.3.1 Process Parameters During CMP

The tribology at the interface is dependent on a wide variety of aspects such as the materials involved at the polishing interface, the normal load, the sliding velocity, and the temperature at the interface. Plots of COF and AE signal dependencies on loads and speeds for low-*k*, SiC, and Cu top layers as obtained during their polishing on a bench-top CMP tester are shown in Fig. 4.3. From the figure it can be seen that there exist conditions of "hydrodynamic contact" (no polishing) at low load and high speeds, light-to-medium polishing at higher load and lower speeds, intense polishing at high load and low speeds, and delamination at very high load and very low speeds. The COF recorded during the polishing experiments reflected the contact conditions (rubbing or floating) and also the impact of the materials being polished (e.g., 0.40 for low-*k* and 0.55 for Cu), and the AE signal reflected regimes of material removal and intensity of surface asperity interaction during polishing (<0.1 for floating, 0.1–0.5 for light polishing, 0.5–1 for intense polishing, and >1 with peaks for delamination). As a result, the optimum CMP conditions, at the highest polishing rate with no delamination, have been determined.

The slurry acts as a lubricant and coolant at the interface during polishing. The lubricant film separates the sliding surfaces and thus reduces friction

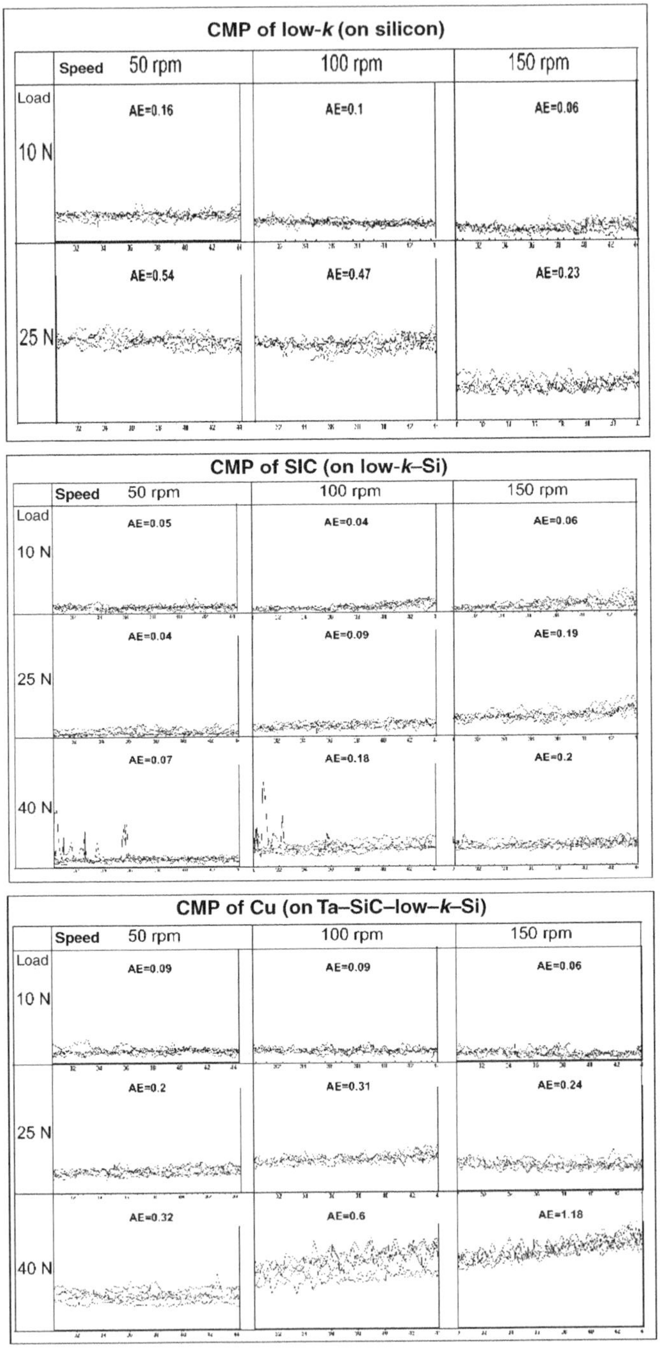

FIGURE 4.3 Coefficient of friction and real-time acoustic emission at different loads and speeds.

between them. During CMP, even though the slurry acts as a lubricating medium, it modifies the wafer surface and the abrasives present in the slurry abrade the chemically modified surface. Thus, the effect of slurry on the tribology is beyond the simple mechanism of lubrication offered by a fluid film. Lu et al. [8] investigated the effect of normal force and pad velocity on the slurry film thickness at the interface and correlated it to the friction coefficient data. Their data indicate that the thickness of the slurry film at the interface decreases with increase in the applied normal force and decrease in the pad velocity. An inverse relation was observed between the slurry film thickness and the coefficient of friction, emphasizing the role of slurry as an effective lubricating medium. The effect of slurry film was also studied by Runnel and Eyman [9]. Their study concluded that the pressure is partially supported by the slurry film at the interface. Thus, we understand that the slurry film and flow pattern significantly affect the frictional characteristics at the interface.

The effect of slurry flow rate on the frictional characteristics has been recently studied by a few research groups [10,11]. The Stribeck curve generated from the friction coefficient data as shown in Fig. 4.4 suggests that the polishing regime does not change. They concluded that the friction coefficient can be modified using the slurry flow rate without changing the polishing regime.

Another important factor that is often not taken into consideration is the process temperature during polishing. Recent researches [12,13] have elucidated the effect of temperature on the coefficient of friction during both copper and ILD CMP by conducting polishing experiments at different pad and slurry temperatures. Sorooshian et al. [12] have attributed the changes in coefficient of friction to the changes in pad properties, which result in an increase in shear force. Conversely, removal rate, surface chemical analysis,

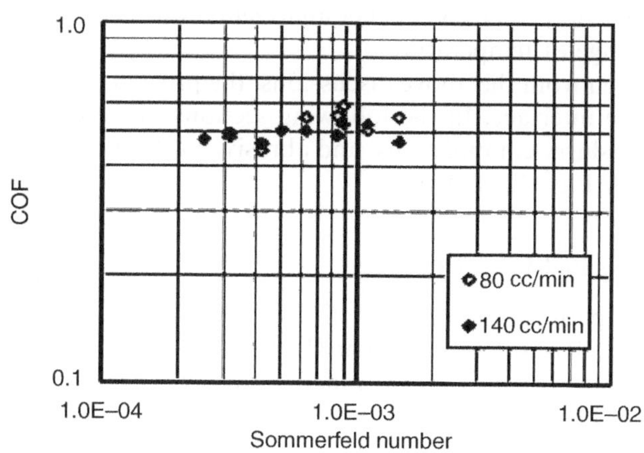

FIGURE 4.4 Stribeck curves generated for different slurry flow rates (from Ref. 10).

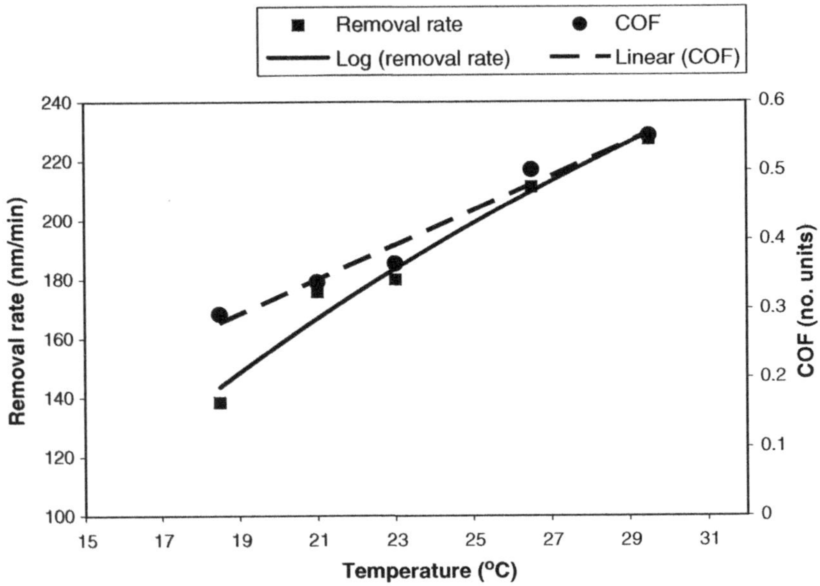

FIGURE 4.5 Coefficient of friction and removal rate versus slurry temperature (from Ref. 13).

and electrochemical studies were carried out by Mudhivarthi et al. [13] to discuss the possible reasons for the increased removal rate and coefficient of friction with increase in slurry temperature. From their data (see Fig. 4.5), it was seen that the coefficient of friction increases almost linearly with temperature. Apart from the surface chemical analysis, the change in viscosity of the slurry can also be a possible additional attribute for the increase in the coefficient of friction [14]. However, to deduce a firm inference, the change in coefficient of friction has to be further analyzed at different process conditions and a wider range of temperature.

As can be seen from the above discussions, the process parameters affect the tribology at the interface. In the following sections, the effect of various pad characteristics, slurry compositions, and abrasive particle characteristics on the tribology of CMP is elucidated.

4.3.2 Polishing Pad Characteristics

The grooves or perforations on the polishing pads have a significant impact on the polishing mechanism and outcome [15,16]. Grooves or perforations on the pad allow an effective slurry flow under the wafer surface and thus are very crucial for an effective CMP process. Philipossian et al. [6] carried out fundamental tribological studies during dielectric CMP on pads with different groove types at various slurry abrasive concentrations. Figure 4.6 presents data

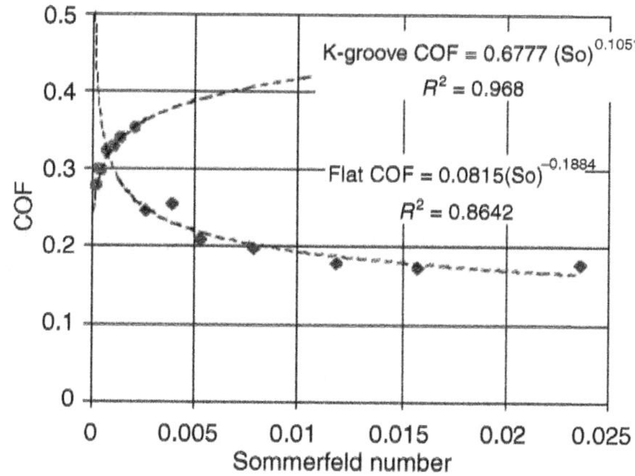

FIGURE 4.6 Values of β (tribological mechanism indicator) describing boundary lubrication for an IC1400 K-groove pad at 25% solids and partial lubrication for an IC1000 flat pad at 12.5% solids (from Ref. 6).

from their work, which show the effect of grooves on the coefficient of friction. Stribeck curves were generated using the friction data for a variety of groove and pad types as presented in Fig. 4.7. From the shapes of individual curves, the authors deduced that some of the pads were polished in partial lubrication regime and some in boundary lubrication regime at lower Sommerfeld numbers and transitioned to the partial lubrication regime. Consistent removal rates and uniformity were observed as long as the polishing regime was in the

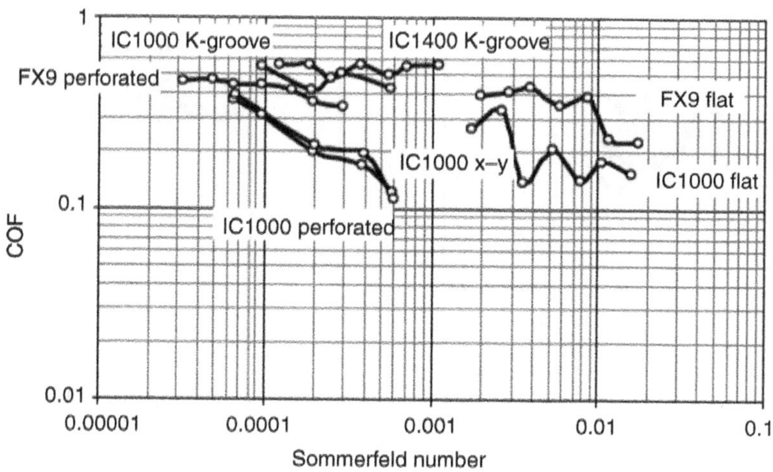

FIGURE 4.7 Stribeck curves for polishing using different pad materials and groove patterns (from Ref. 6).

boundary lubrication regime. However, polishing is aggressive in the boundary lubrication regime (where the particles abrade the wafer surface and a solid contact exists between pad and wafer), which might induce delamination during CMP for next-generation ICs, where mechanically weak low-*k* dielectrics are integrated with copper.

Analyses of Stribeck curves, Preston's coefficient, COF, and tribological mechanism indicator correlating with each other help to understand the polishing mechanisms. Such an analysis not only helps in the process development but also provides useful feedback to the pad development manufacturers.

4.3.3 Slurry Characteristics

Abrasive particles in the slurry provide a majority of the mechanical components during CMP, whereas the slurry chemicals modify the exposed copper surface. Thus, both the slurry chemicals and the abrasive particles play a major role in the abrasion process and the interface tribology. Li et al. [17] have studied the effects of slurry surfactant, abrasive size, and abrasive content on the tribology and kinetics of copper CMP. They generated Stribeck curves as shown in Fig. 4.8. From their results, it was concluded that the effect of slurry abrasive weight percentage had no effect on the tribological mechanisms of polishing, but the slurry size was shown to have a significant effect. They also concluded that the presence of surfactant significantly lowers the coefficient of friction. As per their findings, the removal rate during copper

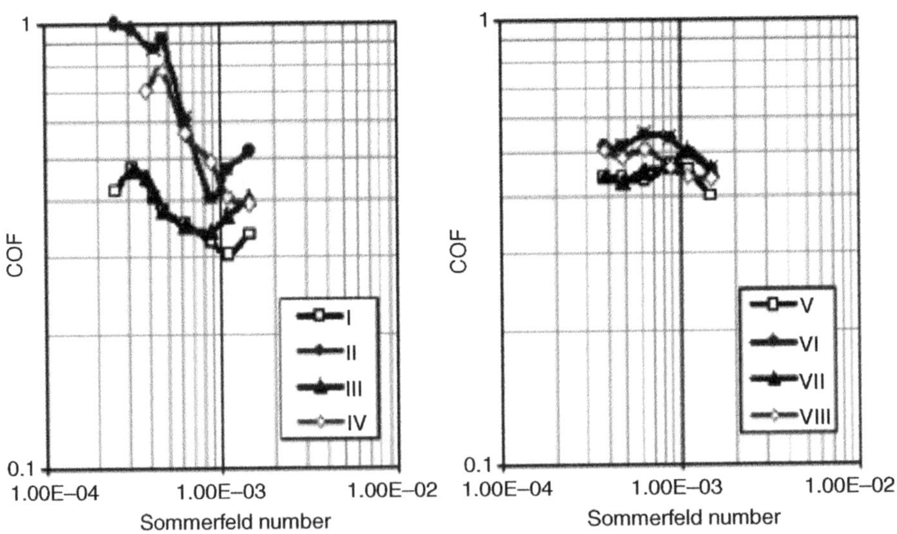

FIGURE 4.8 Stribeck curves for slurries with 13-nm abrasives (left) and 35-nm abrasives (right) (from Ref. 17).

CMP correlated more with the variation in frictional forces (stick–slip) rather than the COF value itself.

Investigations in the past also emphasized the effect of various additives of the slurry on the tribology during CMP [18,19]. It was observed that the presence of a flocculent reduces the surface frictional force (see Fig. 4.9a). Also, the ionic strength of the slurry has a significant impact on the frictional force (shown in Fig. 4.9b). It can be noted from these data that the friction at the surface depends significantly on the chemical interaction at the interface, besides the mechanical aspects.

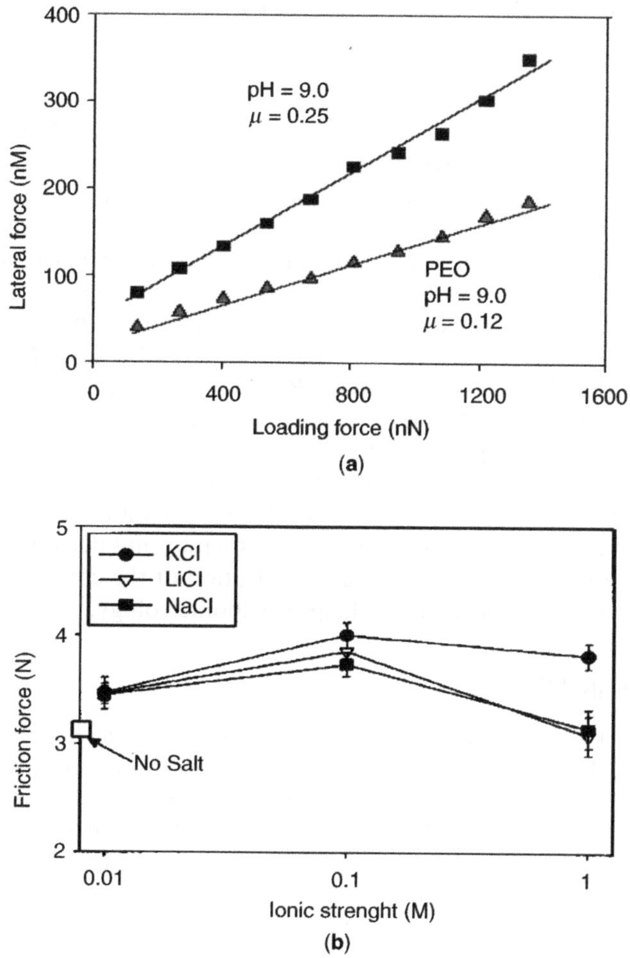

FIGURE 4.9 (**a**) Friction force dependence on flocculant (from Ref. 18). (**b**) Friction force dependence on ionic salts (from Ref. 19).

4.3.4 Wafer Contour Characteristics

The wafer contour determines the area of contact between the wafer and the pad along with the abrasives. Thus, the amount of surface asperity interaction and the particle–wafer interaction also depends on the wafer contour. The fluid film that is in contact with the wafer surface is also dependent on the wafer contour. Thus, the pressure experienced by the wafer at different applied pressures and velocities changes with the shape of the surface. Scarfo et al. [20] conducted polishing tests on wafer samples with concave, convex, and intermediate surface contours and noted that the shape of the wafer affects the coefficient of friction.

4.4 OPTIMIZING PAD CONDITIONING PROCESS

As the surface of a new (unconditioned) polishing pad is, in general, smooth and wets poorly, it does not provide good slurry transport to the pad–wafer interface. Pad conditioning is therefore necessary to open up the closed cells in the polyurethane pad to provide a consistent polishing surface throughout the pad's lifetime. Underconditioned pads are prone to have a glazing effect on their surfaces, resulting in reduction of the surface roughness. Such a reduction in surface roughness reduces the removal rate and increases nonuniformity. Overconditioned pads result in excessive loss of pad material, which dramatically reduces the pad lifetime. Pad replacement becomes necessary owing to two reasons: changes in physical properties of the pad and changes in grove dimensions because of pad wear. Replacing the pad prematurely would increase the cost of consumables and the machine downtime, which affect the throughput of the CMP process and overall process operational costs and thus the productivity index of a manufacturing process. Hence, it is crucial to understand the governing factors of an effective conditioning process. Many aspects of the polishing process such as process end-point detection, evaluation of pad conditioner performance, and optimization of the conditioning process variables can be achieved with the help of tribometrology.

4.4.1 PadProbe™

Gitis et al. [21] demonstrated the effectiveness of a novel noninvasive instrument called PadProbe™, developed by CETR Inc., in monitoring pad surface quality and pad life. PadProbe™ can be installed on rotational, orbital, and linear type CMP polishers (refer to Fig. 4.10). This sensor can monitor pad surface friction and wear *in situ* during the conditioning process. The only requirement is that the sensor should be in direct contact with the pad surface.

PadProbe™ allows either continuous or periodic *in situ*, in-process control of two crucial CMP parameters: pad life (dynamics of pad wear) and pad condition (dynamics and level of pad friction). Data obtained from such a

FIGURE 4.10 PadProbe^{TM} installed on a Strasbaugh polisher (from Ref. 21).

probe can be very effective in a production environment for determining the timing to start and finish *ex situ* pad conditioning, the extent of pad conditioning, the effective pad life based on performance, and the window of optimized pad performance in wafer-to-wafer uniformity.

Pad life is currently conservatively estimated by the number of polished wafers. A direct and real-time measurement of the pad wear will allow the determination of the exact time to replace the pad. Pad surface condition is currently estimated by the results of wafer polishing. The direct real-time monitoring of the pad surface will allow better control of the conditioning process. Pad condition, defined as the coefficient of friction of the pad surface, depends on the contacting materials (wafer, pad, chemicals, by-products), surface roughness (wafer pattern, pad conditioning), relative speed, contact pressure, temperature, presence of water or slurry, its flow rate, distribution (pores, grooves), and viscosity. Therefore, pad condition is the most comprehensive parameter characterizing the state of the pad surface.

Examples of the typical experimental data for wafers polished with *ex situ* pad conditioning, correlating the pad wear (shown as pad thickness change), pad condition, wafer removal rate (RR), and within-wafer non-uniformity (WIWNU), are shown in Fig. 4.11. The experiment included polishing of four groups of wafers, each group consisting of three blanket oxide (TEOS) wafers, with pad preconditioning (T_0–T_1), regular conditioning (T_2–T_3 and T_6–T_7), shortened conditioning (T_4–T_5) between the groups, and without conditioning.

The initial portion of the graphs from the beginning T_0 to moment T_1 corresponds to the preconditioning of a just-installed pad. During this procedure, the pad wear increases and the pad condition also increases and reaches its initial working level, designated as 100%, at time T_1', after which it practically does not change upon preconditioning. The second portion of the graphs from T_1 to T_2, as well as the portions T_3 to T_4, T_5 to T_6, and T_7 to T_8, corresponds to polishing. During these time periods, the pad surface gets clogged with particles and gradually loses its quality, but not its thickness. Therefore, the pad wear stays substantially constant, while the pad condition gradually drops to the level close to 70% of the initial working level. The *ex situ* conditioning (periods T_2–T_3, T_4–T_5, and T_6–T_7 on the time scale) restores the polishing properties of the pad, and the corresponding areas on the graphs are similar to the one described above for the interval T_0–T_1. Thus, the pad wear increases during every conditioning procedure, whereas the pad condition rises during conditioning and then falls during polishing.

Pad surface conditions dictate the RR for the given combination of materials and process parameters. Pad surface conditions also affect WIWNU, which substantially increases during polishing and drops after conditioning. Thus, a strong correlation can be found among pad condition, RR, and WIWNU, which can be verified and established quantitatively through extensive modeling.

Real-time monitoring of the pad thickness and surface conditions allows extended pad use, as long as both these parameters are above the critical threshold, thus increasing pad life. Such an increase in pad life translates into

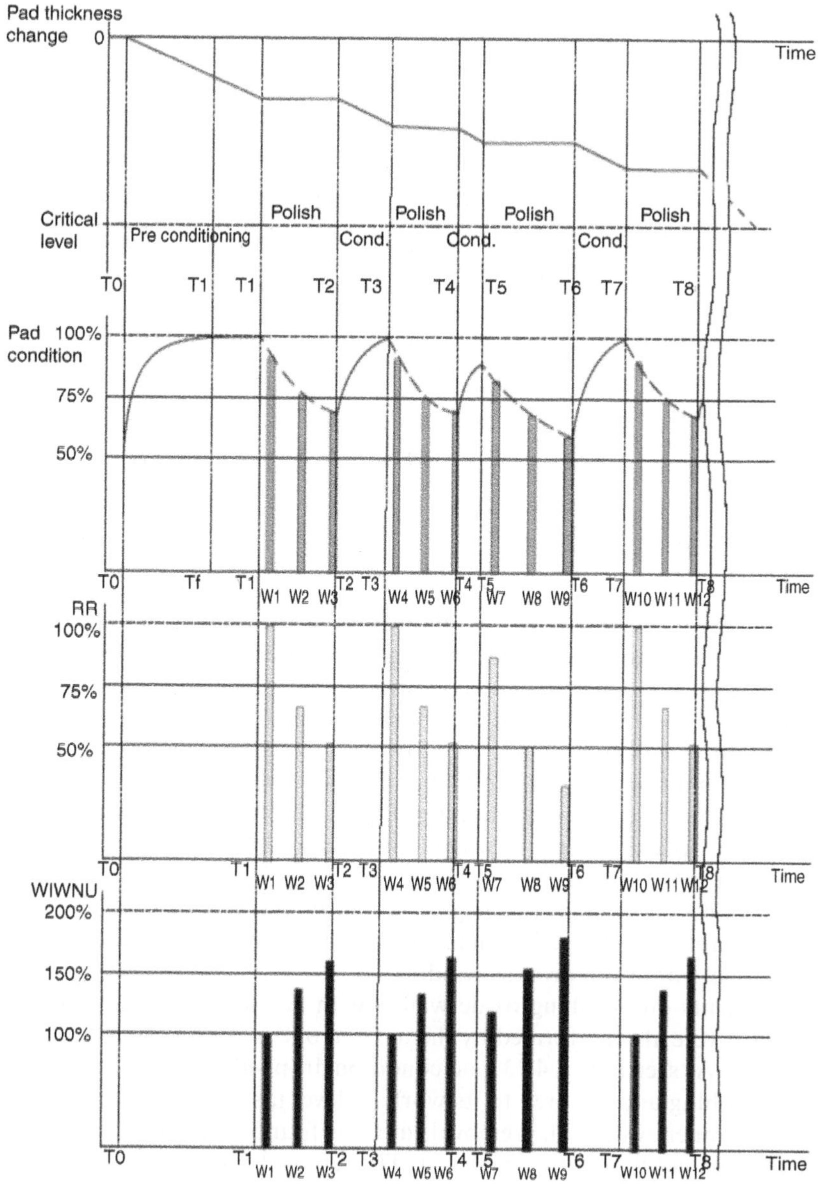

FIGURE 4.11 Correlation of PadProbe™ output with CMP parameters of RR and WIWNU (from Ref. 21).

increased number of wafers polished per pad and less frequent pad replacements [22]. This improves process throughput and reduces cost of consumables and machine downtime by huge margins, all contributing toward an efficient CMP process.

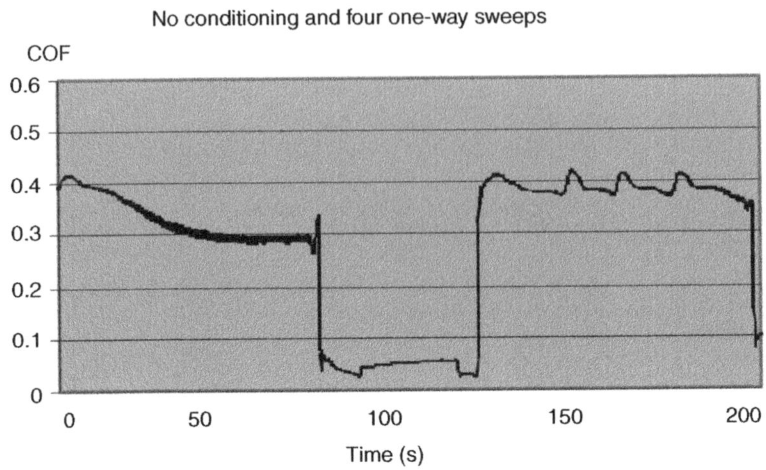

FIGURE 4.12 Pad condition during polishing with (right) and without (left) conditioning (from Ref. 22).

Noninvasive sensors such as PadProbe™ can also provide valuable information to characterize and optimize the conditioning process. Kalenian et al. [22] studied the effect of pad conditioning process with varying number of sweeps. The real-time data from the polishing experiments of oxide wafers with and without *in situ* conditioning on a Strasbaugh polisher are presented below. Both pad condition (*in situ*) and removal rate (postpolish) were measured. During polishing without conditioning (Fig. 4.12, left section of the graph), the coefficient of friction dropped, which reflects pad deterioration, and indeed, a reduction in removal rate was observed. Four one-way full-surface sweeps of the conditioner arm (Fig. 4.12, right section of the graph) were sufficient to maintain stable polishing, characterized by both the pad condition and the removal rate within the process window.

Experiments on polishing oxide wafers with *in situ* conditioning of various duty cycles were also performed by them, from one two-way sweep to four two-way sweeps. As seen in Fig. 4.13, pad condition dropped (as the pad deteriorated) during polishing and restored to its working level (as the pad refreshed) during each conditioner sweep. The removal rate was found to correlate well with the average pad condition during each test. The removal rate was the lowest for the one-sweep test and highest for the four-sweep test. Continuous pad conditioning carried out for 2.5 h with *in situ* measurements of pad wear is presented in Fig. 4.14. Manual pad geometry measurements were done to check its wear. Their results confirmed that the *in situ* pad wear values correspond to the posttest measurements. The coefficient of friction at the pad–conditioner interface gets stabilized after a certain period of conditioning and any further conditioning will not improve the condition of the pad surface [23]. This suggests that excessive conditioning will result only in increased pad wear but not the pad

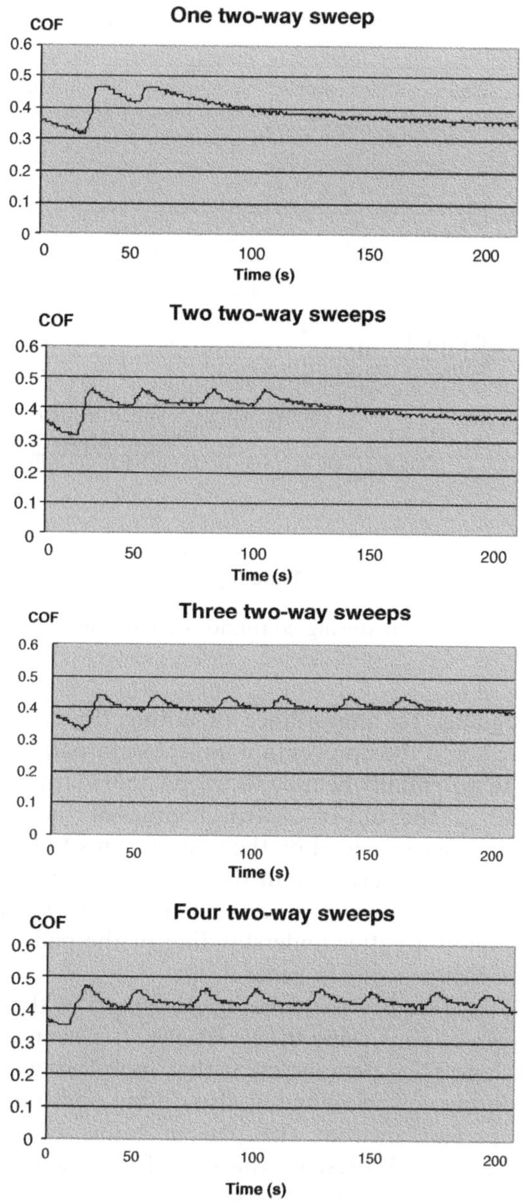

FIGURE 4.13 Pad condition during polishing with periodic conditioner sweeps (from Ref. 22).

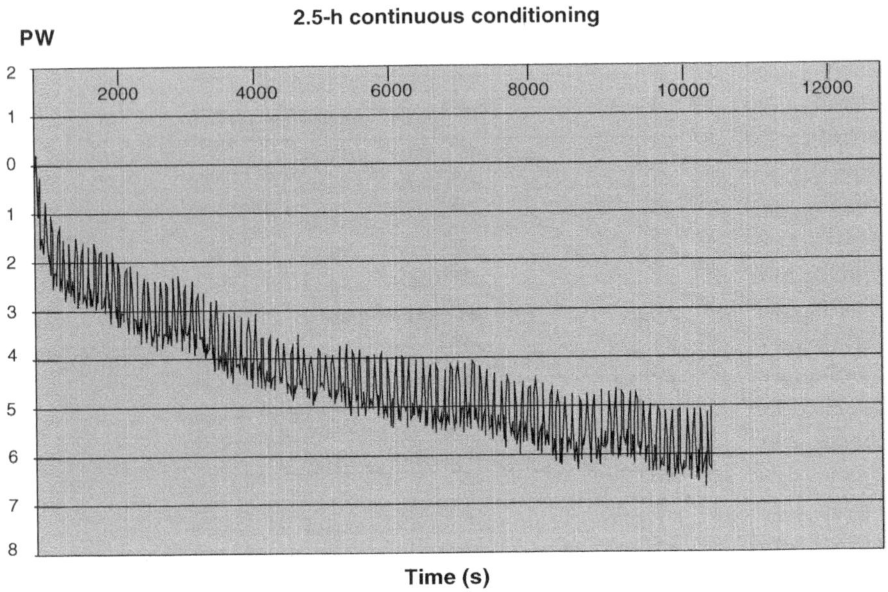

FIGURE 4.14 Pad wear during continuous conditioning (from Ref. 22).

condition. This signifies the importance of noting the end point of the pad conditioning process and thus preventing unnecessary pad wear.

PadProbe™ can potentially be used to trigger tool alarms when pad wear or pad condition fall in the out-of-control regime of the process. The pad condition signal response provided by the sensor seems to be more sensitive to changes in consumable properties than the measured removal rate and serves as an amplification factor for any process changes. This technique can enable the industry to develop a better understanding of the pad conditioning, and, moreover, of the CMP process, in general.

Fang et al. [24] have studied the tribological aspects of the pad conditioning process on two types of pads using the PadProbe™ instrument installed on an Ebara CMP polisher. They demonstrated that monitoring the pad condition (COF), not only during conditioning but also during polishing, can give a good insight into the performance of the pad in terms of removal rate and nonuniformity. Figure 4.15 presents the data from their work, where they monitored the pad condition on two types of pads during batch wafer polishing, with and without conditioning the pad. The pad friction monitored during polishing without conditioning indicated lower pad friction in the 25th wafer, which was found to affect the removal rate and polishing nonuniformity. The friction curves during polishing with conditioning were consistent with respect to the removal rate and nonuniformity. Figure 4.16 shows the pad wear associated with the conditioning and subsequent polishing process. From Fig. 4.16, it can be seen that majority of the pad wear occurs during conditioning.

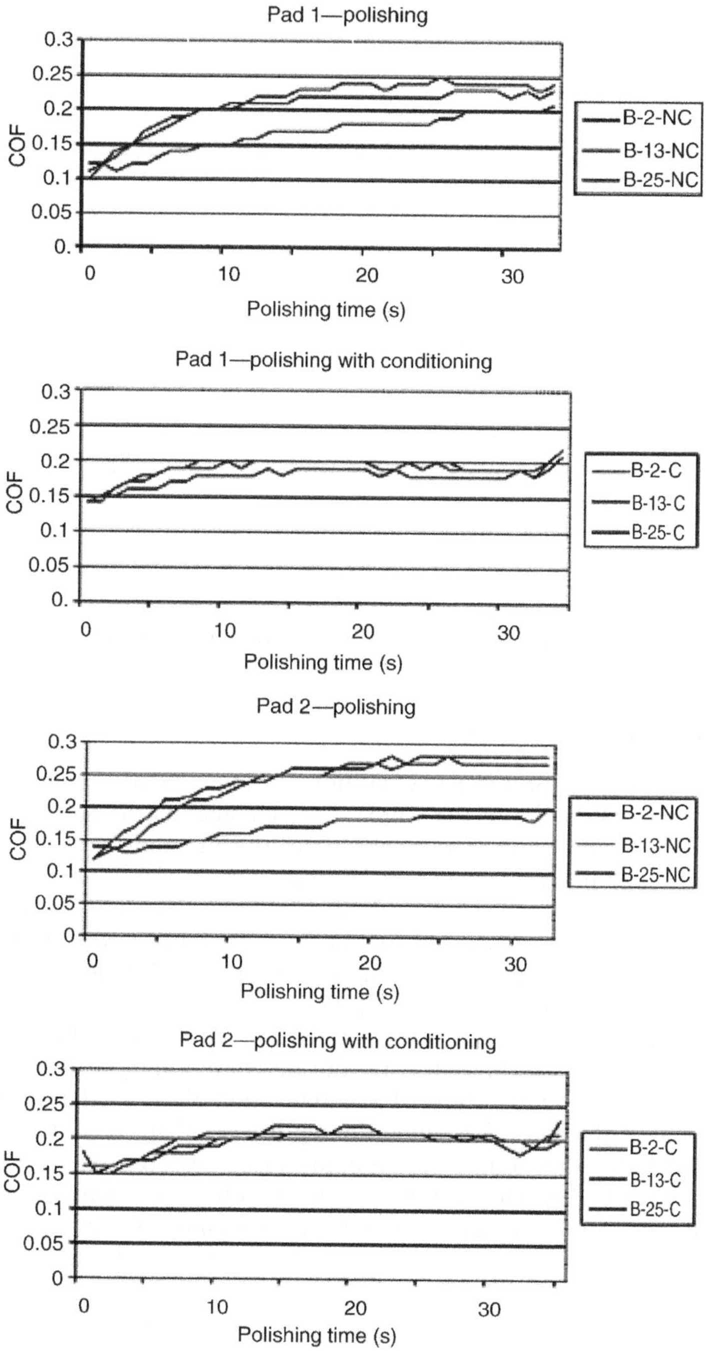

FIGURE 4.15 Pad condition during polishing tests with and without conditioning (from Ref. 24).

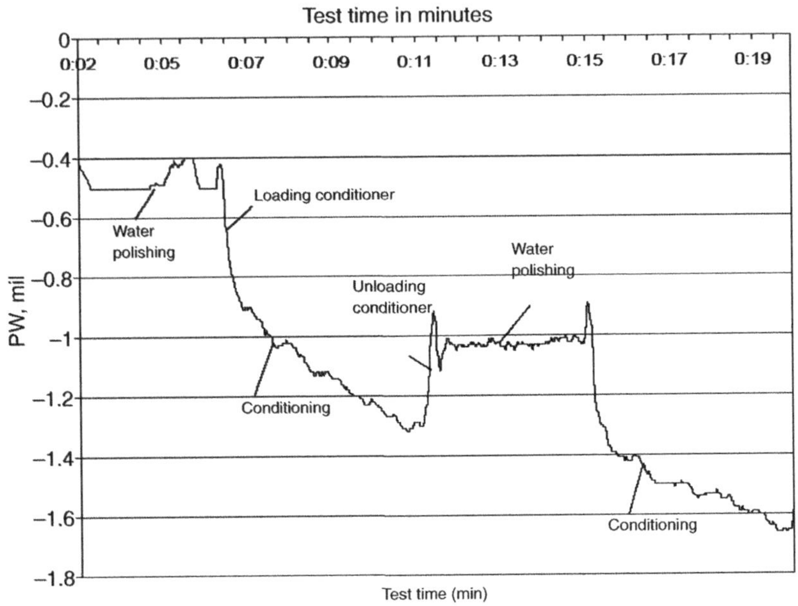

FIGURE 4.16 Pad wear during conditioning and polishing cycles (from Ref. 24).

4.4.2 Effect of Temperature

Another important aspect of pad conditioning is the role of temperature. CMP pads are highly sensitive to the temperature changes as the pad is a polymer material. The following discussion elucidates the effect of temperature on the pad conditioning process while monitoring coefficient of friction and pad wear. Pad conditioning experiments at different temperatures were carried out and the pad–conditioner coefficient of friction along with pad wear was monitored. It was observed that friction between the pad and conditioner increased with increase in conditioning temperature [11].

It can be seen from the real-time friction curves (see Fig. 4.17) that coefficient of friction stabilized faster during conditioning at lower temperatures than at higher temperatures. The stabilization of the coefficient of friction is a measure of the end of the conditioning process [23]. According to this, it can be concluded that longer conditioning processes are needed for full pad conditioning at higher temperatures. Along with the coefficient of friction, the pad wear, an important aspect during pad conditioning, was monitored during the experiments. Pad wear rate was measured using the real-time change in the CMP tester carriage position during these conditioning experiments. Figure 4.18 presents the real-time carriage position (change in thickness of the pad) and the resultant pad cut rate. The pad loss was observed to be high at the lowest temperature and decreased thereafter with increase in conditioning temperature. These observations suggest that the conditioning process is more

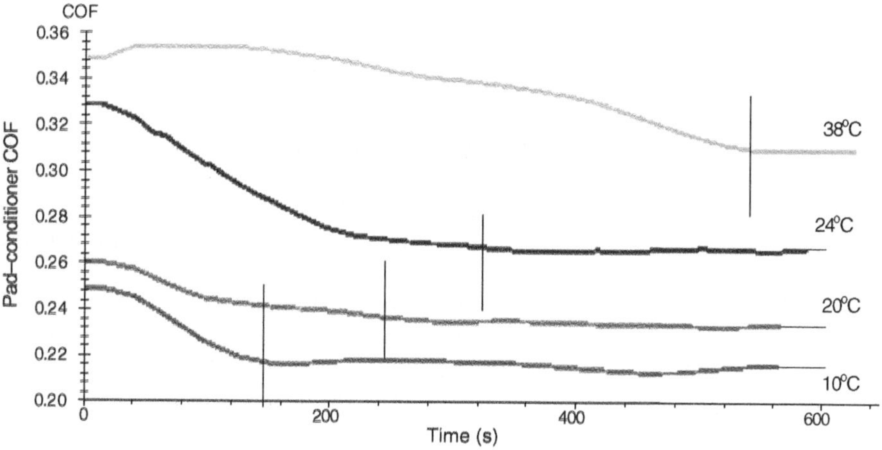

FIGURE 4.17 Coefficient of friction curves during conditioning at different temperatures (from Ref. 11).

aggressive, resulting in higher levels of pad wear at lower temperatures than at elevated temperatures. The reason for such a change in conditioning behavior was attributed to the change in the area of contact and mechanical properties of the pad at elevated temperatures. A possible change in surface hydrolysis of polyurethane pads under mechanical abrasion at elevated temperatures might also have brought about the change in the tribological characteristics during the conditioning process.

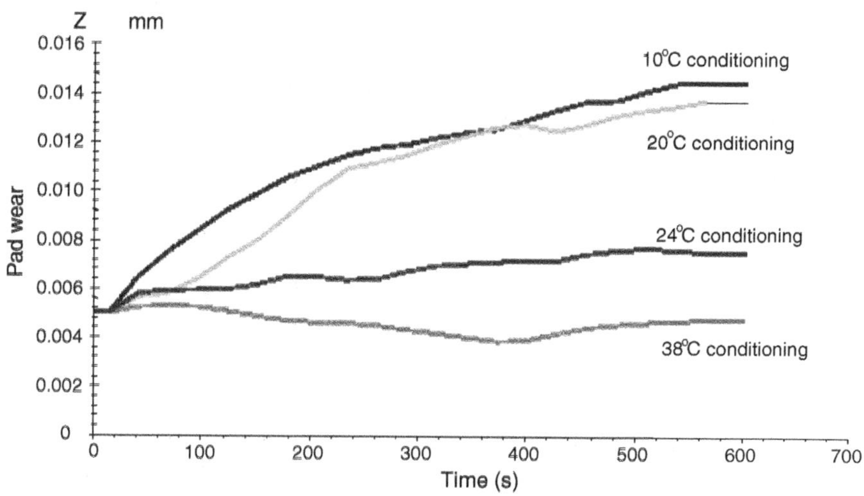

FIGURE 4.18 Pad wear during conditioning plotted versus time (from Ref. 11).

4.5 CONDITIONER DESIGN

The current pad conditioners used by the semiconductor industry result in loss of significant pad material upon conditioning, reducing the pad lifetime. Thus, it is very important to develop new gentle pad conditioners in order to improve the lifetime of the pads. The following section focuses on the usefulness of monitoring tribological aspects to facilitate an efficient pad conditioner design. Aspects that need to be considered for an effective design of pad conditioners are

1. Pad wear (cut) rate, which can be defined as the pad thickness loss over a period of continuous conditioning. It characterizes the abrasiveness, or aggressiveness, of the conditioner.
2. Pad refreshing (dressing) rate, defined as the time taken by a conditioner to change the friction level of a glazed pad to the friction level of a conditioned (unglazed) pad. It characterizes the effectiveness of the conditioner.
3. Pad lifetime.

A goal for conditioner design is to use a good pad conditioner that should be less aggressive (producing low pad wear rate) and more effective (taking short time to bring to and stabilize COF at its conditioned-pad level). Monitoring the friction coefficient and pad wear helps characterize pad conditioners and improvise their design.

Conditioner aggressiveness: The pad wear rate during conditioning indicates the abrasiveness, or aggressiveness, of a pad conditioner. The wear increases linearly with time at a constant wear (cut) rate.

Conditioner effectiveness: The effectiveness of a conditioner can be defined as the inverse of time it takes to regenerate the surface roughness of the pad from its glazed condition, which is brought about during polishing. Figure 4.19 is a

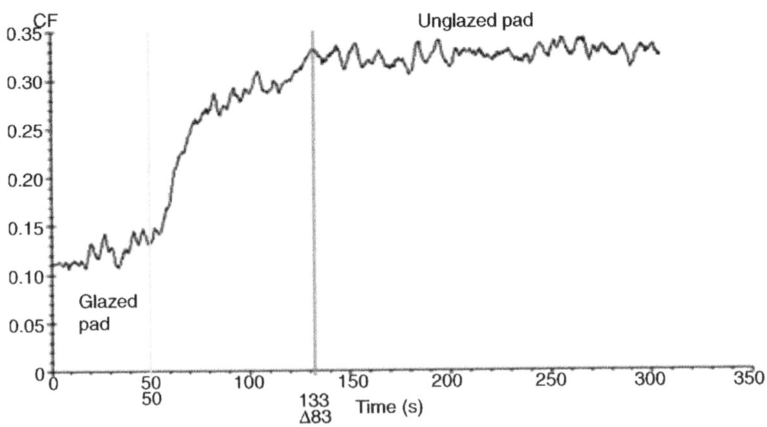

FIGURE 4.19 Graph of pad friction versus conditioning time (from Ref. 26).

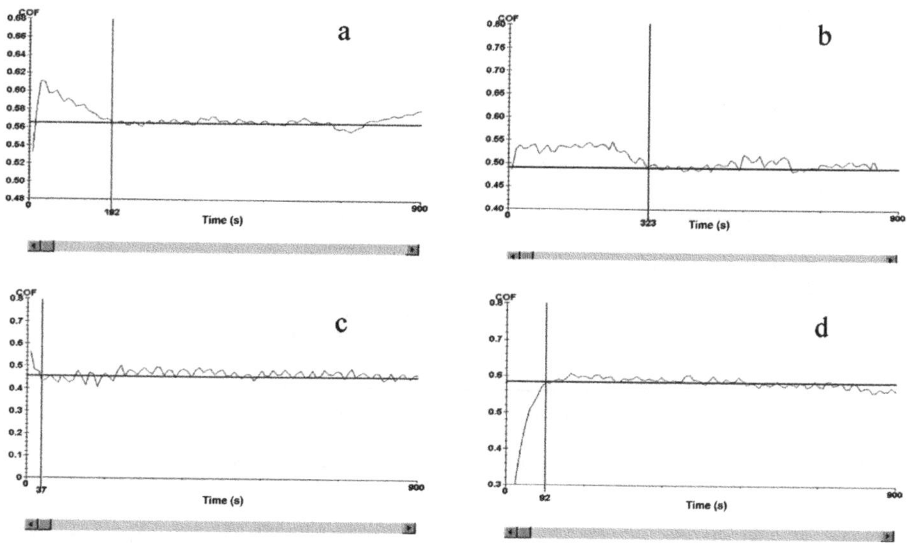

FIGURE 4.20 Comparison of conditioner effectiveness based on coefficient of friction versus time graphs (from Ref. 26).

typical plot of pad–conditioner COF versus time during the conditioning process [26]. The graph shows the variation of coefficient of friction at the pad–conditioner interface, which is very low for the glazed surface. As the surface gets conditioned, that is, as the surface roughness of the pad gets regenerated, the coefficient of friction increases and stabilizes after a certain period of time, indicating the end of the polishing process.

Pad conditioners can be evaluated for their effectiveness by comparing the time required to achieve a stabilized coefficient of friction as shown in Fig. 4.20. It can be seen from Fig. 4.28 that conditioners "a" and "b" take longer times than conditioners "c" and "d" to bring the pad condition to a steady state. In previous results, we have seen that the pad wear is almost linear with time. Thus, the longer a conditioner will take to bring the pad condition up, the higher will be the pad wear and less effective will be the conditioner. Again, pad wear is also dependent on aggressiveness of the conditioner, so care must be taken to manufacture conditioners by optimizing the levels of aggressiveness and effectiveness.

Figure 4.21 proposes a schematic of quadrants, which gives the ideal conditioner design criteria and the conventional conditioner characteristics. Indeed, conditioner manufacturers have been struggling to resolve the trade-off between conditioner effectiveness (pad refreshing rate) and conditioner abrasiveness (pad wear rate). The optimal situation for the conditioner design is shown in the first quadrant of Fig. 4.21, where conditioners will be effective and at the same time gentle on the pads. The abrading characteristics must be optimized by tailoring the shape and hardness of the conditioner asperities.

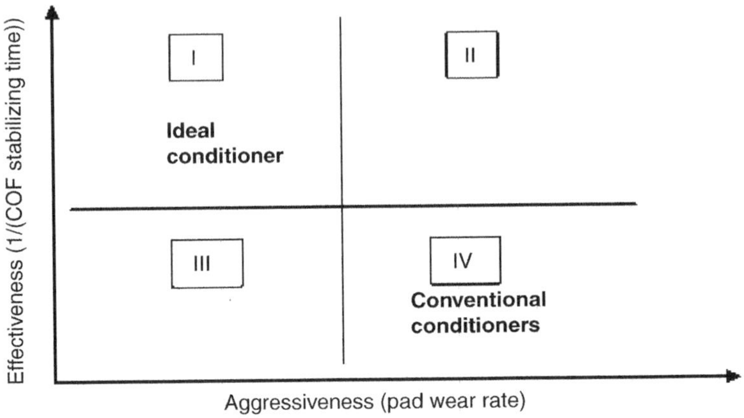

FIGURE 4.21 Schematic quadrants to group pad conditioners (from Ref. 26).

Along with the aggressiveness and effectiveness of the conditioner, the lifetime of the pad conditioner, that is, its durability, should also be taken into consideration in order to evaluate a conditioner. During the conditioning process, even though the conditioner is predominantly harder compared to the pad, blunting of the conditioner still occurs upon several runs, which affects its aggressiveness and effectiveness. This is called aging of the conditioner. Hosali et al. [25] conducted experiments on pad conditioning using two conditioners, 1 (used and dull) and 2 (new and sharp), and monitored pad condition, pad wear, and polishing removal rate. From Fig. 4.22, it can be seen that pad wear with the new conditioner (conditioner 2) was about 2.5 times higher than that with the used conditioner (conditioner 1). Also, as shown in Fig. 4.23, both the pad condition and the removal rate during polishing changed similarly with the conditioner age. The removal rate was about 6% higher and the pad condition was about 20% higher with the new conditioner (conditioner 2).

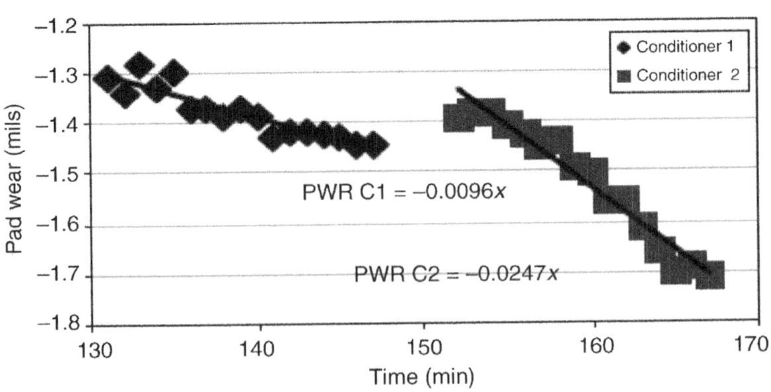

FIGURE 4.22 Pad wear dependence on conditioner age (from Ref. 25).

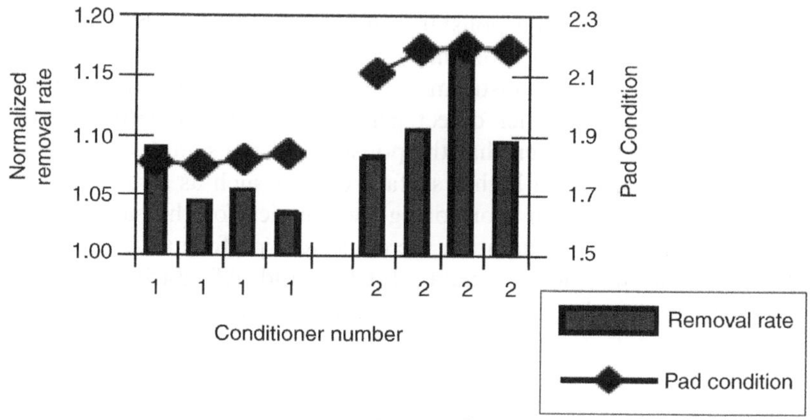

FIGURE 4.23 Removal rate and pad condition dependence on conditioner age (from Ref. 25).

4.6 CMP CONSUMABLE TESTING

CMP consumable quality and performance need to be consistent with the specifications, which is highly critical to have significant process control and yield. In this regard, tribometrology can be utilized to test the consumable quality and functionality.

4.6.1 Slurry Testing

Sikder et al. [27] conducted slurry characterization studies on a CMP bench-top tester. Figure 4.24 shows the COF and AE data collected during the polishing of four TEOS wafer coupons using slurry 1 and slurry 2. The AE and

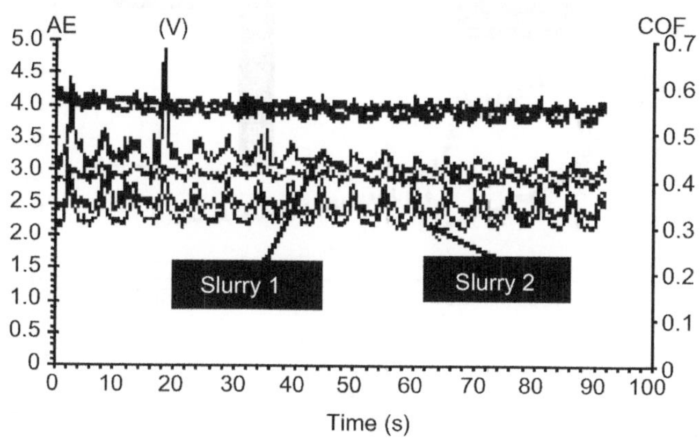

FIGURE 4.24 AE values (marked) and COF during polishing using slurry 1 and slurry 2 (from Ref. 27).

COF signals recorded during polishing show that although both the slurries have the same COF during polishing, they produce different levels of AE signals. Higher values of acoustic may be due to higher intensity of surface asperity interactions or higher defect generation through scratching on the polishing surfaces. It was seen that the polishing of the samples that produced more AE signals resulted in higher surface defects such as scratches. The AE signal is thus very useful in monitoring the slurries for the same batch and batch to batch in order to maintain the consistent performance.

Figure 4.25 shows the raw data of COF and AE during polishing of patterned copper samples with slurry 3 and slurry 4. From Fig. 4.25a it can be

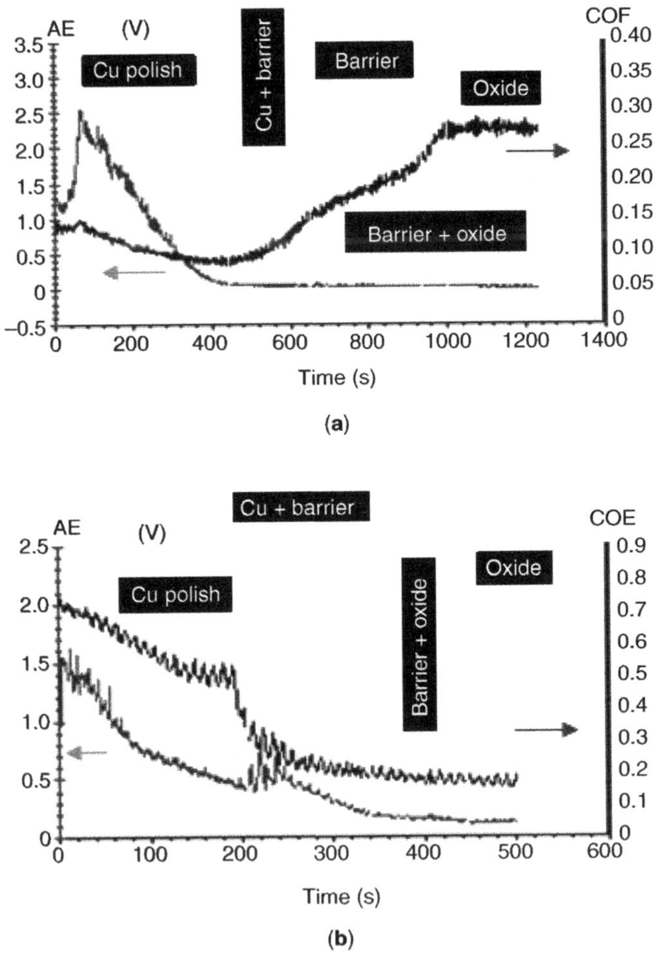

FIGURE 4.25 Complete Cu removal process monitoring with change in COF and AE values: **(a)** Cu polishing with slurry 3, where Cu is removed after 439 s of polishing, and **(b)** Cu polishing with slurry 4, where Cu polishing takes only 200 s [27].

seen that different stages of polishing can be well noted from COF and AE data. Also, it can be noted that COF is high both at the beginning of Cu polishing and at the end of Cu polishing, while polishing with slurry 4 than with slurry 3. Although slurry 4 yielded high removal rate and may be suitable for bulk copper removal during the CMP process of a Cu–TEOS system, the high friction may lead to delamination of a Cu–low-*k* system.

The coefficient of friction and acoustic emission signal can be effectively utilized for functional testing of consumables. An example of such a utility is illustrated in Fig. 4.26. For a quantitative, fast, and inexpensive estimate of the functional properties of two different products from a manufacturer, the slurry samples were tested on a CMP tester equipped with capability for continuous acquisition of the frictional and acoustic signals. The collected signals allow the determination of three important functional parameters, namely, time of complete tungsten removal, time of complete underlayer removal, and intensity of polishing (characterized by the level of acoustic signal).

It was found that six of the seven batches removed both the tungsten and the underlayer before the maximum amount of time allocated to it, whereas one batch (batch 1) took much longer time to remove the tungsten layer (the underlayer did not get fully removed during the test duration). This is consistent with the fact that the slurry gives a low acoustic emission signal, that is, not an aggressive interaction of surface asperities of pad and wafer or wafer

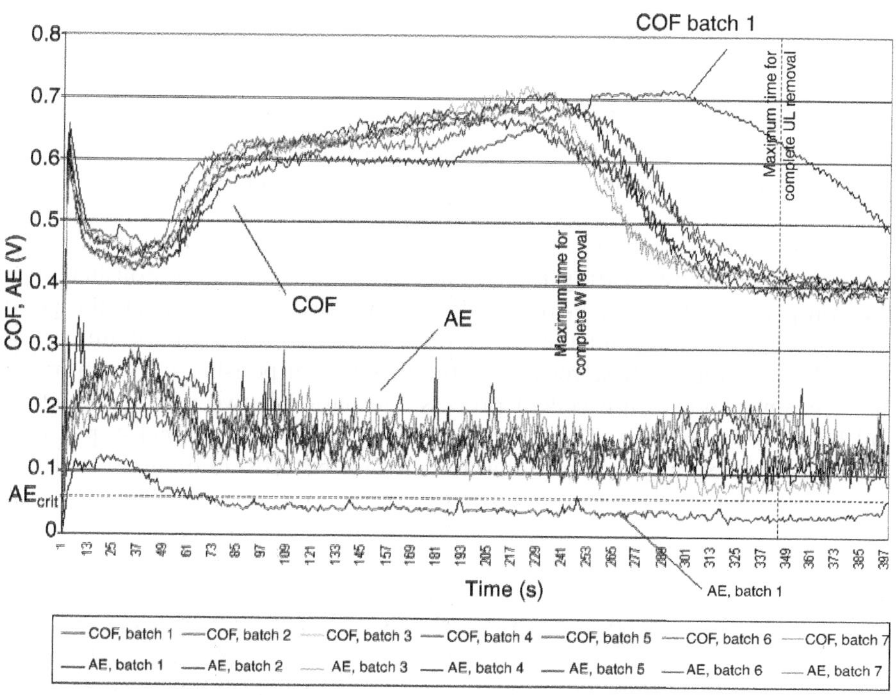

FIGURE 4.26 Quality control testing of tungsten slurry (from Ref. 31).

TABLE 4.1 Screening of Three Lots of Copper CMP Slurry.

No. of Lots/Tests		Bench-top CP-4 Data			Mirra Data
Lot	Test	COF	AE	RR	RR
1	1	0.602	1.34	6.019	
	2	0.628	1.32	6.070	
	3	0.589	1.35	6.172	
	4	0.593	1.31	6.045	
	5	0.617	1.38	6.243	
	Average	0.606	1.34	6.110	6.100
2	1	0.591	1.72	6.152	
	2	0.585	1.75	6.085	
	3	0.631	1.58	6.090	
	4	0.616	1.63	5.986	
	5	0.603	1.87	6.080	
	Average	0.605	1.71	6.079	6.100
3	1	0.533	1.30	5.472	
	2	0.517	1.33	5.514	
	3	0.525	1.29	5.610	
	4	0.546	1.35	5.469	
	5	0.537	1.34	5.531	
	Average	0.532	1.32	5.519	5.500

From Ref. 32.

and abrasive particles. The other batch under suspicion (batch 2), though exhibited slightly rougher acoustic profile and slightly different rate of tungsten removal, still fitted the three criteria and hence got accepted for production.

Another example of slurry prescreening is presented here. Three batches, lot 1 (in spec) and lots 2 and 3 (under suspicion), from a commercial copper slurry were inspected by replicating the conditions of an 8-in. Mirra polisher on a bench-top tester. The bench-top results could differentiate between the good lot 1, the faulty lot 2, and the faulty lot 3 (refer to Table 4.1 for numerical data). Verification of these results on the production polisher fully confirmed the screening data, with excellent process correlation between the production and bench-top machines. Thus, tribometrology collected on bench-top testers can be useful not only in slurry characterization but also in quality control.

4.6.2 Pad Testing

A polishing pad has a significant impact on the performance of the CMP process. It transports the slurry to the pad–wafer interface, impacts the polishing nonuniformity, and affects the global wafer and device planarity. Pads may consist of thin porous closed cell [28], open cell [29], or noncell [30] polyurethane material. The properties of polishing pad can be studied in detail by monitoring the output parameters of a variety of tests. Wear tests (with an upper conditioner either rotating or stationary, no wafer, on the rotating pad) are used to characterize pad wear resistance.

Functional polishing tests (with either *in situ* or *ex situ* conditioning) allow the quantitative estimation of pad functional properties, by the determination of four important functional parameters, namely, time of wafer layer removal (characterizing the removal rate), transition time from one wafer layer to another (characterizing the polishing nonuniformity), acoustic level (intensity of polishing), and pad wear (durability). It can be seen that pad 1 exhibits more intense polishing and faster tungsten removal (Fig. 4.27) and less pad wear (Fig. 4.28), whereas pad 2 exhibits slower tungsten removal (Fig. 4.27) and higher pad wear (Fig. 4.28) [27,31]. The coefficient of friction and acoustic emission curves during tungsten polishing were obtained from the acoustic emission sensor and load sensor that determine both the lateral and normal loads. Pad wear was determined by monitoring the position of the upper carriage. This methodology to determine the performance of polishing pads can be implemented during pad development. This would also help evaluate new pads to be used in fab environments. In the same context, Zantye et al. [32] tested Psiloquest's application-specific pads, which were surface coated with TEOS for different times as a part of the development of application-specific pads. They measured static COF on the pads, dynamic COF (during polishing), and tungsten removal rate. They correlated the data with the surface chemical and mechanical properties.

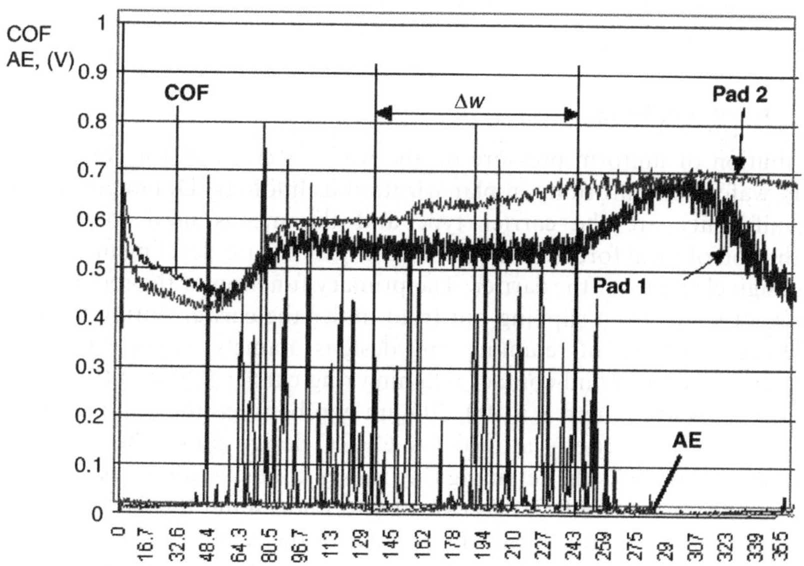

FIGURE 4.27 *In situ* pad friction and AE measurements (from Ref.31).

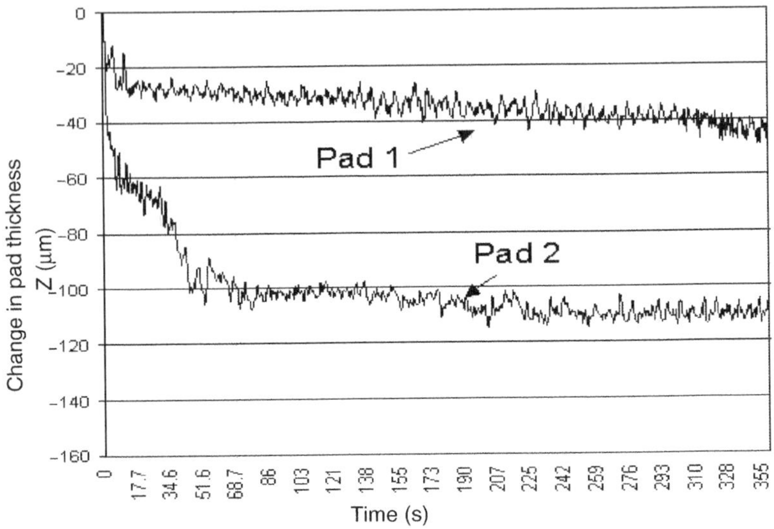

FIGURE 4.28 *In situ* pad wear measurements (from Ref. 31).

Tribometrology in combination with certain other metrology tools such as ultrasound scanning technique can be used to evaluate the quality of the polishing pads. Zantye et al. [33] have investigated the effects of nonuniformities in the pads on the tribological aspects during CMP. They concluded that the friction at the surface changes with a change in the uniformity of the polishing pads, which is measured using the ultrasound scanning technique.

4.6.3 Retaining Rings

Distribution of uniform pressure on the wafer is essential for controlling the within-wafer nonuniformity in planarization technology. Dominant factors for the uniformity are the carrier type and the means used to control the distribution of local forces. The retaining ring and wafer leveling means are two key design elements of the carrier. The primary function of the retaining ring is to prevent wafer from slipping out from under the carrier during the process. There are two types of retaining ring designs, namely, noncontact retaining ring carrier (NRRC) and contact retaining ring carrier (CRRC) (Fig. 4.29). In the case of NRRC, there is a 100–200-μm gap between the retaining ring and the pad, while the carrier is at rest. The CRRC design causes the retaining ring to contact the pad. As the retaining ring makes contact with the pad during the polishing process in CRRC, it is essential to understand the friction behavior, wear characteristics, and chemical resistance of the retaining ring material with respect to the process parameters in order to develop new materials or to apply optimum downpressure on it.

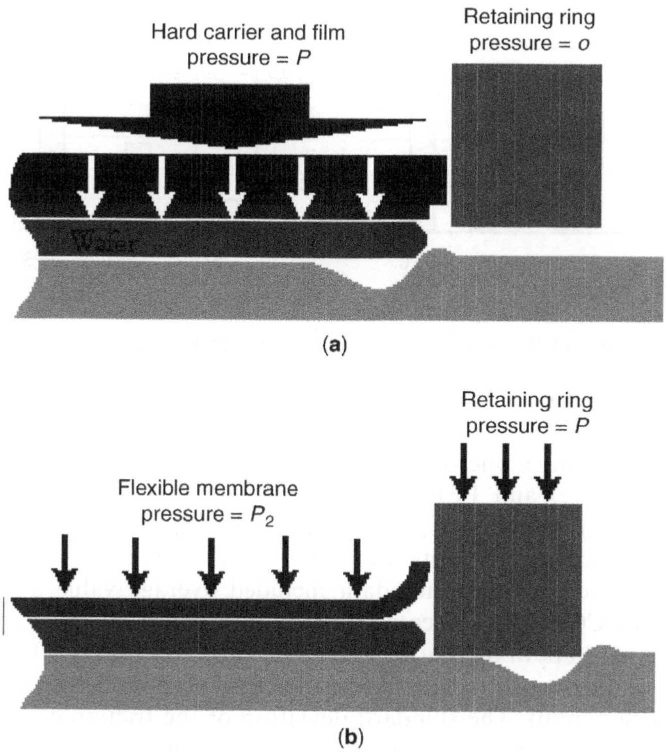

FIGURE 4.29 Schematic of a film-backed noncontact retaining ring carrier (**a**) and of a bladder-backed contact retaining ring carrier (**b**) (from Ref. 34).

Typical industrial plastic selection criteria have focused on pin-on-disk tests (involving plastic sliding over steel) and sand slurry abrasion tests. The CMP environment, however, is very different from these typical industrial tests. In CMP, the retaining ring plastic is subjected to a plastic-to-plastic adhesive force component involving the polyurethane pad, chemical attack from the chemicals in the slurry, as well as an abrasive component associated with slurry particles.

Tribological studies of potential candidate materials for retaining ring applications have been conducted in order to improve the retaining ring performance, thereby reducing defects and inconsistencies during polishing [34,35]. The coefficient of friction of the candidate materials in the presence of different kinds of slurries was measured on a bench-top CMP tester. The difference in weights of the material before and after polishing provided a measure of the removal rate, thus characterizing each material for its tribological performance and wear factor. Several types of plastics, as well as various commercial slurries, were used in the study. Several polymers have

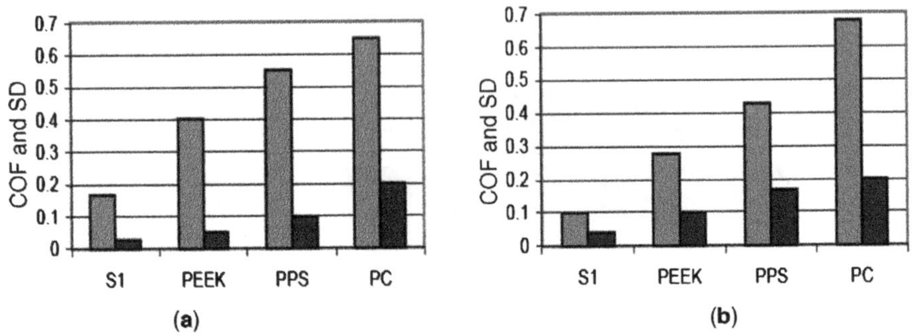

FIGURE 4.30 COF with (**a**) SW-2000 and (**b**) SS-12 slurries (from Ref. 34).

been tested, including numerous unfilled and filled PPS (polyphenylene sulfide), PC (polycarbonate) and PEEK (polyetheretherketone) resins, carbon fiber reinforced (CFR) Arlon, as well as a new proprietary material. The output parameters of the experiments were the wear of the ring materials and coefficient of friction; the latter data included average value and standard deviation. The COF data presented in Fig. 4.30a and b are not very different from each other in both the slurries. The data shown in Figs. 4.30 and 4.31 are the standard deviation of the friction coefficient data averaged over the sampling time (2 min). The standard deviation of the friction coefficient data can be considered a measure of vibration in the system, and thus, a small value is desirable.

Moussa and Quartapella [35] also compared the friction coefficients of various materials, among which they stated an improved performance of the new Arlon material to be applicable for CMP retaining rings. Figure 4.32 shows the coefficient of friction data for various traditional retaining ring materials along with newly researched materials.

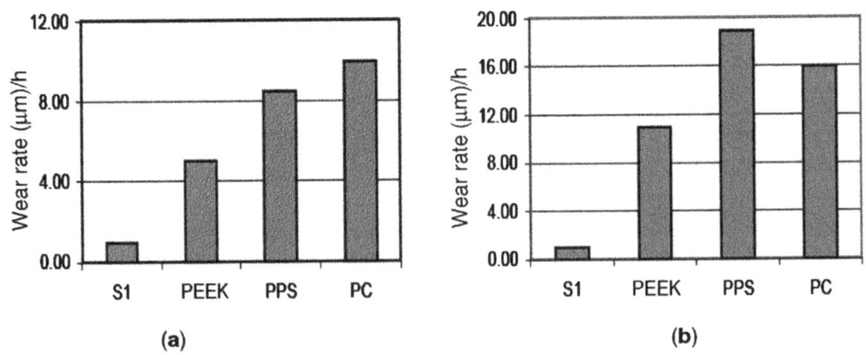

FIGURE 4.31 Wear rate with (**a**) SW-2000 and (**b**) SS-12 slurries (from Ref. 34).

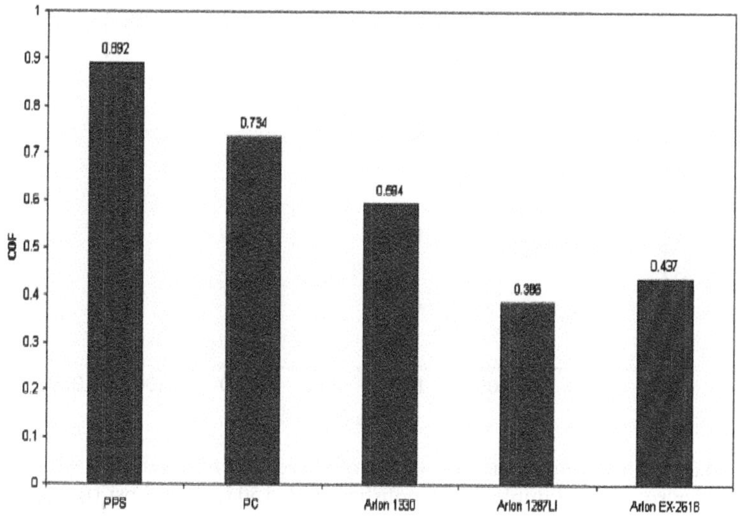

FIGURE 4.32 Average coefficient of friction comparison of all of the materials at the end of the test (28–30 min) (from Ref. 35).

4.7 DEFECT ANALYSIS

4.7.1 Coefficient of Friction and Acoustic Emission Signal

Among the major benefits of the measurement of acoustic and friction coefficients in CMP are that is allows monitoring the intensity of the polishing processes and detecting polishing regimes and conditions when mechanically weak low-*k* materials delaminate (Fig. 4.33). Figure 4.34 shows the AE signal

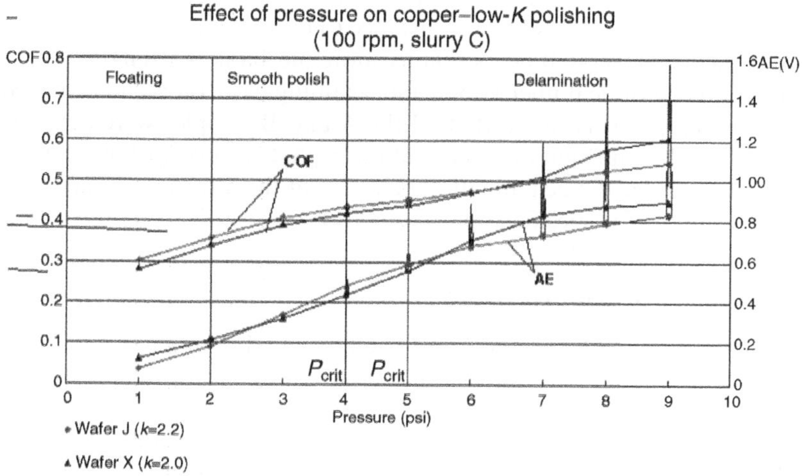

FIGURE 4.33 Coefficient of friction and acoustic emission signal measurements indicating low-*k* delamination (from Ref. 36).

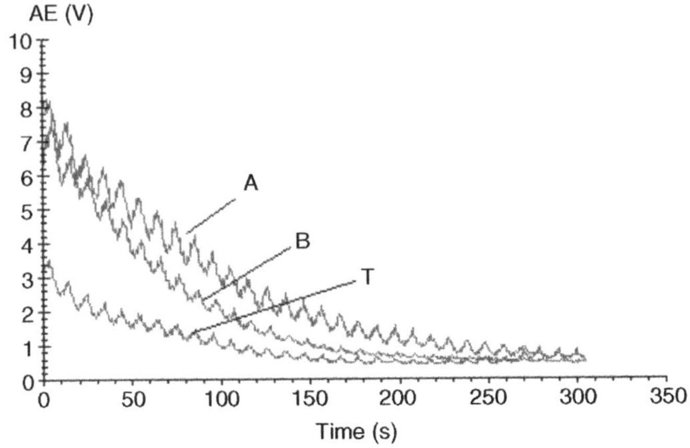

FIGURE 4.34 Variation in the AE signal of three different samples while polishing with no wafer rotation. Samples were polished with 150 rpm at 3 psi (from Ref. 36).

variation while polishing the patterned (MIT 854 test pattern) copper wafers on a bench-top tester. It can be seen in Fig. 4.34 that the highest AE signal is recorded for the sample with low-*k* dielectric A and the lowest signal for the sample with TEOS as dielectric, whereas a slightly smaller value is recorded for the sample with low-*k* B (*k* for B is slightly higher than that for A) when compared with the sample with low-*k* A.

It is prudent to mention that the only difference among the three samples is the underlying ultra-low-*k* ILD film. High AE for samples A and B may be an indication of delamination of Cu from the barrier and ILD interface. Lower amplitude of AE for sample B than for sample A can be seen from the figure indicating that sample B has better resistance to the CMP process. This could be due to the fact that low-*k* B bears slightly more mechanical strength. Analyzing the samples using optical microscope and scanning electron microscope (SEM) (not shown here), it was found that all the A and B samples were delaminated whereas TEOS had no sign of delamination. Signals can be further analyzed in order to characterize the delamination behavior of the ultra-low-*k* materials and distinguish slurries for specific applications [36].

4.7.2 Advanced Signal Processing

Das et al. [37] studied the coefficient of friction signal in order to determine the end-point detection more effectively. As shown in Fig. 4.35, the raw COF data were analyzed and the noise levels were filtered out to more precisely determine the end point of the process. The variance sequential probability ratio test (SPRT) method was adopted to analyze and filter the raw data. The coefficient of friction signal can thus be processed and analyzed to detect the end point more effectively.

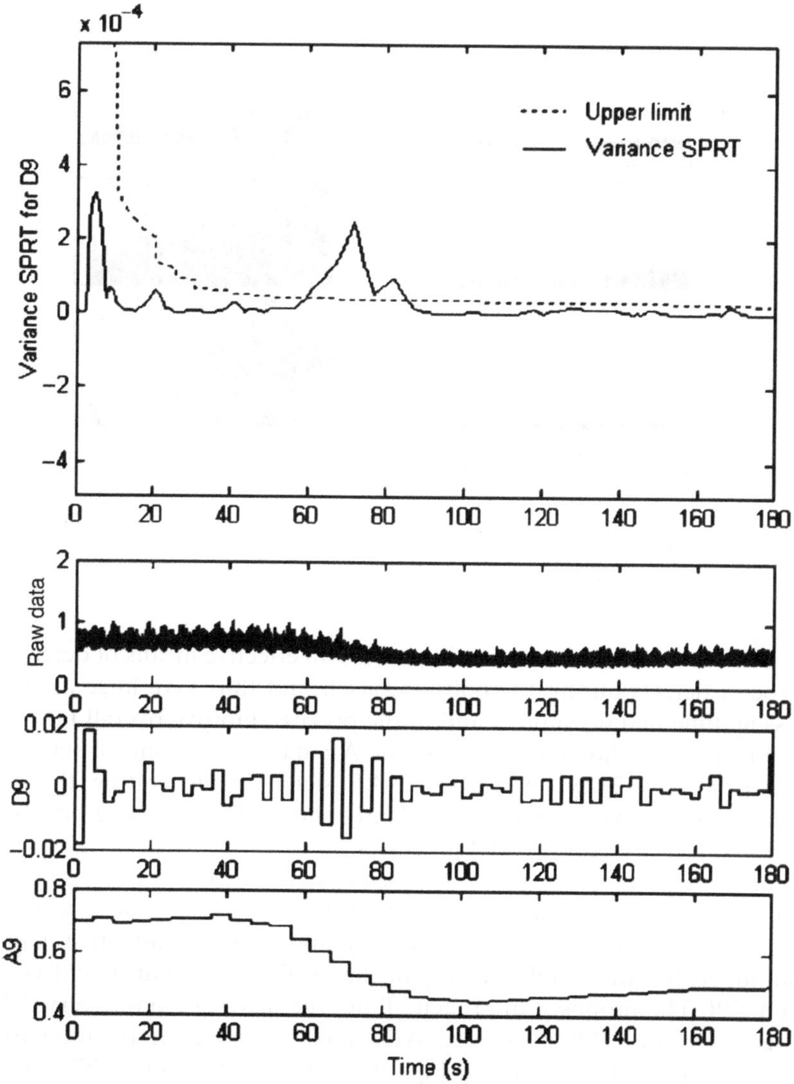

FIGURE 4.35 Variance SPRT for the patterned copper CMP process (from Ref. 37).

Ganesan et al. [38] analyzed the raw AE and COF signals, monitored and recorded during CMP, and filtered them using wavelet-based multiscale analysis and SPRT techniques to detect delamination and end point of copper–low-*k* system during CMP. An offline strategy and moving window-based strategy were implemented on the data collected from two different sources. They demonstrated the use of an online wavelet-based approach of analyzing the AE signal to provide an efficient method to detect delamination during

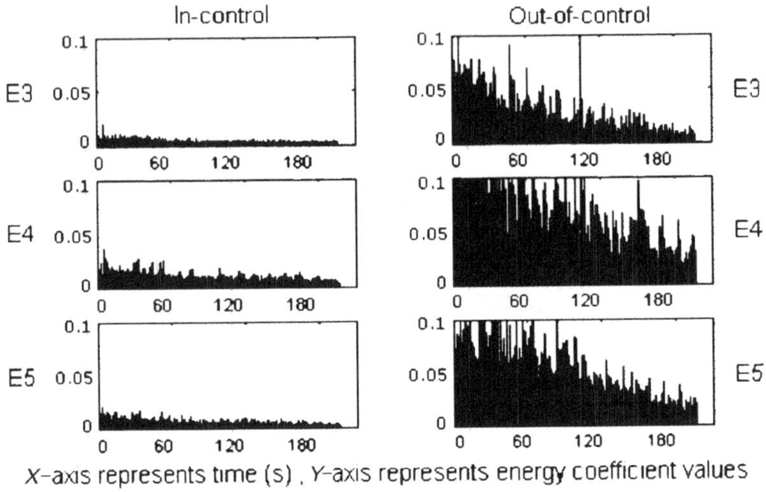

FIGURE 4.36 Energy from online analysis of the in-control and out-of-control AE data (from Ref. 38).

CMP. Such an online strategy would provide an effective means of determining the process defects during the polishing itself and thus save huge amount of processing time and downtime, and in the process improve overall fabrication yield. Figure 4.36 shows the analysis of AE data of in-control and out-of-control processes. They analyzed the AE signal by generating the energy coefficients of the AE signal, which clearly shows the difference between in-control and out-of-control CMP processes.

A further investigation of the AE signals can be carried out by means of advanced digital signal processing as demonstrated by [27]. A debouche digital wavelet filter was employed for signal decomposition. An adaptive matching pursuit algorithm was employed to plot the result in the joint time–frequency domain [39]. The proposed approach allows the investigation of very delicate AE signal features [40]. Processed AE signal results corresponding to two different slurries are shown in Fig. 4.37. Here, "a" and "b" represent performance of slurry 1 and slurry 2, respectively. The bottom of each result plot contains time-domain-filtered AE signal, whereas the frequency domain is plotted on the right-hand side. The inner product of AE signal amplitude in time and frequency domains enhanced by the marching pursuit algorithm produces nonlinear energy maps. These maps are represented in the middle. Here, a harmonic signal is represented by the horizontal lines, whereas the noise-like signal components consist of broken ellipsoids on the joint time–frequency graph. The AE signal corresponding to slurry 1 is rather stochastic in comparison to the AE signal taken from the slurry 2 experiment. This is a signature of the process full of transient behavior indicating potential Cu layer delamination. The AE signal corresponding to slurry 2 is dominated by main

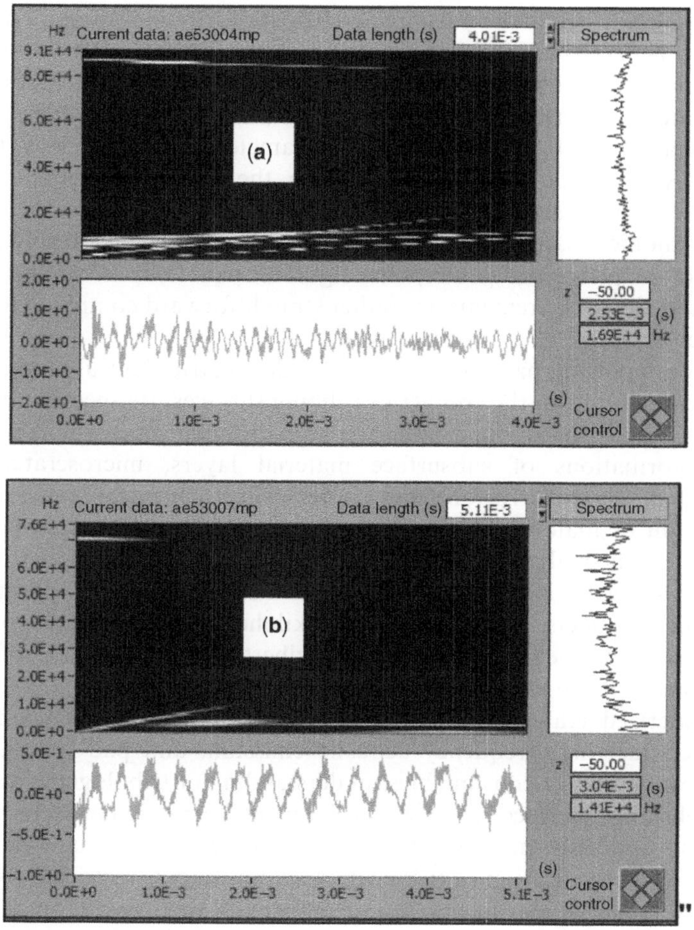

FIGURE 4.37 Joint time–frequency domains of the AE signal for slurry 1 (**a**) and slurry 2 (**b**). The AE signal was filtered by a Debouche 05 wavelet filter. Only middle bands were selected and processed by the marching pursuit joint time–frequency domain algorithm (from Ref. 27).

harmonics indicating the "smooth" CMP process. As previously discussed, no delamination is induced here.

4.8 SUMMARY

Tribometrology is a science of precise, repeatable, and reproducible measurements of friction and wear (material removal) as long as the conditions of the test are accurately replicated. Coefficient of friction, acoustic emission signal, and wear rate constitute the most important set of tribometrology. The friction coefficient between two sliding surfaces either lubricated or unlubricated provides information about the nature of the surface interactions. The tribology at the interface depends, in general, on the nature of the materials sliding against each other. Process parameters influence the tribology significantly. To monitor wear, measuring the linear geometrical wear of each of the rubbing surfaces is preferred. As wafer material removal is on the order of nanometers, its measurements are highly complicated during polishing and are often performed after polishing. The pad wear is on the order of micrometers; its measurements are rather straightforward compared to the thin film removal rate.

Another important parameter to monitor during tribological processes is the acoustic emission from the contact of rubbing surfaces. Its spectrum may have numerous frequencies, corresponding to such different processes as plastic and elastic deformations of subsurface material layers, microscratching and microfatigue, microcorrosion and other electrochemical reactions, and delamination of material layers. At a given speed, higher frequencies reflect the processes on smaller microareas that spread into smaller depths. Thus, the megahertz acoustics is more informative of the specific micro-tribo-processes on tiny microcontacts, in comparison to a kilohertz range reflective of integral characteristics of the interface and a decihertz range reflective of integral characteristics of the entire mechanical system.

Computerized real-time measurements and analysis of the coefficient of friction, contact high-frequency acoustic emission, and pad wear allow the effective evaluation of consumables, understanding of tribological interactions at the polishing interface, process development, dynamic characterization of the polishing process, including rate and nonuniformity of material removal, and so on. The application of tribometrology not only is restricted to research and development departments but also proves very useful in the device production facilities.

QUESTIONS

1. Why is tribology important for the chemical–mechanical planarization process?
2. What are the different kinds of polishing regimes possible during the polishing process? How are they determined? What are the necessary parameters to determine the polishing regime?
3. What is tribometrology? What are its applications during the CMP process?
4. In a fabrication plant, one of the CMP stations encountered inconsistent material removal rate. Given that the slurry is fully functional, how would you use tribometrology to determine the source of inconsistency?

5. Describe the need to monitor the pad conditioning process? What are the consequences of over- or underconditioning? How is the end point of a conditioning process determined?
6. Describe an experimental procedure (name the sensors to be used) to characterize a pad conditioner.
7. List some of the means to detect the generation of defects *in situ* during the CMP process.
8. Name some of the major factors that influence the interface tribology during a CMP process.
9. Consider that you are a polishing pad manufacturer. Describe an experimental procedure to characterize the pad samples that are manufactured in your company.
10. Comment on the applicability of tribometrology in reducing the cost of ownership during a CMP process.

REFERENCES

1. Ludema K. Friction, Wear and Lubrication: A Textbook in Tribology. CRC Press; 1996.
2. Bhushan B. Introduction to Tribology. Wiley; 2002.
3. Bhushan B. Principles and Applications of Tribology. Wiley; 1999.
4. Goryacheva I. Contact Mechanics in Tribology, Solid Mechanics and Its Applications. Kluwer Academic Publishers; 1998.
5. Stribeck R. Characteristics of plain and roller bearing. Zeit Ver Deut Ing 1902;46,1341–1348, 1432–1438, 1463–1470.
6. Philipossian A, Olsen S. Fundamental tribological and removal rate studies of inter-layer dielectric chemical mechanical planarization. Jpn J Appl Phys 2003;42: 6371–6379.
7. Gitis N, Mudhivarthi R. Tribology Issues in CMP. Semiconductor Fabtech is published quarterly by Henley Publishing Ltd, London, UK. 18th ed. 2003. p 125–128.
8. Lu J, Rogers C, Manno V, Philipossian A, Anjur S, Moinpur M. Measurements of slurry film thickness and wafer drag during CMP. J Electrochem Soc 2004;151(4):G241–G247.
9. Runnel S, Eyman L. Tribology analysis of chemical–mechanical polishing. J Electrochem Soc 1994;141:1698.
10. Li Z, Borucki L, Koshiyama I, Philipossian A. Effect of slurry flow rate on tribological, thermal and removal rate attributes of copper CMP. J Electrochem Soc 2004;151(7):G482–G487.
11. Mudhivarthi S, Gitis N, Kuiry S, Vinogradov M, Kumar A. Effects of slurry flow rate and pad conditioning temperature on dishing, erosion, and metal loss during copper CMP. J Electrochem Soc 2006;153(5):G372–G378.

12. Sorooshian J, Hetherington D, Philipossian A. Effect of process temperature on coefficient of friction during CMP. Electrochem Solid-State Lett 2004;7(10):G222–G224.
13. Mudhivarthi S, Zantye P, Kumar A, Kumar A, Beerbom M, Schlaf R. Effect of temperature on tribological, electrochemical, and surface properties during copper CMP. Electrochem Solid-State Lett 2005;8(9):G241–G245.
14. Mullany B, Byrne G. The effect of slurry viscosity on chemical–mechanical polishing of silicon wafers. J Mater Process Technol 2003;132:28–34.
15. Doy T, Seshimo K, Suzuki K, Philipossian A, Kinoshita M. Impact of novel pad groove designs on removal rate and uniformity of dielectric and copper CMP. J Electrochem Soc 2004;151(3):G196–G199.
16. Muldowney G. Modeling CMP transport and kinetics at the pad groove scale. Proceedings of the Material Research Society; 2004. p K5.3.1–K5.3.6.
17. Li Z, Ina K, Lefevre P, Koshiyama I, Philipossian A. Determining the effects of slurry surfactant, abrasive size, and abrasive content on the tribology and kinetics of copper CMP. J Electrochem Soc 2005;152(4):G299–G304.
18. Basim G, Moudgil B. Effect of soft agglomerates on CMP slurry performance. J Colloid Interface Sci 2002;256:137–142.
19. Choi W, Mahajan U, Lee S, Abiade J, Singh R. Effect of slurry ionic salts at dielectric silica CMP. J Electrochem Soc 2004;151(3):G185–G189.
20. Scarfo A, Manno V, Rogers C, Anjur S, Moinpour M. In situ measurement of pressure and friction during CMP of contoured wafers. J Electrochem Soc 2005;152(6):G477–G481.
21. Gitis N, Vinogradov M, Meyman A, Xiao J. PadProbe™ for quantitative control of pad surface conditions and wear. CMP User Group Conference; 2002.
22. Kalenian B, Pautsch B, Sennett B, Gitis N, Vinogradov M. CMP in-situ pad condition monitoring with PadProbe™. CMP-MIC Conference; CA; 2003. p 418–421.
23. Gitis N, Meyman A, Vinogradov M, Faynberg M, Dorfman V. Method and apparatus for monitoring polishing plate condition. US patent 6,702,646,2004.
24. Fang J, Davis K, Gitis N, Vinogradov M. CMP process and consumables evaluation with PadProbe™. Fifth International Conference on Microelectronics and Interfaces. Proceedings of American Vacuum Society, Santa Clara, CA; 2004. p 55–57.
25. Hosali S, Busch E, Vinogradov M, Gitis N. CMP process and consumables evaluation with PadProbe™. Ninth International CMP-MIC Conference; Fremont, CA; 2005. p 115.
26. Sung J, Gitis N, Kuiry S, Vinogradov M, Khosla V, Nishizawa E, Toganoh T. Studies of advanced pad conditioners. Eleventh International CMP-MIC Conference; Fremont, CA; 2006. p 10H.
27. Sikder A, Gitis N, Vinogradov M, Dougela A. In-situ tribological properties monitoring and chemical mechanical characterization of planarization process. Proceedings of 2004 ASME/STLE International Joint Tribology Conference; Long Beach, CA; 2004 Oct 24–27.
28. www.electronicmaterials.rohmhaas.com, Rohm & Haas Electronic Materials, Rohm & Haas Company.

29. http://www.ppg.com/ppgcmp/pads.htm, PPG CMP Consumables, PPG Industries, Inc.
30. http://www.mipox.co.jp/en/, Nihon Micro Coating Co., Ltd.
31. Gitis N, Vinogradov M. Incoming inspection and failure analysis of CMP consumables at the semiconductor fab. VMIC Conference; Santa Clara, CA; 2003.
32. Zantye P, Mudhivarthi S, Kumar A, Obeng Y. Metrology and characterization of application specific chemical mechanical polishing pads. J Vac Sci Technol 2005;23(5):1392–1399.
33. Zantye P, Kumar A, Dallas W, Ostapenko S, Sikder A. Investigation of the nonuniformities in polyurethane chemical mechanical planarization pads. J Vac Sci and Technol B 2006;24(1):25–33.
34. Gitis N, Xiao J, Kumar A, Sikder A. Advanced specification and tests of CMP retaining rings. CMP-MIC Conference; Marina Del Rey, CA; 2004. p 252–255.
35. Moussa R, Quartapella C. Next-Generation materials for CMP retaining rings. VMIC Conference; 2003. p 501.
36. Sikder A, Zantye P, Thagella S, Kumar A, Vinogradov M, GItis N. Delamination studies in Cu–ultra low-*k* stack. CMP-MIC Conference; CA; 2003. p 120–127.
37. Das T, Ganesan R, Sikder A, Kumar A. Online end point detection in CMP using SPRT of wavelet decomposed sensor data. IEEE Trans Semicond Manuf 2005;18(3):440–447.
38. Ganesan R, Das T, Sikder A, Kumar A. Wavelet-based identification of delamination defect in CMP (Cu–low *k*) using nonstationary acoustic emission signal. IEEE Trans Semicond Manuf 2003;16(4):677–685.
39. Kavacevic J, Vetterli M. Wavelets and Subband Coding. Signal Processing Series. Prentice Hall; 1995.
40. Daugela A, Kutomi H, Wyrobek T. Nanoindentation induced acoustic emission monitoring. Ziff Fur Metallkunde 2001;92:1052–1056.

5

PADS FOR IC CMP

Changxue Wang, Ed Paul, Toshihiro Kobayashi and Yuzhuo Li

5.1 INTRODUCTION

The chemical–mechanical polishing or planarization (CMP) process is a complex interplay between the wafer and the consumables involved. The consumables include slurry, pad, conditioner, and so on. During polishing, the pad carries the slurry and delivers it to the wafer surface. It also transmits the normal and shear forces from the polisher to the wafer. Therefore, polishing pad plays a critical role in the CMP process and influences the outcomes such as material removal rate (MRR), within-wafer nonuniformity (WIWNU), wafer-to-wafer nonuniformity (WTWNU), step height reduction efficiency (SHRE), and defect counts.

As the CMP process is a combination of mechanical and chemical processes, the polishing pad has to endure repeated mechanical stress and constant chemical attacks. The applied downforce and polishing-induced friction force cause various levels of compression and abrasion. Slurry components such as oxidizers and acids can react with different components of the pads. Thus the polishing pad must have sufficient mechanical integrity and chemical resistance to survive the rigors of polishing. Polishing pads must balance the needs of hardness, modulus for planarity, and strength to resist wear and tear during polishing. Pads must also be able to survive the aggressive slurry chemistry used in CMP polishing without degrading, delaminating, blistering, or warping, since CMP slurries for the polishing of interlevel dielectric (ILD) oxide layers are usually highly alkaline with pH as high as 11, and CMP slurries for the polishing of metal films are highly acidic

Microelectronic Applications of Chemical Mechanical Planarization, Edited by Yuzhuo Li

with pH as low as 2 or even lower and containing strong oxidizer. Furthermore, in order to carry and deliver slurry efficiently on the pad surface, the pads must be sufficiently hydrophilic. In other words, the pads should have high critical surface tension.

Polishing pads are usually made of polymeric materials such as polycarbonates, nylons, polysulfones, and polyurethanes. The most commonly used pads are made of polyurethanes, which have balanced mechanical properties of strength, hardness, modulus, and excellent chemical stability. In addition, these properties may be readily and precisely controlled during the manufacturing process. The pads can be made to contain a wide range of microstructures, textures, and fillers using various manufacturing processes such as casting, molding, extrusion, web coating, and sintering. Also polyurethanes have the highest critical surface tension values among the above-mentioned polymeric materials [1], which allow polyurethane pads to carry and deliver aqueous slurry on its surface more evenly and effectively.

Pad properties and their effects on the CMP performance have been extensively investigated through polishing experiments as well as theoretical analysis (modeling and numerical simulation) to meet the ever-increasing demand of process reliability and yields. Innovative designs such as fixed abrasive pads, surface-treated pads, and reactive pads have also been reported in open literatures and patents. In this chapter, we will focus on the most commonly used polyurethane pads in terms of their applications in IC CMP. The chapter will be divided into five sections: (1) physical properties and their effects on polishing performance, (2) chemical properties and their effects on polishing performance, (3) pad conditioning and its effect on polishing performance, (4) modeling of pad effects in CMP, and (5) novel designs of CMP pads.

5.2 PHYSICAL PROPERTIES OF CMP PADS AND THEIR EFFECTS ON POLISHING PERFORMANCE

5.2.1 Pad Types

There are four main types of polishing pads currently used for CMP process. They are differentiated by their microstructures and have different physical and mechanical properties. The four types are [1,2] as follows:

Type I: Felts and polymer-impregnated felts

Type II: Porometrics (microporous synthetic leathers)

Type III: Filled polymer sheets (films)

Type IV: Unfilled textured polymer sheets (films)

The key features, properties, and typical uses of these four types of polishing pads are listed in Table 5.1 [1,2].

TABLE 5.1 Key Features, Properties, and Applications of Different Types of Polishing Pads [1,2].

	Type I	Type II	Type III	Type IV
Structure	Felted fibers impregnated with polymeric binder	Porous film coated on a supporting substrate	Microporous polymer sheet	Nonporous polymer sheet with surface macrotexture
Microstructure	Continuous channels between fibers	Vertically oriented open pores	Closed-cell foam	None
Inherent microtexture	High	High	Medium	Low
Slurry holding	Medium	High	Low	Minimal
Examples of commercial pads	Suba™, STT 711™, Pellon™	Politex™, Surfin™, UR100™, WWP3000™	IC1000™, IC1010™, IC1040™, FX9™, MH™	OXP3000™, OXP4000™, NCP-1™, IC2000™
Compressibility	Medium	High	Low	Very low
Stiffness/hardness	Medium	Low	High	Very high
Typical applications	Si stock polish, tungsten CMP	Si final polish, tungsten CMP, post-CMP buff	Si stock, ILD CMP, STI, metal damascene CMP	ILD CMP, STI, metal dual damascene

5.2.2 Pad Microstructures and Macrostructures

The physical and mechanical properties of a polishing pad are determined not only by the chemical composition of the pad materials but also by its microstructures. As described in Table 5.1, each type of pad has its unique microstructure that ranges from large open pores to nonporous solids. Some representative microstructures of pad surfaces are shown in Fig. 5.1.

Cross-sectional SEM images of type III and type IV pads are shown in Fig. 5.2. Cross-sectional SEM images of all the four types of pads have also been reported in other publications [1].

A type I pad has microstructure characterized by nonwoven polyester fibers, partially impregnated with polyurethane to leave open porosity throughout the pad. A type II pad has the most complex microstructure consisting of a porous layer on a supporting substrate similar to the structure of a type I pad. The surface of a type II pad consists of open pores. A type III pad is essentially a closed-cell foam, where the pores are created either by blowing agents or by the addition of microballoons, thus the pad surface has significant texture even prior to conditioning. A type IV pad is non-porous and unfilled and has the simplest microstructure.

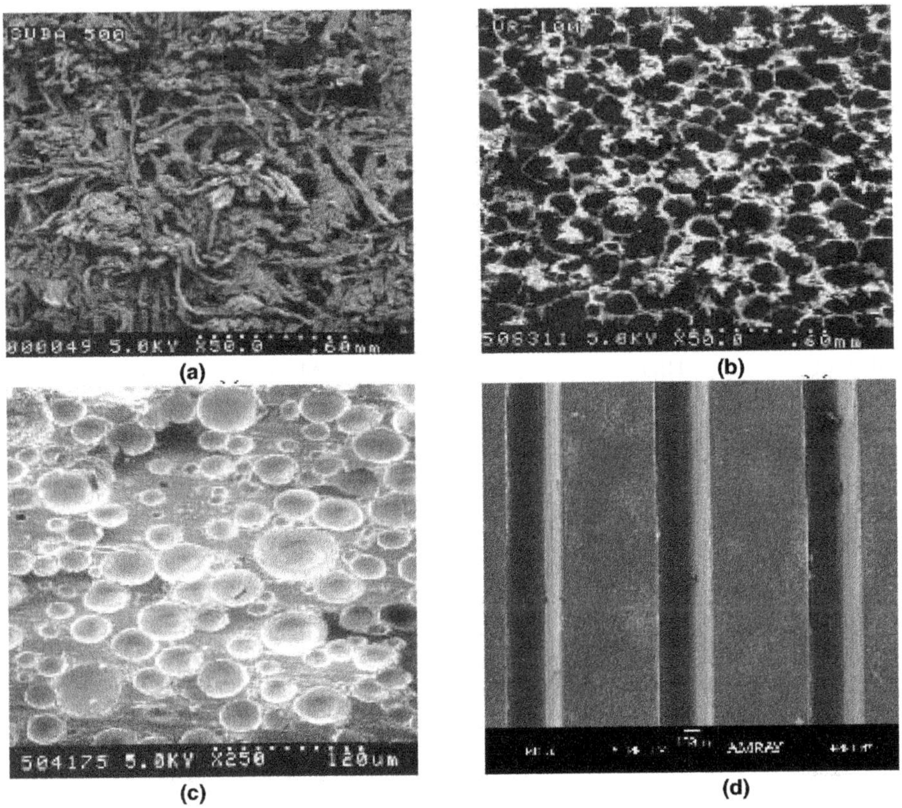

FIGURE 5.1 SEM images of surface of four different types of polyurethane pad [2]: (**a**) type I pad (Suba500^{TM}), (**b**) type II pad (UR100^{TM}), (**c**) type III pad (IC1000^{TM}), (**d**) type IV pad (OXP4000^{TM}, grooved).

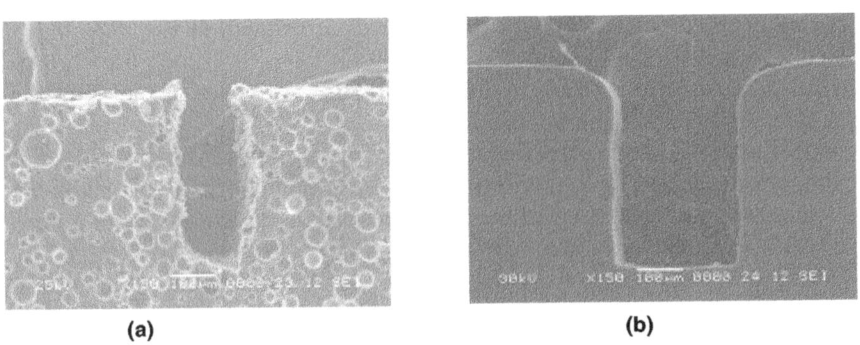

FIGURE 5.2 Cross section of type III and type IV polyurethane pads: (**a**) type III pad (IC1000^{TM}, grooved), (**b**) type IV pad (Mipox NCP pad, grooved).

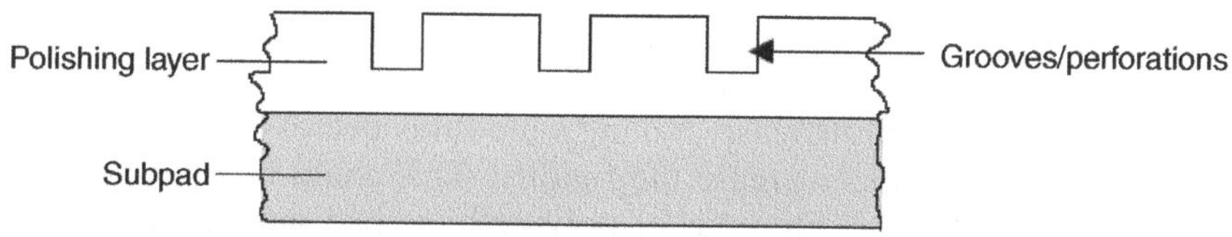

FIGURE 5.3 Schematic cross section of a multilayer polishing pad [2].

Macrostructures of polishing pads are in the form of either vertical perforations through the pad or surface grooves created by embossing, molding, or machining. For most polishing pads, macrostructures as grooves are required in addition to the microstructure for the purpose of better slurry transport. For nonporous type IV pads, grooves are a must owing to a lack of surface microstructures (pores).

Multilayer or stacked pads are commonly used in the CMP processes for better polishing performance such as uniform material removal and good planarization across the wafer surface. A multilayer or stacked pad usually consists of a stiff, hard top layer and a soft, flexible subpad, and possibly some medium layer(s), as shown in Fig. 5.3.

5.2.3 Polyurethane Pad Properties and Control

5.2.3.1 Hardness, Young's Modulus, and Strength Pad physical properties such as density, porosity, hardness, Young's (elastic) modulus, tensile strength, yielding strength, and glass transition temperature (at which the polyurethane softens appreciably) are influenced not only by the composition of the pad materials but also by the thermal history, macrostructure, and conditioning of the pad.

Hardness of polyurethane pads can vary from shore D values of less than 15 (very soft and flexible pads) to greater than 65 (very stiff and rigid pads), and Young's (elastic) modulus of polyurethane pads can vary from a low of 1 MPa (very soft, flexible, and elastomeric pads) to greater than 1 GPa (very stiff and rigid pads). Usually, hardness increases with the increase in Young's modulus.

Hardness and Young's modulus are functions of the pad's polyurethane composition [1]. They are strongly influenced by the types and concentration of the hard and soft polyurethane segments. Typically, increasing the soft segment concentration will lead to the increase in toughness and flexibility but decrease in Young's modulus and hardness. Higher concentration of hard segments improves the property consistency at a high temperature and leads to the increase in stiffness and strength. Thus hardness and Young's modulus can be controlled through the variation and optimization of polyurethane hard-to-soft segment ratios.

Urethane stoichiometry also plays a key role in influencing the polyurethane pad properties. Stoichiometry refers to the ratio of reactive groups (usually diol or diamine moieties) to isocyanate groups. The presence of excess isocyanates will cause side reactions that result in chemical bonding (cross-link) between polymer chains. As the ratio of reactive groups increases, the hardness, tensile strength, and elongation increase. Young's modulus decreases when the ratio increases since cross-links increase the rigidity. This could be used to control the hardness and Young's modulus of polyurethane pad independently [1].

Another important factor that influences polyurethane pad properties is the pad's thermal history (including the temperatures and durations at which the pad is cured). Generally, higher baking temperature and longer baking duration will increase the degree of cure, which leads to an increase in pad hardness, Young's modulus, yielding strength, tensile strength, and glass transition temperature. This is consistent with the fact that a higher level of cure results in the formation of more cross-links between polyurethane chains and a network structure.

5.2.3.2 Pad Porosity/Density Type I, II, and III pads all have pores although their microstructures are different. Type IV pads (noncell, nonporous, solid) do not have native porous microstructure except those generated by the conditioning process. Typical physical properties of IC1000 pad (type III, porous) and IC2000 pad (type IV, noncell, nonporous, solid) are listed in Table 5.2.

Pad porosity is inversely related to its density. Many physical properties of the polyurethan pad are strongly dependent upon its porosity (or density). The hardness and Young's modulus (elastic or storage modulus) of porous pads have a clear linear correlation with the density (or porosity) of the pads [1]. It is obvious that nonporous (noncell) pads have much smaller variability in density and other physical properties compared to porous pads. Nonporous pads have much higher strength, modulus, hardness, and elongation than porous pads.

5.2.3.3 Pad Thickness Typical polishing pads for CMP are about 1.3 mm (0.05 in.) thick. Slightly thicker pads (2.0 mm or 0.08 in. in thickness) have also

TABLE 5.2 Typical Physical Properties of IC1000 and IC2000 Pads [1].

	Value	
Property	IC1000	IC2000
Porosity (# of pores/mm^2)	880 ± 120	0
Density (g/cm^3)	0.748 ± 0.051	1.180 ± 0.002
Hardness (shore D)	52.2 ± 2.5	73.0 ± 1.0
Shear strength (MPa)	51.2 ± 4.1	—
Proportional limit (MPa)	9.1 ± 1.3	(Yielding strength) 33.4 ± 1.4
Tensile strength (MPa)	21.6 ± 2.8	75.0 ± 2.5
Elongation to break (%)	175 ± 20	335 ± 20
Storage modulus (MPa)	310 ± 40	850 ± 61
Loss modulus (MPa)	28.0 ± 4.5	87.0 ± 4.0
Tan delta	0.090 ± 0.005	0.103 ± 0.005

been used for the polishing of sub-0.1-μm devices with smaller feature sizes. Pads with significantly greater thickness (above 5 mm or 0.2 in.) can lead to very poor polishing uniformity due to extremely high stiffness and are rarely used [1].

5.2.3.4 Pad Stiffness/Stacked Pads Pad stiffness is linearly proportional to the product of a pad's Young's (elastic) modulus and the cubic of pad thickness. Usually harder pad has higher Young's modulus; thus a harder pad is also stiffer. Pad stiffness increases much more significantly when its thickness increases because of the cubic relationship between them. Stacked pads combine the top hard/stiff layer and the soft/flexible substrate, which enables the stacked pad to be more flexible, while still maintaining high enough stiffness of the top surface layer.

5.2.3.5 Pad Grooves As discussed above, macrostructures as grooves are usually required for polishing pads for the purpose of better slurry transport. There are many different groove designs such as circular grooves (concentric or spiral), XY grooves, other regular designs (hexagons, triangles, tire-treat-type patterns) or irregular designs (fractal patterns), and a combination of these patterns. Commercial pads typically have circular grooves (concentric K-grooves or spiral grooves), XY grooves, and a combination of these grooves. Groove profile may be rectangular with straight sidewalls, or the groove cross section may be V-shaped, U-shaped, triangular, sawtooth, and so on. The concentric circular K-groove or spiral groove may have its center on- or offset the pad's center. Groove depth, width, and density can be adjusted to tailor a pad's mechanical properties such as stiffness and hardness [2].

5.2.4 Effects of Pad Property on Polishing Performance

Polishing performance is characterized by several parameters that are dependent on the scale of the features being polished. Usually it is divided into three scales: wafer, die, and feature scales. At wafer scale, polishing performance mainly includes material removal rate, polishing nonuniformity, edge effects, macro-scratches, and pad life. At die scale, it includes planarization and defectivity. At feature scale, it includes dishing, erosion, defectivity, and surface roughness. The relationships between polishing performance and pad properties are very complicated and are not fully understood yet. This is because pad physical properties are strongly dependent on each other, and the pad surface topography that changes through conditioning also depends on pad properties. Furthermore, polishing performance depends not only on the polishing pads but also on many other nonpad factors such as slurry, tool (polisher and conditioner), and polishing condition (platen and carrier speed slurry flow rate, back pressure). All these factors make it very hard to distinguish the relationships between pad properties and polishing performance. Some major relationships between pad properties and polishing performance are listed in Table 5.3, and a detailed discussion on some of these relationships follows next.

TABLE 5.3 Relationships between Pad Properties and Polishing Performance [1].

	Polishing Scale			
Pad Property	Wafer	Die	Feature	Conditionability
Density (porosity)	Removal rate, nonuniformity	Defectivity	Dishing, erosion	Yes
Hardness	Macroscratches	Defectivity	Defectivity, roughness, dishing, erosion	Yes
Tensile strength	Pad life			Yes
Abrasion resistance	Pad life			Yes
Modulus (stiffness)	Edge effect, nonuniformity	Planarization		Yes
Thickness	Pad life			
Top pad compressibility		Planarization	Dishing	
Base pad compressibility	Edge effect, nonuniformity	Planarization		
Pad texture (groove)	Pad life, removal rate, nonuniformity, edge effect			
Pad roughness	Removal rate, nonuniformity	Planarization	Dishing, erosion	Yes
Hydrophilicity	Removal rate			Yes

5.2.4.1 ***Pad Roughness Effects*** Pad surface roughness (microstructure as local asperity) comes both from pad conditioning and from porosity within the pad. Rougher pads could lead to higher material removal rate [3]. When a pad is not conditioned during or after polishing, pad surface asperities wear off and become glazed (flattened). This leads to an increase in the area of direct contact between the pad and wafer surface while decreasing the average contact pressure, and the material removal rate drops rapidly [4]. Further, rougher pads can cause more defects such as scratches and residual particles on polished wafer surfaces. With a rougher pad, more asperities (which are in contact with wafer surface) are under higher contact pressure while the total contact area is smaller. Thus particles entrapped within the asperity–wafer contact areas are indented deeper into the wafer surface, and more scratches and pits could be formed when these particles slide with the pad across wafer surface [5]. These scratches and pits on the polished wafer surface make post-CMP cleaning more difficult, and allow more residual particles to stay on the polished wafer surface. For patterned wafer polishing, dishing amount can be reduced by decreasing the pad surface roughness owing to the fact that lower height pad asperity would be more difficult to touch the lower metal part of the patterned features during polishing [6].

5.2.4.2 Pad Porosity/Density Effects Pad pores help retain and transport slurry over the pad surface. As mentioned previously, many pad properties (strength, hardness, stiffness, modulus, etc.) are all related to pad porosity/density; hence it is expected that pad porosity has significant impact on polishing performance. Higher density (or lower porosity) pads produce lower removal rates because of being less effective at transporting enough slurry under the wafer surface [7]. However, they produce better polishing uniformity on the polished wafer surface. The extreme is type IV nonporous (noncell, solid) pad. Since nonporous pads have much higher strength, hardness, and Young's modulus, and have much lower variations in these physical properties than porous pads, more consistent polishing performance can be achieved with nonporous pads compared to porous pads.

When MRR of oxide wafer polishing on porous IC1000 pad and on nonporous NCP pad is compared, it shows that MRR on nonporous NCP pad is slightly lower than that on porous IC1000 pad, and WTWNU of MRR is comparable for NCP and IC1000 pads (see Fig. 5.4). The WIWNU of MRR on NCP pad is much lower than that on IC1000 pad. This demonstrates better

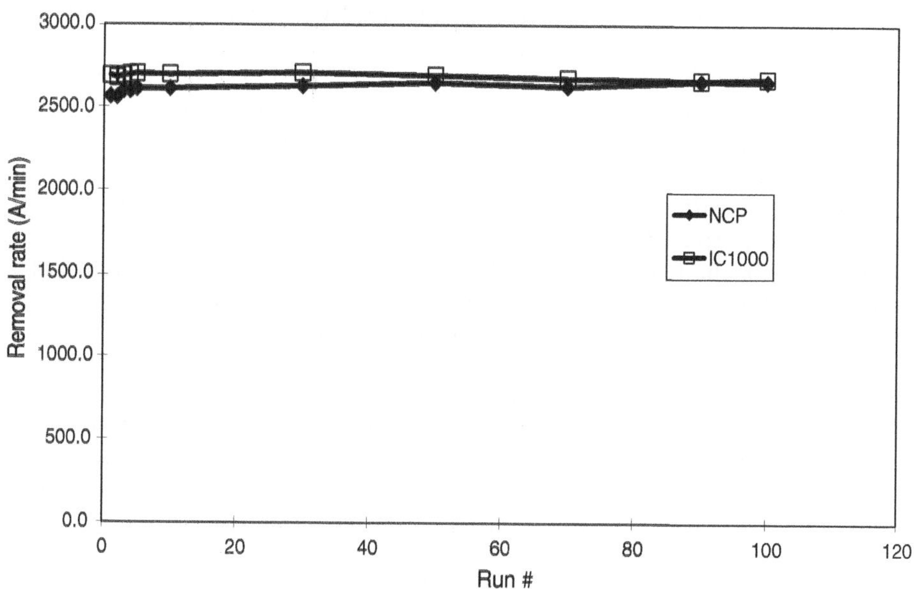

FIGURE 5.4 MRR variation from wafer to wafer (WTWNU): 100 oxide wafers polished on NCP pad (lower line with diamond marker) versus polished on IC1000 pad (upper line with square marker).

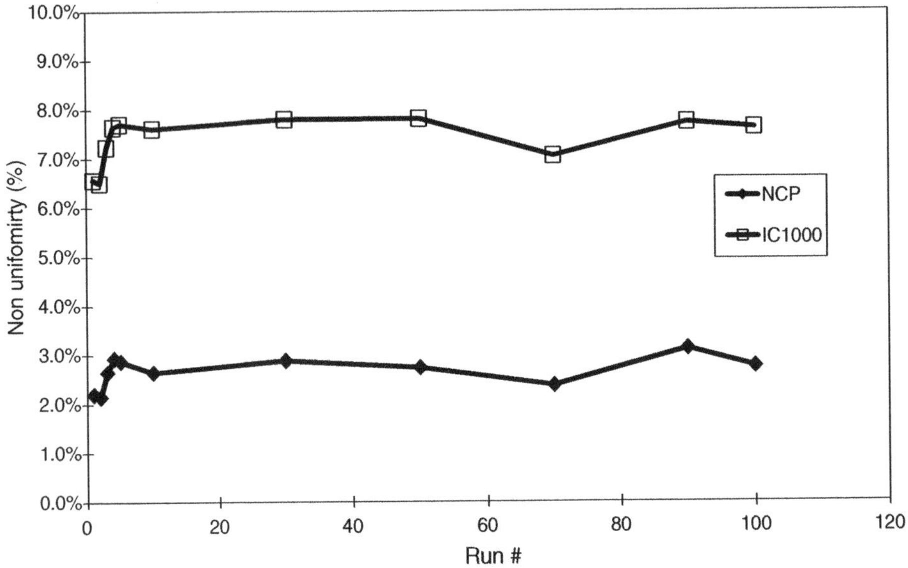

FIGURE 5.5 MRR within wafer variation (WIWNU): 100 oxide wafers polished on NCP pad (lower line with diamond marker) versus polished on IC1000 pad (upper line with square marker).

polishing uniformity across the wafer surface on NCP pad than on IC1000 pad (see Fig. 5.5).

Another benefit of using nonporous pads in polishing is improved planarization of die-scale features, since nonporous pads have much higher Young's modulus and hence much higher stiffness with similar pad thickness, and higher stiffness can help achieve better planarization. Figure 5.6 shows the comparison of patterned wafer profile and step height reduction efficiency for polishing on porous IC1000 pad and nonporous NCP pad.

In Fig. 5.6, the feature profile after polishing on NCP pad is much flatter than that after polishing on IC100 pad. In Fig. 5.6c and d, the lines from top to bottom are for different pattern densities (80%, 60%, 40%, 20%, 10%, respectively) that show the step height versus amount removed. Clearly, polishing SHRE is much higher on nonporous NCP pad than that on porous IC1000 pad. This is also shown in Fig. 5.7 for polishing on porous IC1000 pad and on various nonporous NCP pads.

Since nonporous pads have much higher hardness and strength than porous pads, they have much higher abrasion resistance and lower pad wear rate during polishing; thus they have a much longer pad life under similar polishing and pad conditioning than porous pads. Figure 5.8 shows that NCP pad wear rate is less than one third of that for IC1000 pad under the same pad-conditioning situation. Nonporous pads can also lead to less defectivity, especially for copper CMP polishing. For porous pads, the pores on pad surface can trap the polishing debris and abrasive particles in the slurry. These particles and polishing debris could either scratch the wafer surface (especially when the abrasive particles and polishing debris are harder than the wafer surface materials like the soft Cu, Al films) or leave residual particles on the

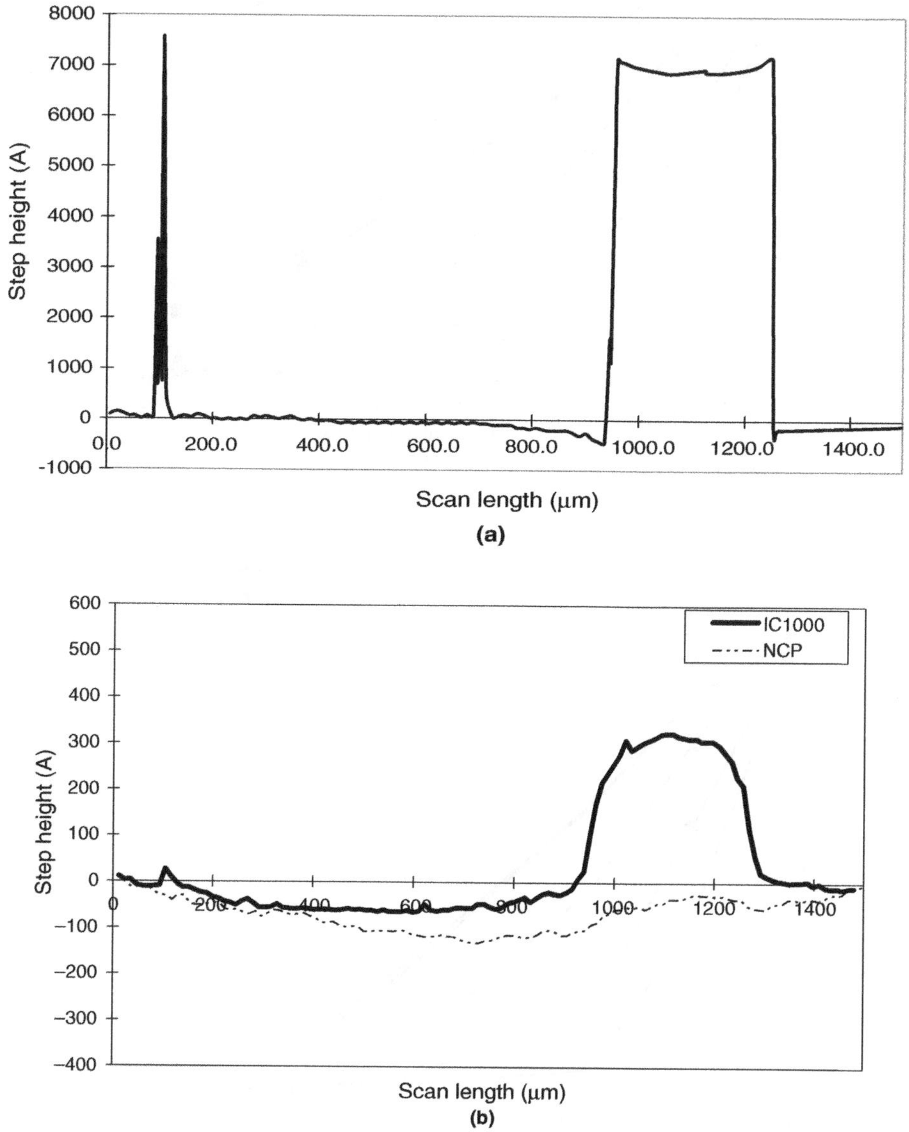

FIGURE 5.6 Comparison of patterned wafer profile and step height reduction efficiency (SHRE) for polishing on IC1000 and NCP pads: (**a**) Wafer profile before polishing, (**b**) wafer profile after polishing, (**c**) SHRE on IC1000 pad, (**d**) SHRE on NCP pad.

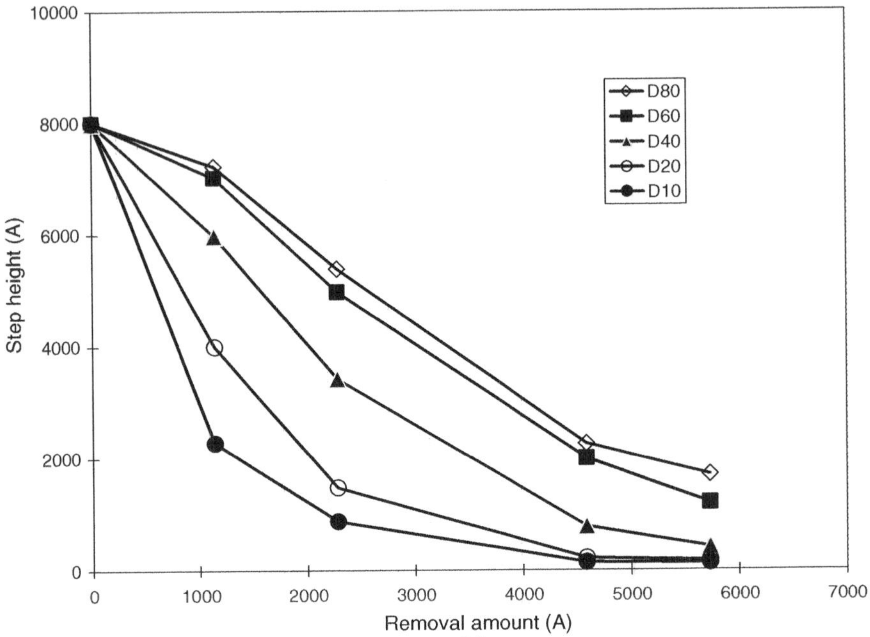

(c)

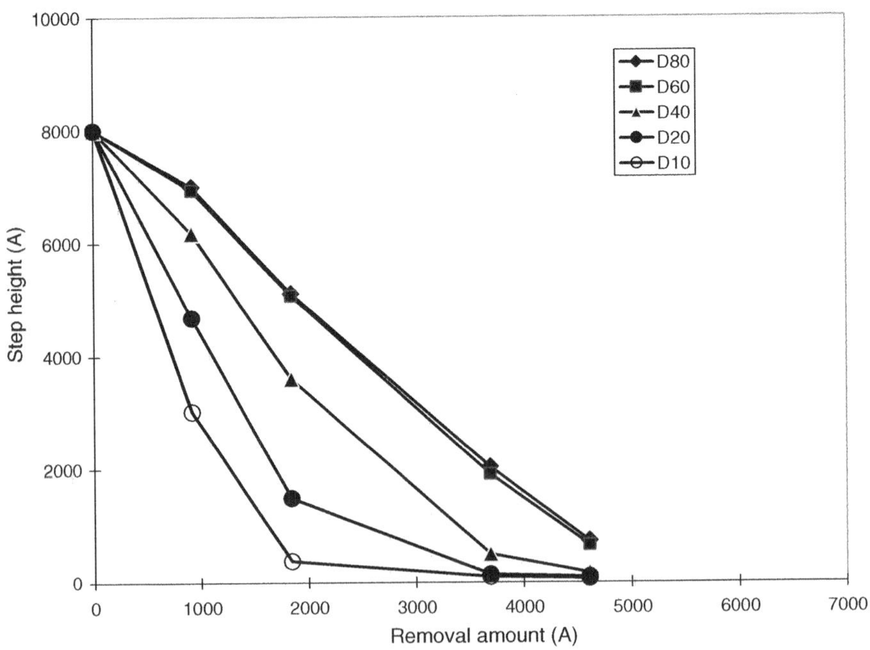

(d)

FIGURE 5.6 (*Continued*)

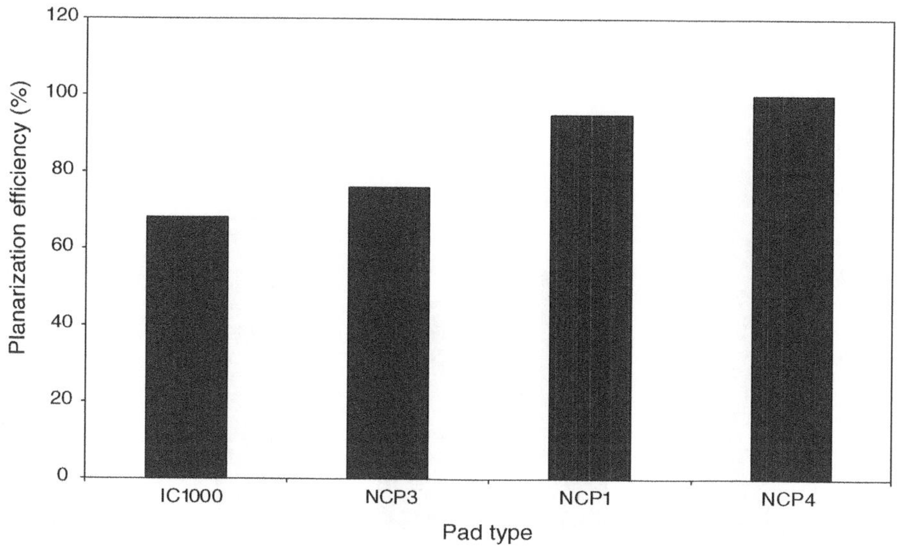

FIGURE 5.7 Comparison of normalized patterned wafer step height reduction efficiency (SHRE) for polishing on IC1000 and various NCP pads.

polished wafer surfaces. The fewer the pores are on a pad surface, the less probable the scratches and residual particles will be on the polished wafer surface. Figure 5.9 shows that the number of defectivity (residual) particles on polished Cu blanket wafer surface is higher with porous IC1000 pad than with various nonporous NCP pads.

Pad temperature influences not only pad physical properties but also the chemistry of the slurry on the pad, especially when reactive metal films are polished, since the metal slurries containing oxidizing and complexing agents are more sensitive to temperature change. Pad temperature increases because

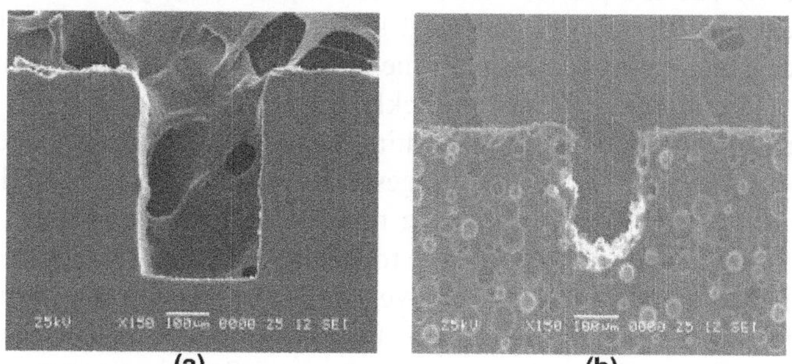

FIGURE 5.8 Comparison of NCP and IC1000 pad wear rates (total conditioning time = 4 h under 5 lbs downforce): (**a**) NCP pad: wear rate 14.3 μm/h, (**b**) IC1000 pad: wear rate 44.0 μm/h.

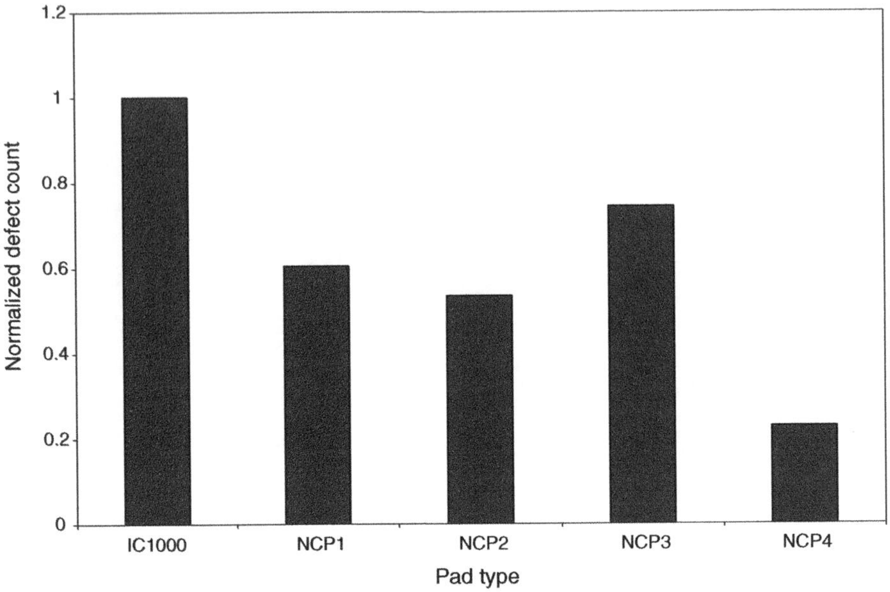

FIGURE 5.9 Normalized defectivity number comparison of Cu blanket wafers polished with first-step Cu slurries on IC1000 pad and various NCP pads.

of the abrasion friction force between wafer and pad surfaces during polishing. The increase in pad surface temperature during polishing is much lower for NCP pad than that for IC1000 pad under various polishing down pressures and table/carrier speeds, especially when 300-mm wafers are polished (see Fig. 5.10a and b). Evidently, in all cases, the temperature increase on NCP (noncell) pad is lower than that on IC1000 porous pad. This is consistent with the fact that the heat capacity of a noncell pad is noticeably higher than that of its porous counterpart. The nonporous NCP pad with grooves on its surface makes it easier for the pad to readily deliver fresh slurry into the pad–wafer contact area and flush away the used slurry from the area than the porous IC1000 pad, which helps to dissipate the heat generated by the friction force in the wafer–pad contact area more quickly.

Although a nonporous pad in polishing has many advantages over a porous pad, it does show some disadvantages. Nonporous pads need longer and harsher conditioning to create surface microstructures (roughness, asperities) along with macrostructures as grooves to better carry and transport slurry over the pad surface so that it can achieve optimum polishing performance. In addition, polishing using harder and stiffer nonporous pads could lead to more severe "edge effects" than when using softer and flexible porous pads.

5.2.4.3 Pad Hardness, Young's Modulus, Stiffness, and Thickness Effects Pad hardness has an influence on many aspects of polishing performance. A harder

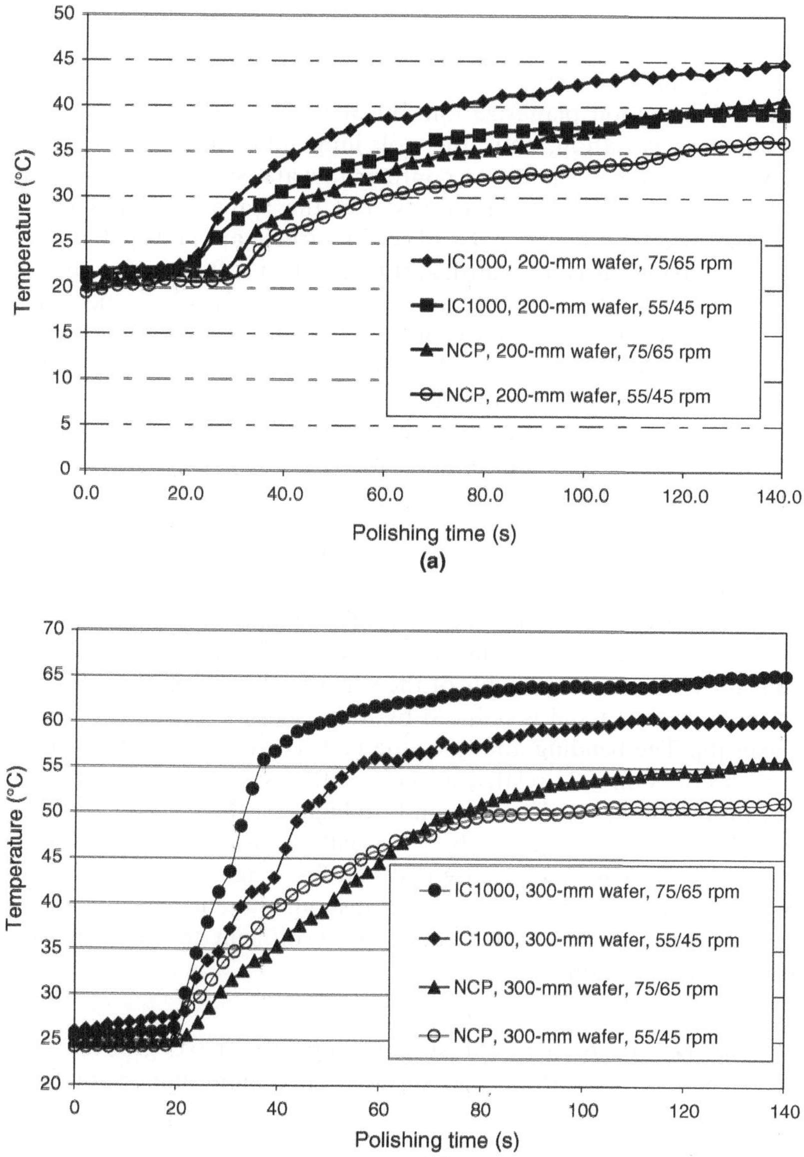

FIGURE 5.10 (**a**) Pad surface temperature profiles during the polishing time (s) of 200 mm blanket oxide wafers using a commercially available silica-based slurry. (**b**) Pad surface temperature profiles during the polishing time (s) of 300 mm blanket oxide wafers using a commercially available silica-based slurry.

pad may lead to higher material removal rate [3], better planarization [8,9], and increased defectivity [8]. With a single-layer polishing pad, it is not easy to achieve both good planarization and polishing uniformity, since hard and stiff pads can achieve good planarization, but it is difficult to conform to wafer surface flatness variation; thus the polishing uniformity is poor. On the contrary, soft and flexible pads can conform to wafer surface flatness variation and achieve good polishing uniformity, but the planarization is very poor because the lower and higher parts of wafer surface are polished at a similar rates. Stacked pads with a top hard, stiff layer and a soft, flexible sublayer can achieve both good planarization and polishing uniformity, since the soft, flexible sublayer enables the pad to conform to wafer surface flatness variation while the top hard, stiff layer still maintains high planarization [10]. The degree of pressure nonuniformity (thus MRR nonuniformity) at a die scale increases with the combination of a stiff hard layer and a thick soft layer, whereas it decreases with the combination of a stiff soft layer and a thick hard layer [11].

Young's modulus (or stiffness) can influence polishing removal rate, planarization, dishing, erosion, and so on. Higher pad stiffness can lead to higher MRR when the pad and wafer touch each other (for a soft pad and low abrasive concentration regime) [12]. The planarization length is highly dependent on the bulk modulus of the pad. Dishing amount is mainly determined by the elastic modulus of the superficial layer of the pad (typically tens of microns thick) and is reduced by increasing the elastic modulus of an open layer [6]. The bending ability of the pad has direct influence on dishing amount since it is directly related to pad's stiffness. Higher bending factor or lower bending ability, which corresponds to higher stiffness, can reduce dishing amounts [13–15]. Increasing pad thickness will increase pad's life (thicker pads last longer), but it will also increase pad's stiffness significantly. This can help obtain better dishing performance (lower dishing amount) while the polishing uniformity deteriorates as stiffer pad is more difficult to conform to wafer surface flatness variation.

5.2.4.4 Pad Groove Effects Most polishing pads have grooves on their surface. Grooves on polishing pad surfaces have many important functions [1,2]:

- To prevent wafer hydroplaning on the pad; thus wafer can be in intimate contact with the pad during polishing;
- To ensure uniform slurry distribution across the pad surface and delivery of slurry to the wafer center in order to maintain uniform polishing from wafer center to edge. This is especially important for larger size wafer (300 mm) and for polishing reactive metals such as Cu, since chemical reaction of slurry with metal layer is as critical to polishing rate as mechanical action;

- To adjust and control both overall and local pad stiffnesses so as to control polishing uniformity and planarization across the wafer surface. Groove design may be varied over the pad surface to tailor the mechanical response and reduce edge effects;
- To provide channels for the removal of polishing debris and heat from the pad surface so as to reduce the possibility of scratch and generation of other defects due to the debris accumulation on the pad surface and to avoid rapid chemical reaction at localized high temperature regions due to frictional and chemical reaction heat;
- To facilitate release of wafers from the pad surface after polishing by breaking the surface suction.

There are many different designs of grooves. An optimal groove design depends on many factors including pad type, materials to be polished, polishing tool (polishers), and even polishing conditions (down pressure, table/carrier speed, slurry flow rate, etc.). Groove width, depth, and density all influence the stiffness of a grooved pad. Grooved pads have lower stiffness than nongrooved pads. As groove density and width increase, pad stiffness becomes less dependent on groove depth and more dependent on the thickness of the remaining ungrooved layer of the pad. But groove depth is the key factor in determining the life of grooved pads, since acceptable polishing performance is possible only until the pad has worn out to the point where grooves are too shallow to prevent hydroplaning, distribute slurry, and remove polishing debris. In order to achieve acceptable pad stiffness as well as long pad life, deep grooves are necessary but sufficient depth of pad must remain to provide enough stiffness [2].

Since grooves on pad surface help in slurry delivery and in lowering the pad stiffness (mainly the top hard/stiff layer in a stacked pad), this could lead to a more consistent material removal rate, better polishing uniformity, and alleviated edge effect. It is shown that better uniformity is obtained throughout the wafer with the embossed politex pad than with regular politex pad because of the presence of grooves on the embossed pad [16]. For different patterns of grooves, combined patterns consisting of spiral and logarithmic grooves were shown to impact on several key attributes of the dielectric and copper CMP processes in terms of slurry retention, hydrodynamic pressure, tribological mechanism, and material removal rates [17]. XY grooves on the pad do not exclusively deliver the assumed outstanding performance compared to the perforated or the K-grooved pad [18]. Grooves on pad surface also change other physical properties of the pad in addition to changing pad stiffness. Mechanical properties (hardness, Young's modulus, etc.) of the pads are shown to be dependent on the orientation of the grooves on the pad [19]. The K-grooved pad, which was the softest, displayed the largest removal rates and average coefficient of friction (COF), whereas the XY-grooved pad, which was the hardest, showed the lowest values of removal rate and COF in ILD CMP [20].

5.3 CHEMICAL PROPERTIES OF CMP PADS AND THEIR EFFECTS ON POLISHING PERFORMANCES

5.3.1 Polyurethane Pad Components

Most commercially available polishing pads are made of polyurethanes that are synthesized from at least three major precursors: long-chain polyol, diisocyanate or isocyanate derivatives, and chain extender [1]. The polyols used in polyurethane manufacturing are mostly polyethers with terminal hydroxyl groups. To obtain certain unique physical or mechanical properties, polyesters with hydroxyl terminals have also been used, though polyester diols are more expensive and the subsequent product tends to be less stable in a basic solution. Almost all polyurethanes are based on either toluene diisocyanate (TDI) or diisocyanato-diphenylmethane (MDI) and its derivatives. MDI has greater chemical flexibility and less toxicity than TDI and is hence preferred. MDI has two reactive isocyanate groups per molecule. MDI with a higher number of functionality groups is being explored with the purpose of controlling the cross-link reaction and the resultant polyurethane properties. The chain extenders are molecules containing groups that can react with isocyanates and link such isocyanates together. The links introduce specialized polymer segments (hard and soft segments) into the polyurethane backbones. These chain extenders include low molecule diols (ethylene glycol, butane diol, etc.), diamines, and water. Other ingredients commonly included in commercial polyurethane manufacturing for polishing pads are catalysts, fillers, and blowing agents (to form pores in the pad) [1].

5.3.2 Polyurethane Property Control by Chemical Components

The choice of polyol, especially its size (molecular weight), flexibility of its molecular structure, and functionality, has significant effect on the properties of the resultant polyurethanes. Varying isocyanates also have major influence on the properties of polyurethanes, since the reaction of di- or polyfunctional isocyanates with polyols forms the polyurethanes. When there is excessive isocyanate with respect to diol, many secondary reactions may occur to create chemical cross-links between chains and network structure in the polyurethanes. Thus pad properties can be controlled and fine-tuned through the control of the stoichiometric ratio of isocyanate to diol.

The molecular structures of polyurethanes vary from rigid cross-linked polymers to linear, highly extensible elastomers. In order to meet a particular requirement or a balance between rigidity and elasticity, the polymers have separate segments or are blended. In general, the soft segments or components are composed of high molecular weight long chain diols in the formation of polyurethanes and are quite mobile, which results in increased flexibility, toughness, and impact resistance. The hard segments or components are stiff oliourethane units mainly composed of reacted isocyanates and chain extender moieties. Hard segments act as pseudo cross-links and control the dimensional thermal stability of polyurethanes. The properties of hardness, stiffness, among others at higher temperatures are controlled by the hard segments. When hard segments "melt" above a certain temperature with missing chemical cross-links, the polyurethanes will become thermoplastic with greatly reduced strength and stiffness (softened) [1].

5.3.3 Chemical Effects on Polishing Performance

The physical properties of the pad are heavily determined by its chemical constituents such as polyols, isocyanates, stoichiometric ratio of isocyanate to diol, and cross-link molecular structures (hard and soft segments). They also have a great impact on the mechanical properties of the pad (strength, hardness, stiffness, Young's modulus, etc.) and on the performance of the CMP process.

CMP is a wet process with DI water and/or polishing slurry present. During polishing, pad is under the attack of water, slurry chemicals, and abrasive particles at elevated temperature due to the friction force among wafer–particle–pad contact. This leads to changes in pad's physical and mechanical properties that influence polishing performance.

When pads are immersed in water, the water molecules penetrate into the pads and reduce their elastic modulus by breaking down the hydrogen bonding between adjacent polymer strands. The reduced elastic and shear moduli of the pad could lead to lower polishing rates. The material removal rate is strongly influenced by the mechanical properties of the pad surface (although the bulk material properties of the pad are nearly unchanged), which may be affected significantly by the time of immersion in water [19,21–23]. In tungsten CMP, polished materials and debris (tungsten), abrasive particles like alumina, oxidizer ion like iodate, and other ions could accumulate in the pad after polishing. The amount of polishing debris accumulated in the pad increases with increased polishing time. The reduction in the isocyanate group and increase in the hydroxyl group after tungsten CMP indicate that a hydrolysis reaction of the isocyanate to form carboxylic acid occurs [24]. In oxide CMP, silica also accumulates in the pad after polishing (material accumulates in the pad pores and grooves). In Cu CMP, polyurethane is fundamentally incompatible with some of the chemicals used, such as hydrogen peroxide. Experiments showed that increase in the hard segment domains on the surface of the polyurethanes upon exposure to the hydrogen peroxide and the polyurethane decomposition (exposed to hydrogen peroxide in Cu slurry) leads to wafer staining and retarded Cu removal rates [25].

Polyurethane pad surface wettability and hydrophilicity also change with the use and conditioning during the CMP process. Conditioning increases pad surface hydrophilicity in addition to refreshing pad surface roughness and removing slurry abrasive particles away from pad surface. Used pads are easier to be wetted and more readily deliver and transport slurry to and under the pad–wafer contact areas, which could lead to better polishing uniformity.

5.4 PAD CONDITIONING AND ITS EFFECT ON CMP PERFORMANCE

It has long been noted that MRR drops rapidly over time (see Fig. 5.11) if the pad is not conditioned during the CMP process [1,4]. This is undesirable because it negatively affects the throughput and the cost of the CMP process. As pad is worn out and its surface becomes glazed during the CMP process (Fig. 5.12), the pad pore size and shape are modified, the direct contact area between wafer and pad surface increases, and the average contact pressure decreases. These and other changes in pad lead to the decrease in material removal rate.

Various pad-conditioning methods have been investigated and adopted to maintain or stabilize the material removal rate. Of these methods, abrasive pad conditioning using diamond-impregnated disks is found to be the most effective. After the installation of each new pad, a series of prepolishing using DI water is performed to break in the pad. The break-in period could last as long as 20–30 min. Before the first monitor or device wafer is polished, several dummy wafers are usually polished using the same slurry that is to be used for the monitor or device wafers. The regular (*in-situ* or *ex-situ*) conditioning process is used during the polishing of these dummy wafers. The purpose of the break-in process is to open up the pores on the pad surface. During the break-in period, there is usually a steady increase in removal rate. Then the removal rate should reach a plateau. It is desirable to reach this plateau within a short period of time. A need for prolonged pad break-in translates to long tool downtime. The purpose of dummy wafer polishing is to let the pad adjust to the slurry environment. The removal rate may go up or down during this phase and then reaches equilibrium. It is also desirable to reach such steady state with as few dummy wafers as possible. If the pad-conditioning process is

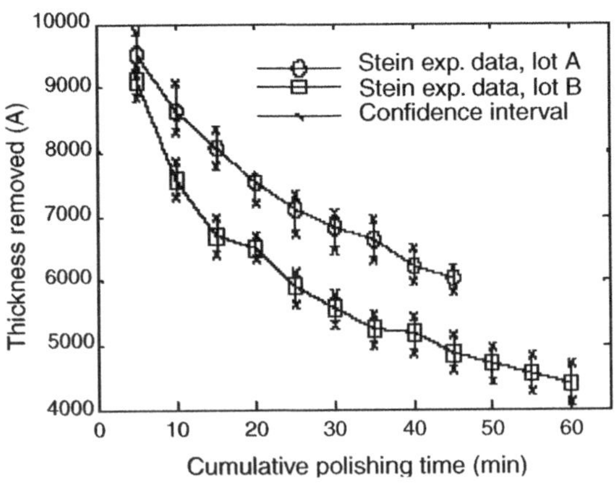

FIGURE 5.11 The oxide thickness removed as a function of cumulative polishing times on the pad. The pads (lot A and lot B) were not conditioned, and cumulative polishing times were obtained using 5 min of polishing [4].

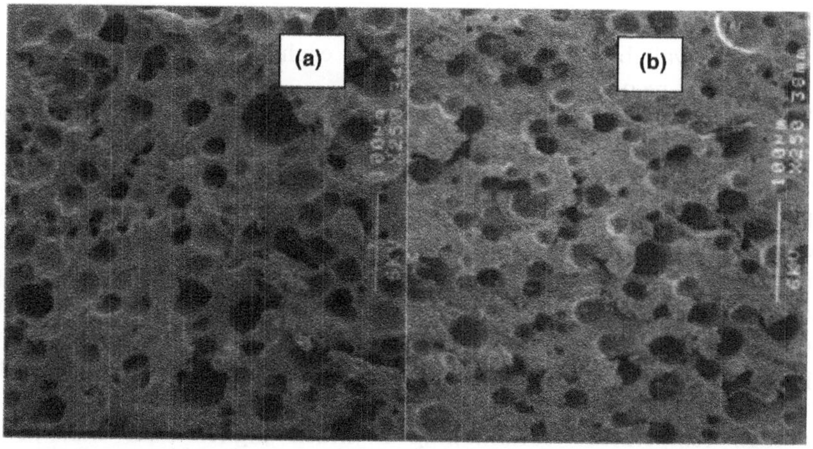

FIGURE 5.12 Scanning electron micrograph of IC1000/Suba IV CMP pad: (**a**) before and (**b**) after polishing [26].

repeated after each wafer polishing, the removal rate profile can be stable for hundreds of wafers [1].

The physical consequence of pad conditioning is to refresh pad surface roughness by reopening the pores [27]. Pad conditioning can be performed during (*in-situ*) or after (*ex-situ*) each wafer polishing. As long as pad asperities are controlled to a consistent height and density, the removal rate remains stable for additional wafers being polished. The disadvantage of an *ex-situ* conditioning process is that it increases, and sometimes doubles, the processing time. On the contrary, an *ex-situ* conditioning process eliminates or significantly reduces the potential side effect caused by the conditioning debris. Currently, there is no well-established conditioning strategy in determining the time intervals to perform pad conditioning (*ex-situ*). The general rule is to use the shortest time required to bring back the removal rate and within-wafer nonuniformity within detectable level.

The phenomenon of material removal rate decay over time during CMP without pad conditioning has been investigated using pad modeling and simulation. Borucki [28] presented a model in which the pad roughness is treated as asperities with randomly varying heights having a common probability distribution function (PDF). The asperity height PDF changes because of the wear caused by the elastic contact between pad asperities and wafer surface under the applied down pressure and pad–wafer relative movement. The effect of abrasive particle on the material removal is ignored since abrasive particles in nanometer scale are much smaller than the asperity heights in micrometer scale. Wang et al. [29] extended the model to include inelastic contact between pad asperity and wafer surface. They also incorporated the effect of abrasive particles on the material removal rate decay [30]. The model correctly predicted the MRR decay trend that matches the experimental results. The models are potentially useful in determining the time window available for polishing between each *ex-situ* conditioning, or the time at which a reconditioning is needed.

There are many factors that can influence the pad conditioning effectiveness. A pad conditioning recipe usually includes down force, rotation speeds of the conditioner (head) and platen (or pad), sweeping speed of the conditioner on the pad, and duration. As the pad wears off in varying rates in different radial zones, it is desirable to program the conditioner to spend different amounts of time in each radial section (from center to edge of the pad) based on the pad wear profile. By implementing such protocol, the pad thickness profile can be better maintained and the WIWNU can be improved [31,32]. Pad conditioning temperature is also an important aspect that could influence the pad surface properties and the CMP performance [33–38]. The removal rate of oxide film after high temperature conditioning is much higher than that at low temperature conditioning. Higher temperature conditioning also makes it easier to remove the slurry residues from pores and grooves of polishing pads [33–35]. Diamond grid shape, size, and numbers, and their positions (alignment) on the conditioner head are also factors that influence the conditioning effects/efficiency.

In addition to opening up and cleaning up the pores, conditioning process may also have impact on other properties of the pad. For harder pads with high modulus, conditioning results in high pad wear. For softer pads with low modulus, conditioning mainly causes material plastic flow and ploughs instead of abrading/wearing that leads to significantly less pad loss. Pad porosity may also change significantly when pad is conditioned. During conditioning, the glazed pad surface is abraded and inner pad pores are surfaced. If there is a vertical gradient in pore density or pore size in the pad materials, the new surface will not have the same pore density and pore size as those found on the pad before conditioning. This may also contribute to removal rate change over time. Therefore, it is desirable for a pad to have uniform pore size and density throughout the pad, both vertically and horizontally. If there is a predictable change in such property, the conditioning process needs to be programmed to counter such nonuniformity.

In principle, the pad break-in and dummy wafer polish process should have brought the pad to a steady state such that each subsequent conditioning will refresh the pad surface with the same physical and chemical properties. For example, the pad can become much more hydrophobic due to the glazing effect (fused polymeric asperities) during the polishing. A pad conditioning step will bring the level of hydrophilicity back to the original level [27,39]. In reality, such a steady state may never be reached as the pad continues to react with the chemicals in the slurry throughout the polishing process. This is particularly problematic for an open-pore pad. The exposure of the pad materials in the lower portion of the pad is much longer than that on the top. By the time they are surfaced, they may exhibit different chemical properties such as hydrophilicity. Therefore, caution must be taken to avoid the use of open-pore pad if the slurry is chemically aggressive that may lead to the formation of a property gradient from the bottom to the top layer of the pad. Because of the complications described above, a steady-state *in-situ* conditioning or an *ex-situ* conditioning with a fixed recipe may lead to severe overconditioning or underconditioning after the pad property starts to drift. Conditioning aggressiveness has significant impact on polishing performance since it controls the pad surface roughness change [27]. Pad surface roughness change and conditioning aggressiveness could be monitored by the surface asperity height distribution through an optical (light white) interferometer profilometer.

During the *in-situ* conditioning process, the conditioner has intimate contact with the slurry chemistry and may react or corrode. The corrosion may lead to a shedding of diamond particles and corroded debris. These large and hard particles can clog the pores on the pad and will cause severe scratches [27,39]. The loss of diamond particles also decreases the conditioning efficiency and polishing performance. Therefore, it is highly desirable to design and improve the chemical resistance of the conditioning materials [40].

5.5 MODELING OF PAD EFFECTS ON POLISHING PERFORMANCE

5.5.1 Review of Modeling of Pad Effects on Polishing Performance

Although extensive work (polishing experiments and theoretical modeling) has been done to investigate the effects of pad properties on polishing performance, the mechanism of pad effects on polishing performance is still not sufficiently understood because of the complicated relationship between pad properties and polishing performance. The major efforts of modeling of pad effects are on how pad physical properties (Young's modulus, viscoelasticity, hardness, stiffness, surface roughness, grooves, etc.) influence removal rate, nonuniformity, dishing, erosion, edge effect, and so on. Modeling on microscratch defect generation due to pad effects during the CMP process has been minimal to date.

The Preston's equation [41] $\mathrm{MRR} = K_p PV$ is the first and most commonly used form for CMP material removal rate. The equation states that (average) MRR is linearly proportional to the product of the applied down pressure and the relative sliding speed between wafer and pad surfaces. Other process effects such as polishing slurry, pad properties, and polishing conditions other than the pressure and rotation speed are all lumped together in the Preston coefficient K_p. However, in many situations, material removal rate deviates from the Preston's equation. The relationship between the MRR and the applied down pressure or the wafer–pad surface relative speed is unusually nonlinear. Zhang et al. [42] proposed an equation $\mathrm{MRR} = K_p \sqrt{PV}$ taking into account the normal and shear stresses acting on the contact area between abrasive particles and wafer surfaces. Shi and Zhao [43,44] gave an expression

MRR = $K_p P^{2/3} V$. Luo and Dornfeld [45] introduced a nonlinear MRR model based on statistical distributions of abrasive particle sizes and considering uniform pad surface roughness (asperity). Fu et al. [12] showed another nonlinear MRR model based on the concept of incomplete and complete contact between the wafer and the pad (beam bending model of pad). Bastawros et al. [46] presented a model considering several saline features of the pad, such as the pad asperity of various amplitudes and frequencies, the local deformation of individual cells, the elastic asperity contact between the wafer and the pad, the multilevel contact evolution at the particle size scale, and the macro asperity scale. These factors lead to several domains of wafer–particle–pad contacts. The MRR dependency on the applied down pressure is different in the three distinct wafer–particle–pad contact regimes. However, these models did not consider the effect of pad surface roughness (asperity height distribution) on polishing removal rate. There are quite a few other publications considering the effect of pad roughness and other pad properties on polishing removal rate [21,47–54].

When the pad is not conditioned during the CMP process, pad surface roughness (asperities) is worn off and becomes glazed, leading to a rapid drop in MRR [4]. Fu and Chandra [55] investigated the effects of pad viscoelasticity on the decay of MRR in a CMP process without considering pad roughness. Borucki [28] presented a dynamic MRR decay model based on the statistical representation of pad surface roughness (asperity height distribution PDF) and elastic contact between wafer and pad asperities, but the effect of abrasive particles on material removal is ignored since abrasive particles in nanometer scale are much smaller than pad asperity heights in micrometer scale. Wang et al. [29] extended the model to consider inelastic deformation of pad asperities in contact with wafer surface. They showed that with this modification and careful selection of pad surface roughness parameters (asperity height distribution PDF), the time-varying (dynamic) MRR model can have significant potential in the prediction of MRR and MRR decay. They further presented another model incorporating the abrasive particle effects and considered that material removal comes from the interaction of the wafer and the active particles entrapped in the pad asperity–wafer contact areas [30]. This model reflects not only the effects of pad properties such as Young's modulus, hardness, asperity density, tip curvature, and height distribution but also the effects of abrasive particle (mean) size and size distribution on the polishing removal rate at a given polishing time.

Polishing nonuniformity (WIWNU and WTWNU) is also investigated through modeling. Fu and Chandra [56] investigated the CMP polishing removal rate WIWNU and edge effect based on elastic pad deformation. They also presented an analytical model for predicting the spatial and temporal distributions of the wafer–pad interface pressure during a CMP process [55]. The spatial distribution correlates with the WIWNU observed in a single wafer, whereas the long-range temporal distribution correlates with the MRR decay and the resulting WTWNU over a batch of wafers. Seok et al. [57]

showed large changes in polishing removal rate near the wafer's edges (edge effects), as often seen in practice, with very uniform removal across most of the wafer surface. Kim et al. [58] in a three-dimensional multiscale elastohydrodynamic lubrication (EHL) contact model showed that the contact force exerted on the wafer surface by asperities is nonuniform across the wafer, which implies nonuniformity in material removal.

The viscoelasticity of a polishing pad has been shown to play an important role in the CMP process [59]. Fu and Chandra [55] in a viscoelastic pad deformation model showed that MRR decay for an unconditioned pad is strongly influenced by the viscoelasticity of the pad. Guo et al. [60] investigated the effects of pad viscoelasticity on dishing and erosion without considering pad roughness (asperity height distribution).

Effects of other pad properties on step height reduction, dishing, and erosion have been investigated through modeling. A relationship that describes the development of dishing with overpolishing time for an arbitrary metal line was given in a physical model presented by Nguyen et al. [61]. This model considers the material removal behavior of two groups of asperities separately (which contain asperities with contact sizes smaller and larger than the line width, respectively). Effects of pad stiffness and bending ability on dishing were shown. Dishing of metal lines is strongly dependent on line width and increases with decreasing pad modulus (higher for soft pads). Higher stiffness and lower bending ability could reduce dishing [13,15,62]. Vlassak [63,64] presented models based on contact mechanics and considered the pad surface roughness (asperity height distribution) using the Greenwood–Williamson model. It was shown that pad roughness enhances dishing of metal lines significantly, but it has only a limited effect on dielectric erosion. As features on the wafer become increasingly narrow, asperity shielding becomes important.

To date, pad effects on micro/nanoscratch generation on polished wafer's surface have not been investigated extensively through modeling. Based on the development of models predicting MRR and MRR decay phenomenon, Wang et al. [65] presented a model considering pad surface roughness (asperity height distribution) as well as slurry abrasive particle size distribution for micro/nanoscratch generation during the CMP process. Particle hardness, particle size distribution, and hardness of wafer surface material together with pad properties such as Young's modulus, hardness, asperity density, tip curvature, and height distribution are all involved in micro/nanoscratch generation probability under a given applied down pressure and wafer–pad rotating speeds. This model is dynamic because it can accommodate pad surface topography evolution (due to pad asperity wear) and particle agglomeration (with debris generated in the polishing process) that changes the slurry abrasive particle size distribution.

Readers are urged to read the above-mentioned publications for model details. The following subsection will give some modeling details of pad effects on polishing performance.

5.5.2 Modeling of Pad Effects on Polishing Performance

5.5.2.1 Pads and Pressure CMP pads have a rough surface. Model studies, fit to experimental profilometer measurements, find that the surface asperities can be characterized as having an average density of $1.2 \times 10^5/\text{cm}^2$, a width of 30 μm (with a standard deviation of 25 μm), and a height variation of 10 μm [47]. In the polishing process, they are pressed against the wafer, pushing abrasive particles from the slurry into the wafer surface. Because of the roughness, only a small fraction of the pad surface is actually in contact with the wafer.

Experimental [66] and theoretical [47] studies of the actual contact area A_C show that during CMP it is only a few tenths of a percent of the total wafer area A_W, and increase with applied pressure P. Stiff pads have a smaller contact area than compliant ones at any given pressure. This has been modeled [47], for pads with Young's modulus E, as Equation 5.1

$$\frac{A_C}{A_W} = \alpha \frac{P}{E} \tag{5.1}$$

and the data from this study can be fitted to give $\alpha = 0.00925$, as shown in Fig. 5.13. Experimental data [66] are in agreement with this fitting.

Equation 5.1 has some important implications. One is that the effective pressure of the pad pushing the abrasive into the wafer depends on Young's modulus of the pad but is independent of the applied polishing pressure. This can be shown as follows: The force applied by the pad is $F = P A_W = P_{\text{eff}} A_C$. Using Equation 5.1 to eliminate A_C leads to the following expression for the average effective pressure:

$$P_{\text{eff}} = E/\alpha \tag{5.2}$$

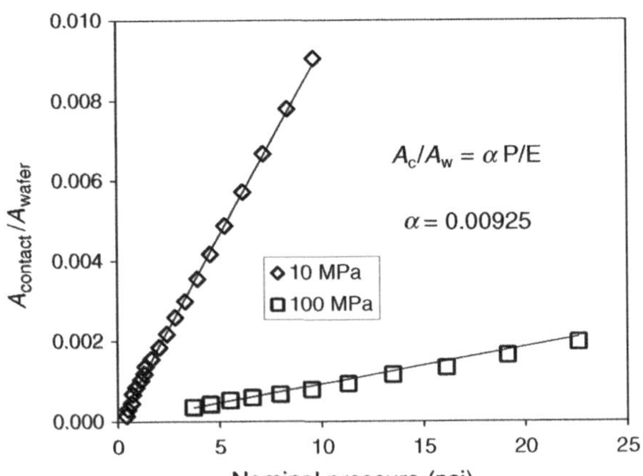

FIGURE 5.13 Fraction of pad surface in contact with the wafer as a function of pressure for two different pads. Data points are from Yu et al. [47]. Lines are fit to Eq. 5.1 with $\alpha = 0.00925$.

Qualitatively, since the contact area increases linearly with applied pressure, the effective pressure is constant for a given pad. Soft, compliant pads have a larger contact area and lower effective pressure whereas hard, stiff pads have a smaller contact area and higher effective pressure. Thus soft pads push abrasive particles against the wafer over a larger area but with less force than hard pads do.

Since the abrasive–wafer surface contact area increases with the force applied, harder pads will remove more material per abrasive than soft pads, but soft pads—with larger contact areas—can remove more material than hard pads at low pressures.

Several early studies [67–69] of CMP focused on the question of whether the CMP process obeyed the Preston polishing equation $R = k_{\mathrm{Preston}}Pv$ or not for the dependence of the polishing rate R on the polishing pressure P and speed v. The experimentally unequivocal answer [69] is that sometimes it does and sometimes it does not. One model [70–72] shows the polishing rate to have the form

$$R = \frac{aPv}{bE + Pv} \tag{5.3}$$

The limiting behavior, for large Pv, is $R = a$. When $bE \gg Pv$, which is the case for hard pads, Equation 5.3 becomes $R = (a/bE)\ Pv$, which identifies the Preston constant as $k_{\mathrm{Preston}} = (a/bE)$. Equation 5.3 thus predicts that hard pads do obey the Preston equation, whereas soft pads do not, except at very low pressures. This is illustrated by Fig. 5.14 for tungsten CMP.

The qualitative explanation of Fig. 5.14 is that as pressure increases, the contact area will be much larger for soft pads than for hard pads, so the removal rate of soft pads will increase faster than the removal rate of hard pads. At low pressures, the mechanical removal step dominates the chemical oxidation step because only a very small fraction of the surface is unoxidized

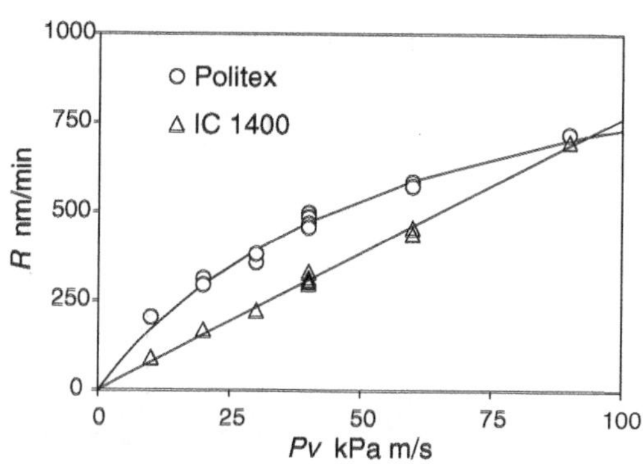

FIGURE 5.14 Removal rate R for tungsten CMP as a function of polishing pressure P and speed v for hard IC1400 and soft Politex pads. Symbols show experimental data from Stein et al. [69]. Solid lines are fit to Equation 5.3.

under these conditions. However, it takes time for the chemical oxidation step to replace the mechanically removed surface film. As the pressure increases, more surface material is removed and the area available for mechanical removal will decrease. In this case, the removal rate will slow down as the system becomes chemically limiting rather than mechanically limiting. This will happen at lower pressures for soft pads than for hard pads. The maximum removal rate will be higher for hard pads than for soft ones, since the effective pressure (and contact area per abrasive particle) will be larger for hard pads. All of this behavior can be seen in Fig. 5.14. The soft Politex pad starts giving higher removal rates than the hard IC1400 pad, but it departs from the Preston dependence as Pv increases. The asymptotic maximum removal rate for Politex is less than that for IC1400. The hard IC1400 pad obeys the Preston equation over the Pv conditions shown and gives higher polishing rates than the soft pad at high Pv. The model prediction [72b] is that as Pv increases, the polishing rate for IC1400 will also slow down to an asymptotic maximum.

Thus the goal of increasing polishing pressure is to increase the contact area of the pad against the wafer surface, while keeping the effective contact pressure constant. Both the contact area and the effective pressure depend on Young's modulus of the pad, with harder pads giving higher effective pressures.

5.5.2.2 ***Pads and Abrasives*** During polishing, material is removed from the wafer surface by abrasive particles that are pressed by the polishing pad onto the surface. A pad with abrasives on its surface is shown [66] in Fig. 5.15.

The removal rate depends on how many abrasive particles on the pad are pressed against the surface. The pad has been modeled [70–72] as a Langmuir

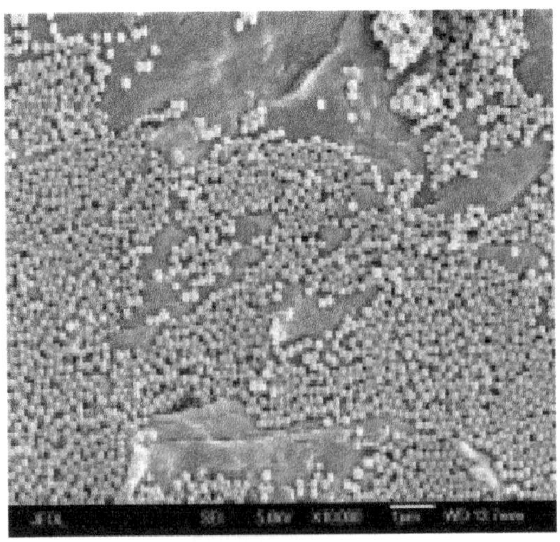

FIGURE 5.15 Section of an IC1000 pad, partially covered with abrasive particles, after CMP with a 12% slurry containing 200-nm abrasives [66].

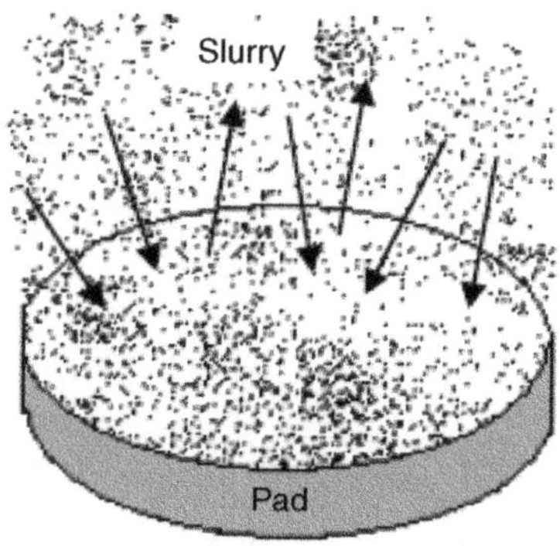

FIGURE 5.16 Pad and slurry with abrasives. Abrasive particles attach to and release from the pad.

surface, with a total site density n_{oPad}. Sites are either empty n_{S} or occupied by abrasive particles n_{A}, so that $n_{\mathrm{oPad}} = n_{\mathrm{S}} + n_{\mathrm{A}}$. Abrasives can attach to and separate from this surface as shown in Fig. 5.16.

The attachment rate r_{on} is proportional to the concentration of abrasives in the slurry [A] and to the number of available sites $r_{\mathrm{on}} = k_{\mathrm{on}}\,[\mathrm{A}]\,n_{\mathrm{S}}$. The separation rate r_{off} depends on the number of occupied sites $r_{\mathrm{off}} = k_{\mathrm{off}} n_{\mathrm{A}}$. At steady state, these rates balance and $k_{\mathrm{on}}\,[\mathrm{A}]\,n_{\mathrm{S}} = k_{\mathrm{off}} n_{\mathrm{A}}$. Combining this expression with that for the total number of sites gives the number of occupied sites n_{A} as a function of the abrasive concentration.

$$n_{\mathrm{A}} = n_{\mathrm{oPad}} \frac{[\mathrm{A}]}{[\mathrm{A}] + K_{\mathrm{Pad}}} \qquad K_{\mathrm{Pad}} = \frac{k_{\mathrm{off}}}{k_{\mathrm{on}}} \tag{5.4}$$

Since the mechanical removal rate is proportional to n_{A}, it is possible to write the overall material removal rate as

$$R = k^* n_{\mathrm{A}} = k \frac{[\mathrm{A}]}{[\mathrm{A}] + K_{\mathrm{Pad}}} \tag{5.5}$$

Figure 5.17 shows the removal rate as a function of abrasive loading %A for tungsten CMP, fit to the prediction of Equation 5.5. The data show a rapid initial rise in removal as abrasive loading is increased, slowing to an asymptotic maximum at high abrasive loading. The model interpretation of this is that at low abrasive loading the pad surface is mostly empty and there is ample open attachment space for abrasive particles from the slurry. Thus the removal rate increases with abrasive loading in this region. At high abrasive loading the pad surface is largely saturated, leaving little attachment space for slurry particles. More concentrated slurry will not significantly change the removal rate in this region.

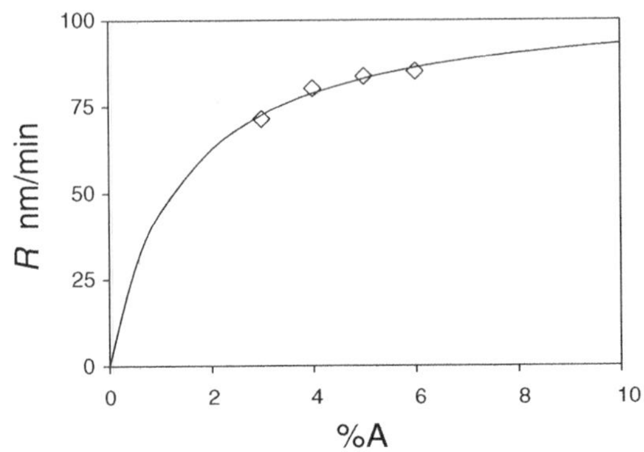

FIGURE 5.17 Removal rate versus weight percent silica abrasive. Symbols are for data from Li et al. [73]. The curve fit to Equation 5.5.

Pad–abrasive particle interaction, as modeled above, appears in several different places. For example, abrasive loading given in weight percentage masks the fact that for a given weight percentage there are more small particles than large ones. Equation 5.6 shows how the relationship between weight percentage %A and particle concentration [A] in particles/cc depends on the abrasive diameter d_A and on the abrasive density ρ_A and the slurry fluid density ρ_f.

$$[A] = \frac{6}{\pi d_A^3}\left[\frac{\%A}{(1-\rho_A/\rho_f)\%A + 100\rho_A/\rho_f}\right] \quad (5.6)$$

Thus the apparent observation that removal rates increase with increasing abrasive loading and decreasing abrasive size, as shown in Fig. 5.18a, is converted to the observation that there are more smaller abrasives than larger ones when the same data are plotted against particle concentration, as shown in Fig. 5.18.b

The data in Fig. 5.18 can be understood as showing a rapid increase in removal rate with increasing concentration at low abrasive concentrations when a large fraction of the pad surface is available, slowing toward an asymptotic maximum removal rate at higher concentrations when the pad is largely saturated by abrasive particles.

The expression in Equation 5.4, and the model it is based on, uses two values to parameterize the interaction between pad and abrasive: n_{oPad} and K_{Pad}. It is reasonable to ask about [76] the effect of abrasive size on each of these. If there is little separation between sites, as appears to be the case shown in Fig. 5.15, then the site density n_{oPad} must decrease with increasing abrasive area, which is proportional to the square of the abrasive diameter d_A. If, however, the mechanical removal rate per particle increases with the particle diameter, then these effects would largely cancel each other. To understand the effect of abrasive diameter on $K_{Pad} = k_{off}/k_{on}$, it is helpful to think about how particle size affects the rates of attachment to and separation from the pad. The

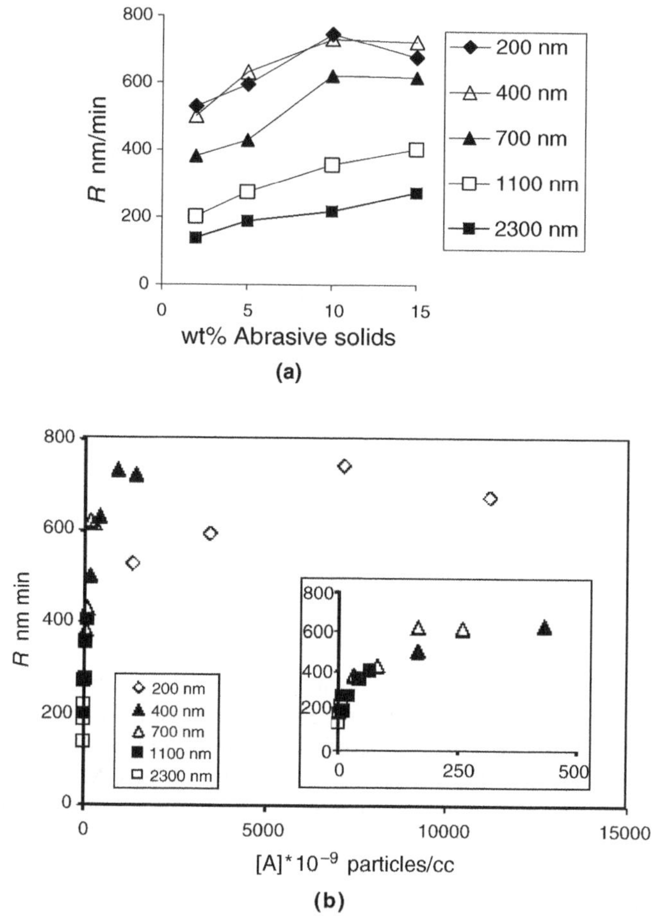

FIGURE 5.18 Tungsten CMP removal rates R versus abrasive loading for five different abrasive sizes. Symbols are for data from Bielmann et al. [74]: **(a)** plotted versus weight percent and **(b)** the same data plotted versus particle concentration, including inset for low-concentration behavior.

attachment rate is based on the collision rate, which is independent of particle size for a well-stirred slurry. Separation, however, depends on the energy from collisions between the attached abrasive particles and the slurry fluid. The energy required for removal and, therefore, the separation rate are inversely proportional to the abrasive mass. Since mass is proportional to volume and volume is proportional to the cube of the abrasive diameter, both k_{off} and therefore K_{Pad} are proportional to d_{A}^{-3}. This implies that a plot of $\log(K_{\text{Pad}})$ versus $\log(d_{\text{A}})$ should have a slope of -3.

The data shown in Fig. 5.18a were used to determine fitting values of K_{Pad} for each different abrasive diameter. A similar procedure was followed for other experimental data representing tungsten, tantalum, and copper polishing

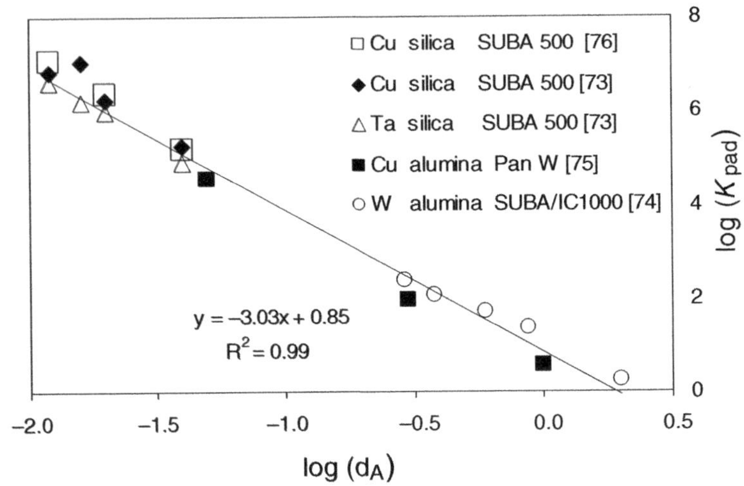

FIGURE 5.19 A log–log plot of K_{pad} versus abrasive diameter d_{A} for different metals and abrasives. Symbols are for experimental data; the solid line is a least squares fit to the data [76].

using either silica or alumina abrasives and using different polishing pads. The resulting set of d_{A} and K_{Pad} pairs were analyzed using a log–log plot. The result, shown in Fig. 5.19, is a slope of −3 as predicted by this model.

Thus this model of the pad–abrasive interactions in CMP can be used to understand how the polishing rate depends on abrasive loading and abrasive size.

5.5.2.3 Pads, Dishing, and Erosion Dishing and erosion are a central problem in CMP of patterned wafers. Polishing of patterned wafers generally begins on an uneven surface left by the various chemical, physical, and electrochemical deposition processes that fill lithographically produced trenches with metal, as shown in Fig. 5.20a. Ideally, CMP would leave the perfectly flat surface shown in Fig. 5.20b. However, to avoid uncertainties in end-point detection and to ensure complete clearance of metal from the ILD, a short overpolishing time is used. Since metal generally polishes faster than ILD, this introduces dishing as shown in Fig. 5.20c. Dishing is found to depend on overpolishing time, on the linewidth λ of the trench, and on the pattern density θ, the fraction of surface covered by metal. In some cases—when metal planarization is not accomplished before the ILD surface is cleared—there is residual dishing even at the clearance end point.

When dishing occurs in a patterned wafer, the ILD beside the dished area polishes faster than the ILD in isolated field regions. The ILD height difference between patterned and field areas, shown in Fig. 5.20, is called erosion. Experimentally, erosion depends on the pattern density.

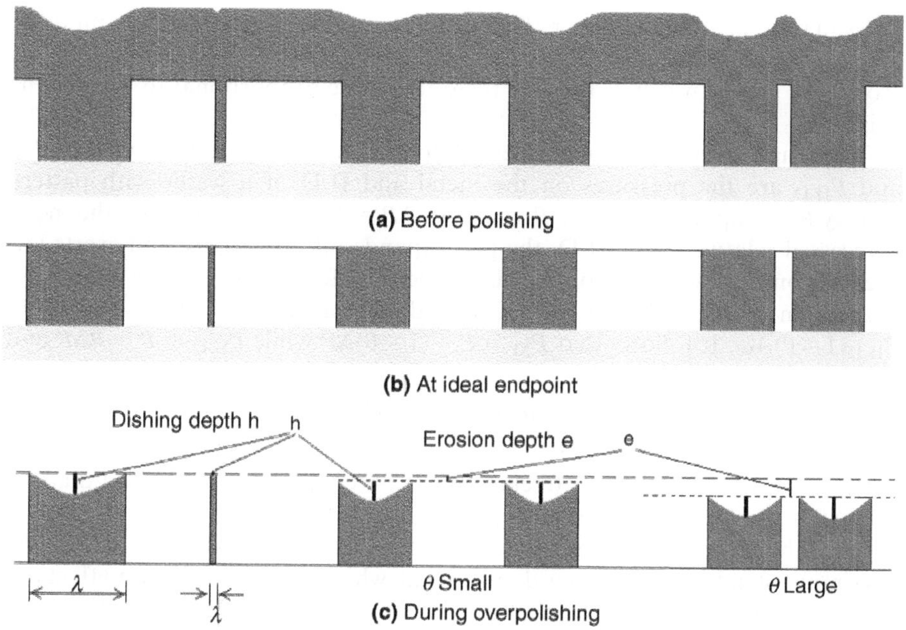

FIGURE 5.20 Surface of a wafer (**a**) before CMP, (**b**) at the ideal endpoint, and (**c**) during overpolishing.

The polishing pad is strongly implicated in both dishing and erosion. A model [77] showing how the pad affects this behavior is based on three fundamental observations: (1) Slurry selectivity leads to metal dishing, (2) Pad expansion into dished regions reduces the pressure and limits dishing, and (3) Pad stretching over the ILD lip in patterned regions leads to higher pressures on the ILD than in field regions and to ILD erosion. Each of these points will be discussed below.

Slurry selectivity σ is commonly defined as the ratio of the polishing rate of blanket wafers at a given polishing pressure and speed, $\sigma = R_M^*/R_{ILD}^*$. Large σ correlates with removal rates that are much faster for metal than for ILD. It is helpful to use $c = (\sigma - 1)/\sigma = (R_M^* - R_{ILD}^*)/R_M^*$, with c taking values between 0 and 1 as an alternate for σ.

Pad expansion into a dished region of height h leads to a pressure difference ΔP between the top and the bottom. This pressure difference is shown in Equation 5.7 as the sum of two terms.

$$\Delta P = \left[\frac{E}{1-\upsilon^2}\right]\frac{h}{\tau} + k_{\text{stretch}}\left[\frac{E}{1-\upsilon^2}\right]\frac{h}{\lambda} = \left[\frac{E}{1-\upsilon^2}\right]\frac{h}{\tau}\left(\frac{\lambda + k_{\text{stretch}}\tau}{\lambda}\right) = \frac{\lambda + b}{a\lambda}h \tag{5.7}$$

The first term is for the conventional Hooke's law—Young's modulus expansion/compression of an elastic pad material with Young's modulus E,

Poisson ratio υ, and thickness τ. The second term is for pad stretching, modeled [61] as a supported beam problem for a material with stretching constant k_{stretch} crossing a gap of width λ. Both of these terms are proportional to the dishing height h.

There are several pressures relevant to polishing of patterned wafers. P_{M} and P_{ILD} are the pressures on the metal and ILD of a wafer with pattern density θ, for applied pressure P and with $\Delta P = P_{\text{ILD}} - P_{\text{M}}$. Because the metal is recessed relative to the ILD, the pad expands into the cavity and exerts less pressure on the metal than on the ILD. The total force F on a wafer of area A_{W} is the sum of pressures on the metal and ILD areas, $F = PA_{\text{W}} = P_{\text{M}}\theta A_{\text{W}} + P_{\text{ILD}}(1-\theta)A_{\text{W}}$. It follows that $P_{\text{M}} = P - (1-\theta)\Delta P$ while $P_{\text{ILD}} = P + \theta\Delta P$ and

$$\Delta P = \left(\frac{c}{1-\theta c}\right)P \tag{5.8}$$

which relates the pressure difference to the pattern density θ. From Equation 5.7, ΔP is also related to the linewidth λ.

With these expressions for the various pressures, it is possible to predict the maximum steady-state dishing depth h_∞ at which the removal rate effects of higher pressure on the ILD balance with those of higher slurry selectivity for the metal. The result is shown in Equation 5.9.

$$\begin{gathered} h_\infty = a\frac{\lambda}{b+\lambda}\frac{c}{(1-\theta c)}P \\ a = \frac{1-\upsilon^2}{E}\tau \quad b = k_{\text{stretch}}\tau \quad c = \frac{R^*_{\text{M}} - R^*_{\text{ILD}}}{R^*_{\text{M}}} \end{gathered} \tag{5.9}$$

In Equation 5.9, h_∞ depends on pressure P, on patterning parameters θ and λ, on slurry parameter c, and on pad parameters a and b, which relate to pad properties E, υ, τ, and k_{stretch}.

In addition to predicting the steady-state dishing depth, the model can be used to predict how dishing changes with time. The result is

$$h(t) = h_\infty + (h_0 - h_\infty)e^{-\kappa(t-t_0)} \tag{5.10}$$

where h_0 is the dishing at clearing time t_0. For an ideal system, $h_0 = 0$. Time constant κ is the ratio of the difference between the time required for blanket metal and ILD wafers to be polished to the steady-state dishing depth h_∞.

$$\kappa = \frac{R^*_{\text{M}} - R^*_{\text{ILD}}}{h_\infty} = \left(\frac{c}{1-c}\right)\frac{R^*_{\text{ILD}}}{h_\infty} \tag{5.11}$$

Equation 5.10 predicts an exponential decay to the final dishing depths, with the decay rate dependent on various parameters described above. This expression for dishing as a function of time, linewidth, and pattern density is compared to experimental data in Fig. 5.21.

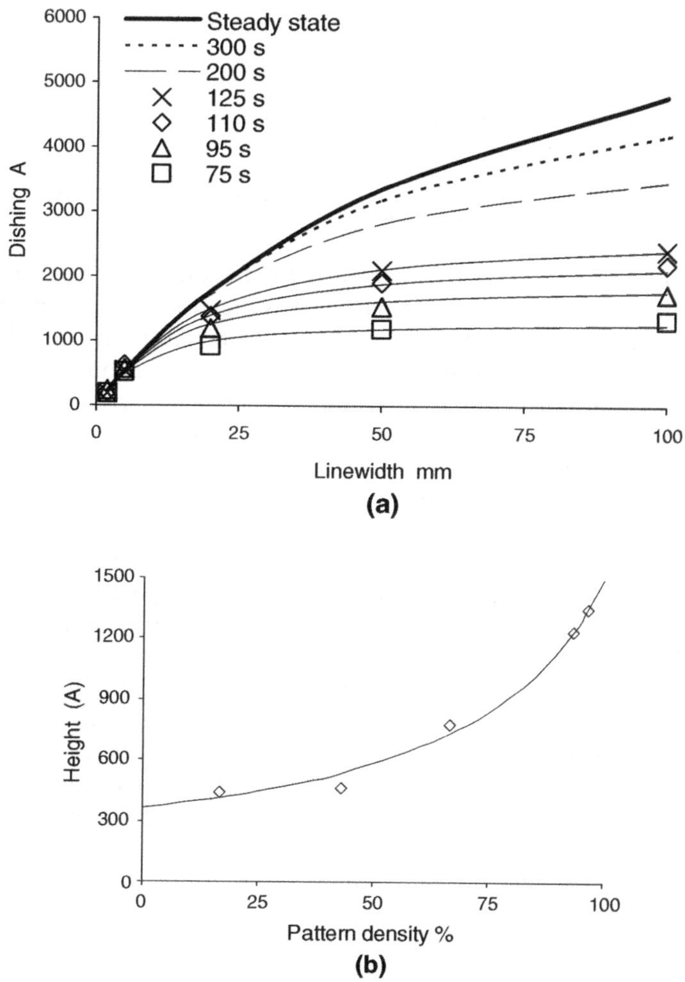

FIGURE 5.21 Dishing as a function of (**a**) linewidth and time and (**b**) pattern density. Symbols are for data from (**a**) Figure 5.18 of Tugbawa [78] and (**b**) Figure 4.12 of Tugbawa [78]. The solid lines are fit to Equations 5.9 and 5.10.

Figure 5.21a shows how dishing increases with time and linewidth toward asymptotic maxima. In time, it approaches the steady-state dishing depths—which are reached sooner for small linewidths than for large ones. As linewidth increases, dishing increases linearly for small linewidths—since the larger gap allows the pad to expand further at any given pressure. But for larger linewidths, the amount that the pad can expand is limited by the stretching term. Thus the amount of dishing is also limited at larger linewidths. Figure 5.21b shows the effect of pattern density on dishing.

Erosion can be qualitatively understood as a result of a pad held up by ILD surrounded by dished metal features. If the ILD covers a broad area, as it would at low pattern density, then the effect is small. If the ILD covers a narrow area, as it would at high pattern density, then it is subject to higher pressure and will erode faster. The $e(t)$ function predicting erosion as a function of time, analogous to Equation 5.10 for dishing, is

$$e(t) = e_0 + R^*_{\mathrm{ILD}}\left(\frac{\theta c}{1-\theta c}\right)\frac{h_\infty - h}{\kappa}h_\infty + R^*_{\mathrm{ILD}}\left(\frac{\theta c}{1-\theta c}\right)(t-t_0) \tag{5.12}$$

where e_0 in the first term is the (generally 0) erosion at clearing time t_0. The second term is a transient that approaches 0 as the dishing height h approaches its steady state. The last term is for continuously increasing erosion. It depends on the pattern density and slurry selectivity and is proportional to the blanket ILD removal rate but is apparently independent of other pad properties. In Fig. 5.22 the model prediction is compared to data for erosion as a function of time and pattern density.

Figure 5.22a shows erosion increasing linearly with time, as predicted by Equation 5.12, at a faster rate for large pattern density than for small pattern density. This is because there is an additional pressure on the ILD surrounded by dished metal, relative to the pressure on a nonpatterned region. The pressure is effectively constant since the dishing is near its steady-state value. When dishing is greater, as is the case for larger pattern density, pressure is higher and erosion is faster. Figure 5.22b shows a steep rise in erosion as pattern density increases. This is because of higher pressures on the smaller ILD regions at high pattern densities than at lower densities.

Thus CMP pads play an important role in dishing of patterned wafers. The model presented here parameterizes CMP pads using Young's modulus, Poisson's ratio, and pad thickness plus an unusual k_{stretch} term. Control of these parameters could lead to improvements in dishing and erosion outcomes during polishing.

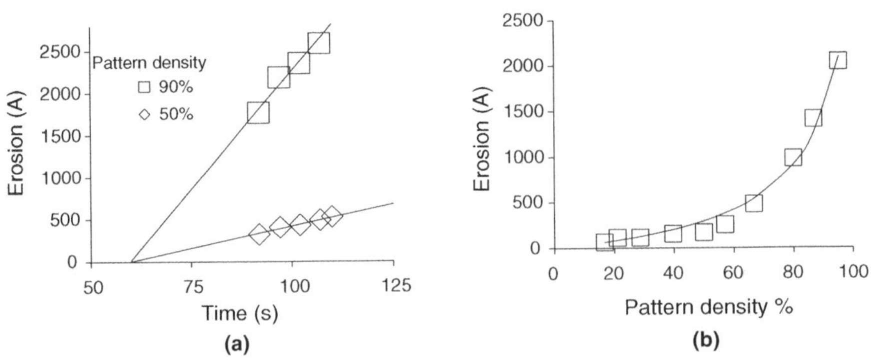

FIGURE 5.22 Erosion as a function of time and pattern density. Symbols are for data from (**a**) Figure 3.14 of Tugbawa [78] and (**b**) Figure 3.16 of Tugbawa [78]. The solid lines are fit to Equation 5.12.

5.6 NOVEL DESIGNS OF CMP PADS

A CMP process is conventionally performed by using slurry (which contains abrasive particles) in combination with a polishing pad. The use of abrasive-particle-containing slurry does lead to some problems. The particles are often the major source of defects such as scratches and particulate residues on polished wafers, even though the polished wafers are supposed to be thoroughly cleaned right after the CMP process. Large and hard particles, even at very low concentration or present in small numbers in the slurry, can lead to scratches and other mechanical defects on a polished wafer surface. The source of these large particles can come from the slurry as a part of normal composition or agglomeration that occurred during storage, transport, additive mixing, and handling. Once they are formed, it is difficult to remove them completely as they can often pass through coarse filters and clog the finer filters. During the polishing, the polishing debris and by-products will interact with these particles and change their surface characteristics and sometimes sizes. Hence it is almost impossible to recycle spent slurries. Furthermore, solid waste disposal is more challenging than solution waste treatment. Therefore, whenever possible, an abrasive-free solution is always preferred over abrasive-containing slurry. For some applications such as oxide, STI, and tungsten CMP, the combination of an abrasive-free solution and a regular pad would not give enough material removal rate under the normal down pressure owing to the mechanical nature of the process and the softness of the regular pads. To overcome this issue, an abrasive-free solution combined with a fixed abrasive pad has been explored and implemented for some applications such as tungsten, oxide, and STI CMP. For a more chemically dominated process such as copper CMP, instead of fixed abrasive pads to enhance the mechanical aspect of the pad, a chemically reactive pad has been investigated to control the surface chemistry among the pad, the wafer, and the chemical solution. In addition, the surface property of the pad can also be modified to meet certain requirements for a given application. In this section, all three types of these pads are described.

5.6.1 Particle-Containing Pads

There are two types of particle-containing pads. The first type holds particles that serve as embedded abrasives. The embedded particles may come off during polishing as a part of normal wear. The second type immobilizes water-soluble particles (WSP) that are slowly released during polishing. The pores left behind by the embedded abrasive or WSP could enhance the conditioning efficiency or possibly eliminate the need for conditioning. In other words, it is an advantage that the particle shedding events continuously and automatically refresh the pad surface [79–81]. Figure 5.23 shows the material removal rate and WIWNU variation with polishing time without pad conditioning for tungsten polishing on fixed abrasive pad [80]. It clearly shows that MRR and WIWNU were quite

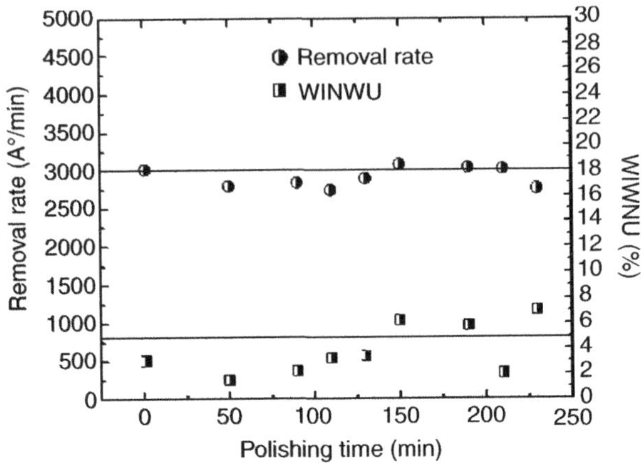

FIGURE 5.23 Removal rate and WIWNU as a function of polishing time without pad conditioning [80].

stable. Compared to conventional polishing using regular pads and abrasive-containing slurry, fixed abrasive pads combined with abrasive-free solution tend to give higher material removal rate for oxide CMP [81]. The increased removal rate could be due to the fact that either the anchored and immobilized abrasives can exert greater pressure on the wafer surface or the self-refreshing effect can reduce the level of pad glazing and enhance the overall polishing efficiency. For pads containing WSP, there is also an increase in removal rate compared to that of a reference pad (IC1000) during oxide CMP [81,85], as shown in Fig. 5.24. In this case, as the WSP have no abrasive function, the increase in removal rate must come from the self-refreshing effect.

In addition to the self-refreshing function, the fixed abrasive pads also give better planarization behavior [79,80,82,83], lower WIWNU and WIDNU

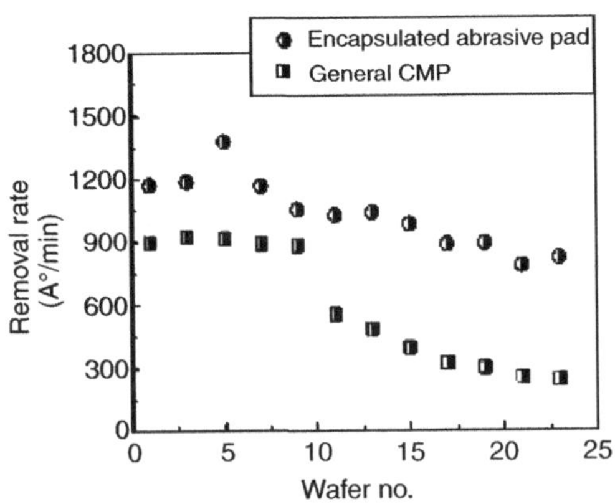

FIGURE 5.24 MRR comparison of oxide wafer polishing on particle-containing pad with DI water and on IC1400 pad with fumed silica slurry at pH = 10.8 [81].

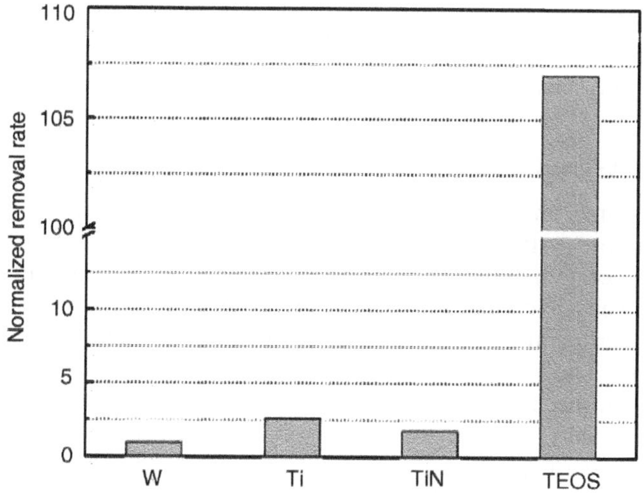

FIGURE 5.25 Material removal selectivity of TEOS, Ti, and TiN to W for polishing on the fixed abrasive pad [80].

(within die nonuniformity) [79,84], good material removal selectivity [80], and lower dishing and erosion [80,83]. Figure 5.25 shows that good material removal selectivity (high TEOS oxide to tungsten, low Ti and TiN to tungsten) can be achieved on a fixed abrasive pad [80]. Figure 5.26 shows the erosion as a function of pattern density after CMP. For fixed abrasive pads, even though

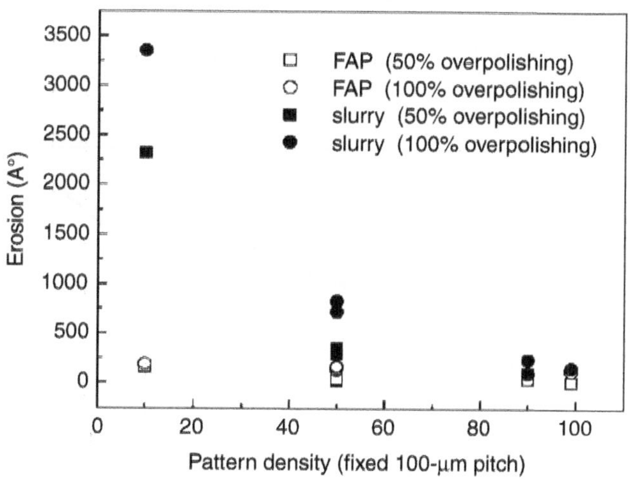

FIGURE 5.26 Erosion as a function of pattern density after CMP. FAP: Polishing with fixed abrasive pad. Slurry: Polishing with abrasive-containing slurry and conventional pad [80].

some suspended particles could still be present because of shedding, the total number of free particles seen by the wafer is far less than a process in which a regular pad and particle-containing slurry are used.

5.6.2 Surface-Treated Pads

During polishing, under mechanical stress and chemical attacks, the chemical and physical properties of the pads can change substantially. For example, hydrolysis of certain functional groups in polyurethane can lead to changes in hydrophobicity/hydrophilicity of the pads. The hot spots on the pad can reach high enough temperature ($>Tg$) so that a polymer may melt locally and form a glazed surface. These chemical and physical changes in the polymer can lead to an alteration in mechanical properties such as strength, stiffness, and hardness, which in turn shifts the polishing performance such as pad life, MRR, and WIWNU. There are two common approaches to overcome this issue. One is to change the chemical composition or the materials of the pad. The second is to increase the level of conditioning and refresh the pad more often. Both methods have disadvantages. There is a long developmental cycle for any change in materials for the pad. The number of variables that can be studied with this approach is limited. Excessive pad conditioning will shorten the lifetime of the pad. One solution is to treat the pads after they have been made. This way, the number of pads with varying properties can be generated in large numbers. Some polymeric pads have been surface modified by depositing a thin film of tetraethylorthosilicate (TEOS) using a plasma-enhanced chemical vapor deposition (PECVD) process. The mechanical and chemical properties of the pad's top surface are a function of the PECVD coating time; for example, the hardness and Young's modulus increase linearly with the duration of the PECVD tetraethylorthosilicate surface coating (Figs. 5.27 and 5.28). Such PECVD-treated pads are hydrophilic and do not require storage in aqueous media during

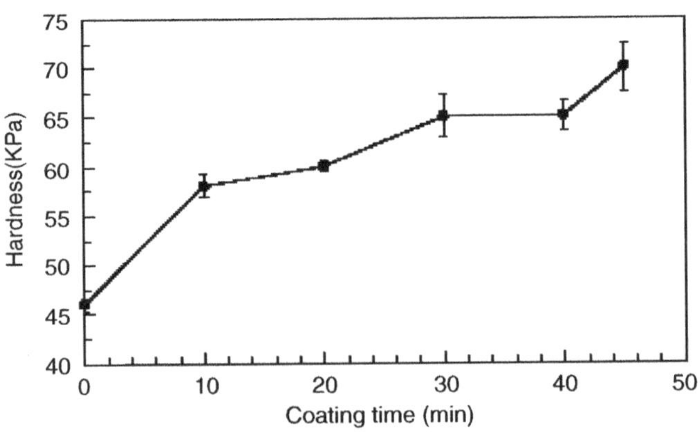

FIGURE 5.27 Hardness as a function of PECVD-TEOS coating time [86].

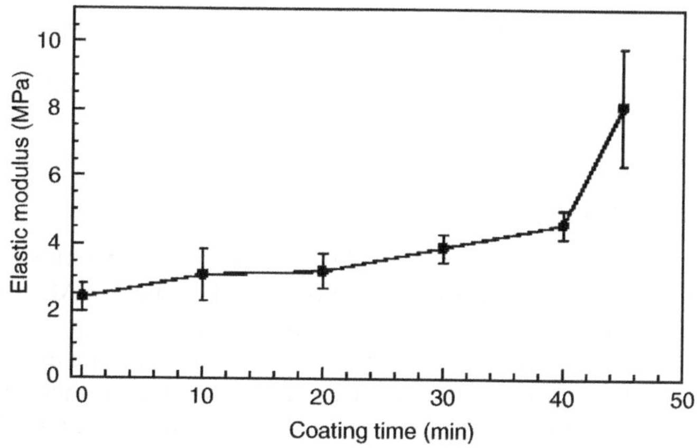

FIGURE 5.28 Elastic modulus as a function of PECVD-TEOS coating time [86].

the idle period. The metal removal rate using such surface-modified polishing pads increases linearly with the PECVD coating time (Fig. 5.29). Thus the use of such surface-modified CMP pads can reduce the cost of ownership by having higher removal rate and enhanced performance [86–88].

Another type of surface-treated pads is the low-shear surface-engineered pads. When used in copper CMP, the pads exhibited vastly improved tribological performance in comparison to the untreated pads. More specifically, these pads gave a much lower COF over a larger range of Sommerfeld numbers compared to untreated pads. The reduction in shear-induced stress and its accumulation within the copper films can significantly

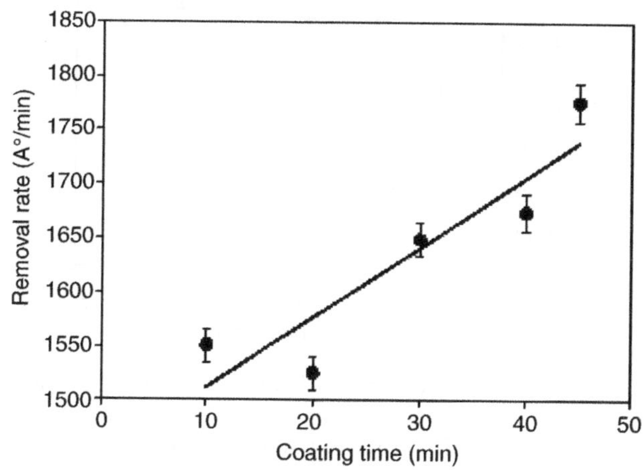

FIGURE 5.29 Correlation of relative blanket tungsten removal rate (W-RR) and PECVD-TEOS coating time [86].

lower the number of defects such as scratches and pits on polished Cu wafer surface. For copper and low-*k* dielectric integration, a low shear polishing has many advantages. It reduces the shear-induced delamination and improves the CMP yield [89].

5.6.3 Reactive Pad

CMP pads are generally designed to play a passive role in terms of chemical reactions that occur on the surface to be polished. For metal CMP, considering the fact that the chemical reactions occurring on the metal surface is critical to the performance of the slurry, a pad that is capable of initiating and controlling various chemical reactions on the metal surface can bring another dimension of flexibility to the CMP process. Furthermore, such a pad can potentially allow a metal CMP process to be conducted at lower down force similar to those of ECMP. With such an added functionality to the pad, a metal CMP process could be simplified. More specifically, the polishing can be accomplished by simply supplying a solution of oxidizer between the pad and the metal film to be polished. The oxidizer is activated only at the sites where pad surface, oxidizer, and metal film meet. The lack of contact for the parts in the low areas on the patterned wafers translates to lower removal rate. The differential removal rate created by such a functional pad surface is able to improve the step height reduction efficiency and prevent the step height from redeveloping [90]. Figure 5.30 shows the comparison of step height changes over polishing time on a reactive pad and on a conventional pad.

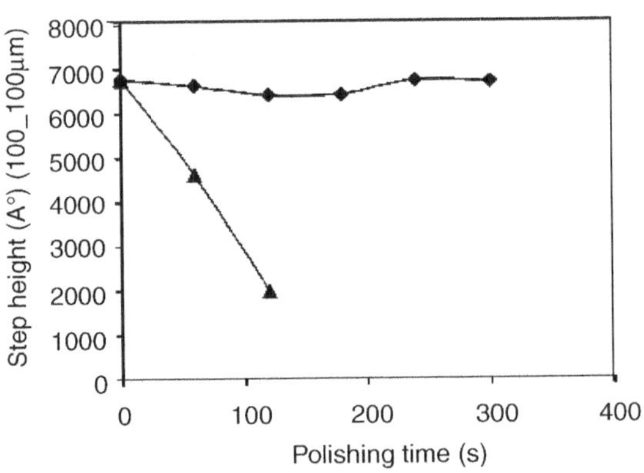

FIGURE 5.30 Step height changes for an 1″ patterned wafer that was cut from an 8″ SKW 6-3 854 wafer on 100-μm copper line in the 50% metal density region. The triangle data were obtained with a reactive pad containing an amine activator for the oxidizer, and the diamond points represent an original pad. The polishing was conducted on a bench-top polisher supplied with a solution of potassium persulfate as the oxidizer.

QUESTIONS

1. Explain why softer pads remove more material than harder pads at low pressures, but why harder pads remove more material than softer ones at high pressures.
2. As the abrasive size increases, the values of K_{Pad} and of k in Equation 5.5 both decrease. Use this information to sketch R versus %A or R versus [A] curves for the polishing rate of an abrasive larger than the 20 nm as shown in Fig. 5.17. Qualitatively explain the difference in polishing rate changes as abrasive concentrations increase at both low and high concentrations.
3. Explain qualitatively why dishing can reach a maximum value for long overpolishing times, while erosion keeps increasing with time.
4. What are the general requirements of polishing pads for IC CMP process? Explain why polyurethane pads are the most commonly used pads in CMP practice. Suggest some alternative materials.
5. What are the differences in microstructures of the four main types of polishing pads? Suggest an additional type.
6. What is the purpose of grooves on a pad surface and how could they influence polishing performance? Suggest some additional designs.
7. What are the main components of polyurethane pads and how the pad properties can be controlled through polyurethane components? Suggest some additional variations.
8. List and explain the advantages and disadvantages of nonporous pad versus porous pad in terms of polishing performance.
9. How would you design a new fixed abrasive pad for copper CMP?
10. How would you design a reactive pad for W CMP?

REFERENCES

1. Oliver MR, editor. Chemical Mechanical Planarization of Semiconductor Materials. New York: Springer; 2004.
2. Muldowney GP, James DB. Characterization of CMP pad surface texture and pad–wafer contact. Mater Res Soc Symp Proc 2004;816:K5.2.1–K5.2.12.
3. Hernandez J, Wrschka P, Hsu Y, Kuan TS, Oehrlein GS, Sun HJ, Hansen DA, King J, Fury MA. Chemical mechanical polishing of Al and SiO_2 thin films: the role of consumables. Electrochem Soc 1999;146(12):4647–4653.
4. Stein D, Hetherington D, Dugger M, Stout T. Optical interferometry for surface measurement of CMP pads. J Electron Mater 1996;25(10):1623–1627.
5. Liang H, Kaufman F, Sevilla R, Anjur S. Wear phenomenon in chemical mechanical polishing. Wear 1997;211:271–279.
6. Kim H, Park DW, Hong CK, Han WS, Moon JT. The effect of pad properties on planarity in a CMP process. Mater Res Soc Symp Proc 2003;767:F2.4.1–F2.4.7.

7. Kim CB, Park SW, Kim JW, Lee WS, Seo YJ. polishing performance of double XY-groove pattern pad for W-CMP application. 207th ECS Meeting, 2006. Abstract #323.
8. Fayolle M, Sicurani E, Morand Y. W CMP process integration: consumables evaluation—electrical results and end point detection. Microelectron Eng 1997;37(38):347–352.
9. Kim SD, Hwang IS, Choi KS. Hard-pad-based CMP of premetal dielectric planarization. Electrochem Soc 2003;150(8):G450–G455.
10. Stavreva Z, Zeidler D, Plotner M, Drescher K. Characteristics in chemical–mechanical polishing of copper: comparison of polishing pads. App Surf Sci 1997;108:39–44.
11. Choi J, Dornfeld DA. Modeling of pattern density dependent pressure non-uniformity at a die scale for ILD chemical mechanical planarization. Mater Res Soc Symp Proc 2004;816:K4.4.1–K4.4.6.
12. Fu G, Chandra A, Guha S, Subhash G. A plasticity-based model of material removal in chemical–mechanical polishing (CMP). IEEE Trans Semicond Manuf 2001;14(4):406–417.
13. Fu G, Li H, Wei D. Understanding the dishing difference between a line and a bond pad in Cu CMP. Electrochem Solid-State Lett 2003;6(12):G143–G145.
14. Guo Y, Chandra A, Bastawros A. Analytical dishing and step height reduction model for CMP with a viscoelastic pad. Electrochem Soc 2004;151(9):G583–G589.
15. Fu G, Chandra A. An analytical dishing and step height reduction model for chemical mechanical planarization (CMP). IEEE Trans Semicond Manuf 2003;16(3):477–485.
16. Zabasajja J, Merchant T, Ng B, Banerjee S, Green D, Lawing S, Kura H. Modeling and characterization of tungsten chemical and mechanical polishing processes. Electrochem Soc 2001;148(2):G73–G77.
17. Doy TK, Seshimo K, Suzuki K, Philipossian A, Kinoshita M. Impact of novel pad groove designs on removal rate and uniformity of dielectric and copper CMP. Electrochem Soc 2004;151(3):G196–G199.
18. Hocheng H, Cheng CY. Visualized characterization of slurry film between wafer and pad during chemical mechanical planarization. IEEE Trans Semicond Manuf 2002;15(1):45–50.
19. Moinpour M, Tregub A, Oehler A, Cadien K. Advances in characterization of CMP consumables. MRS Bulletin October 2002;766–771.
20. Philipossian A, Olsen S. Effect of pad surface texture and slurry abrasive concentration on tribological and kinetic attributes of ILD CMP. Mater Res Soc Symp Proc 2003;767:F2.8.1–2.8.7.
21. Castillo-Mejia D, Gold S, Burrows V, Beaudoin S. The effect of interactions between water and polishing pads on chemical mechanical polishing removal rates. Electrochem Soc 2003;150(2):G76–G82.
22. Li W, Shin DW, Tomozawa M, Murarka SP. The effect of the polishing pad treatments on the chemical–mechanical polishing of SiO_2 films. Thin Solid Films 1995;270:601–606.

23. Lu H, Fookesa B, Obeng Y, Machinski S, Richardson KA. Quantitative analysis of physical and chemical changes in CMP polyurethane pad surfaces. Mater Charact 2002;49:35–44.

24. Moy AL, Cecchi JL, Hetherington DL, Stein DJ. Polyurethane pad degradation and wear due to tungsten and oxide CMP. Mater Res Soc Symp Proc 2001;671:M1.7.1–M1.7.6.

25. Obeng YS, Ramsdell JE, Deshpande S, Kuiry SC, Chamma K, Richardson KA, Seal S. Impact of CMP consumables on copper metallization reliability. IEEE Trans Semicond Manuf 2005;18(4):688–694.

26. Zantye PB, Kumar A, Sikder AK. Chemical mechanical planarization for microelectronics applications. Mater Sci Eng (R) 2004;45:89–220.

27. Lawing AS. Pad conditioning and pad surface characterization in oxide chemical mechanical polishing. Mater Res Soc Symp Proc 2002;732E:I5.3.1–I5.3.6.

28. Borucki L. Mathematical modeling of polish-rate decay in chemical–mechanical polishing. J Eng Math 2002;43:105–114.

29. Wang C, Sherman P, Chandra A. A stochastic model for the effects of pad surface topography evolution on material removal rate decay in chemical mechanical planarization (CMP). IEEE Trans Semicond Manuf 2005;18(4):695–708.

30. Wang C, Sherman P, Chandra A. Modeling and analysis of pad surface topography and slurry particle size distribution effects on material removal rate (MRR) in chemical mechanical planarization. Int J Manuf Technol Manag 2005;7(5/6):504–529.

31. Hooper BJ, Byme G, Galligan S. Pad conditioning in chemical mechanical polishing. Mater Process Technol 2002;123:107–113.

32. Chen CY, Yu CC, Shen SH, Ho M. Operational aspects of chemical mechanical polishing polish: pad profile optimisation. J Electrochem Soc 2000;147(10):3922–3930.

33. Kim NH, Choi GW, Park JS, Seo SJ, Lee WS. Effects of conditioning temperature on polishing pad for oxide chemical mechanical polishing process. Microelectron Eng 2005;82:680–685.

34. Kim NH, Seo YJ, Lee WS. Temperature effects of pad conditioning process on oxide cmp: polishing pad, slurry characteristics, and surface reactions. Microelectron Eng 2006;83:362–370.

35. Kim NH, Ko PJ, Choi GW, Seo YJ, Lee WS. Chemical mechanical polishing (CMP) mechanisms of thermal SiO_2 film after high-temperature pad conditioning. Thin Solid Films 2006;504:166–169.

36. Mudhivarthi S, Gitis N, Kuiry S, Vinogradov M, Kumar A. Effects of slurry flow rate and pad conditioning temperature on dishing, erosion, and metal loss during copper CMP. J Electrochem Soc 2006;153(5):G372–G378.

37. Prasad A, Xiang H, Wang J, Remsen EE. Analysis of pre- and post-conditioned polyurethane CMP pad surfaces as, "a function of conditioning temperature", 210th ECS Meeting, 2006.

38. Tregub A, Ng G, Sorooshian J, Moinpour M. Thermoanalytical characterization of thermoset polymers for chemical mechanical polishing. Thermochimica Acta 2005;439:44–51.

39. McGrath J, Davis C. Polishing pad surface characterisation in chemical mechanical planarisation. J Mater Process Techno 2004;153–154(10):666–673.

40. Yuichi Y, Takaaki K, Shunichi S, Keiichi M, Yasuaki I, Shinji T, Naoki T. The effect of pad conditioning on planarization characteristics of chemical mechanical polishing (CMP) with ceria slurry. Mater Res Soc Symp Proc 2005;867: W3.5.1–W3.5.6.

41. Preston FW. The theory and design of plate glass polishing machines. J Soc Glass Technol 1927;11:214–256.

42. Zhang F, Busnaina A, Ahmadi G. Particle adhesion and removal in chemical mechanical polishing and post-CMP cleaning. J Electrochem Soc 1999;146:2665–2669.

43. Shi FG, Zhao B. Modeling of chemical–mechanical polishing with soft pads. App Phys A 1998;67:249–252.

44. Zhao B, Shi FG. Chemical mechanical polishing in IC process: new fundamental insights. Proceedings of the Fourth International Chemical–Mechanical Planarization for ULSI Multilevel Interconnection Conference, Santa Clara, CA, Feb. 11–12; 1999. pp 13–22.

45. Luo J, Dornfeld D. Material removal mechanism in chemical mechanical polishing: theory and modeling. IEEE Trans Semicond Manuf 2001;14:112–133.

46. Bastawros A, Chandra A, Guo YJ, Yan B. Pad effects on material-removal rate in chemical–mechanical planarization. J Electron Mater 2002;31(10):1022–1031.

47. Yu TK, Yu CC, Orlowski M. A statistical polishing pad model for chemical mechanical polishing. IEEE IEDM Washington DC, Dec 5–8; 1993. pp 865–868.

48. Seok J, Sukam CP, Kim AT, Tichy JA, Cale TS. Multiscale material removal modeling of chemical mechanical polishing. Wear 2003;254:307–320.

49. Zhao Y, Chang L. A micro-contact and wear model for chemical–mechanical polishing of silicon wafers. Wear 2002;252:220–226.

50. Zhao Y, Chang L, Kim SH. A mathematical model for chemical–mechanical polishing based on formation and removal of weakly bonded molecular species. Wear 2003;254:332–339.

51. Qin K, Moudgil B, Park CW. A chemical mechanical polishing model incorporating both the chemical and mechanical effects. Thin Solid Films 2004;446:277–286.

52. Ahmadiz G, Xia X. A model for mechanical wear and abrasive particle adhesion during the chemical mechanical polishing process. J Electrochem Soc 2001; 148(3):G99–G109.

53. Castillo-Mejia D, Kelchner J, Beaudoin S. Polishing pad surface morphology and chemical mechanical planarization. J Electrochem Soc 2004;151(4):G271–G278.

54. Yeruval SB, Park CW, Moudgil BM. Modeling of polishing regimes in chemical mechanical polishing. Mater Res Soc Symp Proc 2005;867:W5.9.1–W5.9.6.

55. Fu G, Chandra A. A model for wafer scale variation of material removal rate in chemical mechanical polishing (CMP) based on viscoelastic pad deformation. J Electron Mater 2002;31(10):1066–1073.

56. Fu G, Chandra A. Wafer scale variation of removal rate in chemical mechanical polishing based on elastic pad deformation. J Electron Mater 2001;30(4):400–408.

57. Seok J, Sukam CP, Kim AT, Tichy JA, Cale TS. Material removal model for chemical–mechanical polishing considering wafer flexibility and edge effects. Wear 2004;257:496–508.

58. Kim AT, Seok J, Tichy JA, Cale TS. A multiscale elastohydrodynamic contact model for CMP. J Electrochem Soc 2003;150(9):G570–G576.

59. Zeng T, Sun T. Size effect of nanoparticles in chemical mechanical polishing—a transient model. IEEE Trans Semicond Manuf 2005;18(4):655–663.

60. Guo Y, Chandra A, Bastawros A. An analytical dishing and step height reduction model for chemical mechanical planarization (CMP) with a viscoelastic pad. J Electrochem Soc 2004;151(9):G583–G589.

61. Nguyen VH, Daamen R, van Kranenburg H, van der Velden P, Woerlee PH. A physical model for dishing during metal CMP. Electrochem Soc 2003; 150(11):G689–G693.

62. Saxena R, Thakurta DG, Gutmann RJ, Gill WN. A feature scale model for chemical mechanical planarization of damascene structures. Thin Solid Films 2004;449:192–206.

63. Vlassak JJ. A contact-mechanics based model for dishing and erosion in chemical–mechanical polishing. Mater Res Soc Symp Proc 2001;671:M4.6.1–M4.6.6.

64. Vlassak JJ. A model for chemical–mechanical polishing of a material surface based on contact mechanics. Mech Phys Solids 2004;52:847–873.

65. Wang C, Sherman P, Chandra A. Scratch generation probability in chemical mechanical planarization. CMP-MIC Feb 2006; pp 471–479.

66. Basim GB, Vakarelski IU, Moudgil B. Role of interaction forces in controlling the stability and polishing performance of CMP slurries. J Coll Interface Sci 2003;263:506–515.

67. Tseng WT, Wang YL. Re-examination of pressure and speed dependences of removal rate during chemical–mechanical polishing processes. J Electrochem Soc 1997;144(2):L15–L17.

68. Luo Q, Ramarajan S, Babu SV. Modification of the Preston equation for the chemical–mechanical polishing of copper. Thin Solid Films 1998;335:160–167.

69. Stein DJ, Hetherington DL, Cecchi JL. Investigation of the kinetics of tungsten chemical mechanical polishing in potassium iodate-based slurries: I role of alumina and potassium lodate. J Electrochem Soc 1999;146(1):376–381.

70. Paul E. A model of chemical mechanical polishing. J Electrochem Soc 2001;148(6): G355–G358.

71. Paul E. A model of chemical mechanical polishing. J Electrochem Soc 2002;149(5): G305–G308.

72. Paul E. Modeling removal rates in chemical-mechanical planarization. Proceedings of the Twentieth International VLSI Multilevel Interconnection Conference VMIC, Sept 22–25; 2003. p 277–283.

72[b]. Paul E. C_MP or $_CM_P$: The Balance in Chemical Mechanical Polishing. Electrochem Solid State Lett 2007;10(7):213–216.

73. Li Y, Hariharaputhiran M, Babu SV. Chemical–mechanical polishing of copper and tantalum with silica abrasives. J Mater Res 2001;16(5):1066–1073.

74. Bielmann M, Mahajan U, Singh RK. Effect of particle size during tungsten chemical mechanical polishing. Electrochem Solid State Lett 1999;2(8):401–403.

75. Thakurta DG, Schwendeman DW, Gutmann RJ, Shankar S, Jiang L, Gill WN. Three-dimensional wafer-scale copper chemical–mechanical planarization model. Thin Solid Films 2002;414(1):78–90.

76. Paul E, Horn J, Li Y, Babu SV. A Model of Pad–Abrasive Interations in Chemical Mechanical Polishing. Electrochem Solid State Lett 2007;10(4):131–133.
77. Paul E, Richardson D, Lerario A. A model of dishing & erosion. Proceedings of the Eleventh CMP-MIC Conference, Feb. 20–23;2006. p 254–283.
78. Gbondo-Tugbawa TE. Thesis, MIT, 2002.
79. van der Velden P. Chemical mechanical polishing with fixed abrasives using different subpads to optimize wafer uniformity. Microelectron Eng 2000;50:41–46.
80. Kim H, Park B, Lee S, Jeong H, Dornfeld DA. Self-conditioning fixed abrasive pad in CMP. Electrochem Soc 2004;151(12):G858–G862.
81. Kim H, Kim H, Jeong H, Seo H, Lee S. Self-conditioning of encapsulated abrasive pad in chemical mechanical polishing. Mater Process Technol 2003;142:614–618.
82. Nguyen VH, Daamen R, Hoofman R. Impact of different slurry and polishing pad choices on the planarization efficiency of a copper CMP process. Microelectron Eng 2004;76:95–99.
83. Simpson A, Economikos L, Jamin FF, Ticknor A. Fixed abrasive technology for STI CMP on a web format tool. Mater Res Soc Symp Proc 2001;671:M4.1.1–M4.1.9.
84. Kulawski M, Henttinen K, Suni I, Weimar F, Mäkinen J. A novel CMP process on fixed abrasive pads for the manufacturing of highly planar thick film SOI substrates. Mater Res Soc Symp Proc 2003;767:F2.11.1–F2.11.6.
85. Charns L, Sugiyama M, Philipossian A. Mechanical properties of chemical mechanical polishing pads containing water-soluble particles. Thin Solid Films 2005;485:188–193.
86. Deshpande S, Dakshinamurthy S, Kuiry SC, Vaidyanathan R, Obeng YS, Seal S. Surface-modified polymeric pads for enhanced performance during chemical mechanical planarization. Thin Solid Films 2005;483:261–269.
87. Zantye PB, Mudhivarthi S, Sikder AK, Kumar A, Obeng Y. Metrology of Psiloquest's application specific pads (ASP) for CMP applications. Mater Res Soc Symp Proc 2004;816:K5.6.1–K5.6.6.
88. Zantye PB, Obeng Y, Mudhivarthi S, Kumar A. Optimization of Psiloquest's application specific CMP pads for commercialisation. Mater Res Soc Symp Proc 2005;867:W3.3.1–W3.3.6.
89. Deopura M, Hwang E, Misra S, Roy PK. Stress characterization of post-CMP copper films planarized using novel low-shear and surface-engineered pads. Mater Res Soc Symp Proc 2005;867:W2.7.1–W2.7.12.
90. Zhang J, Keleher J, Hellring S, Shipp D, Li Y. Reactive pads for metal CMP. Proceedings of 22nd International VLSI Multilevel Interconnection Conference (VMIC), Ferment CA, October 4–6;2005. pp 251–259.

6

MODELING

LEONARD BORUCKI AND ARA PHILIPOSSIAN

6.1 INTRODUCTION

Experimental evidence strongly suggests that material removal in chemical–mechanical polishing (CMP) processes is a result of one or more chemical steps that alter the wafer surface combined with a mechanical step that removes the altered material. Chemical action by itself also removes material by static etching, but generally at a much lower rate than is observed when mechanical action is also present. Similarly, polishing rates observed when a minimally reactive fluid such as water is used instead of slurry are also low. Both chemical and mechanical processes are therefore involved in material removal at commercially practical rates, and the model we describe reflects this dual nature of the process.

In this chapter, we derive and apply to data a simple two-step model that involves a chemical step followed by a mechanical removal step. The model is abstract in the sense that most of the specifics of the slurry composition or of the chemical reaction involved are not given in any detail. Although this appears to be a disadvantage, it is necessary for the application of the model to the analysis of removal rates from proprietary slurries whose compositions cannot be directly investigated. When applied as a compact formula, the model can provide a highly accurate description of removal rate variations as a function of polishing pressure and sliding speed. This then makes it possible to extract the relative contributions of chemical and mechanical processes to removal and to confidently interpolate or extrapolate rates based on the calibration data.

Microelectronic Applications of Chemical Mechanical Planarization, Edited by Yuzhuo Li

When chemical reactions are involved in a process, it is important to know the reaction temperature. In the model described here, reactions at a point on the wafer surface are assumed to be driven by temperature excursions due to contact by passing pad asperities. This is known as flash heating. In systems in which there is dry sliding contact between two rough surfaces, it is known that flash temperatures at asperity contacts can be much higher than the average temperatures of the workpieces involved. In CMP, however, the contact is lubricated and cooled by the slurry, and this needs to be taken into account. In the case of polishing on a rotary tool, it is possible to derive a simple estimate of the mean reaction temperature, and it is this that we use in the chemical part of the two-step model.

Finally, after development of the two-step model and application of the model to blanket wafer polishing, we describe how the model may be applied to patterned wafer polishing and how the latter may differ from blanket polishing.

6.2 A TWO-STEP CHEMICAL MECHANICAL MATERIAL REMOVAL MODEL

We first outline the derivation of the simplest model that can be applied to processes in which both chemistry and mechanics are important. It involves a single rate-limiting chemical step that must occur on the wafer surface before material can be removed by the mechanical process. It is assumed that neither process by itself is capable of removing material at a significant rate. Thus, the model does not directly include static etching. This can be rectified with a three-step model (Question 1).

The derivation of the model follows Langmuir–Hinshelwood style kinetics. The surface at any point in time is assumed to present an array of reaction sites to the slurry, of which a fraction θ_2 has reacted with some component of the slurry, such as an oxidizer, whereas the remaining fraction $\theta_1 = 1 - \theta_2$ has not yet reacted. For example, the unreacted fraction may represent a portion of the surface that has been recently uncovered by mechanical action. The rate r_1 (in MKS units of mol/m^2 s) at which unreacted sites are converted to reacted sites is assumed to follow first-order kinetics; that is, the rate is proportional to the product of the fraction θ_1 of available sites and the molar concentration c of the reactant in the slurry:

$$r_1 = k_1 \theta_1 c \tag{6.1}$$

The proportionality constant k_1 has units of velocity. The mechanical step is assumed to remove only the reacted material at a rate proportional to the fraction of the surface that has been reacted:

$$r_2 = k_2 \theta_2 \tag{6.2}$$

Both r_2 and k_2 have MKS units of mol/m^2 s. A key assumption now is that the chemical and mechanical rates balance, so that

$$r_1 = r_2 \tag{6.3a}$$

$$k_1\theta_1 c = k_2\theta_2 \tag{6.3b}$$

Since $\theta_1 + \theta_2 = 1$, we can eliminate θ_2 from (6.3b) and solve for the fraction of available sites θ_1:

$$k_1\theta_1 c = k_2(1 - \theta_1) \tag{6.4a}$$

$$\theta_1 = k_2/(k_2 + k_1 c) \tag{6.4b}$$

We can now eliminate the surface coverage fractions from the chemical and mechanical rates:

$$r_1 = r_2 = k_2 k_1 c/(k_2 + k_1 c) \tag{6.5}$$

Finally, assuming that none of the mechanically abraded material is redeposited, the material removal rate is

$$\mathrm{RR} = \frac{M_\mathrm{w}}{\rho}\frac{k_2 k_1 c}{k_2 + k_1 c} \tag{6.6}$$

where M_w is the molecular weight of the material being removed (e.g., copper or silicon dioxide) and ρ is the density. Formula 6.6 is the basic two-step removal rate model that we will use. We make an additional assumption throughout that the slurry flow rate is high enough and the surface reaction rate is low enough such that the concentration c of the critical reactant is not significantly depleted across the wafer surface. A more detailed theoretical analysis of transport and reaction [1] suggests that this may usually be the case. The same conclusion can be reached for slurries for which the reactant concentration is known by comparing the rate of consumption required to support the measured removal rate with the known concentration in the fresh slurry to estimate the extent of depletion. A consequence of this assumption is that we can treat $k_1 c$ as a constant and can rename it k_1 without loss of generality. With this change of notation, the basic removal rate model becomes

$$\mathrm{RR} = \frac{M_\mathrm{w}}{\rho}\frac{k_1 k_2}{k_1 + k_2} \tag{6.7}$$

Two limiting cases are of particular interest. When removal is very mechanically limited, $k_2 \ll k_1$, the mechanical rate constant can be neglected in the denominator of (6.7) and we obtain

$$\mathrm{RR} \approx \frac{M_\mathrm{w}}{\rho} k_2 \tag{6.8}$$

In the opposite extreme when removal is very chemically limited, $k_1 \ll k_2$, we obtain

$$\mathrm{RR} \approx \frac{M_\mathrm{w}}{\rho} k_1 \tag{6.9}$$

In order to be specific about the relationship between chemical and mechanical rate constants and polishing variables, we will assume here that the mechanical rate constant is proportional to the frictional power density:

$$k_2 = c_p \mu_k p V \tag{6.10}$$

where μ_k is the kinetic coefficient of friction (COF), p is the polishing pressure, and V is the relative sliding speed between the pad and the wafer. For simplicity we will agree throughout that the pad and wafer rotate counterclockwise at the same rate, so that the sliding velocity V is the same everywhere under the wafer (Question 2). In (6.12), $\mu_k p$ is the lateral force per unit area; multiplying by the velocity gives the rate per unit area at which work is being done by friction. Since some of this work goes into the deformation or abrasion of the pad surface, into pad, slurry, and wafer heating, and into tool vibration rather than into material removal from the wafer, we multiply the frictional power density by an empirical proportionality constant c_p in (6.10) to obtain the fraction that results in material removal. The MKS units of c_p are mol/J. In the mechanically limited extreme, the removal rate is then

$$\mathrm{RR} \approx \frac{M_\mathrm{w}}{\rho} c_p \mu_k p V \tag{6.11}$$

so that (6.10) in this limit leads to Preston's law [2]. There is strong evidence of a correlation between the COF and the removal rate in mechanically limited polishing (Fig. 6.1). We note that various analyses of three-body pad–wafer–slurry particle contact mechanics and adhesion [3–8] lead to other dependencies of the removal rate on pressure and velocity than the one we use here. We choose (6.11) because of its historical relationship with glass polishing.

For the chemical rate constant, we assume that that there is a rate-limiting chemical step (such as copper surface oxidation by hydrogen peroxide in copper polishing) in the polishing process that has an activation energy E. The chemical rate is thermally activated and is assumed to have an Arrhenius form:

$$k_1 = A \exp(-E/kT) \tag{6.12}$$

where the preexponential A will be treated as a fitting constant. The units of A are mol/m^2 s and those of E are J/mol or eV (1 eV $\approx$ 23 kcal/mol). A major difficulty posed by the chemical rate is that the reaction temperature T is not a directly controlled polishing parameter, nor is it easily observable. We will

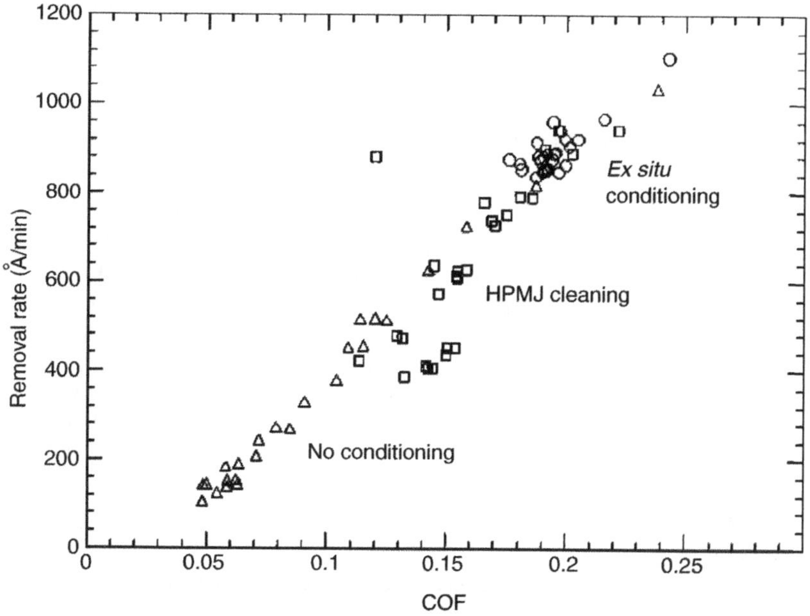

FIGURE 6.1 Relationship between oxide removal rate and COF for three different pad treatments—diamond conditioning, high-pressure microjet cleaning (HPMJ), and no conditioning (from Ref. 9).

presently show how to estimate the mean reaction temperature in a rotary CMP tool.

6.3 PAD SURFACES AND PAD SURFACE CONTACT MODELING

Before discussing the mean reaction temperature, we first review some relevant facts about polishing pads. Images of commercial IC1000™ foamed polyurethane polishing pad surfaces show that they are rough and that pad surface roughness generally exceeds that of the wafer topography that is being polished. Using interferometry or scanning profilometry, a histogram or probability density function (PDF) can be constructed for the distribution of surface heights (Fig. 6.2). The area under the PDF between any two heights is the probability of finding a point on the surface within that range. Thus, the area under the entire PDF is 1. From interferometry or SEM (scanning electron microscopy) images, it may also be seen that pad surfaces are populated with asperities or summits—those points that are higher than immediately surrounding points. Summit heights are statistically distributed and can be described with a summit height PDF, which we denote by $\phi_s(z)$, where the subscript "s" stands for summit. The summit height distribution is shifted to the right relative to the surface height distribution (Fig. 6.2) since

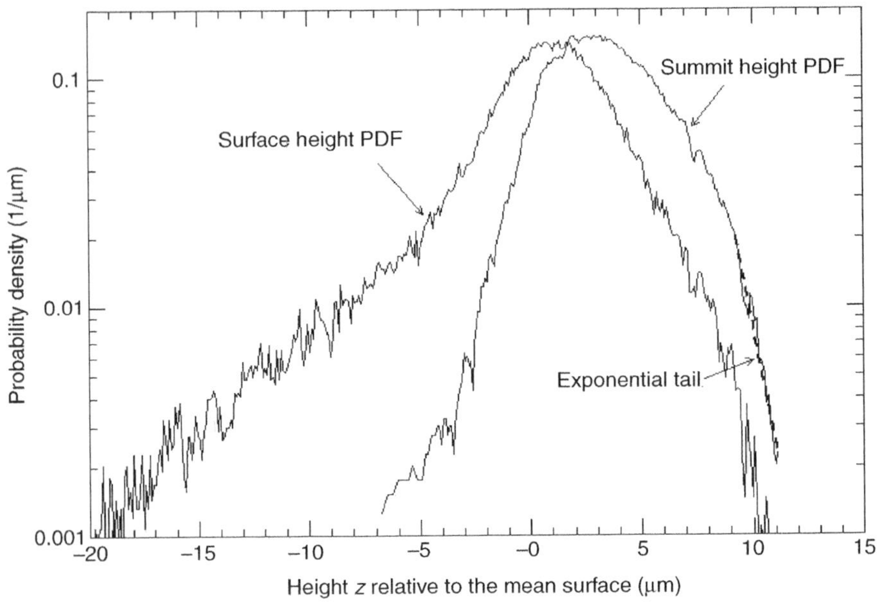

FIGURE 6.2 Surface height and summit height probability density functions for an IC1000 commercial polishing pad.

only the locally highest points are being counted. The area density η_s of the summits can be estimated from pad images and, if the data are sufficiently good, it is possible to estimate the mean summit tip curvature κ_s. (This is not trivial—more information and a critical discussion of the problem of characterizing rough surfaces can be found in [10–13].)

When a wafer is pressed against a pad, only the highest pad summits in the right-hand tail of the distribution actually come into contact. Viewed on a log plot, this critical part of the PDF is often found to decay exponentially for z above some point z_c; that is,

$$\phi_s(z) = B\exp(-z/\lambda) \tag{6.13}$$

for some constants B and λ. This may be partly a result of the diamond morphology or grit size used on the diamond conditioning tool. The parameter λ is a characteristic distance over which the PDF drops by a factor of $1/e$ and therefore is a measure of surface abruptness. In what follows, we will often assume for simplicity that the pad surface has exponentially distributed summit heights in the contacting tail.

We describe next the main elements of the Greenwood and Williamson theory [10,11], which is frequently used to model elastic contact between rough surfaces. In the simplest version of this theory, the load is assumed to be light enough so that deformation of the asperities is elastic and asperity tips can be approximated as nearly spherical. The theory is based on the Hertz solution for the elastic contact problem between a sphere and a half-space [10]. In the Hertzian analysis, deformation of a summit with tip curvature κ_s and height z to some height $d < z$ produces a circular contact area with radius $a = \sqrt{\delta/\kappa_s}$ where $\delta = z - d$. The contact pressure at a distance of r from the center of the area, $r \leq a$, is

$$p(r) = \frac{2E^*}{\pi}\kappa^{1/2}\delta^{1/2}(1 - (r/a)^2)^{1/2} \tag{6.14}$$

where, for a soft pad and hard wafer, E^* is related to the pad Young's modulus and Poisson ratio by $E^* = E/(1 - \nu^2)$. By averaging (6.14) over the circular contact, the mean contact pressure is found to be

$$p_a = \frac{4E^*}{3\pi}\kappa^{1/2}\delta^{1/2} \tag{6.15}$$

The Greenwood and Williamson theory builds on the Hertzian solution. The loads carried by the summits when the wafer is at height $z = d$ are summed up, taking into account their heights and frequencies. The result is that the nominal pressure applied to the wafer is related to its location d by a nonlinear pressure–displacement relationship:

$$p = \frac{4E^*\eta_s}{3\kappa_s^{1/2}}\int_d^\infty (z - d)^{3/2}\phi_s(z)dz \tag{6.16}$$

Similarly, the contact area fraction (the real contact area divided by the nominal area) is

$$A_f = \frac{\pi\eta_s}{\kappa_s}\int_d^\infty (z - d)\phi_s(z)dz \tag{6.17}$$

the mean contact area is

$$\bar{A}_c = \frac{\pi}{\kappa_s}\int_d^\infty (z - d)\phi_s(z)dz \tag{6.18}$$

and the area density of asperities actually in contact is

$$\eta_c = \eta_s\int_d^\infty \phi_s(z)dz \tag{6.19}$$

When the summit height PDF is exponential within the contact region $z > d$, all four of the above integrals can be evaluated explicitly (Question 3). We note that in reality, polyurethane is viscoelastic, loads are usually high enough so that at least some asperities are plastically deformed, and observed contact areas are not generally circular [18]. Thus, the Greenwood and Williamson theory is an imperfect device for describing pad–wafer contact in CMP.

6.4 REACTION TEMPERATURE

We now describe an approach for estimating the reaction temperature. Viewed at a microscopic level, this is a very complex stochastic thermal problem involving the frictional interaction of a relatively smooth surface with a rough one in the presence of a cooling lubricant that contains abrasive particles. Rather than treat this problem fully, we try to capture the most important features while neglecting other potentially interesting ones. The end result will be a simple estimate of the reaction temperature that can be used in a compact model.

From the point of view of a fixed point on the wafer, mechanical removal of the reacted layer occurs during a series of brief encounters with contacting asperities that are randomly spaced in time. When an asperity passes, the wafer surface is momentarily, but rapidly, heated by friction from the transient increase in contact pressure. This rapid temperature rise is called flash heating. Since the mean time between asperity encounters is much larger than the time of contact (Question 4), each flash heating event is followed by rapid cooling of the wafer surface by the surrounding slurry. Because the chemical rate is exponential in the temperature, most of the reaction that regrows the mechanically removable layer occurs during and shortly after the asperity encounters.

When an asperity is in contact with the wafer, the surface of the wafer and the contacting surface of the asperity will have the same temperature if temperature continuity applies, a common assumption in this kind of analysis [14]. Even when an asperity is not touching the wafer, the temperature of the asperity tip and that of the nearest point on the wafer should still be within a few degrees because of cooling provided by the thin ($\sim$20 μm) slurry layer [21]. Thus, the mean temperature of points on the wafer that are immediately over pad summits should be nearly the same as the mean asperity tip temperature. This allows a shift in focus from the wafer to the asperities. It also suggests that the flash temperature experienced by a given point on the wafer depends on the thermal history of the asperity that it has encountered.

Elaborating on this point, we focus on polishing on a rotary tool and assume that on a large scale, the temperature distributions of the pad, slurry, wafer, and other components have reached a steady state. Then a given pad asperity will experience many identical cycles during polishing in which it encounters the wafer leading edge, the contacting tip heats up as the asperity passes under the wafer (if it is in contact), and then the tip cools down again between the wafer trailing edge and the next encounter with the leading edge. The temperature at the leading edge T_{p} can be considered an initial condition during each cycle. On many rotary polishers, T_p can be measured with an IR gun or an IR video camera.

After contact for time τ at real contact pressure p_a and constant frictional power density $\mu_k p_a V$, the asperity tip temperature rise $\Delta T(\tau)$ above the initial condition at the wafer leading edge can be shown to be approximately [14] (Question 5) equal to

$$\Delta T(\tau) = C\tau^{1/2} \tag{6.20}$$

where

$$C = \frac{2\gamma_p \mu_k p_a V}{\sqrt{\pi \kappa \rho C_p}} \tag{6.21}$$

In (6.21), $\kappa\rho C_p$ is the product of the pad thermal conductivity, density, and heat capacity, and γ_p is the fraction of the thermal energy that enters the pad rather than the slurry or wafer, also known as a heat partition factor. In using (6.20 and 6.21), we are ignoring variations in p_a due to wafer tilt and topography, wafer zonal pressure control, and variations in μ_k that may arise as material layers are cleared. The mathematical approximation used to obtain (6.20) applies when the time of contact is small relative to the time required for the thermal energy to reach the platen. For solid polyurethane ($\kappa = 0.22$ W/m K, $\rho = 1.1 \times 10^3$ kg/m^3, $C_p \approx 1.7 \times 10^3$ J/kg K), the thermal diffusivity $D = \kappa/(\rho C_p)$ is about 1.2×10^{-7} m^2/s. At $V = 1$ m/s, the contact time with a 300-mm wafer on a rotary tool is at the most about $t = 0.3$ s, and the diffusion length during this time is $\sqrt{Dt} \approx 190\,\mu$m, roughly consistent with the mathematical assumption. We also note that (6.20) ignores lateral heat loss from the sides of the asperity.

In order to average (6.20) over the wafer surface at any given instant in time, a relation is needed between the locations of points on the wafer and the contact time τ of arriving asperities. To do this, impose a nonrotating polar coordinate system over the pad with the origin coincident with the pad center and the $\theta = 0$ line passing through the wafer center. Then an asperity arriving at (r, θ) on the wafer surface will have entered the wafer at the leading edge at $(r, -\theta_0)$ for some angle $-\theta_0 < 0$, and it will exit at the trailing edge at (r, θ_0). If the rotation rate of the platen is Ω_p rad/s, then the contact time will be

$$\tau = (\theta + \theta_0)/\Omega_p \tag{6.22}$$

If on a rotary tool the wafer has radius r_w and the distance between the wafer center and the pad center is c_w, then the increase in mean temperature averaged over the wafer surface is

$$\Delta\bar{T} = \frac{1}{\pi r_w^2} \iint \Delta T(\tau)\, dA \tag{6.23}$$

$$= \frac{1}{\pi r_{\mathrm{w}}^2} \int_{c_{\mathrm{w}}-r_{\mathrm{w}}}^{c_{\mathrm{w}}+r_{\mathrm{w}}} \int_{-\theta_0}^{\theta_0} C\tau^{1/2} r d\theta dr \tag{6.24}$$

$$= \frac{C}{\pi r_{\mathrm{w}}^2 \Omega_p^{1/2}} \int_{c_{\mathrm{w}}-r_{\mathrm{w}}}^{c_{\mathrm{w}}+r_{\mathrm{w}}} r \int_{-\theta_0}^{\theta_0} (\theta + \theta_0)^{1/2} d\theta dr \tag{6.25}$$

$$= \frac{2^{5/2} C}{3\pi r_{\mathrm{w}}^2 \Omega_{\mathrm{p}}^{1/2}} \int_{c_{\mathrm{w}}-r_{\mathrm{w}}}^{c_{\mathrm{w}}+r_{\mathrm{w}}} r\theta_0^{3/2} dr \tag{6.26}$$

$$= \frac{C}{V^{1/2}} \left(\frac{2^{5/2} c_{\mathrm{w}}^{1/2}}{3\pi r_{\mathrm{w}}^2} \int_{c_{\mathrm{w}}-r_{\mathrm{w}}}^{c_{\mathrm{w}}+r_{\mathrm{w}}} r\theta_0^{3/2} dr \right) \tag{6.27}$$

We have passed from (6.26) to (6.27) by using the fact that $V = c_{\mathrm{w}}\Omega_{\mathrm{p}}$ (Question 2) when the pad and wafer corotate. The expression in parentheses in (6.27) is a function only of the wafer radius and center location; it is a tool-specific parameter that we denote by ζ. Since

$$\theta_0 = \cos^{-1}\left(\frac{r^2 + c_{\mathrm{w}}^2 - r_{\mathrm{w}}^2}{2rc_{\mathrm{w}}} \right) \tag{6.28}$$

the tool parameter ζ can be easily computed once for each tool geometry. We then have

$$\Delta \bar{T} = \frac{\zeta C}{V^{1/2}} \tag{6.29}$$

and therefore

$$\bar{T} = \bar{T}_{\mathrm{p}} + \Delta \bar{T} \tag{6.30}$$

$$= \bar{T}_{\mathrm{p}} + \frac{2\zeta}{\sqrt{\pi\kappa\rho C_{\mathrm{p}}}} \left(\frac{\gamma_{\mathrm{p}}(p_{\mathrm{a}}/p)}{V^{1/2}} \right) \mu_k p V \tag{6.31}$$

where $\bar{T}_{\mathrm{p}}$ is the mean leading edge pad temperature. In (6.31), the middle factor in the mean temperature increment has been singled out for reasons that will be explained shortly. The first factor is specific to a given tool and pad material, and the third factor involves only observable polishing parameters.

Before continuing, we address a technical detail that has been glossed over. In (6.21), the asperity contact pressure p_{a} is not constant over the wafer at any instant in time but varies with the uncompressed height of the asperity and the location of the wafer surface. Thus, we really should have written (6.21) as

$$\Delta T(p_{\mathrm{a}}, \tau) = \tilde{C} p_{\mathrm{a}} \tau^{1/2} \tag{6.32}$$

where $\tilde{C}$ is the part of C that does not involve p_a. The temperature increment at (r, θ) at any instant in time then depends on the height of the asperity arriving there. This complication can be addressed by noting that over one rotational period, (r, θ) should sample pad asperities of all heights, including those that do not contact. Assuming that conditioning is uniform, so that the asperity height distribution does not depend on r, we can then average (6.32) over the entire asperity height distribution, obtaining

$$\Delta T(\tau) = \tilde{C}\bar{p}_a \tau^{1/2} \tag{6.33}$$

The average contact pressure, taken over both contacting and noncontacting asperities (the population for which we wish to estimate ΔT), is defined as

$$\bar{p}_a = \int_d^{\infty} p_a(z - d)\phi_s(z)dz \tag{6.34}$$

where, in the integrand, $p_a(z - d)$ for $z > d$ is the mean contact pressure that results from compressing an asperity from height z to the wafer height d (see Eq. 6.15). Thus, we need only interpret p_a in (6.31) as the average asperity contact pressure in order to address this technical point.

We now return to the factor in parentheses in Equation 6.31. This factor has been singled out because both of the terms in the numerator can potentially depend on V. For example, if the pad is ungrooved, then hydrodynamic pressures can develop under the wafer that increase the ratio p_a/p if suction develops [15] and decrease it if lift develops. The magnitude of the change in contact pressure depends on the sliding speed. In order to simplify the analysis of the model, we will assume throughout that the pad is grooved and that the grooving pattern relieves hydrodynamic pressures enough so that p_a/p is independent of V. Since most polishing processes use grooved pads, this is not a severe assumption.

The analysis of the pad heat partition factor γ_p is more complex. One complicating factor is that the contact between pad asperities and the wafer is not dry; therefore, simple formulas that have been developed for heat partitioning in two-body dry sliding contact [14] do not apply. Because polishing is done with slurry, it is likely that a thin lubrication layer forms between each asperity tip and the wafer surface. Such a layer is predicted to exist by elastohydrodynamic lubrication (EHL) theory [16]. EHL theory further provides compact formulas for estimating the steady-state thickness of the layer. For various choices of material parameters, velocity and asperity compression, the compact formulas suggest that the layer thickness may be on the order of a few nanometers to a few tens of nanometers. The nanolubrication layer should also contain a solid fraction consisting of slurry particles and, in fact, may filter a portion of the particle size distribution [17], points that are not addressed in EHL theory. Active slurry particles, if large enough, should support some of the load and may, in fact, increase the EHL layer thickness. Thus, neither the distribution of lubrication layer thicknesses nor the mean thickness is precisely known.

Nevertheless, lubricated heat transfer to the pad can be analyzed numerically to an extent that is satisfactory for understanding the general form of the dependence of γ_p on V. This can be done using a commercial finite-element software package, for example. In the analysis, a geometrical model is constructed of the lubrication layer and of portions of the wafer and pad asperity. The heights of the computational domains for the pad and wafer must be chosen to be large enough so that the solution shows little geometric sensitivity. We look at heating from the point of view of an observer on the asperity, treating the asperity as stationary and the wafer as a slider with velocity V. In the lubrication layer, shear flow is assumed in which the velocity field matches the pad field at the bottom of the layer and the wafer at the top of the layer. The heat equation

$$\rho C_p\left(\frac{\partial T}{\partial t} + \vec{v}\cdot \vec{\nabla} T\right) = \vec{\nabla}\cdot(\kappa\, \vec{\nabla} T) + Q, \tag{6.35}$$

with an advective term $\vec{v} \cdot \nabla T$ is then solved using thermal parameters ρ, C_p, κ, and a velocity field $\vec{v}$ appropriate for each material. The use of the advective term makes it unnecessary to actually move the portion of the geometry that models the wafer. In the lubrication layer, the frictional power is dissipated via the source term Q in (6.35), where $Q = \mu_k p_a V/h_c$ and h_c is the lubrication layer thickness. At the leading edge of the computational domain, the wafer and slurry are assumed to arrive at ambient temperature. The thermal problem (6.35) with boundary and initial conditions is then integrated over time until the integral average

$$\frac{1}{A_c}\iint_{A_c} \kappa \frac{\partial T}{\partial z}\, dA \tag{6.36}$$

of the heat flux into the pad taken over the contact area A_c stabilizes. The pad heat partition factor γ_p is then the ratio of (6.36) to the total power density $\mu_k p_a V$.

Figure 6.3a shows the calculated heat partition factor γ_p for lubricated sliding contact between copper with a thin Cu_2O layer and polyurethane. The calculations assume a summit height distribution with an exponential tail (6.13) and use the mean asperity contact pressure from (6.34) and the corresponding contact area from the Hertz theory to define the geometry (Question 6). The thermal properties of water are used for the slurry, though for slurries with high solid content, it is necessary to account for the thermal effects of the particles. The pad heat partition factor in Fig. 6.3a is shown as a function of

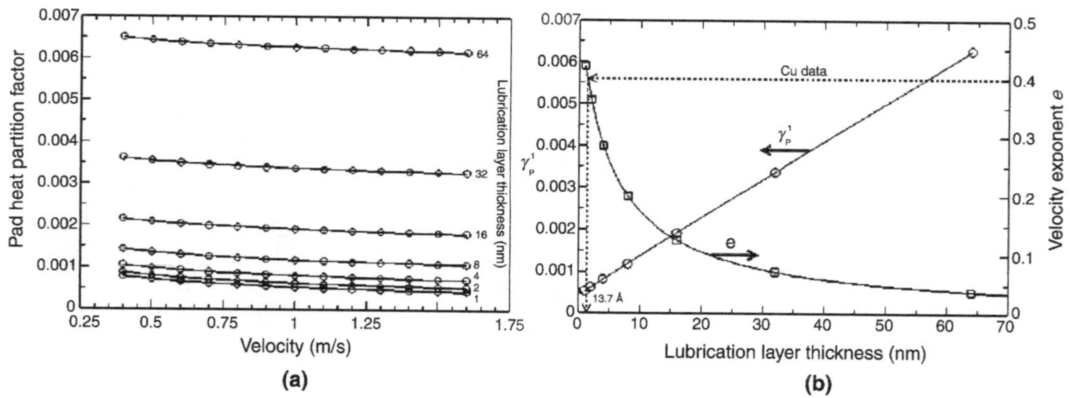

FIGURE 6.3 (**a**) Heat partition factors for lubricated sliding contact of polyurethane against copper with a thin Cu_2O layer. (**b**) Extracted values of γ_p^1 and e.

velocity V for a range of lubrication layer thicknesses. Over the range of parameters considered, γ_p is closely approximated by a power law:

$$\gamma_p = \frac{\gamma_p^1}{V^e} \tag{6.37}$$

for some constants γ_p^1 and e (solid curves in Fig. 6.3a). Extracted values of γ_p^1 and e are shown in Fig. 6.3b. The fact that a power law provides a good description of the finite-element results makes it possible to express the reaction temperature from (6.31) in the simpler form

$$\bar{T} = \bar{T}_p + \frac{\beta}{V^{1/2+e}} \mu_k p V \tag{6.38}$$

where β is defined as

$$\beta = \frac{2\zeta}{\sqrt{\pi \kappa \rho C_p}} \gamma_p^1 (p_a/p) \tag{6.39}$$

Equation 6.38 can be used in several ways. It can be treated as a compact formula for the reaction temperature in which $\bar{T}_p$ and μ_k are measured and β and e are fitting parameters. Alternatively, β and e can be estimated from pad properties ($\kappa \rho C_p$), the tool and wafer geometry (ζ), heat transfer calculations (γ_p^1 and e), and contact mechanics (p_a/p). When considering the changes that occur when moving a process to a larger wafer size or to a different rotary tool, it can also be useful to explicitly segregate out the easily computed tool parameter ζ [20] to estimate the effect of the change in process on the mean reaction temperature.

To summarize the removal rate model, it can be said that consists of a two-step chemical–mechanical model

$$\mathrm{RR} = \frac{M_w}{\rho} \frac{k_1 k_2}{k_1 + k_2} \tag{6.40a}$$

in which the chemical and mechanical rate constants are, respectively,

$$k_1 = A \exp(-E/k\bar{T}) \tag{6.40b}$$

$$k_2 = c_p \mu_k p V \tag{6.40c}$$

and the mean reaction temperature is

$$\bar{T} = \bar{T}_p + \frac{\beta}{V^{1/2+e}} \mu_k p V \tag{6.40d}$$

When used as a compact model for data description, there are five parameters in the model: E, A, c_p, β, and e. Since most polishing tools do not currently measure the COF, it is also possible to define $\tilde{c}_p = c_p\mu_k$ and $\tilde{\beta} = \beta\mu_k$ and to use these as fitting parameters in place of c_p and β. Although this is a practical simplification, it does ignore variations in temperature and rate that occur due to variations in the COF. In tools that do not allow enough access to measure $\bar{T}_{\mathrm{p}}$, it is also sometimes feasible to substitute the ambient temperature T_{a}. We note also that it is usually the case that the mean pad leading edge temperature is linear in the power density and that $\bar{T}_{\mathrm{p}} = T_{\mathrm{a}} + c_{\mathrm{b}}\mu_k pV$ for some constant c_{b} (Fig. 6.4b).

6.5 A POLISHING EXAMPLE

We illustrate the use of the theory in (6.40a)–(6.40d) by applying it to some data from a copper polishing experiment [22]. In the experiment, 100-mm PVD copper wafers were polished with *in situ* conditioning for 90 s at three pressures (1.5, 2, and 2.5 psi) and three sliding speeds (0.31, 0.63, and 1.09 m/s) on a modified bench-top polisher equipped for COF acquisition. The mean pad leading edge temperature was measured with an infrared video camera. Polishing was performed with commercial slurry with hydrogen peroxide as the oxidizer and a low solid fraction of 35-nm colloidal silica particles. Figure 6.4a shows the mean measured removal rates (open circles) and the corresponding mean values of the COF (triangles), both plotted as a function of pV. Differences between repeats in this experiment averaged 195 Å/min, or 5.2 % repeatability error. Figure 6.4b shows the mean pad leading edge temperature, which is close to being linear in the frictional power density $\mu_k pV$.

One of the interesting features of these data is the drop in removal rate that occurs between 10 and 11 kW/m^2. The point left of the transition was measured at 2.5 psi and 0.62 m/s and at the right side at 1.5 psi and 1.09 m/s. There is little change in COF, temperature, or frictional power density $\mu_k pV$ at the transition. The decrease, 2000 Å/min, is about 10 times the experimental error. Similar fluctuations can also occur in silicon dioxide polishing data [19]. Figure 6.5 shows a contour plot of the same removal rate data as a function of p and V separately (solid lines) with contours of constant pV superimposed for comparison (dashed lines). The points on the two sides of the transition are indicated by an "x." Had this copper slurry been perfectly Prestonian, with $\mathrm{RR} = (M_{\mathrm{w}}/\rho)c_p\mu_k pV$, the data in Fig. 6.4 would have been linear in pV within experimental error and the removal rate contours in Fig. 6.5 would have been parallel to the contours of constant pV. The disparity in the directions of the contours becomes more evident at higher velocity. From Fig. 6.5, it is clear that a wide range of rates can be expected at any fixed value of the frictional power density. According to the theory outlined above, Equations 6.40, the origin of this behavior is the flash heating increment (6.40d), and in particular the heat transfer to the pad, since this is the only part of the theory that does not depend just on the pV product.

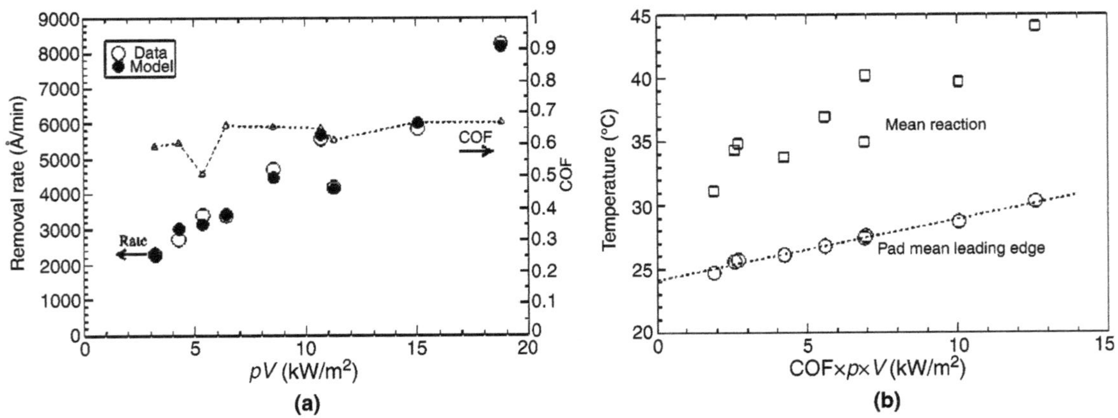

FIGURE 6.4 (**a**) Measured (open circles) and modeled (solid circles) removal rates and COF from a copper polishing experiment. (**b**) Mean pad leading edge temperature and calculated reaction temperature.

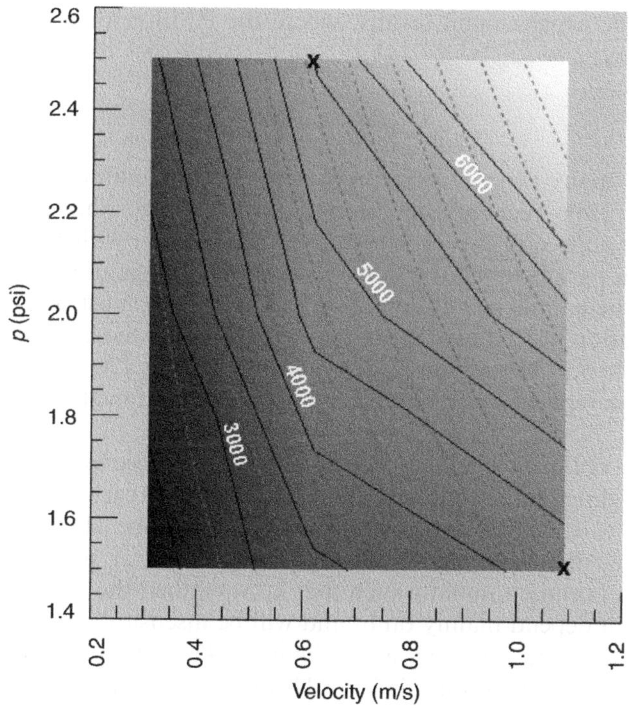

FIGURE 6.5 Contours of removal rate as a function of p and V.

We now describe the application of (6.40) as a compact model. The model contains five parameters, E, A, c_p, β, and e, but there is a problem with using E as a fitting parameter. The problem is that a polishing experiment conducted at room temperature in which the chemical rate constant (6.40b) and the reaction temperature T are not directly measured does not provide enough information to reliably estimate E. In fact, for any assumed value of E over a reasonable range (say 0.25–1.1 eV), optimization of the remaining four parameters in the above theory will produce nearly identical root mean square (RMS) fitting errors [19]. The reason for this is that over a narrow range of temperatures, such as in Fig. 6.4b, a change in E can be compensated for by an opposite change in A, leading to nearly the same value for k_1 and therefore nearly the same removal rate. For copper polishing, one way to address this problem is to perform a dipping experiment in which wafers are immersed at controlled temperatures for fixed but short times in DI water containing the oxidant used in the slurry. The slope of a log plot of k_1 versus $1/kT$ then provides E. For copper polished in slurry that uses hydrogen peroxide as the oxidant, the activation energy is about 0.65 eV [23].

Given E, the above compact model is applied to data by varying the four remaining parameters A, c_p, β, and e to minimize the RMS error between the model and the experimental results, where the RMS error ε is defined for N data points by

$$\varepsilon^2 = \sum (RR_{\text{meas}} - RR_{\text{theory}})^2/N \tag{6.41}$$

There are many ways of carrying out parameter optimization; several are described in [24]. We use the downhill simplex algorithm, which has the advantage of not requiring derivatives of the RMS error with respect to each of the unknown parameters but can have the disadvantage of slow convergence. Since all of the parameters are positive, this constraint must be applied during the minimization procedure. Optimum parameters for the current example are listed in Table 6.1. The minimum RMS fitting error is 170 Å/min, slightly less than the mean replication error. The corresponding removal rates are shown in Figure. 6.4.

When any optimization method is used, it is advisable to be skeptical about whether the global minimum point has actually been reached. This can fail to happen due to experimental noise, which can trap the algorithm at a local minimum, and it can happen if the starting guess is far from optimum. For example, if A is chosen initially such that $k_1 \gg k_2$, then the predicted rate from the model will depend mainly on c_p and will be insensitive to changes in A, β, and e. Data that are scattered, as in Figure. 6.4a, will then appear to fit best with a linear function (Preston's law). However, adjusting the initial guess so that k_1 and k_2 are initially more nearly comparable may produce a significantly different result, as indicated by the solid circles in Figure. 6.4a.

Another potential difficulty is that if the data do *not* show scatter about Preston's law that is significantly in excess of experimental repeatability (indicating a very mechanically limited process), then it will be impossible to estimate the chemical rate parameters A, β, and e with any reliability. In the opposite extreme—a very chemically limited process—it is similarly impossible to estimate c_p. One way to detect these situations is to perform an *uncertainty* or *sensitivity analysis*. A simple uncertainty analysis can be done by varying the optimum parameters up and down one at a time while holding the others fixed until the RMS fitting error changes by a fixed percentage. The size of the resulting interval is then a measure of how well the parameter can be estimated from the data. Table 6.1 lists the range over which each of the optimized parameters can change while increasing the RMS error by less than 10 % in the example. Failure to estimate the chemical rate reliably will show up as an

TABLE 6.1 Optimal Parameters for the Copper Data in Fig. 6.4a with $E = 0.65$ eV.

Parameter	Optimum	Uncertainty (10 % RMS increase)
c_p	7.14×10^{-7} mol/J	6.70×10^{-7}–7.61×10^{-7}
A	5.16×10^{7} mol/m^2 s	5.05×10^{7}–5.27×10^{7}
β	1.18×10^{-3} °C/Pa (m/sec)$^{1-e}$	1.15×10^{-3}–1.21×10^{-3}
e	0.40	0.33–0.46

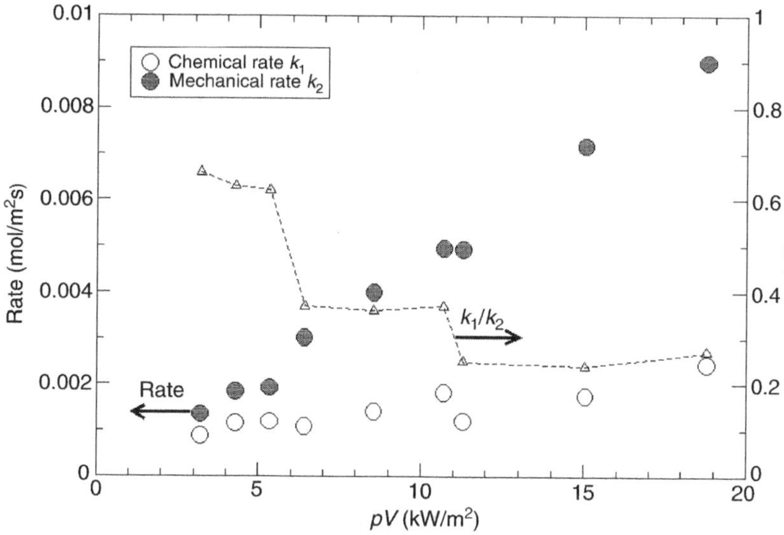

FIGURE 6.6 Chemical and mechanical rate constants and rate ratio for the data in Fig. 6.4.

extremely large uncertainty interval, for example $(0, \infty)$, for at least one of the three chemical parameters.

If the data contain enough evidence to reliably extract all of the model parameters, then (6.40b) and (6.40c) can be used to estimate the mechanical and chemical rate constants, the reaction temperature, and the rate ratio at each operating point. Figure 6.6 shows the extracted rate information for the current example, and Fig. 6.4(b) shows the reaction temperature. The mechanical rate increases nearly linearly with pV, deviations from perfect linearity being due to fluctuations in the COF. Similarly, the chemical rate generally increases with pV but drops when an increase in velocity decreases the fraction of heat transferred to the pad asperities. The process has about equal chemical and mechanical rate constants at low frictional power density but becomes more chemically limited as the power density increases. Since k_1 and the rate ratio are not functions of pV alone, it is sometimes more illuminating to display them using a contour plot similar to the one in Fig. 6.5.

6.6 TOPOGRAPHY PLANARIZATION

Topography planarization differs in several respects from blanket wafer polishing. Topography height variations over the wafer surface induce variations in contact local pressure: high areas have higher contact pressure, and low areas may see little or no contact pressure. At equilibrium, the integral of the contact pressure over the wafer surface must equal the applied load.

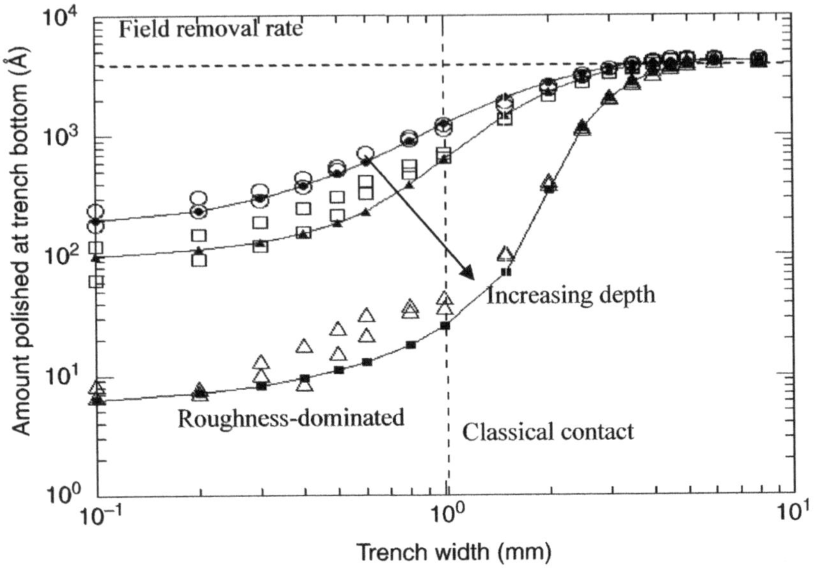

FIGURE 6.7 Amount removed from trench bottoms as a function of trench width and depth.

Because of this, pressures over the wafer surface are coupled, complicating the problem of determining the pressure distribution over patterns. Pressure variations should also affect the heating of asperity summits since the summit temperature should decrease when the summit passes over a low or noncontacting area on the wafer. Finally, as layers are uncovered, the coefficient of friction will in general change, further affecting the temperature distribution on the wafer.

We describe next what happens to trench topography contact mechanics for isolated trenches of different widths and depths in a severely mechanically limited process at constant COF in order to illustrate different behavioral regimes (Fig. 6.7).

For very wide trenches (a large multiple of the pad thickness, depending on the hardness of the pad), the pad conforms easily to both the trench sides and the bottom, and except very close to the sides, the contact pressures and polishing rate are nearly equal. The planarization rate (the rate at which the step height is reduced) is therefore very low, and we could describe the feature as being larger than the pad planarization length.

For somewhat narrower trenches (a small multiple of the pad thickness) that are also deep relative to the pad roughness, the pad initially spans the gap, the entire load is carried by the trench sides, and none is carried by the bottom. The sides therefore polish more quickly than on a blanket wafer, the bottom does not polish at all, and the planarization rate is high. Eventually, the unsupported area of the pad, which projects or bends into the recess, makes contact with the bottom and begins to support some of the load. The bottom then begins to polish and as it does so, the polishing rate gradually drops in the area surrounding the trench. Thus, the planarization rate decreases as the surface becomes flatter. This and the previous case might be termed the classical contact regime since they can be accurately described by classical smooth surface contact mechanics (Fig. 6.7).

As the trench becomes still narrower (less than the hard pad thickness but at least a few micrometers wide), the unsupported surface of the pad no longer bends significantly. If the pad surface were perfectly smooth, then the trench bottom would not polish, as argued above, until the sides were nearly planarized. However, the pad surface is in fact not smooth, so in reality some of the asperities are able to reach into the trench and polish the bottom (Fig. 6.7). The rate at which the bottom polishes then depends mainly on the abruptness of the rough surface. In this regime, which might be called the roughness-dominated regime, the planarization rate is controlled more by the pad asperity height distribution than by pad bending and contact mechanics.

Finally, in the limit of very narrow trench widths, the trench bottom will see contact only from asperities that are both high enough and narrow enough to fit into the trench. All other asperities will simply span the gap between the sides of the feature. This is what might be termed the asperity filtering regime.

We describe a model that combines rough surface contact mechanics with elastic pad mechanics. The theory is capable of describing planarization in both the classical contact and roughness-dominated regimes (see the solid curves in Fig. 6.7). We then introduce a simple approximation for the asperity filtering regime.

Suppose that the height of the wafer surface at (x, y) at time t is $w(x, y, t)$, where we visualize the wafer as being upside down as it is in normal polishing. Let the height of the mean surface of the pad be similarly described by a presently unknown function $\zeta(x, y, t)$. Variations in local pad surface height and asperity height are measured above and below this mean surface. Both the wafer and mean pad shapes are referenced to a fixed plane, such as the top of the platen. There is also a degree of vertical arbitrariness in the description of the wafer surface. If a nominal average pressure $\bar{p}$ is applied to the wafer, then the wafer will be displaced vertically by an amount δ (which is also unknown), so that the local distance between the wafer surface and mean pad surface is $d = w + \delta - \zeta$. From Greenwood and Williamson theory (6.16), the local contact pressure p at (x, y) is

$$p = \frac{4E^{*}\eta_{s}}{3\kappa_{s}^{1/2}} \int_{w+\delta-\zeta}^{\infty} (z - (w + \delta - \zeta))^{3/2} \phi_{s}(z) dz \tag{6.41}$$

The load then balances if

$$\iint p dx dy = \bar{p} A_{w} \tag{6.42}$$

where A_w is the wafer area and the double integral is over the wafer surface. For any particular wafer and pad surface shapes, Equation 6.42 determines the load balancing vertical displacement δ of the wafer. In what follows, we will ignore frictionally induced pitch and bank of the wafer and wafer bending. By applying Greenwood and Williamson theory, we are also ignoring plastic and viscoelastic deformation of the pad, which are complex to account for because of the cyclic nature of encounters between the wafer and fixed points on the pad.

If we knew the mean pad surface shape ζ, then Equations 6.41 and 6.42 would provide the local load balanced solid contact polishing pressure. However, the pad mean surface shape is itself influenced by the local contact pressure. If we knew the pressure distribution, then the displacement field of the pad under the wafer at equilibrium could be found by solving the momentum equations [25]:

$$\frac{\partial \sigma_{ij}}{\partial x_j} = 0 \tag{6.43}$$

where σ_{ij} is the stress tensor and $x_1 = x$, $x_2 = y$, $x_3 = z$. Here, a repeated index implies summation over all spatial dimensions (the Einstein summation convention). Treating each pad layer as a homogeneous and isotropic elastic material, the stress tensor is related to the strain tensor ε_{ij} by

$$\sigma_{ij} = \lambda \varepsilon_{kk} \delta_{ij} + 2\mu \varepsilon_{ij} \tag{6.44}$$

where the Lame constants $\lambda = E\nu/((1+\nu)(1-2\nu))$ and $\mu = E/(2(1+\nu))$ are, in general, different for the pad and subpad. For small strains, the strain tensor is related to the displacements u_i by

$$\varepsilon_{ij} = \frac{1}{2}\left(\frac{\partial u_j}{\partial x_j} + \frac{\partial u_j}{\partial x_i}\right) \tag{6.45}$$

Equations (6.43–6.45) lead to a set of three partial differential equations for the three components of the displacement. These must be solved subject to the constraint that at the pad–platen interface, the displacements are all zero because the pad is glued to the platen. At the mean top surface of the pad, which might, for example, be flat when not loaded, the stress is related to the local contact pressure by

$$\sigma_{33} = -p \tag{6.46}$$

After solving the above elasticity problem, the mean pad surface shape ζ under loading is then the z coordinate u_3 of the displacement at the top of the pad.

Although it is not generally practical to solve the above elasticity problem manually, it is possible to solve it numerically (with the finite element, finite difference, or boundary element method) for layout patterns of limited size. For structures that are long in one direction, such as arrays of long lines, a plane strain problem [25] can be solved instead of the full three-dimensional problem. In the latter case, structures can be analyzed that are on the order of several centimeters wide and that contain hundreds of features [26]. Load balance in these alternative approaches can be done using the nominal pressure $\bar{p}$ on reduced size geometries rather than the full wafer if the geometry size is on the order of the planarization length of the pad.

At first glance there appears to be a dilemma in the above model: We need to know ζ to get p from load-balanced rough surface contact theory and p to get ζ from bulk elasticity theory. A solution to this problem is to apply an iterative numerical method. A very simple approach is to take an initial guess at ζ, say $\zeta = 0$. The local contact pressures p are then calculated from the wafer topography using the load-balanced rough surface model, and these are fed into the elasticity model to obtain a better estimate of ζ. This procedure is then repeated with the new ζ until ζ and p converge. A problem with this idea is that the pad is elastic, so the surface simply oscillates. One remedy is to correct ζ by adding to it only a fixed fraction (called a damping factor) of the pad top surface displacement. When the damping factor is small enough, convergence will eventually be reached and the solution will be independent of the damping factor. The wafer shape is then updated by solving

$$\frac{\partial w}{\partial t} = \mathrm{RR}(p, V) \tag{6.47}$$

where RR(p, V) is the local removal rate at contact pressure p and sliding speed V on the exposed material. Equation 6.47 is most easily solved with an explicit method, such as the Euler or Runge–Kutta method [24].

When polishing multilayer structures, the local removal rate must be calculated with a model that is specific to each exposed material. If the above two-step model is used, then the COF must also be adjusted for the pattern density. We provide an example of pattern correction. Figure 6.8a shows blanket copper polishing data that were collected from an Applied Materials Mirra Mesa 200-mm polisher and modeled using the two-step model. Since COF data are not available from this tool, parameters c_p and β were extracted that implicitly incorporate the COF. Pad temperature data were also collected with an IR gun for both blanket and patterned wafer polishing (Fig. 6.8b) on MIT 854 structures. From the blanket data (Fig. 6.8a), the mean amount of material removed during a 190-s first-step polishing process at 2 psi and 1.63 m/s ($pV = 22{,}500$ W/m^2) would be expected to be about 14,250 Å. In reality, only 8500 Å of copper was removed from the initial 10,500 ECD film on the patterned wafers. The reason for this is shown in Fig. 6.8b, where the slope of the temperature rise during patterned wafer polishing can be seen to be smaller than during blanket wafer polishing. Since $\bar{T}_p = T_a + c_b \mu_k p V$, the thermal data suggest that a reduction in COF has occurred to about 59 % of the blanket

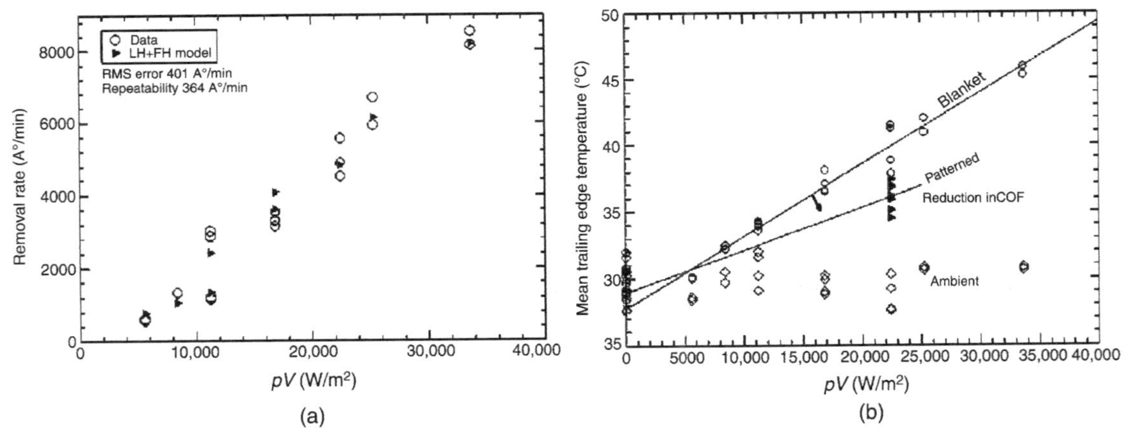

FIGURE 6.8 (**a**) Blanket copper polishing data from a commercial tool. (**b**) Pad thermal data from both blanket and patterned wafer polishing on the same tool.

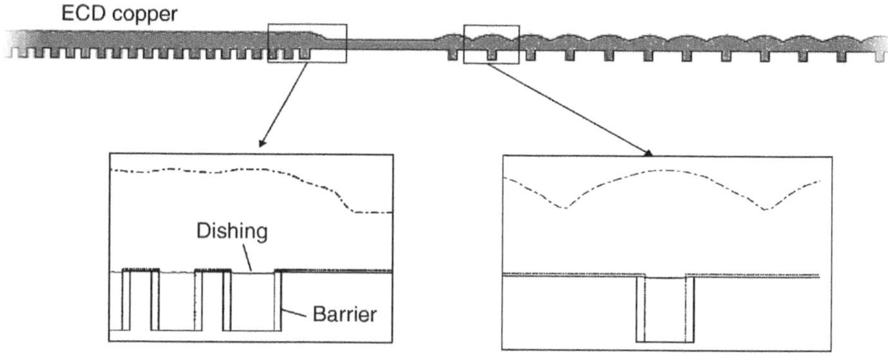

FIGURE 6.9 Simulation of copper polishing and dishing in a large 2D structure (from Ref. 26).

value (the ratio of the two slopes in Fig. 6.8b). Since c_p and β both implicitly incorporate the COF, these parameters therefore must be reduced by a factor of 0.59 to simulate first-step polishing on the patterned wafers. With this correction for the effect of patterning on the COF, the blanket model predicts that 8400 Å of copper should be removed, very close to what was measured. By taking into account the above considerations, it is possible to model planarization, dishing, and erosion using models that have been extracted from blanket wafer data (Fig. 6.9), though this in general requires specialized and somewhat complex software.

When the pad is sufficiently stiff and the area of interest on the wafer has no large-scale topography variations that need to be accommodated by pad bending, the topography polishing model can be greatly simplified by eliminating the solution of the elasticity equations, since in this case it will happen that ζ will be nearly constant. This occurs, for example, when all of the features of interest are in the roughness-dominated and asperity filtering regimes and when there are no variations in topography due to electroplating or other processing steps that are much in excess of a small multiple of the pad roughness standard deviation. The latter is usually on the order of 6–9 μm. If, in addition, the summit height distribution is exponential in the contacting tail, then the load-balanced contact problem (6.41–6.42) on the wafer or on a significantly large pattern area can often be solved explicitly (Question 7). This then leads to a comparatively simple model (6.47) for the planarization rate for features of different sizes and depths.

In the asperity filtering regime, the Greenwood and Williamson theory no longer properly models the local contact pressure since the model contains no notion of asperity width. We describe a simple statistical method for incorporating width effects in two-dimensional polishing. A more sophisticated but more complex approach based on elasticity theory that takes into account asperity shape and the interaction of the asperity with trench structures of similar size can be found in Reference 27. The statistical approach assumes that the summits can be characterized by some approximate measure of width. For a foamed pad, the walls between adjacent pores fundamentally constrain asperity lateral dimensions. By analyzing pad SEM cross sections, a probability density function $\psi(w)$ can be estimated for the wall thickness distribution; that is, $\psi(w)dw$ provides the probability that a wall between two pores at the surface will have a width between w and $w + dw$. If the maximum width that can be accommodated at location x on the wafer is $w_{\max}$, then the proportion of asperities of any height that can make contact at x is

$$\int_0^{w_{\max}} \psi(w)dw \tag{6.49}$$

For example, if there is no constraint on the allowable width at x ($w_{\max} = \infty$), then from (6.49) the proportion of asperities that may contact is 1. If x is at the bottom of a trench of width W, then $w_{\max} = W$ and the proportion increases with the trench width. When this modification is incorporated into the Greenwood and Williamson model, the local contact pressure becomes

$$p = \int_0^{w_{\max}} \psi(w)dw \frac{4E^* \eta_s}{3\kappa_s^{1/2}} \int_d^{\infty} (z-d)^{3/2} \phi_s(z)dz \tag{6.50}$$

Equation 6.50 assumes that the asperity width is not correlated with the height; that is, the joint probability density function for both height and width is separable. Equation 6.50 should be used in place of (6.41) in the topography polishing model when features have sizes in the filtering regime.

QUESTIONS

1. Consider a three-step removal process in which a chemically grown film can be either removed by mechanical abrasion or dissolved by a complexing reaction with rate constant k_3. If the concentration of the complexing agent is c, show that

$$\mathrm{RR} = \frac{M_w}{\rho} \frac{k_1 \cdot (k_2 + k_3 c)}{k_1 + k_2 + k_3 c}$$

   This model allows for static etching.

2. Show that for a rotary polisher, if the pad rotation rate is Ω_p, the wafer rotation rate is Ω_w, and the separation between the pad center and wafer center is c_w, then the relative sliding velocity between the wafer and the pad is

$$\vec{V} = \Omega_p c_w \vec{j} + (\Omega_p - \Omega_w)\vec{k} \times \vec{r}$$

where $\vec{r}$ is a vector to a point under the wafer from the wafer center. It follows that the relative sliding velocity and speed are constant when the pad and wafer corotate.

3. When the contacting population of the pad summit height distribution is exponential, $\phi_s(z) = B\exp(-z/\lambda)$, show using the Greenwood and Williamson theory that when the wafer is at $z = d$, then the pressure–displacement relation is

$$p = \sqrt{\pi}E^* \eta_s \kappa_s^{-1/2} \lambda^{5/2} \phi_s(d)$$

the real contact area fraction is

$$A_f = \pi \eta_s \kappa_s^{-1} \lambda^2 \phi_s(d)$$

and the area density of summits in contact is

$$\eta_c = \eta_s \lambda \phi_s(d).$$

4. Use the summit height PDF in Fig. 6.2 to estimate the number of summits per unit area in contact with the wafer if the total summit density is $10^8/\text{m}^2$ and $d = 9\ \mu\text{m}$. At a sliding speed of 1 m/s, what is the approximate time between contacting asperity encounters at a fixed point on the wafer?

5. This exercise derives formula 6.21 for the flash heating increment. Consider the one-dimensional heat equation

$$\frac{\partial T}{\partial t} = D \frac{\partial^2 T}{\partial z^2}$$

for $z < 0$, where $D = \kappa/(\rho C_p)$ is the thermal diffusivity. Suppose that the heat flux at $z = 0$ is

$$\kappa \frac{\partial T}{\partial z} = H$$

and that the initial temperature is $T = 0$.

(a) Apply the Laplace transform, $\hat{T}(z,s) = \int_0^\infty T(z,t)e^{-st}dt$, to the heat equation and to the boundary condition at the surface and show that

$$s\hat{T}(z,s) = D\frac{\partial^2}{\partial z^2}\hat{T}(z,s) \quad \text{when } z < 0,$$

$$\frac{\partial}{\partial z}\hat{T}(z,s) = \frac{H}{\kappa s} \quad \text{when } z = 0.$$

(b) Solve the differential equation in (a) and show that at the surface, $\hat{T}(0,s) = DH/\kappa s^{3/2}$.

(c) Using a table of inverse Laplace transforms, show that $T(0,t) = (2/\sqrt{\pi})(Ht^{1/2}/\sqrt{\kappa\rho C_p})$. When $H = \gamma_p \mu_k p_a V$, this is the flash heating increment at time t.

6. For a summit height PDF with an exponential tail, show using load balance and formula 6.34 for the expected real contact pressure that

$$\frac{\bar{p}_a}{p} = \frac{2}{3\pi}\eta_s^{-1}\kappa_s\lambda^{-1}$$

Show also that the corresponding contact area is

$$A_c = \frac{\pi}{4}\left(\frac{p}{E^*}\right)^2 \eta_s^{-2}\lambda^{-2}$$

7. Consider an isolated, infinitely long trench of width W and depth S, where the width is in the roughness-dominated regime. Suppose that the structure is polished using a pad with an exponential summit height distribution in the contacting tail.

(a) If the slurry is perfectly Prestonian, show that the planarization rate is governed by

$$\frac{dS}{dt} = \frac{M_w}{\rho} c_p \mu_k \bar{p} V(\exp(-S/\lambda) - 1)$$

(b) Derive the equivalent model for slurries that can be described by the two-step theory.

REFERENCES

1. Thakurta DG, Schwendeman DW, Gutman RJ, Shankar S, Jiang L, Gill WN. Three-dimensional wafer scale copper chemical—mechanical planarization model. Thin Solid Films 2002;414:78–90.
2. Preston F. The theory and design of plate glass polishing machines. J Soc Glass Technol 1927;11:214–256.
3. Zhang F, Busnaina A. The role of particle adhesion and surface deformation in chemical mechanical polishing processes. Electrochem Solid-State Lett 1998;1:184–187.
4. Tseng W, Wang Y. Re-examination of pressure and speed dependencies of removal rate during chemical mechanical polishing processes. Electrochem Soc 1997;144: L15–L17.

5. Shi F, Zhao B. Modeling of chemical mechanical polishing with soft pads. Appl Phys A: Mater Sci Process 1998;67:249–252.
6. Zhao B, Shi F. Chemical mechanical polishing—threshold pressure and mechanism. Electrochem Solid-State Lett 1999;2:145–147.
7. Zhao B, Shi F. Chemical mechanical polishing in IC processes: new fundamental insights. Proceedings of the fourth CMP-MIC;Santa Clara, CA;1999. p13–22.
8. Bastaninejad M, Ahmadi G. Modeling the effects of abrasive size distribution, adhesion, and surface plastic deformation on chemical–mechanical polishing. J Electrochem Soc 2005;152(9):G720–G730.
9. Seike Y, Lee H, Takaoka M, Miyachi K, Amari M, Doi T, Philipossian A. Development of a pad conditioning process for interlayer dielectric CMP using high-pressure micro jet technology. J Electrochem Soc 2006;153(3):G223–G228.
10. Johnson KL. Contact Mechanics. Cambridge University Press; 1985.
11. Greenwood JA. A unified theory of surface roughness. Proc R Soc Lond A 1984;393:133–157.
12. Greenwood JA. Contact of rough surfaces. In: Singer IL, Pollock HM, editors. Fundamentals of Friction: Macroscopic and Microscopic Processes. Kluwer Academic Publisher; 1992. p37–56.
13. Greenwood JA. Problems with surface roughness. In: Singer IL, Pollock HM, editors. Fundamentals of Friction: Macroscopic and Microscopic Processes. Kluwer Academic Publisher; 1992. p57–76.
14. Cowan RS, Winer WO. In:Blau PJ, *et al.*, editors. ASM Handbook. Volume 18, Friction, Lubrication and Wear Technology; 1992. p39–44.
15. Borucki L, Ng S-H, Danyluk S. Fluid pressures and pad topography in chemical–mechanical polishing. J Electrochem Soc 2005;152(5):G391–G397.
16. Szeri AZ. Fluid Film Lubrication Theory and Design. Cambridge University Press; 1998.
17. Ng S-H. Doctoral dissertation. Atlanta (GA): Georgia Institute of Technology; 2004.
18. Muldowney G, Elmufdi CL, Palaparthi R. Proceedings of the 10th CAMP CMP Symposium; Lake Placid, NY;2005; Aug 14–17.
19. Sorooshian J, Borucki L, Stein D, Timon R, Hetherington D, Philipossian A. Trans ASME J Tribol 2005;127:639–651.
20. Rosales-Yeomans D, Borucki L, Doi T, Lujan L, Ichikawa K, Philipossian A. Implications of wafer size scale-up on frictional, thermal and kinetic attributes of ILD CMP. Proceedings of VMIC;2005. p188–193.
21. Sampurno YA, Borucki L, Zhuang Y, Boning D, Philipossian A. A method for direct measurement of substrate temperature during copper CMP. J Electrochem Soc 2005;152(7):G537–G541.
22. Borucki L, Li Z, Philipossian A. Experimental and theoretical investigation of heating and convection in copper polishing. J Electrochem Soc 2004;151: G559–G563.
23. Denardis L. Doctoral dissertation. University of Arizona; 2006.
24. Press WH, Teulkolsky SA, Vetterling WT, Flannery BP. Numerical Recipes in C. 2nd ed. Cambridge University Press; 1992. p994.

25. Landau LD and Lifshitz EM. Theory of Elasticity. 3rd ed. Butterworth-Heinemann; 1986.
26. Im YH, Borucki L, Bloomfield MO, Cale TS. Integrated multistep process simulation using chip-scale structures. Advanced Metallization Conference; Montreal;2003.
27. Vlassak JJ. A contact-mechanics based model for dishing and erosion in chemical–mechanical polishing. Materials Research Society Symposium Proceedings;Materials Research Society,Warrendale, PA;2001.

7

KEY CHEMICAL COMPONENTS IN METAL CMP SLURRIES

Krishnayya Cheemalapati, Jason Keleher and Yuzhuo Li

7.1 INTRODUCTION

Chemical–mechanical planarization (CMP) slurries can be classified into two general categories according to their applications: metal CMP slurry and nonmetal CMP slurry. Two most important metals incorporated into IC chip manufacturing via CMP are W and Cu. Depending upon integration schemes, a metal CMP slurry may or may not carry out the function of removing the film(s) immediately below the overburden metal such as cap, adhesion, and barrier layers. For tungsten CMP, a single-step slurry is typically used to remove both excess tungsten and its barrier/adhesion layer. For copper CMP, after the removal of copper, a subsequent barrier removal step is often required. This chapter reviews the basic requirements for the key components found in common metal slurries.

It is generally understood that a metal CMP slurry chemically modifies the surface to be polished and yields a softer and porous complex layer, which is then removed by mechanical force in the process. Although these metals may differ in their physical and chemical properties, the underlying principle for the design of slurries is the same. A production-worthy metal CMP slurry must address several issues, such as material removal rate (MRR), within-wafer nonuniformity (WIWNU), step height reduction efficiency, dishing/erosion, minimum ILD loss, corrosion, scratching, slurry residue, and other surface defects. Typical metal CMP slurry may contain an oxidant, a chelating agent, abrasive particles, a surfactant, and other additives. These components must work in concert to produce adequate material removal rate, high planarization efficiency, and

Microelectronic Applications of Chemical Mechanical Planarization, Edited by Yuzhuo Li

minimal surface defects. Depending on the electrochemical activity of the metal at a given pH, an additional passivating agent may be required. For example, the presence of an effective passivating agent is absolutely critical to the performance of copper CMP slurry. This topic will be dealt with in detail in the following chapter. The focus of this chapter is on the key components of slurries intended for copper CMP. The basic working principles discussed here are also applicable to other metals such as tungsten and nickel.

7.2 OXIDIZERS

A metal CMP process involves an electrochemical alteration of the metal surface and a mechanical removal of the modified film. More specifically, an oxidizer reacts with the metal surface to raise the oxidation state of the material, which may result in either the dissolution of the metal or the formation of a surface film that is more porous and can be removed more easily by the mechanical component of the process. The oxidizer, therefore, is one of the most important components of the CMP slurry. Electrochemical properties of the oxidizer and the metal involved can offer insights in terms of reaction tendency and products. For example, relative redox potentials and chemical composition of the modified surface film under thermodynamically equilibrium condition can be illustrated by a relevant Pourbaix diagram [1]. Because a CMP process rarely reaches a thermodynamically equilibrium state, many kinetic factors control the relative rates of the surface film formation and its removal. It is important to find the right balance between the formation of a modified film with the right property and the removal of such a film at the appropriate rate.

In this section, the physical and chemical properties of oxidizers commonly found in metal CMP slurries, such as nitric acid, ferric salts, hydrogen peroxide, iodates, permanganates, and chromates, are first described. The focus will then turn to their capability in material removal rate and their corrosion tendency of the metal films of interest.

7.2.1 Nitric Acid

One of the early systems investigated for Cu CMP uses nitric acid (HNO_3) as an oxidizer. HNO_3 can dissociate into proton (H^+) and nitrate ion (NO^{3-}). The presence of proton (H^+) can assist the chemical attack on Cu metal. The nitrate ion at high concentration can oxidize copper to a higher oxidation state. Based on the Pourbaix diagram [1] shown in Fig. 7.1, the oxidized state of copper prefers Cu^{2+} in the low-pH regime. Steigerwald and coworkers [2,3] showed via XPS measurements that no significant native surface film such as copper oxide was formed during the polishing process with a nitric acid system. Any Cu_2O that was present in an *ex situ* analysis may have been a result of the oxidizing NO^{3-} left on the surface after polishing. Therefore, the presence of HNO_3 mainly leads to a direct dissolution of copper into ions.

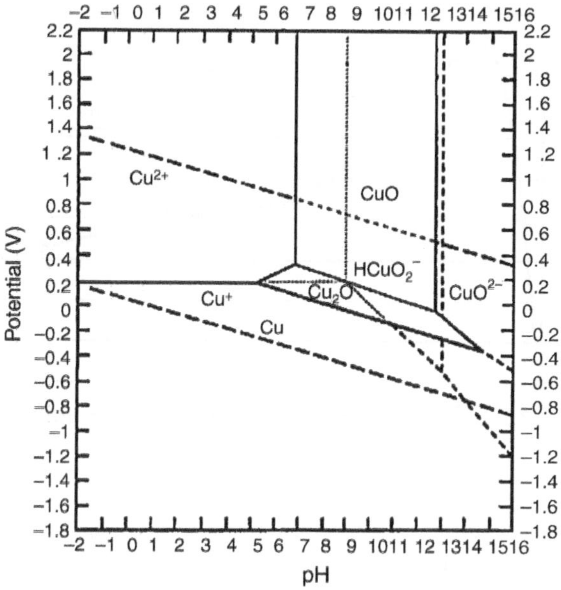

FIGURE 7.1 Pourbaix diagram of copper–water system (from Ref. 1).

Carpio and coworkers [4] supported this hypothesis via a potentiodynamic study of a set of HNO_3-containing slurries. The corrosion currents and potentials under both the static and the dynamic conditions were practically the same. This is consistent with the fact that there was no native copper oxide film formed because of the presence of these slurries. As a matter of fact, the corrosion currents decreased slightly upon abrasion of the copper surface. The contact between the metal surface and the abrasive pad may have limited the mass transport of chemicals to and from the copper surface. This was verified via an AC impedance measurement that showed the importance of the systems' mass transport. It was also concluded that in a dissolution-controlled process, mechanical abrasion would not enhance the chemical corrosion rate or reduce the mass transport of reactants and/or products in the system.

Although HNO_3-containing slurry can yield high material removal rate, the high intrinsic dissolution rate may lead to rapid etching of the copper in the trench. This fast etching action translates to low step height reduction efficiency and severe dishing at the end. Therefore, nitric acid is no longer a preferred oxidizer for advanced copper CMP slurry.

7.2.2 Hydrogen Peroxide

H_2O_2 is one of the most powerful and widely used oxidizers in present-day industrial CMP slurries [5–9]. Hirabayashi and coworkers [10] were among the first to describe a Cu CMP slurry that contains hydrogen peroxide, glycine, and

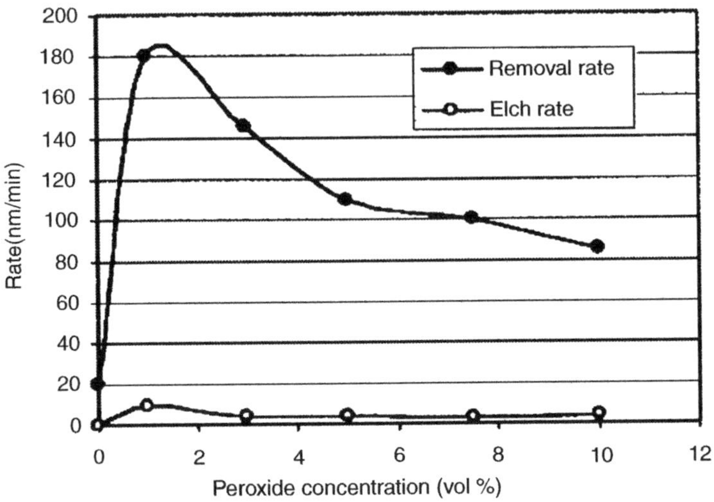

FIGURE 7.2 Material removal and static etch rate of copper as a function of H_2O_2 concentration (from Ref. 5).

silica particles. It was proposed that, at proper pH, a Cu_2O layer is formed by H_2O_2 oxidation. In the presence of an effective complexing agent such as glycine and at low concentration of hydrogen peroxide, the film is porous and easy to be removed with mechanical abrasion. As shown in Fig. 7.2, under these conditions, the material removal rate increases with the increase in hydrogen peroxide concentration (<1 wt %). At higher pH or higher hydrogen peroxide concentration, the film becomes thicker and denser. The film is strong enough to serve as a passivating layer. Under these conditions, the material removal and static etch rates, thus, decrease with the increase in hydrogen peroxide concentration (Fig. 7.2). Such a dual effect of hydrogen peroxide concentration on MRR is schematically illustrated in Fig. 7.3.

As shown in Fig. 7.3, in the so-called etching region, addition of extra effective complexing agent will increase the MRR and reach a peak MRR sooner with lower hydrogen peroxide concentration [11]. Similarly, addition of a passivating agent such as BTA should cause a decrease in MRR and a delay in reaching its peak at a higher peroxide concentration. In the passivation region, an addition of complexing agent will slow down the decreasing trend in MRR. Most copper CMP slurries are formulated with a hydrogen peroxide concentration in the passivation region.

The effect of hydrogen peroxide concentration on tungsten removal rate, however, is not as pronounced as for copper (Fig. 7.4) [12]. This is a direct result of the formation of native oxide film on tungsten surface readily in the presence of hydrogen peroxide. Lim and coworkers studied the effect of hydrogen peroxide on tungsten CMP [13]. It was confirmed that, in the presence of hydrogen peroxide, a native WO_3 formed first, which was then transformed to a much harder WO_2, thus leading to lower static rates. When

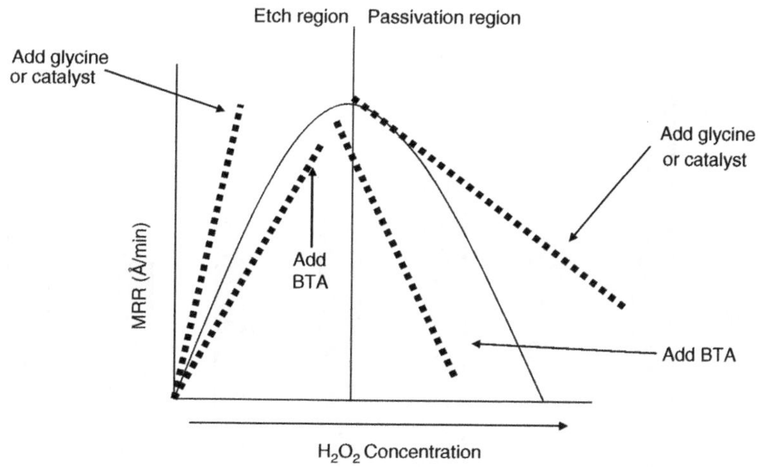

FIGURE 7.3 A schematic illustration of hydrogen peroxide concentration effect on material removal rate of copper with the addition of complexing agent such as glycine and passivating agent such as BTA (from Ref. 11).

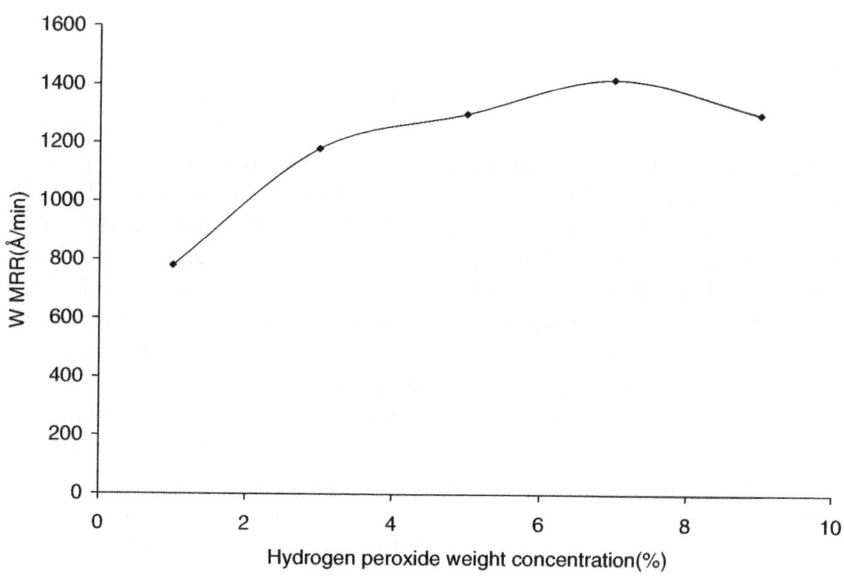

FIGURE 7.4 Material removal rate on 8″ tungsten blanket test wafers as a function of hydrogen peroxide concentration (from Ref. 12).

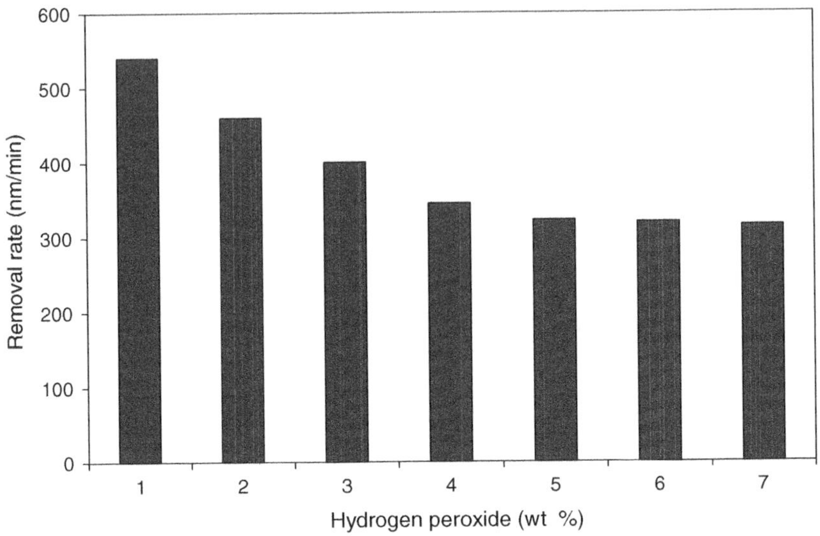

FIGURE 7.5 A correlation between hydrogen peroxide concentration and NiP removal rate for a slurry containing 1 wt % glycine, 5 wt % alumina, and various amount of hydrogen peroxide at pH 4 (from Ref. 12).

hydrogen peroxide and ferric nitrate were used together as mixed oxidizers, the tungsten removal rate was significantly higher than when hydrogen peroxide or ferric nitrate was used separately. The synergistic effect is caused by the Fenton chemistry, which will be described in more detail later in this chapter.

In a computer hard drive, the data are stored on one or more rigid disks coated with magnetic materials. The magnetic film is coated on a planar NiP substrate. NiP film is typically electroplated on an aluminum disk. Before the coating process, the NiP must be planarized. The planarization process is similar to oxide ILD CMP except that the material is a metal. The slurry chemistry is similar to copper or tungsten CMP except that the end point is to form a perfectly planar substrate. Hydrogen peroxide can be used as an oxidizer in NiP CMP as stated in the patent by Jia and coworkers [14,15]. As shown in Fig. 7.5, the removal rate for NiP film follows the same trend as for copper. At higher concentration, the removal rate decreases because of the formation of a native protective NiO film [12].

There have been numerous publications dealing with the mechanism of reaction between hydrogen peroxide and copper film. The electrochemical data generated by Brusic and coworkers [16] suggested that H_2O_2 and glycine promote the formation and removal of a Cu(II) oxide film. Li and coworkers [17] further investigated the formation of the Cu–glycine complex and its effect on the copper film removal rate and static state dissolution rate. It was concluded that the Cu–glycine complex formed during CMP enhanced the decomposition of H_2O_2 to give hydroxyl radical (*OH). As shown in Table 7.1,

TABLE 7.1 Oxidation Potential for Various Oxidizers [18,19].

Oxidant	Oxidative Potential (V)
Fluorine	3.0
Hydroxyl radical	2.8
Ozone	2.1
Hydrogen peroxide	1.8
Potassium permanganate	1.7
Chlorine dioxide	1.5
Chlorine	1.4

hydroxyl radical is a stronger oxidizer than hydrogen peroxide itself [18,19]. The formation of such a strong oxidizer leads to a significant increase in both copper removal rate and static etch rate. Such a decomposition path is not limited to copper ions. As a matter of fact, the decomposition mechanism can be generalized to other transition metal ions. For example, the presence of ferric and ferrous ions in a solution can cause the decomposition of H_2O_2 via a well-known Fenton cycle as shown in Equations 7.1 and 7.2 [20,21]. During copper CMP, the catalytic reactions that lead to hydroxyl radicals involve the Cu(I) and Cu(II) pair.

$$M^{2+} + H_2O_2 \rightarrow M^{3+} + OH^- + {}^*OH \quad (7.1)$$

$$M^{3+} + H_2O_2 \rightarrow M^{2+} + {}^*OOH + H^+ \quad (7.2)$$

As a transient species, hydroxyl radical can react with various components in a slurry. Four of such representative reactions are listed below (Eqs. 7.3–7.6) [22].

$$^*OH + C_6H_6 \rightarrow (OH)C_6H_6 \quad (7.3)$$

$$^*OH + CH_3OH \rightarrow {}^*CH_2OH + H_2O \quad (7.4)$$

$$^*OH + [Fe(CN)_6]^{4-} \rightarrow [Fe(CN)_6]^{3-} + OH^- \quad (7.5)$$

$$^*OH + {}^*CH_2OH \rightarrow HOCH_2OH \rightarrow H_2O + CH_2O \quad (7.6)$$

As shown in Equation 7.3, the first type of reaction is an addition reaction where the hydroxyl radical adds on to an unsaturated compound, aliphatic or aromatic, to form a new free radical. The second is a hydrogen abstraction where a new organic free radical and water are formed (Eq. 7.4). The third involves an electron transfer where an electron donor gets oxidized and the hydroxyl radical gets reduced to hydroxide (Eq. 7.5). The last type consists of two radical species combining into a neutral species (Eq. 7.6).

The practical implication of these radical reactions in metal CMP is multifaceted. First, the presence of any metal impurities can significantly lower the pot lifetime of the slurry. The pot lifetime is often defined as the lifetime of slurry after hydrogen peroxide is added. The presence of metal ions as impurities can cause the decomposition of hydrogen peroxide over time and shorten the pot lifetime of slurry. Various schemes can be employed to overcome this difficulty. One is to ensure that all chemical ingredients contain as little heavy metal ion impurities as possible. Sometimes, this may be an expensive solution due to the cost of high purity oxidizers. The second option is to add a stabilizer into the slurry. As a matter of fact, some of the hydrogen peroxide sources may already contain stabilizer. The function of these stabilizers is to sequest the metal ions and prevent them from directly interacting with hydrogen peroxide. The downside of this approach is that, in some cases, the stabilizer may be too strong and may interfere with the desirable or expected interactions between copper ions and hydrogen peroxide. Therefore, some specific knowledge of the properties of the stabilizer used in the incoming hydrogen peroxide is very useful. The third approach is to use high enough hydrogen peroxide concentration to counter the loss due to such decomposition. In practice, all three approaches are employed in a balanced and moderate fashion.

The presence of impurity metal ions in slurry brought by raw materials will have a direct impact on the slurry performance such as static etch rate and removal rate. For example, as shown in Fig. 7.6, using the same formulation with various sources of hydrogen peroxide, the static etch and polishing rates are quite different [12].

For this reason, all hydrogen peroxide used in the research and production must be well documented and routinely qualified.

The third implication of metal ion mediated formation of radicals is that the metal ions introduced during metal CMP changes the chemical composition of the fluid at the interface between the pad and the wafer. The newly introduced ions may play a significant role in determining the effectiveness of all key components in the original CMP slurry. For example, as shown in Fig. 7.3, the copper ions can be viewed as a catalyst for hydrogen peroxide and can change the static etch and removal rate profiles. Fig. 7.7 shows the addition of copper nitrate to a hydrogen peroxide based slurry; the static etch rate increases significantly upon the addition of copper nitrate [12]. It is important to note that the amount of copper nitrate added appears to be very high, although a back envelope calculation will put the copper ion concentration at 1000 ppm if an 8″ wafer is used during a copper CMP operation with 200 ml/min flow rate and a 500 nm/min removal rate is achieved. We further assumed that all copper removed from the wafer surface was converted in to copper ions and was homogeneously dispersed in the spent slurry. This is very far from truth. The amount of copper ions generated during the CMP operation will stay around the copper surface in the boundary layer until it is diffused out. The thickness of such layer can be as thin as a few microns. The local concentration of copper ions must be at least 100× higher than the bulk concentration.

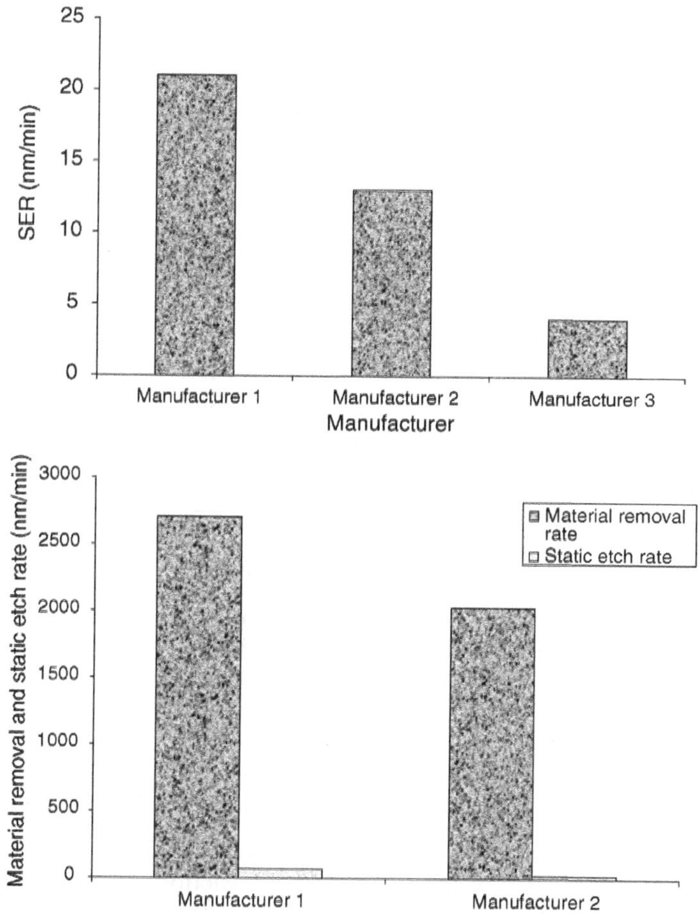

FIGURE 7.6 Material removal and static etch rate of copper with identical formulations incorporating oxidizer from different manufacturers. The total metal impurity is the highest in hydrogen peroxide by Manufacture 1 and the lowest by Manufacturer 3 (from Ref. 12).

As much as the metal ion mediated hydrogen peroxide can cause complications in slurry formulation, one can take advantage of such catalytic effect in designing advanced metal CMP slurry or polishing scheme as well. For example, Li et al. investigated the possibility of encapsulating metal ions inside a supramolecular structure for an abrasive-free system [23]. The use of a reactive pad could also be another alternative where the pad surface is modified using active functional groups such as amino group, thereby substituting the slurry with oxidizer-only solution. The oxidizer-only solution could be a potential solution to overcome slurries with short pot life [24].

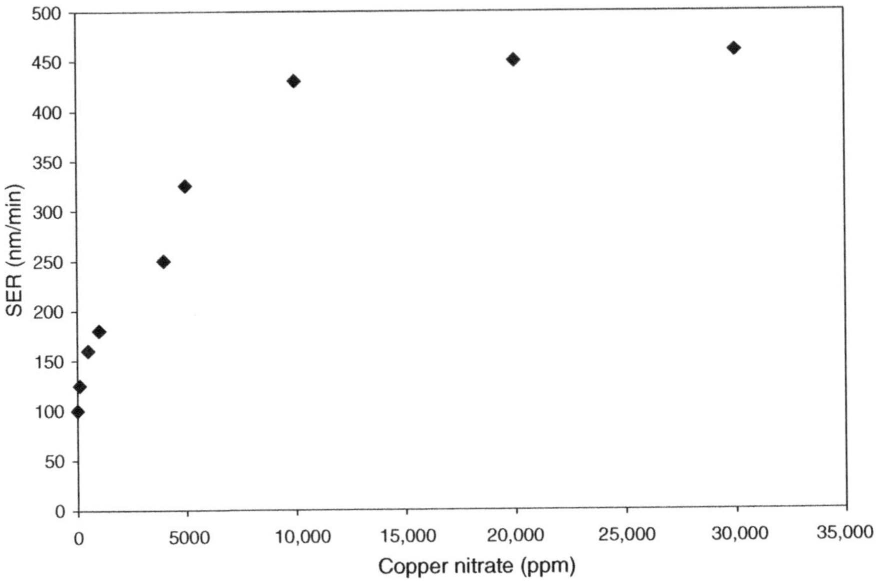

FIGURE 7.7 Correlation between static etch rate and copper nitrate concentration in a solution containing 1 wt % hydrogen peroxide and 1 wt % of glycine with various concentrations of copper nitrate added (from Ref. 12).

7.2.3 Ferric Nitrate

Ferric nitrate was among the first oxidizers used in early tungsten and copper CMP processes. As indicated in the Pourbaix diagram [25] for Fe (Fig. 7.8), the reduction potential is 0.77 V for $Fe^{3+} \rightarrow Fe^{2+}$ reaction. Considering the fact that the reduction potential for $Cu^{+} \rightarrow Cu$ is 0.52 V, ferric nitrate has more than adequate oxidizing power for the Cu CMP process (Eq. 7.7).

$$Fe^{3+} + Cu \rightarrow Cu^{+} + Fe^{2+} \tag{7.7}$$

Owing to the fact that free ferric ions are stable only in the acidic regime (Fig. 7.8), most slurries using ferric ions as an oxidizer are formulated at pH substantially below 4. At such a low pH, the copper surface oxidized via the reaction described in Equation 7.7 will not form any native oxide film. Without the protection from such an oxide, the copper surface is prone to corrode, which results in high static etch rate and practically no planarization efficiency. To provide a balance, therefore, the presence of a corrosion inhibitor is a must for copper CMP slurry.

Babu and coworkers [26,27] investigated the copper dissolution in the presence of Fe^{3+} using a copper rotating disk electrode (RDE). The cathodic reaction was separately studied using a platinum rotating disk electrode, while the overall corrosion process was measured on rotating disks. It was

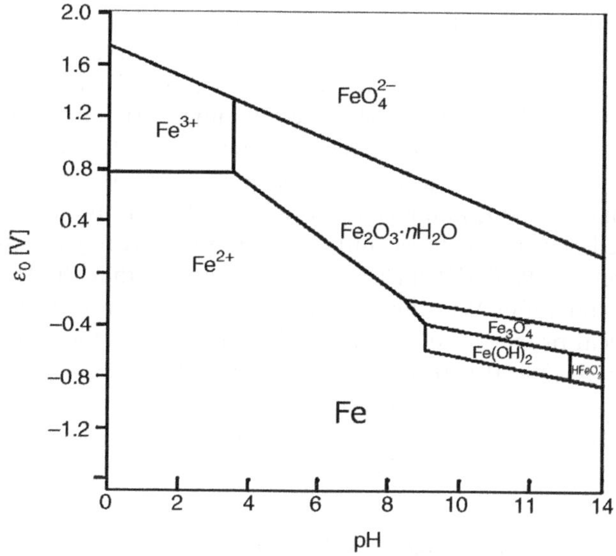

FIGURE 7.8 Pourbaix diagram of Fe–water system (from Ref. 25).

determined that the rate of dissolution is related to the effectiveness of the Fe^{3+} ions reaching the surface. Therefore, external forces such as slurry flow rate and rotating table speed may have an effect on the dissolution rate and material removal rates. The copper static etch and removal rates are usually very high when a ferric nitrate based CMP slurry is used. The high static etch rate leads to drawbacks such as corrosion and dishing. To lower the static etch rate and increase the planarization efficiency, a strong passivating agent such as BTA [28] is usually included in a ferric-based copper CMP slurry. The passivating film formed by BTA molecules is effective in lowering the static etch rate and protecting the trench areas. It is important to point out that, due to severe protonation at low pH, BTA performs more efficiently at pH above 4 (following chapter). Therefore, the BTA concentration required to form an effective passivating layer at a low pH is significantly higher (2–5 times) than that for a slurry at a high pH using hydrogen peroxide as an oxidizer. There is one potential advantage of an acidic copper CMP slurry. As the material removal rate of dielectric oxide is extremely low at pH below 4, a ferric nitrate based low-pH copper CMP slurry usually has a high selectivity of copper over its barrier and dielectric layers underneath.

Ferric nitrate has also been used widely as an oxidizer in W CMP slurries. As a matter of fact, ferric nitrate based slurry was successfully used for tungsten CMP applications long before its application to copper films. Unlike copper, however, tungsten does not usually require the use of inhibitors in the slurries. Tungsten is a much harder metal in comparison to copper. Under oxidizing condition, tungsten has much better passivation characteristics. More specifically, native passivating film such as WO_3 and $FeWO_4$ can be formed very rapidly during CMP [29,13]. In addition to serving as an oxidizer, ferric ions have also been utilized as catalysts or cooxidizers for hydrogen peroxide. Seo and coworkers [29] studied a slurry containing ferric nitrate and hydrogen peroxide and found the slurry gives high removal rates due to synergistic interaction between ferric ions and hydrogen peroxide. It was also found that the formation of $FeWO_4$ film has a passivating effect. Lim and coworkers [13] also compared the use of ferric nitrate based and hydrogen peroxide based mixed oxidizer for W CMP slurries. They concluded that the formation of a much brittle and softer WO_3 film is the key for high removal rates.

The chemical properties of ferric and ferrous ions are the characteristics of transition metals [30]. It is possible that the ferric/ferrous pair can be replaced by other transition metals. However, due to the increasing and overriding concerns on metal ion contamination, the potential use of ferric or any other metal ions in CMP slurry may be limited.

7.2.4 Potassium Permanganate, Dichromates, and Iodate

Potassium permanganate ($KMnO_4$) is a strong oxidizer that is stable in alkaline media. Carpio and coworkers [4] studied the use of $KMnO_4$ in both buffered ($pH = 6.88$) and unbuffered ($pH = 8.76$) systems. Potentiodynamic measurements of the unbuffered systems showed low corrosion current for the unabraded copper and a current increase during abrasion. This implies that the copper surface is passivated in the absence of abrasion and the passivation film is mechanically removed by abrasive action. In general, such a gap between the two currents (in the presence and absence of abrasion) can be correlated with high step height reduction efficiency and is thus desirable for a metal CMP slurry. A reported drawback for unbuffered systems includes the instability of the alumina abrasives at such high pH. The buffered system on the contrary showed a different set of passivation and dissolution characteristics. The electrochemical measurements showed no difference in the corrosion currents between the abraded and the nonabraded systems indicating low oxidation efficiency, thereby resulting in a removal rate less than 500 Å/min. Therefore, the use of such an oxidizer may be limited to high pH slurry.

$K_2Cr_2O_7$ [31] is another strong oxidizing agent that could be employed as an oxidizer in metal CMP. Unlike hydrogen peroxide, the dichromates give a simple relationship between the oxidizer concentration and the material removal rate. More specifically, the dual regimes associated with hydrogen peroxide based copper CMP slurry (Fig. 7.3) was not found for the slurry with dichromates. This is partially due to the fact that the attack on Cu_2O by dichromates is favorable; this converts Cu(II) that is easier to dissolve. For hydrogen peroxide the oxidation from copper(I) to copper(II) is unfavorable. Cu_2O provides the needed passivation. Duquette and coworkers [32] investigated the formation of the passivating film at low pH values for Al

CMP potentiodynamic measurements. Their work showed a large decrease in the corrosion current at 1 wt % of the oxidizer concentration, indicating the formation of a passivating film. AES measurements confirmed the formation of a high oxygen content film with a thickness of 1.5 nm. The addition of 3 wt % alumina to the slurry showed a twofold increase in the removal of copper when compared to the abrasive-free systems. The material removal rate peaked when using 2 wt % oxidizer.

Potassium periodate or periodic acid has also been investigated and used as an oxidizer for W CMP [33,34]. Alumina and periodic acid based slurries for W CMP have been studied in detail by Hetherington and coworkers [35–37]. The impact of process parameters, slurry pH, abrasive and oxidizer concentrations, and also the impact on the process temperature were investigated. The removal rate was found to be a direct function of the oxidizer concentration. In addition to periodic acid, other iodine compounds have also been investigated. Singh and coworkers [38] studied the use of potassium iodate (KIO_3) as an oxidizer in the presence of an inhibitor (BTA) and the use of excess iodide ion (KI) in the intermediate pH regime for copper CMP. It is reported that both BTA and KI played a role in forming the passivating layer. The dissolution of this layer was enhanced with the addition of EDTA as a complexing agent. The formation of the passivating layer on the surface was confirmed by the anodic electrochemical measurement and by XPS measurement. This slurry system works the best in the intermediate pH ranges (pH = 4–6) with 10^{2-} and 10^{3-} M KI and BTA, respectively. Babu and coworkers [39] studied KIO_3 as an oxidizer and found that the copper polishing rate decreased with the increase in the slurry pH suspected because of the formation of passive layer with the increase in the pH, which was also supported by potentiodynamic results. Polishing rate was observed to increase with the oxidizer concentration at pH = 4.0 following the same trend pointed for non-peroxide-based slurries by Vacassy and coworkers [31]. The increase in the removal rate was believed to be due to the increasing iodate concentration formed by copper oxide(s)/$Cu(IO_3)_2$ duplex film.

The presence of KIO_3 in W CMP slurry along with H_2O_2 was found to produce a nonstoichiometric duplex of WO_2/WO_3 leading to low removal rates due to the increasing hardness of the oxide film with the reduction in the oxygen content, and hence the presence of WO_2 in the surface film leads to lower removal rates in case of KIO_3 [29].

The use of iodate and periodic acid as oxidizers for noble metal CMP has also been attempted. Similar to W CMP, a surface oxidation or modification is required for the subsequent removal by mechanical force. For example, the potential use of ruthenium as bottom electrode capacitor for next-generation DRAM devices [40] has been explored. Owing to the fact that a dry-etch process can lead to the formation of toxic RuO_4 [41], the possibility of using CMP to implement Ru has gained interest recently. The studies in this area have indicated that the formation of stable passive layers such as RuO_2 [41,42] and Ru_2O_5 [42] are important steps in the Ru CMP.

7.3 CHELATING AGENTS

During metal CMP, the material removal is a combination of mechanical abrasion and chemical dissolution. The abraded material could remain at the interface, causing issues such as redeposition and increased surface damage. A solution to this problem is to allow the material to dissolve and hence be moved into the exit stream. One mechanism to facilitate the dissolution process is via complexing. For copper CMP, numerous organic and inorganic ligands in aqueous solutions can be used. The formation constants of some complexes are usually known in the literature and can be affected by solution conditions such as pH and temperature. The presence of a complexing agent can also have significant impact on the chemical and physical behaviors of other chemical additives. For example, the passivating film formed on metal surfaces may be altered. The surface adsorption property of various species onto abrasive particles may be changed. This section will survey some of the major complexing agents used in Cu CMP applications.

Metal ions rarely exist in aqueous media alone. They are often strongly associated with a fixed number of molecules or ions called ligands. Water molecules are often the ligands by default, especially in the absence of other stronger ligands in the solution. The ensemble of central metal ion and its ligands forms a complex. The formation of metal complex lowers the free energy of the metal ions and stabilizes the system. As a result, metal ions may have greater solubility at a given pH or temperature. Equation 7.8 illustrates the formation of a complex with a central cation M^{n+}, where $n = 2$ and there are six ammonia ligands. As the ligands are neutral, the overall charge of the complex is the same as the central metal cation. Equation 7.9 shows the chelate effect of ethylenediamine with the generic metal ion of $+2$ charge. For some cases, the ligand may be charged species; the overall charge of the complex may be different from that of the central metal ion.

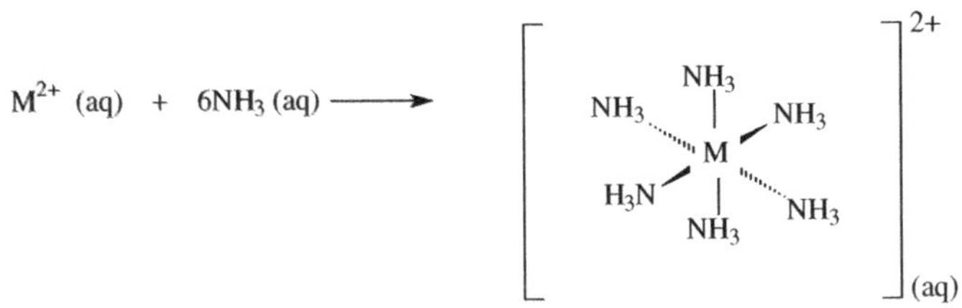

(7.8)

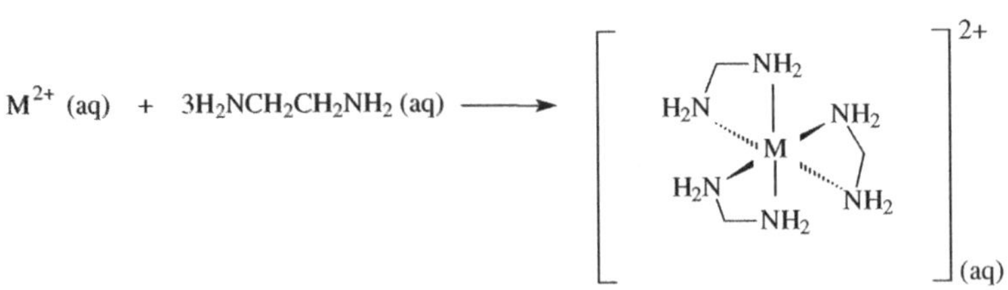

(7.9)

7.3.1 Ammonia

Ammonia (NH_3) is a colorless, pungent gas that is highly soluble in water [43]. A saturated aqueous solution contains 45% ammonia by weight at 0°C (32°F) and 30% at room temperature. Once in aqueous solution, ammonia forms ammonium hydroxide, NH_4OH, which is basic in nature and can dissociate into ammonium ($NH4^+$) and hydroxide (OH^-) ions.

Copper CMP slurries containing ammonium hydroxide have produced highly planar surfaces. The presence of ammonia can weaken the passivating film formed by CuO through a complexation and dissolution process (Eqs. 7.10 and 7.11). The presence of ammonia can also accelerate the oxidation of metallic copper (Eqs. 7.12 and 7.13) [44,45].

$$\mathrm{Cu\cdots O + NH_3} \longrightarrow \left[\mathrm{Cu^{2+}}\begin{matrix}\mathrm{NH_3}\\ \mathrm{O^{-2}}\end{matrix}\right] + \mathrm{H_2O} \longrightarrow \mathrm{Cu(NH_3)^{2+}} + \mathrm{2OH^-} \qquad (7.10)$$

$$\mathrm{Cu(NH_3)2^+ + 3NH_3 \rightarrow Cu(NH_3)_4^{2+}} \qquad (7.11)$$

$$\mathrm{Cu + Cu(NH_3)_4^{2+} \rightarrow 2Cu(NH_3)_2^{+}} \qquad (7.12)$$

$$\mathrm{2Cu(NH_3)_2^{+} + 1/2O_2 + 4NH_3 + H_2O \rightarrow 2Cu(NH_3)_4^{2+} + 2OH^-} \qquad (7.13)$$

The reaction scheme mentioned above has been investigated by Duquette and coworkers [46] using the RDE system. Ammonia can etch copper in the presence of oxidizers by dissolving the oxide film. The dissolution characteristics of the metal upon addition of ammonia into the slurry was related to the mass transport of the $Cu(NH_3)^{2+}$ intermediate away from the metal surface. Three different ammonium-containing compounds, in addition to ammonium hydroxide, have also been studied for their effectiveness. It is clear that the counter ions (hydroxide, nitrate, and chloride) apparently have an impact on the effectiveness of complexation and hence on the removal rate. It was found that ammonium hydroxide resulted in the highest removal rate followed by ammonium nitrate and chloride. A measurement of electrode potentials also followed a similar trend as the removal rate. For example, the difference between the potentials at static and dynamic conditions in the presence of ammonium chloride was much smaller than that for ammonium hydroxide. More specifically, both potentials are low, which strongly indicates that the chloride has a negative effect on the ability of ammonia to form copper complex and dissolve copper oxide on the surface. The presence of chloride in the slurry results in corrosion of the copper surface due to the pitting effect of the anion. Li and coworkers [12] also observed an increased static etch rate with the presence of these anions in the slurry.

7.3.2 Amino Acids

The complexes of amino acids with copper are formed according to the general reaction shown in Equation 7.14 [47]. The formation constants for the copper–amino acid complexes are generally greater than that for copper ammonia complex [48]. Therefore, it is anticipated that amino acids should be more effective than ammonia in assisting the dissolution of oxidized copper during CMP.

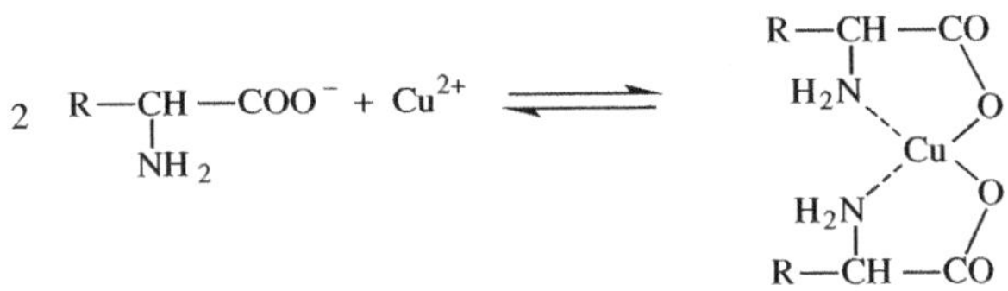

(7.14)

One commonly used Cu CMP model slurry can be traced back to the formulation reported by Hirabayashi and coworkers [10] in which hydrogen peroxide, glycine, and abrasive particles were used. Although the function of hydrogen peroxide as an oxidizer is well known, the role of glycine in Cu CMP was somewhat less clear. In a study reported by Li and coworkers [17], the main role of glycine was defined as a chelating agent to form a complex with the Cu^{2+} ions generated during the polishing process. The complex can catalyze the decomposition of hydrogen peroxide leading to the formation of hydroxyl radicals (*OH). A hydroxyl radical is a much stronger oxidizing agent than hydrogen peroxide itself and thus causes a significant increase in the static etch and material removal rates. The static etch (dissolution) rate of copper was found to be closely correlated with the steady-state *OH concentration. Similarly, the formation kinetics of *OH has also been shown to have a direct correlation with the material removal rate of Cu during the polishing process. Upon the addition of excess Cu^{2+} in the form of $Cu(NO_3)_2$, the material removal rate was further increased as the concentration of *OH was elevated. The relative importance of other factors on the formation and importance of OH* radicals was also studied by Gorantla and coworkers [49]. More specifically, amino acids such as glycine, serine, and cystein were used to deconvolute the specific impact of pH on hydroxyl radical formation since the addition of copper ions causes a decrease in the slurry pH. It was found that the dissolution and polish rates with and without copper ions at $pH = 4$ are essentially the same, suggesting that the amino acid–copper complex may have minimal impact on the formation of hydroxyl radicals or the hydroxyl radicals have negligible impact on the dissolution and removal rate. Considering the fact that, at $pH = 4$, the majority of the amino acid is pronated and unable to form copper–amino acid complex, it is possible that the amount of hydroxyl radicals formed under this condition is essentially unaffected by the presence of extra copper ions. The study does show that, at $pH = 8$, the dissolution and removal rates increase linearly with the increase in copper ions added to a glycine-based copper CMP slurry. At such a high pH, the formation of amino acid–copper complex shall be efficient and so will be its impact on the formation of hydroxyl radicals.

The influence of pH on the roles of amino and carboxylic groups has been studied by Babu and coworkers [50,51]. It was demonstrated that, in the basic regime, the removal rate decreases as the pH increases for a citric acid based slurry due to a competition between hydroxide and citrate for complexation with copper ions. In the acidic regime, amino compounds such as ethylenediamine gives lower removal rate as the pH decreases due to the protonation of the amino groups. For a glycine-based slurry, the removal rate is high throughout the entire pH between 2 and 10. This is a result of a large formation constant for the glycine–copper complex.

The structural influence on the role of complexation with copper ion has been systematically studied by Patri et al. using a set of amino acids that vary the relative distance between the amino and carboxylic functional groups. More specifically, α-aniline, α-amino butyric acid, γ-amino butyric acid, and glycine were employed in the study. It was found that, at low pH, the amino acid with longer carbon chain gives higher removal rate. At high pH, the complexation ability of amino acid decreases with the increase in the distance between the functional groups, which leads to a decrease in removal rate [52].

7.3.3 Organic Acids

The organic acid is a large class of organic compounds. As a matter of fact, amino acids are special cases of organic acids. This section will focus on those organic acids that do not have any adjacent amino moieties. Some of these organic acids can form stable complexes with copper in solution [47]. Similar to amino acids, organic acids that form complex with copper ions typically use their carboxylate group and a nearby electron-donating atom such as nitrogen or oxygen. For example, the hydroxyl and carboxylate groups in 3,4-dihydroxybenzoic acid can serve as an illustration for this type of complexation (Eq. 7.15)

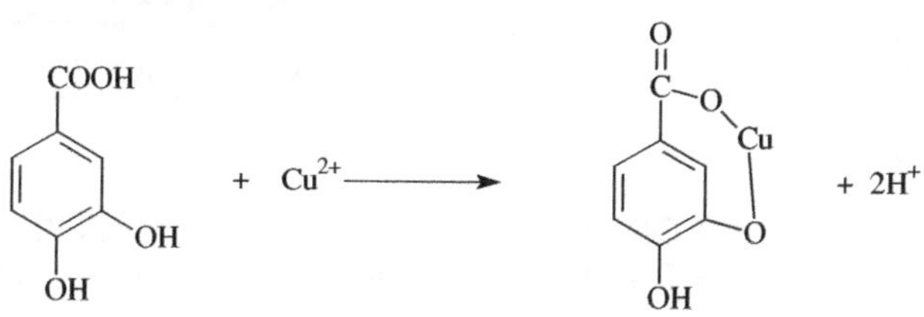

(7.15)

This compound does not complex with copper at a low pH (around 2–3) due to protonation of the carboxylate group. Similar to citric acid, this compound cannot compete effectively with hydroxide at high pH either. Thus, these compounds can only be used at intermediate pH values, around 4–7, in order to achieve high metal ligand binding efficiency. In a similar study, Oehrlein and coworkers [53] showed the effectiveness of phthalate compounds for the chelation of copper. In this study, it was proposed that the oxidation of Cu surface by the H_2O_2 leads to the formation of $Cu(OH)_2$. There are two possible pathways for the hydroxide. It can dehydrate to form the passive CuO film or dissolve to give free copper(II) ions. The outcome is highly influenced by the pH. At low pH of 3–5, direct dissolution is favored. The presence of phthalate does not have any significant impact on the dissolution rate at pH below 3. At pH values between 3 and 5, the dissolution of CuO film will be affected by the presence of phthalate, hence increasing the removal rate.

The use of oxalic acid as a complexing agent has been investigated using polishing and potentiodynamic studies [54]. Maximum removal rates were observed at pH 3. This is consistent with the fact that oxalic acid (HX) had the maximum concentration of X^- at pH = 3.00, which causes complexation with the positive copper surface at the acidic pH condition.

7.3.4 Thermodynamic Consideration and Quantitative Description

The impact of complexing agents on a metal system in the presence of an oxidizer under various pH conditions can be described in a quantitative format such as a Pourbaix diagram. Fig. 7.9 shows a typical shift in Cu–H_2O Pourbaix diagram in terms of the stability and the composition of the species due to the

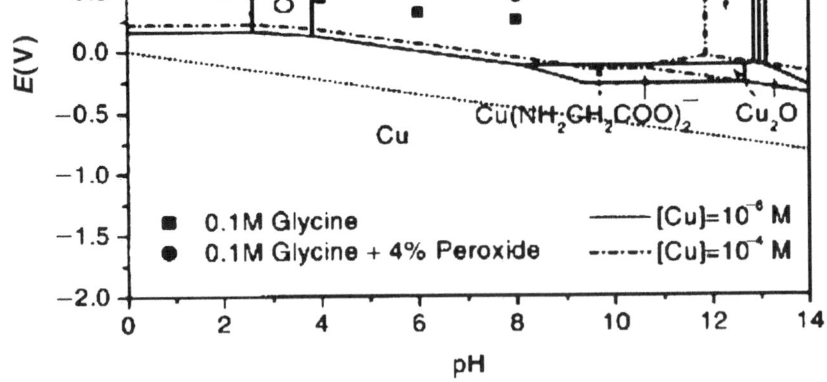

FIGURE 7.9 Potential–pH diagrams of Cu–glycine–H_2O system at two different activities of Cu at the glycine concentration of 0.1 M (from Ref. 55).

presence of a complexing agent [8,55]. The presence of an effective complexing agent significantly extended the soluble regions for copper species to a much higher pH. Readers are highly recommended to refer to the seminal work by Raghavan et al. [55] on the impact of various complexing agents commonly employed in Cu CMP slurries.

7.4 SURFACTANTS [56,57]

7.4.1 Structures and Physical Properties of Surfactants

Surfactants are molecules that have at least one lyophobic moiety, that is, solvent hating part, and a lyophilic moiety, that is, solvent loving part. In case of water as a solvent, lyophobicity means hydrophobicity and lyophilicity translates to hydrophilicity. The molecular structure of a representative surfactant, Triton X-100, is shown in Fig. 7.10.

When exposed to a polar solvent, such as water, the ethylene oxide portion of Triton X-100 becomes lyophilic, whereas the hydrocarbon section becomes lyophobic. The roles of these two moieties will be switched in case of a nonpolar solvent. Thus, the surface active behavior for a given surfactant molecule is governed by the solvent and the conditions of the system. This section will discuss the use of water as a solvent because most CMP slurries are aqueous based.

A characteristic property of surfactant molecules is their tendency to aggregate at interfaces. Examples are adsorptions onto solids and monolayer formation at an air–water interface. Surfactants sometimes create their own interface by forming very small aggregates like micelles or vesicles to remove a portion of their structure from direct contact with a solvent. In case of a micelle formed with a surfactant such as Triton X-100, the hydrocarbon chains are in closer contact in the center and form a hydrophobic microenvironment. The ethylene oxide moieties are exposed to water with much greater frequency. If a hydrophobic species is added into this micellar system, there will be a tendency for the hydrophobic molecules to be concentrated inside a micelle. At low concentration, the micelle system and the added hydrophobic additives can reach a thermodynamic equilibrium, which is often called microemulsion system. At high concentration, the hydrophobic additives form their own separate phase and the surfactant molecules serve only as a decorative layer

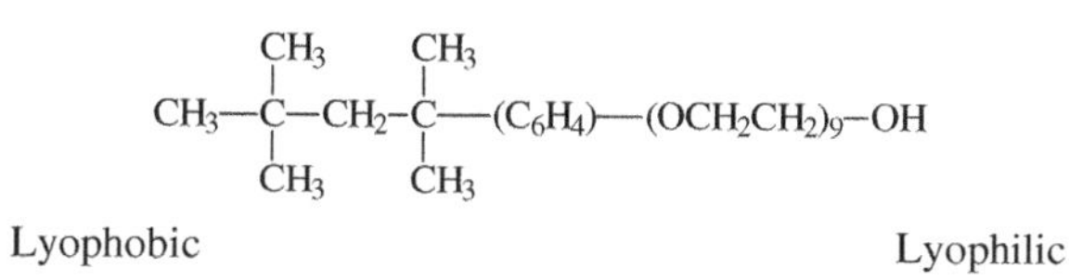

FIGURE 7.10 Structure of Triton X-100 and division between hydrophobic and hydrophilic moieties (from Refs. 56,57).

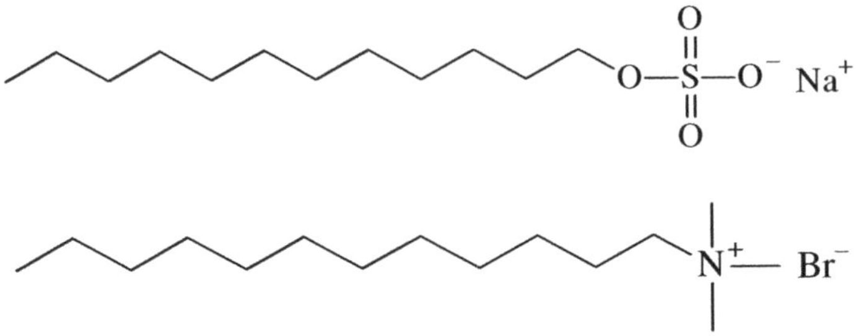

FIGURE 7.11 Structure of a common anionic surfactant, sodium dodecyl sulfate (SDS), and, a cationic surfactant, dodecyl trimethyl ammonium bromide (DTAB) (from Refs. 56,57).

between the hydrophobic phase and aqueous environment. This is a classic case of emulsion.

Surfactants can be classified into four major categories: anionic, cationic, nonionic, and zwitterionic. All ionic surfactants have a tendency to dissociate in an aqueous environment to yield a surface active portion and a corresponding counter ion. In case of anionic surfactants, the surface active portion of the molecule bears a negative charge in a solution. The surface active portion of a cationic surfactant bears a positive charge. The overall charge of a zwitterionic surfactant before and after the dissociation is zero. Some zwitterionic surfactants do not dissociate. Similar to zwitterionic surfactants, the overall charge of a nonionic surfactant remains zero in a solution. Unlike zwitterionic surfactants, there are no charged centers in a nonionic surfactant and the molecules do not dissociate. As examples, the structures of two commonly used anionic and cationic surfactants are shown in Fig. 7.11. After dissociation in an aqueous environment, sodium and bromide are the corresponding counter ions for these two surfactants. The positively or negatively charged "(ammonium or sulfate) head groups" will have a tendency to adsorb onto a negatively or positively charged surface.

A representative zwitterionic surfactant that contains both positive and negative charges in the surface active portion is shown in Fig. 7.12.

Fig. 7.13 is a common example of a nonionic surfactant. Nonionic surfactants tend to adsorb onto surfaces with either the hydrophilic or the hydrophobic locations depending on the nature of the surface.

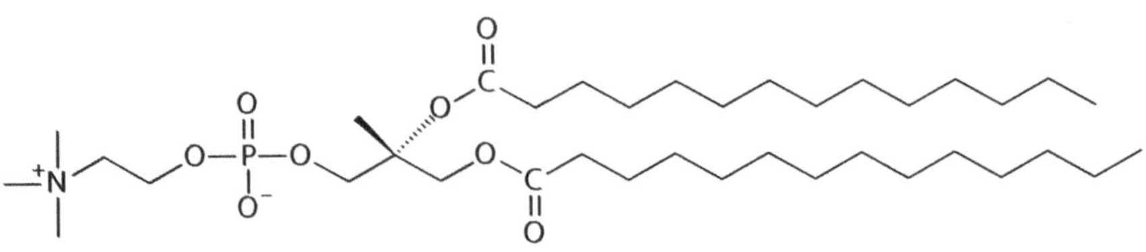

FIGURE 7.12 Structure of a common Zwitterionic surfactant (from Refs. 56,57).

FIGURE 7.13 General structure of a common nonionic surfactant (from Refs. 56,57).

A hydrophobic tail may include such structures as straight-or branched-chain alkyl groups, long-chain alkylbenzenes, alkylnaphthalenes, perfluorinated alkyl groups, polysiloxane groups, and others.

7.4.2 Dispersion of Particles

Dispersion stability is critically important for all CMP slurries. For example, particle–particle aggregation may significantly increase the mean particle size of a slurry and result in surface defects such as scratches and delamination. Severe aggregation can also accelerate the settling of slurry, which can drastically alter the physical and chemical characteristics of a slurry. Therefore, a proper dispersion of metal oxide or abrasive particles in an aqueous environment is fundamentally important for CMP applications [58]. One of the most effective and commonly used methods for the stabilization of a colloidal dispersion involves the use of surfactant.

Colloidal stabilization is a well-studied subject. For most colloidal systems, the so-called DLVO theory [59] works well. Fig. 7.14 shows a typical DLVO potential diagram that describes the primary and secondary energy traps for two particles as well as the energy barrier they must overcome in order to aggregate. Fig. 7.15 shows that large particles have to overcome greater energy barrier before aggregation can occur.

Among many factors that influence the energy barrier between particles in an aqueous environment, the surface charge is a dominating one. Solution properties such as electrolyte strength and pH affect the overall particle charge and, hence, the slurry stability. For example, at low-or high-pH values (significantly away from the isoelectric point for the particles), the surface will carry strongly positive or negative charge. The strong electrostatic repulsion helps the stabilization of these particles. At a pH near the point of zero charge or isoelectric point, the particles have greater tendency to aggregate.

The stability of metal oxide slurries can be enhanced by the addition of surfactants [60]. The formation of a surfactant layer on the particles can modify the particle surface in two ways. It can alter the charge characteristic of the particle, which can stabilize or destabilize the particle dispersion. It can also provide steric barrier between particles, which reduces the flocculation of the particles, thereby increasing the stability of the dispersion [60].

Denoyel and coworkers [61] used a thermodynamic adsorption method to study the adsorption of nonionic and anionic surfactants onto silica, kaolin, and alumina particles. The surfactants formed various structures depending upon their concentrations in the solution. It was found that 2D hemimicelles with low coverage were formed at very low surfactant concentrations. With

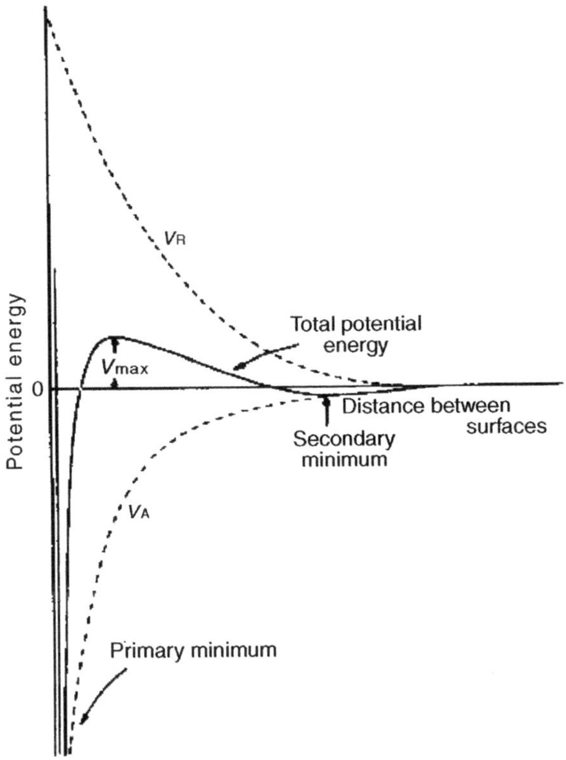

FIGURE 7.14 Schematic of DLVO potential: V_A is the attractive van der Waals potential, V_R is the repulsive electrostatic potential (from Ref. 59).

further increase in the surfactant concentration, micelle-like aggregates were observed. At even higher surfactant concentration, bilayers were formed.

7.4.3 Surface Modification of Wafer Surface

The added surfactant molecules intended for CMP slurry stabilization can adsorb not only onto the abrasive particle but also onto the surface of the wafer to be polished. Depending on the extent of such adsorption, the added surfactant may influence the CMP process in several ways such as change in friction behavior of the slurry, modification of removal rate and selectivity, alteration of defectivity level, and shift in post-CMP profile. In this section the impact of surfactant adsorption on the removal rate, selectivity, and post-CMP cleaning characteristics will be discussed.

Corrosion inhibitors hold an important position in Cu CMP slurry design as they provide the necessary passivation for the low-lying regions. The most commonly used inhibitor in copper CMP slurries is BTA. The formation of the

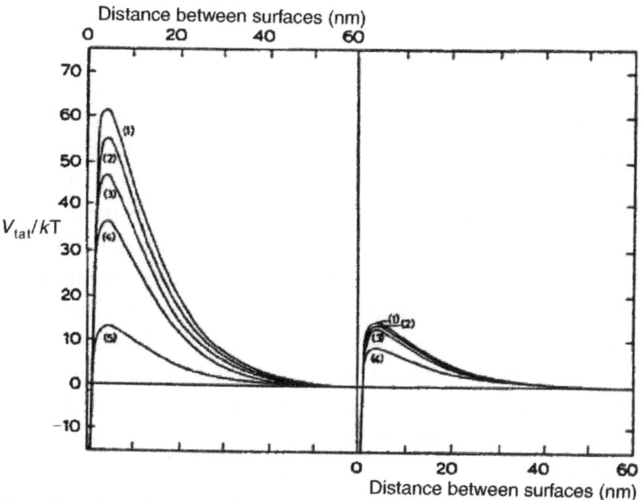

Theoretical curves of total potential energy against the distance of separation of two spherical particles of radii a_1 and a_2 and with equal surface potential (35.86 mV); $A = 5 \times 10^{-20}$ J, $1/\kappa = 10^{-6}\ \mathrm{cm}^{-1}$, $\varepsilon = 78.5$.
(a): $a_1 = 125$ nm, $a_2 = 125$ nm (1), 100 nm (2), 75 nm (3), 50 nm (4), 12.5 nm (5).
(b): $a_1 = 12.5$ nm, $a_2 = 125$ nm (1), 100 nm (2), 75 nm (3), 12.5 nm (4). Source from Parfitt [24]

FIGURE 7.15 A set of representative DLVO potentials for dispersions with various particle sizes. For small particles, $E > 10kT$ is needed to overcome the repulsive barrier (from Ref. 59).

hydrophobic Cu(I)–BTA surface complex functions as a physical boundary between the slurry and the oxidized surface, thereby preventing any unwanted dissolution. However, the use of BTA sometimes introduces scratch-related surface defects and leaves undesirable organic residues [62,63]. It is thus desirable to find alternative passivating agents. Surfactant molecules could also serve as potential corrosion inhibitors. Unlike BTA, the interaction between the surfactant and the oxidized copper surface is electrostatic. Under most Cu CMP slurry operating conditions, the modified copper surface develops a positive charge. Thus, the attraction between a negatively charged anionic surfactant leads to lower static etch rate in comparison to nonionic and cationic surfactants. Hong and coworkers [64] studied the use of ammonium dodecyl Sulfate (ADS) instead of BTA as corrosion inhibitor in electrochemical–mechanical planarization (ECMP). ADS yielded lower corrosion currents in comparison to BTA at similar concentrations through electrochemical measurements. Because of the difference in the passivation mechanisms, the use of surfactants could also eliminate defects related to BTA. The use of mixed surfactants, that is, a combination of an anionic surfactant with surfactants of opposite charges was studied by Li and coworkers [65]. In addition to anionic surfactants, cationic counterparts were introduced to increase the packing density in between the repelling anionic heads. The use of mixed surfactants resulted in even lower dissolution rates in comparison to values obtained from a single surfactant system.

In STI CMP it is important to modulate the selectivity between oxide and nitride in order to effectively stop at the nitride layer and control the oxide dishing. One approach to control the selectivity is to specifically modify the surface of the abrasive or the substrate [66]. Under most STI CMP operating conditions, both silicon dioxide abrasive and substrate filmscarry negative charge, while silicon nitride surface is positively charged. The anionic surfactant added to such systems will specifically adsorb onto the positively charged silicon nitride, thereby reducing the removal rate through the formation of a lubrication layer. Moudgil and coworkers [67] studied the adsorption of sodium dodecyl sulphate (SDS) onto silica and nitride surfaces as a function of pH via zeta-potential measurements. It was concluded that the adsorption of the anionic surfactant onto the nitride surface suppressed the nitride removal rates.

Another application of selective modification of silicon dioxide surface is barrier metal (adhesion layer) CMP. The construction of a copper damascene structure involves the elimination of copper overburden followed by barrier CMP. Barrier slurry encounters a number of materials such as the residue overburden metal, adhesion layer, and the underlying dielectric. In the case of employing of a low-*k* dielectric material as the insulator, an additional layer such as cap is also present. Depending on the integration scheme, the selectivity of barrier slurry is desired to be tunable. During the polishing of barrier metal, in order to minimize the level of erosion and avoid excess loss of dielectric materials, a lower or controlled removal on dielectric material is needed. The use of surfactants could also be extended to barrier CMP quring to their preferential adsorption characteristics based on the surface functionalities and charge. Miller and coworkers [68] investigated anionic, cationic, and nonionic surfactants in order to modulate the selectivity and hence control the erosion of copper damascene structures with silica-based slurries at a pH of 4.00. Under the given conditions, SiO_2 and SiOF dielectric stay predominantly negative. The highest degree of erosion control was achieved with the cationic surfactants owing to the electrostatic attraction between the positively charged molecules and the negatively charged dielectric. The hydrogen bonding present between the nonionic surfactants and the dielectric also helped in achieving control on erosion but to a lesser degree. Anionic surfactants gave the least degree of control over erosion due to the lack of adsorption onto the dielectric surface. Schroeder and coworkers also investigated the use of amphiphilic nonionic surfactants to control the selectivity, erosion, and dishing [69].

The use of surfactants to selectively modify wafer and abrasive particles has also been extended to post-CMP applications. During post-CMP cleaning, surfactants are sometimes included in the cleaning solutions to intentionally wet the wafer film and encapsulate other residues and particles. Itano and coworkers [70] studied the zeta potential of Si, SiO_2, and Si_3N_4 as a function of pH in the presence of both anionic and cationic surfactants. It was shown that particle deposition onto the surfaces was suppressed by modifying the surface charges with the addition of anionic or cationic surfactants. A more detailed discussion on this topic is given in Chapter 11. It is worthwhile to point out that the surfactant molecules added into the polishing slurries and their subsequent adsorption onto the wafer surface not only modify the polishing characteristics but also may have a significant impact on the post-CMP cleaning. Gutmann and coworkers [71] demonstrated the use of anionic surfactants (1% DOWFAX) in the Cu CMP slurries consisting of alumina abrasives and potassium dichromate as oxidizer. These slurries resulted in smooth, low-particulate defect-free copper surfaces without aggressive post-CMP cleaning. It is also important to point out that with the implementation of low-*k* dielectric materials, the hydrophobicity of the wafer surface is significantly different from that of silicon dioxide.

7.5 ABRASIVE PARTICLES

In addition to the chemical component described in the previous sections the mechanical forces are provided by the downpressure, pad, and abrasives. The primary function of abrasive particles is to enhance the mechanical strength of the pad and transmit the downforce to the wafer surface, resulting in an increased removal rate. In addition to the mechanical function, the abrasives serve as adsorption sites to the reaction by-products and polishing debris and assist in their transportation and elimination from the vicinity of the wafer. The debris, if not removed, may lead to unwanted defects such as scratch or corrosion. Therefore, the role of abrasive particles during polishing is multifaceted and dynamic. As a result, both bulk and surface properties of these particles are important in controlling the polishing characteristics such as removal rate and post-CMP surface quality.

A wide variety of materials have been implemented as abrasive particles in CMP processes. They include alumina, silica, ceria, zirconia, titania, and diamond. The effectiveness and suitability of these particles in CMP with particular applications are greatly influenced by their bulk properties (density, hardness, particle size, crystallinity etc.) and the surface properties (surface area, isoelectric electric point (IEP), OH content, etc.). This section will focus on the evaluation of alumina, silica, diamond, and ceria as the major abrasives used for the CMP of metals.

7.5.1 Hardness

Hardness is one of the most influential bulk properties in metal film CMP. Hardness is defined as the resistance of the material to a localized plastic deformation in the form of a small scratch or indentation. The difference in hardness values between the abrasives and the modified substrate film may determine the removal rate during a CMP process. It is important to emphasize here that, for the substrate film, what matters is the hardness of the modified layer that is in direct contact with the abrasive particles. The hardness of such a modified layer may be significantly different from its bulk film hardness. The modification can be accomplished by the oxidizer added into the slurry or the native oxygen in the atmosphere.

One of the most popular techniques used for determining the hardness of a material is the Mohs' scale that consists of a qualitative but an arbitrary hardness index scheme ranging from extremely soft materials (value of 1 Moh) to very hard materials such as diamond (10 Moh). Other techniques that are often employed for measuring hardness of substances are developed by Rockwell [72], Brinell [72], Knoop, and Vickers [73]. Over the years, more quantitative methods such as nanoindentation [74] have been developed. This technique applies a small and a controllable load on to the substrate with a probe. The depth of penetration along with a known geometry of the probe provides an indirect way to measure the area of contact at full penetration, which is then used to determine the hardness. The hardness is determined by the ratio of the total force to the contact area. Table 7.2 lists the bulk hardness of different materials, metal films, and abrasive particles, in both Moh and microhardness scales [75].

It is important to note that the hardness of the modified substrate may not always be less than the native metal surface. One such example is copper oxide, which has a higher hardness value in comparison to native copper. The resultant oxide might lead to unwanted surface defects such as scratch and redeposition. One way to prevent such defects is to direct these polishing debris to the abrasive particles. The function of abrasive particles in this regard, sometimes, is more important than their duty as a force of abrasion.

Li and coworkers [76] investigated the relative importance of abrasive hardness and functional group on the particles for copper and Ta CMP using alumina, silica and diamond particles. It was found that the presence of surface hydroxyl groups is more important than the bulk hardness in determining the Ta removal rate. For copper CMP, given the same chemistry, the abrasive hardness played a more significant role in determining the material removal rate. It is important to point out that, for metal CMP, the material removal rate is not the sole requirement. The final finish or the surface quality of a polished substrate is often just as important as removal rate. Therefore, when considering the benefit

TABLE 7.2 Bulk Hardness Values of Various Substrates and Abrasives [75].

Material	Moh Hardness	Microhardness (kg/mm^2)
Copper	2.5–3.0	80
Tantalum	6.5	230
Tungsten	7.5–8.0	350
Hydrated silica	—	400–500
Silica	6–7	1,200
Copper oxide	3.5–4.0	—
Alumina	9.0	2,000
Diamond	10	10,000

brought by the hardness of abrasive particles, one must not trade surface quality simply for material removal rate. For example, adding a small amount of diamond particles into CMP slurry often increases the material rate significantly because of the extreme hardness of diamond particles. However, the use of diamond abrasive particles can have negative effect on the surface quality where the mechanism of material removal can alter from abrasion to indentation. The transition from abrasion to indentation depends on the relative hardness of the abrasive and the substrate. It is equally important to realize that not all scratches found on copper surface in the presence of diamond particles are caused by the indentation of diamond particles directly. This subject will be discussed more in detail in Section 7.5.6 chapter.

7.5.2 Bulk Particle Density

Density of a substance may be defined as the weight of a substance per unit volume. In principle, the bulk density of agglomerated particles in the slurry can offer an indirect measurement of the abrasive particle hardness. The bulk density of the particle can be calculated by using Equation 7.16, excluding the open pores and voids from the volume calculation, where ρ stands for the specific gravity of the slurry measured using a pycnometer [77].

$$\text{Bulk density} = \frac{\text{Solids(wt \%)}}{\frac{100}{\rho} - (100 - \rho)} \tag{7.16}$$

Babu and coworkers [78] established that the abrasive particle density indeed offered a means for characterizing the hardness of submicron abrasive particles based on the material removal rates. The polishing rates of both Cu and Ta were measured for slurries of submicron-sized alumina particles with varying bulk densities ranging from 3.2 to 3.8 g/cm^3, dispersed in DI water. It was found that the polishing rate increased significantly when the dry powder bulk density exceeded a threshold value.

The bulk density of abrasives also has a direct impact on the slurry dispersion stability. The weight of the abrasives might overcome the repulsive forces among particles required for the stability of the slurry system. For example, the particle settling issue is much more severe for alumina-based slurry than that for silica because of the difference in particle density [79].

7.5.3 Particle Crystallinity and Shapes

In addition to bulk density, particle crystallinity is another physical property that can be related to its hardness and potential effectiveness in providing the mechanical force in CMP. For example, alpha alumina abrasive has a higher hardness, thereby producing higher removal rate in comparison to its gamma crystalline counterpart [80]. As expected, the number of defects is also higher for alpha alumina due to the inherent hardness [81].

Another important factor just as important as the hardness of an abrasive particle that could have a significant impact on the effectiveness of creating indentation on the substrate film is its shape. In principle, irregular and sharp edges of the particle could lead to very high local pressures, thereby causing scratches on the surface. Therefore, similar to hardness, particle shape is another property that has both positive and negative effects on CMP performance. It is desirable to have some degree of sharpness to ensure the material removal rate. However, a balance needs to be achieved in order to avoid additional defects. Li and coworkers investigated the use of plate-like boron nitride particles [82] for metal CMP. The objective of the study was to identify an optimal particle shape and morphology that provide high surface contact area between the abrasive particles and the substrate film and low scratching frequency caused by sharp edges.

Silica can exist in many crystalline forms such as quartz, cristobalite, and tridymite. Fumed silica on the contrary tends to be amorphous, which could be attributed to the fabrication process of the abrasive. The amorphous nature is probably caused by the rapid cooling employed in the process [83]. Colloidal silica, which is usually synthesized via wet chemical methods, is highly amorphous as well. In addition, colloidal silica particles are usually spherical and highly hydrated in nature, which makes them far less likely to cause scratches on metal substrate surface.

Two major classifications for diamond crystals are mono- or single crystalline and polycrystalline. The monocrystalline diamond particles tend to have more uniform surfaces and sharp edges. The abrasiveness of the monocrystalline diamond is mainly governed by its particle size. In the case of polycrystalline diamond, it is sometimes determined by the packing arrangement and the interaction of these single crystals that are related to friability [84]. Friability is defined as the readiness of a substance to crumble and form fine particles or fibers under the application of external pressures.

7.5.4 Particle Size and Oversized Particle Count

In addition to the hardness and shape, particle size is another important property for an abrasive. Three relevant parameters that characterize a CMP slurry are mean particle size, particle-size distribution, and oversized particle count. The particle size modulates the material removal rate, WIWNU, and surface quality for both metal and nonmetal CMP. There are two CMP mechanisms that describe the relationship between particle size and removal rate. For a slurry containing extremely large particles, the indentation mechanism may dominate. More specifically, the material removal rate depends on the indentation volume. In other words, the volume of material removal per particle is directly proportional to the particle size. The net effect is that an increase in particle size leads to an increase in material removal rate. A slurry containing smaller particles will follow the contact-area mechanism where the material removal is determined by the contact area of the particles with the substrate. The contact area increases with the increase in the particle concentration and reduction in the particle size. At a fixed concentration the number of particles increases with the reduction in the particle size.

Mahajan and coworkers [85] studied the impact of abrasive size at different particle concentrations on the oxide removal rate. It was found that the removal rate was a direct function of the particle concentration for monosize abrasives of size 0.2 μm, thereby supporting the contact-area mechanism. The mechanism shifted to indentation for a monodispersed system at 1.5 μm, resulting in reduced removal rates. At 0.5 μm, the removal rate initially increased and then decreased with particle concentration, suggesting a shift in the removal rate mechanism. Particle-size distribution [86] has an equally important effect as the particle size. A larger number of oversized particles in the distribution also cause a shift in the mechanism of material removal. Mahajan and coworkers conducted studies to evaluate the impact of size distribution on oxide removal rates. Baseline commercial slurry was spiked with different concentrations of impurities in the range of 0.5–1.5 μm. The size at different concentrations resulted in removal rates lower than that obtained with the original slurry. Slurry spiked with 1.1 % of 1.5μm particles resulted in a removal rate equal to the baseline slurry, suggesting the predominance of indentation mechanism. Slurries spiked with other concentrations and sizes resulted in a decrease in the removal rate explained by the reduction in the contact area of the abrasives with the oxide substrate.

The impact of particle size also follows similar principles in metal CMP. Lu and coworkers [87] confirmed the contact-area mechanism for material removal in copper CMP with silica abrasives. It was observed that the material removal rate increased with the increase in the specific surface area of the abrasives in the slurry. Bielman and coworkers [88] conducted a similar investigation and found that the removal rate of tungsten decreased with the increase in alumina particle size from 0.29 to 2 μm at 4.5% slurry concentration.

The slurry particle size can have a dramatic impact on the surface quality of the polished surface. A large number of so-called oversized particles tend to give high scratch counts on the polished wafers. There is no general agreement on the specific cutoff for the classification of oversized particles. However, it is generally agreed that particles that are larger than 1.0 μm will certainly scratch most surfaces during CMP. It is also possible that particles that are greater than 0.5 μm can cause scratch-related defects. This accidentally coincides with the size limit (0.5 μm) for one of the most popular optical particle counters [89]. Remsen and coworkers [90] studied the impact of oversized particles on fumed silica slurries and their correlation with scratches on silicon dioxide films. It was found that the count had a direct correlation with the number of scratches observed on the substrate. The equivalent diameter of scratch-causing particles was found to be 0.68 μm.

7.5.5 Particle Preparation

Abrasive particle properties can be attributed to the method or process used to generate the particles. The surface properties can be altered or modified by varying the physical conditions (temperature, pressure, time) or chemical composition of the initial reactants. This section will survey the various preparation techniques for silica, alumina, and diamond.

Fumed silica is traditionally produced by vapor phase hydrolysis of silicon tetrachloride in a hydrogen–oxygen flame [83]. Figure 7.16 shows the reaction sequence for this process.

The combustion process creates silicon dioxide primary particles that then condense to form the desired aggregates. The resultant aggregates are ultimately sintered together and have a chain length of 0.2–0.3 μm. These aggregates become mechanically entangled to form larger agglomerates. The particle size that lies in the submicron range depends on the process.

Unlike fumed silica, colloidal silica is produced by a wet method. In addition, colloidal silica can be produced in spherical shape with uniform size distribution. Stober and coworkers [91] developed a system that allows controlled growth of silica spheres. The reaction occurs through the hydrolysis of alkyl silicates and the condensation of silicic acid in alcohol solution with an ammonia catalyst to control the morphology. This method can produce uniform spheres in suspension that range from less than 0.05 to 2 μm in diameter. Tabatabei and coworkers [92] showed that the size and morphology could also be controlled by the reaction kinetics and also the molar ratio of the reagents. The design and fabrication of microfluidic chemical reactors [93] for the synthesis of colloidal silica particles were reported by Khan and coworkers, in which different reactor configurations and flow types were employed to yield varying sizes of the colloidal silica particles.

Alumina is traditionally formed via thermal dehydration of aluminum hydroxides [94,95]. The final size and crystallinity of the alumina abrasives largely depend on the temperature and time of the thermal treatment process. It has been reported that the total reaction conversion occurs at a temperature of 1500 K. Technical grades of calcined alumina are commonly used for smelting, ceramics, and abrasive particles. Other common forms of alumina produced are fused and white tabular alumina. Fused alumina is produced by melting calcined alumina at a high-temperature furnace for extended time periods. White tabular alumina is composed of large well-developed crystals of

$$2H_2 + O_2 \longrightarrow 2H_2O$$

$$SiCl_4 + 2H_2O \longrightarrow SiO_2 + 4HCl$$

$$SiCl_4 + 2H_2 + O_2 \xrightarrow{1800^{\circ}C} SiO_2 + 4HCl$$

FIGURE 7.16 Synthesis of fumed silica via vapor phase hydrolysis of silicon tetrachloride in an oxygen flame (from Ref. 83).

α-alumina, which is generally produced in high-temperature furnaces. By elevating the temperature of dehydrated alumina, the system is allowed to reach a temperature close to fusion. The formation of alumina via furnace heating methods has been shown to be the most successful for both small-scale and industrial-scale applications.

The first artificial diamonds were synthesized by Henri Moissan in 1893 by heating charcoal at high temperatures with iron in a carbon crucible in an electric furnace, in which an electric arc was struck between carbon rods inside blocks of lime. The first commercially successful synthesis of diamond was produced on December 16, 1954, by Hall at General Electric, using an elegant "belt" apparatus [84]. Modern manufacturers of synthetic diamonds use the same method discovered by Hall. A mixture of graphite and a catalyst, typically nickel, is subjected to a pressure of approximately 1,000,000 lb/in^2. and a temperature of 1800 °C for a period of approximately 1 h during which the diamond crystals nucleate at many sites in the mixture. The mixture is then cooled and the pressure is reduced to atmospheric pressure. The diamond crystals are then separated from the remaining graphite and nickel using an acid wash after which they are sorted by shape, size, and impurities. The larger diamonds are used for sawing concrete, granite, and marble. Smaller diamonds are used in grinding wheels and, more recently, are used as abrasive particles for various applications. Polycrystalline compact diamond (PCD) is diamond particles sintered together using high pressure high temperature (HPHT) technology. In addition to HPHT techniques, chemical vapor deposition (CVD) technique has also attracted attention as a synthetic method for diamond and diamond-like films.

Other than single crystalline and polycrystalline diamonds, cluster diamond (CD) [96] is another variety that contains a core and an overgrown region containing a plurality of diamond crystallites extending outward from the core. Some CD, which consists of ultrafine diamond particles with a mean particle size of 5 nm, shows excellent lubricating ability [97].

7.5.6 Surface Properties

The abrasive content is usually defined by weight percentage. At a constant weight percentage of the abrasives, smaller particle size translates to larger total surface area. At a certain point, the surface property of a particle becomes important enough to compete with its bulk properties in influencing the application outcome. For CMP slurries, as the abrasive particles are often smaller than or close to 100 nm, the surface properties are definitely influential on the CMP performance. In some cases, they may exert more influence on the CMP outcome than their bulk properties such as hardness, density, crystallinity, shape, and size. Relevant surface properties include charge and functionalities on the abrasive surface.

The nature of the charges developed on abrasive surface when introduced into solutions depends upon the solution pH, ionic strength, and the surface functionalities of the abrasive. For example, in the case of silica and alumina, the amount of negative or positive charge depends on the relative number of hydroxyl groups on the surface, which in turn depends on the solution properties mentioned above and also on the manufacturing process. Surface charge of abrasive governs the interactions among the abrasives and also between the particles and the substrate. The interactions among the abrasives determine the stability of the dispersion as explained in the earlier section. The interaction between the abrasive and the surface functionality plays a key role in determining the removal rate in some cases. Such an interaction also plays a role in forming or eliminating surface defects. For example, the debris or the by-products formed during the CMP process must find their way out of the vicinity of the wafer. Otherwise, they may lead to unwanted surface defects such as scratches or pits. The adsorption of these by-products onto the abrasives offers a platform for an efficient exit. The effectiveness of such surface adsorption depends heavily on the functional groups and charges on the abrasive surface.

In a CMP slurry, all chemical additives have an opportunity to reach an adsorption–desorption equilibrium with abrasive surface. During polishing, the slurry is in direct contact with the wafer and the polishing by-products. The added surface adsorption onto wafer surface and the interaction with the by-product can disrupt the equilibrium established before CMP. It is worthwhile to point out that, compared to the total surface area of the abrasive particles, the wafer surface area is small. However, the impact of polishing debris and by-products are difficult to ignore. For example, Li and coworkers [98] have found that the effective concentration of BTA can decrease significantly due to the introduction of copper ions during CMP. The positively charged complex formed between BTA and copper ions has a much greater tendency to adsorb onto negatively charged silica surface.

In addition, to aid the transport of polishing debris, the hydroxyl groups on the abrasive particles, especially for silica, have a significant impact on the polishing rate and removal rate selectivity. For example, Li and coworkers [76] performed a mass-balance study to determine the amount of copper adsorbed onto the abrasives with a model CMP slurry. It was found that the amount of copper in the form of oxide adsorbed onto the silica (45%) was approximately five times higher in comparison to alumina (8%). The removal rate of copper, however, was higher with the alumina abrasives, thereby showing a larger impact of the abrasive hardness on the removal rate under the experimental conditions. Similar studies in Ta CMP resulted in the adsorption of 75% of the abraded metal onto the silica surface, further proving the importance of the surface functional groups toward the material removal. The specific interaction of the surface groups with Ta substrate makes it a favorable candidate for Ta CMP.

In general, silica surface has a higher number of hydroxyl groups (4 per nm^2) [99] than alumina, ceria, titania, and diamond. As a matter of fact, the number of functional groups on diamond particles is very small without surface treatment. Therefore, silica has an advantage in accommodating polishing debris or by-product. Diamond particles are the best materials to use if the function of transporting polishing by-products and debris is placed upon the abrasive particles. This is why the addition of silica particles can help the reduction of surface scratches. This is also consistent with the fact that some abrasive-free solutions (without the benefit of abrasive particles) give more surface scratches than their counterparts that contain the same chemistry except a small amount of extra silica particles.

7.6 PARTICLE SURFACE MODIFICATION

The surface properties of the particles can be modified according to the requirement. Surface modification of particles could be divided into three major categories according to the procedures employed. These are covalent modification, surface adsorption, and encapsulation.

Covalent modification involves the substitution of the reactive surface groups present on the abrasives using covalent reactions. Babu and coworkers [76] investigated the impact of surface modification for silica particles on Ta CMP. As discussed earlier, the surface hydroxyl groups are responsible for the interaction between the particles and the surface to be polished. By replacing the free hydroxyl groups with the capped $-OCH_3$ groups, the Ta removal rates decreased significantly.

Another method of surface modification involves surface adsorption. The driving force for surface adsorption may involve hydrogen bonding, electrostatic charge interaction, hydrophobic–hydrophobic interaction, among others. A simple example is the adsorption of surfactants onto silica abrasives in order to increase the stability of the dispersion. The advantage of surface adsorption method is the simplicity. It does not involve lengthy or complicated chemical reactions and work-up procedures compared to the covalent method. The disadvantage is that desorption can readily occur unless there is a sufficient amount of the same materials in the solution to maintain the equilibrium. Therefore, the modification is usually not permanent. In addition, sometimes, it may not be possible or practical to keep substantial amount of the adsorbate in free solution. It is generally true, however, that a polymeric surfactant tends to adsorb onto a surface and yield a much stronger adsorbed layer due to higher equilibrium constant. Unlike monomeric adsorbate, polymer molecules interact with the surface through multiple point of contacts. Another variation of surface adsorption method is to physically attach smaller particles onto a larger core particle. One such example is the coating of organic latex core particles with many smaller silica particles. This yields composite particles consisting of an organic core coated by an inorganic shell due to the electrostatic attraction [100,101] (Fig. 7.17). The organic core allows the assembly to have a lower effective hardness and be compressible under the load. It has been demonstrated that the compressibility and lower hardness lead to a reduction in surface defects. The organic particles can be selected from thermoplastic resins such as

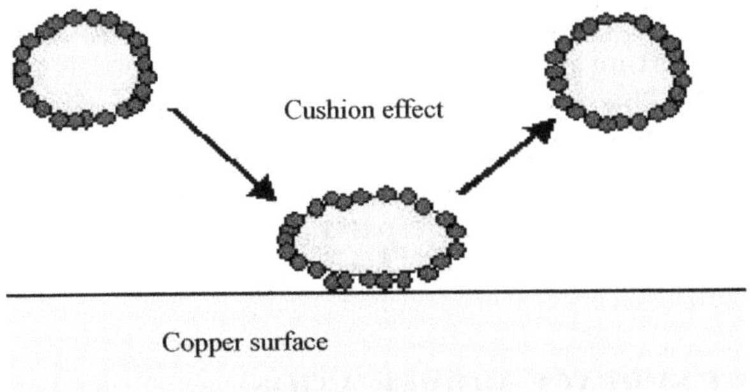

FIGURE 7.17 Polymer particles deforming under the pressure providing a cushioning effect. Polymer particles coated with colloidal silica particles (from Ref. 101).

polystyrene and styrene-based copolymers, phenoxy resins, and polyolefins and thermosetting resins such as phenol and urethane resins [100]. Yet, another example of such surface adsorption assembly is the inorganic mixed-abrasive system. A mixed-abrasive system consists of inorganic composite particles of different sizes coated onto each other with the larger abrasive forming the core. This approach avails the advantages of the individual abrasives while mitigating their disadvantages. For example, alumina abrasives could be coated with silica abrasives. The hardness of the alumina would still contribute toward the removal rate, and the surface coating with softer silica abrasives would prevent the surface defects. Hegde and coworkers [102,103] demonstrated this concept using the electrostatic interaction between the individual abrasives at a given operating pH.

Particle encapsulation is another technique that can be used for surface modification that involves the complete coating of a particle with polymer [104]. The net effective hardness in this case tends to be less than that of the core. Particle encapsulation in CMP, hence, could be used to reduce surface defects of the surface to be polished. It should be noted here that excessive coating of the particle could result in reduced abrasiveness and in problems with dispersion stability.

7.7 SOFT PARTICLES

The preceding discussion has been focused on the hard inorganic particles such as metal oxides. The reason for employing hard abrasives in CMP is due to a long standing assumption that a certain degree of hardness is required to indent and abrade the modified surface during CMP. The progress in the synthesis of fumed silica, alumina, and ceria particles that are substantially smaller than 100 nm [105–107] have prompted the industry to use smaller and smaller abrasive particles to meet the ever-increasing demand in defect reduction. A logical question to ask is how much hardness do these nanoparticles still posses? How important is it for those particles to have the original bulk hardness possessed by their large counterpart? The success with the core-shell assembly illustrated in Fig. 7.17 is certainly encouraging for those who explore the potential of using even softer materials for copper CMP applications. As a matter of fact, the softest materials that have been explored are from the abrasive-free system for copper CMP [108]. Within this category, some versions of the abrasive-free systems (AFS) do not contain any particles at all. One disadvantage of such a system is its inability to clear the copper puddles, especially when the CMP process is used to remove overburden copper at a higher metallization level (>3). This is the reason that some of the improved versions became almost abrasive-free systems (AAFS) or diet abrasive systems (DAS). In a similar approach, an abrasive-free system that contains pure nonabrasive organic particles or supramolecular structures has also been reported [23]. An extension of such a supramolecular system is the use of pure organic abrasive particles. The use of these organic abrasive particles for copper CMP has been investigated by Cheemalapati et al. [109–111]. The key findings from this study will be discussed in details in the following section.

7.8 CASE STUDY: ORGANIC PARTICLES AS ABRASIVES IN Cu CMP

7.8.1 Particle Characterization

In this study Cheemalapati and coworkers demonstrated the usefulness of hydrophilic organic particles as abrasives in formulating copper CMP slurries. The particles are melamine-based resin particles functionalized with surface amino groups. Because of the presence of these amino groups, the native particles are basic in nature (IEP > 8). The particles are usually stabilized using polymeric anionic surfactant during synthesis. As shown in Fig. 7.18, the stabilized particles show negative charge throughout the tested pH range. This is a strong indication that the incorporation of such anionic surfactant is almost permanent or irreversible. A benefit of having high zeta potential across a wide pH range may translate to a wider window for formulation and less sensitivity to pH variation. The resultant slurry was reasonably stable and had comparable mean particle size and particle-size distribution in comparison to other conventional abrasive particles such as silica and alumina. The narrow particle size of the resultant set of particles is shown in Fig. 7.19.

7.8.2 Material Removal Rate and Selectivity

The organic abrasives were incorporated into a copper CMP slurry that contains an oxidizer such as hydrogen peroxide, a complexing agent such as glycine, and a passivating agent such as BTA. After a set of initial optimization on slurry formulation using a bench-top polisher and a 1″ copper disk, promising slurries were selected based on their removal rates (>5000 Å/min) and static etch rates (<200 Å/min). Figures 7.20 and 7.21 show the data

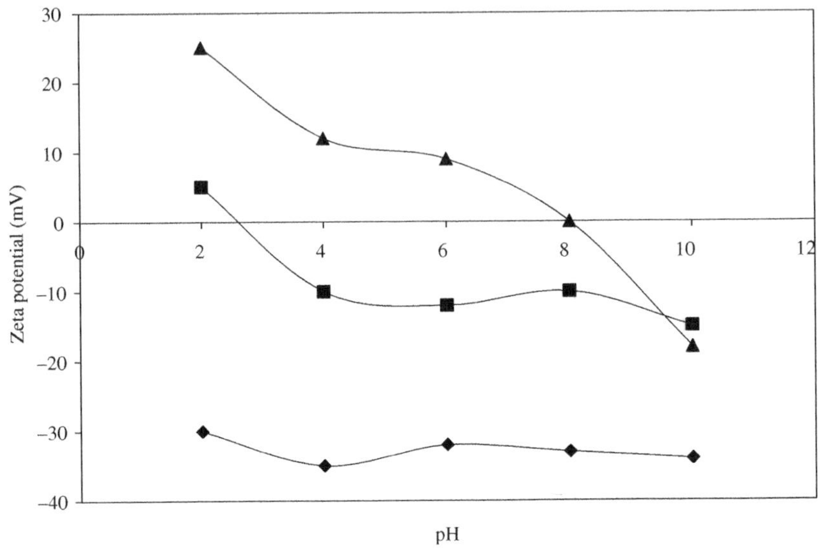

FIGURE 7.18 Zeta-potential data obtained from Brookhaven's Zeta Pals Electrokinetic Analyzer showing negative charge over a wide pH range (from Ref. 110).

obtained on 8-in. copper blanket wafers using a rotary polisher (Strasbaugh nHance) and a representative slurry that contains organic abrasive particles. As a comparison, the data from a silica-based slurry is also included in the figure. The slurry containing organic particles gave a high removal rate at all

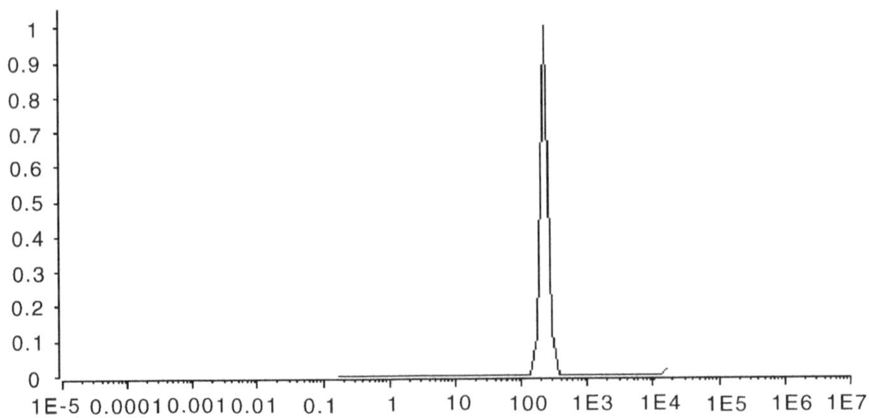

FIGURE 7.19 Particle sizing data obtained on a high-angle dynamic light scattering instrument (ALV HPBS) for slurry sample prepared using pure organic particles ready for CMP. Note: x axis denotes the radius of the particle in nanometers and y axis denotes normalized relative number of a particular particle size (1 denoting the mean particle size) (from Ref. 110).

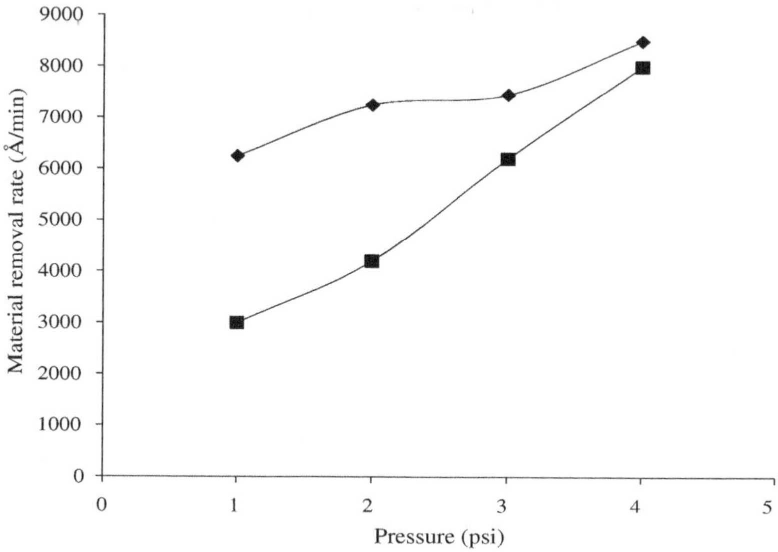

FIGURE 7.20 Removal rate versus downforce on 8″ Cu blanket test wafers polished at 75/65 rpm table/carrier speed and 200 ml/min slurry flow rate on Strasbaugh *n*-Hance polisher. Diamond data points indicate the removal rate values with the organic particles, and square data points denote removal rate values for silica particles, both polished under identical formulation and abrasive concentration (from Ref. 110).

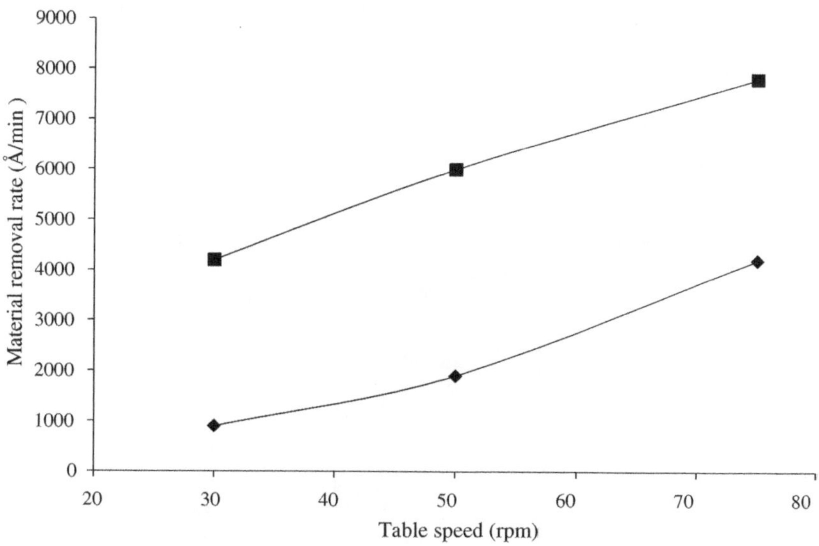

FIGURE 7.21 Removal rate versus table speed on 8″ Cu blanket test wafers polished at 2 psi down force and 200 ml/min slurry flow rate on Stasbaugh *n*-Hance polisher. Square data points indicate the removal rate values with the organic particles, and diamond data points denote removal rate values for silica particles, both polished under identical formulation and abrasive concentration (from Ref. 110).

TABLE 7.3 Removal Rate of 8″ Blanket Test Wafers of Different Substrates at Similar Process Conditions with the Organic Abrasive Slurry [109].

Wafer substrate	MRR (Å/min)
Cu	6920
Ta	5
Oxide	28
Black diamond	9
TEOS	20
TaN	5

downforce settings in comparison to silica-based slurry with similar chemistry. Furthermore, the organic-particle-based slurry gave a removal rate above 4000 Å/min even at much lower platen speed (<30 rpm). This behavior is unexpected if one considers the softness of organic particle. The low sensitivity toward downforce is believed to be a direct result of the compressibility nature of the organic abrasive particles. As the particles are readily compressible, the variation in downpressure may have little impact on the total contact area between the abrasive particles' wafer. These soft abrasives may also reduce pad wear and extend the lifetime of a polishing pad.

As the abrasive particles are specifically designed to interact with copper surface, the removal rate on the barrier and dielectric materials are very low (Table 7.3). This will allow longer overpolishing without incurring significant erosion.

The surface quality of the polished Cu substrate is shown in Fig. 7.22. The polished copper surface has an average surface roughness of 6 Å without any signs of corrosion or pitting. This is consistent with the fact that the slurry is well balanced between the chemical and mechanical strengths. This is

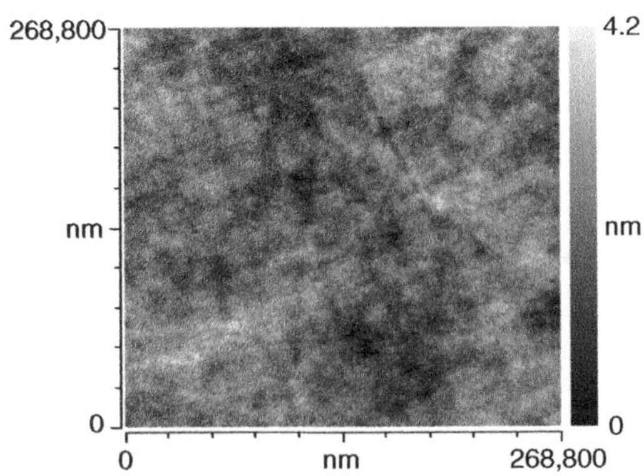

FIGURE 7.22 Representative surface quality image of Cu blanket test wafer polished with a slurry consisting of organic abrasive particles. Surface roughness RMS value of 0.6 nm was achieved (from Ref. 109).

quite different from the abrasive-free systems. The surface of a wafer polished with an AFS usually shows signs of strong chemical attacks.

7.8.3 Step Height Reduction Efficiency and Overpolishing Window

The performance of a copper slurry is measured by its ability to planarize large features such as 100 μm lines with 50% metal density. As shown in Fig. 7.23, the reductions of copper thickness and step heights are parallel on a standard SEMATECH 854 patterned testing wafer. This translates to nearly 100% step-height reduction efficiency. The finial dishing value after clearing of overburden copper (80 s) is close to 200 Å. A long overpolishing window was also observed with only an increase of 300 Å in dishing for a 45 s overpolish. This is particularly useful as overpolishing is often required for incoming wafers with large within wafer nonuniformity in terms of copper deposition. It is remarkable that more than 50% (45 s/80 s × %) overpolishing causes only a little increase in dishing.

7.8.4 Summary on the Organic Particles

This study demonstrated that, by incorporating copper complexation functionality onto the abrasive, it is possible to achieve high copper removal rate and high selectivity over other dissimilar materials. The removal rate and selectivity are dominated by the interaction particle surface functional groups and substrate surface. The importance of abrasive particle hardness becomes secondary or minor. Furthermore, the softness and compressibility of these organic particles provide unique properties to the slurry such as low sensitivity toward polishing downforce and long overpolishing window. Compared to an abrasive-free system (AFS), the organic particles do serve as a soft abrasive to clear the copper puddles.

7.9 CONCLUSIONS

The four key components of a typical CMP slurry, that is, oxidizer, complexing agent, surfactants, and abrasives have been highlighted in this chapter. Although each component has its own role to play during a CMP process, their effective concentrations or effectiveness in performing a particular function are highly interrelated and dynamic. A critical phenomenon that has the maximum impact on all components is surface adsorption or surface–surface interaction. The sources of surfaces include abrasive particles, pads, polishing debris, and wafer. A strong surface adsorption can alter the effective concentration of that component in the solution. A competitive surface adsorption can change the polishing dynamics and removal rate selectivity. Therefore, the surface properties of the particles are important characteristics of an abrasive. The surface adsorption behavior of each chemical additive should be carefully

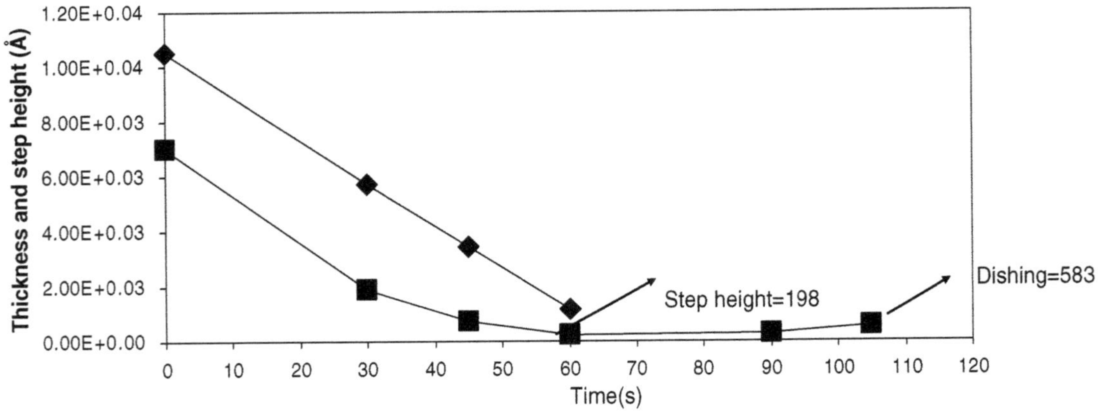

FIGURE 7.23 The overall average overburden copper thickness and step height of 100 μm copper in the 50 % metal density region on a SEMATECH 854 patterned wafer (from Ref. 109).

considered in formulating a slurry. The bulk properties of the abrasive, such as hardness and size, are also important as they determine the mode of interaction with the surface that is being polished.

QUESTIONS

1. List some of the key differences between the following:
   (a) Metal and nonmetal CMP slurries.
   (b) Cu and W CMP slurries.
   What are the main reasons for the difference in each case?
2. How will you design a slurry for a noble metal? List all the items you would take into consideration.
3. What steps would you take to design CMP slurry for low-*k* applications?
4. What are some of the key variables impacting the WIWNU of a CMP process?
5. What key monitoring parameters in your view should be required for a slurry at production scale?
6. Explain the requirement differences among ILD, metal, and MEMS CMP slurries.
7. In general, do you want a CMP process to be mechanically limited or chemically limited? Explain.
8. Differentiate metal and nonmetal CMP slurries in terms of their Prestonian behavior.
9. Explain the importance and limitation of electrochemical studies in CMP slurry.
10. Explain why the requirement of CMP slurries become more stringent with the advancement in technology node.

REFERENCES

1. Pourbaix. Atlas of Electrochemical Equilibria in Aqueous Solutions. Houston, TX: National Association of Corrosion Engineering; 1974.
2. Steigerwald JM, Murarka SP, Duquette DJ, Gutmann RJ. Advanced metallization for devices and circuits: sciences, technology, and manufacturing. Mater Res Soc Symp Proc 1994;337:133–138.
3. Steigerwald JM, Murarka SP, Gutmann RJ. Chemical Mechanical Planarization of Microelectronic Materials. New York: John Wiley & Sons, Inc.; 1995.
4. Caprio R, Farkas J, Jairath R. Initial studies on copper CMP slurry chemistries. Thin Solid Film 1995;266:238–244.
5. Du T, Tamboli D, Desai V, Seal S. Mechanism of copper removal during CMP in acidic H_2O_2 slurry. J Electrochem Soc 2004;151G231–G235.
6. Wang L, Doyle F. Mater Res Soc Sympo Pro 2003;767:F6.5.1–F6.5.10.

7. Lu J, Garland JE, Petti CM, Babu SV, Roy D. Relative roles of H_2O_2 and glycine in CMP of copper studied with impedance spectroscopy. J Electrochem Soc 2004;151(10):G717–G722.
8. Aksu S, Wang L, Doyle FM. Effect of hydrogen-peroxide on oxidation of copper in CMP slurries containing glycine. J Electrochem Soc 2003;150(11): G718–G723.
9. Ein-Eli Y, Abelev E, Starosvetsky D. Electrochemical aspects of copper chemical mechanical planarization(CMP) in peroxide based slurries containing BTA and glycine. Electrochem Acta 2004;49:1499–1503.
10. Hirabayashi H, Higuchi M, Kinoshita M, Kaneko H, Hagasaka N, Mase K, Oshima J. Copper-based metal polishing solution and method for manufacturing semiconductor device. US patent 5,575,885. 1996 Nov 19.
11. Li Y. Copper slurry developments. CMP for ULSI Multilevel Interconnection-Short Course. Marina Del Rey, CA; 2003.
12. Unpublished results. Dr. Yuzhuo Li andcoworkers. Clarkson University, Potsdam, NY.
13. Lim G, Lee J-H, Kim J, Lee H-W, Hyun S-H. Effect of oxidants on the removal of tungsten in CMP process. Wear 2004;257:863–868.
14. Jia K. Inventor; Colloidal silica slurry for NiP disk polishing.US patent 6,149,696. 1997 Apr 18.
15. Streinz CC, Neville M, Grumbine SK, Mueller BL, Composition and method for polishing rigid disks. US patent 6,015,506. 1997 Nov 6.
16. Brusic V, Scherber D, Kaufman F, Kistler R, Streinz C. Chemical–mechanical planarization. Proc Electrochem Soc 1997;96–22:176–185.
17. Hariharapurthiran M, Zhang J, Ramarajan S, Keleher JJ, Li Y, Babu SV. Hydroxyl radical formation in H_2O_2–amino acid mixtures and chemical–mechanical polishing of copper. J of Electrochem Soc 2000;147(10):3820–3826.
18. Hage R, Iburg JE, Kerschner J, Koek HH, Lempers E, Martens RJ, Racheria US, Russe WW, Suarthoff T, Vliet M, Warner JB, Wolf L, Krijnen B. Effect of manganese catalysts for low-temperature bleaching. Nature 1994;369:637–639.
19. Thompson KM, Spirito M, Griffith WP. Mechanism of bleaching by peroxides. Part 4—kinetics of bleaching of malvin chloride by hydrogen-peroxide at low pH and its catalysis by transition metal-salts. J Chem Soc Farad Trans 1996;92:2535.
20. Suguwara H, Toma Y, Takabe T, Yokoi K, Bleaching compositions. US patent 4,756,845. 1988 Jul 12.
21. Kissa E, Dohner JM, Gibson WR, Strickman D. J Am Oil Chem Soc 1991;68:532.
22. McMurry JE. Organic Chemistry. Brooks Cole; 2003.
23. Keleher J, Rushing K, Zhao J, Wojtczak B, Li Y. Supramolecular abrasive-free system for Cu CMP. Proc Mater Res Soc 2003;767:F6.1.1–F6.1.11.
24. Li Y. Reactive pads for metal CMP. 10th International Symposium on CMP. NY: Lake Placid; 2005.
25. Wikipedia. The free encyclopedia. Available at, http://en.wikipedia.org/wiki/Image:Pourbaix_Diagram_of_Iron.svg#filelinks.

26. Luo Q. Chemical mechanical polishing of thin copper films [dissertation]. Potsdam (NY): Clarkson University; 1997.
27. Luo Q, Campbell DR, Babu SV. Proc Electrochem Soc Interconnect Con Metall 1998;97–31:73–83.
28. Luo Q, Babu SV. Dishing effects during chemical mechanical polishing of copper in acidic media. J Electrochem Soc 2000;147(12):4639–4644.
29. Seo Y-J, Lee W-S. Effect of different oxidizers on the W CMP performance. Mater Soc and Engg 2005;B118:281–284.
30. Cotton FA, Wilkinson G, Gaus PL. Basic Inorganic Chemistry. 3rd ed. New York; John Wiley & Sons Inc.;1995.
31. Paul E, Kaufman F, Brusic V, Zhang J, Sun F, Vacassy R. A model for Copper CMP. J Electrochem Soc 2005;152(4):G322–G328.
32. Kallingal CG, Duquette DJ, Murarka SP. An investigation of slurry chemistry used in chemical mechanical planarization of aluminum. J Electrochem Soc 1998; 145:2074–2081.
33. Small RJ, Laurence M, Maloney DJ, Peterson ML. Chemical–mechanical composition and process. US patent 6,635,186. 2003 Oct 21.
34. Small RJ, Laurence M, Maloney DJ, Peterson ML. Chemical–mechanical composition and process. US patent 6,117,783. 2000 Sep 12.
35. Stein DJ, Hetherington DL, Cecchi JL. Investigation of the kinetics of tungsten chemical mechanical polishing in potassium-iodate based slurries (role of alumina and potassium iodate). J Electrochem Soc 1999;146(1):376–381.
36. Stein DJ, Hetherington DL, Cecchi JL. Investigation of the kinetics of tungsten chemical mechanical polishing in potassium iodate based slurries (role of colloid species and surface chemistry). J Electrochem Soc 1999;146(5):1934–1938.
37. Stein DJ, Hetherington DL, Guilinger T, Cecchi JL. Insitu electrochemical investigation of tungsten electrochemical behavior during chemical mechanical polishing. J Eletrochem Soc 1998;145(9):3190–3196.
38. Lee SM, Mahajan V, Chen Z, Singh RK. Chemical mechanical planarization in IC devices, manufacturing III. Proc Electrochem Soc 2000;99–37:187–192.
39. Li Y, Babu SV. Chemical mechanical polishing of copper and tantalum in potassium iodate based slurries. J Electrochem Soc 2001;4(2):G20–G22T.
40. AoyamaYamazaki S, Imai K. Ultra-thin Ta_2O_5 Film capacitor with Ru bottom electrode. J Electrochem Soc 1998;145:2961–2964.
41. Lee W-J, Park H-S, Lee S-I, Sohn H-C. Effect of ceric ammonium nitrate (CAN) additive in HNO_3 solution on the electrochemical behavior of Ruthenium for CMP processes. J App Electrochem 2004;34:119–125.
42. Yun S-K, Lee J-D, Orui K, Nojo H, Yoon B-U, Hong C-K Cho H-K, Moon J-T. Development of Ru CMP slurry and its application to node-separation of RIR RIR capacitor. Semiconductor. R&D Samsung Electronics Co. Ltd.
43. Atkins PW. Physical Chemistry. 6th ed. W.H. Freeman and Company; 1997.
44. Halpern J. Kinetics of dissolution of copper in aqueous ammonia. J Electrochem Soc 1953;100:421.
45. Zembura Z, Piotrouski A, Kolenda J. J App Electrochem 1990;20:365.

46. Steigerwald JM, Duquette DJ, Murarka SP, Gutmann RJ. Electrochemical potential measurements during the chemical-mechanical polishing of copper thin films. J Electrochem Soc 1995;142(7):2379–2385.

47. Snoeyink VL, Jenkins D. Water Chemistry. New York: John Wiley & Sons, Inc.; 1980.

48. Hancock RD, Martell AE. Metal Complexes in Aqueous Solutions (Modern Inorganic Chemistry). 1st ed. Springer; 2005.

49. Gorantla VRK, Matijevic E, Babu SV. Amino acids as complexing agents in chemical–mechanical planarization of copper. Chem Mat 2005;17:2076–2080.

50. Patri UB, Aksu S, Babu SV. Role of functional groups of complexing agents in copper slurries. J Electrochem Soc 2006;153(7):G650–G659.

51. Gorantla VRK, Goia D, Matijević E, Babu SV. Role of amine and carboxyl functional groups of complexing agents in slurries for chemical–mechanical polishing of copper. J Electrochem Soc 2005;152(12):G912–G916.

52. Patri UB. Role of slurry chemicals in chemical mechanical planarization of copper [dissertation]. Potsdam (NY): Clarkson University; 2005.

53. Hernandez J, Wrschka P, Oehrlein GS. Surface chemistry studies of copper chemical mechanical planarization. J Electrochem Soc 2001;148(7):G389–G397.

54. Gorantla VRK, Babel A, Pandija S, Babu SV. Oxalic acid as complexing agent in CMP slurries of copper. Electrochem Solid State Lett 2005;8(5):G131–G134.

55. Tamilmani S, Huang W, Raghavan S. Potential-pH diagrams of interest to chemical mechanical planarization of copper. J Electrochem Soc 2002;12:G638–G642.

56. Evans DF, Wennerström H. The Colloidal Domain: Where Physics, Chemistry, and Biology Meet. 2nd ed. Wiley-VCH; 1990.

57. Tardos TF. Applied Surfactants: Principles and Applications. John Wiley & Sons; 2005.

58. Kim QT-Y, Park E-K. Fabrication of ceria slurry and its stability on aqueous phase. The 2nd PacRim International Conference on Planarization CMP and its Application Technology, 2005; vol. 261.

59. Parafitt GD. Dispersion of powders in liquids. 3rd ed. Applied Science. London; 1981.

60. Yu X, Somasundaran P. Colloids Surf A 1994;89:277.

61. Denoyel R, Rouquerol J. Thermodynamic (including microcalorimetry) studies of the adsorption of nonionic and anionic surfactants onto silica, kaolin and alumina. Colloid Interface Sci 1991;43(2):555–572.

62. William MH, Leeuwee M, Roberson M, Bentley P. Removal of CMP and and post-CMP residue from semiconductors using supercritical carbon-dioxide process. US patent 7,064,070. 2006 Jun 20.

63. ESC Inc. Cu-Based Interconnect Post-CMP Cleaning-Technology Update. CMPUG meeting. 2003.

64. Hong Y, Roy D, Babu SV. Ammonium dodecyl sulphate surfactant for as a potential corrosion inhibitor for electrochemical planarization of copper. Electrochem Solid State Lett 2005;8(11):G297–G300.

65. Bundi D, Cheemalapati K, Duvvuru V, Li. Y. Novel passivating system in copper CMP (Mixed surfactant system). CAMP (Clarkson University) Meeting May 2005. Can andigua, NY; 2005.

66. Park J-G, Katoh T, Lee W-M, Jeon H, Paik U. Surfactant effect on oxide to nitride removal selectivity of nano-abrasive ceria slurry for chemical mechanical polishing. Jap Journ App Phy 2003;42:5420–5425.
67. Bu K-H, Moudgil BM. Colloidal silica based high selectivity shallow trench isolation (STI) chemical mechanical polishing (CMP) slurry. Proceedings of MRS. Symposium W. Spring; 2005.
68. Miller AE, et al. 9th International Symposia On CMP. NY: Lake Placid; 2004.
69. Schroeder D.J. et al., CMP method utilizing amphiphilic nonionic surfactants. US patent 6,936,543. 2005 Aug 30.
70. Itano M, Kezuka T, Ohmi T. Minimization of particle contamination during wet porcessing of si wafers. J Electrochem Soc 1995;142:971–978.
71. Lee BC, Wang B, Duquette DJ, Gutmann RJ. Synthesis of model alumina slurries for damascene patterning of copper. Proc. of Materials Research Symposia Symposium M. Spring; 2001.
72. Schaffer JP, Saxena A, Antolovich SD, Sanders TH Jr, Warner SB. The Science and Design of Engineering Materials. 2nd ed. McGraw Hill; 1999.
73. Callister WD Jr, Materials Science and Engineering: An Introduction. 4th ed. New York: John Wiley & Sons Inc.; 1997.
74. Chowdhury S, Barra E, Laugier MT. Hardness measurement of CVD diamond coatings on SiC substrates. Surf Coatings Tech 2005;193:200–205.
75. Petrovic I, Mathur S, Planarization Composition. US patent 20,060,283,093. 2006 Dec 21.
76. Li Y, Zhao J, Wu P, Lin Y, Babu SV, Li Y. Thin Solid Films 2006;497:321–328.
77. Ramarajan S. Chemical–mechanical planarization of copper/tantalum for microelectronic applications [dissertation]. Potsdam (NY): Clarkson University; 2000.
78. Ramarajan S, Li Y, Hariharaputhiran M, Babu SV, Her YS. The role of alumina particle density in chemical mechanical planarization of copper, tantalum, and tungsten disks and films. J CMP On-Chip Interconnect IMIC 2000;1(1):28–38.
79. Luo Q, Campbell DR, Babu SV. Stabilization of alumina slurry for chemical mechanical of copper. Langmuir 1996;12:3563–3566.
80. Industrial products-alumina abrasives. Available at http://www.metallographic.com/Industrial%20Products/Alumina.htm.
81. Hegde S, Babu SV. Study of surface charge effects on oxide and nitride planarization using alumina/ceria mixed abrasive slurries. Electrochem Solid State Lett 2004;7(12):G316–G318.
82. Li Y. Particle innovations in copper CMP slurry development, impact of hydrophobicity, hardness, and functionality. Solid State IC Tech 2004;1:508–513.
83. Barthel H, Heinemann M, Stintz M, Wessely B. Particle size of fumed silica. Chem Eng Tech 1998;21:745–52.
84. Prelas MA, Popovici G, Bigelow LK. Handbook of Industrial Diamond. Marcel DeKiler: New York; 1998.
85. Mahajan U, Bielman M, Singh RK. Chemical Mechanical Polishing Fundamentals and Challenges. In: SV Babu, S Danjluk, M Krishnan, M Tsujimara, ed. Materials Research Society Proceedings, Vol. 566.

86. Basim GB, Adler JJ, Mahajan U, Singh RK, Moudgil BM. Effect of particle size of chemical mechanical polishing slurries for enhanced polishing with minimal defects. J Electrochem Soc 2000;147(9):3523–3528.
87. Lu Z. Investigation of slurry systems in metal and dielectric chemical–mechanical polishing [dissertation]. Potsdam (NY): Clarkson University; 2004.
88. Bielmann M, Mahajan U, Singh RK. Effect of particle size during tungsten chemical–mechanical polishing. Electrochem Sol State Lett 1999;2(8):401–403.
89. Particle–Sizing Systems Inc. Accusizer size-range. Available at http://www.pssnicomp.com/acculab.htm.
90. Remsen EE, Anjur S, Boldridge D, Kamiti M, Li S, Johns T, Dowell C. Analysis of large particle count in fumed silica slurries and its correlation with scratch defects generated by CMP. J Electrochem Soc 2006;153(5):G453–G461.
91. Stöber W, Fink A, Bohn E. J Coll Interf Sci 1968;26:62–69.
92. Tabatabaei S, Shukohfar A, Mirhabibi A. Experimental study of the synthesis and characterization of silica nanoparticle via sol-gel method. J Phys Conf Series 2006;26:371–374.
93. Khan SA, Gunther A, Jensen KE. Langmuir 2004;20:8604–8611.
94. Gitzen WH. Alumina as a Ceramic Material. USA: The American Ceramic Society; 1970.
95. Wefers K, Misra C. Oxides and hydroxides of aluminium. Alcoa Laboratories technical report1987.
96. Davies GJ. et al. Growth of diamond clusters. US patent 6,676,750. 2004 Jan 13.
97. Hanada K, Mayuzumi M, Nakayama N, Sano T. Processing and characterization of cluster diamond dispersed Al–Si–Cu–Mg composite. J Mat Proc Tech 2001;119(1–3):216–221.
98. Dhane S. Investigation of surface adsorption behavior of bta in a model copper CMP slurry [Masters thesis]. Potsdam (NY): Clarkson University; May 2006.
99. Zhuravlev LT. Langmuir 1987;3:316–318.
100. Uchikura K. et al. Aqueous dispersion for chemical mechanical polishing and chemical–mechanical polishing process. US patent 20,010,008,828. 2001 Jul 19.
101. JSR Micro CMP slurries. Available http://www.jsrmicro.com/pro_CMP_slurry.html.
102. Jindal A, Hegde S, Babu SV. Electrochem Sol St Lett 2002;5(7):G48–G50 (2002).
103. Jindal A, Hegde S, Babu SV. J Electrochem Soc 2003;150(5):G314–318.
104. Wang Y, Pfeffer R, Dave R. Polymer encapsulation of fine particles by a supercritical antisolvent process.
105. Wu S, Xie Y, Cheng G. J Biomed Engg 2006;23(2):362–365.
106. Kim J, Paik U, Jung Y-G, Park J-G. Jap J App Phys 2002;41:4509–4512.
107. Murray PG, Coy DC. The emerging role of nanocrystalline ceria in microelectronic polishing applications. NSTI Nanotechnology Conference and Tradeshow. 2005

108. Kondo S, Sakuma N, Homma Y, Goto Y, Ohashi Ni, Yamaguchi H, Owada N. Abrasive-free polishing for copper damascene interconnection. J Electrochem Soc 2000;147(10):3907–3913.

109. Cheemalapati K, Li Y, Tang K, Bian G. Organic particles for copper CMP at low down force. Proceedings of 9th International CMP for ULSI Multilevel Interconnection Conference. 2004. P23–27.

110. Cheemalapati K, Chowdhury AR, Li Y, Tang K, Bian G. Novel organic particles for copper CMP at low down force. Materials Research Society Symposium Proceedings, Vol. 816. 2004. PK 1.7.

111. Li Y, Choudary A, Tang K, Bian G, Cheemalapati K. Nonpolymeric organic particles for chemical mechanical planarization. US patent 7,037,351. 2006 May 2.

8

CORROSION INHIBITOR FOR Cu CMP SLURRY

SURESH KUMAR GOVINDASWAMY AND YUZHUO LI

Copper chemical–mechanical planarization (Cu CMP) is a rapidly growing segment in the fabrication process of today's semiconductor devices [1]. Copper CMP slurry typically contains an oxidizer that chemically converts the metal film for easy removal, abrasive particles that enhance the abrasiveness of the pad, a complexing agent that enhances the solubility of the abraded metal/metal oxide, a passivating agent that protects the lower lying areas, a pH regulating agent, and a surfactant [2]. The main function of the passivating agent is to protect the copper film from aggressive chemical attack that may lead to isotropic dissolution of the copper film. With the protection of such passivating agent, ideally, only the copper film in the protruded area is selectively removed by mechanical force, thus yielding a step height reduction. It is commonly observed that a passivating agent in a Cu CMP slurry will lower not only the static etch rate but also the removal rate. As a strong corrosion inhibitor, benzotriazole (BTA) is the most commonly used passivating agent in Cu CMP slurry. A vast number of publications and reports have been devoted to the use of BTA in metal CMP slurry [3–16]. The study of BTA analogs for CMP application, however, has been relatively limited in open literature. In this chapter, the role of corrosion inhibitor in Cu CMP slurry is first described. As a case study, the relative effectiveness of some tetrazoles such as 5-aminotetrazole monohydrate (ATA), 5-phenyl-1H-tetrazole (PTA), and 1-phenyl-1H-tetrazole-5-thiol (PTT) as corrosion inhibitor in Cu CMP slurry will be discussed.

Microelectronic Applications of Chemical Mechanical Planarization, Edited by Yuzhuo Li

8.1 THERMODYNAMIC CONSIDERATIONS OF COPPER SURFACE

A proven strategy for Cu CMP slurry formulation involves three common stages: (1) selection of an oxidizer–complexing agent pair that significantly softens the copper film; (2) selection of an effective passivating agent that can prevent the film from isotropic dissolution; and (3) introduction of abrasive particles into the above solution. There have been many successful combinations for each stage and overall formulations. The discussion in this section will also follow the same analogy. First, the thermodynamic behavior of copper surface under oxidizing condition will be examined. The influence of a representative complexing agent on these surface events will then be described. The impact of a passivating agent on the thermodynamic behavior will also be discussed.

In the presence of an oxidizer, the pH dependence of copper dissolution and passivation can be easily illustrated using a Pourbaix diagram. As shown in Fig. 8.1, at acidic pH, the oxidation of Cu to Cu^{2+} is favored. In a near neutral pH system, the formation of cuprous and cupric oxide is thermodynamically favored. Under this condition, the copper surface is passivated by the oxide film. At higher pH, depending on the copper ion activity, copper oxides may be dissolved to form $HCuO_2^-$ and CuO_2^{2-} [17]. To illustrate the importance of copper ion concentration, the potential-pH diagram of Cu–H_2O system at copper ion activities of 10^{-4} and 10^{-6} is depicted. In addition, the potential influences of two commonly seen oxidizers, hydrogen peroxide and hydroxylamine, are also indicated in the figure.

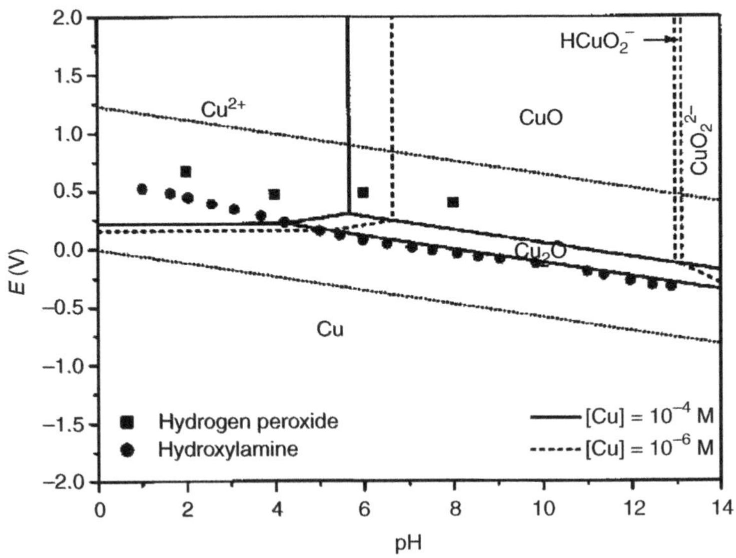

FIGURE 8.1 Potential-pH diagram for Cu–water system (from Ref. 17).

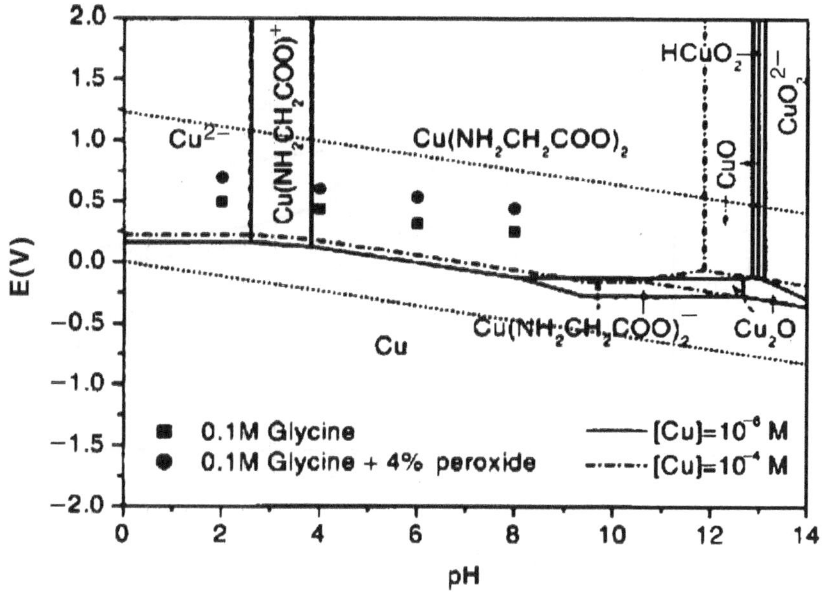

FIGURE 8.2 Potential-pH diagram for Cu–glycine–water system (from Ref. 17).

As mentioned earlier, a working Cu CMP slurry usually contains a complexing agent as well as an oxidizer. The dissolution regime for copper in the presence of a complexing agent typically expands at the expense of passivating region. As shown in Fig. 8.2, the presence of a representative complexing agent, glycine (LH, NH_2CH_2COOH), causes an expansion of the soluble copper region and a retreat of the copper oxide area. Tamilmani et al. reveals that water soluble CuL_2 has a very high stability constant [17]. Pourbaix diagram indicates that copper–glycine complexes are thermodynamically stable in a wide pH range of 3–12. The Pourbaix diagram also shows that, under thermodynamically controlled conditions, the copper film will freely dissolve into copper ions in the presence of hydrogen peroxide and glycine at a wide pH range. As the dissolution is usually isotropic and leads to practically no planarization effect, this event is not desirable for CMP applications. In order to formulate a CMP slurry with planarization potential, a passivating film must be constructed on such a softened copper surface to protect the trench or recessed areas.

As far as passivating film is concerned, in principle as shown in Fig. 8.1, the native copper oxide structures formed in the presence of a strong oxidizer under certain pH conditions can serve the purpose for many applications. For CMP purposes, such a hard surface film does not usually lead to any meaningful material removal under the mechanical forces exerted by a polishing pad. Therefore, there is a need to build a soft passivating layer on a copper film softened by the complexing agent. The most commonly used

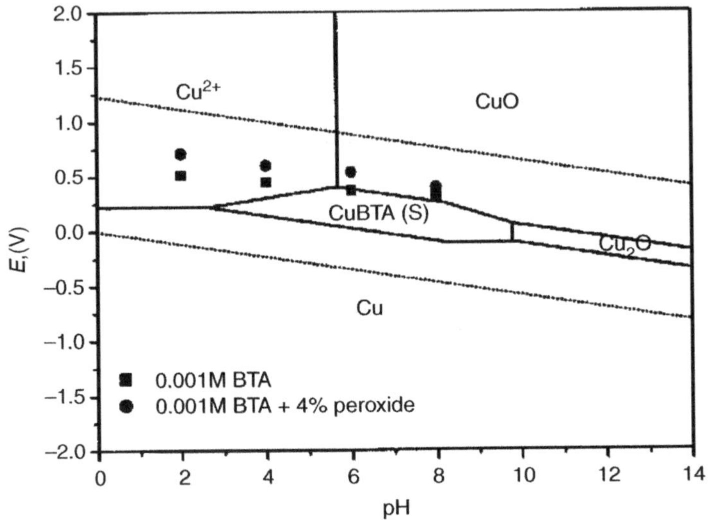

FIGURE 8.3 Potential-pH diagram for Cu–BTAH–water system (from Ref. 17).

passivating agent for this purpose is benzotriazole. Fig. 8.3 shows the Pourbaix diagram of copper surface in the presence of BTA.

It is important to point out that the Pourbaix diagram shown in Fig. 8.3 does not contain any complexing agent. The boundaries indicated in the diagram will likely change when a complexing agent is introduced. Furthermore, one should be reminded that the CMP process is rarely thermodynamically controlled. Many kinetic factors such as diffusion, viscosity, and reagent availability may dominate the polishing outcome. A thermodynamic analysis of such a complex surface should serve only as a general guideline at best.

8.2 TYPES OF PASSIVATING FILMS ON COPPER SURFACE UNDER OXDIZING CONDITIONS

As described earlier, the native copper oxide structures formed in the presence of an oxidizer under certain pH conditions can prevent the free dissolution of copper into copper ions. In addition to such oxide structures, there are at least three other types of layers that can accumulate on the oxidized copper surface to inhibit copper corrosion:

- salt layer
- surfactant layer
- hydrophobic complex stack

It has been reported that some ionic species may accumulate at the surface because of static charge interactions. At high local concentration, the diffusion of these species may be limited; hence the dissolution of copper is inhibited. Various phosphate salts are examples of this type. A more detailed discussion of this type of corrosion inhibition can be found in Chapters 10 and 11, which deal with passivation films in electrochemical planarization (ECP) and electrochemical–mechanical polishing (ECMP). The second type of corrosion-inhibiting film may be built with a passivating agent that is capable of forming a complex with copper. Unlike the complexing agent described earlier, the passivating agent–copper complex does not lead to rapid dissolution of copper ions. Instead, it attracts more passivating agent to adsorb onto the complex and then to the passivating agents themselves. Eventually, a thin film of passivating agent is formed that completely inhibits the copper corrosion. BTA is an excellent example of this type. The third type of corrosion-inhibiting agent (typically surfactant) is attracted to the copper surface by static charges. Unlike the salts, these surfactant molecules tend to stack into monolayer or double layers in accordance with their phase behavior. The focus of this section is on the characteristics and applications of hydrophobic passivating agent. A brief discussion on the use of surfactant as a corrosion inhibitor will also be included.

Many commercial and developmental Cu CMP slurries contain BTA as a corrosion inhibitor. In a representative Cu CMP slurry, a combination of hydrogen peroxide (H_2O_2) and a complexing agent is used to oxidize and soften the copper surface. Without any passivating agent, such a solution can give high copper removal regardless of the involvement of any abrasive particles. The material removal using such a solution is, however, mostly isotropic. In other words, the step height reduction efficiency is practically zero when using such a polishing solution because the softened copper surface can be significantly disrupted or removed even with the weakest mechanical force including the shear force impinged by the fluid flow. In the presence of a dedicated passivating agent such as BTA, the softened film is somewhat protected and hardened. The art of slurry formulation is to balance the need for protection in the lower lying areas and the need for removal at higher or protruded areas. More specifically, a proper combination of BTA as passivating agent and a complexing agent can balance the need to have low static etch rate (in the absence of mechanical abrasion) and a high polishing rate (in the presence of mechanical abrasion) [18]. For a CMP solution containing glycine and hydrogen peroxide, addition of BTA results in a significant reduction in the Cu removal rate because of the formation of Cu–BTA complex on the copper surface. For example, Deshpande et al. showed that BTA acted as corrosion inhibitor and decreased the dissolution rate [19]. They also showed that the inhibition efficiency of BTA was enhanced by an increase in BTA concentration as well as the presence of hydrogen peroxide. This is consistent with the fact that the passivation film has two key components. The first is a complex layer between BTA and oxidized copper. The second is a hydrophobic layer stacked with BTA molecules as shown in Fig. 8.4.

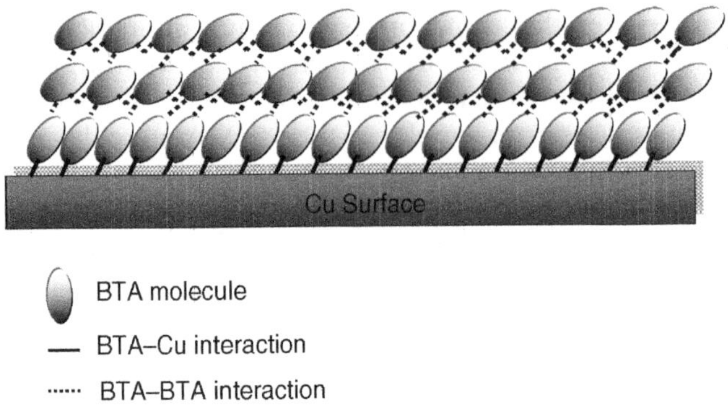

FIGURE 8.4 Schematic representation of a BTA passivating layer formed on copper.

An investigation by Luo and Babu on a ferric nitrate based CMP slurry showed that copper polishing rate drops significantly in the presence of BTA [20]. According to the study, at low BTA concentration (about 0.001 mol/dm^3) and moderate oxidizer concentration (about 0.05 mol/dm^3 of $Fe(NO_3)_3$) the Cu–BTA film may not be strong enough to withstand shear stress during polishing either mechanically or hydrodynamically, leading to insufficient protection. For a slurry that contains higher concentration of BTA ($\gg$0.005 mol/dm^3), the passivating efficiency is significantly improved, which leads to a lower copper dishing value. In a study by Tsai et al. it was shown that corrosion rate in a solution of 5 wt% of HNO_3 declined by two orders of magnitude after the addition of 0.1 wt% of BTA [21]. Addition of corrosion inhibitor also improves the surface roughness of copper surface as the inhibitor reduces the copper corrosion such as pitting corrosion that takes place on the surface. As a copper oxide film is never fully developed under the conditions described above using nitric acid and ferric nitrate as oxidizers, BTA must form the first complex layer with an oxidized surface that is rich in copper ions.

Steigerwald et al. reported that the Cu–BTA passivation film was almost 20 nm thick after a 10-min immersion in a solution at pH 2 [22]. Cohen and coworkers also studied the stoichiometry, thickness, and chemical composition of the Cu–BTA using *in situ* ellipsometry and *ex situ* X-ray photoelectron spectroscopy [13]. The authors reported that film grown on Cu_2O and bare Cu under oxidizing conditions are on the order of 5–40 Å thick and the chemical composition of this layer is mostly Cu^{+1}–BTA. Similar to the schematic view portrayed in Fig. 8.3, Walsh et al. suggests that the BTA film is composed of a monolayer that is in direct contact with the copper film and a multilayer built on top of the monolayer [6]. They reveal that in the monolayer, BTA molecular plane is oriented within 15° of the surface normal. In the multilayer, the molecular plane is tilted by about 40° from the plane of copper surface.

Notoya et al. showed that BTA exhibited the highest inhibition efficiency at pH 6 [23]. This is consistent with the fact that, to form both complexing and multilayer effectively, the BTA molecules must be neutral. BTA would not be effective if the molecules are protonated (under acidic condition) or deprotonated (under extreme basic condition).

Besides BTA, a range of other chemicals have also been studied as corrosion inhibitor in Cu CMP solution/slurry. Sekar and Ramanathan studied hydrazine as an inhibitor for Cu CMP in nitric acid based slurry [24]. They reported that material removal rate and static etch rate decreased with the addition of hydrazine. They also noticed that the addition of hydrazine to the slurry improves the surface roughness of the polished copper surface. Du et al. used 3-aminotriazol as corrosion inhibitor for Cu CMP slurry based on hydrogen peroxide–glycine system [25]. The result of their study showed that the addition of amniotriazol suppresses both static etch rate and material removal rate of copper. In the X-ray photoelectron spectroscopy (XPS) analysis, authors reveal that addition of aminotriazol corrosion inhibitor suppresses the oxide formation on the copper surface. It is possible that the surface adsorption of amniotriazol on the copper surface prevents the normal growth of copper oxide. Hu et al. showed that citric acid could be used as corrosion inhibitor in 3 vol% of HNO_3 solution [26]. It is found that the addition of citric acid reduces the material removal rate and improves the planarization efficiency for Cu CMP. Using a potentiodynamic polarization study, the authors showed that citric acid inhibits copper corrosion in HNO_3 solution. They suggested that the passivation layer consists of a nonnative citrate complex film that inhibits etching. Considering the fact that citric acid is commonly used as a complexing agent that promotes dissolution of copper, the formation of such passivating layer under such a circumstance is unique. Lee reported the inhibiting effect of imidazole on copper corrosion in HNO_3 solution using potentiodynamic study [27]. The imidazole was shown to act as an effective inhibitor to prevent Cu corrosion. Cu–imidazole complex film is simultaneously formed with the Cu oxide in the presence of imidazole that reduces the copper corrosion.

A surfactant is commonly used as a dispersing agent in CMP slurry for abrasive particle stabilization. It is important to point out that the wafer surface is also available for surfactant molecules to adsorb. The net result of such surface adsorption may function as a passivating film. Depending on the nature and the concentration of surfactant, the adsorption may result in a monolayer, a double layer, or an array of hemimicelles (Fig. 8.5). Also, depending on the operating pH of the slurry, the copper surface may be positively or negatively charged. As the isoelectric point of copper oxide surface is about 6, the surface may exhibit slight positive charge in a solution that is below pH 6. At high pH, the surface may be slightly negatively charged. It is logical to expect that a surfactant with charge opposite to that of the copper surface should be more effective in serving as a passivating agent because of electrostatic attraction between the surfactant molecule and copper surface. Hong et al. investigated the performance of anionic, cationic, and nonionic surfactants as corrosion inhibitors at various slurry pH values [28]. They showed that slurry containing anionic surfactant drastically suppresses

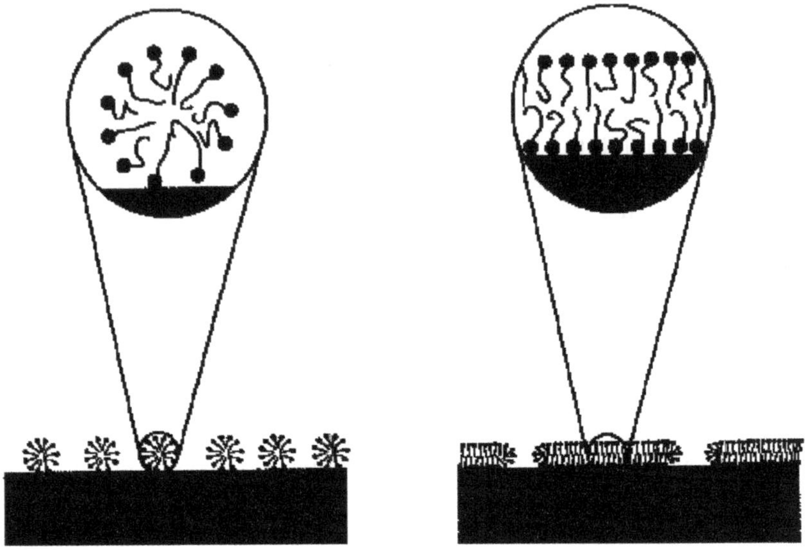

FIGURE 8.5 Surface adsorption of surfactant [29] (from Ref. 29).

the copper etching in the pH range of 2–8. For cationic surfactant, the suppression of copper corrosion was effective in the pH range 2–3 and in the pH greater than 6. It was concluded that nonionic surfactants did not show significant corrosion-inhibiting characteristics in CMP slurry. This is consistent with the charge analysis described above.

Our recent investigation on the use of surfactants as potential corrosion inhibitors showed that the electrostatic attraction between the charged surfactant molecules and the copper surface may not be the only criterion for forming an effective passivation layer. For example, a low concentration of cationic surfactant can also form an effective passivating film on copper surface at a pH where the copper surface is clearly positive such as 3–5 [30]. An answer to such a puzzle can be traced back to the counterion effect. More specifically, the counterion of a surfactant may play a significant role in forming a film on copper surface. For example, the bromide ions in CTAB may bridge the gap between the two positively charged centers on copper surface and cationic surfactant molecule. It is also important to note that the packing density of the surfactant layer could be low in general because of the charge repulsion among surfactant head groups with the same charge. It is easy to understand that when a single surfactant is used, because of charge repulsion, the protection of the metal film by such a surfactant system may be inadequate. When a mixed surfactant system is employed, the charge repulsion among surfactants may be minimized, which leads to a better and more tightly packed passivating film. It was shown by Yuzhuo Li and coworkers that a mixed surfactant system containing anionic and cationic surfactants in the molar ratio of 4 : 1 with total surfactant concentration of 0.058 wt % could reduce the copper static etch rate from over 100 nm/min to less than 10 nm/min for a CMP solution containing 2 wt % of H_2O_2 and 1 wt % of glycine at pH 5 [31]. The optimum molar ratio between the cationic and anionic surfactants is a function of copper surface charge density, which is related to pH and other environmental factors. In general, a greater copper surface charge density should translate into a lower demand on the availability of anionic countersurfactant.

As mentioned earlier, BTA forms an effective passivating layer on copper surface relying on the hydrophobic nature of the molecule. One of the problems associated with such a passivating mechanism is that the introduction of copper ions can lead to the destruction of such a film by increasing the water solubility of these complexes. For the mixed surfactant system, however, the increase in static etch rate is minimal upon introduction of copper ions. This is because the passivating film formed is through charge–charge interaction. Unlike compounds such as BTA, the surfactant passivating film that is based on static charge interaction is not likely disturbed by the introduction of copper ions. This may translate into an advantage during slurry formulation. The potential disadvantage of a surfactant-based passivating film is its durability against shear flow during polishing. After all, the film (usually mono- or double layer of surfactant molecules) may be too thin to withstand the level of shear force during polishing.

8.3 EFFECT OF pH ON BTA IN GLYCINE–HYDROGEN PEROXIDE BASED Cu CMP SLURRY

Figures. 8.6 and 8.7 show the static etch rate (SER) and material removal rate (MRR) of Cu as a function of slurry pH with 1 wt% of glycine, 3 wt% of hydrogen peroxide, 3 wt% of SiO_2, and 1 mM of BTA. The static etch and

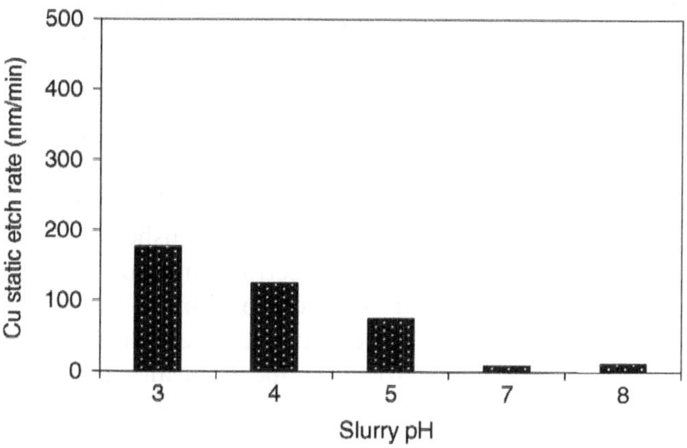

FIGURE 8.6 Effect of CMP solution pH on the Cu SER in CMP slurry containing 1 wt% of glycine, 3 wt% of H_2O_2, 3 wt% of SiO_2, and 1 mM of BTA.

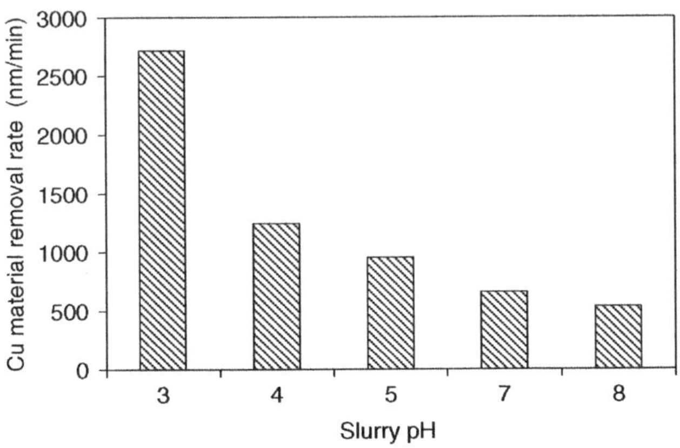

FIGURE 8.7 Effect of CMP solution pH on the MRR of copper in CMP slurry containing 1 wt% of glycine, 3wt% of H_2O_2, 3 wt% of SiO_2, and 1 mM of BTA.

material removal rates of copper are lower at higher pH. There are two synergistic effects that lead to such a trend. At higher pH the copper oxide formed under oxidizing condition is stronger and provides stronger passivating effect. This is consistent with the thermodynamic property of copper as indicated by Pourbaix diagram [17]. At the same time, within the pH range shown in the figures, BTA is less protonated at higher pH. In another words, at acidic pH the formation of Cu–BTA passivation film is less effective because of increased water solubility of protonated BTA [19]. This is a key reason why many copper CMP slurries are formulated at pH 5 or above. It is important to point out that there is a trade-off for high pH slurry as BTA and BTA–copper clusters have low solubility in water. It often leads to the formation of large particles that may cause scratches during polishing and leave organic residues that are difficult to clean. It is also worthwhile to mention that the copper static etch rate obtained on slurry prior to polishing may not reflect the true picture for the copper surface during polishing. As soon as the polishing starts, copper ions are introduced into the solution. During the polishing, the local concentration of copper ions can be extremely high. Such a high concentration of copper ions can lower the pH to below 3. At such a low pH, the effectiveness of BTA as a passivating agent can be severely diminished [32].

Owing to the integration of low-*k* dielectric materials, the IC industry is moving toward using acidic CMP slurries for bulk Cu CMP to minimize delamination at the metal–low-*k* dielectric interface [19]. Cu CMP process is thus becoming more chemically active than mechanically driven processes [33]. BTA, however, as mentioned earlier is known to have poor corrosion-inhibiting efficiency under acidic conditions and cause polishing debris aggregation at high pH. A poor inhibition of copper surface may lead to surface defects such as dishing, corrosion, and erosion [34,35]. Therefore there is a need for identifying a passivating agent that forms effective protective film on copper surface under acidic condition.

8.4 EVALUATION OF POTENTIAL BTA ALTERNATIVES FOR ACIDIC Cu CMP SLURRY

Organic heterocyclic compounds containing azole nucleus are known to display corrosion-inhibiting characteristics for copper and its alloys [36–40]. Various studies have been reported on the application of such compounds. However, detailed studies using such compounds as a passivating agent in Cu CMP is limited. The corrosion inhibitors selected for this study have an amino group, a phenyl group, or a thiol group attached to the parent tetrazole moiety. It is generally known that a tetrazole moiety is capable of forming strong chemical interaction with various metal surfaces including copper. Within this class of compounds, 5-phenyl-1H-tetrazole is an interesting contrast to benzotriazole. Although the tetrazole and phenyl moieties resemble the overall structure of BTA, the flexibility of phenyl group in PTA may lead to a film with different properties from BTA as BTA is a rigid planar molecule. It is reported that the solubility of BTA is higher at pH values below 4 and above 8. For PTA, the solubility is low throughout the acidic region and is higher at pH above 5 [41,42]. It is obvious that a corrosion inhibitor would form a more effective passivating film if its solubility in water is low at the pH it is being used. It is such property that serves as a driving force to form the film on copper surface through hydrophobic–hydrophobic interaction. For BTA, the best operating pH range is 4–7. For PTA, the range extends well into acidic region (<4). Furthermore, as the molecule has a flexible moiety, the film may have different packing density than its rigid counterpart. The molecular structure of heterocyclic tetrazole compounds that were investigated is shown in Fig. 8.8 along with BTA structure.

The static etch rates of Cu were measured using a set of CMP slurries that contain 1 wt% of glycine, 3 wt% of H_2O_2, 3 wt% of silica abrasive particles, and various concentrations of tetrazole. The pH of the CMP slurry system was selected based on the ratio of the material removal rate to the static etch rate (MRR/SER). At pH 3, ATA- and PTA-containing slurry yield low SER and high MRR/SER, whereas for PTT containing slurry pH 4 was observed. The results from the SER measurements are summarized in Fig. 8.9, which clearly indicate that addition of tetrazole inhibitor decreases the static etch rate of copper. There is a sharp decrease in SER with the addition of 0.5 mM of PTA. ATA and PTT, however, did not significantly alter the static etch rate. This suggests that the passivation film formed by PTA on copper surface is different from the one formed by ATA or PTT. The flexible phenyl ring present in PTA must play a significant role in determining the hydrophobicity of the molecule and its tendency to form an effective passivating layer on copper.

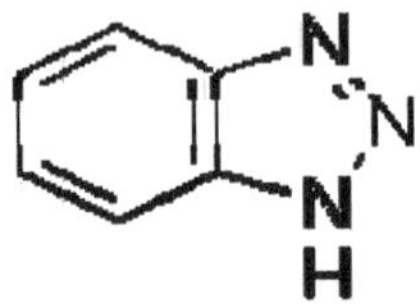

Benzotriazole (BTA)

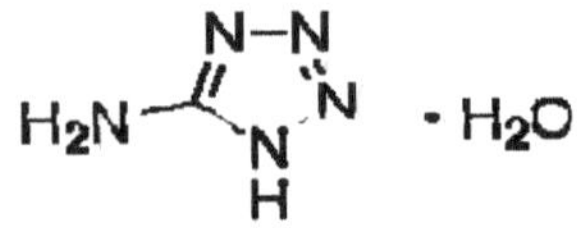

5-Aminotetrazole monohydrate (ATA)

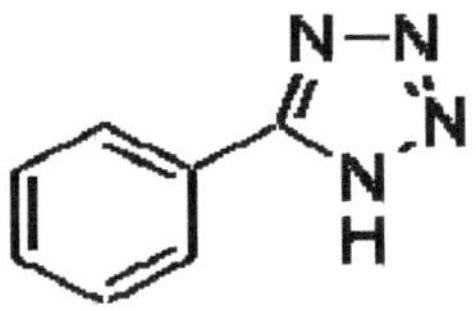

5-phenyl-1H-tetrazole (PTA)

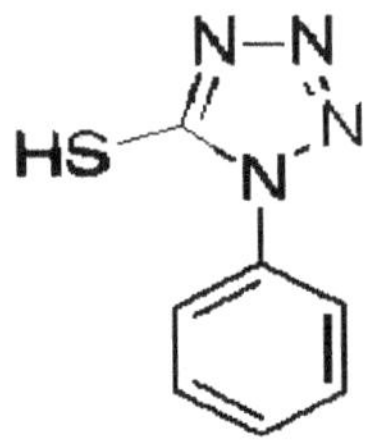

1-phenyl-1H-tetrazole-5-thiol (PTA)

FIGURE 8.8 Molecular structure of BTA and heterocyclic tetrazole compounds (ATA, PTA, and PTT).

During CMP, the temperature between the polishing pad and the wafer is elevated due to friction and chemical reactions [43,44]. The elevated temperature may have an impact on the surface adsorption behavior of the passivating agent. The passivating film may become thinner due to an

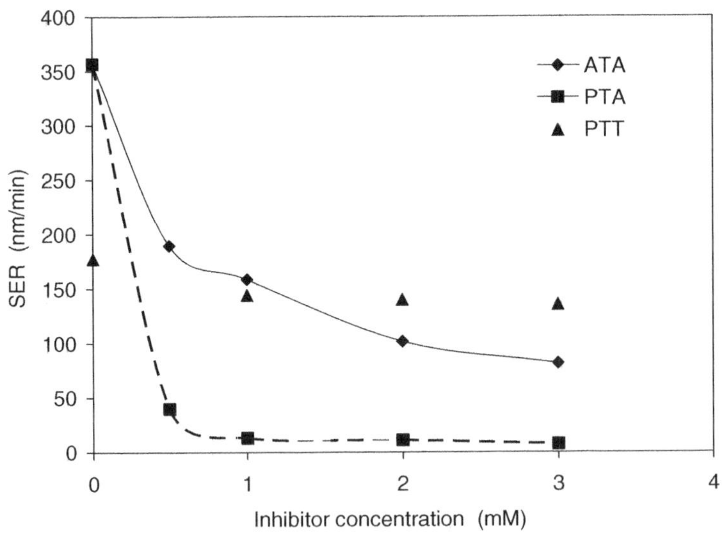

FIGURE 8.9 Static etch rate of copper in a solution containing 1 wt% of glycine, 3 wt% of H_2O_2, 3 wt% of silica and various concentrations of inhibitors (ATA and PTA at pH 3 and PTT at pH 4).

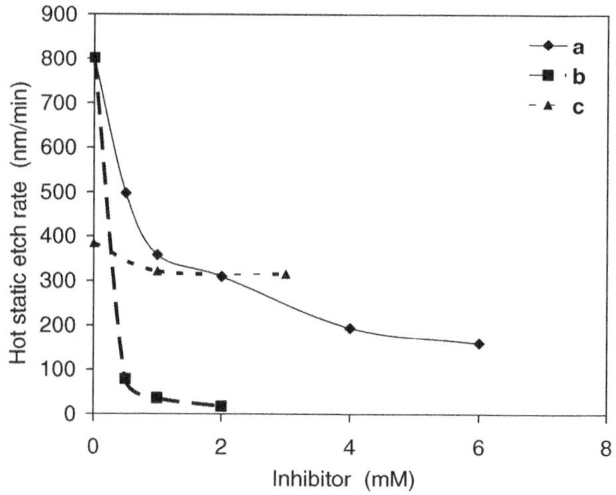

FIGURE 8.10 Hot static etch rate of copper in a solution containing 1 wt% of glycine, 3 wt% of H_2O_2, 3 wt% of silica, and various concentrations of inhibitors (**a**) ATA and (**b**) PTA at pH 3 and (**c**) PTT at pH 4.

increased desorption. A reduced passivating efficiency can lead to a higher level of dishing and corrosion during CMP. To properly characterize a slurry, therefore, its static etch rate should be measured at such an elevated temperature. Fig. 8.10 shows the "hot" (40 °C) static etch rate of copper in the presence of 1 wt% of glycine, 3 wt% of H_2O_2, and 3 wt% of silica particles with tetrazole inhibitors. The results indicate that hot static etch rate (HSER) follows the same trend as SER at room temperature. The ratio of HSER/SER is on the order of 2–3 at low inhibitor concentration and decreases with the increase in inhibitor concentration [45].

Similar to BTA, the tetrazoles may form passivation film on the copper surface in two steps. During the first step, tetrazole molecules are adsorbed onto the copper surface to form a strong complexing monolayer. Once Cu-tetrazole monolayer is formed, tetrazole–tetrazole interaction would be needed to form multiple layers of adsorption. The driving force for the further adsorption can be hydrophobic–hydrophobic, hydrogen bonding, or copper ions bridged complexes. In case of ATA, the ATA–copper complex may have enough solubility in water. The first monolayer of ATA–copper complex does not adhere to the copper surface strongly enough to serve as an anchor for subsequent adsorption. In addition, as the intermolecular ATA–ATA attraction may be weaker than that with water molecules, there is little driving force to form a thick ATA–ATA layer. In case of PTA, electron-rich phenyl group may form π–π stacking among phenyl rings, which leads to a thicker subsequent PTA layer. The hydrophobicity of the phenyl ring may also promote the further deposition of PTA molecules onto PTA layer. For PTT, due to the strong affinity of a thiol group (–SH) toward copper, a strong anchor layer must have been formed on the copper surface [45]. The phenyl

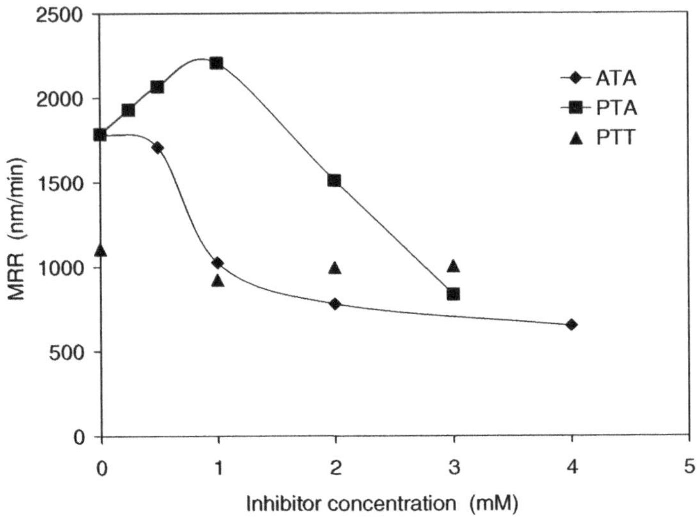

FIGURE 8.11 Material removal rate of copper obtained using slurry containing 1 wt% of glycine, 3 wt% of H_2O_2, and 3 wt% of silica with various concentrations of inhibitors (ATA and PTA at pH 3 and PTT at pH 4).

group in PTT could then promote the subsequent adsorption and stacking. The static etch rate measurement for PTT-containing slurry suggested that the passivating layer is significantly weaker than that made of PTA. The results strongly suggest that the PTT film formed on the surface is not tightly packed [46].

The copper material removal rates obtained using the slurries described above are shown in Fig. 8.11. The MRR for ATA- and PTT- containing slurries share the same trend as their corresponding static etch rates. This is consistent with the observations for all BTA-containing slurries. The MRR for PTA-containing slurry, however, shows a maximum at about 1 mM of the inhibitor concentration. This phenomenon strongly suggests that, under certain circumstances, the passivating film formed by PTA on copper surface may assist the copper removal. In another words, at 1 mM or below, PTA provides dual functionality—complexing and passivating. At concentration above 1 mM, PTA resumes normal behavior as a passivating agent [47].

Fig. 8.12 shows the static etch rate of copper as a function of slurry pH in CMP solution with 1 wt% of glycine and 3 wt% of H_2O_2. Without any passivating agent, the static etch rate increases with the decrease in pH. With PTT or ATA, the static etch rate shows a minimum at pH = 4–5 indicating that a slightly stronger passivating film can be formed at such pH. For a PTA-containing system, the SER of copper decreased with the pH of CMP solution. This is significantly different from that found for BTA. For BTA systems, an increase in pH in this range leads to an increase in passivating efficiency [46,47]. These trends have been correlated with the water solubility of each passivating agent (see Chapter 7 for details).

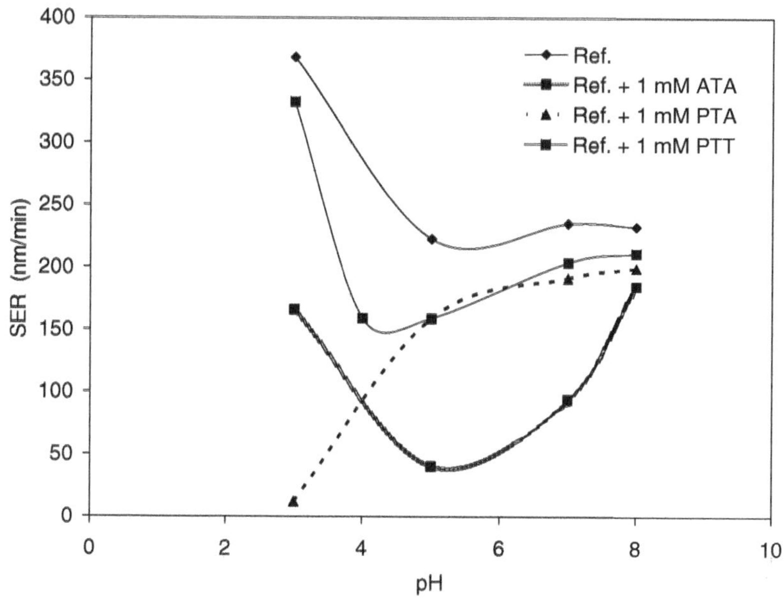

FIGURE 8.12 Static etch rate of copper obtained using slurry containing 1 wt% of glycine, 3 wt% of H_2O_2, and 3 wt% of silica + 1 mM inhibitors (ATA, PTA, and PTT) at various pH.

The material removal rates were measured using slurries containing 1 wt% of glycine and 3 wt% of H_2O_2 with silica particles at different slurry pH. As shown in Fig. 8.13, the MRR for the reference slurry and slurries containing ATA and PTT follow a similar trend as their SER. The removal rate decreases as the pH increases from 3 to 5. For PTA- containing slurry, the trend is reversed in relation to its SER. This created a unique situation, that, at pH = 3, the MRR/SER becomes extremely large. It was reported by Paul and Vacassy that a large inhibitor with good blocking power would also offer a large cross section for mechanical removal [48]. PTA that offers a good passivation for copper corrosion at low pH also helps in high material removal rate during polishing because of a large cross section for mechanical removal.

8.5 ELECTROCHEMICAL POLARIZATION STUDY OF CORROSION INHIBITORS IN Cu CMP SLURRY

Potentiodynamic polarization measurement is an effective electrochemical technique in characterizing film formed on a metal surface under various redox conditions. The electrochemical polarization measurements were

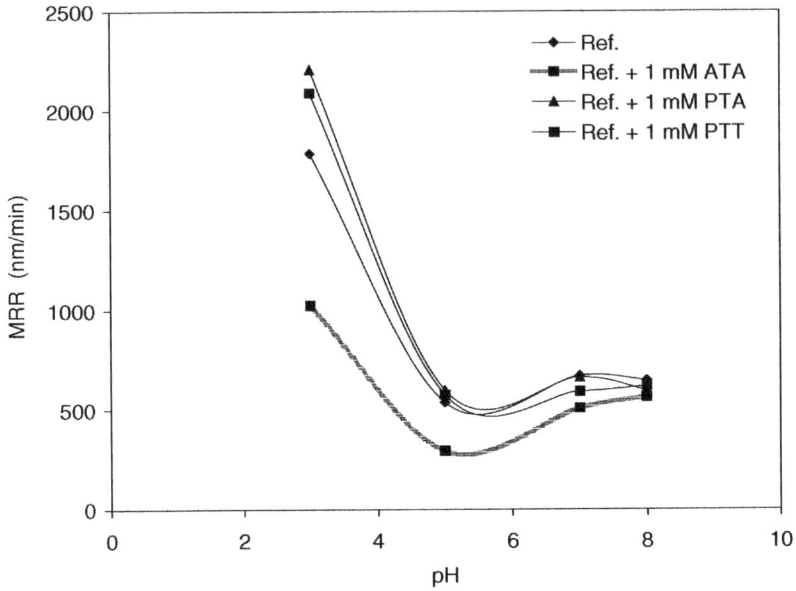

FIGURE 8.13 Material removal rate of copper obtained using slurry containing 1 wt% of glycine, 3 wt% of H_2O_2, and 3 wt% of silica + 1 mM inhibitors (ATA, PTA, and PTT) at various pH.

performed using potentiostat and galvanostat in an electrochemical cell [46]. The polarization curves for copper in the presence of CMP slurry containing 3 wt% of silica, 1 wt% of glycine, 3 wt% of H_2O_2, and 1 mM of tetrazole inhibitor (ATA and PTA at pH 3 and PTT at pH 4) are shown in Fig. 8.14. In the presence of ATA, PTA, and PTT, the cathodic and anodic curves were shifted to the higher potential region, which indicates that tetrazole compounds form a protective film on the copper surface. Similar to BTA, the polymeric Cu-tetrazole passivation film increased the barrier for the transport of ionic species to and from the copper surface. The values of corrosion potential (E_{corr}) and corrosion density (I_{corr}) were estimated from the polarization curves. Without any inhibiting agent the I_{corr} value was 2649 $\mu A/cm^2$. As shown in Table 8.1, the corrosion potential increased and corrosion current density decreased with the increase in inhibitor concentration, which indicates the formation of a denser passivation film. Although all three compounds exhibit corrosion inhibition and the performance of the tetrazole can be ranked as PTT < ATA < PTA, the presence of PTA brought an I_{corr} value as low as 251 $\mu A/cm^2$, an indication that PTA acts as an excellent inhibitor in Cu CMP slurry. Despande et al. reported I_{corr} of 8 $\mu A/cm^2$ and E_{corr} of 336 mV (Ag/AgCl) for a solution composition of 0.1 M of glycine, 5% hydrogen peroxide, and a high BTA concentration of 0.01 M for solution pH of 2 [19].

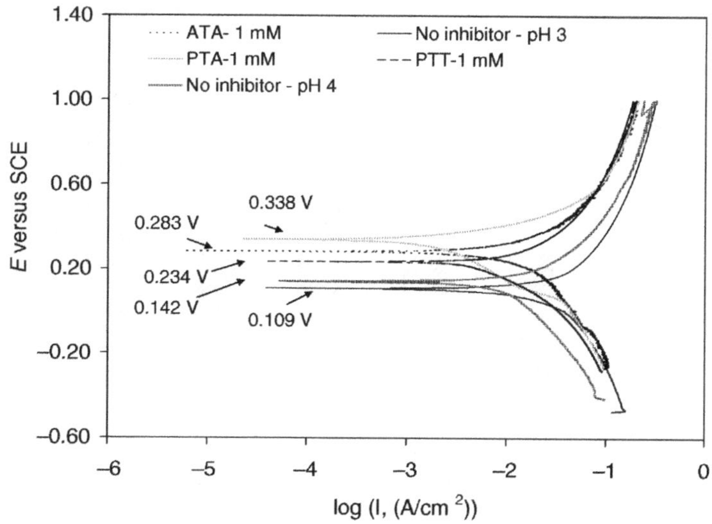

FIGURE 8.14 Effect of various inhibitors on the polarization behavior of copper in a slurry containing 3 wt% of silica, 1 wt% of glycine, and 3 wt% of H_2O_2 (at pH = 3 for no inhibitor, PTA and ATA; at pH = 4 for no inhibitor and PTT).

TABLE 8.1 E_{corr} and I_{corr} Estimated from the Polarization Studies of Copper in Various Slurries.

Slurry	E_{corr}, V	I_{corr}, μA/cm^2
3% Silica + 1% glycine + 3% H_2O_2 (Ref.)	0.109	2649
Ref. + 1 mM ATA at pH 3	0.283	1193
Ref. + 1 mM PTA at pH 3	0.338	348
Ref. + 1 mM PTT at pH 4	0.234	1532

8.6 HYDROPHOBICITY OF THE SURFACE PASSIVATION FILM

Owing to the nature of the passivating film, as shown in Fig. 8.4, the passivating layer that is in direct contact with slurry should be relatively hydrophobic. Contact angle measurement on the copper surface exposed to CMP solution can reveal the presence and the extent of a hydrophobic film formed on the surface. Fig. 8.15 shows the contact angle measured for various inhibiting systems. High contact angle value indicates more hydrophobic nature of the film whereas low contact angle value indicates less hydrophobic nature of the film formed on the copper surface. Among the samples, Cu surface exposed to PTA has the largest contact angle, which suggests that Cu–PTA film is more hydrophobic compared to other films. Presence of phenyl group in PTA molecule increased the hydrophobic nature of the film. Also, if the orientation of adsorbed phenyl group is parallel to the copper surface, then

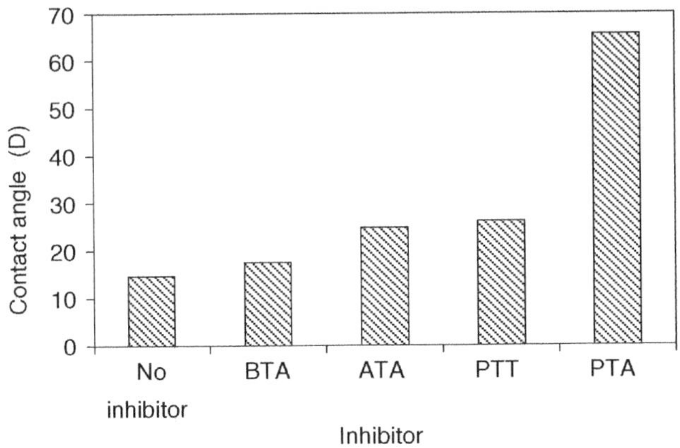

FIGURE 8.15 Contact angle measurement on copper surface exposed to CMP solution containing 1 wt% of glycine and 3 wt% of H_2O_2 (for no inhibitor, BTA, PTA, and ATA, at pH = 3 and for PTT at pH = 4).

this structure could favor large contact angle. Low contact angle of Cu–PTT film suggests that even though PTT has phenyl group in its molecular structure, either the film formed is not continuous or the PTT does not have absolute flat orientation but is tilted relative to the copper surface and hence the film is less hydrophobic in nature. Cu surface exposed to ATA-containing solution shows low contact angle as ATA does not have hydrophobic functional group. In case of BTA, even though it has hydrophobic aromatic group in its molecular structure, it has low contact angle. This could be because the BTA molecule does not form a strong Cu–BTA passivation film at low-pH slurry and the orientation of aromatic group is not parallel but is tilted relative to the copper surface [6]. As the inhibitor molecules are adsorbed onto the copper surface to form copper-inhibitor film (through amino group for BTA and ATA, phenyl group for PTA, and thiol group for PTT) the presence or hydrophobic nature of the functional group present in the molecule acts as an additional layer of protection that shields the underlying copper from corrosive species present in the CMP solution/slurry.

8.7 COMPETITIVE SURFACE ADSORPTION BEHAVIOR OF CORROSION INHIBITORS

As shown in Fig. 8.4, to form an effective passivating film on copper, a corrosion inhibitor must adsorb onto the surface. The adhesion to the copper surface must be strong enough to protect against the shear force impinged by slurry flow. The coordinate bond found in a typical copper(I) or copper(II) complex is usually sufficient to serve as an anchor. The static charge attraction between a surfactant and a charged surface may be too weak to perform such function. In order to completely stop the corrosion current or the transport of ionic species, there should be additional layer(s) of inhibitor molecules stacked on top of the anchor layer. An effective passivating film must have strong adhesive forces among and within these layers. A weak link anywhere in the film will lead to an insufficient passivation or local corrosion.

At molecular level, the presence of electron-donating nitrogen, oxygen, and sulfur atoms, and multiple bonds in the organic inhibitor molecule usually facilitates adsorption on the metal surface [49]. More specifically, the electronic density on the donor atoms may determine the strength of adsorption. The steric hindrance effect may have an impact on the packing density of the passivating film [50,51]. On the copper surface exposed to CMP slurry, the inhibitor molecules must compete with other species such as ions and surfactants [52]. There are many useful techniques available for the purpose of characterization of the chemical and physical nature of the passivating film *in situ* or *ex situ* [6,13,53–56]. For detailed discussions, the reader can refer to books listed in reference numbers [57,58]. A simple technique that can semiquantitatively estimate the surface adsorption tendency and quickly screen for potential inhibitors is described below.

The technique requires that the exact amount of corrosion inhibitor in a model CMP slurry or solution be determined without significant interference from the other chemical components. For example, the exact concentration of BTA in a solution can be determined using its UV absorption intensity against a calibration curve. As the determination is done using a peak at 260 nm for BTA, there should be no other species in the system that absorbs strongly in the same region. To simulate the presence of copper surface, a certain amount of copper particles with known specific surface area is added to the model CMP slurry or solution. The added copper particles should be small enough to provide large enough total surface area and large enough to allow easy separation. For the study described below, particles with an average of 80 μm in size are used. The particles can be pretreated to simulate the fact that the surface is oxidized or not. An adequate amount of time (usually less than a minute) is given to allow the added particles to reach equilibrium with BTA in the solution. A series of UV/Vis absorption spectra were then obtained as a function of Cu particles added. The level of decrease in BTA concentration in the solution as a result of added copper particles indicates the adsorption tendency for BTA onto copper surface. As a case study, Fig. 8.16 shows the effect of oxidized copper particles on the residue concentrations of ATA, PTT, PTA, and BTA. It is clear that the total amount of ATA adsorbed onto the same copper surface is significantly less than PTT, PTA, and PTT. As described earlier, both PTT and ATA do not yield strong enough passivating film on copper surface. The copper particles' adsorption experiment implies that the ATA molecules do not form thick enough stacking layers on top of the anchor layer. For PTT, although significant thickness of stacking layer may have formed on the anchor layer, the stacking may have very low packing density or high porosity.

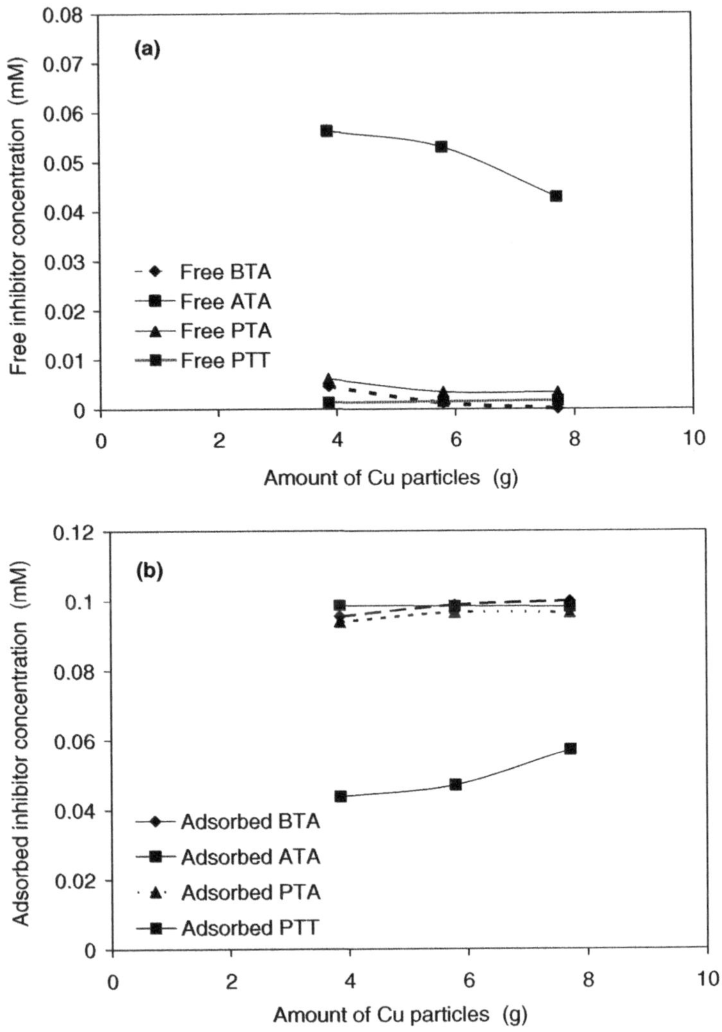

FIGURE 8.16 Concentration of (**a**) free inhibitor in solution and (**b**) adsorbed inhibitor onto Cu particles after surface adsorption in various amounts of copper particles.

As mentioned earlier, chemisorption of an organic inhibitor is the main driving force for the formation of the anchor layer. The presence of lone pair electrons on the nitrogens in all three tetrazole molecules gives rise to the tendency to form such a complex with the oxidized copper species on the surface. The subsequent adsorption of molecules onto the surface is determined by the solubility of these molecules in water and attraction between themselves such as hydrogen bonding, π–π interaction, and hydrophobic–hydrophobic interactions. If the solubility is high and the intermolecular interaction among inhibitor molecules is weak, only the anchor layer will form. If the solubility is low and the intermolecular interaction among inhibitor molecules is high, a multilayer will occur. The solubility profiles of the inhibitors are illustrated in Fig. 8.17 [41,42,59,60]. It is thus possible to predict or explain the passivating behaviors of potential corrosion inhibitors according to their water solubility. For PTA and PTT, similar to BTA, the presence of phenyl groups may increase the tendency to form stacking layer [40]. Unlike PTT, PTA, and BTA, ATA does not have any hydrophobic moiety that leads to a high water solubility. This results in ATA forming monolayer or very thin film on copper surface [61].

The water solubility profiles of corrosion inhibitors are also very helpful in predicting the optimum pH for the CMP slurry to give the best performance and design matching post-CMP cleaning solution. For example, according to Fig. 8.17a, BTA- containing slurry will yield the best results at pH 4–7 in terms

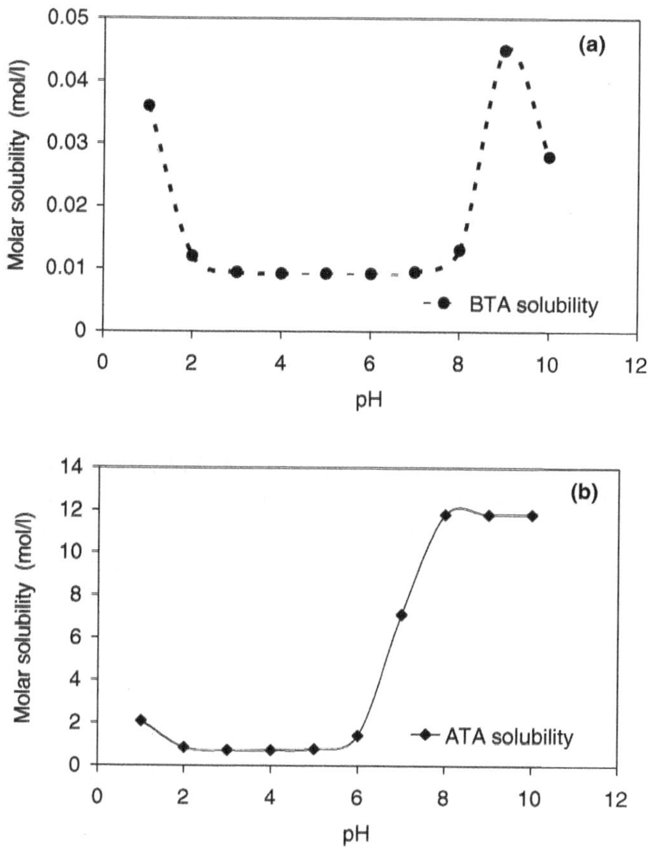

FIGURE 8.17 Inhibitor molar solubility data for (**a**) BTA, (**b**) ATA, (**c**) PTA, and (**d**) PTT at various pH values (from Refs. 41,42,62,63).

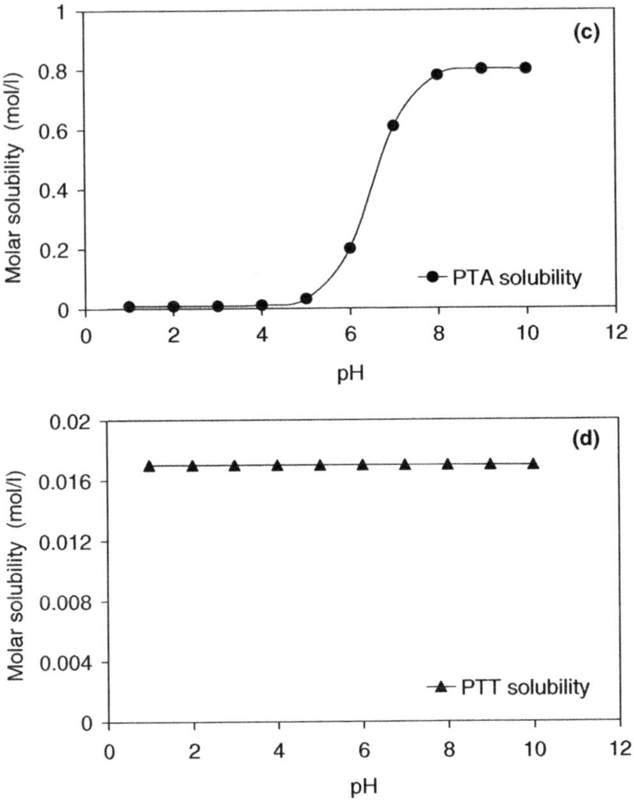

FIGURE 8.17 (*Continued*).

of its corrosion-inhibition property. The post-CMP cleaning solution should be formulated at pH below 3 or above 8. For PTA, using a similar analogy, the CMP slurry should be formulated at pH 2–4 and its post-CMP cleaning solution should work well at pH above 7. For PTT, however, owing to its low solubility across the pH range, an effective post-CMP cleaning solution may be difficult to formulate at any pH.

8.8 SUMMARY

Corrosion inhibitors are an important chemical constituent of Cu CMP slurry. The critical performance of a slurry such as static etch rate, material removal rate, and post-CMP defect counts are greatly influenced by the concentration and the structure of the inhibitor used in the slurry. BTA is proven to be an effective corrosion inhibitor for Cu CMP especially in nonacidic medium.

Significant progress has been made in the development of CMP slurry in identifying and investigating alternate chemical additives for CMP slurry. Continuous improvement in the development of slurry will find the acceptance of CMP in new applications such as Cu–low-*k* integration. The investigation of new corrosion inhibitors such as tetrazoles for copper CMP process reveals that the SER and MRR of copper are strongly affected by functional groups attached to the tetrazole ring. Experimental results clearly indicate that tetrazole compounds ATA, PTA, and PTT inhibit copper corrosion more effectively at acidic pH. This is in sharp contrast to the commonly used BTA molecule. The result reveals that the inhibiting efficiency of the tetrazole inhibitor in Cu CMP increases in the following order—PTT < ATA < PTA. Unlike BTA, PTA shows an excellent inhibiting characteristic while maintaining a steady material removal rate over a respectable concentration range. Studies suggest that hydrophobicity of the passivation film formed on the copper surface influences the barrier properties of the film and hence the inhibiting efficiency of the tetrazole inhibitor. The inhibiting efficiency of inhibitor decreases with the temperature and the rate of decrease is in the following order—PTA < PTT < ATA. This study reveals that the surface adsorption behavior depends strongly on the solubility of the inhibitor. It is revealed that strong surface adsorption tetrazole additives such as PTA that can form thick passivation film on copper surface could be used as an effective corrosion inhibitor for Cu CMP slurry at acidic pH.

QUESTIONS

1. What is the role of corrosion inhibitor in metal CMP slurry?
2. List some of the corrosion inhibitors other than BTA that are used in CMP slurry formulation.
3. What is the pH range in which the BTA has good corrosion-inhibitor property in CMP slurry? What are the reasons for BTA having good corrosion-inhibitor property in that pH range?
4. Name some of the functional groups that favor surface adsorption on copper metal.
5. What is the effect of temperature on the passivating property of a corrosion inhibitor during CMP process?
6. What do you expect in a potentiodynamic polarization measurement if you have corrosion inhibitor in a CMP slurry, and what are the reasons for expecting such behavior?
7. What are the factors that favor monolayer or multilayer passivation layer formation by a corrosion inhibitor on copper surface?
8. What is the effect of corrosion inhibitor on the hydrophobicity of copper surface? What is the effect of change in hydrophobicity on the material removal rate and static etch rate of copper?

9. What is the effect of corrosion inhibitor on the post-CMP cleaning?

10. What are the selection criteria for a chemical substance to be used as a corrosion inhibitor in a CMP process?

REFERENCES

1. Steigerwald JM, Murarka SP, Gatmann RJ. Chemical Mechanical Planarization of Microelectronic Materials. New York: John Wiley & Sons Inc.; 1996.
2. Hariharaputhiran M, Zhang J, Ramarajan S, Keleher J, Li Y, Babu SV. Hydroxyl radical formation in H_2O_2–amino acid mixtures and chemical mechanical polishing of copper. J Electrochem Soc 2000;147(10):3820–3826.
3. Luo Q, Campbell DR, Babu SV. Stabilization of alumina slurry for chemical-mechanical polishing of copper. Langmuir 1996;12:3563.
4. Luo Q, Campbell DR, Babu SV, Proceedings of the 1st International VMIC Specialty Conference on CMP Planarization; 1996 Feb; Santa Clara CA;1996. P145.
5. Wang MT, Tsai MS, Liu C, Tseng WT, Chang TC, Chen LJ, Chen MC. Effects of corrosion environments on the surface finishing of copper chemical mechanical polishing. Thin Solid Films 1997;518:308.
6. Walsh J, Dhariwal H, Gutierrez A, Finneti P, Muryn C, Brookes N, Oldman R, Thomton G. Probing molecular orientation in corrosion inhibition via a NEXAFS study of benzotriazole and related molecules on Cu(100). Surf Sci 1998;415:423.
7. Notoya T, Poling G. Corrosion (Houston) 1976;32:216.
8. Brusic V, Frisch MA, Eldridge BN, Novak FP, Kaufman FB, Ruch BF, Frankel GS. Copper corrosion with and without inhibitors. J Electrochem Soc 1991;138:2253.
9. Tommesani L, Brunoro G, Frignani A, Monticelli C, Dal Colle M. On the protective action of 1,2,3-benzotriazole derivative films against copper corrosion. Corros Sci 1997;39:1221.
10. Carpio R, Farkas J, Jairath R. Initial study on copper CMP slurry chemistries. Thin Solid Films 1995;266:238–244.
11. Thierry D, Leygraf C. Simultaneous raman spectroscopy and electrochemical studies of corrosion inhibiting molecules on copper. J Electrochem Soc 1985;132:1009.
12. Rubim J, Gutz IGR, Sala O, Orville-Thomas WJ. Surface enhanced Raman spectra of benzotriazole adsorbed on a copper electrode. J Mol Struct 1983;100:571.
13. Cohen SL, Brusic VA, Kaufman FB, Frankel GS, Motakef S, Rush B. X-ray photoelectron spectroscopy and ellipsometry studies of the electrochemically controlled adsorption of benzotriazole on copper surfaces. J Vac Sci Technology 1990;A8(3):2417.
14. Poling GW. Reflection infra-red studies of films formed by benzotriazole on Cu. Corrosion Sci 1970;10(5):359.
15. Chadwick D, Hashemi T. Adsorbed corrosion inhibitors studied by electron spectroscopy: Benzotriazole on copper and copper alloys. Corrosion Sci 1978;18:359.
16. Fox PG, Lewis G, Boden PJ. Some chemical aspects of the corrosion inhibition of copper by benztriazole. Corrosion Sci 1979;19:457.

17. Tamilmani S, Huang W, Raghavan S, Small R. Potential-pH diagrams of interest to chemical mechanical planarization of copper. J Electrochem Soc 2002;149(12): G638–G642.
18. Hang YK, Eom DH, Park JG. Electrochemical Society Meeting; Sep 2–7; San Francisco, CA; 2001.
19. Deshpande S, Kuiry SC, Klimov M, Obeng Y, Seal S. Chemical mechanical planarization of copper: role of oxidants and inhibitors. J Electrochem Soc 2004;151:G788.
20. Luo Q, Babu SV. Dishing effects during chemical mechanical polishing of copper in acidic media. J Electrochem Soc 2000;147(12):4639–4644.
21. Tsai T-H, Yen S-C. Localized corrosion effects and modifications of acidic and alkaline slurries on copper chemical mechanical polishing. Appl Surf Sci 2003;210(3–4):190.
22. Steigerwald JM, Murarka SP, Gutmann RJ, Duquette DJ. Chemical processes in the chemical mechanical polishing of copper. Mater Chem Phys 1995;41:217.
23. Notoya T, Satake TM, Ohtsuka T, Yashiro H, Sato M, Yamauchi T, Schweinsberg DPPaper 076 (https://www. umist.ac.uk/corrosion/JCSE), International Symposium on Corrosion Science in the 21st Century, UMIST, Manchester, UK, July 6–11, 2003.
24. Surya Sekhar M, Ramanathan S. Characterization of copper chemical mechanical polishing (CMP) in nitric acid–hydrazine based slurry for microelectronic fabrication. Thin Solid Films 2006;504(1–2):227–230.
25. Du T, Luo Y, Desai V. The combinatorial effect of complexing agent and inhibitor on chemical–mechanical planarization of copper. Microelectron Eng 2004;71(1):90–97.
26. Hu TC, Chiu SY, Dai BT, Tsai MS, Tung I-C, Feng MS. Nitric acid based slurry with citric acid as an inhibitor for copper chemical mechanical polishing. Mat Chemi Phys 1999;61(2):169–171.
27. Lee W-J. Inhibiting effects of imidazole on copper corrosion in 1 M HNO_3 solution. Mat Sci Eng 2003;AC348:217.
28. Hong Y, Patri UB, Ramakrishnan S, Roy D, Babu SV. Utility of dodecyl sulfate surfactants as dissolution inhibitors in chemical–mechanical planarization of copper. J Mater Res Soc 2005;20(12):3413.
29. Holberg L, Jonseen B, Kronberg B, Lindman B. Surfactants and Polymers in Aqueous Solution. 2nd ed. Wiley; 2003.
30. Govindaswamy S, Cheemalapati K, Li Y. Evaluation of Surfactant as Corrosion Inhibitor in Copper Chemical Mechanical Planarization. Unpublished results 2007.
31. Zhao J, Bundi D, Cheemalapati K, Duvvuru V, Li YA Non-BTA Based Novel Post CMP Clean Solution. CMP-MIC.2005.
32. Govindaswamy S, Wu Z, Li Y. Evaluation of Novel Chemical Additive as an Inhibiting Agent in Copper CMP. CMP-VMIC.2006.
33. Gorantala VRK, Goia D, Matijević E, Babu SV. Role of Amine and Carboxyl Functional Groups of Complexing Agents in Slurries for Chemical Mechanical Polishing of Copper. J Electrochem Soc 2005;152:G912.
34. Wrschka P, Hernandez J, Oehrlein GS, King J. Chemical Mechanical Planarization of Copper Damascene Structures. J Electrochem Soc 2000;147:706.
35. Kertit S, Essoufi H, Hammouti B, Benkaddour MJ Chem Phy 1998;95:2072.

36. Yan CW, Lin HC, Cao CNElectrochim Acta 2000;45:2815.
37. Zucchi F, Trabanelli G, Fonsati M. Tetrazole derivatives as corrosion inhibitors for copper in chloride solutions. Corros Sci 1996;38:2019.
38. Essoufi H, Kertit S, Hammouti B, Benkaddour MBull Electrochem 2000;16:2005.
39. Ravichandran R, Rajendran N. Electrochemical behaviour of brass in artificial seawater: effect of organic inhibitors. Appl Surf Sci 2005;241:449.
40. Mihit M, El Issami S, Bouklah M, Bazzi L, Hammouti B, Ait Addi E, Salghi R, Kertit S. The inhibited effect of some tetrazolic compounds towards the corrosion of brass in nitric acid solution. Appl Surf Sci 2006;252:2389.
41. Substance identifier no. 18039-42-4,SciFinder Scholar (2006), Calculated using advanced chemistry development (ACD/Labs) software V8.14 for solaris (1994–2006 ACD/Labs).
42. Substance Registry No. 95-14-7,SciFinder 2006, Calculated using advanced chemistry development (ACD/Labs) software V8.14 for solaris (1994–2006 ACD/Labs).
43. Zantye PB, Kumar A, Sikder AK. Chemical mechanical planarization for microelectronics applications. Mater Sci Eng 2004;45:89.
44. Lu H, Obeng Y, Richardson KA. Applicability of dynamic mechanical analysis for CMP polyurethane pad studies. Mater Charact 2002;49(2):177.
45. Tan YS, Srinivasan MP, Pehkonen SO, Chooi SYM. Effects of ring substituents on the protective properties of self-assembled benzenethiols on copper. Corrosion Sci 2006;484:840.
46. Govindaswamy S, Li Y. Investigation of tetrazole based corrosion inhibitor for Cu CMP. Unpublished results 2007.
47. Govindaswamy S, Li Y. Investigation of 5-pheny-1-H-tetrazole as corrosion inhibitor in chemical mechanical planarization of copper film. Unpublished results 2007.
48. Paul E, Robert V. Mat. Res. Soc. Symp. Proc. Vol. 767 © 2003 Materials Research Society. PAGE: F1.2.1-F1.2.6
49. Bouklah M, Benchat N, Hammouti B, Aouniti A, Kertit S. Thermodynamic characterisation of steel corrosion and inhibitor adsorption of pyridazine compounds in 0.5 M H_2SO_4. Mater Lett 2006;60(15):1901.
50. Bentiss F, Traisnel M, Lagrenée MJ. Appl Electrochem 2001;31:41.
51. Riggs Jr. OL. Corrosion Inhibitor, C. C Nathan, Ed., NACE International Houston, TX, 1981.
52. Vracar L-M, Drazic DM. Adsorption and corrosion inhibitive properties of some organic molecules on iron electrode in sulfuric acid. Corros Sci 2002;44:1669.
53. Lalitha A, Ramesh S, Rajeswari S. Surface protection of copper in acid medium by azoles and surfactants. Electrochim Acta 2005;51(1):5.
54. Aramaki K, Kiuchi T, Sumiyoshi T, Nishihara H. Surface enhanced Raman scattering and impedance studies on the inhibition of copper corrosion in sulphate solutions by 5-substituted benzotriazoles. Corr Sci 1991;32(5–6):593.
55. Ito M, Takahashi M. IR reflection-absorption spectroscopic study of benzotriazole on copper. Surface Science 1985;158(1–3):609.
56. Szöcs E, Vastag GY, Shaban A, Konczos G, Kálmán E. Investigation of copper corrosion inhibition by STM and EQCM techniques. J App ElectroChem 1999;29:1339.

57. Holmerg K, editor. Handbook of Applied Surface and Colloid Chemistry,Volume 1–2. John Wiley & Sons; 2002.
58. Brundle CR, Evans CA Jr,Wilson S, editors. Encyclopedia of materials characterization—surfaces, interfaces, thin films.Elsevier; 1992.
59. Substance Registry No. 86-93-1,SciFinder 2006, Calculated using advanced chemistry development (ACD/Labs) software V8.14 for solaris (1994–2006 ACD/ Labs).
60. Substance Registry No. 4418-61-5, SciFinder 2006, Calculated using advanced chemistry development (ACD/Labs) software V8.14 for solaris (1994–2006 ACD/ Labs).
61. Govindaswamy S, Li Y. Surface adsorption behavior of CMP corrosion inhibitors on Cu surface. Unpublished results 2007.

9

TUNGSTEN CMP APPLICATIONS

JEFF VISSER

9.1 INTRODUCTION

Tungsten chemical–mechanical planarization (CMP) is one of the earliest forms of planarization techniques for integrated circuit (IC) manufacturing following the induction of oxide CMP. Unlike oxide CMP that produces a planar surface to address the depth-of-focus issue, tungsten CMP enables the construction of a functional structure. More specifically, tungsten CMP removes the excess tungsten and barrier layer, leaving contact plugs filled with tungsten and lined with the barrier layer (usually a combination of titanium and titanium nitride). These now planarized contacts, also known as vias, are the building blocks for metallization interconnects. In addition to the differences in function between tungsten and silicon dioxide CMP, the material removal mechanism is also dissimilar. Oxide CMP is a mechanically dominated process that removes the higher topologies at a higher rate than those in recessed areas [1]. The chemically assisted dissolution of silicon dioxide often plays a minor role unless the process is performed at extremely high pH [2]. In comparison, tungsten (W) CMP is a chemically driven process. Think of W CMP as a carpenter would do when refinishing a painted piece of wood. First, the paint remover is applied, allowing a layer of paint to de-adhere from the surface. The now bubbled layer of paint can be easily wiped away with a clean cloth. This two-step process is repeated until all of the paint has been removed or only small amounts of residue remain; W CMP operates in the same way. More specifically, an oxidizer is added to the slurry to soften the tungsten surface. The modified tungsten layer is subsequently removed by mechanical force through the actions of abrasive particles and the polishing pad. This

Microelectronic Applications of Chemical Mechanical Planarization, Edited by Yuzhuo Li

two-step process is then repeated until the tungsten layer is completely removed or the process is manually stopped. This mechanistic view of the CMP process is called the Kaufman theory because F. Kaufman, who wrote the foreword for this book, was the first to propose such an approach [2]. Commonly used oxidizers include hydrogen peroxide, periodic acid, and ferric salts [3]. Since the implementation of W CMP process and the initial publications on CMP in the early 1990s [2,4–6], there have been vast number of publications on this topic. Readers are referred to some of those excellent reviews [7–9]. The focus of this chapter is on some of the practical aspects of W CMP, including those issues that practitioners face in consumable selections for defect reduction. More specifically, the basic application of tungsten CMP including the general requirements and a simple polishing flow is first introduced. Some representative defects that may be generated during W CMP will be discussed. Finally, various processing design options and suggestions will be explored.

9.2 BASIC TUNGSTEN APPLICATION, REQUIREMENTS, AND PROCESS

9.2.1 Basic Applications of Tungsten CMP

The primary and initial events that occur on the tungsten surface upon contact with W CMP slurry are a series of oxidation reactions. These reactions can be best described using a Pourbaix diagram with the corresponding chemical components, mainly the metal, the oxidizer, and other active ingredients that can influence the thermodynamics of these oxidation reactions. It is important to point out that as CMP is a dynamic process, the chemical reactions described in the Pourbaix diagram may never reach their thermodynamic equilibrium. Regardless of this, a Pourbaix diagram offers tremendous amount of information on the chemical reactions involved and their possible impact on the CMP outcome. Using hydrogen peroxide as an example, a simplified $W–H_2O_2$ Pourbaix diagram is shown in Fig. 9.1. As most of the W CMP processes are performed in an acidic regime, the Pourbaix diagram predicts the formation of a passive WO_3 layer. As this passivation layer is removed, the pads are slowly stained with a brownish-yellow color indicating the presence of both a tungsten IV oxide and a tungsten VI oxide. Unlike copper CMP, the formation of an effective passivating film by tungsten on its own eliminates the need to include an extra passivating agent for simple tungsten film CMP. If there is any reason to include an anticorrosion agent in the slurry, the requirement must be related to the need to protect the tungsten structures in a particular environment such as when patterned with other metals that may lead to additional electrochemical reactions [10].

Similar to oxide CMP, a basic W CMP process uses a polisher equipped with a polishing pad, a cleaning unit, and a slurry distribution system. Owing to the need for an oxidizer, the slurry distribution system also includes a premixing sequence. As the lifetimes of these oxidizers are usually short, the

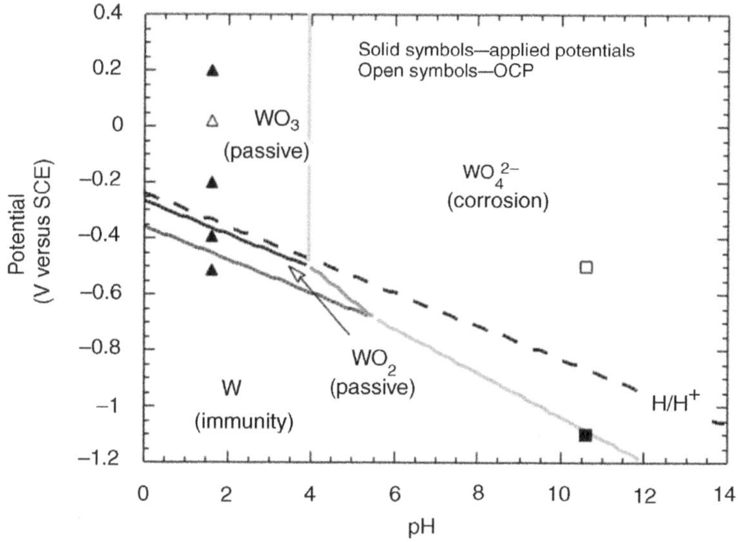

FIGURE 9.1 Pourbaix diagram for tungsten calculated at 10^{-6} M WO_4^{2-} (from Ref. 11).

pot lifetime of the slurry may be significantly shorter than that of oxide CMP slurry [1].

As mentioned in the introduction, the basic application of a tungsten CMP process is in constructing a set of tungsten vias as part of the interconnect network on a wafer. Figure 9.2 shows a basic contact layer stack after tungsten deposition and before W CMP. STI stands for shallow trench isolation (for more details see Chapters 12 and 13), BPTEOS for boron phosphorous tetraethyl orthosilicate, poly for polycrystalline silicon, substrate for Si (silicon) wafer with slight P- or N-type doping, and barrier layer stands for a combination of Ti and TiN. The goal of a W CMP process is to remove all excess tungsten above the Ti/TiN barrier layer along and the barrier layer itself. The ideal end result is shown in Fig. 9.3, where the dielectric material BPTEOS is exposed and the tungsten studs are formed. The minimum requirement for such a process is that the loss of tungsten due to dishing be less than 0.05% of its total height [12]. A typical wafer stack before CMP consists

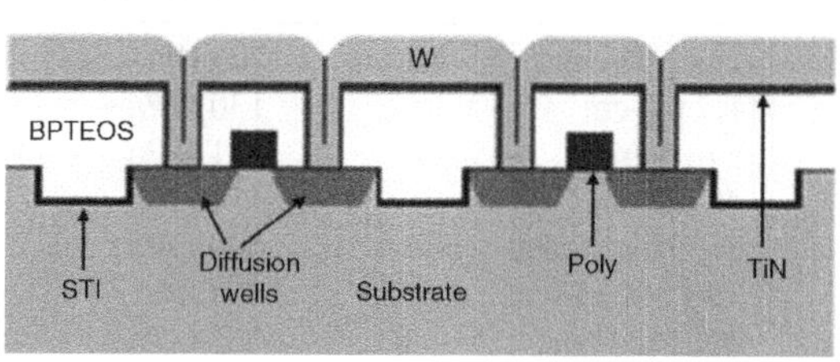

FIGURE 9.2 Post-W deposition (basic contact layer stack diagram).

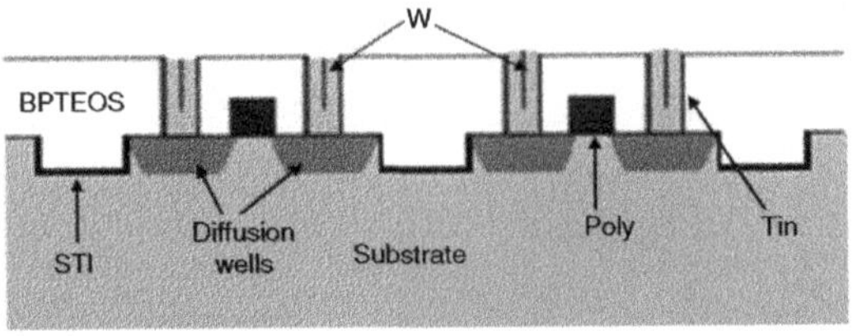

FIGURE 9.3 Post W CMP, left two contacts show minor plug recession, right two contacts show minor plus protrusion.

of 3500 Å of excess tungsten, 250 Å of a combination of Ti and TiN, and 5000 Å trench height of tungsten. Table 9.1 lists the physical and mechanical properties of the key chemical ingredients commonly found in a wafer stack that is encountered by the W CMP process. It is important to point out that these properties may vary significantly depending upon the processes that yielded the film stacks and subsequent treatment such as annealing and mechanical processes prior to W CMP. Similarly, the specifications given on the film stack may also change depending upon the overall integration scheme.

The circuit shown in both Figs. 9.2 and 9.3 also illustrates a basic CMOS (complementary metal oxide semiconductor) inverter. From left to right, the circuit flows from VDD (power supply voltage) into the PMOS (positive-channel metal oxide semiconductor) and then pulls up the output of the PMOS to near VDD, giving a low input on the polysilicon gate. Just the opposite happens from right to left; the circuit flows from GND (ground) to the NMOS (negative-channel metal oxide semiconductor), subsequently pulling down the output of the NMOS to GND, giving a high input on the polysilicon gate. To effectively allow carrier flow through the transistors, the transistors need an efficient low-resistance path to VDD, GND, and other transistors. Tungsten plugs provide the path from the metal interconnect lines (aluminum, copper, or, in rare cases, tungsten) through the tungsten plugs to the appropriate location in the circuit.

Tungsten CMP, to a lesser extent, has also been utilized to construct tungsten lines as a part of the interconnect via damascene or dual-damascene

TABLE 9.1 Physical Properties of Pre-WCMP Wafer Stack.

Material	Electrical Resistivity	Young's Modulus	Density
Aluminum	2.65 μΩ cm	70 GPa	2700 kg m^{-3}
Copper	1.7 μΩ cm	130 GPa	8920 kg m^{-3}
Tungsten	5 μΩ cm	411 GPa	19,250 kg m^{-3}
Titanium	40 μΩ cm	116 GPa	4507 kg m^{-3}
Titanium nitride	30–70 μΩ cm	600 GPa	5430 kg m^{-3}
Silicon	100,000 μΩ cm	47 GPa	2330 kg m^{-3}
Polysilicon	4000 μΩ cm	165 GPa	2330 kg m^{-3}
SiO_2	10^{12} Ω m	66 GPa	2650 kg m^{-3}

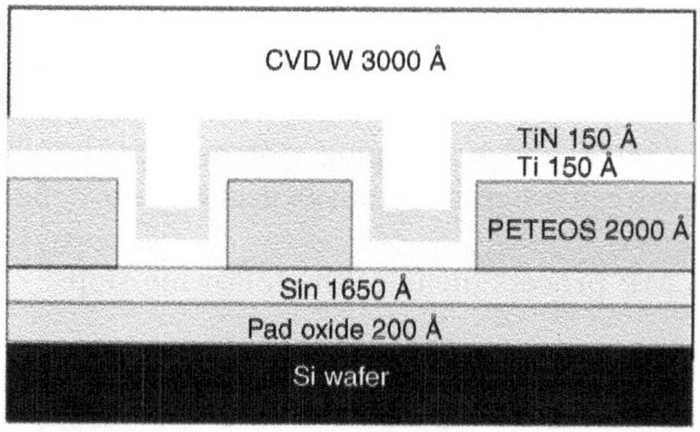

FIGURE 9.4 A cross section of a typical testing wafer in which tungsten lines are used as an interconnect (from Ref. 13).

processing. With damascene, the process yields a recessed metal line within the dielectric layer shown in Fig. 9.4. A dual damascene gives a simultaneous recessed line and vertical contact. Figure 9.5 shows a typical testing wafer for tungsten plus or Vias. The benefits of using tungsten over copper in such a scheme are mainly process related, including minimized contamination risk, decreased risk of corrosion, lower cost of consumables, and possibly higher yield. Copper on the contrary has many circuit-related gains such as decreased resistance and hence decreased signal processing delays.

9.2.2 Basic W CMP Requirements and Procedures

From a practical point of view, similar to any other CMP process, W CMP is required to have a polisher equipped with polishing pad, a cleaning unit, and a slurry distribution system with a premixing sequence. The polisher is usually composed of a polishing table, a slurry dispense and dispersion apparatus, a pad conditioning unit, an assembly designed to hold the wafer while polishing occurs (head), and a robotic unit to bring the wafer into and out of the system. The

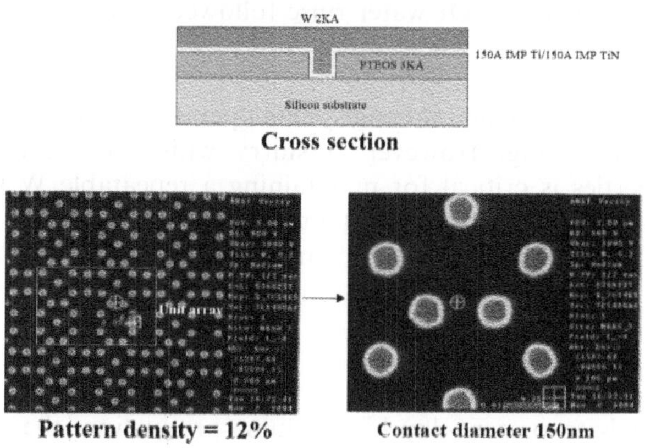

FIGURE 9.5 A typical testing wafer for tungsten plugs (from Ref. 13).

cleaning unit contains a cleaning solution port, a DI (deionized water) port, one or two physical brush stations, a megasonic clean, and a dry cycle. The slurry distribution system is composed of a mixing vessel, multiple chemical inputs, a short-term storage tank, and a pumping system with the ability to maintain sufficient head pressure in order to sustain an accurate slurry flow at the polishing tool dispense.

A routine W CMP process includes the following steps:

1. Loading wafer into the polishing tool.
2. Robotically Loading the wafer onto the polishing head.
3. Preconditioning the pad or breaking-in the pad (if new pad).
4. Prepolishing to saturate the pad with the slurry using dummy wafers.
5. Loading and polishing the device wafer.
   (a) A pressure ramping segment provides the delay required between initial wafer polishing and full pressure, alleviating wafer tension and minimizing the chance of wafer fracture.
   (b) Main polishing is done at full pressure, speed, and oscillation rate.
   (c) DI water buffing on main pad, initial stage of wafer cleaning, and flushing the consumables further diluting the etchant.
6. Moving the head to a secondary platen if needed.
   (a) DI water buffing on a soft buffing pad to minimize surface microscratching and further minimize wafer slurry contamination.
   (b) Unloading the polishing head and moving to an unloading station or cleaning unit.
7. Robotically loading the wafer onto the cleaning stations.
   (a) Station 1 of the cleaner uses a megasonic bath to aid in breaking the bonds formed between the slurry and the wafer.
   (b) Station 2 of the cleaner uses brushes to scrub both the front side and backside of the wafer while using a cleaning agent at a much higher pH.
   (c) Station 3 of the cleaner also uses brushes to scrub both the front side and backside of the wafer repeating the actions from station 2.
   (d) Station 4 runs a DI water rinse followed by a combination of spin and heat dry.

The slurry distribution and mixing system operates in the background of the actual wafer processing. However, a slurry with consistent physical and chemical properties is critical for maintaining a repeatable W CMP process with steady yield. Two of the most important parameters to monitor are etchant (oxidizer) and solid (abrasive particle) concentrations.

9.3 W CMP DEFECTS

From early trials to present-day high-volume production, there have always been defect concerns associated with W CMP, such as scratches and residues

TABLE 9.2 Basic W CMP Defects.

Defect Class	Defect Type	Defect Class	Defect Type
Residuals	Visual residual tungsten	Particle	Surface
	Micro residual tungsten		Imbedded
	Visual residual barrier layer	Contact deformation	Popped
	Micro residual barrier layer		Cored
	Residual chemical		Underetched
	Residual pad debris		Overetched
Scratches	Chatter	Polish nonuniformity	Edge fast
	Micro		Center fast
	Small		Edge/center fast
	Macro		Saddle removal
	Comets		Wedge removal
	Handler		Hot spot removal
	Circular	Miscellaneous	Miscellaneous

that interfere with both photoalignment and wafer yield. Table 9.2 gives a more detailed list of possible W CMP defects. As geometries shrink, some of the defects once considered nuisance become unacceptable. For example, in the 1990s chatter marks and small scratches were considered nuisance defects to the level of 300 microscratches per wafer at ~0.5-μm technology node. Today, the tolerance level for scratch is zero. To reach this goal, tremendous efforts have been made to reduce inherent film defects, eliminate pattern density concerns, optimize scribe orientation, lower wafer edge profile limitations, and increase metrology repeatability and reliability.

As indicated in Table 9.2, scratches can be divided into many categories from chatter marks to macroscratches. Chatter marks are a series of small scratches shown in Fig. 9.6. They are usually less than 5 μm in length and tend to skip much like a rock across the surface of a lake. In some cases, a single depression can be classified as a chatter mark, though others might classify it as a large pit. Pitting can come from corrosion during ammonia cleaning, a surface particle indentation, or even embedded particles ripped out during polishing. Chatter marks are mainly caused by large surface particles skipping on the wafer from

FIGURE 9.6 Chatter mark.

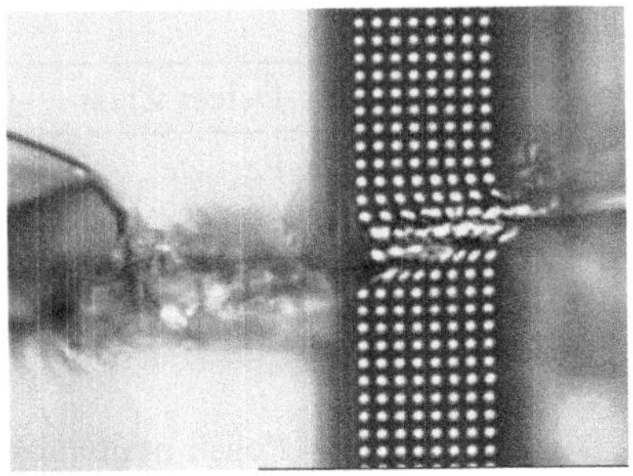

FIGURE 9.7 Macroscratch.

peeled tungsten, scribe, and insufficient or overly aggressive pad break-in, dried slurry flaking, slurry delivery tube shedding, or even tungsten deposition clamps. Macroscratches are greater than 50 μm in length and 0.5 μm in width and are usually caused by embedded particle rip-off, diamond loss from conditioner, and tool debris. As shown in Fig. 9.7, macroscratches can cause significant damage to a wafer. All die that comes in contact with a macroscratch will fail. Therefore, most wafers that suffer from macroscratches are scrapped by fabs to prevent further propagation of defects in downstream processing steps.

Another common defect found on a polished wafer is tungsten coring shown in Fig. 9.8. Tungsten coring occurs when the tungsten etchant is able to seep into the seam created during tungsten deposition. The level of coring depends on the nature of the etchant used in the slurry and on the contact time. The true cause of a coring defect is not tungsten CMP but the formation of the contact and deposition of both tungsten and barrier layers.

Residual tungsten and barrier layer are also viewed as defects (Fig. 9.9). Residual W and barrier layers are usually caused by insufficient polishing that can be attributed to inadequate end-point detection and high within-wafer nonuniformity (WIWNU) due to process drift or poor optimization of polishing parameters. Insufficient polishing times are usually solved by the implementation of end-point detection. For more detailed analysis of tungsten defects, see Chapter 17.

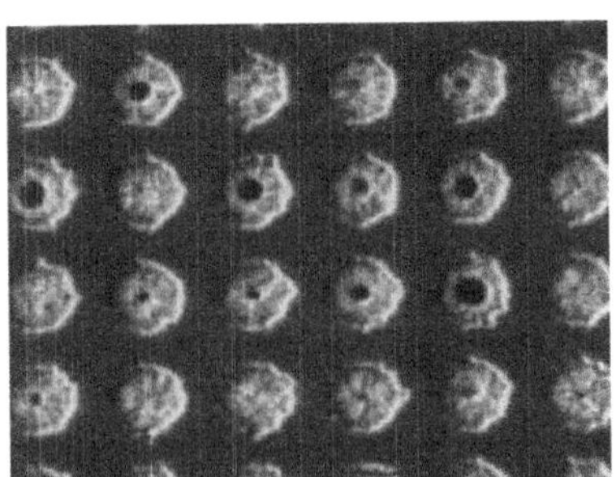

FIGURE 9.8 Coring and plug recession.

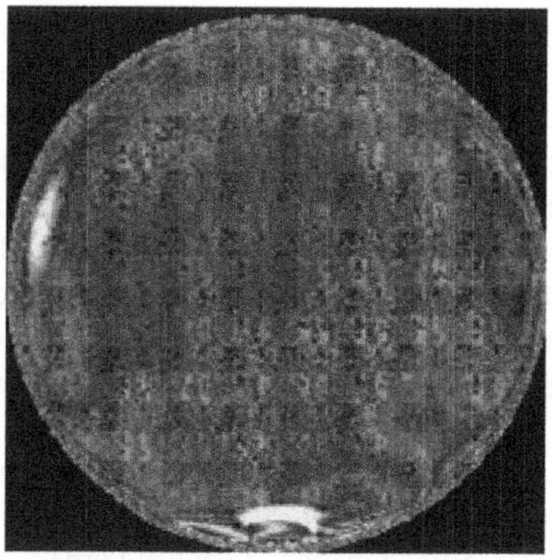

FIGURE 9.9 Residual W and barrier materials.

9.4 VARIOUS W CMP PROCESSING OPTIONS

9.4.1 Basic Considerations

In order to prepare for a W CMP process, a tool set (slurry distribution and mixing, a polisher, and a cleaning system) is usually first identified as the minimum requirement. The consumables (slurry, pads, conditioners, and a specific head membrane or a carrier film) are then selected. Depending on the technology node or wafer type to be polished, a polishing sequence is designed and tested. Based on the polishing outcome, the process is repeatedly optimized. In some cases, the consumables are reselected and optimized. In addition to objectives that are specific to the wafer type and technology node, the first level of polishing requirement for W CMP usually includes material removal rate, removal rate selectivity on related films, within-wafer nonuniformity, surface quality, dishing, and erosion (Fig. 9.10). Defect inspections and investigations usually occur after these basic requirements are met unless the wafer suffers severe and massive delamination or damage. To facilitate the initial consumable evaluation and process optimization, instead of production wafers, one may want to use standard testing blanket

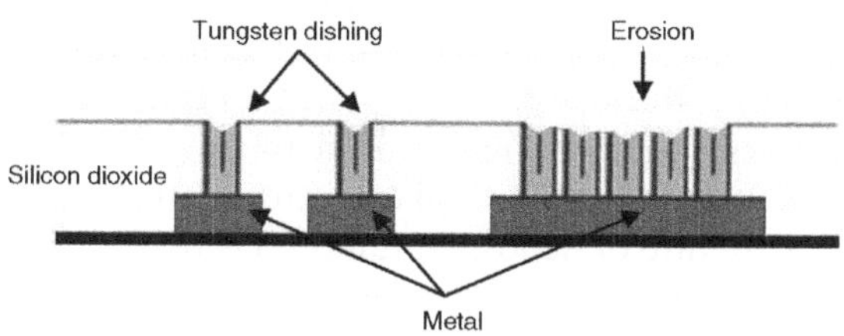

FIGURE 9.10 A schematic illustration of dishing on two individual vias (left) and an array of five vias (right).

and patterned wafers that are available commercially. In addition to the obvious cost advantage, the results from these testing wafers can be communicated within the CMP community in general or between vendor and user without confidentiality and incompatibility concerns.

The standard testing patterned wafers should represent three aspects of the production line: wafers at mean pattern density, wafers at the maximum allowed pattern density, wafers at a minimum pattern density, and combinations of the above three. The patterned wafers should also represent a full cross section of the product line, taking into account the number of backend layers (metal layers) and the type of tungsten process to be polished, contacts (plugs), damascene, or even dual damascene. The damascene processes with tungsten are rather rare, given the resistivity of tungsten (5.65 μΩ cm) compared to aluminum (2.65 μΩ cm) and copper (1.67 μΩ cm), so the likelihood of qualifying such a process remains low. Some manufacturing facilities may prefer to use a representative selection of the production wafer within their own facility as short loop wafers. A short loop wafer is simply a shortened version for the full production wafer to gain quick and relatively inexpensive feedback on process performance. Figure 9.11 shows a basic representation of an SKW floor plan for a tungsten line process evaluation or development. The benefit of designed test wafers comes from the extensive amount of information that can be gathered from one wafer. Generally, designed test wafers have a wide range of pattern densities and plug sizes, most of them also include sections with varied linewidths to evaluate damascene processing. Finally, by using designed test wafers full design of experiments (DOEs) can be performed more efficiently, decreasing the process development time significantly.

The entire tungsten CMP operation can be broken into four separate stages (tungsten polishing, barrier layer removal, oxide buffing, and post-CMP cleaning); the first two steps may be carried out together without a physical stop. In this section, we shall focus on the removal of excess tungsten from the surface. In order to fully understand the polishing characteristics of tungsten, we must first examine the physical and chemical properties of the material and their dependence on the preparation methods. More specifically, we need to know more about the incoming tungsten wafer in terms of its preparation method and history. Generally, the tungsten layer is deposited by chemical vapor deposition (CVD) that can produce profiles of various thickness across the wafer, depending on deposition tool and process [14]. For example, the tungsten thickness profile of some incoming wafers may vary significantly from thick center, thick edge, thick edge and center, thin center and edge, and so on. Each of these profiles should be understood to properly design the tungsten CMP polishing and ensure optimum planarization efficiency.

The crystalline structure of deposited tungsten is body-centered cubic (BCC) with varying grain size based on the deposition temperature. The tungsten grain size directly affects the polishing rate [8]. For example, a tungsten film that was deposited at 500°C may give a removal rate of 2200 Å/min. Under the same polishing conditions, a film that was deposited at 300°C may have a removal rate

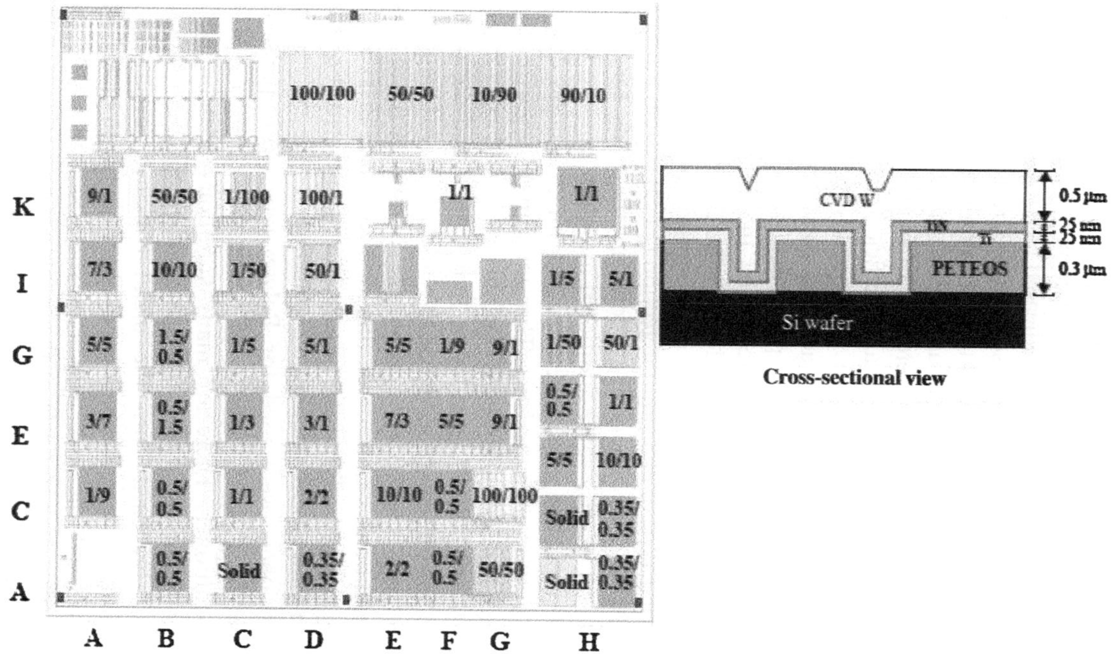

SKW5-3 mask floor plan

FIGURE 9.11 SKW tungsten line test wafer mask floor plan (from Ref. 13).

of 2800 Å/min [8]. Since most production facilities have a range of product lines with various technology nodes that may require their own corresponding deposition recipes, the incoming wafers for CMP operation may have significantly different physical and chemical properties that require a separate optimization process for polishing conditions. A familiarization with those various deposition processes and conditions will help to minimize duplicated effort in the CMP process design. From a production point of view, it is critical to identify an appropriate tool set that is dedicated to a particular product line. The tool qualification process must represent an accurate cross section of the product that will be polished at that facility or at a minimum be correlated to that product, and for this, most facilities use blanket tungsten wafers.

Most tungsten slurries have both an etchant or oxidizer and a passivating agent or mechanism to improve planarization efficiency. To a certain extent, hydrogen peroxide serves as an etchant that quickly oxidizes the tungsten surface, and its subsequent product WO_3 acts as a passivating agent to prevent any further oxidation of the tungsten. The protruded areas are quickly polished away, whereas the recess areas are protected by the passivating layer. The planarization efficiency is strongly dependent on the combination of pad and slurry selected. When selecting a pad, it is important to know whether the pad supports the particular type of end-point detection (EPD) system built into the polisher. Some EPD systems require windows in the pads for refraction measurements and will require some additional attention prior to testing the pads. Not all pad window designs are the same and not all window designs will be optimum for all process characteristics. The same can be said with respect to subpads that are used to create additional flexibility in the pad as a form of fine-tuning the planarization efficiency, wafer nonuniformity, and die nonuniformity. For more specific information related to pads, please see Chapter 5.

Once the pad has been selected, the appropriate conditioner for that pad will also be needed. Conditioners range from diamond rings and disks to bristled brushes, depending on the pad and the process that is intended. For tungsten CMP, the most common type is a diamond conditioner. In the early years of W CMP, brush conditioners were used for Polytex-type (soft) pads. Besides switching to a harder pad, the choice of conditioner also shifted to the diamond-based conditioner. There are many variations of diamond conditioners depending on their metal base and diamond arrangements. Some disks have the diamonds brazened onto the metal backings, others have the diamonds bound in nickel to metal backings, others bind with nickel onto the metal backings and coat the disk with a polymer to protect the bonding material, and yet others lead to CVD of the diamonds in diamond to increase diamond retention and chemical resistivity. Each technology has its own advantages and disadvantages. Brazened diamonds have a tendency to fracture; nickel bonding tends to erode with the tungsten slurries, thus loosing diamonds; and even the small diamond particles bounded via a CVD process could pose particle concerns during conditioning. Some of the newer conditioners are manufactured by individually placing diamond particles to provide even cut rates and uniform diamond wear. The improvement on backing materials has also been reported, such as changing from the common metal alloys to newer ceramics to improve disk flatness and minimize corrosion from the aggressive slurries [15].

In addition to overall material removal rate and selectivity, within-die and within-wafer nonuniformities are the two most important factors that could significantly affect throughput and wafer yield. For example, high wafer nonuniformity requires extensive overpolishing that can lead to increased dishing and erosion as well as prolonged process time. High within-die nonuniformity could lead to various residue issues. In general, the within-die nonuniformity is pattern dependent and within-wafer nonuniformity is not pattern dependent. Most tool parameters play a small role in controlling within-die nonuniformity as compared to their larger role in controlling overall wafer nonuniformity. For example, a polishing head with better zone control (edge, middle, and center) can significantly improve within-wafer nonuniformity but helps little in improving within-die uniformity. Slurry may have an impact on both types of nonuniformities.

9.4.2 Barrier Polishing

Once the tungsten polishing process has been understood, the barrier polishing needs to be examined. Within a tungsten module, titanium nitride (TiN) is generally used as the barrier layer; the TiN is usually deposited by a CVD process. Depending on the quality of the CVD process used, the barrier layer directly influences both the electrical characteristics and the polishing performance (parameters) of both tungsten and TiN. It is not uncommon to vary the barrier layer thickness to help improve contact resistance. Any significant change in relative thickness ratio between tungsten and TiN will require a reevaluation of the tungsten slurry for material removal rate selectivity. For example, a typical W CMP slurry may have a ratio of ~20:10:1 for $W{:}TiN{:}SiO_2$ to ensure a minimum overpolishing on the barrier and loss of oxide due to erosion. It is important to point out that the variation in barrier thickness may have an impact on the exposure of seam at the center of tungsten plugs. A thicker barrier layer translates to a slightly higher ratio for the plugs and leads to a greater chance of opening the seam that allows corrosive slurry to leak in. Ultimately, the slurry will cause defects such as coring.

9.4.3 Oxide Buffing

Oxide buffing is used to remove a couple dozen nanometers of oxide in order to make the tungsten plugs slightly protruded. During such a process, the tungsten plug recess or dishing is eliminated and oxide array erosion is minimized. Furthermore, by employing such an oxide buff, defects can be reduced, yield can be increased, and contact resistance can be reduced as well. Oxide buffing is usually performed with moderate downforce and speed

(3–4 psi, ~40 rpm) using the same slurry as used for oxide polishing. For an oxide buffing, there are multiple implementation schemes. Depending on the number of platens available on the tool, the oxide buffing can be performed either on the second or on the third platen. One possible scheme involves a main tungsten polishing on the first platen using a hard pad, light tungsten and barrier polishing on the second platen using a hard pad, and oxide buffing on the third platen using a soft pad. This is very similar to a typical three-step copper polishing. Another possibility is to use the first platen to finish the tungsten and barrier removal. The second platen with a soft pad is used for oxide buffing. The objective is to balance the need for lower defect count and shorter processing time and lower consumable usage.

9.4.4 Post-W CMP Cleaning

Wafer cleaning starts with a DI buff, usually on an extremely soft pad to remove the bulk of the slurry debris, and a thin layer of the oxide ~10 Å. During this initial stage of cleaning, some surface defects and pitting can also be minimized. The wafers are then put through megasonic clean and a couple of brush clean stations. The most commonly used chemical for the cleaning process is ammonia. Although ammonia is very effective in neutralizing the polishing slurry and undercutting the adhesion between abrasive particles and oxide surface, a high pH solution sometimes does cause defects such as corrosion and pitting. Therefore, alternative cleaning solutions have been sought and evaluated.

9.5 OVERALL TUNGSTEN PROCESS (VARIOUS PROCESSING DESIGN OPTIONS AND SUGGESTIONS)

9.5.1 W CMP Process Controls

Similar to oxide and STI CMP processes, in the W CMP process also a critical polishing parameter is the pressure exerted on the wafer during polishing. The term pressure is also expressed as downforce (DF) or downpressure. It is important to point out that downforce refers to the force applied from the back of the wafer, which can be converted to downpressure after a correction of the wafer size. Among all the factors also known as "knobs" in the semiconductor industry (downpressure, table speed, head speed, backpressure, slurry flow rate, etc.), downpressure is the most effective variable in controlling the removal rate according to the Preston equation. Unlike copper CMP, which has lowered the operating pressure to below 2 psi to avoid delamination and damage to the low-*k* dielectric films, there has been less demand to reduce the downforce in W CMP. Today, W CMP still routinely operates at 3–4 psi or higher. The higher tolerance to increased pressure is due to the dielectric material used in conjunction with the W plugs. Even when softer dielectrics are used, such as densified BPTEOS and densified BPSG, an additional thin film of USG is added as a buffer for better W:dielectric selectivity. The drive for a low-pressure copper process is aimed at lowering the dishing on large copper structures, which leaves almost no room for correction by sacrificing a portion of the dielectric materials. For tungsten plugs, the level of dish is usually less severe and there is room to correct such recess by removing a thin layer of the oxide.

On the chemistry side, the key knobs are pH, concentration of oxidizer, and abrasive content, given a relatively developed slurry formulation. Among these factors, pH has the most profound impact on removal rate as it determines how deep the oxidization process can penetrate into the tungsten film. A deeper penetration of the tungsten film allows higher removal rate provided that there is sufficient mechanical force to remove the softened film. The mechanical forces include downpressure, abrasive particles, and pad. In other words, at any given pH and oxidizer concentration, there is a removal rate limit. Within the limit, the removal rate follows the Preston equation. If the mechanical force is above this limit, scratch-type defects will be present. If the mechanical force is severely below what is needed to remove all the softened fill, corrosion-type defects will dominate the surface.

In a rotary platform, the polishing head rotates toward the platen or the table. This is to ensure that the linear velocity of every single point on the wafer during the polishing remains the same. In order to do so, there is also an offset between the two speeds. The ratio of the two speeds is largely determined by the size difference between the platen and the head. This ratio can be used as a knob to tune the within-wafer nonuniformity. Another important parameter for WIWNU control is the backpressure. On an older tool, there are usually a limited number of zones. On a newer tool, the backpressure can be individually applied to the polishing head in more than six zones. The relative backpressure on these zones can be monitored and regulated during polishing. Sometimes, the monitoring and controlling system can be linked to the end-point detection unit that allows real-time adjustment and automatic correction for the next wafer.

9.5.2 Platen Temperature Control

As W CMP is a combination of chemical and mechanical actions, the temperature between the pad and the wafer will rise during polishing because of mechanical friction and chemical reactions. Without a precise control of platen temperature, the removal rate from wafer to wafer will not be consistent. For example, as shown in Fig. 9.12, the removal rates increase significantly and then reach a relative plateau for a typical slurry using a platen that is not temperature controlled.

By controlling the platen temperature, the time to polishing rate saturation becomes much faster, in most cases with no variation at all. It is important to realize that higher table temperature may have negative effects on the polishing performance, such as increased within-wafer nonuniformity and higher dishing and corrosion. For a given temperature range, between 15 and 35°C, there is a predictable and steady increase in the material removal rate. At temperatures significantly higher or lower than this range, the pad adhesion to the platen

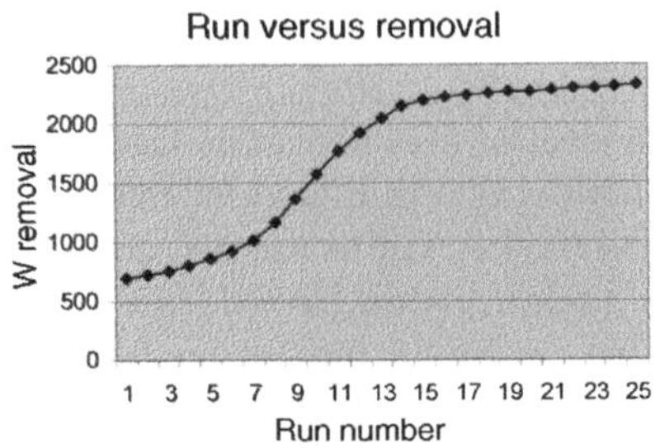

FIGURE 9.12 Removal rate profile for a tungsten process conducted on a platen without temperature control.

may become a problem. At temperatures slightly lower than 15°C, there is an indication that the removal rate is relatively independent of the temperature, and the slurry usage can be minimized.

9.5.3 Slurry Selectivity

To achieve the desired CMP outcome, it is necessary to optimize the material removal rate ratio between metal (tungsten plus barrier) and dielectric. Such a ratio is obviously a function of slurry composition and processing condition. Some of the early W CMP processes used ratios as high as 50:1 (tungsten:oxide). Based on the thickness of tungsten and barrier in a typical stack, the absolute removal rate for tungsten is desired to be in the range of 2000–5000 Å. Depending on the integration scheme and slurry capability, such a removal rate ratio (tungsten:oxide) can now be anywhere from 1:1 to 50:1, though with most slurry the ratio tends not to exceed 20:1. In order to decide an optimum ratio, many factors must be considered. For example, a higher tungsten-to-oxide removal rate ratio often means higher absolute tungsten removal, which leads to decreased slurry usage and higher throughput (cycle time). A lower oxide removal means that the polishing can stop on SiO_2 more effectively. At the same time, a high tungsten removal rate usually leads to increased wafer nonuniformity, increased intradie nonuniformity, increased edge polishing (dependent on tool head selection and process parameters), and increased risk of higher contact recession (thus increased contact resistance). If the increase in tungsten removal rate is accomplished via higher chemical reactivity from the slurry, such a process has a high potential to give greater wafer surface defects such as corrosion and pits.

9.6 CONCLUSIONS

Similar to copper and STI CMP, tungsten CMP constructs vias and sometimes lines as a part of the interconnect in IC fabrication. The ability to form these microstructures enables the multilevel interconnect network that shortens the processing delay time and IC performance. Due to the chemical nature of tungsten, a slurry for W CMP usually does not have to contain passivating agent other than oxidizers such as hydrogen peroxide. The slurry pH and abrasive content are the major factors that influence the W CMP outcome in terms of removal rate and selectivity. In order to reduce surface defects after CMP, a number of parameters must be optimized, such as temperature, removal rate selectivity, post-CMP buffing, and post-CMP cleaning.

QUESTIONS

1. The current slurry formulation used for W CMP consists of fumed silica abrasive particles, ferric nitrate, and hydrogen peroxide as the key components. Can you think of other key components that could be used to develop additional W CMP slurries?
2. The post-W CMP cleaning with ammonia in the past has caused wafer surface defect issues. Is it possible for you to determine alternative cleaning chemistries for the post-W CMP process? Explain why you think they will provide a more efficient process.
3. The tungsten:oxide selectivity is vital to achieve a longer overpolishing window. Do you think it is possible to add an additive to the W slurry to protect the oxide surface to enhance the overpolishing window? If so, what type of additive would you recommend? Keep in mind that this additive should not hinder the W material removal rate.
4. What is the best regulation point for W CMP platen temperature? Explain why.
5. Is it possible to conduct barrier polishing on a separate platen using dedicated slurry, similar to copper CMP? If yes, explain the process and both the positive and negative effects.
6. Write the Preston equation for W CMP polishing. Explain how each processing variable is used with respect to W polishing.

REFERENCES

1. Murarka SP, Steigerwald JM, Gutmann RJ. Chemical Mechanical Planarization of Microelectronic Materials. Wiley VCH; 1997. p129.
2. Zantye PB, Kumar A, Sikder AK. Chemical mechanical planarization for microelectronics applications, Materials Science and Engineering: R: Reports, 2004;45(3–6):89–220.
3. Kaufman FB, Thompson DB, Broadie RE, Jaso MA, Guthrie WL, Pearson DJ, Small MB. Chemical–mechanical polishing for fabricating patterned W metal features as chip interconnects. J Electrochem Soc 1991;138(11):3460–3465.

4. Seo Y-J, Lee W-S. Effects of different oxidizers on the W-CMP performance. Mater Sci Eng B 2005;18(1–3):281–284.
5. Stein DJ, Hetherington D, Guilinger T, Cecchi JL. In situ electrochemical investigation of tungsten electrochemical behavior during chemical mechanical polishing. J Electrochem Soc 1998;145(9):3190–3196.
6. Elbel N, Neureither B, Ebersberger B, Lahnor P. Tungsten chemical mechanical polishing. J Electrochem Soc 1998;145(5):1659–1664.
7. Fayolle M, Sicurani E, Morand Y. W CMP process integration: consumables evaluation—electrical results and end point detection. Microelectron Eng 1997; 37–38:347–352.
8. Li Y, Cheemalapati K, Wang C, Burkhard C, Jun W, Dodaka I, Atsushi H, Toshihiro K, Hozumi K. Correlation between pad structures and CMP performance. VMIC IMIC;2006. p 444–450.
9. Ireland PJ. High aspect ratio contacts: a review of the current tungsten plug process. Thin Solid Films 1997;304:1–12.
10. Ganguly U, Krusius JP. Novel compensation chemical metal polishing for low dishing and high global planarity for ultra-planar die applications in micro-optics and micro-electro-mechanical systems. Thin Solid Films 2004;460:306–314.
11. A study on the correlation between electrochemical corrosion and chemical mechanical polishing performance of W and Ti film. Microelectron Eng. Forthcoming. Corrected proof available online: 2006 11 Oct.
12. Lillard RS, Kanner GS, Butt DP. The nature of oxide films on tungsten in acidic and alkaline solutions. Los Alamos (NM): Materials Corrosion and Environmental Laboratory, Materials Science and Technology Division, Los Alamos National Laboratory.
13. Oliver MR. Chemical–Mechanical Planarization of Semiconductor Materials. Springer; 2004. p247.
14. Website: http://www.testwafer.com/.
15. Website http://www.diamonex.com/products_cmp.htm.

10

ELECTROCHEMISTRY IN ECMP

JINSHAN (JASON) HUO

10.1 INTRODUCTION

Electrochemical polishing, or electropolishing, is conventionally used for producing shiny surfaces where mechanical polishing is difficult to perform. Examples include components with complicated surfaces, decorative items, and other special applications. For microelectronic fabrication, planarization is emphasized in addition to surface smoothness. Hence, the term electrochemical planarization (ECP) is used throughout this chapter/book.

Surface polishing can be achieved under certain conditions of electrochemical dissolution, which is a reverse process of electroplating (EP). A simple electrochemical cell is shown in Fig. 10.1. Two metal (e.g., Cu) bars are immersed in an electrolyte. A voltage is applied between the two bars. The one connected to the positive pole of the power supply is anode. The other one is cathode. The positive potential applied to the anode may pump out electrons from copper atoms on the anode surface. As a result, copper dissolution may occur in certain electrolytes. Conversely, copper deposition may occur on the cathode. That is, copper electroplating results when the working electrode is chosen to be cathode, and copper dissolution is accomplished when the working electrode is chosen to be the anode.

For EP, the biggest challenge is void-free gap filling. This is realized by adding additives such as accelerator, suppressor, and leveler into a plating solution. Among them, accelerators have a function of catalyzing charge transfer. These rapidly diffusing small electroactive molecules accumulate onto the trench bottom as the copper deposition process goes on. As a result, the current density on the trench bottom becomes higher and higher than that on

Microelectronic Applications of Chemical Mechanical Planarization, Edited by Yuzhuo Li

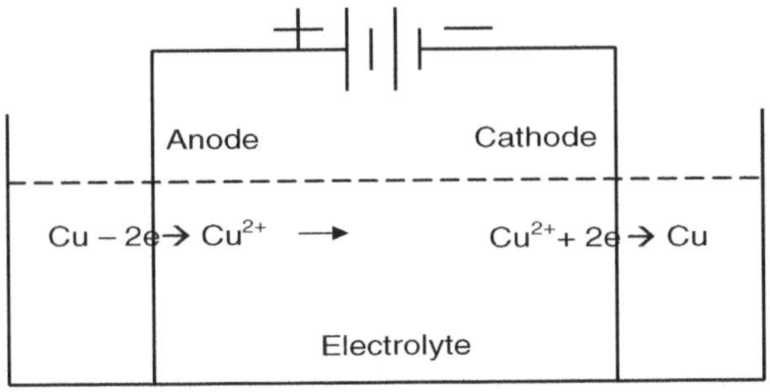

FIGURE 10.1 Schematic of an electrochemical cell.

the top flat surface. Therefore, a bottom-up filling is realized. This bottom-up filling effect also leads to an unflat surface topography (see Fig. 10.2), which is a challenge for the following polishing process.

For ECP, the biggest challenge is planarizing large features. Electrochemical dissolution, the reverse process of electroplating, does not necessarily produce a smooth surface. Indeed, a smooth surface can be produced only at certain potential ranges in certain electrolyte solutions.

Electrochemical polishing can be used for metallic materials including metals, alloys, and conductive metallic compounds to get a smooth and shiny surface. Bulk metal materials are normally polycrystalline, which are constructed by the repetition of identical structural units (crystalline cells) in space. The crystal periodicity is disrupted at grain boundaries and metal surfaces. However, their macroproperties are isotropic if the crystal grains are randomly orientated.

Thin films deposited by various techniques usually have preferential orientations, which have a significant influence on the electrochemical behavior of the material. For example, electrodeposited copper films on silicon wafers may have most crystal grains orientated in such a way that their (111) planes are parallel to the surface. Cu(111) planes have a higher dissolution rate than Cu(100) planes in the active potential region of an electrochemical process.

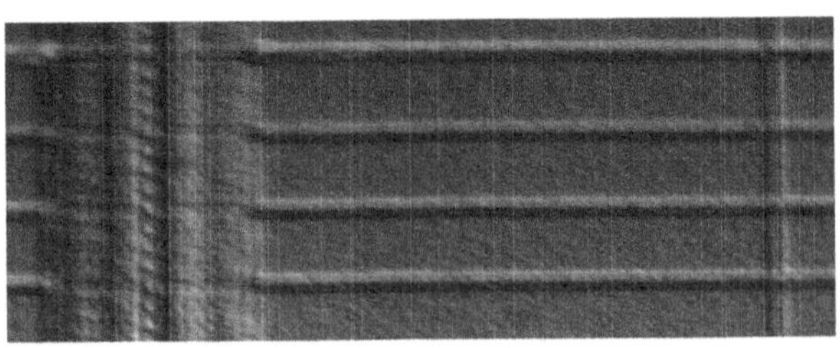

FIGURE 10.2 Surface topography of an electroplated copper film.

10.2 PHYSICAL AND CHEMICAL PROCESSES IN ELECTROCHEMICAL PLANARIZATION

Electrochemical dissolution is associated with current flow between anode and cathode. The dissolution of an atom from the anode surface involves (1) electrochemical reaction, (2) mass transport, and (3) charge transfer. First, the metal atoms on the anode surface are ionized (or oxidized) under the applied anodic potential:

$$M - ne \rightarrow M^{n+} \tag{10.1}$$

where M is a metal atom, e is an electron, and n is the number of electrons transferred in the electrochemical reaction. Then, with or without chelating with a complexing agent, M^{n+} ions move away from the anode surface into the electrolyte solution. The M^{n+} ions may reach the cathode surface and recombine with electrons (reduction). The mass transport process (ions move in an electrolyte) makes current flow in the solution possible. Charge transfer includes electron transport through conductive wire, electron transfer at electrode surfaces (oxidation and reduction), and charge carrier (ions) transport in the solution. To understand these processes, one has to know the electrode/electrolyte interface structure, where the most in an electrochemical system complicated and important physical and chemical processes occur.

10.2.1 Electrode/Electrolyte Interface

In an anode material (usually metal, e.g., Cu), there is a communal sharing of outer-shell electrons among all the atoms. The atoms are bonded together by the electrostatic attraction between the positively charged ions and negatively charged freely moving electrons. The free electrons and atoms at the solid surface have higher energy states. Some of the atoms on the metal surface may lose electrons to become ions at one moment. The ions may also recombine with electrons and become atoms at another moment. Depending on the electronic structures, it is easier for some metals (such as sodium) to lose electrons (be ionized) than others (such as platinum).

When a metal is in contact with an electrolyte, a solution or melt that consists of charged species—ions, the ionization process may be promoted due to the following reasons [1,2]:

1. Metal ions cannot move in the metal electrode but can move through the solution, producing electric current in the solution under an applied potential.
2. Electrons can move freely in solid metal (electric current in a metal) but cannot survive in the solution and will quickly recombine with positive ions.
3. Water dipoles and negative ions in the solution may drag the surface metal ions into the solution.

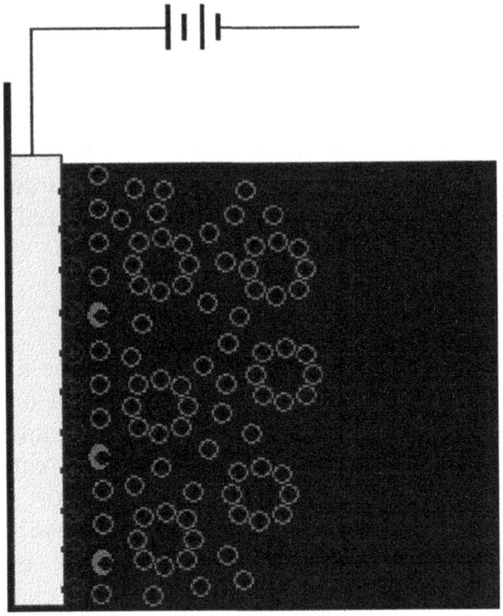

FIGURE 10.3 Electrode/electrolyte interface.

As more surface atoms ionize, more extra electrons remain inside the solid metal. The electrostatic attraction between these electrons and metal ions adsorbs the ions onto the metal surface.

If a positive potential is applied to the metal, as shown in Fig. 10.3, the ionization of the surface atoms will be promoted, and thus more metal ions will be produced at the surface. In the solution, water molecules, positive ions (cations), and negative ions (anions) drift around. The adsorbed layer of positive metal ions attracts nearby water dipoles in a preferential direction. The negative ions in the solution near the anode surface are also attracted toward the surface. The adsorbed fixed layer and the negative ion layer (Fig. 10.3) together are the so-called electrical double layer. Details about the double layer are available elsewhere [3]. Electrochemical reactions and mass transport for further electrochemical dissolution occur and pass through this double layer.

10.2.2 Electrochemical Reaction

Electrochemical reaction (e.g., $Cu - 2e \rightarrow Cu^{2+}$) proceeds when the applied potential (E) on the anode is higher than the formal potential $E^{0\prime}$ [4]:

$$E^{0\prime} = E^0 + \frac{RT}{nF} \ln \frac{\gamma_{ox}}{\gamma_{re}} \tag{10.2}$$

where E^0 is the *standard redox potential* of the anode material (e.g., 0.34 V for Cu/Cu^{2+}), $R = 8.317$ J/mol K is the gas constant, T is the temperature in kelvins (K), $F = 9.64853 \times 10^4$ C is the Faraday constant, and γ_{ox} and γ_{re} are the activity coefficients of the oxidation and reduction products, respectively. From Equation 10.2 we can see that formal potential $E^{0\prime}$ and thus the electrochemical reaction depends on

1. anode material, which determines E^0;
2. electrolyte, which determines γ_{ox} and γ_{re}; and
3. temperature.

10.2.3 Mass Transport

According to the above discussion, the metal ions produced under applied potential may dissociate from the anode surface and get into the electrolyte solution due to the electrostatic attraction from polar water molecules and anions in an electrolyte. Driven by the electric field between anode and cathode, all cations move toward the cathode and all the anions move toward the anode. The ion motion driven by electric field is called migration, as shown with white arrows in Fig. 10.4.

In the case of Fig. 10.4, cations Cu^{2+} and H^+ can easily be reduced on the cathode and thus consumed. But anions PO_4^{3-} and OH^- are not consumed. Continuing the migration of PO_4^{3-} and OH^- toward the anode results in higher concentration of these ions near the anode; that is, concentration gradients of the ions are built up. The concentration gradients can drive these ions to move from higher concentration area toward lower concentration area, that is, opposite to the migration direction. The ion motion driven by concentration gradient is called diffusion, as shown with black arrows in Fig. 10.4. The current flow through the closed circuit is a result of the net transport of charge carriers in the x direction (from anode to cathode) [4].

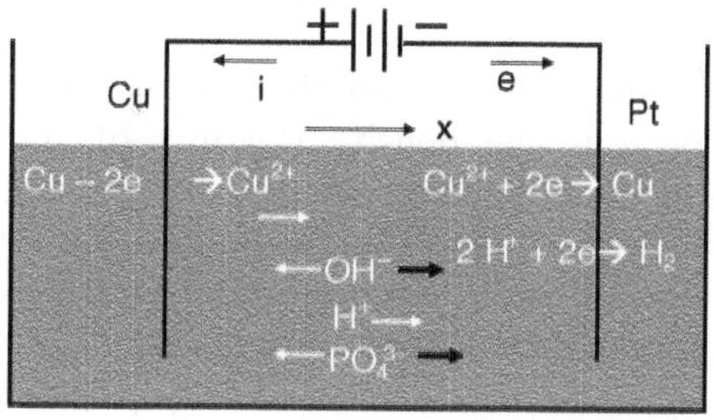

FIGURE 10.4 Mass transport and current flow in an electrochemical cell with phosphoric acid solution. Migration: positive ions (Cu^{2+} and H^+) migrate toward the cathode and negative ions (PO_4^{3-} and OH^-) migrate toward the anode. Diffusion: PO_4^{3-} and OH^- ions diffuse in the direction opposite to migration, driven by their concentration gradient resulting from the migration.

The third form of mass transport is convection driven by pressure. When forced circulation exists in electrolyte, convection may be the dominant form of mass transport. Thus, in general, a flux J_j (mol/s cm) of species j may occur due to the above three types of mass transport mechanisms. The flux can be described by the Nernst–Planck equation [5]

$$J_j = -D_j \frac{\partial C_j(x)}{\partial x} - \frac{z_j F}{RT} D_j C_j \partial \frac{\phi(x)}{\partial x} + C_j\, v(x) \tag{10.3}$$

where C_j, z_j, and D_j are, respectively, the concentration, number of valence electrons, and diffusion coefficient of species j, ϕ (x) is the electrical potential at position x, and $v(x)$ is the velocity of solution in the x direction. If j is an ionic species, then the flux J_j is equivalent to the current density. Thus, the current (i, C/s) for a solution flow through a cross-sectional area A normal to the axis of mass flow is

$$i_j = i_{\mathrm{d},j} + i_{\mathrm{m},j} + i_{\mathrm{c},j} = J_j z_j F\, A \tag{10.4}$$

where $i_{\mathrm{d},j}$ is the diffusion current, $i_{\mathrm{m},j}$ is the migration current, and $i_{\mathrm{c,j}}$ is the convection current, which represent the contribution of diffusion, migration, and convection, respectively.

In a real electrochemical system, convection is usually introduced by such means as rotating electrode, stirring, or other forced circulation. In any case, the electrolyte moves relative to electrode surfaces. Due to the mechanical friction between electrolyte solution and electrode surface, a velocity $v(x)$ variation exists. The velocity of solution flow is generally a constant (v_∞) in bulk solution (far from the electrode surface and the wall of solution container) and decreases while approaching the solid surfaces [6]. The solution flow velocity $v(x) = 0$ at solid surface ($x = 0$). A *hydrodynamic* (or *Prandtl*) *boundary layer* is defined as [6]

$$\delta = \frac{5x}{\sqrt{\mathrm{Re}_x}} \tag{10.5}$$

where $\mathrm{Re}_x \equiv \rho v_\infty x/v$ is the Reynolds number (in the range of 10^5–3×10^6 for flow over flat plate), ρ is the density (kg/m^3) of the solution, x is the characteristic length in the flow direction, and v is the viscosity (kg/s m) at x. Inside the boundary layer, convection is limited, and the mass transport mechanisms are mainly diffusion and migration. Outside the boundary layer, mass transport is dominated by forced convection.

10.2.4 Anodic Polarization Curve and Conditions for Electrochemical Planarization

Electrochemical dissolution depends on many factors such as applied anode potential, electrolyte, temperature, interelectrode distance, disk rotation speed, and electrolyte convection. As stated earlier, ECP occurs only under certain conditions. To find the right conditions for ECP, rotating disk electrode (RDE) is used to measure the anodic polarization curve of an anode material in an electrolyte. A simple setup for the measurement includes a rotator, a RDE (anode), a cathode, a potentiostat, and a computer. A schematic of the setup is shown in Fig. 10.5. Different than a regular power supply, a potentiostat can apply a voltage between working electrode (anode in the case of ECP) and counter electrode (cathode in the case of ECP) such that the potential of the working electrode can be program controlled relative to the reference electrode. The potential of the working electrode and the current between working and counter electrodes can be controlled in various forms by software. The technique used to measure anodic polarization curves is *linear sweep voltammetry* (LSV).

Depending on the anode material, electrolyte, and other factors, the polarization curves can have various shapes. A typical polarization curve of copper in phosphoric acid solution is shown in Fig. 10.6. At low potential (close to zero), electrochemical reactions of low redox potentials such as

$$H^+ + e \rightarrow H \quad (E^0_{H^+/H} = 0\ V) \tag{10.6}$$

may occur. As the potential increases, when the anode potential is high enough for

$$Cu - 2e \rightarrow Cu^{2+} \quad (E^0_{Cu/Cu^{2+}} = 0.34V) \tag{10.7}$$

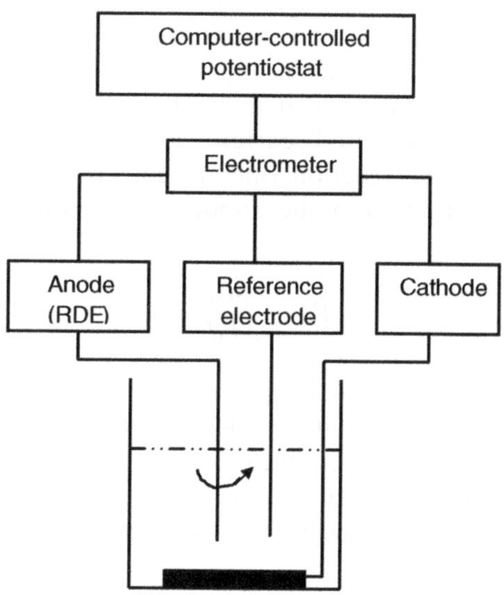

FIGURE 10.5 Schematic of the setup for anodic polarization curve measurements.

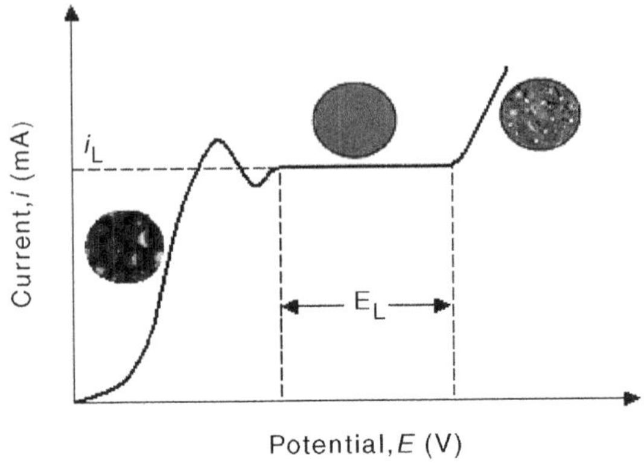

FIGURE 10.6 Polarization curve and surface topography after anodic dissolution in different potential ranges.

to occur, the anodic current largely increases with the increase of *overpotential* [9]:

$$\eta = E - E^{0'} \tag{10.8}$$

$$i = \frac{\eta}{R_p} \tag{10.9}$$

where E is the applied potential on the anode and R_p is the *polarization resistance*. In other words, in the low potential region, the measured current is linearly proportional to the electrode potential. In some electrolytes, there may be two slightly different linear regions with different polarization resistances, as shown in Fig. 10.6. The two linear regions are very likely correspond to different electrochemical reactions such as (10.6) and (10.7).

Assuming that the measured current is contributed only by anodic dissolution and its reverse reaction (metal deposition) on cathode, one can calculate the removal rate from the anode (or deposition rate on cathode) surface [4,10]

$$R_d = \frac{M_W}{2\,FdA} i \tag{10.10}$$

where R_d (μm/s) is the removal (or deposition) rate, M_W (g) and d (g/cm^3) are, respectively, the molecular weight and density of the removed (or deposited) material, i (A) is the measured current, F is the Faraday constant, and A (cm) is the polished (or deposited) surface area. For Cu ECP in phosphoric acid solution, Equation 10.10 has been proved to agree well with experimental measurements [10].

For polycrystalline materials, the atoms at grain boundaries and different crystal planes have different energy states. Thus, the actual formal potentials

($E^{0\prime}$) for all the different locations of an anode surface are different than that calculated from Equation 10.2. Hence, the actual overpotentials (η) at grain boundaries and different crystal planes are different. As a result, these different areas exhibit different dissolution rates according to Equations 10.8–10.10. This phenomenon is called crystallographic etching. Therefore, anodic dissolution in the linear region produces a rough surface (see Fig. 10.6) due to crystallographic etching.

With continuing increase in anode potential, the current–potential relationship deviates from the linear relationship. As the potential continues to increase, the increase in the current slows down until it reaches a maximum and then decreases to a minimum point, and finally increases to the limiting current plateau. This is a transition from kinetics of electrochemical reaction domain to mass transport domain. This region may have a very different shape depending on electrolytes, potential scan rate, and other factors. Details of this region are discussed elsewhere [7,8].

Further increase in anode potential gets into limiting current plateau range (E_L in Fig. 10.6). In this region, the potential is so high that the electrochemical reaction is faster than mass transport; that is, Cu^{2+} ions produced on the anode surface in a unit time are more than those that mass transport can remove from the anode surface into the bulk solution. As a result, a Cu^{2+} ion concentrated layer is developed inside the Prandtl boundary layer [4]. The concentrated Cu^{2+} layer is called Nernst layer or diffusion layer, which has a thickness of [13]

$$\delta_c = 1.61 D^{1/3} \nu^{1/6} \omega^{-1/2} \qquad (10.11)$$

The current in this region, termed as limiting current (i_L), depends on the diffusion and migration of Cu^{2+} ions from the anode surface to the bulk solution across the anodic layer. The value of the limiting current can be influenced by many factors, as shown in the following equation [11,12]:

$$i_L = 0.62\, n\, F\, A\, D^{2/3} \nu^{-1/6} \omega^{1/2} C^* \qquad (10.12)$$

where D is the diffusion coefficient, C^* is the Cu^{2+} concentration at the electrode surface, $\omega = 2\pi$ rps (s^{-1}, rps stands for revolution per second) is the angular velocity of the rotating disk, v is the kinematic velocity, and A is the area of anode surface. These factors affect i_L through their effects on the diffusion layer.

Experimental study indicated that the mass transport limiting species in copper–phosphoric acid solution is the so-called acceptor (water molecules), which diffuses into the diffusion layer and facilitates Cu^{2+} removal [14]. In some cases, a salt film of metal ion complexes can form as an anodic layer to control the mass transport processes [15–17]. Mass transport limiting species can also be metal ions, which diffuse and migrate through anodic layers (ion-concentrated diffusion layer and/or salt film) into the bulk solution [17].

Anodic layers are proved to be the key for electrochemical planarization, which can be achieved under mass transport controlled conditions.

Increasing anode potential beyond the limiting current plateau range may cause gas bubbles to form on anode surface. The gas bubbles break anodic layers and thus mass transport controlled condition. Hence, the current–potential relation becomes linear again. Gas bubbles produced on anode lead to a rough surface [4].

10.3 MECHANISMS AND LIMITATION OF ELECTROCHEMICAL PLANARIZATION

ECP can be achieved by three mechanisms: ohmic leveling, diffusion leveling, and migration leveling. Understanding these mechanisms can help us understand the limitation of ECP in microelectronic applications. Understanding ECP mechanisms can also help us understand other electrochemistry involved polishing techniques such as electrochemical–mechanical polishing (ECMP).

10.3.1 Ohmic Leveling

As discussed earlier, electrochemical dissolution proceeds under dynamic conditions when the applied anode potential is relatively low; that is, the process is dominated by electrochemical reactions, and potential and current obey Ohm's law, as expressed in Equation 10.9. Under these conditions, surface planarization can be achieved by electrochemical dissolution if the interelectrode distance and anode surface roughness have the same order of magnitude. As illustrated in Fig. 10.7, the distance (l) between the anode and

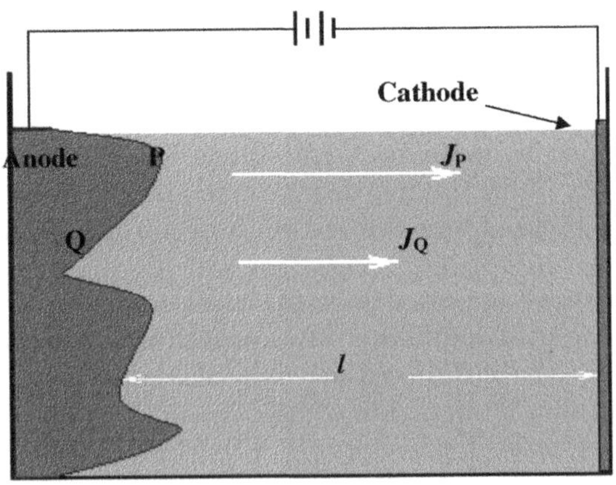

FIGURE 10.7 Schematic of current density between the cathode and a rough anode.

the cathode varies due to the roughness of the anode surface. Hence, the current density (J) varies along the anode surface since [4]

$$J = \frac{V}{\rho_e l} \tag{10.13}$$

where V is the voltage between the electrodes and ρ_e is the polarization resistivity of the electrolyte between the anode and the cathode. From Equations 10.10 and 10.13, one can get *ohmic leveling* effect, which is the difference of the removal rate between the protruding point (P) and the recessed point (Q) [4]

$$\Delta R_d = R_d(\mathrm{P}) - R_d(\mathrm{Q}) = \frac{MV}{nFd} \frac{2b}{\rho_e(l^2 - b^2)} \tag{10.14}$$

where b is the amplitude of the anode surface undulations. Equation 10.14 indicates that the ohmic leveling effect depends on the surface profile and interelectrode distance. Larger surface roughness (b) and closer electrodes can produce a better leveling effect, which can be significantly high when l is very small [4]. In addition, smaller resistivity (ρ_e) of the electrolyte solution and higher voltage (V) between the electrodes also facilitate ohmic leveling. However, the planarization under dynamic conditions or ohmic control cannot produce a shiny surface due to crystallographic etching effect.

10.3.2 Diffusion Leveling

Under mass transport control, an anodic layer (salt film or ion-concentrated layer) forms on the anode surface. The electrochemical dissolution process is then controlled by the diffusion and migration of mass transport limiting species in the anodic layers. In the case of Cu in concentrated phosphoric acid solution, it is said to be acceptor (water molecules) diffusion controlled. Since any arbitrary surface profile can be obtained by superimposing a series of sine waves, it is reasonable to simplify the anode surface profile to a single sine wave of wavelength "a" and amplitude "b." Figure 10.8 shows a sinusoidal anode surface profile with anodic layer in the case of Cu in concentrated phosphoric acid solution; that is, the mass transport limiting species is acceptor. Cu^{2+} and acceptor concentration profiles are also illustrated in Fig. 10.8. The surface profile can be mathematically expressed as

$$y = b \sin\left(\frac{2\pi}{a} x\right) \tag{10.15}$$

Wagner [18] calculated the acceptor concentration gradient ($\partial C/\partial y$) and Cu dissolution rate (R_d) along the sinusoidal anode surface according to Fick's second law with the following assumptions: (1) steady-state, (2)

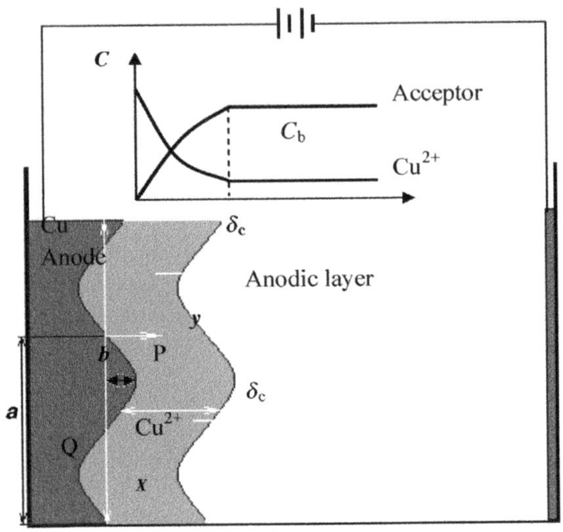

FIGURE 10.8 Sinusoidal anode surface profile with anodic layer and concentration profiles.

$b \ll a \ll \delta_c$, and (3) boundary conditions shown in Fig. 10.8. The calculation showed that [18]

$$\left(\frac{\partial C}{\partial y}\right)_{y=b\,\sin(\frac{2\pi}{a}x)} = B\left[1 + 2\pi\frac{b}{a}\sin\left(\frac{2\pi}{a}x\right)\right] \tag{10.16}$$

$$R_{\mathrm{d}} = C\left(2\pi\frac{b}{a}\right)\sin\left(\frac{2\pi}{a}x\right) \tag{10.17}$$

where $B = (\partial C/\partial y)_{\mathrm{avg}}$ is the average concentration gradient at the anode surface. Equations 10.16 and 10.17 suggest that the concentration gradient, the driving force of diffusion, versus location x has the same shape as the geometric profile of the anode surface. Thus, the variation of Cu dissolution rate (R_{d}), directly proportional to diffusion rate, also has the same shape as the geometric profile of the anode surface; that is, the dissolution rate at the peak area R_{d}(P) is higher than the dissolution rate at valley area R_{d}(Q). Hence, a planarization effect can be achieved due to the diffusion rate difference between protruding and recess areas. The planarization effect is named as diffusion leveling, which can be derived from Equation 10.17 [4]:

$$\Delta R_{\mathrm{d}} = R_{\mathrm{d}}(P) - R_{\mathrm{d}}(Q) = 4C\pi\frac{b}{a} \tag{10.18}$$

Equation 10.18 indicates that short-wavelength profiles are planarized faster than long-wavelength profiles.

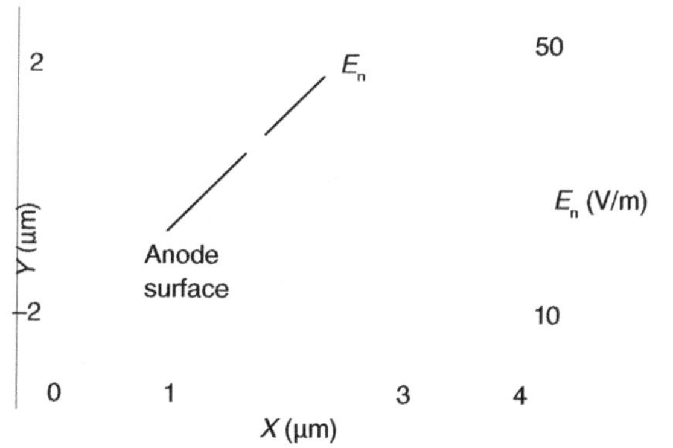

FIGURE 10.9 Sinusoidal anode surface and normal electrical field profiles.

10.3.3 Migration Leveling

As stated earlier, mass transport limiting species can be metal ions in some cases. An example is copper in 1-dydroxyethylidene diphosphonic (HEDP) acid solution [10,17,19]. In this case, a salt film $(Cu\text{-}HEDP)_n$ forms on Cu anode surface [20]. Cu^{2+} ions are the mass transport limiting species, which diffuse and migrate from Cu anode surface through the salt film into the bulk electrolyte solution.

Migration is ion motion driven by electric field (E_n). Computer simulation has been performed to study the leveling effect driven by migration [4,10]. In a 2D model, a sine waveform is located at the center of a Cu anode (3 mm thick and 300 mm long) surface. The interelectrode distance is 100 mm. Electrolyte resistivity is 15 Ω cm. The simulation was performed using boundary element simulation software. The simulation data indicate that the amplitude of normal electric field (E_n) varies along the sinusoidal surface (see Fig. 10.9). E_n–X has a shape similar to the surface profile Y–X. For a given amplitude (b) of the sine wave, the difference of the normal electric field between the peak point P and the recessed point Q, $[E_n(P) - E_n(Q)]$, increases as the wavelength (a) decreases (i.e., the frequency f increases), as shown in Fig. 10.10. The results suggest that Cu^{2+} ions at peak areas have stronger migration-driving force than those at valley areas. Thus, the planarization effect due to the migration rate difference between protruding and recess areas, named as migration leveling, can be expressed as

$$\Delta R_d = \frac{\alpha q}{2m}[E_n(P) - E_n(Q)]t \tag{10.19}$$

where α is a constant, q and m are the charge and mass of Cu^{2+}, respectively, and t is time.

Migration leveling and diffusion leveling effects expressed in Equations 10.18 and 10.19 are plotted in Fig. 10.10, which shows that both migration

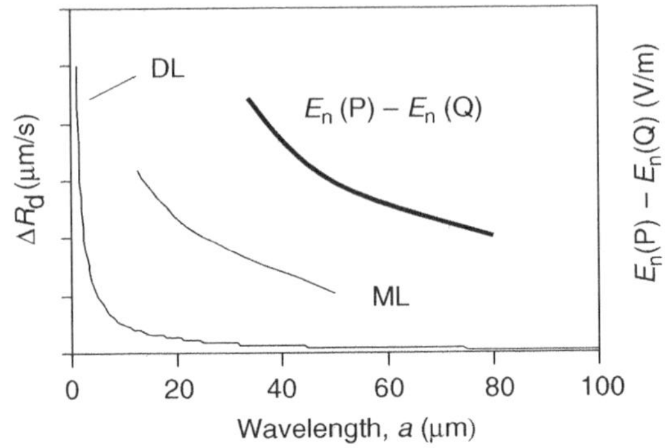

FIGURE 10.10 Migration leveling and diffusion leveling effects and peak–valley normal electric field difference versus wavelength for a given amplitude of the anode surface profile.

leveling and diffusion leveling have a strong dependence on the wavelength of the anode surface profile for a given amplitude. Shorter wavelength has better leveling effects than longer wavelength. Mechanically polished metal surfaces usually have sharp surface undulations (i.e., short wavelength, Fig. 10.11a) and

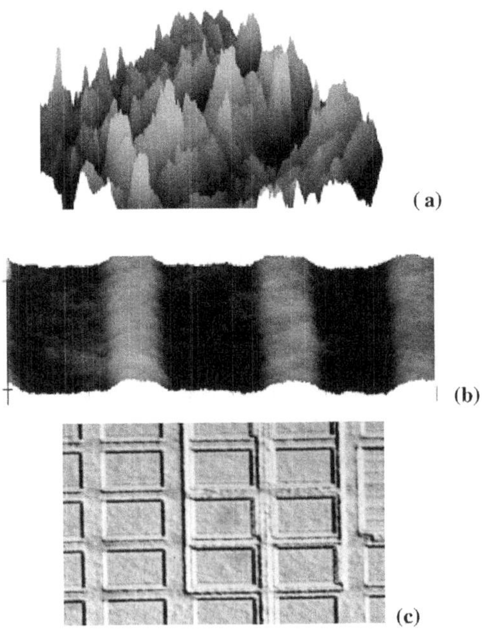

FIGURE 10.11 Surface profiles: (**a**) mechanically polished Cu surface (short wavelength), very good ECP effect; (**b**) electroplated Cu film (longer wavelength), less ECP effect; (**c**) electroplated Cu film (flat step, infinite wavelength), zero ECP effect.

thus can be easily planarized by ECP. Electroplated Cu films for microelectronic applications usually consist of gentle undulations (longer wavelength, Fig. 10.11b) and flat steps (infinite wavelength, Fig. 10.11c). Planarization effect by ECP is small or zero on these features. To achieve the desired planarization efficiency of these features, modified polishing techniques such as ECMP are explored and developed on the basis of ECP.

10.4 *IN SITU* ANALYSIS OF ANODIC/PASSIVATION FILMS

The formation of an anodic film (ion-concentrated diffusion layer or salt film) depends on electrolyte and processing conditions. The characteristics of the film are critical to the planarization efficiency of ECP and ECMP. Hence, the analysis, especially *in situ* analysis of anodic films, is highly valuable. A few techniques can be used for the purpose.

10.4.1 Impedance Measurement

An anodic film may have a different resistivity than the bulk solution of an electrolyte. Thus, a simple method to detect the existence of an anodic film is impedance measurement. Figure 10.12 illustrates a setup for impedance measurement. One can first measure the impedance (Z_0) between the anode (WE) and the reference electrode (RE) in active region (i.e., at a low DC potential, refer to anodic polarization curve in Fig. 10.6), and then measure the impedance (Z_1) in the mass transport limited region (i.e., at a DC potential in the limiting current plateau of the anodic polarization curve). In both measurements, a low-frequency (e.g., 1 Hz) small AC signal (e.g., 5 mV) is applied together with the DC signal. Keep in mind that the formation of an anodic film takes time [10]. Comparing Z_0 and Z_1, one can tell the existence of an anodic film.

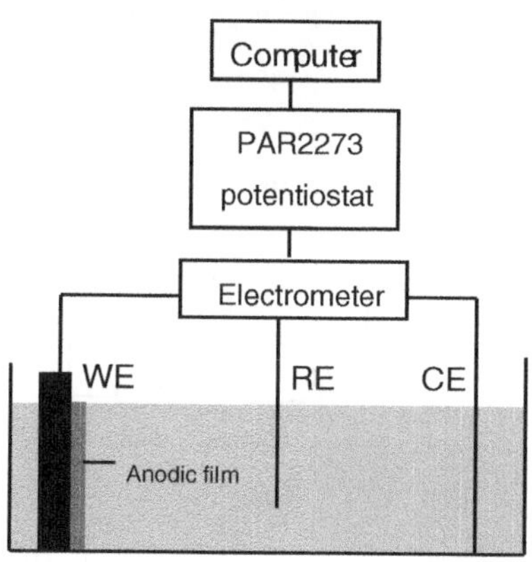

FIGURE 10.12 Influence of electrolyte on ECMP planarization efficiency.

10.4.2 Electrochemical Impedance Spectroscopy

Although simple impedance measurement can tell the existence of an anodic film, electrochemical impedance spectroscopy (EIS) can obtain more information about the electrochemical processes. In general, the anode/electrolyte interface consists of an anodic film (under mass transport limited conditions) and a diffuse mobile layer (anion concentrated), as illustrated in Fig. 10.13a. The anodic film can be a salt film or a cation (e.g., Cu^{2+}) concentrated layer. The two layers (*double layer*) behave like a capacitor under AC electric field. The diffuse mobile layer can move toward or away from anode depending on the characteristics of the anode potential. The electrical behavior of the anode/electrolyte interface structure can be characterized by an equivalent circuit as shown in Fig. 10.13. Impedance of the circuit may be expressed as

$$Z = R_1 + \frac{R_2}{1+\omega^2C^2R_2^2} - j\frac{\omega C R_2^2}{1+\omega^2C^2R_2^2} = Z' + jZ'' \tag{10.20}$$

where

$$Z' = R_1 + \frac{R_2}{1+\omega^2C^2R_2^2}$$

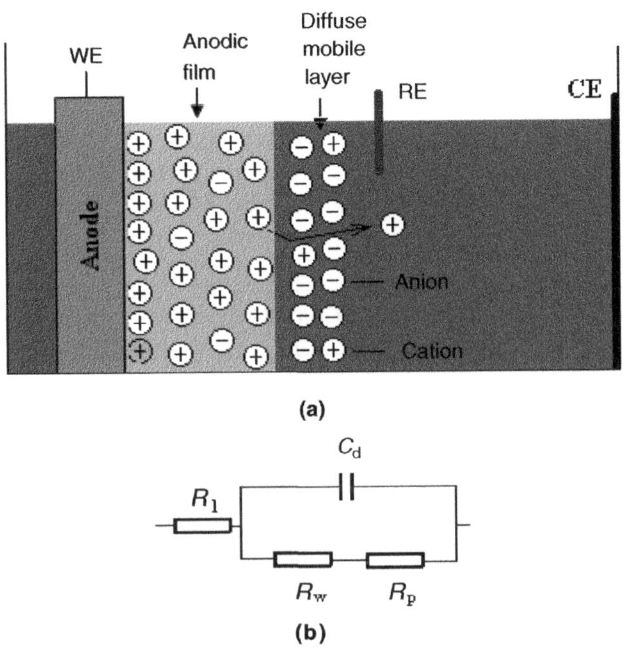

FIGURE 10.13 Illustration (**a**) and equivalent circuit (**b**) of an anode/electrolyte interface structure.

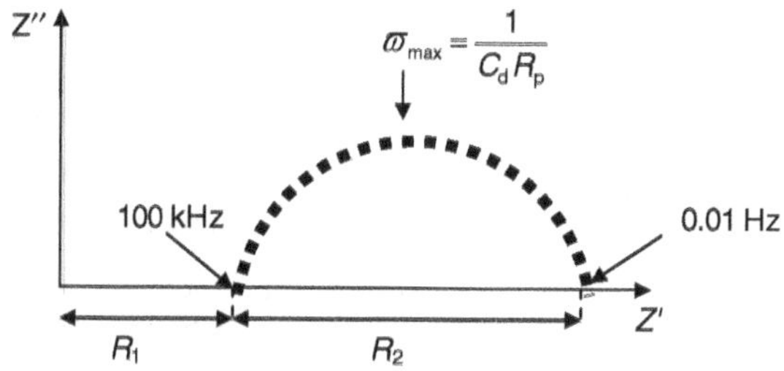

FIGURE 10.14 Impedance spectrum of the circuit in Fig. 10.17b.

$$Z'' \frac{\omega C R_2^2}{1 + \omega^2 C^2 R_2^2}$$

$R_1 = R_c + R_e + R_f$, $R_2 = R_p + R_w$, and R_c, R_e, R_f, R_p, and R_w are the electrical resistances of connection wires, electrolyte, anodic film, polarization, and cation diffusion, respectively. When an AC signal with frequency sweep is applied to the circuit, an impedance spectrum as shown in Fig. 10.14 can be obtained. From the spectrum, one can get the values of parameters such as R_f, C_d, and R_w. These parameters indicate not only the existence of an anodic film but also other important information of the electrochemical processes occurring at the anode/electrolyte interface. The correlation between these parameters and the electrochemical processes is not fully understood yet. However, some applications of this technique with detailed discussion have been reported [1,2,17].

10.4.3 Ellipsometry

Electrochemical impedance measurement is convenient and usually no extra instrument is needed since contemporary potentiostat, which is normally used in an ECP or ECMP system, tends to possess EIS function. However, the thickness of an anodic film cannot be easily determined from the measured impedance or impedance spectra because the electrical resistivity of the film is usually unknown.

A technique that can possibly measure the thickness of an anodic film is ellipsometry [21–23]. Figure 10.15 illustrates the setup of an ellipsometer and how is works. A light beam of known polarization reflects and passes through a sample. The polarization change between the incident and reflective beams, expressed as $\tan \Psi$—the amplitude (of electrical filed, E) change upon reflection and Δ—the phase shift, is measured. Since

$$\tan \Psi e^{i\Delta} = \frac{E_{rp} E_{is}}{E_{ip} E_{rs}} \qquad (10.21)$$

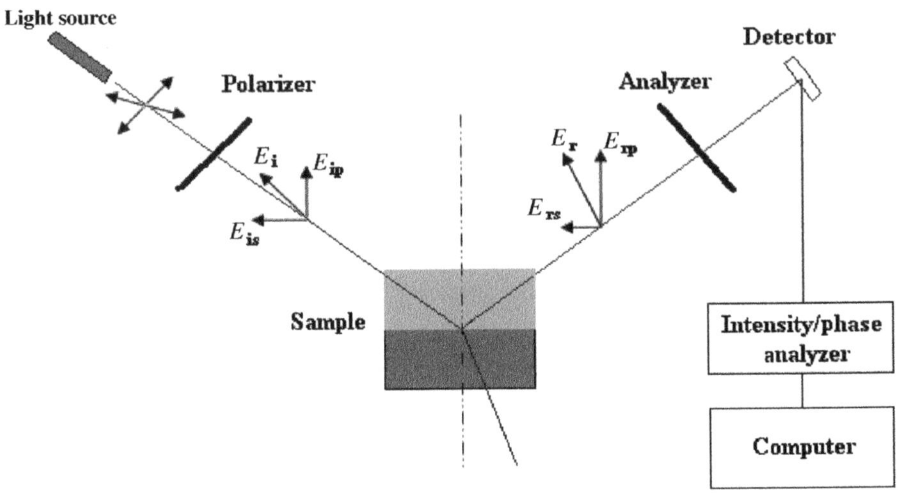

FIGURE 10.15 Schematic of the ellipsometry setup.

is a function of N ($= n + ik$), λ, L, and θ, the measured data together with wavelength (λ) and angles (θ) of the incidence and reflection can be used to determine the optical constants (n and k) and thickness (L) of a thin film. Here $n = c/v$ (speed ratio of light in vacuum and the sample material) is the refraction index; $k = (\lambda/4\pi)\alpha$ is the extinction coefficient; α is the absorption coefficient; and E_{is}, E_{ip}, E_{rs}, and E_{rp} (see Fig. 10.15) are the electric field (E) components in the directions perpendicular (s) and parallel (p) to the plane of incidence, which describe the polarization state of a light beam.

Ellipsometry can measure films from subnanometer to a few micrometers, depending on material properties and wavelength of the light source. It has been widely used for thin film measurement in various applications, from biology to semiconductor, and from solid/solid to solid/liquid interfaces [24,25]. Ellipsometer with electrochemical cell for *in situ* thin film analysis is available from J.A. Woollam Co., Inc. and has been used in the research on electrochemical deposition [26]. However, *in situ* measurement of anodic films is more challenging because the films are usually metal complexes with unknown optical properties and difficult to verify with other *ex situ* techniques.

10.5 MODIFIED ELECTROCHEMICAL POLISHING APPROACHES

To achieve better planarization efficiency than ECP, different modified ECP approaches have been explored. These approaches, including ECP-DI water technique [27], sectorial cathode ECP [28], membrane-mediated ECP [29], and contact ECP (i.e., ECMP) [30,31], work by different mechanisms and exhibit different performances.

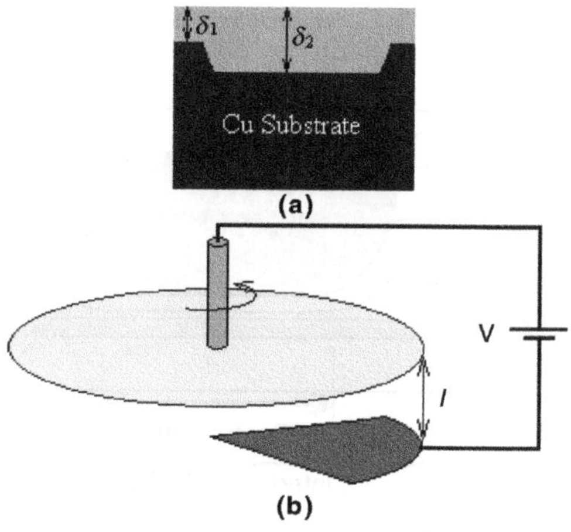

FIGURE 10.16 Flat anodic film (**a**) produced by ECP with ultraclose electrodes and sectorial cathode (**b**).

From the above discussion about planarization mechanisms, we can see that better planarization efficiency for large features and flat steps can be achieved if a planar anodic film can be produced. In this case, as shown in Fig. 10.16, the anodic film at protruding area (δ_1) is thinner than that at recess area (δ_2). Thus, Cu^{2+} diffusion and migration from Cu surface through the anodic film to electrolyte have a shorter distance at protruding than that at recess; that is, higher Cu removal rate can be achieved at protruding than that at recess. The planarization efficiency is attributed to anodic film thickness variation and is independent of surface feature (wavelength of the surface profiles). However, to achieve the best planarization efficiency, the interelectrode distance (l) needs to be very small (challenging for large size wafers). For this purpose and for good electrolyte circulation, a sectorial cathode was used (Fig. 10.16) and certain planarization efficiency was achieved on wafer coupons [28].

Another interesting approach is membrane-mediated ECP, as shown in Fig. 10.17 [29]. In this approach, a half electrochemical cell consists of cathode, electrolyte, and membrane and moves on the surface of a wafer. The membrane allows cations (e.g., Cu^{2+}) to pass through, whereas electrolyte is retained inside the cell. DI water is supplied between the wafer and the half cell and functions as lubricant, solvent, and medium for cation transport. The thickness of the water film between the membrane and the wafer varies with the wafer surface profiles if the membrane is flat. When a voltage is applied between a wafer and the cathode, ionization (e.g., $Cu \rightarrow Cu^{2+} + 2e^-$) occurs on the anode (wafer). Cations (e.g., Cu^{2+}) dissolve in DI water and move (migration—driven by the electric field between anode and cathode, and diffusion if a cation concentration gradient exists in the water film) through the membrane into the half cell. The water film at protruding areas of a wafer is thinner than that at recess areas. Hence, cation migration and diffusion

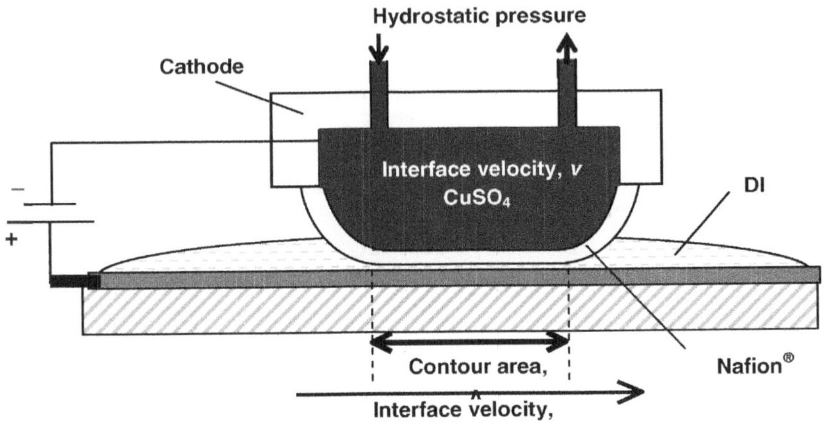

FIGURE 10.17 Illustration of membrane-mediated ECP.

distance through the water film at protruding is shorter than that at recess, which results in different removal rates. Therefore, planarization is achieved.

Considering performances, throughput, and feasibility, ECMP is the best approach. The cross section of an ECMP setup is shown in Fig. 10.18 [32]. The setup consists of a perforated conductive pad, a perforated insulating layer, a cathode, and a power supply. A voltage is applied between the conductive polishing pad and the cathode. When a wafer contacts and rotates against the conductive polishing pad, a positive potential is applied to the metal film (e.g., Cu) on the wafer; that is, the metal film becomes anode once it contacts and moves against the conductive polishing pad. When an electrolyte is supplied between the anode and the cathode, electrochemical processes such as

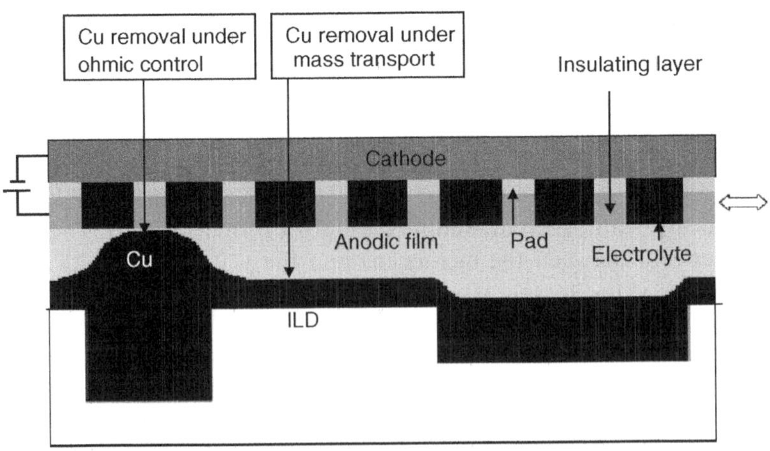

FIGURE 10.18 Illustration of ECMP.

electrochemical reactions, mass transport, and current flow, as well as mechanical abrasion, can occur simultaneously. Hence, the process is called electrochemical mechanical polishing. For low downforce, the process parameters and electrolyte can be set such that metal removal is driven by electrochemical processes whereas planarization efficiency is promoted by mechanical abrasion. The best planarization efficiency, which is an issue for ECP, can be achieved when

1. an anodic film forms (the film can be formed chemically, i.e., by chemical reaction, or electrochemically, i.e., electrochemical processes proceed under mass transport limiting conditions);
2. the anodic film has mechanical properties such that it can be easily removed by mechanical abrasion at low downforce and the film above protruding areas of the metal surfaces can be entirely removed;
3. an electrolyte exhibits low limiting current (i_L) and low polarization resistance tan ϕ (see Fig. 10.19).

These conditions mainly depend on the electrolyte. However, the thickness and surface profile of the anodic film, which strongly affect planarization efficiency, also depend on the ECMP pad.

When all the above conditions are met, as shown in Fig. 10.18, protruding and recess areas are in different electrochemical conditions. The protruding area directly contacts electrolyte and electrochemical dissolution occurs under ohmic control; that is, its dissolution rate (R_d) is determined by the polarization

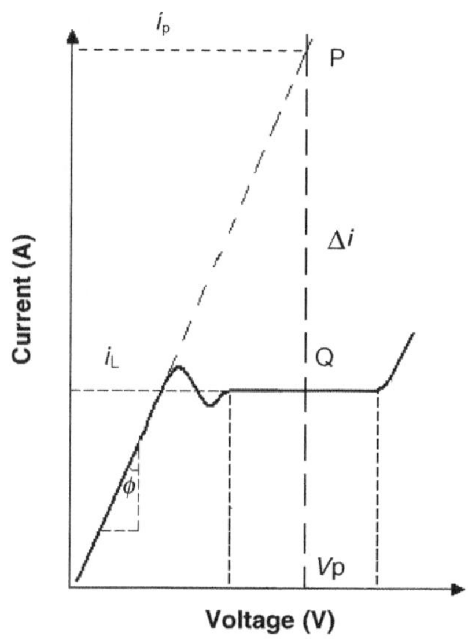

FIGURE 10.19 Influence of electrolyte on ECMP planarization efficiency.

resistance tan ϕ. With applied voltage V_p,

$$R_d(P) = A\, i_p \tag{10.22}$$

where A is a constant. In comparison, the recess areas are covered by the anodic film. Thus, the dissolution rate is determined by the limiting current i_L (refer to Fig. 10.19). With applied voltage V_p,

$$R_d(Q) = A\, i_L \tag{10.22}$$

The planarization ability can be expresses as

$$\Delta R_d = R_d(P) - R_d(Q) = A(i_p - i_L) \tag{10.23}$$

Therefore, a good electrolyte should produce an anodic film that gives low limiting current i_L and low polarization resistance tan ϕ (thus high i_p) for good planarization ability [32]. Additionally, an electrolyte should also meet the requirements for other performances such as corrosion and post-ECMP cleaning.

QUESTIONS

1. What are the differences between the surface profiles of electroplated Cu/Si wafers and mechanically polished Cu bulk materials?
2. What is an electrolyte? What are the requirements for an ECP electrolyte?
3. What electrolyte properties and process parameters can affect the formation of anodic layers?
4. What could be the results of multiphase metal ECP?
5. Why is it difficult for conventional ECP to achieve satisfactory planarization of electroplated Cu/Si wafers?
6. Compare three different ECP approaches with ECMP. What are the planarization mechanism, advantages, and disadvantages of each technique?

REFERENCES

1. Robinson RA, Stokes RH. Electrolyte Solutions. 2nd ed. London, UK: Butterworths Publications Limited; 1959. p 2.
2. Solartron Analytic. Understanding electrochemical cells. Technical report 17. 1997
3. Bard AJ, Faulkner LR. Electrochemical Methods Fundamentals and Applications. 2nd ed. New York: Wiley; 2001. p 12.

4. Huo J, Solanki R, McAndrew J. Electrochemistry: New Research. New York: Nova Science Publishers, Inc.; 2005. p 83.
5. Bard AJ, Faulkner LR. Electrochemical Methods Fundamentals and Applications. 2nd ed. New York: Wiley; 2001. p 138.
6. Incropera FP, DeWitt DP. Fundamentals of Heat and Mass Transfer. 4th ed. New York: Willey; 1996. p 294–351
7. Huo J, Solanki R, McAndrew J. Electrochemical polishing of copper for microelectronic applications. Sur Eng 2003;191:11–16.
8. Dmitriev VA, Rzhevskaya EV, Periodic phenomena in the anodic dissolution of copper in phosphoric acid. Russ J Phys Chem 1961;35:425–429.
9. Koryta J, Dvořák J, Kavan L. Principles of Electrochemistry. 2nd ed. New York: Wiley; 1993.
10. Huo J. Electrochemical planarization of copper for microelectronic applications. .Doctoral dissertation. Oregon: Oregon Health and Science University; 2004.
11. Koryta J, Dvořák J, Kavan L. Principles of Electrochemistry. 2nd ed. New York: Wiley; 1993. p 134–139
12. Paunovic M, Schlesinger M. Fundamentals of Electrochemical Deposition. New York: Wiley; 1998. p 73–97
13. Bard AJ, Faulkner LR. Electrochemical Methods Fundamentals and Applications. 2nd ed. New York: Wiley; 2001.
14. Landolt D. Review article fundamental aspects of electropolishing. Electrochim Acta 1987;321:1–11.
15. Grimm RD, West AC, Landolt D. AC impedance study of anodically formed salt films on iron in chloride solution, J Electrochem Soc 1992;1396:1622–1629.
16. Sridhar N, Dunn DS. In situ study of salt film stability in simulated pits of nickel by Raman and electrochemical impedance spectroscopies. J Electrochem Soc 1997;14412:4243–4253.
17. Huo J, Solanki R, McAndrew J. Study of anodic layers and their effects on electropolishing of bulk and electroplated films of copper. J Appl Electrochem 2004;343:305–314.
18. Wagner C. Contributions to the theory of electropolishing. J Electrochem Soc 1954;101:225–228.
19. Huo J, Solanki R, McAndrew J. Electrochemical planarization of patterned copper films for microelectronic applications. J Mater Eng Perform 2004;134:413–420.
20. Fang JL, Wu NJ. Determination of the composition of viscous liquid film on electropolishing copper surface by XPS and AES. J Electrochem Soc 1989;136: 3800–3803.
21. Christensen TM. Physics of Thin Films Lecture Notes. Available at http://www.uccs.edu/˜tchriste/courses/PHYS549/549lectures/opticalchar.html. Viewed on, 2006 Nov 20.
22. Gonçalves D, Irene EA. Fundamentals and Applications of Spectroscopic Ellipsometry. Available at http://www.scielo.br/scielo.php?pid = S0100-4042200 2000500015&script = sci_arttext Viewed on 2006 Nov 20.
23. The University of Texas at Arlington. Ellipsometry. Available at http://www.uta.edu/optics/research/ellipsometry/ellipsometry.htm. Viewed on 2006 Nov 20.

24. Arwin H. Spectroscopic ellipsometry and biology: recent developments and challenges. Thin Solid Films 1998;313–314:764–774.
25. Hilfiker JN, Thompson DW, Hale JS, Woollam JA. Thin Solid Films 1995;270:73–77.
26. Hilfiker JN, Glenn DW, Heckens S, Woollam JA. J Appl Phys 1996;79:6193–6195.
27. Tsujimura M. VMIC Conference Proceedings; 2004. p 267–274.
28. Huo J, Solanki R, McAndrew J. A new electroplanarization system for replacement of CMP. Electrochem Solid-State Lett 2005;82:C33–C35.
29. Mazur S, Jackson CE, Foggin G. WIITC 2005.
30. Ong P, Naujok M, Knarr R, Chen L, Moon Y, Neo S, Salfelder J, Duboust A, Manens A, Lu W, Shrauti S, Liu F, Tsai S, Swart W. Economikos L, Wang X, Sakamoto A, et al. IITC 2004.
31. Yamada K, Abe D, Fukaya K, Shimada M, Kobayashi N, Kondo S, Tominaga S, Namiki A. IITC 2005.
32. Huo J. 2006 CMP-MIC Short Course Lecture; Fremont; 2006 Feb 20. p 161–180.

11

PLANARIZATION TECHNOLOGIES INVOLVING ELECTROCHEMICAL REACTIONS

LAERTIS ECONOMIKOS

11.1 INTRODUCTION

Copper started replacing aluminum and tungsten as an interconnect and via material in silicon-based devices in 1990s owing to its superior conductivity and electromigration resistance [1,2]. Copper is poorly etched by plasma and wet-etch methods; therefore, a dual damascene process was introduced for Cu removal and planarization. Following etching of features into the underlying dielectric, diffusion barriers such as Ta and TaN are deposited to prevent migration of copper atoms to active devices. Copper is then deposited by an electrolytic process followed by chemical–mechanical planarization (CMP) to remove the excess Cu. The electrodeposition of copper for interconnects has been well established, and several researchers have studied the impact of additives on topography formed across lines with various widths and pattern densities [3,4]. Despite the fundamental understanding of electrodeposition mechanisms, the plated copper topography across various pattern densities has not changed over the years. As Fig. 11.1 shows, Cu is recessed on wide lines whereas "momentum plating" is observed on high pattern density areas. A local topography of about 4000 Å is typical within a die.

CMP was first developed at IBM East Fishkill in 1983 and is being used for copper planarization since the late 1990s. An important characteristic of CMP is that it planarizes Cu across the wafer with least dependence on pattern densities compared to other methods. An inadequate planarization will lead to

Microelectronic Applications of Chemical Mechanical Planarization, Edited by Yushuo Li

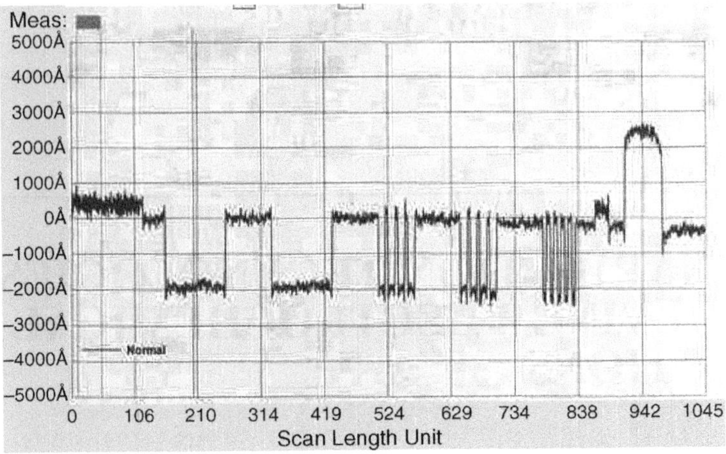

FIGURE 11.1 Plated Cu topography across pattern densities.

copper puddles formed in subsequent layers creating shorting defects. Furthermore, depth-of-focus limitations of lithography could create patterning problems [5,6]. As can be seen by the expression below, planarity requirements become more important as technology moves beyond the 90-nm node.

$$\rho = \frac{\lambda}{(\mathrm{NA})^2}$$

$$\mathrm{NA} = n \sin \alpha$$

where ρ is the depth of focus, NA is the numerical aperture, n is the refractive index, 2α is the light collection angle of lens, and λ is the wavelength.

The chemical component of CMP slurry creates porous unstable oxides or soluble surface complexes. The slurries are designed to have additives that initiate the above reactions. The mechanical component of the process removes the above-formed films by abrasion. In most planarization systems the mechanical component is the rate-limiting step. As soon as the formed porous film is removed, a new one is formed and planarization proceeds. Therefore, the removal rate is directly proportional to the applied pressure. To achieve practical copper removal rates, pressures greater than 3 psi are often required. These pressures should not create delamination, material deformation, or cracking on dense or relatively dense dielectrics used in silicon microfabrication on conventional dielectrics. However, the introduction of porous ultra-low-*k* (low dielectric constant) materials will require a low downpressure (<1 psi) polishing to maintain the structural integrity of the device [7–9]. It is expected that dielectrics with *k* value less than 2.4 will require a planarization process of 1 psi downpressure or less when they are introduced to production. It is expected that this process requirement will become even more important for the 45-nm technology node [10].

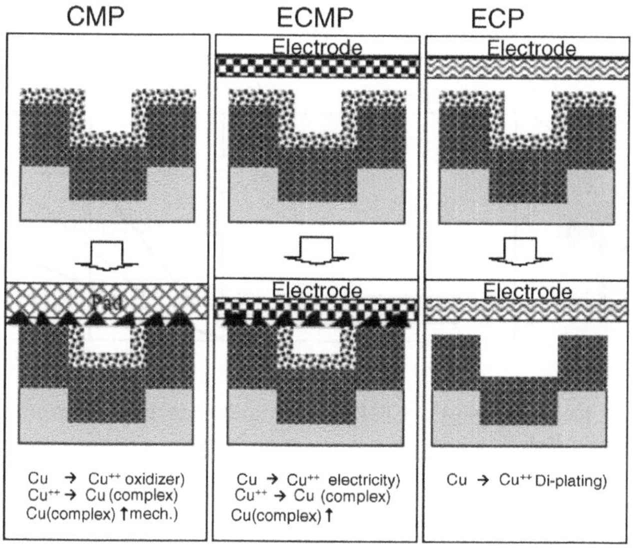

FIGURE 11.2 Schematics of three planarization technologies (CMP, ECP, and ECMP).

To meet the copper planarization requirements imposed by ultra-low-*k* materials, two trends in CMP process developments were reported in recent publications. Some slurry suppliers have started developing slurries capable of achieving high removal rates of bulk copper at pressures of 2 psi with copper clearance at 1 psi. In the second approach, low downpressure copper removal is achieved by incorporating voltage-activated electrochemical reactions in CMP process [11–13]. Figure 11.2 shows the schematics of three planarization technologies, chemical–mechanical polishing (CMP), electrochemical polishing (ECP), and electrochemical–mechanical polishing (ECMP). In this chapter, as a background to the discussion on full-sequence ECMP process, the working principles of three relevant planarization processes (CMP, ECP, and ECMP) are first described and compared. The advantages and challenges of ECMP are discussed in light of its potential to address the major concerns brought by the need to implement low-*k* dielectric materials in integrated circuit (IC) manufacturing.

11.2 CMP

The principal steps of Cu polishing are shown in Fig. 11.3 as (1) chemical reaction: Cu oxidized by oxidizer, $Cu \Rightarrow Cu^{++}$; (2) chemical reaction: oxidized Cu becomes Cu complex, $Cu^{++} \Rightarrow$ Cu (complex); and (3) mechanical reaction: Cu complex is removed by slurry and pad. Removal rate is governed by Preston's law: $RR = k_1 P_p P_s$, where, RR is the removal rate (nm/min); k_1 is the coefficient of Preston's Law; P_p is the polishing pressure (kPa); and P_s is the polishing speed (m/s).

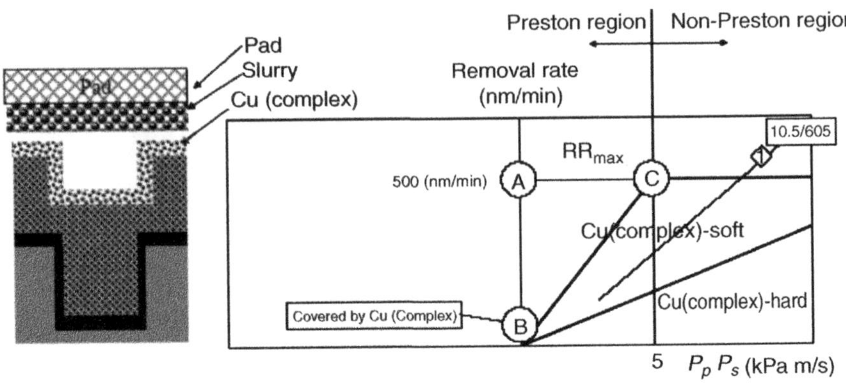

FIGURE 11.3 Key aspects of a CMP process and their impact on material removal rate (MRR) (from Ref. 14).

In the acidic region, just after the Cu surface is first oxidized by chemical oxidizer, the oxidized Cu surface immediately forms a Cu (complex) with the available complex agent. This Cu (complex) is then removed mechanically. This represents the competing reaction between chemical reactions (1) and (2), and mechanical reaction (3), in accordance with Preston's law. Point A in Fig. 11.3 shows the maximum removal rate determined by the chemicals used in reaction (1). Point B shows the minimum removal rate determined by the chemicals used in reaction (2). Point C is the intersection of the maximum removal rate and Preston's curve. The example data used in Fig. 11.3 are

$$\text{removal rate} = 605\,\text{nm/min at } 3\,\text{psi} \times 0.5\,\text{m/s} = 10.5\,\text{kPa m/s [14]}.$$

The removal rate is governed by Preston's law if $P_p \times P_s$ is situated to the left side of point C. The left side of point C is called the Preston region and the right side is the non-Preston region. The removal rate curve depends on the hardness of Cu (complex). The removal rate of hard Cu (complex) is lower than that of soft Cu (complex).

11.3 ECP

Electropolishing is a process of controlled material removal and smoothening of the surface of a conductive material by anodically polarizing it in the presence of certain electrolytes. It is also referred to as "reverse plating" or "super passivation." Cu electropolishing has been studied since the early 1990s as a replacement to CMP by a number of researchers [15–18]. The electrolytes typically used for electropolishing metals are phosphoric acid based [19]. Some electrolytes may also use hydrofluoric acid (HF), sulfuric acid (H_2SO_4),

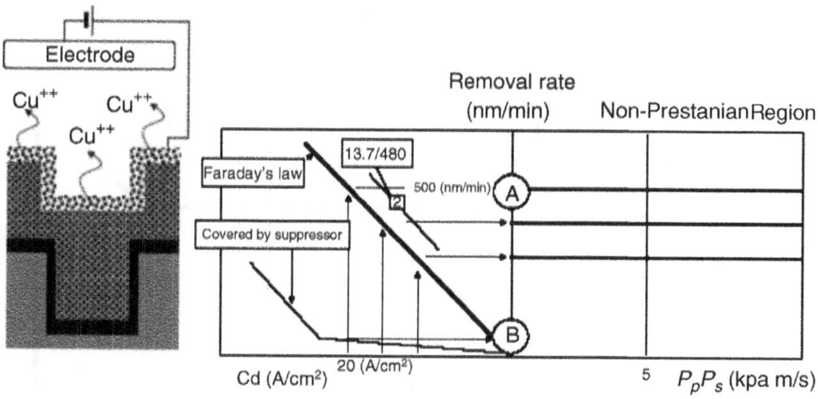

FIGURE 11.4 The working principle of ECP based on an illustrative case in which the removal rate of 480 nm/min is obtained at 13.7 mA/cm^2 (from Ref. 13).

chromic acid (H_2CrO_4), and nitric acid (HNO_3) [12]. The underlying principle of ECP is shown in Fig. 11.4. ECP is the reverse reaction of plating. It is thus referred to as deplating. Here, the removal rate is governed by Faraday's law. Chemical reaction when Cu is oxidized by electricity, $Cu \Rightarrow Cu^{2+}$. Removal rate is governed by Faraday's law RR = k_2 Cd, where RR is the removal rate (nm/min); k_2 is the coefficient of Faraday's law; and Cd is the current density (A/cm^2).

A schematic of an apparatus used for electropolishing a copper-coated silicon wafer is shown in Fig. 11.5 [20]. As can be seen, the electrochemical cell comprises the anode (the positive terminal), which is the copper surface, the cathode (the negative terminal), which could be made of platinum, and a reference electrode, all connected to a DC power supply and immersed in an electrolyte.

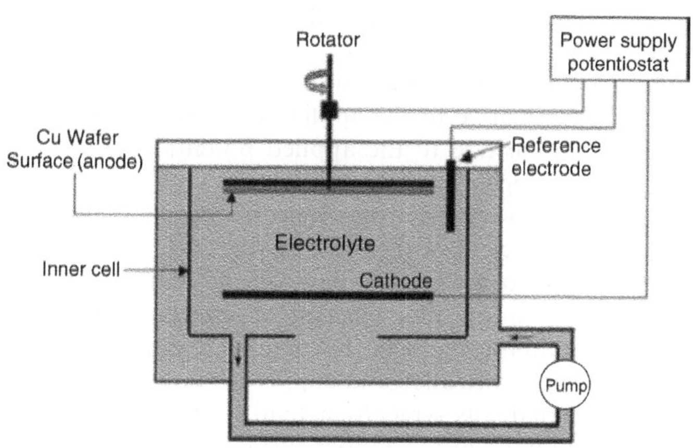

FIGURE 11.5 Schematic of electropolishing apparatus (from Ref. 20).

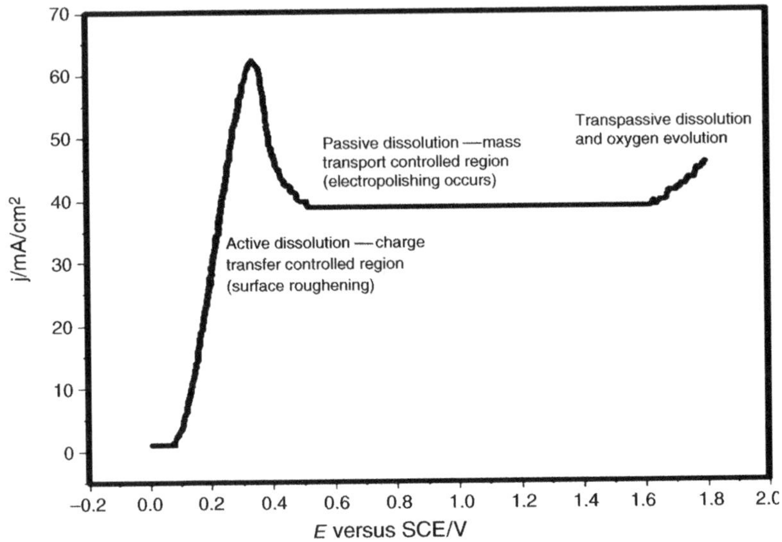

FIGURE 11.6 Anodic potential versus current density polarization curve (from Ref. 20).

Figure 11.6 shows the current–voltage behavior during the electropolishing of copper in a 13.3 M phosphoric acid + 17.7 M water solution with an electrode rotation speed of 240 rpm [20].

When a voltage is applied, active dissolution of the copper surface occurs and the current density increases rapidly until it reaches a maximum, which indicates the beginning of the formation of a passivation film. Surface kinetics or charge transfer is the rate-determining step during this regime. The copper surface is dull in appearance with a rough surface after active dissolution, as only preferential crystallographic sites with high surface energy are etched.

With further increase in potential, the current density drops and reaches a plateau where it is insensitive to the applied potential. This value of current density where it is insensitive to the applied potential is called the limiting current density, and it depends on various factors such as solubility of copper in the phosphoric acid based electrolyte, electrolyte concentration, temperature, and diffusion coefficient of the limiting reactant. In this plateau region, the copper surface is covered by a passivation film, which is constantly removed at the passivation film–electrolyte interface and grown at the copper–passivation film interface. In this plateau regime, the electropolishing process is mass-transfer controlled and copper is removed at a very slow rate. The limiting current plateaus typically extend over a range of 1 V. Visually, one can see the brightening of the copper surface during electropolishing and the smoothness of the surface. At higher potentials, an increase in the current density from the limiting value is observed and this is due to the gas evolution reaction. Electropolishing has several advantages including the nonelectrical contact, smooth finish, corrosion-resistant surface, minimal contamination, and ease in end-point detection.

However, electropolishing is pattern sensitive and this may limit its applications. The passivation film or diffusion layer thickness is minimal above narrow features as the diffusion layer profile is unaffected by the copper surface profile. Therefore, the diffusion flux that corresponds to the removal rate is large for these features and planarization of these features is possible. For wide features, the diffusion layer profile follows the copper surface profile, the removal rate is low, and this leads to a conformal copper removal and inadequate planarization.

The theories proposed to explain the formation of passivation film are salt-film mechanism and acceptor mechanism [21]. In the salt-film mechanism, the assumption is that during the active dissolution regime, the concentration of metal ions (in this case, copper) in solution exceeds the solubility limit and this results in the precipitation of a salt film on the surface of copper. The formation of the salt film drives the reaction forward, where copper ions diffuse through the salt film into electrolyte solution and the removal rate is determined by the transport rate of ions away from the surface. As the salt-film thickness increases, the removal rate decreases. In the acceptor mechanism, it is assumed that the metal-ion products remain adsorbed onto the electrode surface until they are complexed by an acceptor species like water or anions. The rate-limiting step is therefore the mass transfer of the acceptor to the surface. Recent studies confirmed that water may act as an acceptor species for dissolving copper ions [22].

Electrochemical removal rates can be correlated with the current densities between 10 and 100 mA/cm^2 [23]. The limiting current density for a mass-transfer limited process is given by the Levich equation [20]

$$J_{\text{lim}} = 0.62\, n\, \mathrm{F}\, D_{\mathrm{H}}^{2/3} w^{1/2} v^{-1/6} C_{\mathrm{H}} / s_{\mathrm{H}}$$

where n is the stoichiometric number of electrons in the reaction, F is the Faraday constant, D_{H} is the diffusion coefficient of water, w is the angular rotational frequency, v is the kinematic viscosity, C_{H} is the water concentration, and s_{H} is the number of water molecules associated with each dissolving Cu^{2+} ion.

Substituting for constants, the copper removal rate in μm/min is given by

$$\text{Rate} = 441\, D_{\mathrm{H}}^{2/3} w^{1/2} v^{-1/6} C_{\mathrm{H}}$$

The copper removal rates can be varied from 200 to 1700 nm/min. The removal rate is varied by altering the rotational speed and the water concentration. However, as mentioned, pattern sensitivity limits its application to planarization of integrated circuits.

11.4 ECMP

ECMP relies on a combination of mechanical abrasion and electrochemical removal of the copper film from the wafer to achieve planarization. It combines the anodic dissolution of copper with the planarization capability of a pad to achieve topography leveling. It thus overcomes the biggest hurdle of Cu electropolishing—its inability to planarize features with variable defect density. In conventional electropolishing, the removal rate of the protruded and recessed areas are approximately the same; thus removal of Cu follows the initial topography, and planarization efficiency is low. In ECMP as shown in Fig. 11.7, a passive thin film is formed across the Cu surface. The passive film on the protruded areas is removed by the mechanical abrasion of the polishing pad allowing electrochemical dissolution of copper. In the Cu recess areas, not in direct contact with the polishing pad, the passive film is intact, preventing electrochemical dissolution of copper. Thus planarization of areas with different pattern densities having features of variable dimensions can be achieved [24].

ECMP planarization efficiency depends greatly on the electrolyte chemistry as opposed to electropolishing where planarization is mass-transfer controlled within the boundary layer formed on the surface to be polished [20,25]. The electrolyte used in ECMP is H_3PO_4 based, containing additives that can form a complex on the Cu surface. Electrolytes based on H_2SO_4 or HNO_3 have also been reported. The passive film formed must be thick enough to prevent current to pass through in the recessed areas and at the same time soft enough to be removed by the mechanical abrasion of the polishing pad at low

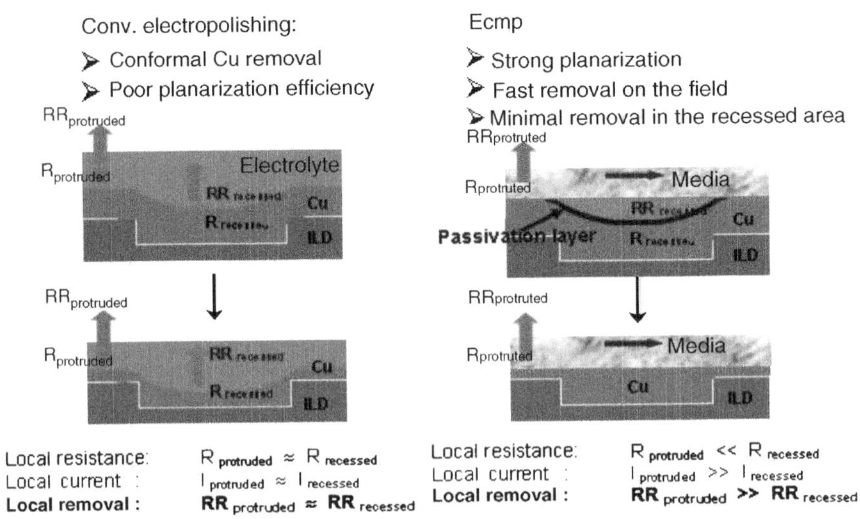

FIGURE 11.7 An illustration of ECMP polishing mechanism in relation to a conventional electropolishing process (from Ref. 25).

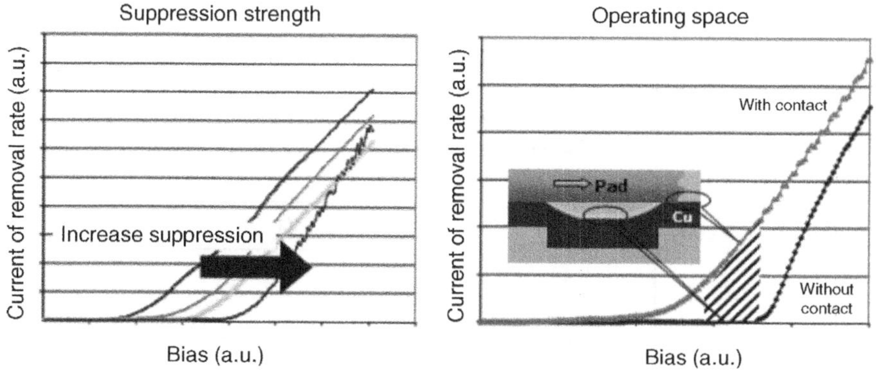

FIGURE 11.8 The suppression strength and operating space for representative electrolyte chemistry.

downpressures (< 0.5 psi). Figure 11.8 defines the operating voltage range for the electrolyte. At low voltages the copper removal rate in contact area (protruded region) is low and not practical. The removal rate in the noncontact area (recessed region) is zero. As the applied voltage is increased, the removal rate in the contact area increases. The chemistry's operating process window is defined at the lower end as the voltage necessary to achieve copper removal rate is greater than 2000 Å/min. At the upper end the voltage has almost zero removal rate in the noncontact region. As the voltage increases, the suppression strength of the electrolyte in the recessed areas is reduced significantly.

The removal amount on the wafer is proportional to the charge flowing through the polishing cell, in accordance with Faraday's law of electrolysis. The linear relationship is also in agreement with the understanding of the ECMP electrochemical reaction: Copper is oxidized into Cu^{2+} ions and the reaction releases two electrons for each atom removed from the wafer surface.

$$Cu \rightarrow Cu^{2+} + 2e^-$$

The Cu removal is voltage controlled and independent of applied downforce as shown in Fig. 11.9.

In order to control the removal profile on the wafer, a multizone cathode is used. Each of the three zones of the cathode is biased independently by a separate power supply as shown in Fig. 11.10. A five-zone cathode has also been reported to better control the wafer removal rate profile. Figure 11.11 shows that zone 1 is responsible for 60% of the amount of removed copper up to 100 mm radius; the remaining 40% is coming from zone 2. The copper removal of the last 50 mm of the wafer is coming from a combination of the three zones. The net removal is a linear combination of the removal from each zone.

In conventional CMP, end-point detection relies on *in situ* sensors such as eddy current sensors [26] to measure the remaining copper film thickness. In

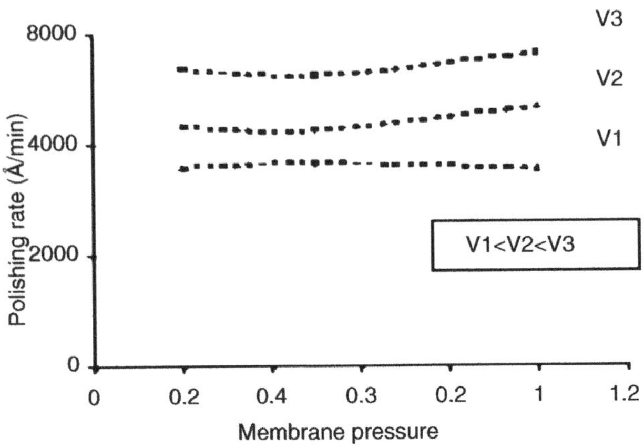

FIGURE 11.9 Removal rate versus downpressure (from Ref. 25).

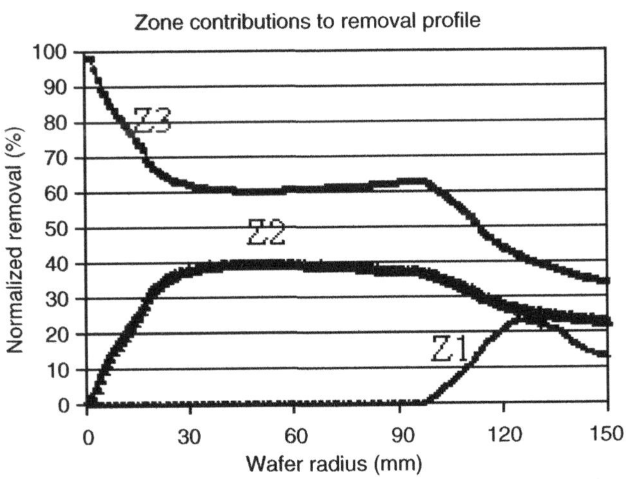

FIGURE 11.10 Contribution from each zone to the removal profile on the wafer.

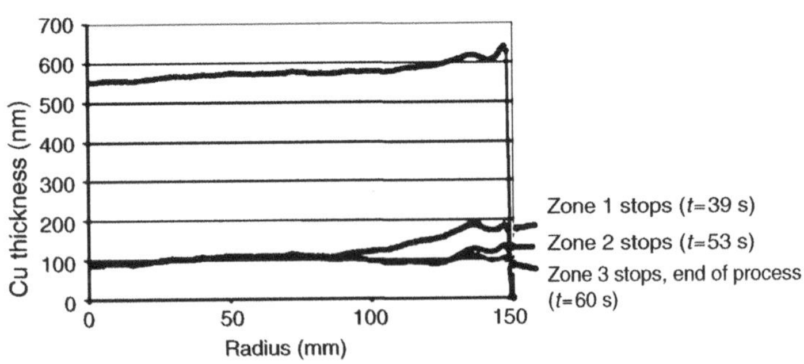

FIGURE 11.11 Modulation of copper profile across wafer using a multizone cathode (from Ref. 25).

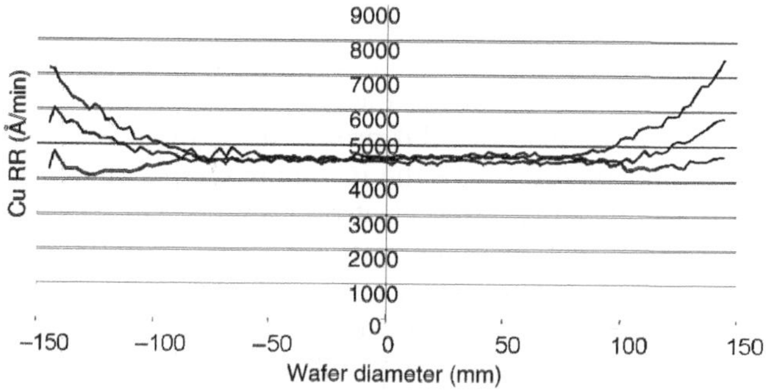

FIGURE 11.12 Cu removal profiles across the wafer modulated by zone charges in an ECMP process (from Ref. 25).

ECMP, end point is based on the amount of charges delivered to each zone, which follows Faraday's law. There is a linear relationship between the charge required by each zone and the amount of copper removed. By modulating the charge on each zone as shown in Fig. 11.12, a desirable removal profile can be obtained.

Manens et al. [27] reported that charge-based end-point detection coupled with advanced process control (APC) can be used to adjust the amount of copper removed and compensate for the large variations in incoming Cu thickness. Figure 11.13 shows wafers with various incoming Cu thicknesses ranging from 3500 Å to more than 6000 Å, which simulates, in an extreme fashion, the variations that can be observed from wafer-to-wafer and lot-to-lot

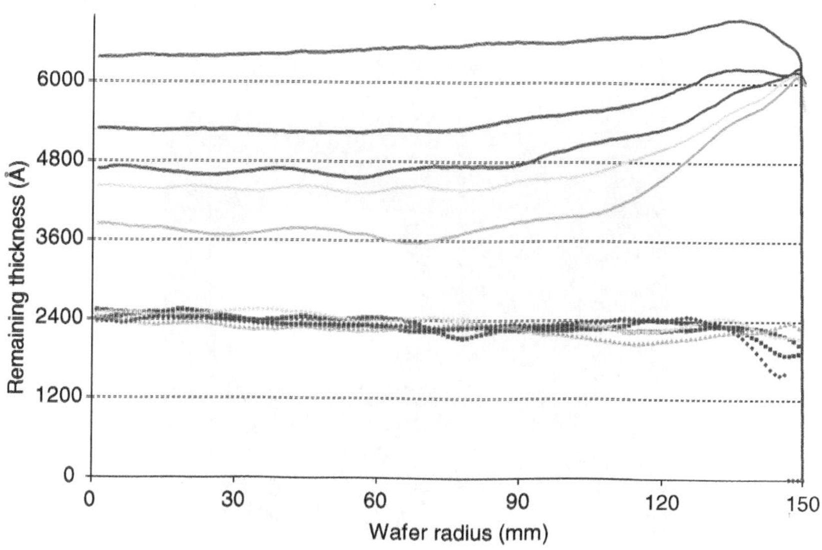

FIGURE 11.13 Prepolished (solid) and Postpolished (dotted) wafer thickness profiles with end-point control (from Ref. 27).

due to variations in copper seed deposition and electrochemical deposition profile and thickness. A noncontact, eddy-current-based metrology unit measures the incoming Cu thickness, and the data are fed forward to adjust the multizone removal to achieve a flat profile at the target thickness of 2400 Å after electropolishing. In Fig. 11.13 the incoming copper thickness nonuniformity is reduced significantly after electropolishing. The thickness range is reduced from >2000 Å before planarization to less than 500 Å after electropolishing. The preplanarization mean copper thickness range is 2800 Å or a variation of 80% among wafers. Postelectropolishing copper thickness is reduced to less than 4%, showing the ability of the end-point algorithm to accurately evaluate the required charge for each zone based on incoming profile. This approach differs greatly from the APC in conventional CMP, in which profile control requires real-time *in situ* thickness measurement to correct for profile changes [28]. Furthermore, adjusting the pressures in a multizone polishing head in real time is limited by the delays encountered in pressure controllers.

ECMP process is typically non-Prestonian; therefore, the control "knobs" of conventional CMP, such as downpressure and platen speeds, cannot be directly used to modify the polishing profile. The control knobs of ECMP are mainly the charges applied to each zone. Different voltages are used for simultaneous completion of the applied charge by each zone.

In most ECMP applications a perforated polishing pad with holes through it is used. Thus, electrical contact between the Cu surface that is contacted to the anode and the cathode located under the polishing pad is achieved via the electrolyte [29]. In the applied materials ECMP reflexion platform, the wafer is contacted to the anode located at the center of the pad as shown in Fig. 11.14.

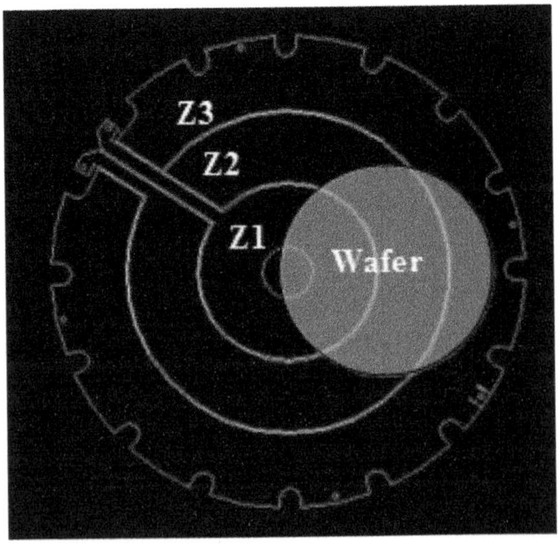

FIGURE 11.14 Schematic of multicathode–wafer–anode contact in AMAT ECMP relexion polisher (from Ref. 25).

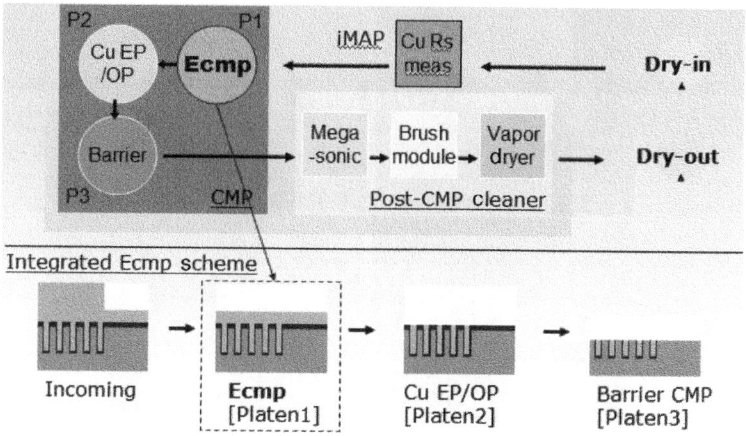

FIGURE 11.15 A schematic view of a hybrid-ECMP system.

The design shown in Fig. 11.14 requires that copper film always remain across the wafer throughout the polishing as only the edge of the wafer contacts the anode. As soon as the copper at the wafer edge becomes too thin, resistance at the anode increases significantly leading to "copper pullouts," which in turn result in complete discontinuity between the anode and the cathode. Therefore, this design is intended for only bulk copper removal by electromechanical means. In one ECMP system shown in Fig. 11.15, electrochemical–mechanical planarization is used to remove the bulk Cu down to approximately 2000 Å of the remaining copper. Conventional CMP is used to clear the remaining copper and barrier. Because the remaining copper film is still removed with conventional slurry, the ECMP reflexion platform is also referred to as a hybrid-ECMP design.

In this mode of operation, ECMP leaves a uniform Cu film across pattern densities and allows conventional CMP to operate at low pressures less than 2 psi. The remaining Cu film after ECMP is measured by the focus ion beam (FIB) technique [25]. The film thickness is quite uniform across the wafer regardless of the pattern structures. For example, the thickness of the film over some representative structures (0.18-μm array, 0.25-μm array, 10-μm line, and 50-μm bond pad) is within a range of 200 Å or less, as seen in Fig. 11.16.

Figure11.13 shows that the target remaining copper thickness can be achieved for various incoming copper thicknesses without compromising wafer uniformity. This in turn implies that ECMP can achieve planarization with reduced copper overburden compared to conventional planarization. Indeed, the planarization efficiency of the process is 100% (copper thickness removal:step height removal = 1:1) as shown in Fig. 11.17 [25]. As shown in Fig. 11.18 [30], thick-plated copper film often leads to stress-induced voids in multilevel interconnects. Figure 11.19 shows the schematic of a stress-induced void caused by copper shrinkage during the copper annealing process.

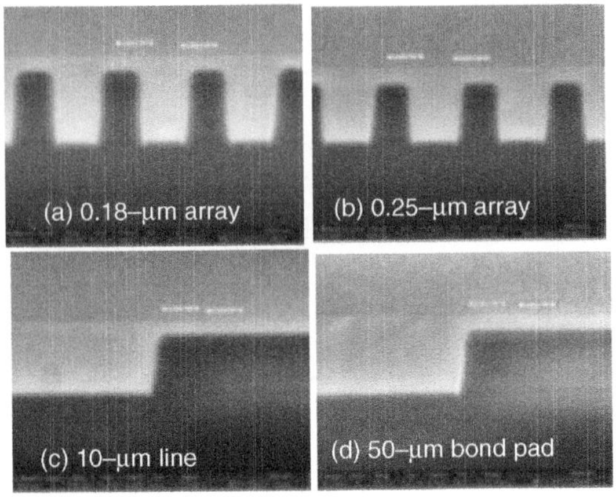

FIGURE 11.16 FIB images of remaining Cu film over different pattern structures post-ECMP process [25] where the film thickness is (**a**) 1000 Å over 0.18 μm, (**b**) 1100 Å over 0.25 μm, (**c**) 1100 Å over 10 μm, and (**d**) 1200 Å over 50 μm structures.

Conventional CMP typically requires an overburden of copper of at least 4000 Å due to low planarization efficiency. An overburden of 1000 Å is sufficient for ECMP to achieve planarization without compromising performance in terms of dishing, erosion, and defectivity. A stress migration test was carried out at 250 °C for 333 h on a structure with metal level 1 and level 2 lines with thickness in the range of 0.14–0.16 μm. An increase in resistance by more than 10% is considered a fail. The reduction in electroplated copper thickness can reduce the number of fails in stress migration test as shown in Table 11.1.

The copper-clearing step is a conventional planarization process. This step introduces most of the topography on the wafer. The two main reasons are the dependency of the copper removal rate on the wafer pattern density and the fact

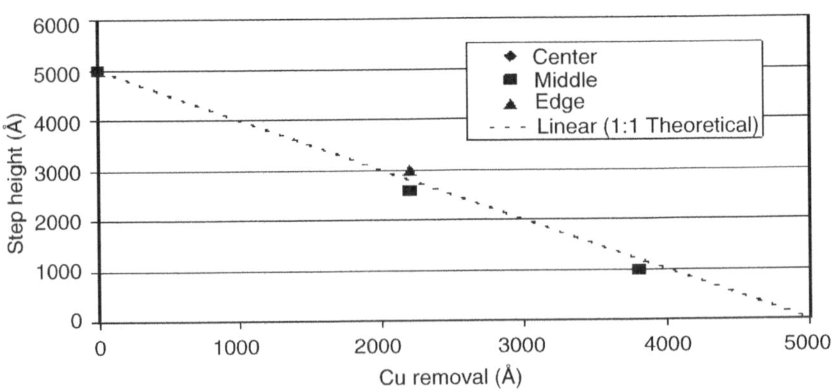

FIGURE 11.17 ECMP planarization efficiency at removal rate of 6000 Å/min on 100 × 100 μm lines (from Ref.25).

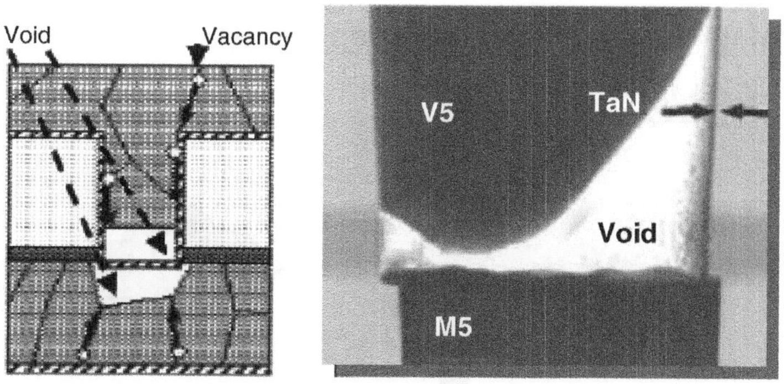

FIGURE 11.18 Stress-induced void formed in Cu via interconnect, Cu is separated from barrier film (TaN) (from Ref. 30).

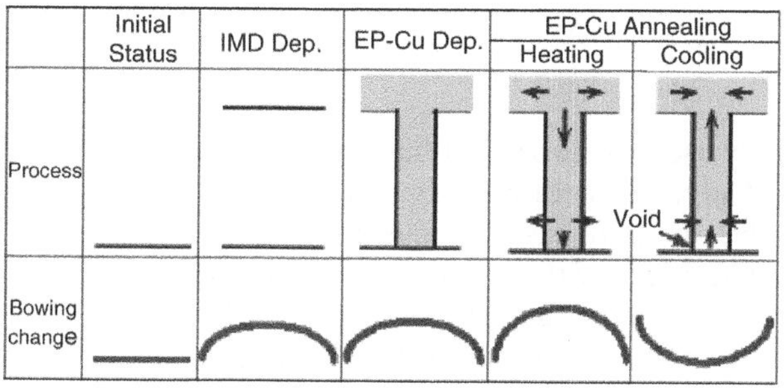

FIGURE 11.19 Schematic diagram showing formation of stress-induced void caused by large stress change and shrinkage of copper during annealing (from Ref. 30).

that most commercial copper-clearing slurries have high selectivity over the barrier materials. Areas of the die with high pattern density will polish faster than lower pattern density areas. In these areas, copper will erode with reference to the field area owing to slurry selectivity. This mechanism leads to large topography variations within the die. The barrier removal process typically "corrects" the topography introduced by the copper-clearing step. This correction depends on

TABLE 11.1 Stress Migration Fails as a Function of Electroplated Copper Thickness.

ECP Cu Thickness, Å	Failed Dies/Total Dies
7000, (baseline)	28/152
4000 (reduced Cu thickness enabled by ECMP)	19/152

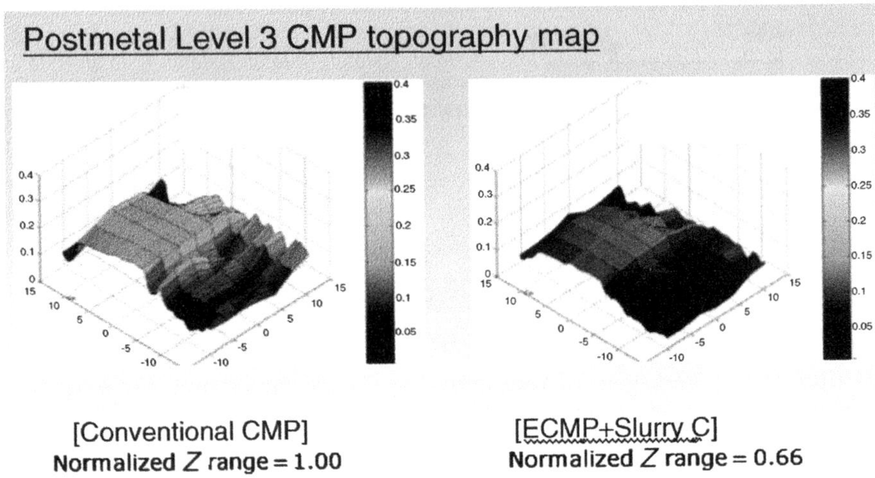

FIGURE 11.20 Final topography comparison between wafers after a conventional and an ECMP planarization process, where slurry C is used in the barrier removal (from Ref. 25).

the barrier slurry used. A slurry that has high removal rate of dielectric film and relatively low Cu removal rate tends to have a better topography correction compared to the one that has a smaller ratio of dielectric removal rate to Cu removal rate. The effect of the barrier slurries on the wafer final topography at multiple wiring levels has been discussed in detail by Ong et al. [31].

It is obvious that the final topography after copper clearing is influenced to a large extent by the properties of the Cu slurry used, and it can be reduced by the barrier removal process. Depending on the selection of Cu clearing and barrier slurries, some of the advantages of hybrid ECMP, such as reduced dishing/erosion and low polishing scratches and polishing residues, can be compromised. Figure 11.20 shows the topography across die of wafers polished with a conventional planarization process and a hybrid-ECMP process at metal level 3. The maps were generated from the lithographic tool that uses a laser interferometer to adjust the focus plane across the die.

11.5 FULL SEQUENCE ELECTROCHEMICAL–MECHANICAL PLANARIZATION

The electrochemical removal of both bulk and remaining copper is referred to as full-sequence ECMP (FS-ECMP), and it has been investigated by many researchers [20,32]. In FS-ECMP the entire wafer has to be contacted to the anode throughout the polishing process. A conductive polishing pad provides a path for current to flow even when copper islands are formed and allows for residual copper to clear. The research focus has been on developing polishing pads with high conductivity that will not scratch copper. In one approach the pad is immersed in an electrolyte bath in the same manner as in hybrid-ECMP design. In another approach, electrolyte is supplied through the pad [33]. In the first design the wafer cannot be rinsed, and the electrolyte will be carried over to a different platen where barrier polishing is conducted. Electrolyte has to be rinsed off on this platen before barrier polishing starts. In the second design, wafer rinsing is done on the barrier removal platen as is done in conventional CMP. The electrolyte chemistry does not have to be fully compatible with the barrier slurry chemistry. In this approach the pad design is quite complex; its manufacturability has to be demonstrated along with its stability in a production environment.

Electrochemical polishing in DI (ECP-DI) is another approach for "full copper removal." It involves removal of Cu only in the areas that come in contact with an ion-exchange film [32]. Cu is removed through an electrochemical reaction between OH ions in DI water and Cu atoms. Both electrodes are contacting the wafer front face via an ion-exchange film. One electrode is used to remove copper and the other is used to supply electricity. DI water is supplied to the interface of the wafer–ion exchanger as shown in Fig. 11.21. Current flows from the supply electrode to the Cu surface of the wafer and the processing electrode through DI water. Multiple units of those shown in Fig. 11.21 can be used to increase the total area of the processing electrodes and thus the removal rate. High removal rates of more than 8000 Å/min were reported. A linear relationship between removal rate (up to 370 nm/min) and current density (up to 0.9 Å/cm^2) is shown in Fig. 11.22. Figure 11.23 shows normalized planarity performance for an array of 100-μm copper line with 50% metal density.

Unlike in the ECMP process described earlier, the initial conductivity of the liquid between the wafer and the pad is not required. As a matter of fact, a conductive solution actually lowers the planarization efficiency as shown in Fig. 11.24. Similar to ECMP, high planarization efficiency can be obtained at

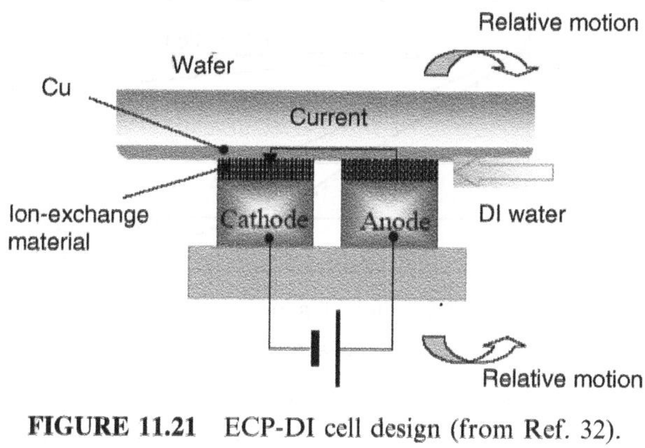

FIGURE 11.21 ECP-DI cell design (from Ref. 32).

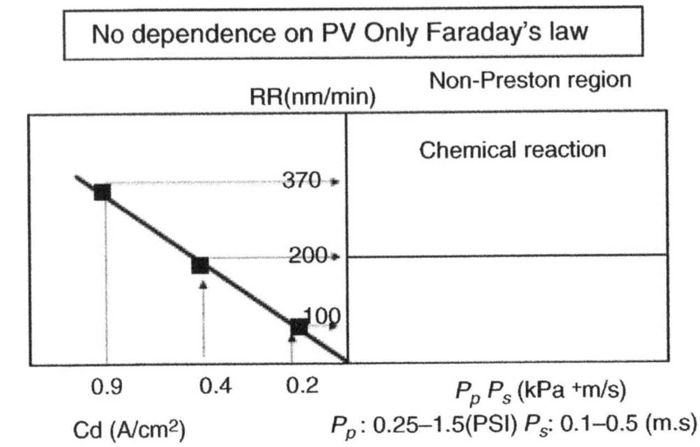

FIGURE 11.22 Linear relationship between removal rate and current density in an ECP-DI process. The removal rate is independent of pressure and linear velocity of the platen (from Ref. 32).

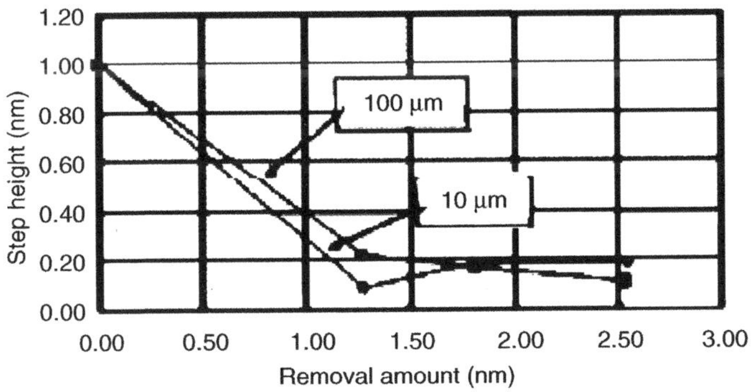

FIGURE 11.23 Planarization performance on the 10- and 100-μm copper lines with 50% metal density for an ECP-DI polishing (from Ref. 32).

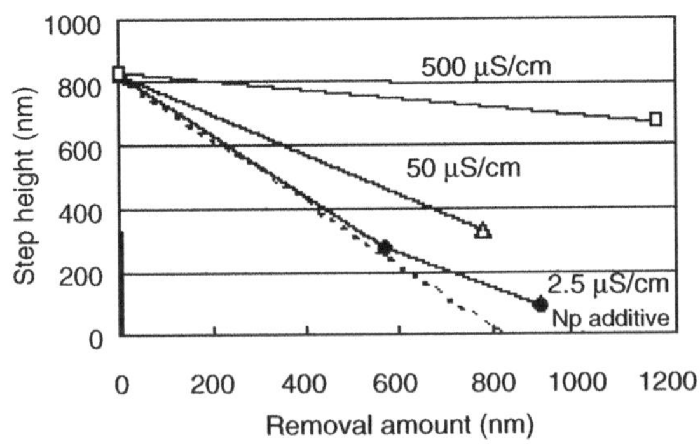

FIGURE 11.24 Dependency of step height reduction efficiency on electrical conductivity of the liquid used for ECP-DI polishing (from Ref. 32).

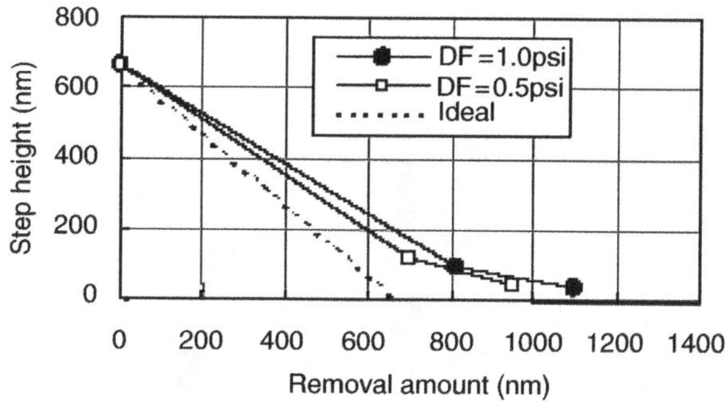

FIGURE 11.25 Correlation between planarization efficiency and downforce used during an ECP-DI polishing (from Ref. 32).

low downpressure. As shown in Fig. 11.25 the planarization efficiency actually improves when the downforce is smaller.

During an ECP-DI process, the electrical conductivity of the water coming in contact with the ion exchanger is increased because the protruded areas of the wafer will come in contact with the ion exchanger more often and will have high removal rate. In comparison, the recessed areas are farther away from the ion-exchange material. They will experience lower material removal rate. The planarization is thus accomplished.

With this approach, all Cu can be removed as the wafer contact is done through a series of supply/processing electrodes contacting the whole wafer surface. The ion-exchange film will increase the resistance; thus, high current densities need to be used to maintain the removal rate. This in turn might cause Cu pitting or formation of other defects. Defectivity data were not reported for the ECP-DI process [32].

A novel conductive carbon pad reported by Kondo et al. [33] shows capability to remove all Cu on an orbital 300-mm CMP system. The pad consists of a top carbon layer as an anode, an intermediate insulating layer, and an underlying cathode sheet. More than 100 electrocells were fabricated within the pad as shown in Fig. 11.26. Cu removal follows the same mechanism described in Fig. 11.7, with removal rates as high as 4000 Å/min obtained at a current of 30 Å. Defectivity data were not reported with this setup. Segmentation of the cathode will allow one to optimize the process to achieve postpolishing wafer uniformity for those with various incoming plating profiles. Similar to the mechanism described before, the passivation film formed in the recessed areas prevents the copper film from being etched; thus planarization can be achieved.

Another fully conductive pad design is shown in Fig. 11.27. The pad is fully immersed in electrolyte solution as the polymeric perforated pad in the hybrid-ECMP system. The fully conductive pad is made of polymer as the matrix, with embedded conductive particles serving as conducting medium to carry the

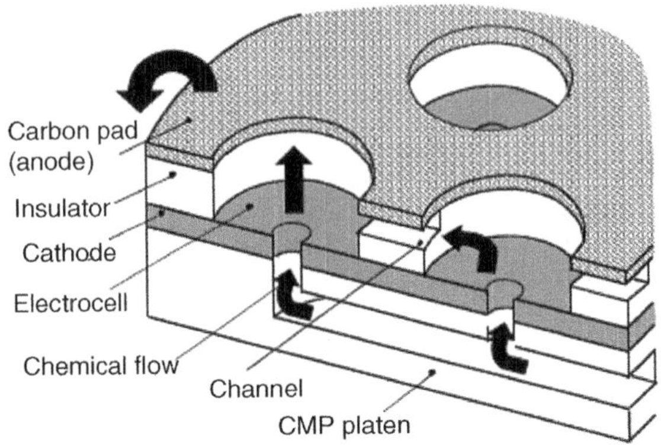

FIGURE 11.26 Novel carbon pad with fabricated electrocells for ECMP (from Ref. 33).

voltage to the wafer surface during polishing. The pad is perforated to provide the conductive path for the metal to dissolve in ionic forms into the solution.

Figure 11.28 is the schematic of an ECMP platen that carries a conductive pad. The counterelectrode (cathode) is in the back of the conductive pad. It can be segmented into three or more zones similar to the platen for the hybrid-ECMP process.

The copper removal mechanism is the same as in the hybrid process shown in Fig. 11.7. To achieve good planarization, the polishing rate in the trench area must be lower than that in the field area. The passivation provided by the electrolyte has to be strong enough to prevent dishing in the trench area and at the same time should not be too strong to resist the mechanical removal in the field area by the pad abrasion. The electrochemical behavior of such an electrolyte is shown in Fig. 11.29.

The upper curve shows the response of electric current to applied voltage in the field area (electrode rotates on the pad). The lower curve shows the same response in the trench area (no rotation). There are four regions where different electrochemical phenomena take place [34].

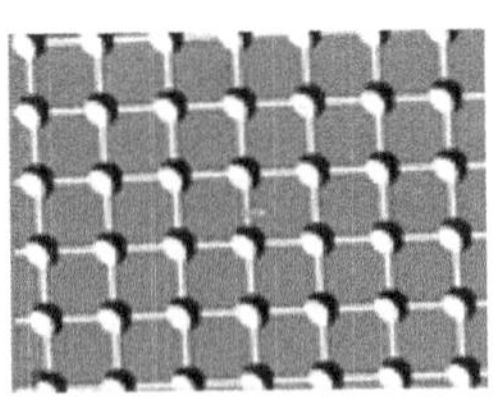

FIGURE 11.27 Picture of a fully conductive pad, conductive polymer matrix with through pad perforations (from Ref. 34).

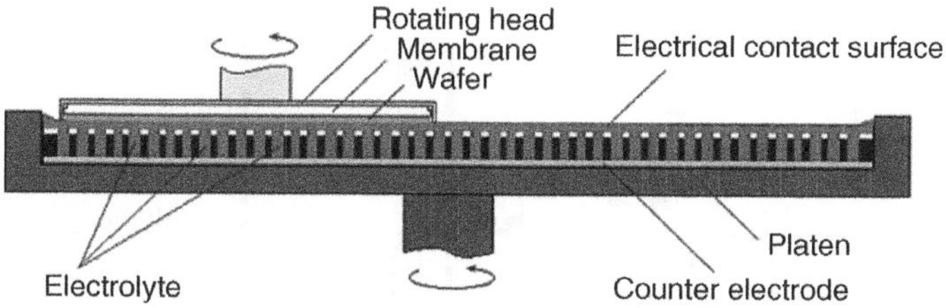

FIGURE 11.28 Schematic of a platen carrying the conductive pad (from Ref. 34).

Region I. Minimal etching takes place in this region. Owing to the passivation provided by electrolyte, the copper dissolution is kept at minimal, especially compared to the traditional electropolishing.

Region II. Cu_xO is suspected to form a film surface that provides additional passivation in addition to that due to the electrolyte. However, the Cu_xO film might be of a loose structure that can be easily removed. This is seen in the upper curve when the Cu_xO film is removed due to abrasion as Cu electrode rotates against the pad.

Region III. When Cu electrode rotates, the decrease in electric current with increasing voltage in the first half of the region indicates a stronger passivation on the Cu surface. Also the significant current difference between rotation and nonrotation of the Cu bar leads to a good planarization as a result of the rate difference between the field and the trench area on a patterned wafer. The main polishing phase is located in this region. A smooth surface finish is also expected from polishing using these parameters.

Region IV. Rapid gas evolution occurs in this region due to electrolysis of water to oxygen. A rougher surface is expected if polishing is conducted in this region.

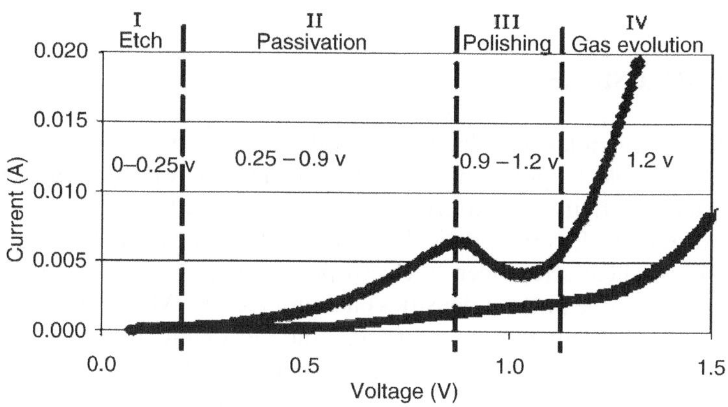

FIGURE 11.29 Electrochemical behavior of a representative electrolyte on copper in cell tests, simulating the polishing rate in the field area (upper curve) and that in the trench area (lower curve) (from Ref. 34).

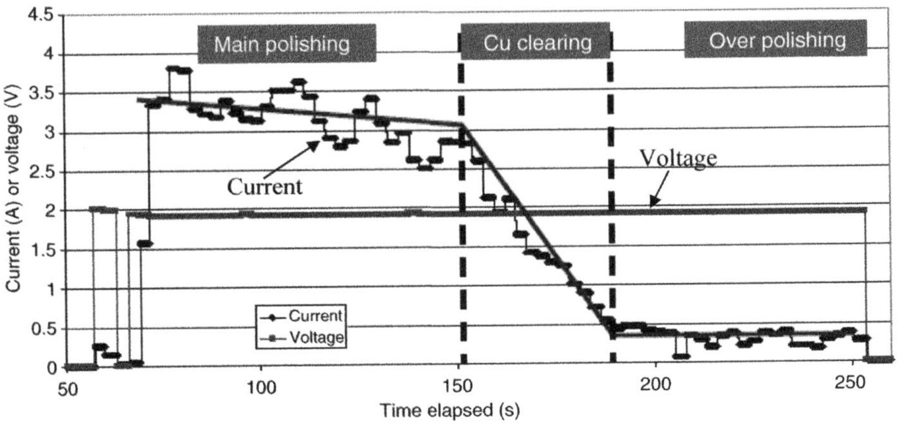

FIGURE 11.30 A typical electric current trace for Cu-clearing process on a patterned wafer, showing drop in current upon Cu clearance (from Ref. 34).

In the hybrid process described earlier, the electric current remains constant during polishing as there is a continuous film of copper across the wafer throughout the process. Figure 11.30 shows the current during polishing on the platen shown in Fig. 11.29 for the full-sequence process. Initially, the current is stable until the Cu film starts to break and expose the barrier. Cu islands are formed across the wafer and the current drops significantly. After the copper residue is completely removed, the current can flow only through the copper pattern that is passivated by the electrolyte, and thus it is much smaller than the initial current. Hence, the end point can be defined by the current. In the hybrid process, the end point is defined by the charge since the goal of the process is to remove a certain amount of copper.

11.6 CONCLUSIONS

The general principles of electrochemical planarization technologies were introduced. Cu electropolishing is a proven method for local planarization; however, its sensitivity to pattern density may prevent it from been used in ULSI processing. Electrochemical–mechanical planarization combines the benefits of low shear stress seen in electropolishing and the ability to planarize structures of variable pattern densities encountered in CMP. For this and other obvious reasons, tremendous industrial and academic interests have been focused on the research and development of ECMP as planarization technology with low shear stress and low dishing/erosion for wafers that contain ultra-low-*k* dielectric materials. Efforts have been revolved around the electrolyte chemistry development and pad design to enable full removal of copper film by electrochemical means.

ECP in DI water shows promise in planarization performance. However, owing to poor electrical conductivity of the ion-exchange film to the wafer, it has to resort to a high current during polishing, which may lead to defects such as copper pullouts. Until this issue is addressed, the ECP-DI approach will be difficult to implement for copper polishing process at industrial level.

ACKNOWLEDGMENTS

Several people in the IBM Semiconductor Research and Development Center including the alliance partners helped in reviewing this chapter for clarity and consistency and provided valuable input. The author would also like to thank the Applied Materials team who has been working on electrochemical–mechanical planarization for their continuous support during the development of the hybrid ECMP as well as for providing their inputs on the recent developments for full-sequence ECMP. In addition, the author gratefully acknowledge to the kind contribution from Manabu Tsujimura of Ebara on the comparison of ECP, ECP-DI, ECMP, and CMP.

QUESTIONS

1. What are the major advantages and potential disadvantages of ECMP versus conventional CMP?
2. To fully take advantages of what ECMP can offer, all copper and the barrier layer should be removed. What would be the principal challenges in accomplishing this?
3. In addition to the design for electrical contacts with the wafer described in this chapter, what would you do if you were designing and implementing an ECMP process for copper?
4. In addition to the pad design described in this chapter, what would you do if you were designing and implementing an ECMP process for copper?

REFERENCES

1. Edelstein DC, et al. Full copper wiring in a sub-0.25-μm CMOS ULSI technology. IEDM. 1997.
2. Muraka SP. Multilevel interconnections for ULSI and GSI era. Mater Sci Eng R 1997;19:87–151.
3. West AC, Mayer S, Reid J. Electrochem Solid State Lett 2001;4:C50–C53.
4. Dini JW. Electrodeposition of Copper in Modern Electroplating. 4th ed. New York: John Wiley & Sons, Inc.; 2000.
5. Licata TJ, Colgan EG, Harper JME, Luce SE. Interconnect fabrication processes and development of low-cost wiring for CMOS products. IBM J Res Develop 1995;39:419–435.

6. Zantye PB, Kumar A, Sikder AK. Chemical mechanical planarization for microelectronics applications. Mater Sci Eng R 2004;45:89–220.
7. Mosig K, Jacobs T, Brennan K, Rasco M, Wolf J, Augur R. Integration challenges of porous ulta-low-*k* spin-on dielectrics. Microelectron Eng 2002;64:11–24.
8. Lanckmans F, Brongersma SH, Varga I, Poortsmans S, Bender H, Conard T, Maex K. A quantitative adhesion study between contacting materials in Cu damascene structures. Appl Surf Sci 2002;201:20–34.
9. Fayolle M, Passemard G, Louveau O, Fulabla F, Cluzel J. Challenges of back end of the line for sub 65 nm generation. Microelectron Eng 2002;70:255–266.
10. Ong P, Ponoth S, Economikos L, Sakamoto A, Hong DH, Chae M, Rhee SH, Nicholson LM, Landers W, Werking J, Li B, Chen F, Sankaran S. Cu CMP with direct polish on ultra-low-*k* dielectric film ($k \sim 2.4$) for 45 nm technology node. ICPT; Oct 2006.
11. Goonetilleke PC, Roy D. Electrochemical–mechanical planarization of copper: effects of chemical additives on voltage controlled removal of surface layers in electrolytes. Materials Chem Phy 2005;94:388–400.
12. Dettner PP. Electrolytic and Chemical Polishing of Metals. Tel Aviv, Israel: Ordentlich; 1987.
13. Sato S et al. Newly developed electrochemical polishing process of copper inlaid in frangile low-*k* dielectrics. IEDM. 2001. p 4.4.1–4.4.4.
14. Pallinti J et al. An overview of stress free polishing of Cu with ultra-low-*k* ($k < 2.0$) films. IITC. 2003.
15. Contolini RJ, Bernhardt AF, Mayer ST. Electrochemical planarization for multilevel metallization. J Electrochem Soc 1994;141:2503–2510.
16. Contolini RJ, Mayers ST, Graff RT, Tarte L, Bernhardt AF. Electrochemical planarization of ULSI copper. Solid State Technol 1997;10:155–161.
17. Duboust A, Wang Y, Neo S, Chen LY. End point detection for electrochemical–mechanical polishing and electropolishing processes.U.S. Patent 20030 136684. 2003 Jul 24.
18. Chang SC, Shieh JM, Huang CC, Dai BT, Li YH, Feng MS. Microleveling mechanisms and applications of electropolishing on planarization of copper metallization. J Vac Sci Techol B 20:2002;2149–2153.
19. Padhi D, Yahalom J, Gandikota S, Dixit G. Planarization of copper thin films by electropolishing in phosphoric acid for ULSI applications. J Electrochem Soc 2003;150:10–14.
20. Suni II. Cu planarization for ULSI processing by electrochemical methods: a review. IEEE Trans Semiconduct Manuf 2005;18:341–349.
21. Vidal R, West AC. Copper electropolishing in concentrated phosphoric acid. J Electrochem Soc 1995;142:2689–2694.
22. Du B, Suni II. Mechanistic studies of copper electropolishing in phosphoric acid electrolytes. J Electrochem Soc 2004;151:C375–C378.
23. West AC, Deligianni H, Andricacos PC. Electrochemical planarization of interconnect metallization. IBM J Res and Dev 2005;49(1):37–48.
24. Sato S, Yasuda Z, Ishihara M, Komai N, Ohtorii H, Yoshio A, Segawa Y, Horikoshi H, Ohoka Y, Tai K, Takahashi S, Nogami T. Newly developed

electrochemical polishing process of copper as replacement of CMP suitable for damascene copper inlaid in fragile low-*k* dielectrics. IEEE. 2001.

25. Economikos L, et al. Integrated electrochemical–mechanical planarization (Ecmp) for future generation device technology IITC. 2005.
26. Bennet D et al. Real-time profile control for improved copper CMP. Solid State Technol Jun 2003.
27. Manens A, Miller P, Kollata E, Duboust A. Advanced process control extends ECMP process consistency. Solid State Technol Feb 2006.
28. Bennet D et al. Real-time profile control for improved copper CMP. Solid State Technol Jun 2003.
29. Butterfield PD, Chen LY, Hu Y, Manens AP, Mavliev R, Tsai SD, Liu FQ, Wadensweiler R. Conductive polishing article for electrochemical–mechanical polishing. U.S. Patent 20 040 134792 2004 Jul 15.
30. Park BL. Mechanism of stress-induced voids in multilevel Cu interconnects. IITC. Jun 2002.
31. Ong P, Economikos L, Hong DH, Chae M, Quon R, Grunow S, Dipaola D, Siddiqi S, Liegl B, Ponoth S, Tseng W.-T, Ticknor A, Fang R, Kulkarni D, Lagus M, Matusiewicz G, Angyal M, Watts D. Design influence on CMP-induced topography at chip and wafer scales over multiple levels. ICPT. Oct 2006.
32. Wada Y, Noji I, Kobata I, Kohama T, Fukunaga A, Tsujimura M. The enabling solution of Cu/low-*k* planarization technologu. IITC. Jun 2005.
33. Kondo S, Tominaga S, Namiki A, Yamada K, Abe D, Fukaya K, Shimada M, Kobayashi N. Novel electrochemical–mechanical planarization using carbon polishing pad to achieve robust ultra–low-*k*/Cu integration IITC. Jun 2005.
34. Jia R, Wang Y, Wang Z, Tsai S, Diao J, Mao D, Karuppiah L, Chen L. ECS Fall. 2005.

12

SHALLOW TRENCH ISOLATION CHEMICAL MECHANICAL PLANARIZATION

YORDAN STEFANOV AND UDO SCHWALKE

12.1 INTRODUCTION

Since the creation of the first integrated circuit (IC) in 1958 by Jack Kilby at Texas Instruments, microelectronic devices have witnessed unprecedented development. During the last almost 50 years, transistors have continued to become smaller, faster, and more reliable, following the prodigious Moore's law [1] according to which IC complexity doubles every 18 months. As scaling down continued, the planarity requirements imposed by lithography, deposition, etching, and cleaning processes became more and more severe [2,3]. In the mid-1980s, IBM researchers started investigating the possibilities of chemical–mechanical planarization or polishing (CMP) [3] as a technique for achieving local as well as global planarity of standard silicon wafers. CMP was not a novel discovery as metal and glass polishing have been known to mankind for centuries [4]. Still, the introduction of CMP in IC manufacturing was not a straightforward process because of the different layer thicknesses, process windows, planarity requirements, and cleanliness standards. The cleanliness of the process in particular was an issue, considering the fact that CMP is inherently dirty and the semiconductor industry had tried to keep wafers as clean as possible for the past 2 decades. Nevertheless, CMP and post-CMP cleaning processes and their consumable optimization allowed the introduction of polishing in a number of microelectronic applications.

Microelectronic Applications of Chemical Mechanical Planarization, Edited by Yuzhuo Li

Initially, CMP was used in the back end of line (BEOL) as an enabling technology for achieving more complex multilevel metallization schemes with reduced RC delays. The first to benefit from it were the interlevel dielectric (ILD) oxide and the tungsten vias [3,5]. The two polishing processes are indicative of the possibilities that CMP offers. The first planarization is a "blind" process in which polishing stops on the initial material, simply reducing topography and overall film thickness. In the case of tungsten CMP on the contrary, the polishing stops on an underlying patterned stop layer, patterning indirectly the polished layer as a negative imprint of the former—a type of inlaid patterning or "damascene" process. The abilities to achieve both global planarity and structure thin films are the basis of the continuous interest that CMP has experienced. As the technique matured, it found a number of applications in microelectronics. Nowadays CMP is used, among other things, for ILD, Cu/AlCu/W damascene, borophosphosilicate glass (BPSG), polysilicon, shallow trench isolation (STI), and replacement metal gate planarization [3,6]. Its applications have been extended to front end of line (FEOL), where requirements are much stricter than in BEOL but the benefits are certainly worth the effort. If BEOL CMP has allowed an increase in vertical IC complexity (number of interconnect lines), FEOL CMP and especially STI CMP have helped horizontal complexity (number of elements per area) to reach the present-day ULSI levels of several hundred million elements per square centimeter.

The purpose of this chapter is to give an overview of the application of CMP for shallow trench isolation, which is the present-day state-of-the-art device isolation technology. STI CMP is without a doubt one of the toughest planarization tasks, and it has been a process driver for CMP development for many years because of its strict planarity requirements. Looking at the planarization, in the case of STI, is helpful in understanding the different aspects and problems that are associated with the planarization process in general: polishing rates, selectivity, pattern density dependence of polishing rates, process and design optimization, slurry and pad selection, post-CMP cleaning, and so on.

This chapter is organized in the following way: Section 12.2 discusses the reasons for the transition from the older local oxidation of silicon (LOCOS) device isolation to STI; Section 12.3 describes the STI process module, showing the standard processing steps for creating trench device isolation; Section 12.4 analyzes the CMP step in more detail; Section 12.5 illustrates the different design optimization methods for improving post-CMP planarity; and, finally, Section 12.6 gives an outlook to the future development and challenges facing STI CMP.

12.2 LOCOS TO STI

Before discussing shallow trench isolation, a brief summary should be made about its predecessor—the local oxidation of silicon [7]—as an illustration of

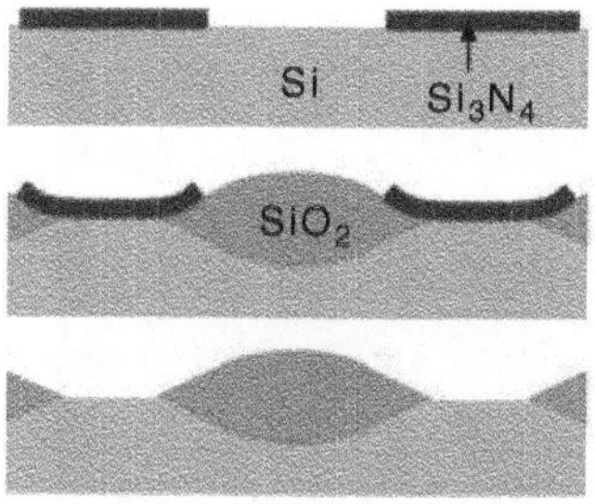

FIGURE 12.1 Schematic cross sections showing the major LOCOS fabrication steps.

how different a device isolation technology is without a global planarization technique and of the numerous advantanges CMP brings.

LOCOS [8,9] relies on thermal oxidation of parts of the silicon surface for device isolation. It has been the main isolation technique for many years because of its simplicity and low cost. A standard process flow for manufacturing LOCOS proceeds as follows (Fig. 12.1). Starting with a blanket-preimplanted wafer, windows are opened with lithography and etching in a deposited nitride layer on the silicon surface in places where the isolation areas are to be created. A long high-temperature wet oxidation follows, which oxidizes the open silicon areas according to the following:

$$Si(s) + 2H_2O(v) \rightarrow SiO_2(s) + 2H_2(v) \tag{12.1}$$

The selective oxidation is based on the much lower diffusion rates of oxygen atoms through silicon nitride compared to silicon dioxide. Because of this difference, the nitride layer is an efficient mask for oxidation. Since the formation of a certain amount of SiO_2 consumes approximately. 44% silicon as the oxidation process takes place, the oxide front moves deeper below the original silicon surface, separating the active silicon areas by oxide isolation areas. After nitride removal, LOCOS isolation is complete (Fig. 12.2) and processing continues with threshold voltage implantation and gate stack formation.

Thermal oxidation is both the reason for the success of this technology and the cause for its shortcomings and eventual replacement by STI. On the one hand, oxidation makes LOCOS easy to manufacture because isolation areas are created with a single oxidation step. On the other hand, the geometry is not easily controllable. First of all, 56 % of the oxide is situated above the silicon surface where it does not serve as isolation, but still creates additional topography of several hundred nanometers on the wafer surface. As devices scale down, more advanced lithography tools are required for imprinting smaller and more precise images on the surface. These tools, however, are applicable only for very planar surfaces because of their limited depth of focus. Therefore, they are not compatible with LOCOS isolated wafers, projecting

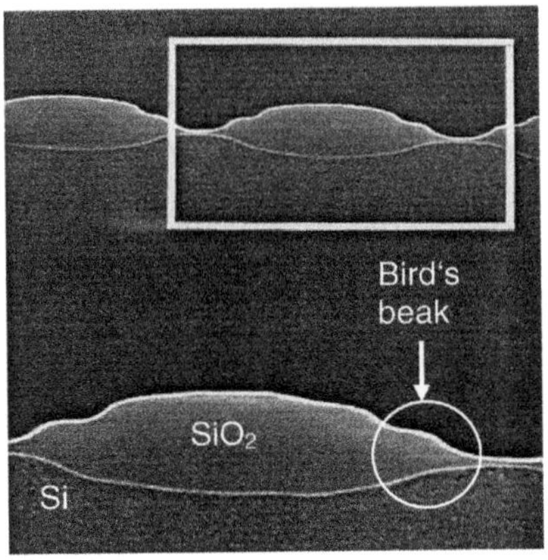

FIGURE 12.2 REM cross section showing a wafer with LOCOS isolation (above) and zoom-in (below).

unsharp images on the surface in consecutive lithography steps (for example, the critical gate lithography with the gates running from isolation to isolation area through active areas). A second important disadvantage of LOCOS is the lateral encroachment [8,9] of the oxide into the active area below the nitride layer, which leads to effective reduction in the active area dimensions by forming a "bird's beak" at the active-isolation transition. This transition, which is typically 100–200 nm in length even for advanced LOCOS integration schemes (and sometimes several times of that), limits the packing density of LOCOS isolated ICs. Thirdly, there is uncontrollable thinning of the oxide in small isolation areas [10] that additionally reduces the isolating properties of LOCOS for high-density designs. There are also additional limitations of using LOCOS, including doping encroachment, susceptibility to latch-up, and oxidation-induced stress in the substrate [8–10].

All of these problems have led to the need for an isolation technology with better control over geometry dimensions for improved scalability. Such a technology is shallow trench isolation, which relies on reactive ion etching (RIE) and chemical vapor deposition (CVD) oxide filled trenches as isolation rather than areas formed by oxidation. In the 1990s, STI gradually replaced LOCOS as the state-of-the-art device isolation technology for ULSI ICs, thanks to the development and maturing of the CMP process. Of course, LOCOS was not simply left behind and forgotten. Like every technology that has been developed and applied over such a long period of time, LOCOS continued its development, and various modifications allowed its application to further technology generations than initially expected. Although initially there were doubts about the scalability of the technology beyond 1 μm, ingenious approaches such as polybuffered LOCOS [11] and polybuffered recessed LOCOS [12] extended it even to 0.25 μm [10]. Still, complexity for advanced LOCOS schemes increased dramatically and did not offer any significant advantages over STI. The latter isolation technology presented a more direct approach for meeting the different requirements for sub-half-nanometer nodes, and therefore was the technology of choice for new generations such as the 65-nm technology node that entered production in 2005.

12.3 SHALLOW TRENCH ISOLATION

As explained above, the basic advantage of STI over LOCOS is the improved control of the former over the geometry of the isolation area. Here, a standard process flow for manufacturing STI with direct planarization is presented. The important considerations for each step are mentioned along with typical process and geometry parameters. Additional process steps required for various optimization techniques are added to this process flow depending on the specific technique. These steps are described in Section 12.4.

A schematic overview of the STI fabrication module [9,10,13] is shown in Fig. 12.3. Manufacturing starts with a thin pad oxide growth and nitride deposition. The nitride serves as a hard mask for etching, and the oxide is a stress reducer. Typical values for the thicknesses of the pad oxide and the nitride are 15–25 and 100–150 nm, respectively. The thickness of the nitride layer is determined by the desired post-CMP step height between the active and isolation areas [13]. Lithography defines the isolation areas in which the nitride and oxide are then etched with RIE. Next, isolation trenches with depth of 250–700 nm are etched into the substrate with chloride-based (e.g., $SiCl_4$) plasma. The pad oxide is underetched slightly. The trench sidewalls are chosen to be at 80–86° to the horizontal [14]. The angle must be sufficiently big in order to avoid the reduction in trench depth for small isolation spaces, but not too big so that optimal trench fill and trench corner properties can be

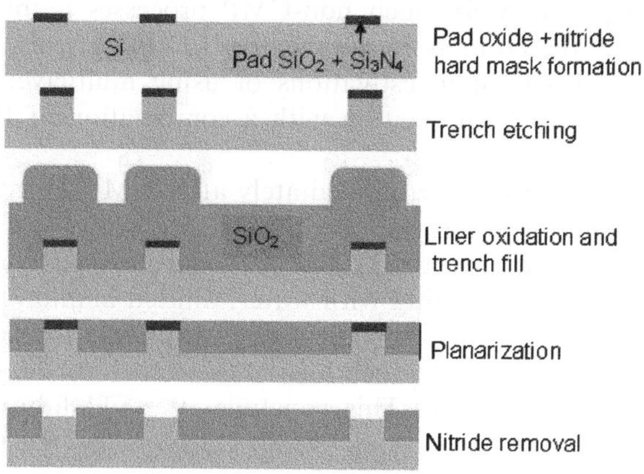

FIGURE 12.3 Schematic cross sections showing the major STI fabrication steps.

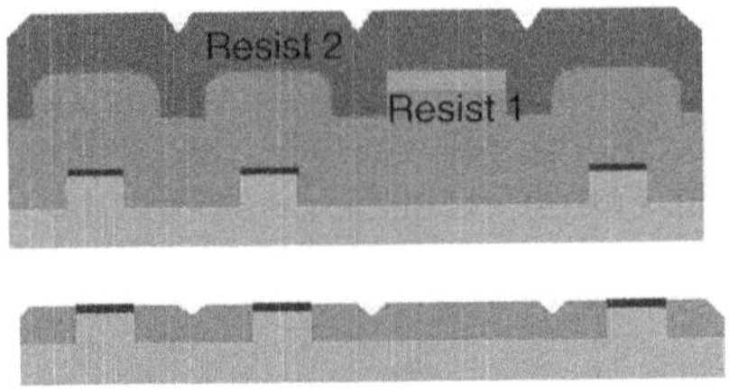

FIGURE 12.4 Schematic cross section of a wafer with block resist-RIE for non-CMP planarization before (above) and after RIE (below).

achieved [13]. After trench etching, a multipurpose liner oxidation is performed [14]. It reduces RIE damage to the trench walls [15], rounds the bottom trench corners thus reducing mechanical stress, and rounds the top trench corner for improved device performance [16]. Liner oxide thickness is in the range of 20–50 nm. On top of the thermal oxide, a thicker oxide is deposited [14]. The total trench oxide thickness is higher than the combined trench depth plus nitride layer thickness, sometimes as much as twice as high. Oxide deposition technique is chosen in such a way as to allow conformal void-free filling of high-aspect-ratio trenches. Tetraethylorthosilicate–ozone (TEOS:O_3) and high-density plasma (HDP) oxides are commonly used [14,17,18]. A high-temperature densification is performed in order to lower the etch rate of the deposited oxide to values close to those of thermal oxide [13,19]. This is important for avoiding trench oxide loss at the end of the STI module during pad oxide etching and cleaning procedures before gate oxidation.

CMP comes next. Its goal is to planarize and pattern the oxide [14]. The CMP step is discussed in more detail in the following two chapters. It should be mentioned here that CMP is not the only option for planarization. More cost-effective solutions have been studied and used, such as multilayer resist processes and spin-on glass in combination with RIE (Fig. 12.4). However, the obtained global planarity for such non-CMP processes is inferior to CMP planarity. They are therefore more suitable for ILD planarization than for STI [20]. There have also been investigations of using multilayer resist/spin-on material techniques for planarization with a combination of RIE and CMP [21].

Cleaning must always follow immediately after CMP. A good post-CMP cleaning is very important for removing all slurry and polished material residues from the wafer surface [22]. The procedure is usually a combination of brush and megasonic cleaning with water, diluted ammonium hydroxide, and diluted HF [23]. Finally, the nitride is stripped with hot phosphoric acid while trying to prevent as much loss of the trench oxide and damage to the silicon surface as possible. This concludes the STI fabrication module (Fig. 12.5).

Looking at the process steps that are responsible for the geometry of the active and isolation areas, it is easy to see the extent to which STI scalability

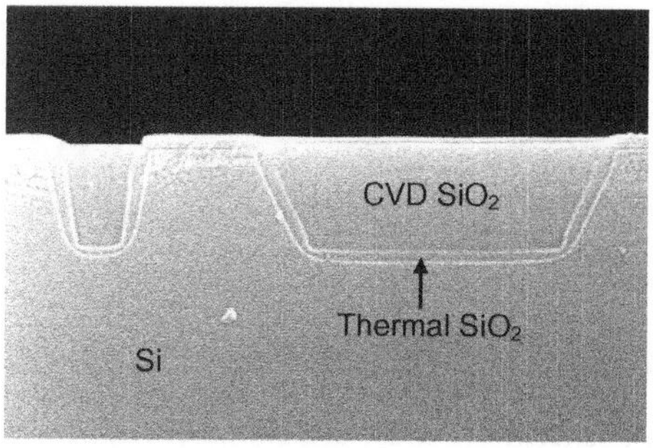

FIGURE 12.5 REM cross section of a wafer with STI.

better as compared to LOCOS scalability. Lithography defines the top lateral dimensions of the trenches and therefore can be scaled with each technology node. RIE defines the depth and controllable sidewall angle. Finally, CMP patterns the oxide laterally and eliminates vertical topography. As additional advantages of STI, the additional degree of latch-up immunity, the lower junction capacitance, and the reduced substrate stress can be mentioned.

12.4 THE PLANARIZATION STEP IN DETAIL

Mastering the polishing step is recognized as a major difficulty in STI formation [13]. The stringent planarity requirements also make it one of the most demanding CMP tasks. STI has been a process driver for the development of planarization equipment, consumables, and process parameters for years. For successful planarization, there are a number of facts to consider—polishing is done very close to the active surface, and process windows are quite tight. On the one hand, the oxide must be completely removed from the active areas in order to allow nitride stripping gate oxide growth. On the other hand, however, reaching the silicon surface during polishing must be avoided at all costs as it leads to serious degradation and even failure of device performance. Fulfilling both requirements is a difficult task as pattern densities inevitably vary across the die and polishing is not uniform.

As shown in Fig. 12.6, the task of STI CMP is to remove the oxide on the top of the active areas and stop on the nitride layer before reaching the silicon

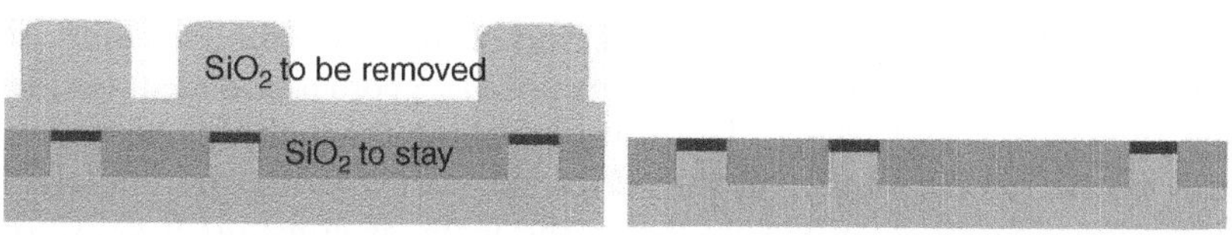

FIGURE 12.6 Schematic cross section of a wafer before and after CMP, ideal case.

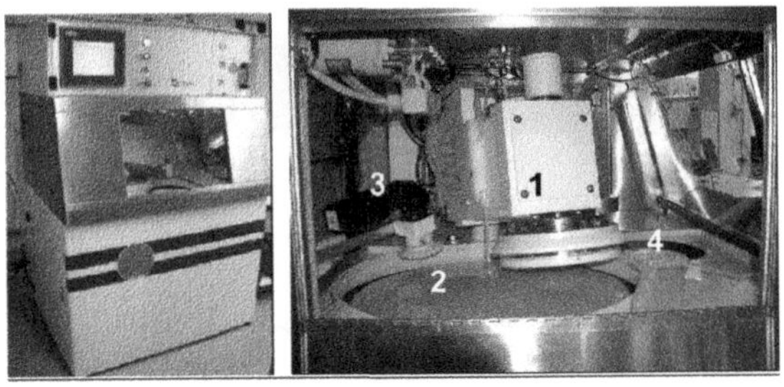

FIGURE 12.7 6EC polisher (left) and work chamber (right) with overarm and carrier (1), polishing table and pad (2), pad conditioner (3), and loading station (4).

surface. Polishing can be performed on either a rotary or a linear CMP tool. Fig. 12.7 shows an example of a small research-type rotary polisher. A basic fumed-silica slurry (pH $\geq$ 10) with KOH as stabilizer is commonly used, although polishing with ceria slurries and acidic slurries is also possible [24]. The exact mechanism of polishing is not yet fully understood. Water is the chemical agent in the slurry, which provides the chemical action by softening the silicon oxide surface. Water breaks Si–O bonds and forms weaker Si–OH bonds:

$$\Xi \mathrm{Si} - \mathrm{O} - \mathrm{Si} \Xi + \mathrm{H_2O} \leftrightarrow \Xi \mathrm{Si} - \mathrm{OH} \tag{12.2}$$

This reaction aids the mechanical abrasion by the silica nanoparticles. The silicon nitride stop layer is also slowly abraded according to the following reaction (for KOH stabilized slurry):

$$\mathrm{Si_3N_4} + 3\mathrm{H_2O} + 6\mathrm{KOH} \leftrightarrow 3\mathrm{K_2SiO_2} + 4\mathrm{NH_3} \tag{12.3}$$

Polishing rates with standard processing parameters are in the 100–200 nm/min region for oxide. Polishing times are therefore normally in the order of several minutes. Oxide-to-nitride selectivity depends significantly on the composition of the slurry, the deposition technique, and the thermal treatment of oxide and nitride and polishing conditions. Minimum and maximum values for commercially available slurries are several to one and several hundred to one, respectively.

At first, an ideal CMP process will be considered for simplicity. Next, the nonidealities of real-life planarization will be discussed along with the effects they have on planarity. An ideal process (Fig. 12.8) is characterized by the following ideal features: polishing pad and wafer are infinitely stiff, slurry selectivity is infinitively high, and wafer pre-CMP geometry is ideal (oxide and nitride thicknesses and trench depth are constant throughout the wafer and there is no nanotopography on the wafer surface). During the planarization process, the pressure distribution across the wafer is constant because of the stiffness of the pad and wafer. This guarantees uniform polishing rates of all

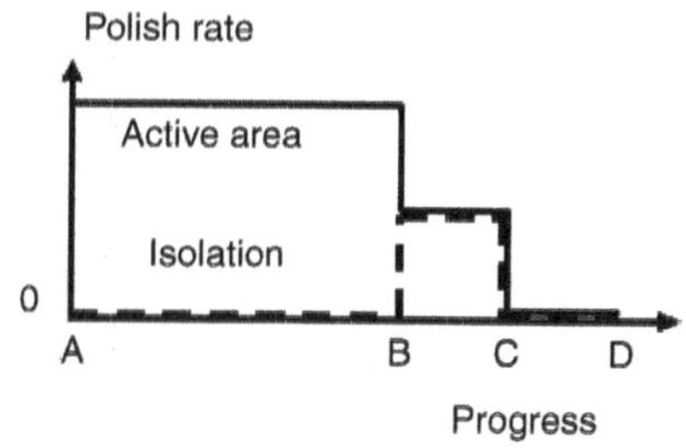

FIGURE 12.8 Diagram of idealized STI CMP polish rates in active and isolation areas.

elevated areas throughout the process. Because no pad compression occurs, the polish rate in the isolation areas (low features) is zero until step height is reduced to zero (Fig. 12.9, A and B). As shown in Fig 12.8, planarization rate, defined as

$$\text{Planarization rate} = \text{elevated feature polish rate} - \text{low feature polish rate} \tag{12.4}$$

equals the polish rate of elevated features. When the initial step height between elevated and low areas is reduced to zero (B), the planarization phase ends and afterward the polishing rates everywhere on the wafer are constant—there is uniform blanket film removal (B and C). Planarization rate is zero and the polishing rate in active areas is reduced because of the increased pattern density (100 %) and consequently decreased pressure. When polishing reaches the nitride layer (C), polishing rates are reduced to zero and polishing stops (C and D). Overpolishing in case of longer polishing times is not possible as the process is self-arresting once end point is reached and again, due to pad stiffness, no further reduction of the trench oxide occurs. The infinitively high slurry selectivity guarantees no thinning of the stop layer. After planarization, a perfectly planar surface is realized. The only topography left at the end of the STI module is the vertical step in the active-isolation transition after nitride removal, equal to its thickness. This is a very ideal polishing progress.

In reality, the planarization process significantly deviates from the ideal case. The polishing pad properties have without a doubt the most pronounced effect on the quality of planarization. The pad has certain compressibility because of which polishing rates are not averaged over the wafer, but over smaller areas. Removal rates vary within the die and across the wafer. Areas with higher pattern densities are polished slower than areas with lower pattern densities [25] (Fig. 12.9). This leads to reaching end point (nitride exposure) in some areas earlier than in others. In the STI CMP process, there is zero tolerance for the remaining oxide on top of the nitride because it would prevent subsequent nitride strip. Therefore, an overpolishing step is required for exposing the nitride in slow polishing areas, during which nitride and oxide

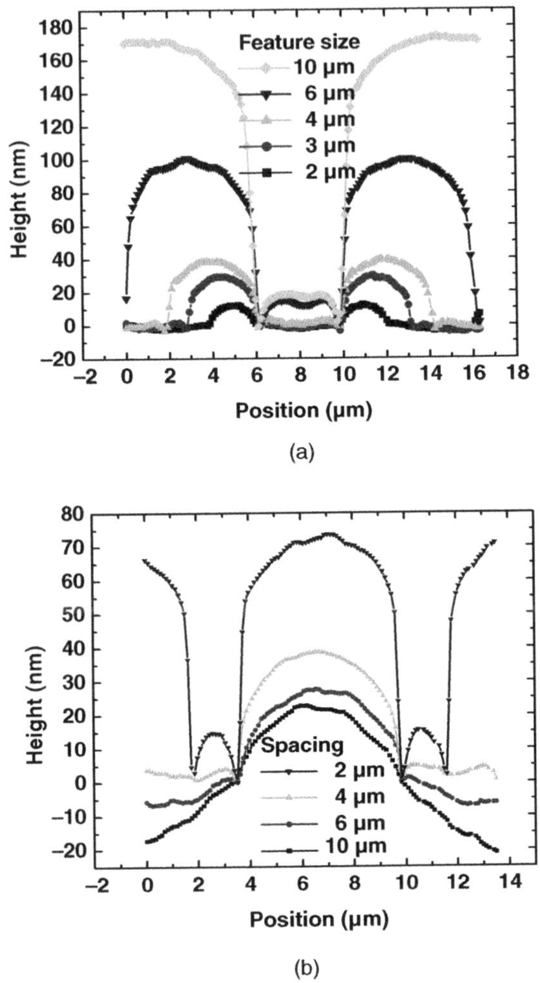

FIGURE 12.9 AFM post-CMP measurements illustrating pattern density dependence of polish rates for different patterns on the same wafer: (a) different feature size, same spacing and (b) different spacing, same feature size.

thinning [26] occurs in faster polishing areas due to the limited selectivity of real STI slurries.

Additionally, on a more local level between two structures, again due to pad compressibility, polishing of the isolation area starts earlier than in the ideal case (Fig. 12.10, B), before the initial step is reduced to zero. This slows down the planarization rate after this point (B and C) and, generally, planarity is not reached at the moment the nitride layer is exposed (C). During the over-polishing step, the polishing rate of the active areas is reduced because of the nitride stop layer, but is nevertheless different from zero, especially near the

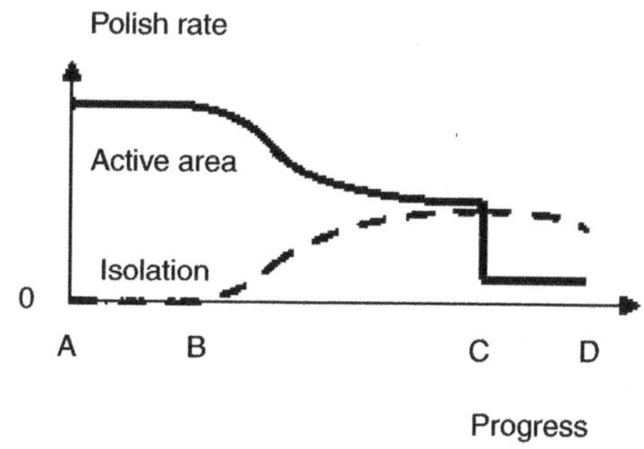

FIGURE 12.10 Diagram of real STI CMP polish rates in active and isolation areas.

corners of the active area. Pad compressibility allows polishing to continue in the isolation area with polishing rates that are highest and farthest away from the active areas. As a result, bowl shaping of the oxide occurs—dishing as well as erosion of the stop layer [26] that is enhanced along the periphery of the active areas (Figs. 12.11 and 12.12). The larger the distance between the active areas, the more pronounced these effects are and the closer the polishing rate in the low area is to the one in the elevated area. Furthermore, pre-CMP wafer nonidealities such as surface nanotopography [27] and nonuniformities in trench depth and layer thicknesses contribute to additional deviations of the polishing rates across the wafer.

Center-to-edge (CTE) polishing uniformity is also an issue that must be addressed in real-life planarization. Rotation velocities of the wafer and table, back pressure, and retainer ring pressure are all parameters that affect the polishing rates for different distances from the center (Fig. 12.13). Different ratios of the angular velocities of the wafer and table, for example, result in different linear velocities of the polished surface relative to the polishing pad [24]. Increasing retainer ring pressure leads to reduced polishing in the periphery of the wafer. Backpressure distribution has probably the most pronounced effect on polishing rate distribution, but even slurry flow rate can influence CTE uniformity. Therefore, the right combination of parameters for

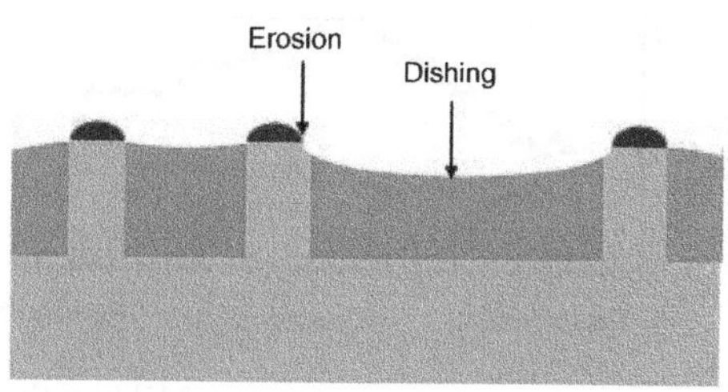

FIGURE 12.11 Schematic cross section of a wafer after CMP showing oxide dishing in large isolation areas and nitride erosion in active area corners next to them.

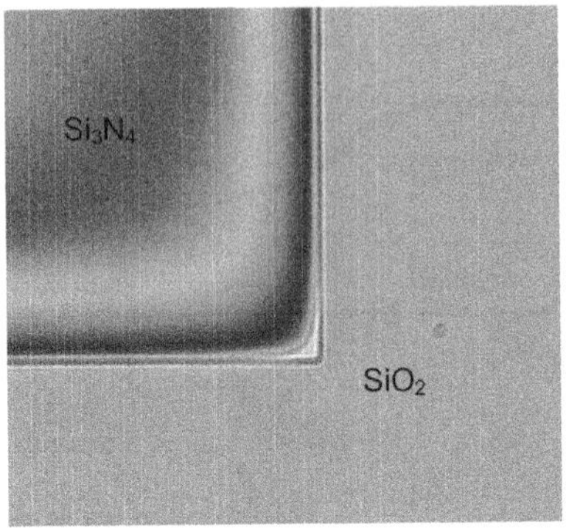

FIGURE 12.12 Optical microscope image of post-CMP wafer surface, showing oxide dishing and nitride erosion.

a certain set of tools and consumables must be determined with test runs for each individual selection of tools, consumables, and wafers.

What do oxide thickness nonuniformities and nitride erosion lead to? As in the case of LOCOS, topography complicates lithography. Recess of the trench oxide also causes gate wraparound and oxide thinning at the active area corner that create a parasitic low threshold voltage corner device in parallel with the main MOSFET. This device has a negative effect on performance, causing a "kink effect" [28] in the MOSFET transfer characteristics (Fig. 12.14) and adversely increasing subthreshold leakage. Nitride erosion is very dangerous as device functionality is very sensitive to damage of the silicon surface. If the nitride is eroded and such a damage occurs during CMP, severe performance degradation and even failure can be expected. One more potential risk must be mentioned, the risk of scratching during polishing. Agglomerations of slurry

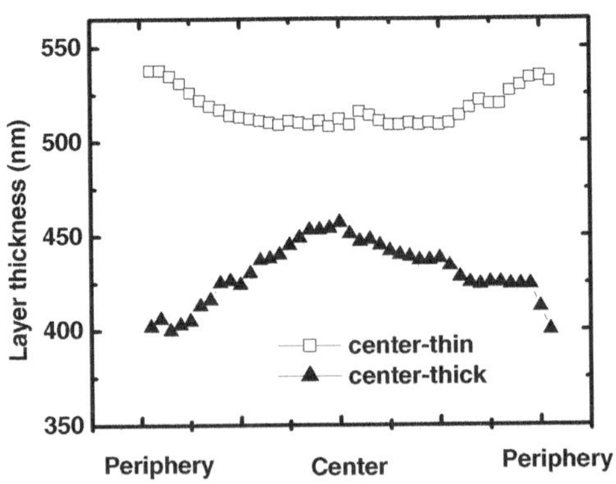

FIGURE 12.13 Ellipsometric measurement of post-CMP film thickness on two wafers, polished with different process parameters, showing center-to-edge polish rate nonuniformity.

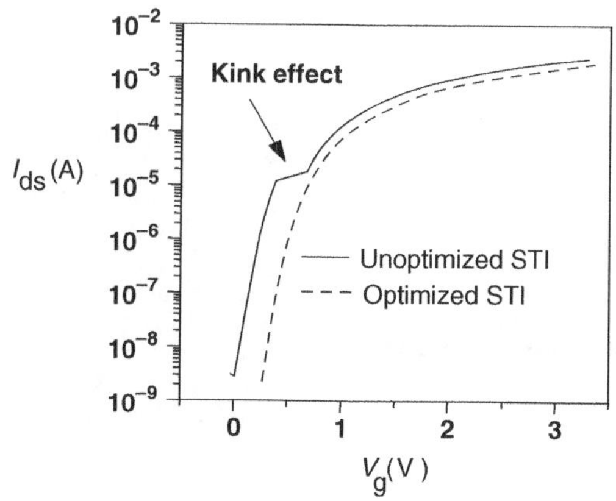

FIGURE 12.14 Transfer characteristics of MOSFETs with (straight line) and without (dashed line) gate wraparound.

particles, foreign particles, or simply abraded material can damage the stop layer and the underlying silicon surface.

Several CMP end-point detection (EPD) methods [29] have been developed for determining the correct moment to stop polishing in order to minimize dishing and erosion. All of them sense changes in certain characteristics of the wafer surface before and after the nitride stop layer is exposed by polishing. Both *in situ* and in-line optical techniques (interferometry, reflectometry, and ellipsometry) are implemented in industry. The friction method, which relies on detecting changes in the polished table motor current, is another well-established technique. More recent developments are the acoustic and electrochemical detection systems. Chemical analysis is done directly on the material being polished or the by-products of the reactions during CMP. In the case of STI CMP, reaching end point is registered by the ammonia molecules or ammonium ions in the slurry on the table when the nitride layer is reached, as long as the slurry itself does not contain ammonia as a stabilizer.

Post-CMP topography is commonly evaluated by profilometry and atomic force microscopy (AFM). Both techniques are suitable for dishing measurements with the latter having superior resolution. Additionally, when equipped with an electrical measurement system, AFM is capable of detecting excessive nitride erosion with nanometer resolution (Fig. 12.15) [30]. Except for process control, the detailed dishing and erosion AFM data can be successfully implemented for calibration of CMP simulation tools.

There are a number of aspects to be considered for improving STI CMP, including consumable and polishing parameter optimisations, and all of them have been investigated during the past years. For example, since polishing pads are the main cause of polishing nonuniformity, it makes sense to produce and implement harder pads. Here, a compromise must be made in choosing the right hardness because with increasing hardness, surface damage also increases.

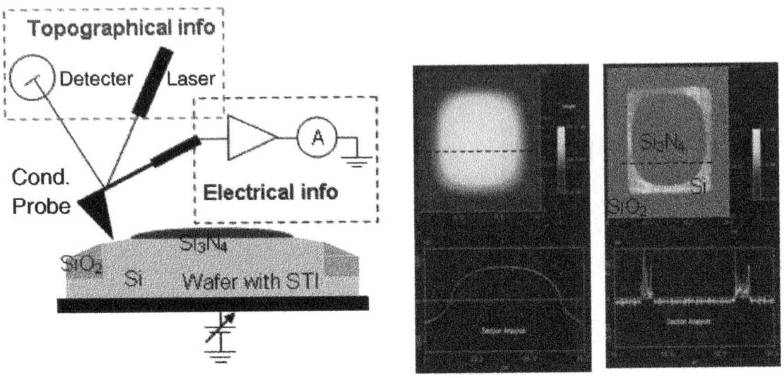

FIGURE 12.15 Erosion C-AFM measurement schematic setup (left), topographic AFM scan (center), and electrical AFM scan (right), illustrating excessive nitride erosion detection with C-AFM. Topographic AFM does not offer a reliable means of detecting erosion in case of smooth transitions from one material to the next. The electrical measurement "recognizes" the exposed nitride area by the greatly enhanced currents at the eroded active area corner (brighter area in electrical scan).

Another possibility is to optimize polishing parameters. Experiments have shown that dishing and erosion are proportional to the pressure applied to the wafer during polishing [24]. Lower pressure reduces pad compression, and higher velocity of the pad relative to the wafer reduces the pad relaxation effect. Therefore, lower pressures and higher rotation velocities have been used at the expense of decreased overall polishing rates and slightly increased within-wafer nonuniformity. Other researchers have used two-step processes in order to keep both polishing performance and planarity at reasonable levels: a first step with low pressure and high platen speed for planarization and a second one with high pressure and low platen speed for uniform blanket film removal [32]. Slurry developers have produced more advanced slurries. Slurry selectivity has increased significantly over the years. Whereas the first STI slurries have had selectivity in the range of 3–4:1; nowadays they are up to several hundreds to 1 [31].

Even with all process parameters and material optimisations, it is usually not possible to reach acceptable levels of post-CMP planarity and therefore additional optimization of the wafer to be polished itself is needed [25]. The next section discusses several different optimization techniques.

12.5 OPTIMIZATION TECHNIQUES

The optimization techniques are performed on mask design or processing level and approach planarity problems from different directions. Some of them reduce the pre-CMP topography, others reduce density variations or nonuniformity dimensions with respect to the planarization length, and still others provide protection against overpolishing. All of these processes have their benefits as well as disadvantages, both of which are discussed below. On the downside, optimization techniques have a common disadvantage compared to direct nonoptimized CMP—all of them add in one way or another to complexity and cost.

12.5.1 Dummy Active Area Insertion

In dummy area insertion, optimization is performed on design level. Since dishing and erosion are enhanced around large isolation areas, one way to eliminate the former is by avoiding the latter. In large isolation areas, additional active areas can be introduced [13] (Fig. 12.16). These additional dummy areas are not required for the IC functionality, but serve only as supporting blocks for CMP. The optimization results in equalized and maximized pattern density across the die. Planarity improvement is significant [25]. Additionally, from the manufacturing point of view, the technique is very attractive since the additional optimization work is performed only on design level and no additional processes or steps are introduced. On the contrary, the dummy structures have a negative effect on performance. They couple capacitively to the rest of the circuit, leading to higher RC delays. For minimizing this effect, complex design software is needed that analyzes the information on the above-lying layers in order to position the dummies optimally. The dummy active area insertion technique is a standard state-of-the-art optimization and is commonly used in industry.

12.5.2 Patterned Oxide Etch Back

Another established optimization technique is patterned oxide etch back [14]. Its purpose is to remove most of the oxide in active areas prior to CMP. An additional lithography step is performed in order to mask the isolation areas

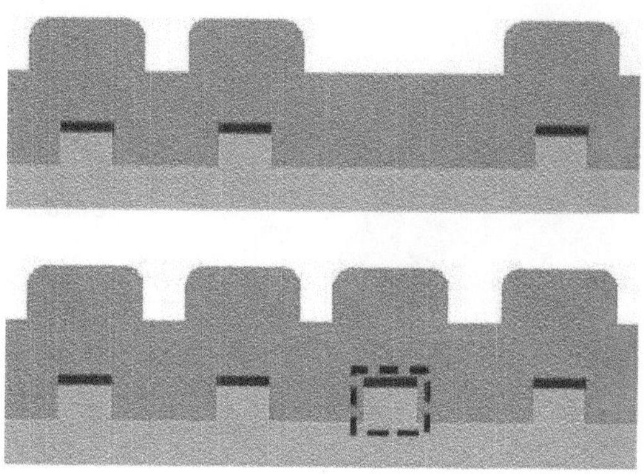

FIGURE 12.16 Schematic cross section of a wafer without optimization (above) and with dummy active area insertion optimization (below).

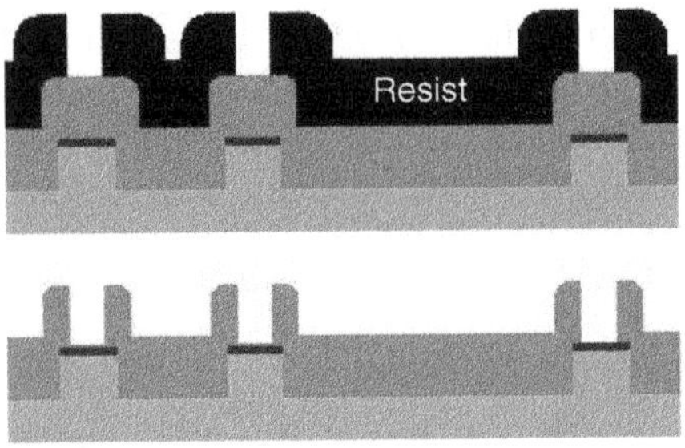

FIGURE 12.17 Schematic cross section of a wafer with reverse oxide etch back after lithography (above) and before CMP (below).

(Fig. 12.17). The additional mask required for OE is a reversed copy of the STI mask with reduced structure size. RIE then removes the oxide in the exposed areas, stopping on the nitride. After RIE of the oxide and resist removal, the remaining nonplanarity at the periphery of the active areas is easily removed with CMP. The planarity results are excellent (Fig. 12.18), and unlike in the case of dummy insertion, planarity improvement is achieved without a negative impact on performance. The main disadvantage of the technique is the increased process complexity due to the additional lithography and RIE steps.

12.5.3 Nitride Overcoat

The goal of the technique is to minimize dishing by physically protecting isolation areas. The optimization consists of simply depositing a thin nitride layer (20–40 nm depending on trench geometry, slurry selectivity, etc.) on top of the oxide before planarization [13,25,32] (Fig. 12.19). The effect is based on the different polishing rates of oxide and nitride. When CMP starts, the pressure exerted in the elevated areas is higher than that in the lower ones.

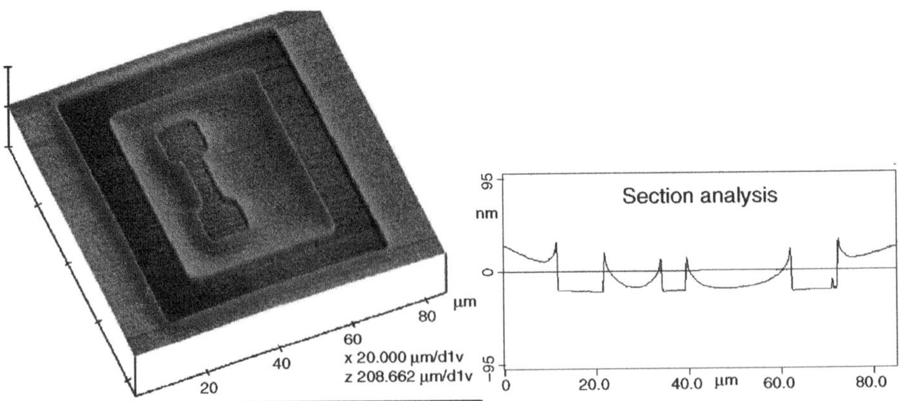

FIGURE 12.18 AFM scan (above) and cross section across the dashed line (below) of a transistor area on a wafer with oxide etch back after planarization and nitride strip.

FIGURE 12.19 Schematic cross section of a wafer with continuous nitride overcoat.

Therefore, the nitride layer is removed faster on the former. When polishing of the oxide starts in the elevated areas, the lower areas are still protected by some remaining nitride, whose polishing rates are much lower, and by the time it is removed, the oxide overburden is already polished away. The planarization results are good for most circuit designs. There are, however, certain limitations of using this technique when the design contains active areas of size that is comparable to the planarization length of the process (several hundred micrometers). In this case, the nitride on top of the active area simply cannot be removed faster than the one in the isolation areas, and a planarization optimization effect cannot be achieved. Considering process complexity, there is one additional nitride layer deposition. However, there is added processing cost of up to 40 % due to the increased polishing time, required to remove the nitride layer [25]. This technique is also used in industry, sometimes in combination with dummy area insertion.

The nitride layer could alternatively be structured with additional lithography and etching steps in such a way as to leave the protective overcoat only in isolation areas (Fig. 12.20) [33,34]. In this case, planarity is further improved at the expense of the increased process complexity. Using a high-selectivity slurry in combination with a patterned nitride overcoat allows complete elimination of dishing in large isolation areas (Fig. 12.21).

12.5.4 EXTIGATE

A novel isolation technique based on STI is the extended trench isolation gate technology (EXTIGATE) [35,36]. Although EXTIGATE was not originally invented as a CMP optimization technique, its implementation nevertheless has an effect on the planarization process. The main distinctive feature in the EXTIGATE process flow arises from the fact that the gate stack is produced *before* isolation fabrication (Figs. 12.22 and 12.23). Among the advantages of this process flow are the better control of the active-isolation transition and better $n+$ $p+$ dual work-function technology. Regarding the CMP process,

FIGURE 12.20 Schematic cross section of a wafer with structured nitride overcoat.

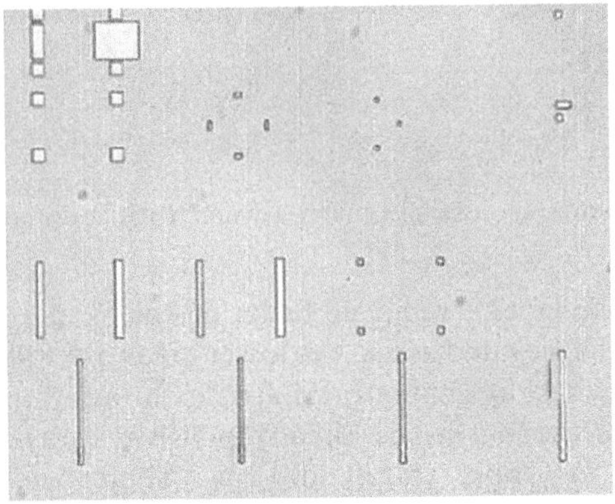

FIGURE 12.21 Optical microscope picture of wafer surface with a patterned nitride overcoat after planarization and nitride strip. Notice the complete elimination of dishing (color uniformity in the green isolation area).

the fact must be considered that the nitride stop layer is situated a couple of hundred nanometers above the silicon surface. In case of erosion of the stop layer, the top of the polysilicon gate is polished. This, however, does not have any effect on device performance. Therefore, although no improvement in planarity arises from it, EXTIGATE brings the advantage of a significantly enlarged process window of the planarization process.

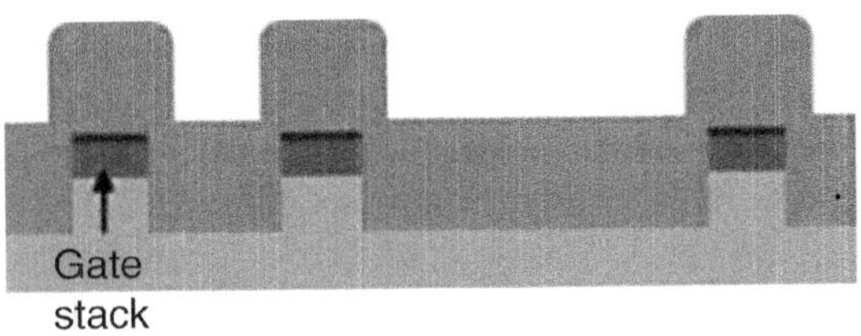

FIGURE 12.22 Schematic cross section of a wafer with EXTIGATE.

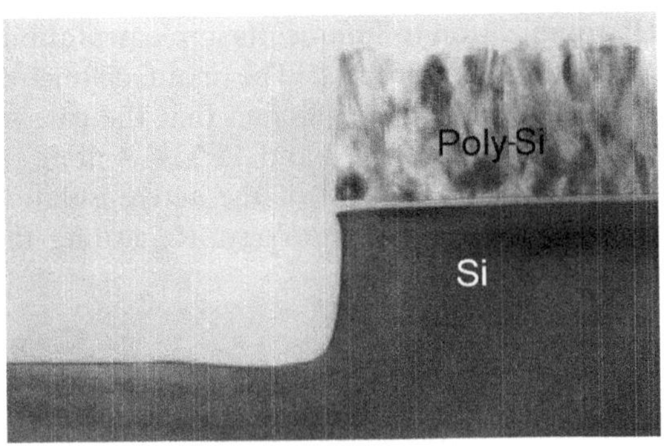

FIGURE 12.23 REM cross section image of a wafer with EXTIGATE. The mini "bird's beak" can be seen at the active area corner.

FIGURE 12.24 Schematic cross section of a wafer with selective oxide deposition.

12.5.5 Selective Oxide Deposition

A simple and effective topography reduction can be achieved by optimizing the trench fill process. Selective oxide deposition (SELOX) is based on the different deposition rates of CVD oxide on silicon and silicon nitride for a modified ozone-activated TEOS process. With such a process, the deposition in the isolation trenches occurs at much higher rates than in the active areas, resulting in a reduction in and eventually almost full elimination of the initial step height (Figs. 12.24 and 12.25). Selectivity of deposition rates on silicon and nitride of 5:1 has been reported [37]. The obvious benefit of SELOX is that topography is significantly reduced even before CMP. In this case, no additional optimization is needed for improving planarity. What is more, there is practically no increase in process complexity compared to the basic STI module.

The selective oxide deposition technique has been additionally developed for BEOL planarization of ILD films, where the difference in deposition rates of ozone/TEOS on PECVD oxide and TiN is exploited [38].

12.5.6 Polysilicon-Filled Trenches

With this optimization technique, topography is reduced *after* the planarization step. The STI trenches are filled with polysilicon instead of SiO_2 [39]. After trench, etching, the first patterned hard mask is stripped and a new continuous oxide/nitride layer is deposited for electrical isolation between the substrate and the trench polysilicon. After CMP, the dishing in isolation areas is counteracted by a high-temperature oxidation step, which oxidizes the upper part of the remaining polysilicon in the trenches. As the volume of the SiO_2 is

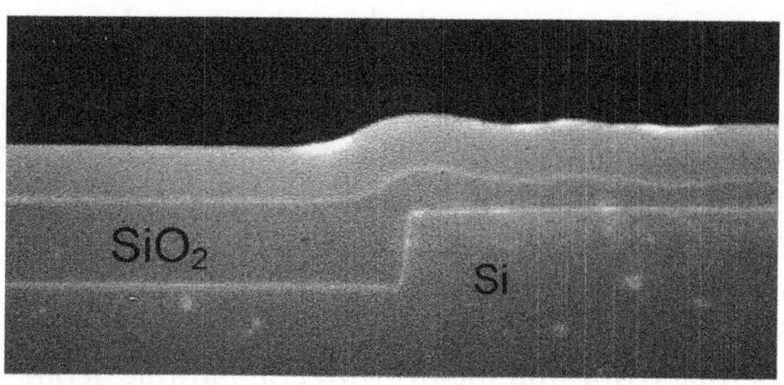

FIGURE 12.25 REM image of a wafer after selective oxide deposition.

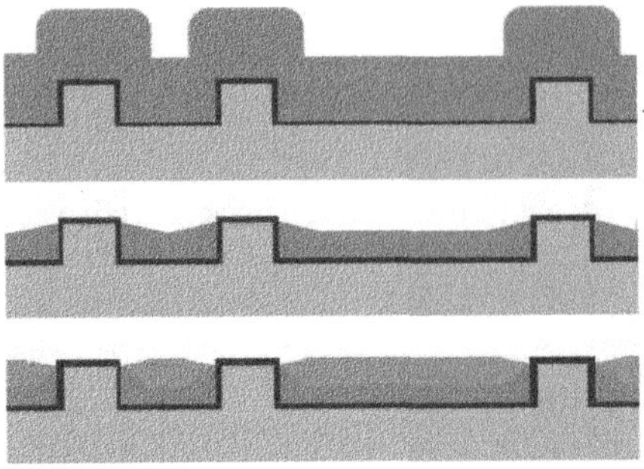

FIGURE 12.26 Schematic cross sections showing the major steps in STI formation with polysilicon-filled trenches.

approximately twice as high as that of the consumed silicon, by controlling the time of the oxidation, the top of the trench-fill stack can be raised to the desired level (Fig. 12.26).

12.6 OUTLOOK

STI CMP has become an indispensable process in state-of-the-art semiconductor technology, accounting for 8–9% of the $2 billion CMP applications market. It will continue to provide the needed trench isolation scalability for future technology generations. The achieved level of planarity improves continuously, but the planarity requirements also become more and more stringent. The future development of the process will be driven first of all by meeting those planarity requirements and also by the manufacturers' desire to decrease the cost of ownership of the process.

Advanced consumables are needed for avoiding complicated and costly optimization techniques and replacing them by direct CMP. Reduction in consumable consumption is also desired, as slurry and polishing pad costs alone account for more than 50% of COO. Better control and automation techniques are required for the improved stability of the polishing process, most important of which are more accurate and reliable end-point detection systems. Further improvements in throughput and post-CMP cleaning procedures as well as defect density reduction are also desired.

The introduction of silicon-on-insulator (SOI) technologies in manufacturing as a way of decreasing parasitic leakages and capacitances in MOS devices will alleviate the planarization process because of the reduced trench depth and consecutively reduced initial step height.

Emerging front-end CMP applications can gain from the experience of STI CMP. The replacement metal gate process, for example, comprises two CMP damascene steps (one oxide and metal) with comparable geometry dimensions. The know-how in terms of consumables, tools, process parameters, and optimization techniques can be easily transferred from the well-established state-of-the-art device isolation planarization to the novel gate formation planarization.

ACKNOWLEDGMENTS

A part of this work was funded by the German Federal Ministry of Education and Research (BMBF) under the KrisMOS project (01M3312C). The authors would like to thank Lorraine Rispal, Christian Brueckner, and Fatih Cilek for their help with CMP and AFM measurements; Dr. Klaus Haberle, Karin Boye, Rudolf Heller, Gisela Hess, Gudrun Mueller, and Gerhard Tzschoeckel for wafer processing at the IHT clean room facility; and H.C. Starck GmbH and, in particular, Dr. Gabriele Hey for the supplied STI CMP slurries.

QUESTIONS

1. Suggest a formula linking a certain technology node to the maximum allowable post-CMP nonplanarity for this node.
2. Which optimization technique do you think gives the best planarity results? Why?
3. Evaluate the scalability of the different optimization techniques.
4. What do you think is the major limiting factor for post-STI CMP planarity at present? How about in 10 years?
5. Compare STI CMP module to other planarization modules. Where would you place it in terms of complexity, planarity requirements, and cost?

REFERENCES

1. Moore G. Moore's law. http://www.intel.com/technology/silicon/mooreslaw/
2. Wei CS, Fraser DB, Wu AT, Paunovic M, Ting CH. The use of selective electroless metal deposition for micron size contact fill. IEDM Technical Digest; Dec 1988. p 446–449.
3. Landis H, Burke P Cote W, Hill W, Hoffman C, Kaanta C, Koburger C, Lange W, Leach M, Luce S. Integration of chemical–mechanical polishing into CMOS integrated circuit manufacturing. Thin Solid Films 1992;220:1–7.
4. Brown N, Cook L. The role of abrasion in the polishing of metal and glass. The science of optical polishing; OSA Technical Digest; TuB-A3-1, Apr 1984.
5. Ali I, Roy S, Shinn G. Chemical mechanical polishing of interlayer dielectric: a review. Solid State Technol 1994;37(10):63–70.
6. Chatterjee A, Chapman RA, Dixit G, Kuehne J, Hattangady S, Yang H, Brown GA, Aggarwal R, Erdogan U, He Q, Hanratty M, Rogers D, Murtaza S, Fang SG, Kraft R, Rotondaro ALP, Hu JC, Terry M, Lee W, Fernando C, Konecni A, Wells G, Frystak D, Bowen C, Rodder M, Chen I-C. Sub-100 nm Gate length metal gate

nmos transistors fabricated by a replacement gate process. IEDM Technical Digest; 1997. p 821–824.

7. Appels JA, Kooi E, Paffen MM, Schatorjé JJH, Verkuylen WHCG. Local oxidation of silicon and its application in semiconductor-device technology. Phil Res Rep 1970;25:118–132.
8. Oldham WG. Isolation technology for scaled MOS VLSI. IEDM Technical Digest; Dec 1982. p 216–219.
9. Bryant A, Haensch W, Mii T. Characteristics of CMOS device isolation for the ULSI Age. IEDM Technical Digest; Dec 1994. p 671–674.
10. Fazan PC, Mathews VK. A highly manufacturable trench isolation process for deep submicron DRAMs. IEDM Technical Digest; Dec 1993. p57–60.
11. Han Y, Ma B. Isolation process using buffer layer for scaled MOS/VLSI. ECS Extended Abstracts 1984;84(1):98.
12. Shimizu N, Naito Y, Itoh Y, Shibata Y, Hashimoto K, Nishio M, Asai A, Ohe K, Umimoto H, Hirofuji Y. A poly-buffer recessed LOCOS process for 256Mbit DRAM cells. IEDM Technical Digest; Dec 1992. p 279–282.
13. Chatterjee A, Ali I, Joyner K, Mercer D, Kuehne J, Mason M, Esuivel A, Rogers D, O'Brien S, Mei P, Murtaza S, Kwok SP, Taylor K, Nag S, Hames G, Hanratty M, Marchman H, Ashburn S, Chen I-C. Integration of unit processes in a shallow trench isolation module for a 0.25 μm complementary metal-oxide semiconductor technology. J Vac Sci Technol 1997;B15(6):1936–1942.
14. Nandakumar M, Chatterjee A, Sridhar S, Joyner K, Rodder M, Chen I-C. Shallow trench isolation for advanced ULSI CMOS technologies. IEDM Technical Digest; Dec 1998. p 133–136.
15. Pyi S-H, Yeo I-S, Weon D-H, Kim Y-B, Kim H-S, Lee S-K. Roles of sidewall oxidation in the devices with shallow trench isolation. Electron Device Lett 1999;20:384–386.
16. Chang CP, Pai CS, Baumann FH, Liu CT, Rafferty CS, Pinto MR, Lloyd EJ, Bude M, Klemens FP, Miner JF, Cheung KP, Colonell JI, Lai WYC, Vaidya H, Hillenius SJ, Liu RC, Klemens JT. A highly manufacturable corner rounding solution for 0.18 μm shallow trench isolation. IEDM Technical Digest; Dec 1997. p 661–664.
17. Nag S, Chatterjee A, Taylor K, Ali I, O'Brien S, Aur S, Luttmer J, Chen I-C. Comparative evaluation of gap-fill dielectrics in shallow trench isolation for sub-0.25 μm technologies. IEDM Technical Digest; Dec 1996. p 841–844.
18. Curtis TO, Pye JT, Poreda JT. APCVD TEOS: 03 advanced trench isolation applications. 9th ed. Semiconductor Fabtech; 1999.
19. Lee HS, Park MH, Shin YG, Park T-S, Kang HK, Lee SI, Lee MY. An optimized densification of the filled oxide for quarter micron shallow trench isolation (STI). Symposium on VLSI Technology; Technical Digest; June 1996. p 158–157.
20. Neureither B, Binder F, Fischer E, Gabric Z, Koller K, Rohl S. Resist etch back as a manufacturable low cost alternative to CMP.Proceedings of the Eleventh International VLSI Multilevel Interconnect Conference; 1994. p151–157.
21. Davari B, Koburger CW, Schulz R, Warnock JD, Furukawa T, Jost M, Taur Y, Schwittek WG, DeBrosse JK, Kerbaugh ML, Mauer JL. A new planarization technique using a combination of RIE and chemical mechanical polish (CMP). IEDM Technical Digest; Dec 1989. p 61–64.

22. Wolf S, Tauber RN. Silicon Processing for the VLSI Era, 2nd ed. Lattice Press; 2000, Vol. 1, Process Technology.
23. Roy SR, Ali I, Shinn G, Furusawa N, Shah R, Peterman S, Witt K, Eastman S. Post chemical–mechanical planarization cleanup process for interlayer dielectric films. ECS 1995; 312(1):216–226.
24. Steigerwald JM, Murarka SP, Gutmann RJ. Chemical mechanical planarization of microelectronic materials John Wiley & Sons, Inc. ; 1997.
25. Chatterjee A, Kwok SP, Ali I, Joyner K, Shinn G, Chen I-C. Chemical mechanical planarization (CMP) process windows in shallow trench isolation for advanced CMOS. Electrochem Soc Proc 1996;96–22:219–227.
26. Smekalin K. CMP dishing effects in shallow trench isolation. Solid State Technol 1997;40:1987–1994.
27. Boning D, LeeB. Nanotopography issues in shallow trench isolation CMP. MRS Bulletin; Oct 2002. p 761–765.
28. Bryant A, Haensch W, Geissler S, Mandelman J, Poindexter D, Steger M. The current-carrying corner inherent to trench isolation. Electron Device Lett 1993; 14(8):412–414.
29. Berman M, Bibby T, Smith A. Review of In Situ & In-Line Detection for CMP Applications. Semiconductor Fabtech, 8th Edition, 1998, pp. 267–274.
30. Stefanov Y, Ruland T, Schwalke U. Electrical afm measurements for evaluation of nitride erosion in shallow trench isolation chemical mechanical planarization.MRS Proceedings; 2005. Vol. 838E.
31. Derbyshire K. Making CMP Work. Semiconductor Magazine 2002;3(7):40–53.
32. Boyd JM, Ellul JP. A one-step shallow trench global planarization process using chemical mechanical polishing. ECS Proc 1996;95–5:290.
33. Sarlet G, Morthler G, Baets R. Dummy-free STI concept for mixed-signal applications. IMEC Newsletter; No. 25, Oct 1999. p 3–4.
34. Stefanov Y, Cilek F, Endres R, Schwalke U. Alternative optimization techniques for shallow trench isolation and replacement gate technology CMP. Proceedings of PacRim-CMP; 2005. p 51–57.
35. Schwalke U, Kerber M, Koller K, Jacobs H. EXTIGATE: The Ultimate Process Architecture for Submiron CMOS Technologies. Transactions on Electron Devices. 1997; 44(11):2070–2077.
36. Nakabayashi T, Uehara T, Segawa M, Ukeda T, Yamanaka M, Yamada T, Arai M, Yabu T, Yamashita K, Kobayashi S, Murakami T, Saeki M, Okuyama H, Kanda A, Ogura M. A novel 0.25 μm CMOS technology for 6. 82 μm^2 6-Tr. SRAM cell with elevated trench isolation and line-and-space shaped gates (ETILS). IEDM Technical Digest; Dec 1995. p 1011–1013.
37. Elbel N, Gabric Z, Langheinrich W, Neureither B. A new STI process based on selective oxide deposition. Symposium on VLSI Technology, Digest of Technical Papers; June 1998. p 208–209.
38. Fischer E, Gabric Z, Neureither B, Spindler O, Graßl T. Global planarization by selective deposition of ozone/TEOS. VMIC Conference Proceedings; 1995. p 247–253.
39. Cheng J-Y, Fu T, Chao TS. A Novel shallow trench isolation technique. Jap J App Phys 1997;36:1319–1324.

13

CONSUMABLES FOR ADVANCED SHALLOW TRENCH ISOLATION (STI)

CRAIG D. BURKHARD

13.1 INTRODUCTION

Chemical–mechanical polishing (CMP) has enabled the fabrication of advanced devices by preparing planar dielectric layers and forming functional structures such as copper line, tungsten vias, and isolated silicon microstructures. As shown in the previous chapter and Fig. 13.1, there are two competing approaches for lateral isolation of a microstructure of silicon. They are termed LOCOS (local oxidation of silicon) and STI (shallow trench isolation). LOCOS isolation has been used for technologies down to 0.25 μm since it is a very manufacturable and well-understood process. The LOCOS method, however, does cause undesirable "bird's beak" structures [1]. During the LOCOS isolation process, a thick oxide is grown in the field region of MOSFETs. Some oxidant diffuses under the edges of the nitride where unwanted oxide grows. The growth beneath the nitride decreases as oxidant moves inward under the nitride edge. Therefore, a gradually tapering oxide wedge merges into the pad oxide, which can then raise the nitride edges [2]. The shape of the field oxide as it penetrates under the nitride has been given the name bird's beak as shown in Fig. 13.1.

One of the main motivations for the development of the STI process is to eliminate the bird's beak phenomenon. With STI, the field oxide is well embedded into the Si and is clearly distinctive from the active-area regions. It allows very narrow active-area pitches and a higher device-packing density.

Microelectronic Applications of Chemical Mechanical Planarization, Edited by Yuzhuo Li

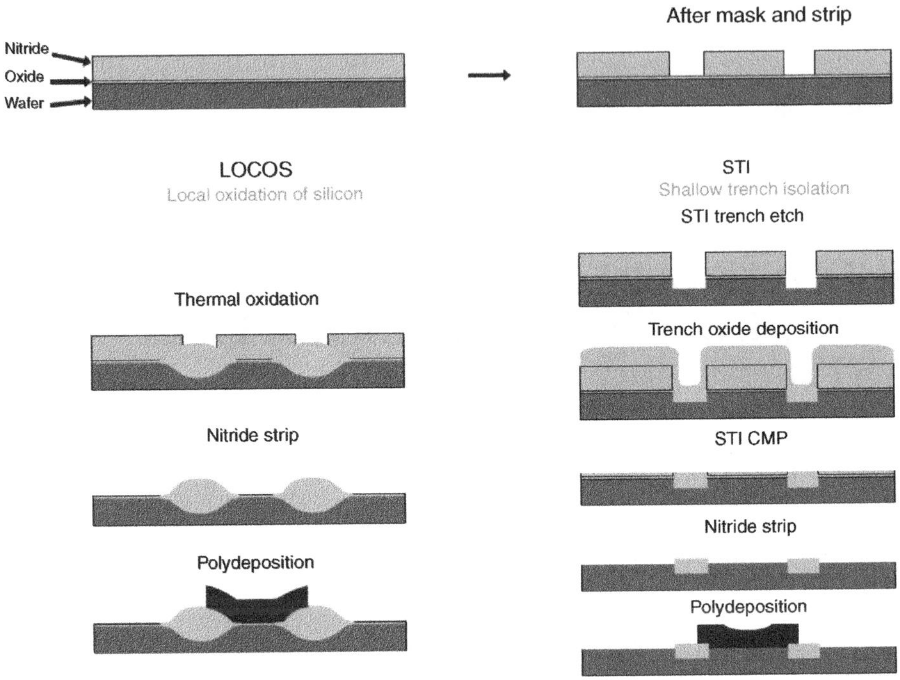

FIGURE 13.1 Illustrations of the various processing steps for local oxidation of silicon and shallow trench isolation (from Ref. 7).

The advantage of STI over LOCOS becomes more pronounced as the device-critical dimensions are reduced [3]. However, the incorporation of STI structures is not without difficulties and drawbacks such as increased leakage current of devices as compared to the LOCOS structure, which is a particular problem for DRAM devices [4–6]. Defects left by an STI process such as dislocations, dishing, and microscratches are the main sources of current leakage. In this chapter, a brief introduction to the technology will be given. The focus of this chapter will then turn to the working principles and selection guidelines for consumables that minimize the level defects present during an STI CMP operation.

As shown in Fig. 13.1, a CMP step is required in the STI process. There are two possible approaches to implementing the CMP step STI—direct and indirect. In a direct STI process, the CMP process is applied directly on a wafer right after the trench oxide deposition (Fig. 13.1). The CMP process removes the overburden oxide as well as the topography created during trench oxide deposition over the features with various pattern densities across the dies and the wafer. For some slurries, especially the conventional oxide slurries, these topographies are a challenge. To overcome this difficulty, a pre-CMP step is sometimes implemented in which a reverse mask is applied onto the film and the oxide in the raised area is preferentially etched. After the etching step, the

CMP step is used to planarize and remove overburden oxide. This is also called indirect STI process. The requirements for direct-polish STI CMP processes include minimal within-die (WID) oxide and nitrite ranges, limited nitride loss, limited oxide dishing, low defectivity, and excellent nonuniformity, both within wafer (WIW) and wafer to wafer (WTW) [8].

13.2 REPRESENTATIVE TESTING WAFERS FOR STI PROCESS AND CONSUMABLE EVALUATIONS

Figure 13.2 illustrates the cross section of a representative patterned wafer design for the evaluation of a direct STI CMP process or consumables such as slurry and pad. An ideal STI process should remove all overburden oxide and stop at the silicon nitride layer without any dishing and nitride loss, as shown in Fig. 13.3. Figure 13.4 is a representative STI patterned wafer layout [9].

The key requirements for an acceptable STI slurry include adequate removal rate (> 3000 Å/min) on oxide, desirable selectivity of oxide over nitride (between 200:1 and 400:1), high planarization efficiency across the die regardless of the patterned density, low scratch count, and low particle residue. The typical operating downpressure for an STI process is 3–5 psi. The typical platen and carrier speeds of a rotary tool are 100 and 60 rpm, respectively. The

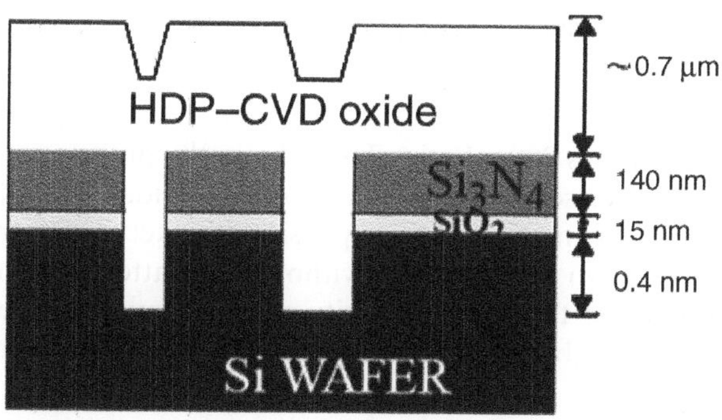

FIGURE 13.2 A cross-sectional view of a representative patterned wafer for STI process and consumable evaluation.

FIGURE 13.3 A cross-sectional view of an ideally polished STI patterned wafer.

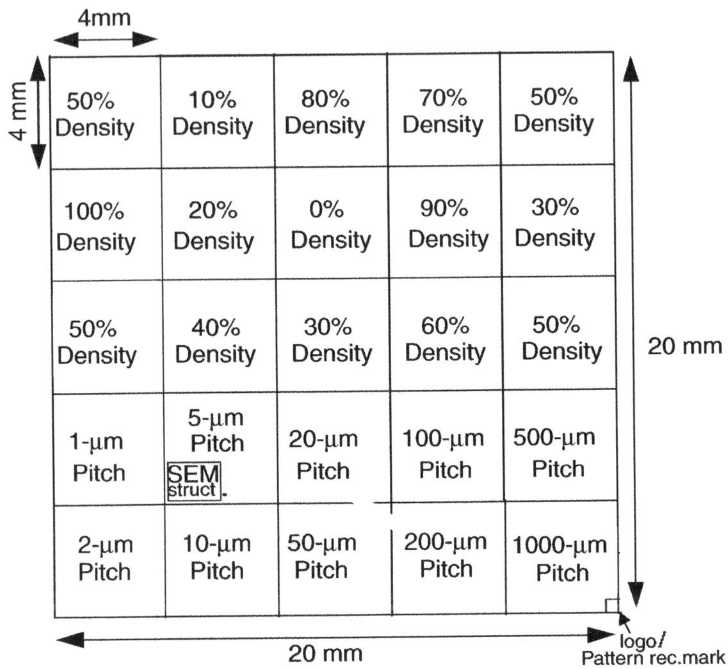

FIGURE 13.4 A schematic layout of a die on a representative patterned wafer for STI CMP process and consumable evaluation.

typical slurry flow rate is 200 ml/min. To evaluate the process consumables for material removal rate, selectivity of oxide over nitride, scratch counts, and particle residue, one can use a less expensive blanket wafer with the same material stack as shown in Fig. 13.2 (without the patterned structures). It is important to point out that the selectivity of oxide over nitride obtained on blanket wafers may not be portable to patterned wafers [8,10]. To evaluate the

TABLE 13.1 Representative Target Metrics for STI and ILD.

Metric Description	Target (Å)[a]
Nitride loss	< 25
Nitride range	< 30
Oxide dishing (50-µm trench)	< 45
Oxide dishing (500-µm trench)	< 200
WIWNU[b]	< 300–400
WIDNU/TIR (die-scale)[b]	< 200–300

[a]These parameters apply to both STI and ILD applications.
[b]Numbers are structure dimension dependent.

planarization capability of an STI process or consumable, the patterned wafers with a range of patterned densities and pitch width as shown in Fig. 13.4 must be used. In order to properly evaluate the defect levels, a patterned wafer with known defect count should be used [9]. Table 13.1 shows the planarization requirements continuing to advance to the subangstrom length scale [11].

13.3 EFFECTS OF ABRASIVE TYPES ON STI SLURRY PERFORMANCE

There are two major types of abrasives used for STI CMP. One is silica based, which resembles those used for silicon oxide dielectric CMP. The other is ceria based, which is similar to those in glass polishing. Silica-based slurries typically give low oxide-to-nitride removal rate selectivity and hence often lead to high nitride losses at the end. In addition, a silica-based slurry tends to polish the structures with different densities at varying rates, which translates to poor within-die planarity [8]. Ceria-based high-selectivity slurries (HSS) exhibit high oxide-over-nitride removal rate selectivity, which leads to a low oxide dishing and nitride loss [12]. In addition to those slurries containing single type of particles, a mixed particle approach has also been considered. One such example is the use of composite particles comprised of ceria and silica [13]. Li and Burkhard also reported the use of photochemical method for the synthesis of ceria-coated TiO_2 particles for STI CMP [14]. One advantage of using such composite particles is related to their overall density that is significantly lower than that of pure ceria particles. Unlike ceria particles, low-density composite particles are less likely to settle.

When polishing silicon dioxide films, it is often observed that on a per abrasive particle basis, ceria polishes planar surfaces significantly more effectively than silica [15]. For example, as shown in Fig. 13.5, the polishing rate for planar silicon dioxide is higher with slurry containing 0.5 wt% of ceria than that for silica-based slurries containing 13 wt% of silica.

For structured surfaces, ceria slurry behaves differently. More specifically, when planarizing a structured surface such as interlevel dielectric (ILD), silica-based slurries polish the high areas at a higher rate than the trench areas. This is consistent with what Preston equation would predict. For ceria-based slurries, however, this Prestonian effect is insignificant. Sometimes the initial polishing rate on the field can be lower than that on the trench [15]. This low removal rate on the field is directly influenced by the type of ceria as well as the additive in the slurries. Even though ceria does not give high planarization efficiency, the selectivity toward nitride is better than silica [15]. In addition, the polishing characteristics of ceria slurries are also sensitive to the nature of the oxide film being polished. Therefore, the variation in incoming wafers in deposition conditions may lead to varying polishing results [16,17].

The reverse-Prestonian behavior for ceria-based slurries has been reported by Merricks et al. [18] who investigated the STI CMP and ILD CMP processes.

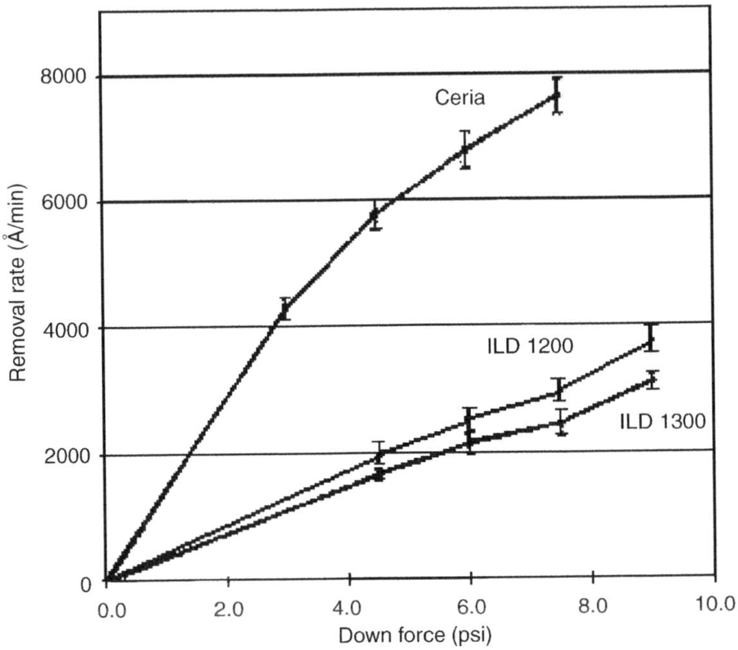

FIGURE 13.5 Oxide removal rate for two silica slurries with 13 wt% of silica (ILD 1200 and ILD 1300) compared to a ceria slurry with 0.5 wt% of ceria under the same polishing conditions (from Ref. 15).

One of the ceria-based slurries used for this study showed that the removal rate decreased as the pressure and velocity ($P \times V$) increased, which is reverse-Prestonian. As industry strives to reduce the CMP process time, the reverse-Prestonian behavior could translate to advantages to obtain higher removal rates at lower downpressures. Silica-based slurries generally require higher downforce to achieve the desired removal rate. Figure 13.6 shows several different ceria-based slurries containing various additive combinations.

From Fig. 13.6, it can be seen that the ceria-based slurry with the combination of additives A and B showed the strongest indication of the reverse-Prestonian behavior. However, the other three slurry combinations showed a Prestonian behavior.

During the STI CMP process, the particle interaction with the wafer surface plays a significant role in the planarization of the wafer topography. Burkhard [19] investigated the chemical interactions between the pad and the wafer surface with ceria slurry. It was observed from the polishing results that some ceria slurries exhibit a reverse-Prestonian behavior. The oxide removal rate decreased as the polishing downforce increased for one of the STI slurries. When the polishing was performed on an orbital platform, this trend was not observed.

Kim et al. [20] studied the effectiveness of CMP on 0.18-μm complementary metal-oxide semiconductor (CMOS) shallow trench isolation using CeO_2-based

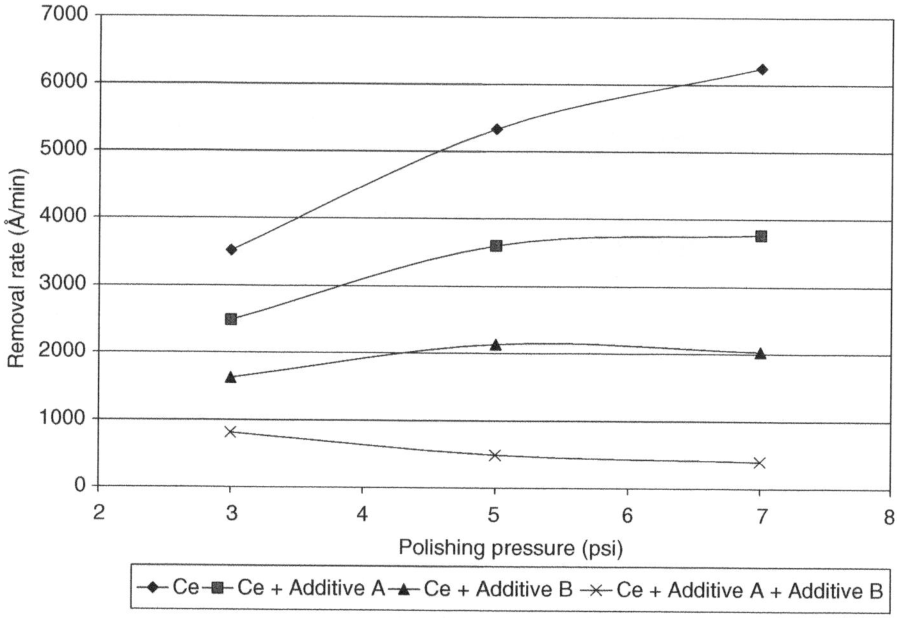

FIGURE 13.6 Effect of additives on reverse-Prestonian behavior.

slurry that has high silicon dioxide over silicon nitride selectivity. The wafer stack consisted of 50 Å oxide layer (pad oxide) and low-pressure chemical vapor deposition (LPCVD) silicon nitride (pad nitride). CMP operation was performed on a rotary-type polisher using unstacked polyurethane pads and two types of slurries, ceria-based HSS and silica-based low-selectivity slurry (LSS). The thickness of the remaining dielectrics and the step height profile at wide pattern pitches (10/10 μm) were monitored. When HSS was used, the removal rate selectivity of CVD oxide over silicon nitride was relatively insensitive toward the two mechanical parameters [(downforce) × (table rpm)] as shown in Fig. 13.7. The selectivities observed were consistently greater than 100, regardless of the processing conditions.

The variation in the field oxide recess was examined on six consecutive lots (25 wafers per lot) polished with HSS and LSS. It was observed that the HSS provided a more consistent field oxide recess from lot to lot, wafer to wafer, and within wafers as shown in Fig. 13.8.

It is not uncommon that the oxide:nitride removal rate selectivity is insensitive toward processing parameters. The additives present within the slurry have a greater influence on the selectivity than the processing parameters for STI CMP. For example, Choi et al. [21] evaluated the selectivity, uniformity, and field oxide dishing during ceria-based HSS CMP using a rotary-type CMP tool. The nitride-to-oxide selectivity was obtained by determining the material removal rates of LPCVD nitride and PE-TEOS oxide blanket wafers. Even though nitride and oxide removal rates depend on

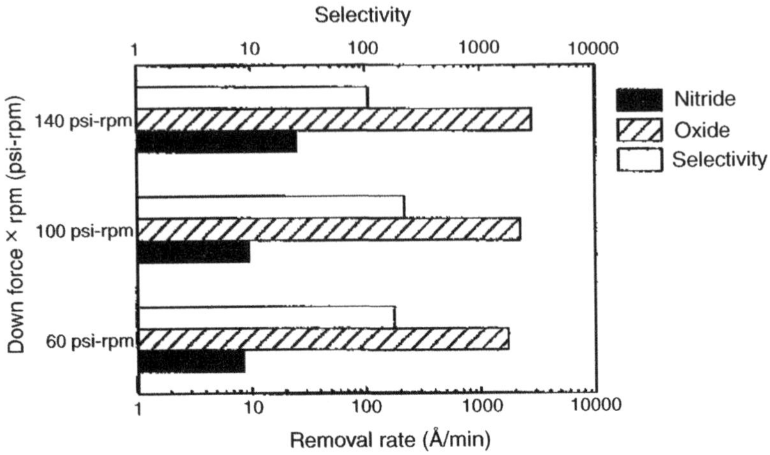

FIGURE 13.7 Removal rate and HDP-CVD oxide/SI_3N_4 selectivity versus (downforce) × (table rpm) obtained on a rotary-type polisher using high-selectivity slurry.

the mechanical parameters, it was observed that the selectivity was constantly over 100:1 regardless of the process conditions.

Martin et al. [12] reported a direct STI CMP process using HSS for the production of 0.13-μm technology microelectronic devices. Figure 13.9 shows the SiN thickness variations after various comparable STI processes that include two reverse-mask processes, two direct-polishing processes using Klebosol silica slurry, and two direct-polishing processes using ceria HSS. HSS leads to significantly better silicon nitride uniformity after polishing.

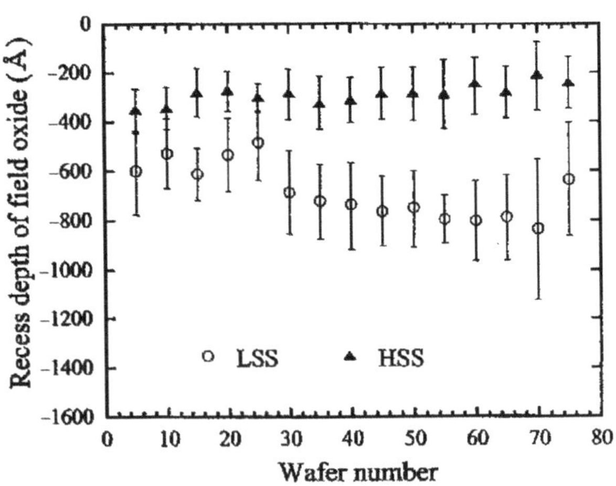

FIGURE 13.8 Relative thicknesses (with respect to zero overpolishing) of the remaining field oxides measured from six successive lots after being polished by HSS and LSS, respectively.

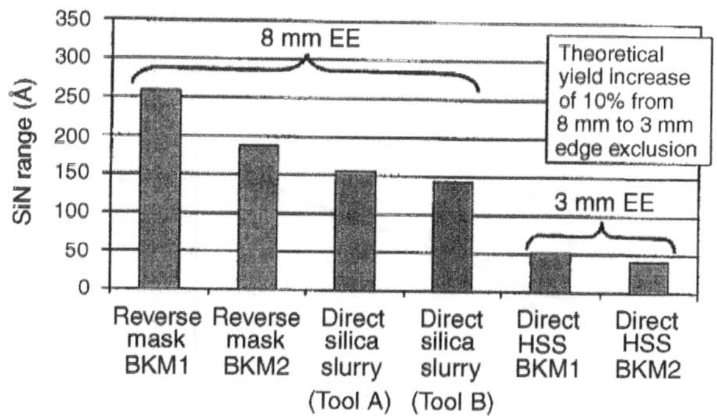

FIGURE 13.9 Postpolishing silicon nitride ranges show greatly improved uniformity for the direct-polishing processes using HSS. This improvement is realized with smaller (3 mm) edge exclusion than was used in the other processes (8 mm).

The scanning electron microscope (SEM) in Fig. 13.10a shows the representative results after HSS BKM 2. The SEM image in Fig. 13.10a was taken just after CMP and illustrates minimal dishing and SiN loss. The SEM in Fig. 13.10b was taken after the silicon nitride strip. It shows that the isolated oxide is more than 500 Å above the level of the active silicon that will become the transistor. It is critical to keep the dishing low and the step height over the active silicon consistent for the sake of proper device performance.

A slurry performance may be heavily influenced by the selection of other consumables. For example, it has been demonstrated that a concentric grooved pad enhanced the performance of a high-selectivity slurry (Rohm & Haas—XSHD3562) [22]. The pad allows the nitride film to be "truly a polishing stop layer." In this study, different combinations of rotary and linear polishing platforms in addition to conventional and high-selectivity polishing slurries were evaluated. It was determined that the best results were achieved

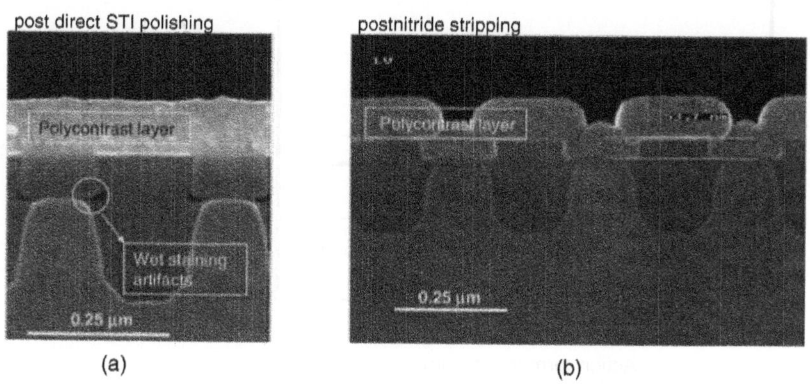

FIGURE 13.10 (a) SEM cross-sectional view of a wafer just after polishing with HSS BKM 2 shows minimal SiN loss and minimal dishing. (b) SEM cross-sectional view of wafer polished with HSS BKM 2 just after SiN strip shows a nominal and consistent step height.

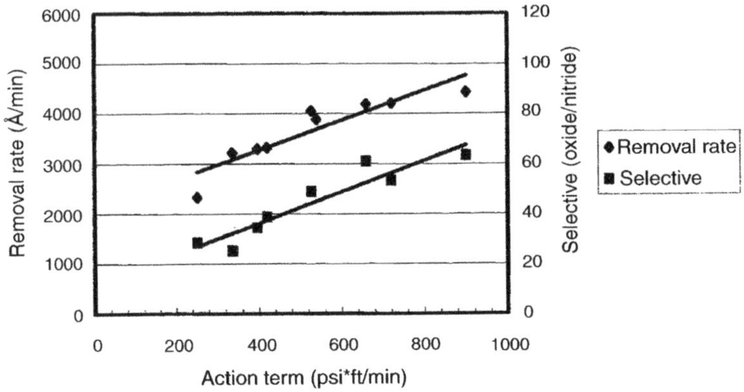

FIGURE 13.11 Oxide removal rate and selectivity (oxide:nitride) as a function of the action term. The action term is defined as the head pressure times the linear velocity.

with a linear planarization platform and a highly selective slurry. A less than 150 Å of oxide dishing in the field region was achieved with a complete removal of oxide from isolated active regions 10–4000 μm in size and a less than 1000 Å total nitride thickness variation.

Hansen et al. [23] examined the major factors that influence the (oxide:nitride) selectivity. These results were obtained in order to determine the optimum processing conditions for maximizing the (oxide:nitride) selectivity. It is important to point out that the oxide removal rate and selectivity (oxide:nitride) has a good correlation with the action term (downforce × linear velocity) as shown in Fig. 13.11. The nitride removal rate and nonuniformity as a function of action term is shown in Fig. 13.12. The average nitride removal rate is 83 Å/min and is

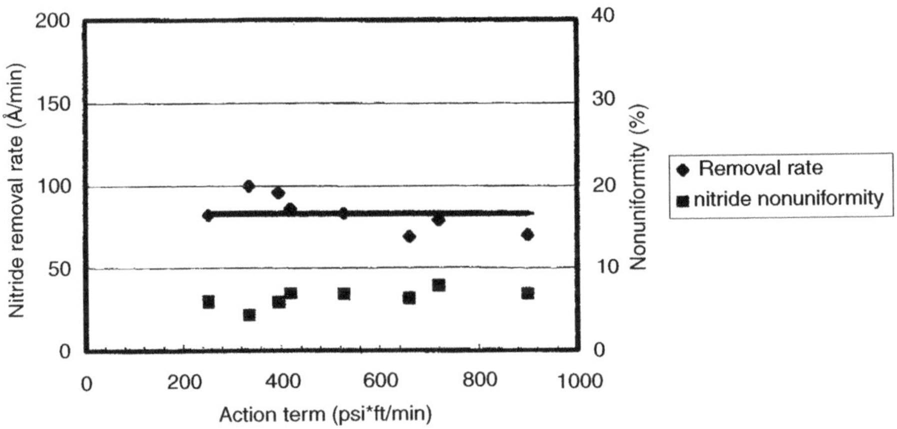

FIGURE 13.12 Nitride removal rate (closed diamonds) and nitride nonuniformity (closed squares) as a function of the action term. The solid line is the average removal rate with a value of 83 Å/min.

independent of the action term, which indicates that as the action term increases, the selectivity also increases with a similar slope as that of the removal for oxide. To polish the device wafers, a process recipe with the highest selectivity was chosen from the initial experimental results. A downforce of 5 psi and a linear velocity of 180 ft/min were selected as the process conditions to polish device wafers. It was observed that the time to planarize the device wafer was less than 60 s. With this decrease in time compared to the typical planarization time, the cost of ownership (COO) could ultimately be reduced.

13.4 EFFECTS OF CHEMICAL ADDITIVES TO OXIDE: NITRIDE SELECTIVITY

At the operating pH (4–6) of most STI slurries, the oxide surface is negatively charged and the nitride surface is slightly positively charged. As shown in Fig. 13.13, a negatively charged surfactant or polyelectrolyte should enhance the selectivity of oxide:nitride.

Kang et al. [24] investigated the effect of anionic surfactants on the oxide:nitride selectivity. As shown in Fig. 13.14, due to static repulsion, the addition of a low molecular weight anionic surfactant did not have any

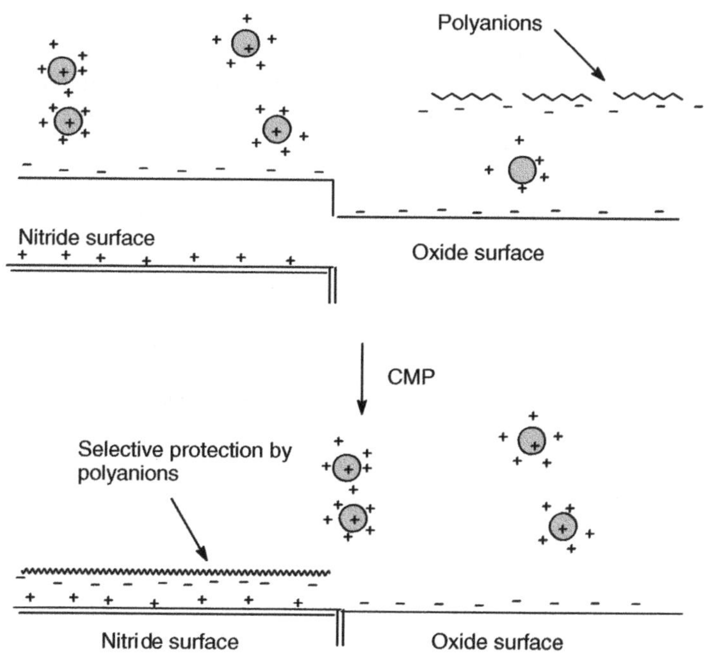

FIGURE 13.13 A typical chemical interaction between a surfactant and the STI wafer surface.

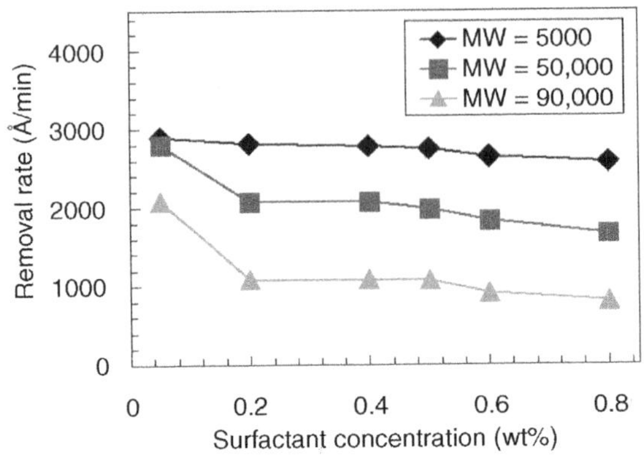

FIGURE 13.14 Oxide film removal rate versus anionic surfactant concentration.

significant impact on the removal rate of oxide. It was also observed that the oxide removal rate decreased with increasing molecular weight, especially at concentrations greater than 0.2 wt%. It is possible that the accumulation of polymeric surfactant on the oxide film may have altered the fluid dynamics of the surface and lead to a minor lubrication effect. Fig. 13.15 shows a graph of the nitride removal rate versus the surfactant concentration.

The nitride removal rate severely decreased with decreasing molecular weight and became stable beyond the surfactant concentration of 0.2 wt%, which was similar to what was observed for the oxide removal rate. With increasing surfactant concentration, the slurry that contained the surfactant with the lowest molecular weight had a larger nitride removal rate than that with the surfactant with the highest molecular weight. Based on the data shown in Figures 13.14 and 13.15, the oxide-to-nitride removal rate selectivity can be calculated. The selectivity increased approximately from 9:1 to 18:1 with

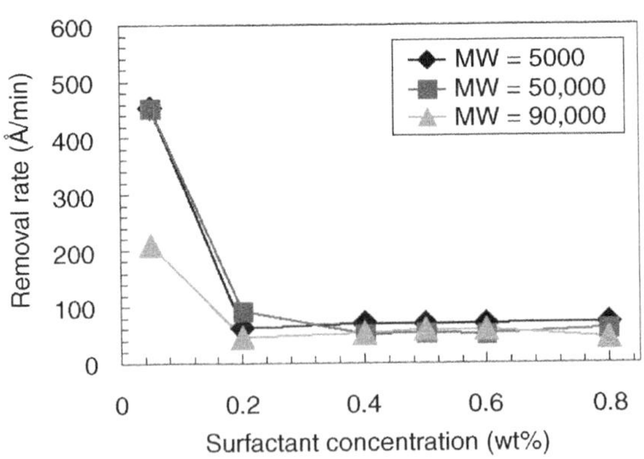

FIGURE 13.15 Nitride film removal rate versus surfactant concentration.

increasing surfactant concentration for the molecular weight of 90,000. For the surfactant with molecular weight of 5000, the selectivity increased approximately from 6:1 to 35:1.

The use of surface adsorption of polyelectrolyte to modulate the removal rate and selectivity has also extended to achieve the so-called self-stopping polishing. Self-stopping implies that the removal rate will decrease automatically when the polishing reaches a certain planarity or material film. Mueller et al. [11] investigated the functions of self-stopping behavior using an inhibiting polymeric additive. The dissolved polymer is capable of adsorbing onto the surface to be polished. Unlike small molecules, the surface adsorption kinetics for a polymer is influenced not only by the functional groups in the polymer backbone but also by the length of the polymer or the number of repeating functional units in the polymer as shown in Fig. 13.16.

The desorption process for a polymer is usually much slower than that for a small molecule. As shown in Fig. 13.16, the polymer molecule may be in equilibrium between fully desorbed and fully adsorbed via a "zipper" process. Under tribological conditions, the absorbed polymer will lubricate the surface and therefore prevent material removal under the applied shear. This will take place until polishing conditions surpass the engineered "threshold" shear force. At low shear force, the removal rate (RR) will be minimal. As the threshold is obtained, the RR response will significantly increase and follow a typical pressure–velocity (*PV*) response as further force is applied.

If the applied shear to the wafer is less than the threshold value, this mechanism can be used effectively to enhance the planarization efficiency. The localized pressure on the topography will be effectively larger than the threshold, which will selectively remove material at a high rate from the

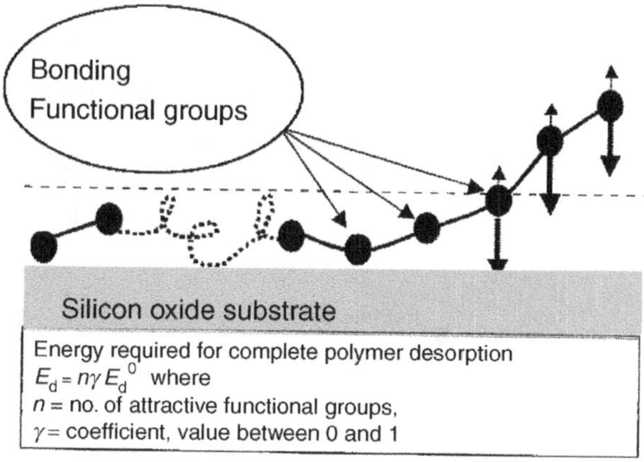

FIGURE 13.16 Diagram illustrating the kinetic "zipper" effect of a polymeric additive in CMP slurry that functions to stabilize the adsorption from removal under certain applied forces.

topography. As planarity is obtained and the pressure becomes uniform across the wafer, below the threshold, the removal rate will approach zero.

In a conventional direct STI polishing step, nonuniform clearing of the oxide over nitride results in higher dishing. Low-density active areas will be planarized and cleared sooner than the larger, dense active areas. Therefore, the low-density areas will experience overpolishing. By using self-stopping slurry, much lower dishing can be obtained. The final result is a more planar surface, and less residual step height, with less remaining overburden. Target step height is more easily obtainable with loss of overburden or downarea, and the step height range across the 10–90% density range is lower.

The effectiveness of a wider range of chemical additives on the oxide:nitride selectivity has been investigated by Srivivasan et al. [25]. Fig. 13.17 shows the polishing rates of silicon nitride with various additives. It was observed that the slurries containing peroxides were unstable. Additives such as urea, ammonium sulfate and glycerol, or ethanol did not sufficiently reduce the silicon nitride polish rate. The nitride rate did not decrease when ammonium hydroxide was added. It is possible that this is a pH effect more than the effect caused from the ammonium concentration. Among all the additives investigated, glycine caused the greatest reduction in nitride removal rate. The influence of other amino acids on nitride and oxide polish rates and the resulting selectivity have

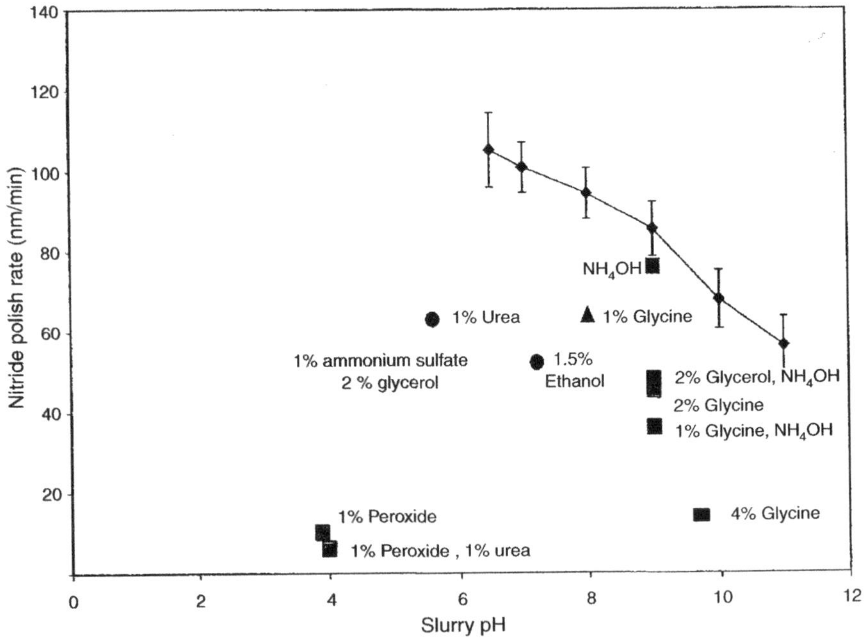

FIGURE 13.17 Removal rates of nitride blanket films with 4 wt% of cerium oxide. Particle diameter of 440 nm, 5.6 psi with ICI400/Suba IV pad. 5% of ceria, SRS-231 slurry. pH adjusted with KOH unless otherwise indicated.

TABLE 13.2 Polish Rates and Selectivity for Ceria Slurry with Additives. All Slurries were 4 wt% of Ceria at pH = 10 Except where Indicated.

Additive, wt%	Nitride Polish Rate, nm/min	Oxide Polish Rate, nm/min	Selectivity (oxide:nitride)
No additive	66 ± 4	489 ± 9	7
1% Glycine	34 ± 1	547 ± 30	16
4% Glycine	13 ± 2	420 ± 30	32
8% Glycine	6 ± 1	410 ± 10	68
4% Glycine (pH = 9.5)	6 ± 1	417 ± 32	70
4% L-Proline	~1	536 ± 10	536
1% L-Proline	12 ± 1	487 ± 19	41
1% L-Arginine (pH = 10.8)	2 ± 1.9	23 ± 3	12
1% L-Lysine	5 ± 3	71 ± 3	14

also been reported. The nitride removal rates using 4% ceria slurries containing amino acids such as glycine, proline, arginine, and lysine are shown in Table 13.2.

As can be seen in Table 13.2, while the removal rates for oxide films were relatively unaffected, the removal rates for the silicon nitride films decreased from approximately 66 to 13 nm/min in the presence of 4% glycine and to ca. 1 nm/min with 4% proline. For arginine and lycine, as the removal rates of oxide are also affected, the overall selectivity is not significantly high and useful. In addition, as shown in Fig. 13.18, the effect of glycine concentration is more dramatic for the first 1% by weight. The selectivity has a relatively linear behavior with glycine concentration since the oxide rate remains very high with a slow decreasing trend.

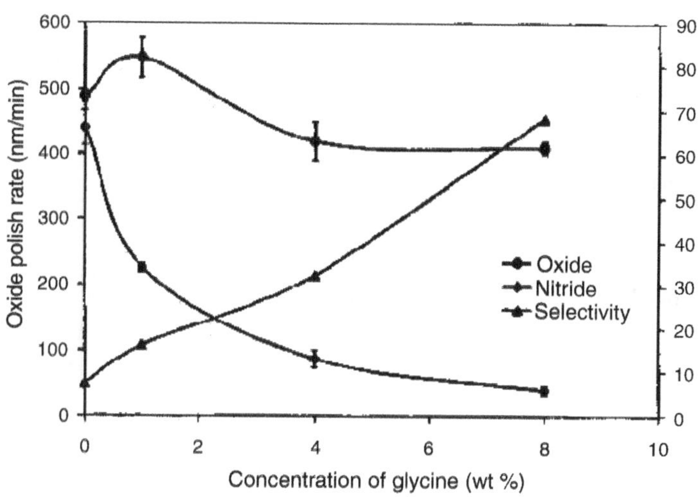

FIGURE 13.18 Polish rates and selectivity of oxide and nitride as a function of glycine concentration. Polishing conditions are 5.6 psi, 30 rpm, and 4 wt% of ceria.

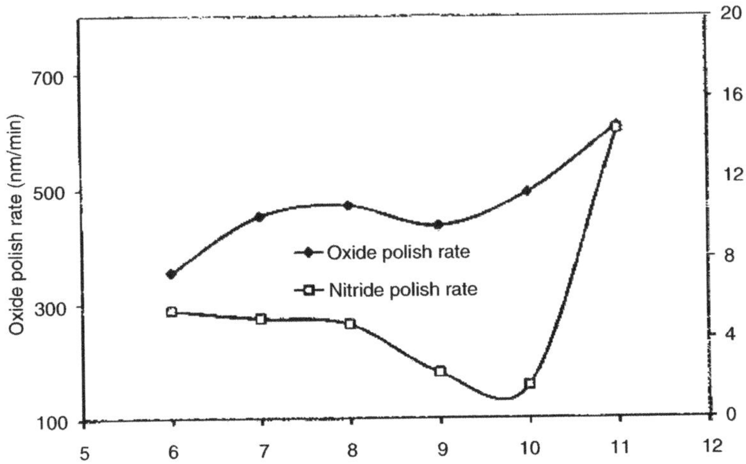

FIGURE 13.19 Removal rates of TEOS oxide and CVD silicon nitride with proline ceria slurry over pH range of 6–11. Ceria was 1 wt% with 2 wt% of L-proline.

Oxide-to-nitride selectivity for a 1–4% glycine–ceria slurry ranges from 16 to 70. The proline–ceria slurry is capable of achieving an even higher selectivity of over 500:1. The pH of the proline–ceria slurry has a significant influence on the selectivity. Fig. 13.19 shows the removal rates of oxide and nitride films for different values of the slurry pH.

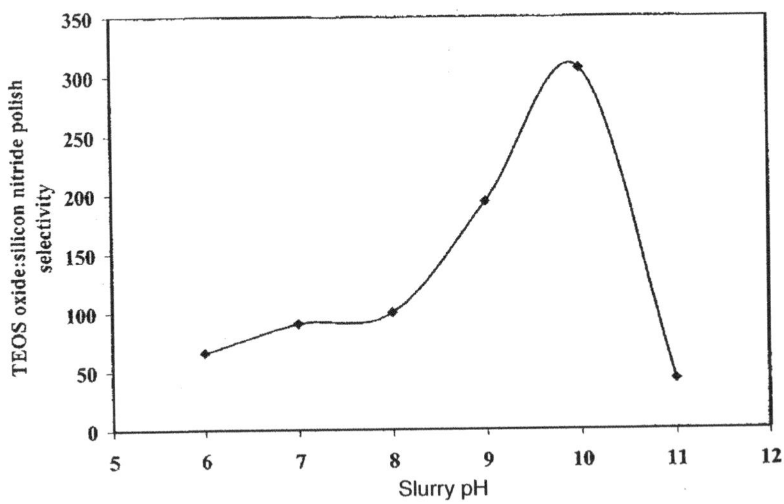

FIGURE 13.20 Selectivity of oxide to nitride of a 2% proline in 1% ceria slurry. Polishing conditions are 6 psi pressure, 40 rpm, and IC1000 on Suba IV.

The slurry in Fig. 13.19 consists of 2 wt% of L-proline and 1 wt% of ceria with pH adjusted to the desired value with either HCl or KOH. The nitride rate is relatively constant from pH 6–8 while the oxide rate gradually increases with pH. The nitride removal rate drops dramatically to a rate of roughly 1 nm/min as the pH increases from 8 to just below 10. The nitride rate rapidly increases from pH 10 to its highest value (with proline additive) at pH 11. Fig. 13.20 shows the selectivity of oxide to nitride in the proline–ceria slurry that has a linear behavior over the pH range of 6–8, has a very large rise from pH 8 to just below pH 10, and then falls quickly from pH 10 to pH 11.

13.5 EFFECT OF SLURRY pH

The pH of a slurry has a profound influence on its colloidal stability and CMP performance. Strong correlations have been established between the particle isoelectric point (IEP) and the optimal pH for slurry stability. The general rule is that the slurry is more stable at a pH that is away from the IEP, so the zeta potential of the particles is greater than ± 20 mV. The focus of this section is on the influence of pH on the slurry performances such as material removal rate and defectivity. In order to examine the impact of slurry pH on these two important performance features, we first take a closer look at the interaction between abrasive particles and the surface to be polished. There is a vast amount of literature on the interaction between abrasive particles and silicon dioxide surface [26]. The discussion below will focus on the interaction between ceria abrasive particles and the silicon dioxide surface to be polished. The basic principles and conclusions can be easily extended to other pairs of abrasive particles and surfaces.

Interactions between the ceria abrasives and the oxide surface have been investigated using both the chemical and the instrumental approaches. Suphantharida and Osseo-Asare [27] used zeta-potential measurements, silicate adsorption, and polishing experiment to investigate the role of ceria abrasives–SiO_2 surface interaction. To determine the effect of pH on the surface charges, the zeta potentials of abrasive particles were measured (Fig. 13.21). The points of zero charge (pzc) or isoelectric point is at pH 6.0 for ceria and pH 1.5 for silica. These values are consistent with those reported by others [28,29].

Silicate ions of varying concentrations were added to ceria dispersions in order to study the electrostatic interaction between ceria particles and silicate ions. It was observed that the zeta potential of ceria becomes less positive or more negative. With an increase in Na_2SiO_3 concentration, the isoelectric point shifts to lower pH values. At high silicate concentration, the zeta potential of ceria particles follows a trend similar to that of silica. This is a clear indication that the silicate ions adsorb onto the ceria particles [30,31].

The adsorption isotherms for the silicate–ceria system were obtained for slurries with pH ranges from 2 to 12 (Fig. 13.22). With an increase in silicate concentration, the total amount of adsorption onto ceria also increases. The adsorption curves reach a maximum at a pH of about 9.

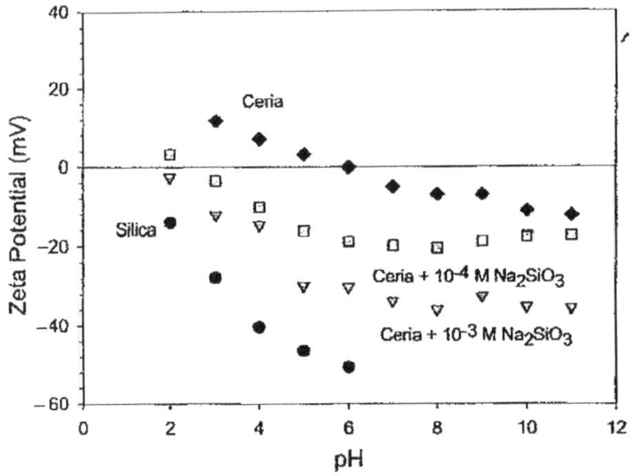

FIGURE 13.21 Effect of pH on the zeta potential of ceria (90%) and silica particles in the absence and presence of silicate ions.

An instrumentation approach was explored by Abiade et al. [32] who investigated ceria-silica interactions using atomic force microscopy (AFM) and SEM. Based on these studies, a model for silica polishing with ceria particles is illustrated in Fig. 13.23.

Figure 13.23a shows the first mechanism, which is mainly chemical in nature. The silica removal rate is accelerated by the ceria–silica interactions, which results in the improved dissolution of the silica substrate during polishing. Figure 13.23b shows the second mechanism, which is based on physicochemical particle–surface interactions in which the ceria–silica bonding does not result in direct modification of the silica substrate but enhances the

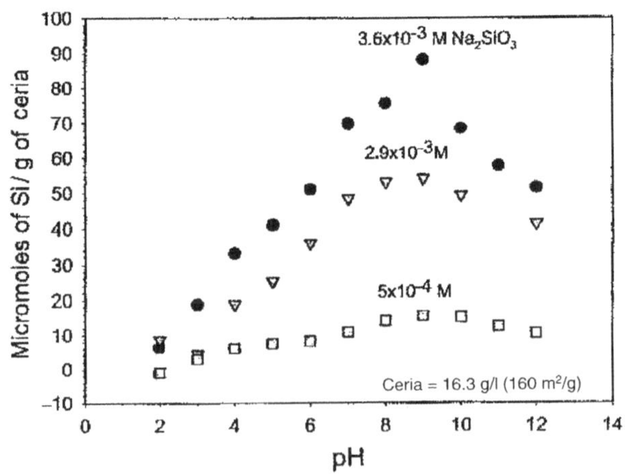

FIGURE 13.22 Effect of pH on the adsorption of silicate ions onto ceria (99.9%) particles.

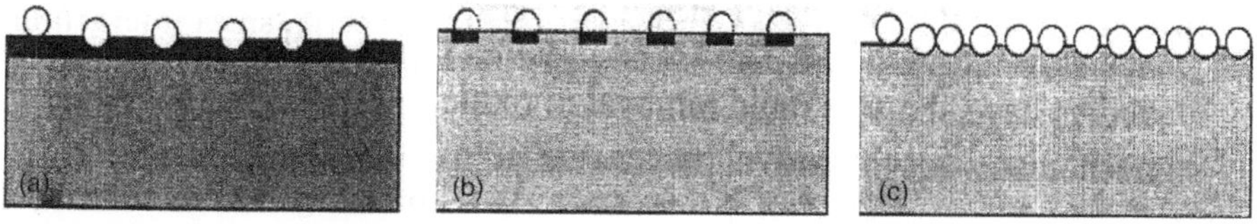

FIGURE 13.23 Schematic depiction of possible ceria–silica removal mechanisms (a) primarily chemical, (b) surface chemical, and (c) enhanced mechanical.

polish rate by increasing the frictional forces. The dissolution of the surface layer is represented by the shaded region and the particle–substrate contact area is represented in Figure 13.23a and b, respectively. Figure 13.23c is based mainly on mechanical interactions, and it would dominate in regimes wherein the slurry is unstable and/or a significant number of abrasive particles are present in the polishing process.

During CMP, various interparticle, surface, and environmental effects give rise to specific interactions that are strongly dependent on the relative distance they are from one another. The summation of the long- and short-range interactions is the sum of the forces (F_{total}) acting on an AFM tip or particle [33].

$$F_{\text{total}} = F_{\text{electrostatic}} + F_{\text{van der Waals}} + F_{\text{chemical}} \tag{13.1}$$

where $F_{\text{electrostatic}}$ refers to the long-range interactions related to charge on the opposing surfaces, $F_{\text{van der Waals}}$ refers to interactions of fluctuating dipoles, and F_{chemical} refers to the short-range chemical bonds. Interaction forces are measured by monitoring the displacement of the AFM cantilever as it comes in contact with the surface. A hysteresis is observed in the approaching and attracting force versus distance curves for strongly adhering surfaces. The force versus distance curves are shown in Fig. 13.24 for the individual pH conditions (2.5, 4.5, 6.0, and 10.5). The force was normalized to the particle radius.

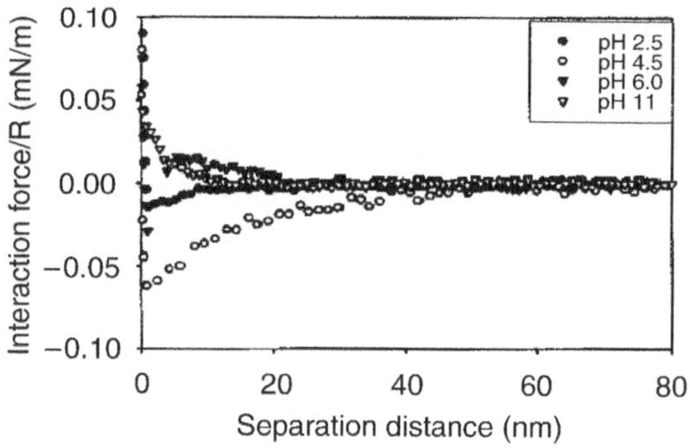

FIGURE 13.24 Force measurements for silica particles and ceria thin film as a function of pH.

The pH of the slurry plays an important role in the ceria–silica interaction. Fig. 13.24 shows that at pH 4.5, a significant decrease in the interaction force was observed compared to the other pH values that were evaluated. At pH of 2.5, 6.0, and 11, the interaction force was relatively consistent over the entire separation distance range. At pH = 2.5, a weak, short-range attraction occurs at separation distances less than 5 nm. This is consistent with the fact that the isoelectric point for silica and ceria are approximately at pH 2 and 6, respectively. A nearly neutral silica surface does not have any strong static interaction with positively charged ceria. At pH values between the two individual IEP, electrostatic attraction should dominate. At pH 4.5, the dominant interaction between the silica particle and the ceria thin film is electrostatic. This measurement agrees well with the measurements of zeta potential and friction force for the ceria–silica system [34]. At pH 6.0, a moderate attraction is also present. The ceria–silica electrostatic interactions are entirely repulsive in highly basic pH conditions. Therefore, a pH between the two IEP values may be favorable for polishing, and a pH that is significantly outside the IEP window may be favorable for post-CMP cleaning.

13.6 EFFECT OF ABRASIVE PARTICLE SIZE ON REMOVAL RATE AND DEFECTIVITY

In essence, STI CMP and ILD CMP are in a similar category. They are both required to remove certain amount of silicon dioxide film from the wafer surface. They are both dominated by mechanical actions. Therefore, the particle characteristics of the slurry are very important. In general, the material removal rate is proportional to the particle size. In a study reported by Park et al. [35], over a wide range of experimental conditions the oxide removal rate decreased with decreasing abrasive size. As abrasive particles play a significant role in determining the overall removal rate and surface quality, it is important to examine the relative importance of the mean particle size and particle size distribution.

The mean particle size is typically represented by the so-called D50, which indicates when the cumulative distribution of particles for a particular particle size reaches 50%. The particle size distribution is typically represented by the difference between D50 and D99. The larger the gap between D50 and D99, the greater the particle size distribution. In a representative study by Chandrasekaran et al. [36], the effects of ceria particle size and size distribution on the removal rate of dielectric materials were carefully elucidated.

The dependence of oxide removal rate on the tail particle size (D99) and mean particle size (D50) are shown in Fig. 13.25a and b, respectively.

As shown in Fig. 13.25, the oxide rate is almost independent of slurry D99 under all pressure conditions. The rate for slurry with D99 = 0.14 did show a drop in removal rate. This is the result of a change in D50 for the slurry. As shown in Fig. 13.25, the oxide removal rate shows a linear relationship with D50 under various downpressures. The polish rate slope also increases with increasing

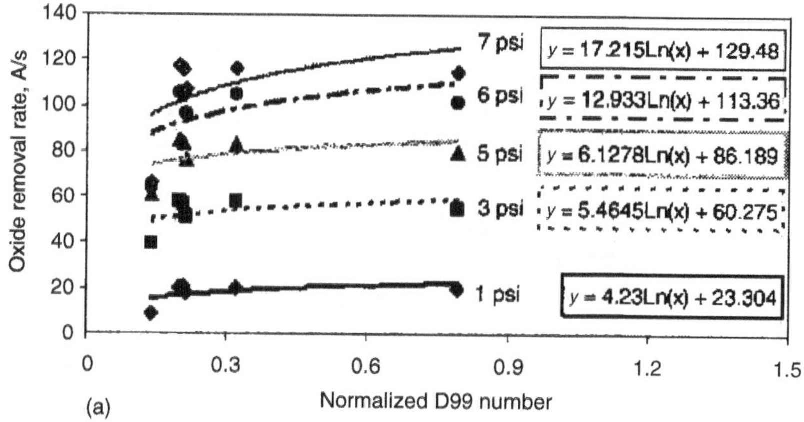

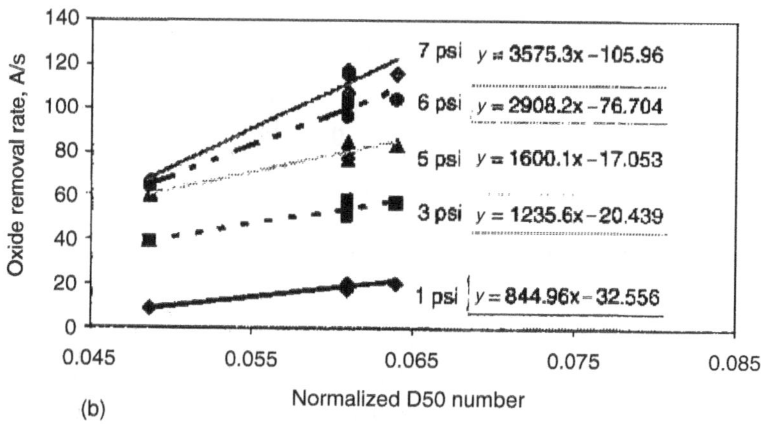

FIGURE 13.25 Variation in oxide removal rate under various downforces with (a) normalized D99 and (b) normalized D50, describing slurry particle size qualitatively.

pressure [36]. A similar study was performed to determine the dependence of nitride removal rate on the normalized D99 and D50 particle size. Similar to oxide removal, the increase in nitride removal rate is insignificant with the increase in D99 for the slurry. The nitride removal rate does respond to the slurry D50 linearly [36]. Therefore, the overall conclusion is that both oxide and nitride removal rates are determined or directly correlated with the D50 of a slurry.

For CMP, material removal rate is only one of the many performance metrics. In most cases, the defect count is much more critical to the overall yield of a CMP process. As shown in Fig. 13.26a and b, the defect count has a stronger correlation with the D99 of a slurry than the D50.

On the basis of the studies described above and reported in the literature, the particles present in a CMP slurry can be classified into two groups: (1) The

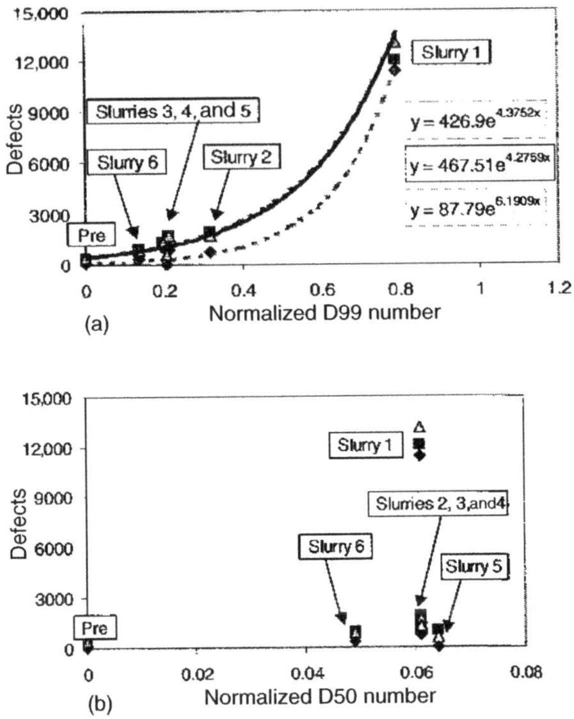

FIGURE 13.26 Variation in defects with D99 particle size for various slurries at a downforce of 3 psi.

particles that have sizes near the mean particle size. These particles are primarily responsible for the overall removal rate; (2) The particles that have sizes above D99; These particles are referred to as oversized particles or large particles in the slurry. These are the particles that can cause defect to the wafer surface. During polishing on a rotary platform tool, the wafer surface comes into contact only with a small percentage of the particles present in the slurry. As the concentration of oversized particles in the slurry is very small to begin with, these large particles must be very effective in generating surface defects.

It is important to point out that a slurry with small particles may have the so-called secondary particle size due to the formation of agglomerations. Depending on the analytical techniques, the primary particles may appear to be absent. For example, as large particles give significantly greater intensity than small particles when using dynamic light scattering method, an intensity-based particle sizing report may show only the secondary particles whereas the small primary particles are faded in the background. As described earlier, a strong correlation has been established between the large particle count and surface defects. Tateyama et al. [37] investigated the effect of abrasive particle size on scratch formation by preparing various slurries with different particle size distributions of ceria abrasives. The effects of the mean secondary particle size

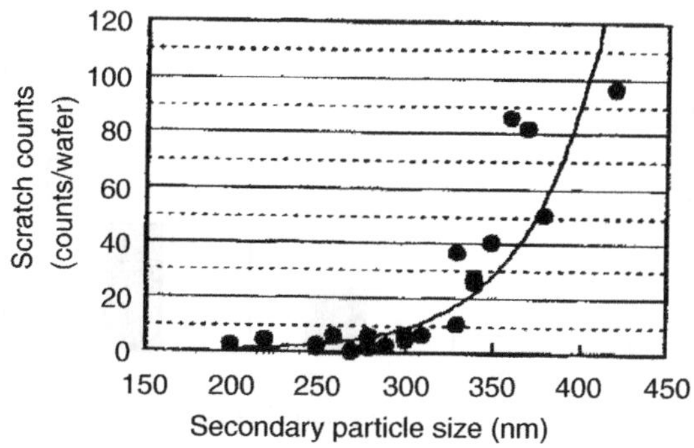

FIGURE 13.27 Dependence of scratch counts on secondary particle size.

on the number of scratches formed on the blanket HDP oxide wafer is shown in Fig. 13.27..

From this study it was observed that the secondary particle size has a significant effect on the number of scratches. The scratch count rapidly increases from approximately 10 to 100 counts if the mean particle size is greater than 340 nm. Therefore, this suggests that the main cause of the scratch formation is possibly the uncontrolled large secondary abrasive particles.

It is important to point out that the particle size effect sometimes can be masked by the presence of other additives such as surfactants. For example, larger particles may require high surfactant concentration to stabilize the slurry. Furthermore, as larger particles tend to have smaller total surface area, they will take up fewer surfactant molecules and leave greater amount of free surfactant solution. Regardless of the abrasive size, the viscosity may increase at high surfactant concentration. A passivation layer may form as a result of surfactant molecules adsorbed onto the oxide and nitride surfaces during polishing.

Stoke's law of resistance can be used to explain the decreasing removal rate with increasing surfactant concentration. The local viscosity near the film surface increases as the amount of adsorbed surfactant increases. The surfactant layer then acts as a passivation layer, which prevents the abrasive from coming into contact with the film surface. If taking kinematical energy into account, a larger particle can travel through the fluid more freely than a smaller particle [38]. The removal rate is thus strongly influenced by contact probability of the abrasives on the film surface [39]. Therefore, the changes in the removal rate of ceria slurry as a function of surfactant concentration and its dependence on the abrasive size can be attributed to the abrasive motion behavior in the surfactant adsorption layer on the surface to be polished. The rheological performances of these slurries are similar to Newtonian performance where the dispersion stability is mainly determined by electrostatic potentials and selective adsorption. The viscosity is lower for the ceria suspension without the surfactant than that with the surfactant.

The most persistent and commonly seen defects in ILD CMP and STI CMP are scratches, chatter marks, and pits. They are related to processing

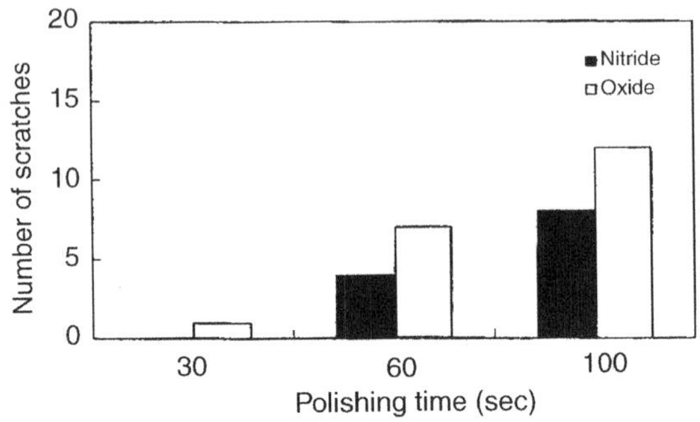

FIGURE 13.28 Scratch counts with polishing time; defects were measured on blanket wafers. Scratch counts increase with polishing time for both films.

parameters as well as slurry characteristics. It is important to point out that, in practice, the level of defects is also heavily related to the CMP and post-CMP cleaning processing parameters. For example, Choi et al. [21] reported that HSS tends to generate more scratches during CMP. This is directly related to the fact that ceria in HSS are harder particles. The study also indicated that the number of scratches increases with the increase in polishing time as shown in Fig. 13.28. This implies that the overpolishing window must be kept minimal not only to limit the amount of dishing and nitride loss but also to reduce the level of scratches. An initial evaluation for defects such as scratches, chatter marks, and pits is usually carried out using blanket oxide wafers. After polishing, large scratches can be seen visually. Small and shadow scratches and chatter marks require specialized defect tools to count. It is sometimes helpful to use dilute HF to highlight the scratched area.

Another practical difficulty in characterizing the correlation between slurry properties and defect level is that the absolute number of these oversized particles is very less in relation to the overall number of particles in the slurry. As shown in Fig. 13.29, the number of large particles that is capable of causing defects is in the range of 2.0×10^6 to 9.0×10^6. A typical slurry with mean size of 130 nm has an average of 1.1×10^7 oversize particles $>0.5\,\mu m$ and 2.3×10^{16} total abrasive particles.

There are very limited numbers of techniques that can be directly used to investigate the formation and the properties of these large particles. For example, practically all ensemble techniques such as dynamic light scattering for particle sizing, pH measurement, zeta-potential measurement, and most of the friction measurements are not capable of detecting the level of oversized particles commonly seen in a qualified CMP slurry. Some model studies doped the slurry with large particles at levels that are several orders of magnitude higher than the minimum defect-causing concentration. These particles have been identified as micron-sizde, hard, irregularly shaped aggregates of abrasive particles.

Oversized particles that develop during the manufacturing process are typically filtered prior to being shipped to the vendor. But once the slurry

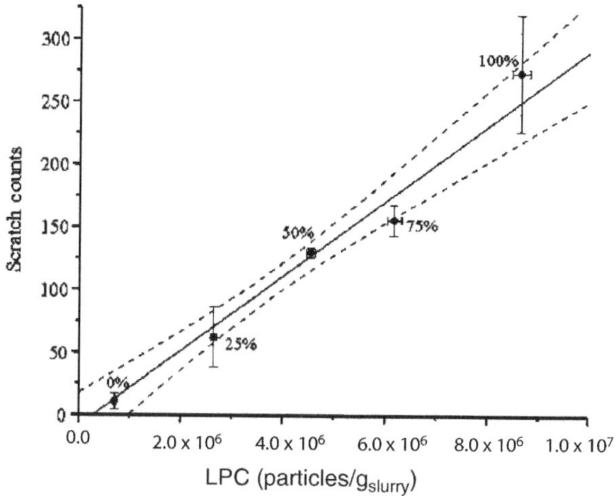

FIGURE 13.29 Correlation between scratch counts and LPC determined for particles with diameter $\geq$0.469 μm ($SO_3 + SO_5$) with mixtures of slurry A (0 wt% of slurry B) and slurry B. Weight percentage of slurry B for the mixtures is labeled at the corresponding data point. The weighted linear regression fit to the data set and 95% confidence limits are represented by the solid line and dashed lines, respectively. Error bars correspond to ±1 standard deviation (from Ref. 40).

leaves the slurry manufacturer, several factors can influence the development of oversized particles. First, oversized particles can occur during shipping and handling. This includes temperature variations (extreme cold or warm conditions) and particle collisions during transportation. Second, agglomerations could occur during the blending of chemical additives prior to use. The addition of additives could cause a change in the surface charge of the particles. Finally, with time, the particle size may change due to settling of the particles. Point-of-use filtration at the polisher is used in order to prevent these oversized particles from coming into contact with the wafer surface.

The removal of large particles and agglomerates is the main objective of filtration. Filtration is also used to eliminate any other contaminants that possibly have been introduced by improper slurry handling or have been collected from within the distribution system. The major criterion of filtration is to remove the oversized particles that could be defect causing without altering the slurry performance. Filtration will be further discussed in Chapter 18.

An alternative method to filtration has been considered in order to reduce the large particles and agglomerates. Li and co-workers [41] investigated a novel slurry conditioner to reduce the oversized particle count present in CMP slurries. This study focused on the correlation between the oversized particle counts in STI CMP slurries and their influence on the wafers after polishing. A representative oversized particle count reduction after the treatment with the conditioner is shown in Fig. 13.30.

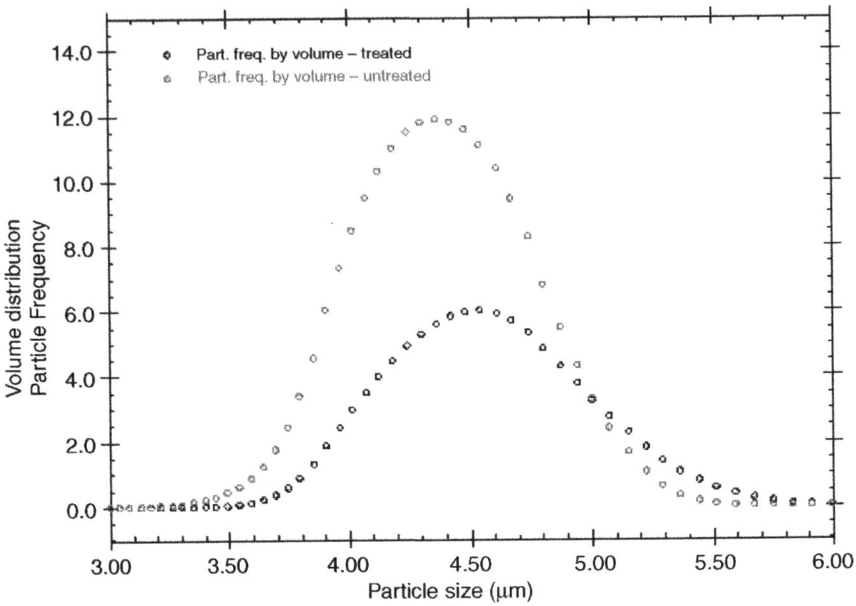

(a)

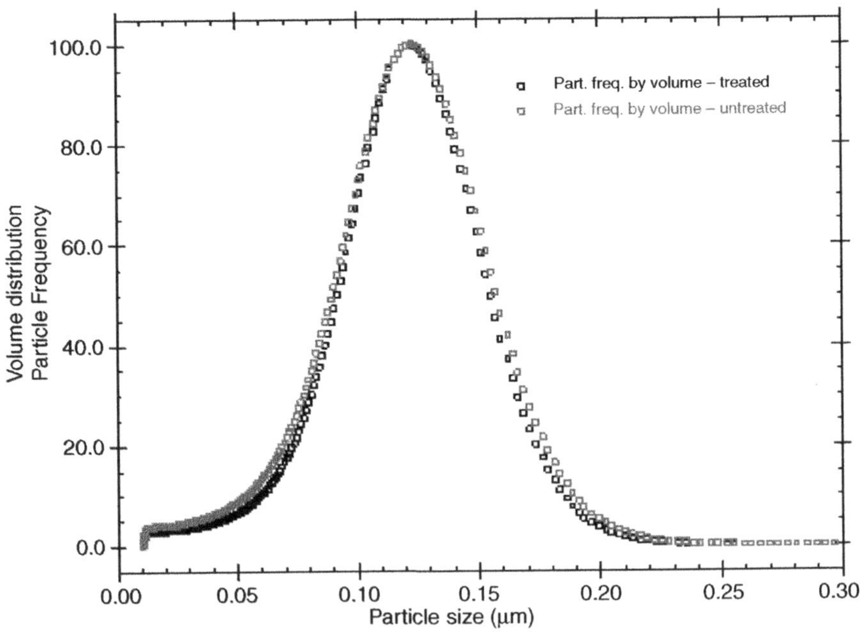

(b)

FIGURE 13.30 Representative changes in the large (a) and small (b) particle size volume distributions before and after treatment with the imation CMP slurry conditioner.

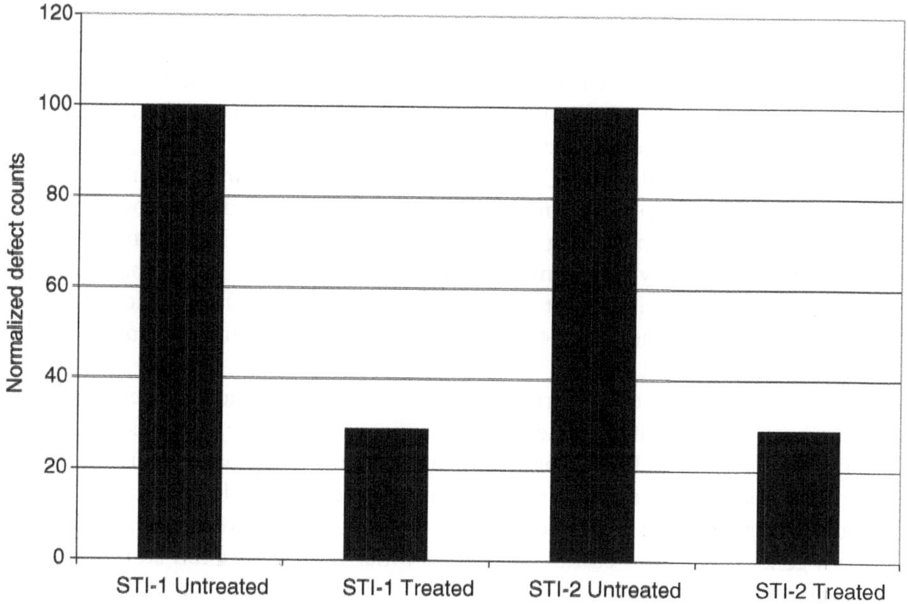

FIGURE 13.31 Representative defect counts for STI wafers polished with imation slurry conditioner. STI-1 is a ceria-based slurry and STI-2 is a silica-based slurry.

Figure 13.30a shows a significant change after treatment with the imation CMP slurry conditioner in the oversized particle volume distribution, where the mean particle size of 120–125 nm was unaffected as shown in Fig. 13.30b. Therefore, the imation CMP slurry conditioner is capable of reducing the number of oversized particle count without influencing the mean slurry particle size and size distribution.

A graph of the normalized defect counts versus the STI slurries, both untreated and treated, is shown in Fig. 13.31. The bars in the graph represent 20 wafers of polishing.

It was observed that the defect counts for the treated ceria-based and silica-based slurries were significantly reduced because the total number of oversized particles in contact with the wafer surface during polishing was lowered.

13.7 CONCLUSION

As STI is becoming vital to the reduction in the feature size, it is important to develop a further understanding of the STI CMP process. STI CMP has some similarities and several differences with oxide CMP.

Similar to oxide CMP, STI CMP is a mechanically dominated process. Therefore, the material removal rates are mainly controlled by the abrasive type, content, and size. Furthermore, the source of defects and their formation mechanism are analogous to oxide CMP. The post-CMP cleaning process for oxide ILD is similar to that of STI. One key difference is that STI polishing stops on the nitride layer with significant areas covered by oxide features. The abrasive particles for STI CMP need to be removed not only from the oxide but also from the nitride as well.

Unlike oxide CMP, STI CMP forms a microstructure that requires an effective polishing stop at the nitride layer and minimized dishing on the oxide (for direct STI). The polishing stop can be accomplished by a control of the removal rate selectivity (oxide:nitride). The typical additive component for STI ceria slurries is polyanions that form a protective film on the nitride surface in order to minimize the nitride material removal rate and to increase the overpolishing window.

Compared to silica, ceria particles polish the planar surfaces more effectively. It has been shown that some ceria slurries exhibit the reverse-Prestonian behavior on certain polishing platforms and not for silica-based slurries. Ceria-based slurries often contain additives that are specifically designed to alter the surface charge on the abrasive particle and/or wafer surface, whereas silica slurries generally do not contain additives other than dispersion stabilizers and pH-adjusting agents.

QUESTIONS

1. Considering the fact that the polishing requirements become more and more stringent as discussed in Section , suggest some changes in the current CMP process that could obtain lower WIWNU, nitride loss, and oxide dishing across the pattern density?
2. The reverse-Prestonian behavior has several benefits in the CMP process. Recommend some possible chemical components that could yield reverse-Prestonian behavior, and provide reasons why you choose these components?
3. During STI and oxide CMP, defects can come from sources such as oversized particles, polishing debris, diamond particles from the pad conditioner, and particles trapped in the polishing pad. In addition to the techniques discussed in this chapter, could you think of additional solutions for the reduction of these defects?
4. Considering the IEPs of ceria particles, oxide, and nitride surfaces, explain the potential advantages and disadvantages of conducting STI CMP at pH = 4 to 10.

REFERENCES

1. Oliver M, Hall R, Osgood RM, Jr., Parisi J, Warlimont H, editors. Chemical–Mechanical Planarization of Semiconductor Materials. Germany:Springer-Verlag Berlin Heidelberg; 2004.

2. Wolf S, Tauber RSilicon Processing for the VLSI Era: Volume 1—Process Technology. 2nd ed. California: Lattice Press; 2000.
3. Lin M, Chang C-Y, Liao DC, Wang B, Henderson AImproved STI-CMP Technology for micro-scratch issue. Proceedings of the CMP-MIC; 1999.
4. Chen, CIEDM Tech Dig 1996;837.
5. Fazan, PIEDM Tech Dig 1993;57.
6. Geissler, SIEDM Tech Dig 1991;91.
7. Jarrell Steve, Personal Communication.
8. Bonner BA, Iyer A, Kumar D, Osterheld TH, Nickles AS, Flynn D. Development of a direct polish process for shallow trench isolation modules. CMP–MIC; 2001.
9. Internet Web site: http://www.testwafer.com
10. John Givens, Personal communication.
11. Mueller B, Lawing AS, Flanagan P, Yu C, Lane S, Huynh D. Challenges for CMP slurries in next generation STI and dielectric applications. Proceedings AVS First International Conference on Mircoelectronics and Interfaces, Santa Clara, CA; 2000.
12. Martin A, Spinolo G, Morin S, Bacchetta M, Frigerio F, Bonner B, McKeever P, Tremolada M, Iyer A. The development of a direct-polish process for STI–CMP. Mater Res Soc Symp Proc 2003;767: F5.10.1.
13. Lu Z, Lee S-H, Gorantla RK, Babu SV, Matijevic E. Effect of mixed abrasives in the chemical mechanical polishing of oxide films. J Mater Res 2003;18: 2323.
14. Li Y, Burkhard C. The charaterization of advanced STI CMP process and consumables. Proceedings of AVS Fifth International Conference on Microelectronics and Interfaces (ICMI'04).
15. Lee S-I, Kim C-I, Kim H, Kim J-H, Nam C-W, Kim S, Kim C-T. Proceedings 1997 CMP-MIC Conference, 163, IMIC, Tampa;1997.
16. Oliver M, Evans D, Hetherington D, Sten D, Stevens J, Hosali S.Proceedings 1999 CMP-MIC, 383, IMIC, Tampa;1999.
17. Stein D, Hetherington D, Oliver M, Hosali S, Evans D, Her B.MRS, Paper M3. 8; San Francisco:2001.
18. Merrick D, Santora B, Her B, Frink S. An Investigation of ceria-based slurries exhibiting reverse-prestonian behavior. Proceedings 11th International Symposium on Chemical Mechanical Planarization; 2006.
19. Burkhard C. Characterization of advanced shallow trench isolation (STI) CMP Processes and consumables. Ph.D. Thesis; 2006.
20. Kim S, Hwang I, Park H, Rhee J. Chemical mechanical polishing of shallow trench isolation using the ceria-based high selectivity slurry for sub-0.18 μm complementary metal-oxide-semiconductor fabrication. J Vac Sci Technol B 2003;20: 918–923.
21. Choi K-S, Lee S-I, Kim C-I, Nam C-W, Kim S-D, Kim C-TApplications of ceria-based high selectivity slurry to STI CMP for Sub 0. 18 μm CMOS technologies. CMP-MIC Conference;1999. p307.
22. Singer P. CMP developers take aim at STI applications. Semiconductor International; 1998. p40.

23. Hansen D, Wu C, Lu HB. Characterization of a shallow trench isolation slurry using a multi-head CMP tool. Proceedings Third International CMP-MIC Conference at Santa Clara; CA Feb 19–20;1998. pp72–75.
24. Kang H, Katoh T, Lee M, Park H, Paik U, Park J. Effect of molecular weight of surfactant in nano ceria slurry on shallow trench isolation chemical mechanical polishing (CMP). Jpn J Appl Phys 2004;43:L1060.
25. Srivivasan R, America WG, Her Y-S, Babu SV. Ceria based slurries for STI planarization. CMP-MIC; 2000. p148.
26. Choi W, Lee S-M, Singh R. Further investigation of effects of ph on silicon dioxide chemical mechanical polishing (CMP). The Electrochemical Society 204th Meeting; 2003.
27. Suphantharida P, Osseo-Asare K. In: Opila RL, Reidsema-Simpson C, Sandaram K, Seal S, Huff H, Suni II, editors. Chemical Mechanical Polishing (V). Pennington, NJ: The Electrochemical Society;2002. PV-2002-1, p257.
28. Parks G. The isoelectric points of solid oxides, solid hydroxides, and aqueous hydroxo complex systems. Chem Rev 1965;65:177.
29. Ray KC, Sengupta PK, Roy SK. Ind J Chem1979;17A:348.
30. Osseo-Asare K, Supthantharida P. In: Bautista RG, Mishra B, editors. Rare Earths and Actinides: Science, Technology and Applications. Warrendale, PA: TMS; 2000. p139.
31. Osseo-Asare K, Suphantharida P In: Opila RL, Reidsema-Simpson C, Sandaram KB, Seal S, editors. Chemical Mechanical Polishing (IV). Pennington, NJ: The Electrochemical Society; 2001. PV-2000-26, p222.
32. Abiade JT, Jung S, Yeruva S, Singh RK. Investigation of ceria-silica interactions during STI polishing. CMP-MIC; 2005. p335.
33. Garcia R, Perez R. Surf Sci Rep, 2002;47:197.
34. Abiade JT. In: Duane SB, Johann WB, Ara P, Greg S, Ingrid V, editors. Advances in Chemical–Mechanical Polishing. Warrendale, PA: Materials Research Society Symposium Proceedings; v816 2004, p K9.5 and Abiade JT, Choi W, Singh RK.Journal Mater Res forthcoming.
35. Park H, Jung J, Park J, Shin J, Ryu C-H. Effects of abrasive size and surfactant concentrations in ceria slurry for shallow trench isolation CMP. Proceedings Ninth International CMP-MIC Conference at Santa Clara, CA; 2005. p357.
36. Chandrasekaran N, Taylor T, Sabde G. Effect of ceria particle-size distribution and pressure interactions in chemo-mechanical polishing (CMP) of Dielectric Materials. Mater Res Soc Symp Proc 2003;767: F3.2.
37. Tateyama Y, Hirano T, Ono T, Miyashita N, Yoda T In: Opila RL, Reidsema-Simpson C, Sandaram KB, Seal S, editors. Chemical Mechanical Polishing (IV). Pennington, NJ: The Electrochemical Society; 2001. PV-2000-26, p297.
38. Katoh T, Kang HG, Paik U, Park JG Jpn J Appl Phys 2003;42:1150.
39. Hoshino T, Kurate Y, Terasaki Y, Susa KJ. Non-Crystalline Solids 2001;283: 129.
40. Remsen E, Anjar S, Boldridge D, Kamiti M, Li S, Johns T, Dowell C, Kasthurirangan J, Freeney P. Analysis of large particle count in fumed silica

slurries and its correlation with scratch defects generated by CMP. J Electrochem Soc 2006;153(5)G453–G461.

41. Li Y, Burkhard C, Serafin M, Nelson N, Olmsted R, Hu F. The impact of a novel slurry conditioner on copper and STI performance. VMIC 2006; Fremont CA, Sep 25–27, 2006.

14

FABRICATION OF MICRODEVICES USING CMP

GERFRIED ZWICKER

14.1 INTRODUCTION

This chapter describes further applications of chemical–mechanical polishing (CMP) beyond those used in classical microelectronic production. Since its success as an enabling technology for integrated circuit (IC) manufacturing, CMP has also found application in fabricating sensors, actuators, microoptical devices, read/write (R/W) heads for hard-disk drives, and other miniaturized structures. The fabrication of these microdevices is described by terms or acronyms like micromachining, microelectromechanical systems (MEMS), or microsystems technology (MST). In this chapter, "microfabrication" is used to cover the different terms and is defined as the fabrication of devices with at least some of their dimensions in the micrometer range [1]. The term "systems" stands for a combination of sensors, actuators, and processing units that have a higher level of functionality than the single devices. Technologies for system integration, microsystem packaging, or chip stacking (3D integration), which often employ CMP, are also included in this chapter.

The manufacturing technologies and materials used for microfabrication are in many cases derived from semiconductor technology, but independent methods have also emerged [2–4]. CMP was originally developed to overcome processing issues in microelectronics manufacturing, such as planarization of interlayer dielectric materials. The technology was then extended to enable the manufacturing of structures that were impossible or difficult to build, such as copper structuring via damascene. Comparably, CMP was used initially in

Microelectronic Applications of Chemical Mechanical Planarization, Edited by Yushuo Li microfabrication for simple planarization tasks and is now offering new solutions [5], for example, for multilevel constructions like micromotors, gears, or micromirrors.

In the following, a short introduction of the key manufacturing methods for microsystems will be given and some application-specific mass-produced devices will be presented. Subsequently, the requirements, similarities, and differences of CMP processes used for microfabrication will be compared to those used in microelectronics manufacturing. CMP tools appropriate for microfabrication and consumables and suitable for manufacturing of MEMS will be listed and discussed. With the help of detailed examples, the use of CMP for the manufacture of some typical MEMS devices will be shown in the last part of this chapter. For that purpose, the fabrication of an integrated pressure sensor in bulk micromachining, the production of an angular rate sensor in poly-Si surface micromachining, and the buildup of a micromirror array with copper as a sacrificial layer will be discussed.

14.2 MICROFABRICATION PROCESSES

Microfabrication has emerged from microelectronics manufacturing and is using its proven processes and process sequences. Additionally, specific methods have been developed to fabricate mechanical, electrical, optical, or sensor structures, which are characteristics of microfabrication. In order to stay within the scope of this book, only top-down methods, that is, the manufacture of smaller structures with higher functionality from larger structures by the use of subtractive methods, will be discussed. Bottom-up methods, which create larger structures by ordered arrangement of small units (molecules, nanoparticles), are still in their infancy and mainly employed for biosensors.

Three top-down manufacturing methods are currently used in microfabrication:

- Bulk micromachining: Microfabrication of three-dimensional features such as membranes, cavities, and so on by anisotropic dry or wet etching into the bulk of substrate materials like silicon, quartz, or others.
- Surface micromachining: Buildup of features using layer-by-layer deposition on the surface of a substrate, structuring of polysilicon or metal layers by dry etching to form combs, levers, wheels, and so on, and release of features from surface by undercutting of sacrificial layers employing selective isotropic wet, dry, or vapor-phase etching processes.
- Micromolding (LIGA[1]): Patterning of three-dimensional resist structures by high-aspect-ratio lithography, fabrication of a mold by electroplating

[1]Lithographie, Galvanoformung, Abformung; German acronym for lithography, electrodeposition, and molding.

of thick metal, and subsequent resist removal. Metal structure can be the final product (e.g., gear wheel) or is used for micromolding of, for example, polymers.

Most of the manufacturing steps using the above-described three methods are directly adopted from semiconductor technology, but some techniques for pattern transfer had to be evolved to accommodate greater film thickness. For example, deep reactive ion etching (DRIE) had to be developed for the anisotropic dry etching of thick poly-Si layers and electroplating is often deployed for the deposition of thick metal films. CMP can be viewed as an enabling process technology when smooth or planarized surfaces are needed.

14.3 MICROFABRICATION PRODUCTS

About 15 years ago, the mass fabrication of microdevices such as pressure sensors, microvalves, and scanning tunneling microscope (STM) tips was started. Soon, more complex structures like accelerometers, angular rate sensors (gyros), scanning mirrors, micromirror arrays, microspectrometers, R/W heads for hard-disk drives, ink-jet print heads, and RF microrelays, to name only a few, have been developed and are produced in large quantities using microfabrication. An incomplete list of the fields of application of microdevices includes automotive, information technology, telecommunications, medical and biomedical, household appliances, environmental control, aerospace, and defense.

In addition to the term MEMS, categories like microoptoelectromechanical systems (MOEMS), radio-frequency MEMS (RF-MEMS), and MEMS for medical or biomedical application (BioMEMS) have been established. The market for microfabricated devices is still in its infancy and is growing with rates comparable to the first years of microelectronics. According to one market research report, the sales of microfabricated systems was $12 billion in 2004 and is expected to grow with a CAGR of 16% to $25 billion in 2009 [6]. Recent extensions of fields of application are consumer, entertainment, and homeland security. Upcoming new devices are MEMS microphones, micro-energy sources, micropumps, chip coolers, and micromachined wafer probes, and this will definitely not be the end of new developments.

Although more related to microelectronics and chip packaging, chip stacking or 3D integration will also be discussed in this microfabrication chapter. In 3D integration, thinned wafers or chips are stacked and wired, for example, by so-called vertical interconnects. By stacking and connecting 2–8 chips, complex systems with a high packing density are formed, which can be mounted into discrete packages or even directly into mobile devices.

In order to be protected against environmental interference, microelectronic or microfabricated chips must be enclosed into specific packages. For some micromechanical sensors, cap wafers with etched cavities are attached hermetically onto the device wafers using anodic, eutectic, or glass frit bonding.

To form effective bonding between these wafers, very smooth surfaces are required, which are produced by means of CMP. After the wafer bonding and singulation of the dies, the fragile structures are protected, sometimes even under high-vacuum conditions, and can then be integrated into the system.

14.4 CMP REQUIREMENTS IN COMPARISON WITH IC FABRICATION

The advancements of microelectronics with its increasing device performance and decreasing structure dimensions, which recently fell below the 100-nm mark, follow the path given by Moore's law. In contrast, microfabrication deals with a broad range of structure dimensions between submicrometers and millimeters [7]. The main developments in CMP are traditionally connected to those in advanced IC manufacturing. In recent years, however, the CMP equipment and consumable communities have also paid close attention to microfabrication. It is recognized that MEMS-specific CMP processes require dedicated consumables [8] and optimized tool sets [9] due to their diversity in structure dimension and materials to be processed.

The differences and similarities in the requirements of CMP for microelectronics manufacturing and microfabrication are summarized in Table 14.1. This comparison gives an overview of the parameters relevant to CMP and is intended to serve as a guideline for CMP practitioners, consumable suppliers, equipment manufacturers, and others. Polishing tools and consumable sets capable of handling MEMS-specific tasks will be discussed at the end of this section. The following is a list of these terms for comparison.

Layers: Typical materials for which CMP processes originally have been developed for microelectronic applications include various types of silicon dioxide such as thermal oxide, TEOS, HDP, BPSG, and other B- or P-doped oxide films. These films are used for various isolation purposes including interlevel dielectric (ILD), intermetal dielectric (IMD), or shallow trench isolation (STI). In addition, n- or p-doped poly-Si, which is a semiconducting material used as capacitor electrode material for DRAMS or gate electrode for MOS applications (CMOS as well as power MOS devices), also has to be polished. Metals for which CMP processes have emerged over the last 10–15 years are W for vertical interconnects (vias) and most importantly Cu as a low-resistivity replacement for aluminum interconnects, employed in the damascene or dual-damascene processing scheme. Other metals that are required for future nonvolatile memories are noble metals like Pt or Ir for which CMP processes have been explored.

Microfabrication to a large extent relies on microelectronic processing technologies. Therefore, all aforementioned materials have found their way into various microsystem applications. Beyond that, other metals to be planarized like Ni or alloys like NiFe are regularly used for applications such as hard-disk drives. In microfabrication, the spectrum of materials is extended by including

TABLE 14.1 Comparison of CMP Requirements for Microelectronics Manufacturing with Microfabrication.

Parameters	Microelectronics	Microfabrication	Comparison	Comments
Layers	Various SiO_2, poly-Si, W, Cu, noble metals	Various SiO_2, poly-Si, Si, various metals, ceramics, polymers	Great variety	Upcoming materials: PZT, magnetic alloys
Typical layer thickness	~0.1–2.0 μm	~0.2–50 μm	Larger	
Required CMP removal	~0.1–1.5 μm	~0.1–20 μm	Larger	
Required polishing rate	0.2–0.5 μm/min	>0.5 μm/min	Higher	New consumable sets, rougher processes required. Alternative: grind/polish sequence
Nonuniformity	<2%	1–5%	Relaxed	Exception: functional layer defines device properties. dependency on deposition NU
Edge exclusion	Down to 1.5 mm	Typ. 6 mm	Relaxed	Due to use of older equipment generations
CMP tool throughput	20–50 wafers/h	<10 wafers/h	Relaxed	
Pattern width	Down to 65 nm	μm–mm	Larger	
Lithography	Optical, 193 nm, DOF 0.5 μm	Optical, g- or i-line, DOF 5 μm e-beam, X-ray, DOF 5–50 μm	Specific	
Required planarity	<100 nm	>100 nm	Relaxed	Exception: optoMEMS: $<\lambda/10$
Topography	Maximum 1 μm	1–10 μm	Larger	
Dishing	<50 nm	>100 nm	Relaxed	Problem: mm-scale patterns
Erosion	<50 nm	>100 nm	Relaxed	Depending on pattern density
Roughness	Some nm	10 nm	Relaxed	Exception: wafer bonding ≈0.5 nm
Particle size	>32 nm	Pattern-size dependent	Relatively uncritical	

(Continued)

Table 14.1 (*Continued*)

Parameters	Microelectronics	Microfabrication	Comparison	Comments
Particle density	$<0.02/cm^2$	Die-size dependent	Relatively uncritical	Exception: wafer bonding $<1/cm^2$
Surface contamination	$<1 \times 10^{10}/cm^2$	CMOS: $<5 \times 10/cm^2$ Mechanical devices: product specific	Uncritical	Wafer bonding and BioMEMS need "clean surface"
Substrates	Si, SOI, III–V semiconductors	Si, metal, glass, quartz, ceramics	Greater variety	
Substrate size	200–300 mm	100—max 200 mm	Smaller	Production: lower unit numbers
Substrate thickness	Production: 500–800 μm Finishing: down to 50 μm	production: 200–700 μm Finishing: down to 20 μm	Specific	

various ceramic or polymeric materials. For example, some polyimide or hardened photoresists that are suitable for CMP processes have been reported to serve as a sacrificial layer [10]. As a matter of fact, a large number of new materials have been tested in laboratories for a wide range of new applications [11]. A recent development is the use of micropatterned lead zirconium titanate (PZT) for actuators due to the piezoelectric property of the material.

Layer thickness: The typical film thickness used by the microelectronics community lies in the range between about 0.1 and 2.0 μm, which is relatively small in comparison to microfabrication, where layers of up to 50 μm or more have to be deposited and structured. In future, film thickness in microelectronics will continue to decrease with increasing number of stacked layers, whereas in microfabrication the range of thickness will increase in both large and small vertical dimensions.

Required CMP removal: The difference in film thickness is strongly connected to the amount of material that has to be removed by CMP. In certain cases, material in the range of 10 μm thickness or more has to be abraded. The material removal at this magnitude is often necessary for microfabrication. One such example is ILD planarization on the terminal metal level due to topographical and depth-of-field requirements set by postfab processing [7]. Other examples include planarization of optical waveguides, where doped oxide strips with a certain refractive index have to be embedded in thick oxide layers or metal plugs serving as chip feedthroughs used in chip stacking (3D-integration), where metal layers with a thickness on the order of the plug diameter have to be removed [12]. Removing a very thick layer has consequences on the requirement of polishing rate.

Required polishing rate: In microelectronics CMP, a removal rate of several hundred nanometers per minute is often adequate and desirable in order to precisely control the process. In microfabrication, a removal rate of more than 0.5 μm/min is often a minimum in order to achieve the CMP targets within reasonable processing times. This demand for higher removal rate can be fulfilled by either increasing the aggressiveness of slurry chemistry or changing the mechanical components of the processing conditions such as higher downforce, or greater chuck and/or platen speeds. In addition, when possible, the insertion of a pre-CMP sequence consisting of rough/fine grinding may also be considered. Some grinding processes can have removal rates greater than 100 μm/min with high homogeneity, depending on the material to be polished. Afterwards, CMP removes the grinding damage and produces a surface with desirable smoothness.

In order to keep the polishing rate constant during the long CMP steps, *in situ* conditioning of the polishing pad using diamond wheels is often a characteristic of a microfabrication process. To control the process more precisely, an end-point detection system that can measure the remaining layer thickness is extremely useful.

Nonuniformity (NU): Generally, the NU specifications in microelectronics are more stringent than those in microfabrication in order to achieve high yield.

Mechanical structures are more robust with respect to thickness variations, that is, the NU requirements are relaxed. As an exception, NU specs could be much more stringent when functional layers of the microsystem have influence on device properties (see, for example, Section 14.5.2). The final layer NU after CMP strongly depends on the NU of the layer deposition process, although homogeneity deviations can partly be compensated by adjustment of the polishing process (backside pressure, chuck/platen speeds). Some strain in deposited layers might lead to increased bow and warp of the substrates. This has to be taken into account when optimizing the CMP process as wafer bow can only partly be flattened by the downforce of the polishing chuck against the pad. Concave surfaces to be polished can be compensated by backside pressure; however, there is the danger that strongly bowed wafers are lost during the polishing process due to slipping of the wafer underneath the retaining ring.

Edge exclusion: In order to increase the number of dies on a wafer, the edge exclusion zone, that is, the area at the edge of the substrate with undefined thickness or pattern specifications, is reduced continuously. For 300-mm wafers, the edge exclusion specifications for IC applications are well under 3mm and are about to hit the 1.5-mm mark. The minimum edge exclusion width depends significantly on wafer geometry (edge roll-off) and chuck design. This can partly be controlled by advanced chuck concepts such as zone-controlled chucks. Microfabrication today relies mainly on older manufacturing equipment with typical 100–150mm wafer size and edge exclusion of 5–6mm.

Throughput: Due to the high productivity demands in microelectronics, the typical throughput of the manufacturing tools lies between 20 and 50 wafers/h. Larger vertical dimensions in microfabrication ease the throughput requirements to $\leq$10 wafers/h.

Pattern width: Following Moore's law, the pattern dimensions in microelectronics design are decreased by a factor of 0.7 in every 2–3 years and will reach the 65-nm node in production in 2007. In contrast, the pattern dimensions in microfabrication are much larger and lie in the range between several micrometers and several millimeters. Planarization of larger patterns poses a major challenge, as the planarization lengths typically obtained during IC CMP are on the order of several millimeters and are only sufficient for small structure widths. The work on the adjustment of existing CMP models to include larger features has started recently [13]. New consumables such as fixed abrasive pads are promising in solving the problems.

Lithography: In order to precisely resolve the nanometer structures in microelectronics, various enhancement techniques have been applied to the current optical exposure tools that are equipped with deep UV light (193nm wavelength). These enhancement techniques include phase-shift masks and immersion lenses (putting a liquid between final lens of the stepper and the wafer). The trade-off for the high resolution of modern steppers is an extremely small depth of focus (DOF) that is around 0.5µm over a typical field size of

20mm × 20mm. This was one of the main reasons for the introduction of planarization by means of CMP to the IC industry at the beginning of the 1990s. Microfabrication, on the contrary, can rely on much more robust 1:1 optical proximity lithography or g- or i-line steppers with DOF values of up to 5 μm. The challenge of microfabrication for certain devices is a very strong CD control relative to the larger pattern dimensions. For certain high-resolution applications, microfabrication can use e-beam, ion-beam, or X-ray lithography with relaxed DOF of 5–50 μm.

Required planarity: The planarity requirements to fulfill the extremely strong DOF demands in microelectronics can only be achieved by employing CMP for planarization. In microfabrication, CMP is used when high topography has to be smoothed, usually with much relaxed planarity requirements. One exception is the manufacturing of optical microsystems (optoMEMS, MOEMS), where the surface planarity demands as a rule of thumb are typically less than a tenth of the wavelength ($<\lambda/10$).

Topography: Due to the smaller layer thickness values in microelectronics, the topography evolving from patterning processes is typically <1 μm, which can be planarized easily with proven CMP processes. The topography in microfabrication can be much larger due to the thicker structured layers underneath. A planarization process for a topography of up to 10 μm requires thicker sacrificial material to be deposited, a soft ramp of the downforce in order to avoid sticking of protruding structures in the pad with the associated danger of wafer loss, longer polishing times and, in the case of larger lateral dimensions, a larger planarization length of the consumable set/process. The danger of pattern delamination is increased due to shear forces being effective on protruding features and has to be taken into account by employing optimized glue layers.

Dishing: A difference in the hardness of materials to be polished such as those during metal CMP might lead to dishing, that is, a larger abrasion in the middle of larger structures due to the sinking of the deformable pad. This might lead to a variation in the cross-sectional area of metal lines and is specified to be <50 nm for advanced microelectronic circuits [14]. This requirement is being constantly revised and getting stronger for each generation of technology node. Dishing is much more severe at microfabrication due to the larger lateral dimensions up to the millimeter scale. It can be compensated by intentionally designed support pillar structures, also known as cheesing. The maximum allowable dishing depends on the application and has to be smaller for microoptical devices than for some mechanical structures.

Erosion: Dielectric erosion depends on pattern density and has to be kept below 50 nm in microelectronic array structures [14]. Erosion can also occur in microfabrication when pattern arrays have to be planarized. Due to larger vertical and lateral dimensions, the requirements are relaxed compared to microelectronics.

Roughness: CMP of silicon oxide leads to roughness values (R_{rms}) well below 1 nm depending on the abrasive used. The roughness of some metals can be on the order of several nanometers for some CMP processes. This is sufficient for today's minimal structure dimensions of 65nm. The requirements might get tighter in the future when the resistance increases in small copper lines due to the fact that electron scattering at the polished surface is no longer tolerable. In microfabrication, larger values are allowed with the exception of wafer bonding, where smooth surfaces with $R_{rms} \approx 0.5$ nm are required. This is also true for optical structures for MOEMS devices.

Residue particles: In CMP for microelectronics applications, residue particles larger than half of the minimal linewidth, that is, ≥ 32 nm for the 65-nm node, are considered as yield-limiting defects ("killer defects"). Most sensitive surface particle counters today using UV laser illumination claim to resolve 30-nm particles (under well-defined test conditions) [15], which suffices for current device generations. In microfabrication, the residue particle size requirements depend on pattern size, but can be regarded as relatively uncritical as long as the surface is thoroughly cleaned with a well-developed post-CMP cleaning process.

The post-CMP cleaning process can be the same as in microelectronics because the planarized surfaces for microfabrication are relatively robust against the physical forces of brush scrubbing or megasonic cleaning. Depending on the materials exposed, chemical additives such as diluted ammonia, diluted ammonia/peroxide mixtures (SC1), weak organic acids, or surfactants can help to improve particle removal. Sensitive structures, for example, tilting mirrors or suspended moving masses, are typically released at the end of the process flow and are therefore not subject to the shear forces of CMP and post-CMP cleaning.

Residue particle density: Advanced post-CMP cleaning processes typically fulfill the killer defect density requirements of $<0.02/\text{cm}^2$ for microelectronic applications [14]. The defect density in microfabrication is relatively uncritical with the exception of wafer bonding, where a particle density $<1/\text{cm}^2$ is necessary for reliable bonds.

Surface contamination: Residual metals on the wafer surface after CMP and post-CMP cleaning have a large impact on device properties in microelectronics. Depending on the stage of manufacturing (e.g., transistor definition versus formation of interconnect levels), the surface contamination requirements vary. For the earlier manufacturing steps (e.g., STI CMP), metal contamination should be $<1 \times 10^{10}\text{cm}^2$. If microsystems with on-chip CMOS control circuits (integrated MEMS) are to be manufactured, contamination requirements depend on the generation of the employed technology, but are typically relaxed compared to advanced ICs ($\sim <5 \times 10^{10}\text{cm}^2$). The requirements for mechanical devices are product specific but can be regarded as uncritical. Wafer bonding processes and bio- or biomedical chips generally need clean, fresh surfaces. Critical contaminant can be organic residues after CMP and post-CMP cleaning, as they can influence the sticking of subsequently deposited films. Organic residues include corrosion inhibitors after Cu CMP (BTA), surfactants in the slurries, and adsorbed hydrocarbons from the environment after prolonged process flow interruptions.

Substrates: The substrates in microelectronics are mainly Si wafers. For mobile applications, silicon-on-insulator (SOI) wafers increasingly replace bulk Si wafers and for very specific high-frequency applications, III–V compound semiconductors (e.g., GaAs) are used. The majority of substrates in microfabrication are Si wafers, but metal, glass, and ceramic substrates are also common. Particularly when using glass, quartz, and ceramic wafers in CMP processes, it has to be taken into account that they are brittle and easy to break. The situation is worse when the material is also under stress induced by deposited layers. For applications where the backside of the wafer has to be structured (e.g., in bulk micromachining), double-side polished substrates are employed.

Substrate size: In microelectronics production, large wafers with diameters of 200 or 300mm are used for economical reasons. The discussion of an introduction of 450-mm wafers started recently [14]. Typical substrate sizes in microfabrication are 100–150mm, which is mainly due to the use of older generation manufacturing equipment. The smaller surface area is still adequate for a small-scale microdevice production, compared to the chip industry. Only a few MEMS fabs are using 200-mm Si wafers today, mainly due to integration with advanced CMOS circuits that are fabricated only on this larger wafer size. Sometimes, in proof-of-concept or feasibility studies only coupons have to be polished, where the achievement of good nonuniformity can be quite demanding [16].

Substrate thickness: Today, most of the chips are thinned after manufacturing, for advanced mobile applications even down to 50µm, by using backside grinding or etching. Nevertheless, the substrate thickness during manufacturing lies between 500 and 800µm due to an increasing risk of wafer breakage with thinner substrates. The substrate thickness is a compromise between material cost and mechanical yield. Very thin wafers that are backside polished for stress relief after grinding have to be attached to a support plate in order to reduce risk of breakage. In microfabrication, thinner substrates are sometimes used to lower the costs of substrate materials, for example, single-crystalline $LiTaO_3$ or $LiNbO_3$ (thickness down to 200µm), for surface acoustic wave (SAW) filter applications. Final Si wafer thinning by grinding and polishing to 5µm for 3D integration (chip stacking) has been reported recently [17].

Equipment: CMP tools for microfabrication are not covered by the large equipment manufacturers as it is still regarded as a niche market. Therefore, the supply situation for MEMS CMP tools consists not only of older, used polishers for small wafer sizes but also of specialized, highly MEMS dedicated polishers and cleaners from smaller companies. The bandwidth varies between simple tabletop polishers for R&D and fully automated CMP cluster tools for production. As the process requirements are equal to or even exceed those of microelectronics, simple polishers or refurbished older generation CMP equipment often cannot meet the demands of the planarization process for microfabrication.

European suppliers are Logitech (United Kingdom) and Presi (France). These suppliers concentrate on tabletop polishers and stand-alone tools for sample sizes between coupons and 200mm. The Mecapol series from Presi and the CDP and CP series of Logitech are manually loaded polishers for delayering and planarization, which need additional post-CMP cleaning equipment. In contrast, the German tool manufacturer Peter Wolters offers, with their PM200, a modular CMP cluster tool, which can be configured with up to two independent polishers, two cleaners, and automatic wafer handling capabilities. The cluster can handle all wafer sizes between 100 and 200mm and is capable for *ex situ* and *in situ* conditioning. Due to its modularity, the cluster can be upgraded from a MEMS R&D tool to a dedicated system for high-volume MEMS production. For end-point control, torque can be measured by means of the motor current of the polishing table.

In Asia, the Korean company G&P Technology offers stand-alone polishers for MEMS under the GNP POLI brand. In the United States, Strasbaugh offers two CMP tools capable of meeting microfabrication needs, the nSpire R&D CMP system (model 6EC) and the nTegrity production CMP system (model 6DS-SP), both for 100–200mm wafer size. The wet-out systems need additional cleaning modules. *Ex situ* and *in situ* pad conditioning is included. Various kinds of polishing heads as well as optical and motor current end-point systems are available. Links to some equipment manufacturers are provided in the reference section [18].

Materials: Over the years, the consumable suppliers, especially the slurry manufacturers, have developed dedicated materials for the needs of MEMS and microfabrication. Often, slurries optimized for microelectronics manufacturing or modifications also do a good job in the world of MEMS. Based on the key ingredients used for dielectric CMP - alkaline dispersions of silica abrasives- or metal CMP solutions of abrasives (silica, alumina) with oxidizers, inhibitors, complexing agents, and surfactants, custom-specific polishing slurries can be formulated for microfabrication. Polishing pads are generally the same as used for glass polishing, wafer manufacturing, or microelectronics planarization.

Table 14.2 gives an overview of various pads and slurries, suitable for microfabrication. The compiled list is based on the information obtained from some consumable manufacturers as well as the experience accumulated at the author's lab as of Spring 2006. Besides the large suppliers like Cabot Microelectronics, Fujimi, and Rohm & Haas, smaller companies are concentrating on custom-specific developments such as Kemesys in France as well as suppliers of abrasives or dispersions, for example, Degussa AG or H.C. Starck, both from Germany. Links to some of these suppliers are provided in the reference section [19].

14.5 EXAMPLES OF CMP APPLICATIONS FOR MICROFABRICATION

CMP has been used for the mass production of microsystems for a number of years now. Among many representative applications, a limited number of examples in which CMP processes were employed have been selected here to illustrate the use of this technology and related issues. These examples cover

TABLE 14.2 Pads and Slurries for CMP of Characteristic Materials in Microfabrication.

Material	Pads	Reference	Slurries	Reference	Remarks
Si oxides (SiO_2, TEOS, PETEOS, doped oxides)	IC1000™ Series, Politex™ Prima	Rohm & Haas	Klebosol® 1501-50, Klebosol® 1508-50	Rohm & Haas	
			EP-MD-8052, EP-MD-8073	Cabot	Fumed silica, for stack heights $\leq 1\,\mu m$
			EP-MD-8080, EP-MD-8090	Microelectronics	Colloidal silica, for larger stack heights
			Planerlite-4000 Series	Fujimi	
			Levasil 50CK/30%, Levasil 50CN/30%, Levasil 100K/30%	H.C. Starck	
			SS 25 from Cabot	Fraunhofer ISIT	
SiN	Suba™ 500, Politex™ Regular	Rohm & Haas	Nalco® 2360	Rohm & Haas	
Al_2O_3	MH Series	Rohm & Haas	Eminess UltraSol™ M5D4 (main) Eminess UltraSol™ 500 (final)	Rohm & Haas	
Saphire	IC1000™ Perforated, Suba™ 1250, Politex™ Regular	Rohm & Haas	$6\,\mu m/3\,\mu m$ diamond (1) $1\,\mu m$ diamond (2) UltraSol 500 (3)	Rohm & Haas	
SiC	IC1000™ Perforated, Suba™ 1250, Politex™	Rohm & Haas	Nalco 2360	Rohm & Haas	
Silicon	Suba Series, SPM pads	Fraunhofer ISIT	Glanzox series	Fujimi	

(*Continued*)

Table 14.2 (*Continued*)

Material	Pads	Reference	Slurries	Reference	Remarks
			Levasil 50/50%, Levasil 100/45%, Levasil 200/ 30%-WALM	H.C. Starck	
	Fixed abrasive pads	VTT, 3M			
Polysilicon	IC1000™ Series, Politex™ Prima, WWP3000	Rohm & Haas	Klebosol® PL 1509-35, Klebosol® PL 1509-12, Klebosol® PL 1618	Rohm & Haas	
			SS 25 from Cabot	Fraunhofer ISIT	
			Planerlite-6000 Series	Fujimi	
			EP-MP-8010	Cabot Microelectronics	For stack heights ≤1 μm
			EP-MP-8188		For larger stack height removal
Germanium	Suba™ X, Politex™ embossed	Rohm & Haas	Nalco® 2360	Rohm & Haas	
SiGe	Politex™ Regular, SPM3100	Rohm & Haas			
Metal			Levasil 50CK/30%-V1, Levasil 50CK/30%-V2	H.C. Starck	Abrasive raw material for metal slurry formulation
Copper	IC1000™ Series, VisionPad™ Series, Politex™ Prima	Rohm & Haas	EPL2361	Rohm & Haas	
			EP-MH-870, EP-MH-883	Cabot Microelectronics	For stack height ≤5 μm removal
			EP-MM-8042, EP-MM-8043		High rate slurry, for stack heights ≥5 μm
			Planerlite-7000 Series	Fujimi	

			iCue® Slurries from Cabot	Fraunhofer ISIT	
Tantalum, TaN, TiN	IC1000™ Series, Politex™ Regular	Rohm & Haas	CUS Series, LK Series, SSA Series	Rohm & Haas	
Tungsten	IC1000™ Series, Politex™ Series, FBP Series	Rohm & Haas	W2000 from Cabot	Fraunhofer ISIT	
NiFe	IC1000™ Series, Politex™ Prima, WWP3000	Rohm & Haas	MSW1500/2000	Rohm & Haas	
ZnCu	IC1000™ Series	Rohm & Haas	MSW1500	Rohm & Haas	
Compound semiconductors			Compol Series	Fujimi	
			Levasil 50/50%, Levasil 100/45%	H.C. Starck	Abrasive raw material for slurry formulation
AlN	Suba™ 500	Rohm & Haas	Nalco® 2370	Rohm & Haas	
GaAs	Suba™ X embossed, Suba™ 500, OPC6350, SPM3100	Rohm & Haas	Nalco® 2360	Rohm & Haas	
InP	Suba™ 500, Suba™ X, SPM3100	Rohm & Haas	Nalco® 2360 w/bleach	Rohm & Haas	
$LiNbO_3$	Suba™ 1250, Politex™ embossed	Rohm & Haas			
Glass	MH Series, Suba™ X, GS	Rohm & Haas	Ceria based (main) Nalco® 2360 (final)	Rohm & Haas	
Polyimide			EP-MI-8005	Cabot Microelectronics	Alumina-based slurry
			EP-MI-8006		Alumina based, low defects

aspects such as bulk and surface micromachining, dielectric and poly-Si CMP, as well as the employment of metal CMP for the planarization of sacrificial layers and the formation of damascene structures.

14.5.1 Case Study I: Integrated Pressure Sensor

The term integrated sensor describes the technology when MEMS structures and signal processing circuits are combined on the same chip. For the fabrication of an integrated pressure sensor, a CMOS circuit for signal processing had first been produced on the wafer front side. As shown schematically in Fig. 14.1, a silicon nitride membrane structure on the front side has to be opened subsequently by a selective silicon wet etching process from the wafer backside using KOH. The size of the membrane is defined by the size of the opening in the SiN_x etch mask and the thickness of the substrate, taking into account that anisotropic KOH etching of a $\langle 100 \rangle$ Si surface does not lead to perpendicular sidewalls but to the formation of $\langle 111 \rangle$ crystal planes with an angle of 54.7° to $\langle 100 \rangle$. Due to a finite etch rate selectivity between $\langle 100 \rangle$ and $\langle 111 \rangle$ planes, a mask undercut has to be expected.

The lithography for the definition of the SiN_x etch mask opening and the subsequent RIE step requires a polished wafer backside. This can be accomplished by the use of double-side polished wafers that, however, show some disadvantages like handling and chucking problems with standard semiconductor manufacturing equipment. Therefore, the CMOS circuit has been processed using standard 150-mm wafers with unpolished backside. Afterwards, the substrates were thinned to the required thickness by grinding and subsequently smoothed by CMP.

After front-side protection of the finished CMOS circuits by means of grinding tape or photoresist, the wafers were ground from 675 to 503 μm thickness using a two-step grinding process with #300 and #2000 grinding wheels. As the wafer thickness defines the size of the membranes, the uniformity of the grinding process must not exceed a total thickness variation (TTV) of 3.0 μm in the case of this example. A subsequent two-step Si CMP process using colloidal silica with an additive in order to enhance polishing

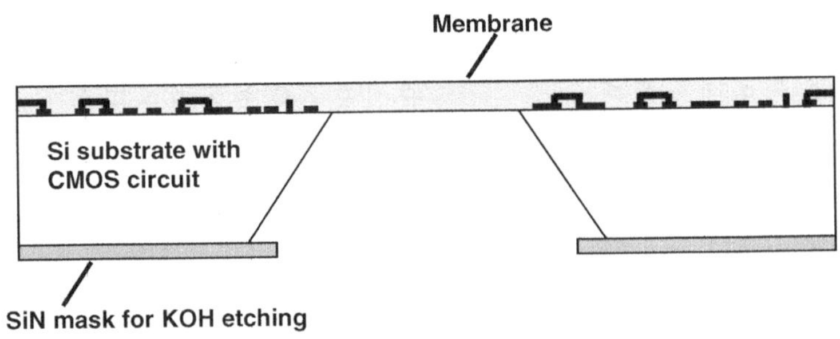

FIGURE 14.1 Schematic of integrated pressure sensor in bulk micromachining.

rate, which had originally been developed for Si wafer manufacturing, removed 3.0–3.5 μm of silicon from the wafer backside. The CMP step slightly decreased the excellent mean TTV = 2.1 μm after grinding to TTV = 2.5 μm, measured over 50 wafers, which still meets the spec.

The removal of about 3 μm by CMP led to a sufficient clearing of the subsurface damage of the grinding process, indicated by minimal undercut variation of the following KOH anisotropic etch. The necessary damage layer removal of about 3 μm by polishing corresponds well with the results from other groups developing technologies for wafer thinning, chip stacking, and packaging [20].

Polishing and cleaning quality has been controlled with a laser surface particle counter (Surfscan) by using the haze measurement option, which is a measure of surface microroughness. After removal of the front-side protection, deposition and structuring of the SiN_x mask, and exposure of the pressure sensor membranes by KOH etching, the wafers have to be sealed hermetically with cap wafers using wafer bonding. Finally, the chips are separated by wafer sawing.

In addition to integrated pressure sensors, the described technology for the formation of membranes by bulk micromachining can also be used for the fabrication of integrated airflow sensors, MEMS microphones, and other related devices. The represented grinding/polishing sequence can also be applied for the thinning of completed device wafers for very thin packages, chip stacking or 3D integration with substrate thickness well below 100 μm. In all cases, the grinding-induced subsurface damage of the crystalline material can be successfully removed by CMP. Besides conventional CMP, the use of fixed abrasive pads to smooth and planarize ground substrates for, for example, thick-film SOI, capping of bulk-micromachined silicon, or surface preparation for bonding of different substrates, has been investigated [21].

14.5.2 Case Study II: Poly-Si Surface Micromachining and Angular Rate Sensor

One of the earliest applications of CMP for microfabrication was the pioneering work at Sandia with the planarization of polysilicon for poly-Si surface micromachining [22]. The motivation for CMP was the elimination of topography caused by the stacking of several structured layers. The Sandia Ultraplanar, Multilevel MEMS Technology (SUMMiT^{TM}) fabrication process with up to four CMP-planarized mechanical layers has been successfully used to build MEMS gear systems, planar pressure sensors, and other devices employing high-aspect-ratio micromachining.

Figures 14.2 and 14.3 show closeups of poly-Si micromachines from Sandia Labs, manufactured by using several poly-Si CMP processes. Figure 14.2 depicts the gear wheels of a 10:1 transmission. Due to CMP, the gears are completely planar. Figure 14.3 shows the alignment clips for positioning the gears. The two layers of gears are clearly visible. More examples that show where the five-layer surface micromachining process has been employed are

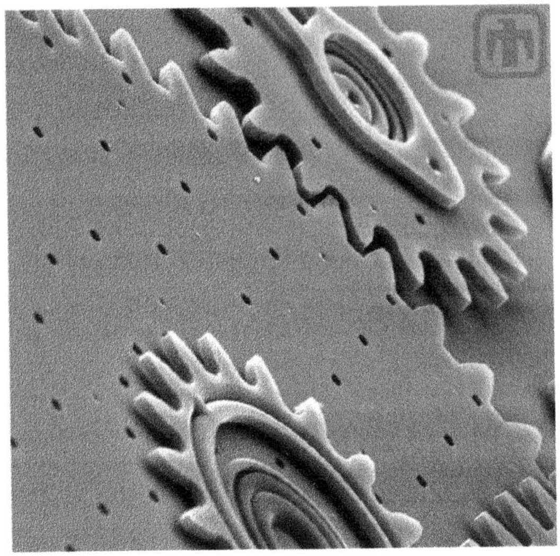

FIGURE 14.2 Poly-Si gear wheels of a 10:1 transmission, fabricated with the SUMMiT™ process. Courtesy Sandia National Laboratories, SUMMiT™ Technologies, www.mems.sandia.gov.

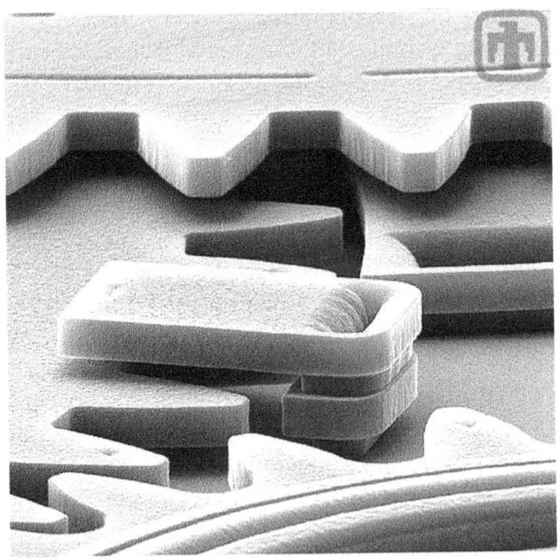

FIGURE 14.3 Alignment clips for positioning the gears. The several poly-Si layers of the SUMMiT™ process are clearly visible. Courtesy Sandia National Laboratories, SUMMiT™ Technologies, www.mems.sandia.gov.

found on Sandia's Web site: www.mems.sandia.gov. Investigations of a South Korean group on the optimization of slurries for polysilicon CMP with respect to selectivities and dishing have been published recently [23].

By using the example of an angular rate sensor (gyro), developed and fabricated at the author's lab, the requirements of a CMP process for polysilicon micromachining will be discussed in detail in the following. Gyros are employed in growing quantities for automotive applications like GPS navigation systems or antiskid and rollover protection systems, and also in cameras for image stabilization or in hard-disk drives for read/write head protection.

One of the various possible concepts for the measurement of angular rates is the use of a rotational Coriolis force sensor. A photograph of the gyro structure is shown in Fig. 14.4. The sensor consists of a ring/disk structure that is performing circular vibrations around an axis perpendicular to the picture plane with frequency f_{drive}. The stimulation takes place capacitively by applying an alternating drive voltage at the comb-like drive structures, which are shown in detail in Fig. 14.5. An angular movement around a second axis, which lies in the ring/disk plane, leads to a deflection of the vibrating structure perpendicular to both axes due to the Coriolis force. The ring/disk structure starts to stagger with a frequency f_{sense} and the amplitude is measured capacitively. In order to increase sensitivity, the system is excited under resonance conditions. The final sensor has to be packaged under vacuum conditions in order to obtain a high quality factor Q. Wafer-scale packaging is accomplished by bonding the sensor wafer against a structured wafer with cavities, which are filled with a structured getter material in order to keep high vacuum over the complete lifetime of about 15 years.

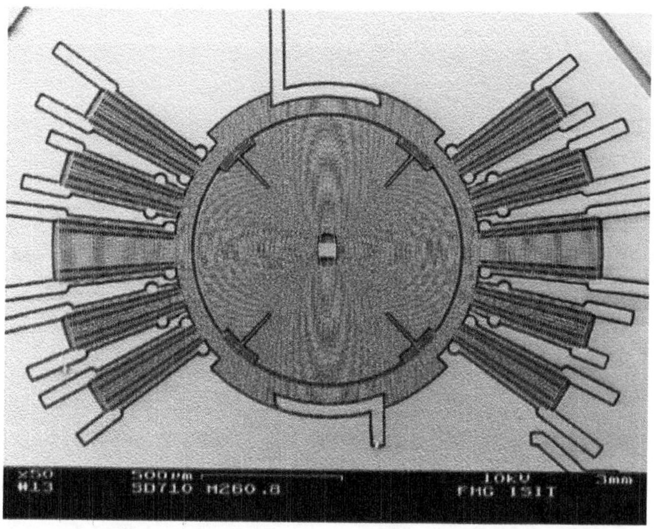

FIGURE 14.4 Structure of the angular rate sensor (gyro), fabricated in poly-Si micromachining.

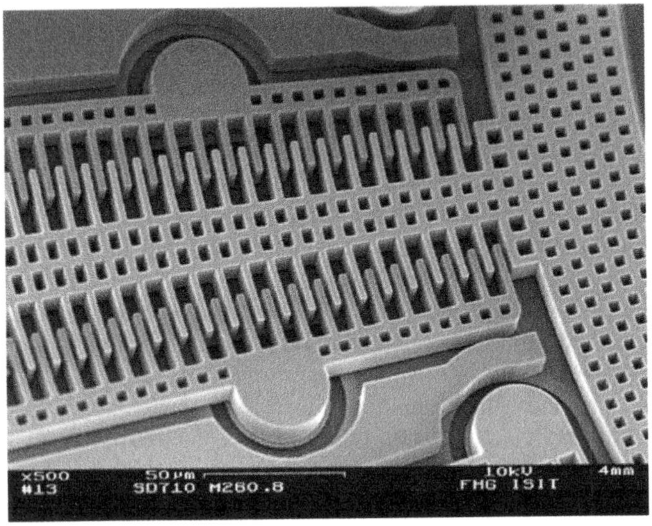

FIGURE 14.5 Detail of the drive structure of the angular rate sensor (gyro).

The vibrating ring/disk structure as well as the drive mechanism consists of 11-μm-thick poly-Si, which has been structured by deep RIE and released from the sacrificial oxide layer underneath by HF vapor phase etching. For the deposition of the thick poly-Si, a modified epitaxy deposition process (EPI poly) has been used [24]. However, as can be seen in Fig. 14.6, the deposition process leads to a rough poly-Si surface with $R_a \approx 100\,\text{nm}$. For the removal of underlying topography, the surface has to be planarized by CMP in order to

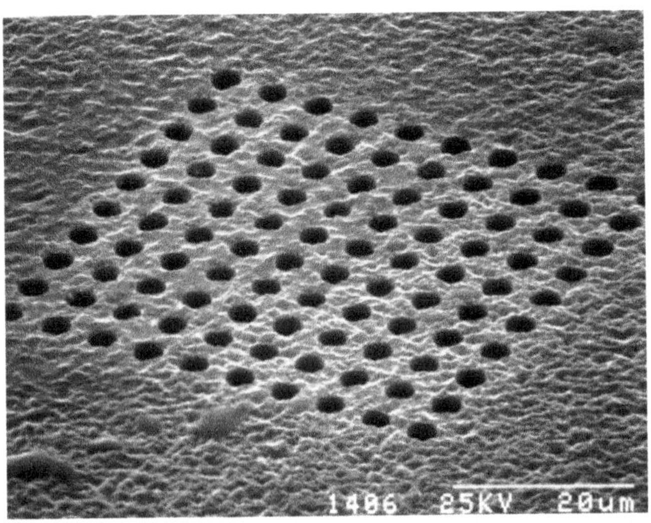

FIGURE 14.6 Roughness of a 10-μm-thick EPI-poly layer.

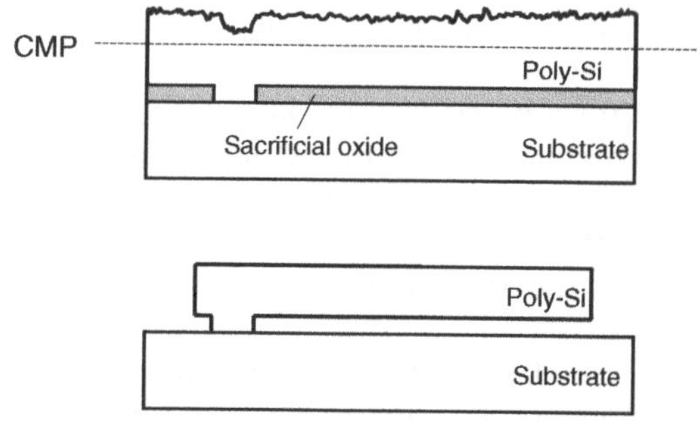

FIGURE 14.7 The roughness of the EPI-poly layer can be planarized by using CMP. After removal of the sacrificial oxide layer by vapor phase etching, the sensor structures are released.

enable high-resolution lithography and etching. Figure 14.7 shows schematically the roughness removal by means of CMP and the released structures after HF vapor-phase sacrificial oxide etching. The polysilicon thickness after CMP and its homogeneity over the wafer is very critical, as it has direct influence on sensitivity, measurement range, and total yield.

Simulations of the angular rate sensor have proved that the required specifications of the gyro can only be fulfilled when the difference between f_{drive} and f_{sense} is less than 300 ± 125 Hz. The dependence of drive frequency and sense frequency on EPI poly-Si thickness is shown in Fig. 14.8. f_{drive} is nearly

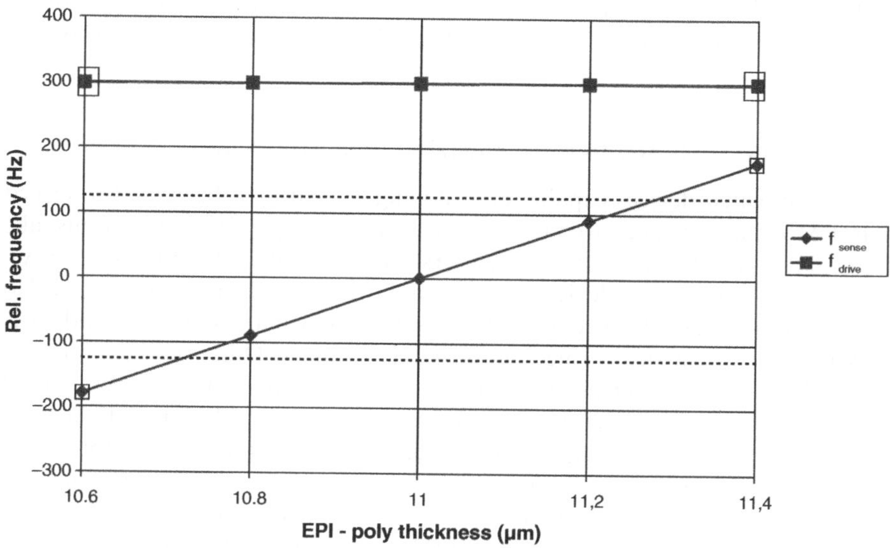

FIGURE 14.8 Simulation of angular rate sensor: drive and sense frequencies versus EPI-poly thickness. Specifications of the gyro can only be fulfilled for $\Delta f = 300 \pm 125$ Hz; that is, EPI-poly thickness after CMP has to be 11.0 ± 0.27 μm.

independent of the thickness of the ring/disk sensor structure, whereas f_{sense} shows a linear increase with EPI thickness. The specified frequency difference can only be achieved for a polysilicon thickness of 11.0 μm within a thickness range of ±270 nm. This means that the requirements of the deposition and polishing process are quite ambitious. The nonuniformity of the 11.0-μm-thick poly-Si layer after CMP has to be in the range of ±2.5% (3σ), which is much more demanding than the typical nonuniformity specs of ±2.0% (1σ) in microelectronics.

The poly-Si thickness of the EPI-poly deposition process is a compromise between underlying topography, processing time, and nonuniformity of the deposition and polishing processes. A reasonable start thickness is 14 μm; that is, 3 μm has to be removed by CMP. For the poly-Si CMP process, a fumed-silica based alkaline slurry has been used which achieved a removal rate of about 500 nm/min and leads to a final roughness of $R_a \approx 0.3$–0.5 nm. The thickness is measured by using an adapted reflectometer that, however, requires a smooth surface. Therefore, the control wafers had to be prepolished before they could be used for removal rate determination.

The CMP task is a so-called blind polish process; that is, the process has to be terminated at a given thickness. In our case, it is assumed that the EPI-poly thickness is repeatable from wafer to wafer. The polishing time is determined with control wafers and preset. Every wafer of a production batch is measured after CMP. If the thickness results run out of spec, the polishing time has to be adjusted for the following wafer (closed-loop control). By combining uniformity profiles of EPI-poly deposition and poly-Si CMP, sensor structures with a final thickness of 11.0 μm and an absolute thickness distribution of ≈80 nm (1σ) are obtainable. The optimized CMP process itself shows a nonuniformity of <2% on 150-mm wafers.

14.5.3 Case Study III: Infrared Digital Micromirror Array

Microfabricated devices capable of switching optical light beams, also termed MOEMS, have gained attention about a decade ago for applications like optical fiber switches, microscanners, or digital micromirror arrays. Particularly, the latter devices have found widespread application in video projection systems for office presentations, home cinema, and very recently, the replacement of classical projectors in movie theaters. Other applications, for example, for head-up displays on auto windscreens are in development. In all cases, arrays of small micromirrors are used for the modulation of light.

As the optical mirrors require planar surfaces, CMP is regularly employed in the manufacturing sequence. Following are two examples of light modulators, based on arrays of tilting micromirrors. The first example is a commercially available digital micromirror device (DMD) from Texas Instruments (TI) for video projectors, using one CMP step for micromirror fabrication. The second example is a micromirror array for switching IR light, which needs three CMP steps in the manufacturing process flow.

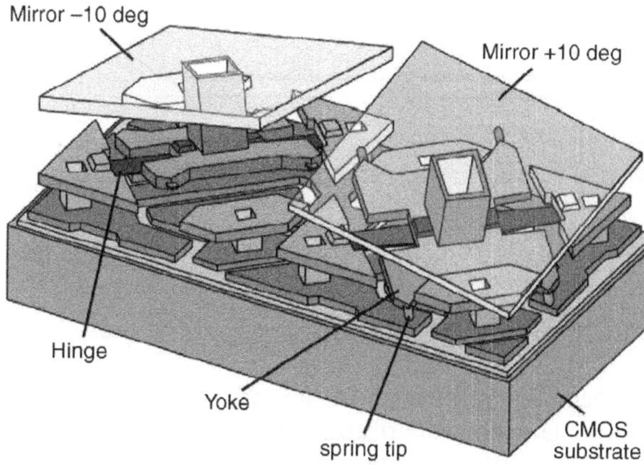

FIGURE 14.9 Illustration of two tilted micromirrors of TI's digital micromirror device.

Figure 14.9 shows a cut through the TI device displaying the design of the MEMS-based DMD. The mirrors can be tilted by electrostatic forces by means of address and bias voltage applied to the mirrors and yokes. The structures are manufactured by a sequence of appropriate deposition and structuring processes. TI uses hardened photoresist as a sacrificial material, which is later removed by a dry etch step to form the air gaps to release up to 1.3 million $16 \mu m \times 16 \mu m$ mirrors. The only CMP step is the planarization of a thick oxide film, deposited over the metal-2 level of the underlying CMOS device, which mainly consists of SRAM address circuits.

Apart from photoresist or silicon oxide, metals can also be used as a sacrificial layer. In the following, a metal surface micromachining technology for the manufacturing of a 256×256 digital micromirror array of an integrated infrared imaging system using Ni as a structural material and Cu as a sacrificial layer will be presented. The process flow, which has been performed at the author's CMP clean room, in all consists of three planarization steps and gives a nice synopsis of CMP applications in microfabrication [25].

The 64k, $80 \mu m \times 80 \mu m$ sized tilting mirrors are built on the top of a CMOS-based control ASIC. In order to reduce the topography of the underlying metallization/passivation structures, a 2.5 μm-thick PECVD oxide film is first deposited on the ASIC. An ILD oxide CMP step based on Klebosol^{TM} 30N50 colloidal silica slurry is used for planarization. In order to connect the ASIC with the deflection electrodes above (see Fig. 14.10), vias have to be etched into the planarized dielectric film. Then, a copper metal stack including a TaN barrier has to be deposited and a two-step Cu damascene CMP process has to be performed. As this process is equivalent to Cu damascene in microelectronics fabrication, standard Cu CMP slurries can be used.

The wedge-shaped deflection electrodes and the 7-μm-high posts for the suspension of the mirrors consist of Ni, fabricated by using 3D electroplating. The complete wafer surface is then covered with a Cu seed layer that has been

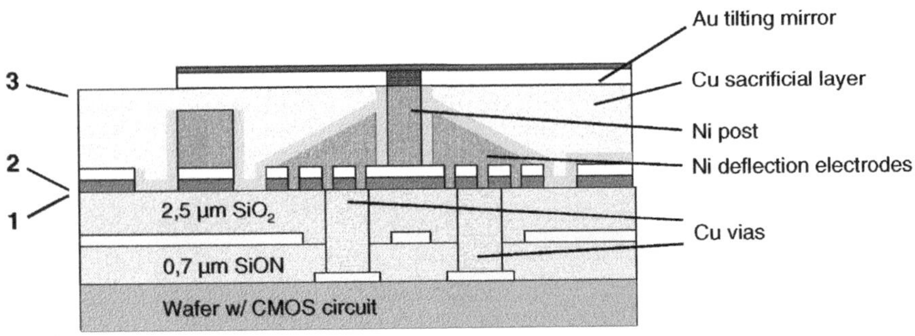

FIGURE 14.10 Schematic cut-through of the micromirror device. Three CMP steps have to be performed: (1) oxide planarization of CMOS passivation, (2) Cu damascene of vias, and (3) CMP of thick Cu sacrificial layer until the Ni posts are exposed.

reinforced by Cu electroplating to 10µm thickness. The third CMP step removes and planarizes the Cu layer until the Ni posts are exposed. This Cu CMP process has to fulfill the requirements of high selectivity to nickel in order to avoid Ni recess and very low copper dishing between the nickel posts in order to achieve flat mirrors. Results fulfilling these specifications have been obtained by using iCue™ 5003 Cu slurry from Cabot Microelectronics. Due to the chemical composition, the slurry shows an intrinsically high selectivity to nickel and the optimized process had a copper removal rate of >500nm/min. Although the nickel posts have distances of 100µm, a dishing between the posts of only <100nm could be measured by using white light interferometry. This value fulfills the specifications of the device and could be obtained using a hard K-grooved IC1400™ polishing pad from Rohm & Haas.

On top of the exposed nickel posts, a gold film has been deposited and structured to form the micromirrors. Subsequently, a wet etch process selective to gold and nickel clears the underlying copper sacrificial layer. This step releases the mirrors and allows them to tilt by means of electrostatic forces by applying a control voltage at the deflection electrodes. Figure 14.11a and b shows some mirrors of the array and a closeup view of the cell structure, which depict mirrors, wedged electrodes and posts for suspension of the torsion hinges.

Besides the application of micromirror arrays, nickel surface micromachining with a copper sacrificial layer is a technology that can be used for various microfabrication concepts. Only recently, the method has been applied for the construction of capacitive RF switches for antenna impedance matching in multiband mobile phones [26].

In contrast to the above example of high selectivity, which uses a low removal rate of nickel, other MEMS-related applications require higher Ni removal rates. Using alumina-based slurries originally developed for tungsten CMP, the addition of oxidizers like H_2O_2 can lead to a fourfold increase in Ni polishing rate [27,28].

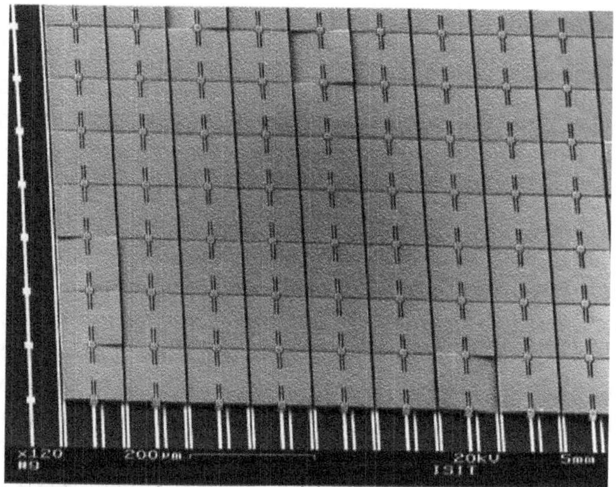

(a)

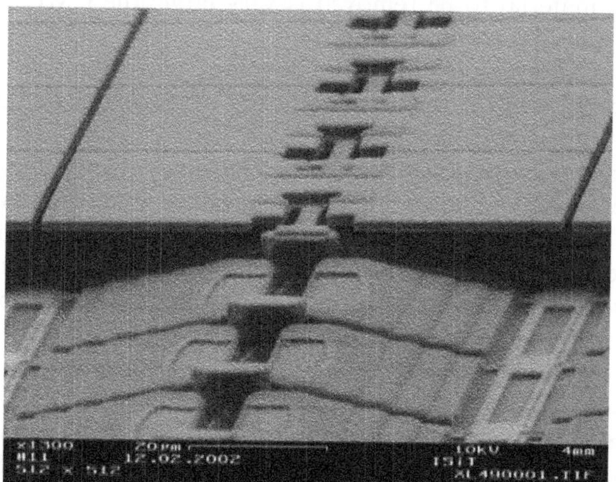

(b)

FIGURE 14.11 (a) Micromirror array, fabricated in metal surface micromachining. Three CMP steps have been employed in the manufacturing sequence. (b) Closeup view of micromirror device. Gold mirror plates and nickel wedged deflection electrodes are clearly visible.

14.5.4 More Representative Applications

Beyond the few examples described in the preceding sections, CMP has many more applications in microfabrication. As described in Section 14.5.1, 3D-integration by chip stacking recently gained interest. One possible concept for vertical interconnection of thin chips is the etching of through-vias and a subsequent filling with a metal like W or Cu. Using the damascene concept, overburden metal can be removed by a metal CMP planarization process [12,29]. As the via diameters and distances of 20–100 μm are relatively large, the CMP process requirements are comparable to those in microfabrication.

For many years, CMP processes have been employed for the manufacturing of R/W leads for the hard-disk industry. The construction of advanced recording systems for high-density data storage is described in E.E. Fullerton and coworkers' report [30]. In a very nice overview, C.M. Pitcher from Seagate disclosed that at least four CMP steps are necessary for modern R/W heads [31]. The alloys NiFe or CoNiFe which, up to now have not been used in microelectronics, need to be polished during the formation of the lower shield, shared pole, and top pole of the head construction. Customized slurries for NiFe, based on alumina abrasives and adjusted at pH values between 3 and 5, have been developed in order to fulfill the very stringent roughness requirements of below 1.5 nm. Recent developments for CoNiFe contain H_2O_2 as an oxidizer and have to be operated at higher pH (5–8) [31]. The fourth CMP step is the planarization of the oxide overcoat layer. In all cases, the amount of material to be removed is very large; the CMP steps therefore belong to the microfabrication category.

CMP of the combination of NiFe together with SU-8 as an insulating material has been used for the manufacturing of electromagnetically actuated MEMS optical microswitches [32].

Although not very successful in microelectronics due to the formation of microscratches, CMP of aluminum will certainly find its way into microfabrication. The structuring of interdigitated electrodes for gigaherts surface acoustic wave (SAW) devices using Al damascene has been reported [33]. Recently, a method for the fabrication of transparent plastic displays has been proposed, where hot embossing into PMMA, metal (Al) sputtering, and subsequent metal CMP are used for the structuring of metal lines [34]. In this publication, an Al CMP process with 1:1 selectivity to PMMA based on alumina abrasive and phosphoric acid, citric acid, and H_2O_2 aqueous solution has been optimized and applied to fabricate 10-μm Al lines embedded in the transparent polymeric material.

14.6 OUTLOOK

It can be predicted that more applications of CMP for microfabrication will be developed in the future. Wafer bonding gains more and more attention, especially for thin- and thick-film SOI wafers and for wafer-scale packaging. For direct bonding, polishing processes for Si or SiGe are already available. Anodic bonding or glass frit bonding generally rely on planarized surfaces. New materials like SiC or diamond layers will enter the microfabrication arena and will become a challenge for CMP engineers. Cheaper materials such as various polymers, ceramics, or glass, as well precious metals like gold or platinum, are already under consideration for microfabrication. Sooner or later, the CMP engineer will be confronted with the question: Can you polish this?

QUESTIONS

1. What are the main differences between CMP processes for microelectronics versus microfabrication?
2. Which chemical property is necessary for a layer material in order to be suitable as sacrificial material when fabricating a surface micromechanical system?
3. What are the most common methods for end-point control in CMP and which of them can be used in CMP for microfabrication?
4. What are the alternative methods for removing very thick layers in microfabrication? Compare these methods with CMP.
5. You have to planarize a SiN layer for the fabrication of a microdevice. What kind of slurry would you try for first CMP tests?
6. Pressure sensors use the principle of deflected membranes. How can this deflection be measured?
7. What are the main reasons for employing CMP in poly-Si surface micromechanics?
8. Design a process flow for the fabrication of an optical waveguide consisting of a doped oxide strip with square cross section embedded in and covered by an undoped oxide by using CMP.

REFERENCES

1. Madou MJ. Fundamentals of Microfabrication. Boca Raton: CRC Press; 2002.
2. Maluf N. An Introduction to Microelectromechanical Systems Engineering. Norwood: Artech House; 2004.
3. Kovacs GT. Micromachined Transducers Sourcebook. New York: McGraw-Hill; 1998.
4. Liu C. Foundations of MEMS. Prentice-Hall; 2005.
5. Moy AL, Hetherington DL. CMP processing issues for MEMS fabrication technology. Proceedings of the 11th CMP-MIC; 2006. p335–342.
6. NEXUS Market Analysis for MEMS and Microsystems III; 2005–2009.
7. Camiletti L. CMP revisited for the MEMS/Foundry Era. MRS Symp Proc 2003;767:205–208.
8. Stein D. CMP technologies for MEMS device fabrication. SEMI Technical Symposium, Semicon West; 2002.
9. Trotha LV, Moersch G, Zwicker G. Advanced MEMS fabrication using CMP. Semicond Int 2004;27(9):54–56.

10. Hornbeck LJ. Digital Light Processing^{TM}: A new MEMS-based display technology. Texas Instruments.www.dlp.com.
11. Rhoades RL. Snapshot of CMP technology evolution for non-traditional applications. Proceedings of the 11th CMP-MIC; 2006. p343–346.
12. Lefevre P, Qiao S, Ina K, Sakai K, Tamai K, Van Calcar P, Moll A. Copper CMP for 3D chip with through wafer vertical interconnects. Proceedings of Ninth CMP-MIC; 2004. p433–439.
13. Tang B, Boning D. CMP modeling and characterization for polysilicon MEMS Structures. MRS Symp Proc 2004;816:209–216.
14. ITRS Roadmap. 2005 ed.http://public.itrs.net.
15. See for example, Surfscan SP2.www.kla-tencor.com.
16. Lee H, Miller MH, Bifano TG. CMOS chip planarization by chemical–mechanical polishing for a vertically stacked metal MEMS integration. J Micromech Microeng 2004;14:108–115.
17. Kobayashi K. Accretech wafer thinning technology. Sixth International Workshop on Thin Semiconductor Devices; Munich; 2005.
18. www.logitech.uk.com, www.presi.com,www.peter-wolters.com, www.gnptech.com, www.strasbaugh.com.
19. www.cabotcmp.com, http://electronicmaterials.rohmhaas.com, www.fujimiinc.co. jp, www.kemesys.com Degussa:https://www1.sivento.com/wps/portal/p3,www. hcstarck. com.
20. Kröninger W, Schneider L, Wagner G. Creating stable and flexible chips for thin packages. Semicond Int 2004;2712:59.
21. Kulawski M, Luoto H, Suni T, Weimar F, Mäkinen J. Integration of CMP fixed abrasive polishing into the manufacturing of thick film SOI substrates. Mater Res Soc Proc 2005;867:111–116.
22. Sniegowski JJ. Chemical–mechanical polishing: enhancing the manufacturability of MEMS. SPIE Proc 1996;2879:104–115.
23. Jeong M-K, Park S-M, Jung J-W, Jeong H-D, Kim H-J. Consumable approaches of polysilicon MEMS CMP. Proceeding of the Second PacRim-CMP; 2005. p297–300.
24. Lange P, Kirsten M, Riethmüller W, Wenk B, Zwicker G, Morante JR, Ericson F, Schweitz JA. Thick polycristalline silicon for surface-micromechanical applications: deposition, structuring and mechanical characterization. Sens Actuators A 1996;54:674–678.
25. Reimer K, Engelke R, Witt M, Wagner B. 16k infrared micromirror arrays with large beam deflection and tenth millimeter pixel size. SPIE Proc 1999;3878:272–280.
26. Lisec T, Huth C, Shakhray M, Wagner B. Surface-micromachined capacitive RF switches with high thermal stability and low drift using ni as structural material. Proc MEMSWAVE 2004:C33–C36.
27. Vijayakumar A, Du T, Sundaram KB, Desai V. The application of chemical mechanical polishing for nickel used in MEMS devices. MRS Symp Proc 2004;816:203–208.
28. Park S-W, Seo Y-J, Choi G-W, Kim N-H-, Lee W-S,Chemical mechanical polishing performance of nickel for MEMS application. Proceedings of the Second PacRim-CMP; 2005. p230–233.

29. Lu J-Q, Rajagopalan G, Gupta M, Cale TS, Gutmann RJ. Planarization issues in wafer-level three-dimensional (3D) integration. MRS Symp Proc 2004;816:217–228.
30. Fullerton EE, Margulies DT, Moser A, Takano K. Advanced magnetic recording media for high-density data storage. Solid State Technol 2001;449:87–94.
31. Pitcher CM. A history of nickel-iron (NiFe) CMP as used in the magnetic head industry. Proceeding of the Eighth CMP-MIC; 2003. p550–557.
32. Gatzen HH, Obermeier E, Kohlmeier T, Budde T, Ngo HD, Mukhopadhyay B, Farr M. An electromagnetically actuated bi-stable MEMS optical microswitch. Proceedings of the 12th International Conference on Solid State Sensors Actuators and Microsystems, Transducers '03; 2003. p1514–1517.
33. Holland AS, Reeves GK, Leech PW. Planarisation of patterned aluminum/diamond surfaces for SAW devices. MRS Symp Proc 2003;767:217–222.
34. Cha N-G, Yu Y-S, Kang Y-J, Park C-H, Park J-G, Characterization of the aluminum and PMMA CMP (chemical mechanical polishing) for transparent plastic display. Proceedings of the Second PacRim-CMP; 2005. p239–244.

15

THREE-DIMENSIONAL (3D) INTEGRATION

J. JAY MCMAHON, JIAN-QIANG LU AND RONALD J. GUTMANN

15.1 OVERVIEW OF 3D TECHNOLOGY

Three-dimensional (3D) integration with interstrata vias has the potential of improving system performance while providing a platform for heterogeneous integration. Performance improvement in 3D integrated circuits (ICs) is mainly due to the reduction of interconnect length, which decreases interconnect delay and power consumption [1,2]. Small form factor is achieved in 3D ICs due to the stacking of active device layers one on top the other. A path for heterogeneous integration is realized if this stacking is done in a fashion that is back-end-of-line (BEOL) compatible, because interconnecting devices using fully fabricated wafers can reduce process incompatibilities.

Approaches to 3D integration are often divided into die-level approaches and wafer-level approaches. Most often, the heterogeneous integration and small form factor advantages are obtained with die-level approaches; in addition, high performance and low manufacturing cost can be achieved with wafer-level approaches.

CMP has permeated many aspects of IC processing but is particularly a dominant factor in BEOL interconnection. Copper damascene patterning has become the standard methodology for building a large number of interconnect layers in the vast majority of high-performance ICs. In 3D ICs, CMP is also critical in backside thinning, surface roughness reduction, and, in particular, feature topography control. Topography mitigation can particularly be a critical

Microelectronic Applications of Chemical Mechanical Planarization, Edited by Yuzhuo Li

aspect for wafer-level 3D because most wafer-bonding approaches require a high degree of surface planarity prior to the bonding step. This chapter aims to provide an introduction to the wide variety of 3D integration approaches under research and emphasizes the critical role of CMP in each of these approaches.

15.2 FACTORS MOTIVATING RESEARCH IN 3D

Three main factors motivating research in 3D will be discussed: reduction in form factor, heterogeneous integration, and enhancement of performance.

15.2.1 Small Form Factor

In recent years, a clear trend in consumer products has been the integration of functionality into existing technology. Reducing form factor has been an enabling factor allowing the integration of global positioning systems (GPS), digital photography, music, and video into the cell phone and automotive platforms [3]. Furthermore, research in miniaturized actuators and distributed sensor networks has drawn considerable interest in biomedical, aerospace, and telecommunication industries [4,5]. System-on-chip (SoC), multichip module (MCM), and system-in-package (SiP) technologies all reduce form factor by unifying integrated circuits at various levels, as illustrated in Fig. 15.1 [6–9]. SoC technology integrates circuit functionality during wafer-level fabrication and provides reduction in packaging cost; however, SoCs are the most

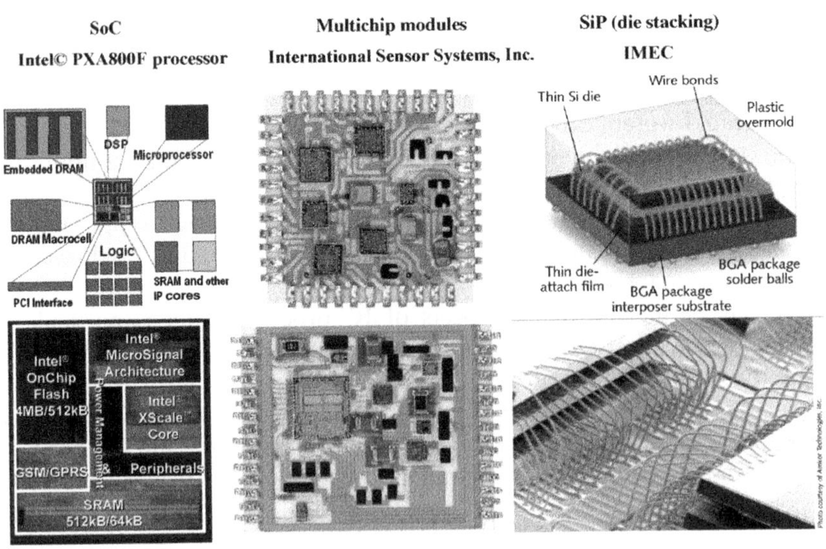

FIGURE 15.1 Representative SoC, MCM, and SiP approaches for reducing form factor and increasing functionality through integration (from Ref. 6–9).

restrictive approach in terms of fabrication because similar fabrication steps must be used to realize different functions. Some multichip module approaches offer higher density interconnection than conventional through-hole printed circuit board (PCB), surface mount, and chip-on-board (COB) technology by interconnecting singulated die using substrate-level interconnects. SiP technologies (some of which are referred to as die-to-die 3D or stacked-die 3D) combine the advantages of MCMs and SoCs by vertically or horizontally integrating die in a single package with wire bonds or solder bumps as interdie interconnects. This approach offers the highest integration flexibility and reduction in form factor among these three approaches but requires handling multiple singulated die. Although stacked die with peripheral wire bonds are common in cell phones today, applications requiring low parasitics use through-die interconnect and flip-chip assembly.

15.2.2 Heterogeneous Integration

As mentioned above, the integration of functionality has become an increasingly important component of commercial product development in recent years. No clear quantifiable metric exists for defining how restrictive a fabrication approach is in terms of integration, but stacked-die 3D (such as SiP), die-to-wafer 3D, and wafer-level 3D offer more integration capability than SoC. An approach demonstrated by Gann [10] assembled singulated known-good-dies (KGDs) in a pseudowafer for wafer-level processing, before being singulated and stacked. These stacked dies were then edge polished to expose interconnects, and interlayer interconnections were processed on the edge. This approach was demonstrated by assembling a stacked computer with 48 layers, including processor, interconnection layers, 10 DRAM memory layers, and 32 flash memory layers. A different layout was used to fabricate a vertical cavity surface emitting laser (VCSEL) array and detector at the layer level [10]. A schematic representation of the pseudowafer along with pictures of individual layers and a stacked assembly is shown in Fig. 15.2. This approach demonstrates that heterogeneous integration can lead to small form factor solutions, with layers from different processes assembled together to create a functional multicomponent system.

Wafer-level approaches demonstrating the integration of microelectromechnical systems (MEMS) with active electronics have been reported using less common processes as early as 1995 [11–13]. Gianchandani et al. demonstrated processes involving the integration of active ICs with an electrochemical etch stop and bulk micromachined accelerometers with a boron impurity etch stop. This was achieved by anodically bonding these structures to a glass wafer with metal interconnect, counter electrodes, and shielding. This early demonstration of interwafer interconnection, that the active electronics were unaffected by the bulk etch and that the structural integrity of the accelerometers was maintained throughout the process, was an important step in wafer-level heterogeneous integration. Other early approaches to integration of MEMS were also

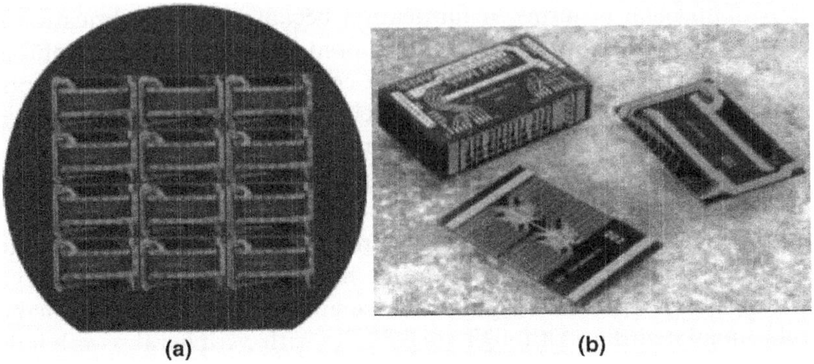

FIGURE 15.2 (**a**) A schematic representation of assembled pseudowafer; all 12 die can be KGD. (**b**) Two single-layer dies and one assembled stack with edge connections shown (from Ref. 10).

reported in 1995 [12,13], and a later review paper summarized the advantages of wafer bonding [14].

15.2.3 Performance Enhancement

As gate delays are reduced due to a variety of improvements including scaling, mobility enhancement through channel strain, high-*k* gate dielectrics, and geometrically advanced gate structures such as FinFETs, interconnect delay is predicted to play a more dominant role in determining the overall performance of future ICs. Since the interconnect delay is proportional to the RC time constant, both the effective resistance and capacitance of the interconnect are the factors determining this component of the overall delay. The RC delay is proportional to the length squared or, with repeaters, proportional to length.

The RC delay improvement offered by 3D interconnect has been the subject of a number of studies [1,2], but recently Bamal et al. [15] have discussed the impact of two-lamina 3D and five-lamina 3D and compared them to the following future interconnect options: (1) scaled Cu/low-*k*, (2) unscaled Cu/low-*k*, (3) wafer-level package (WLP) LC lines (reverse-scaled Cu lines), (4) optical interconnect, and (5) carbon nanotube interconnect. The benchmarking circuit modeled in this study was a scaled inverter connected to a fan out of four scaled inverters using each interconnection approach. The resulting analysis showed that stacking of five lamina in a 3D fashion was the most effective approach for the reduction in interconnect delay for 1-mm global interconnects, as ICs are scaled from the 130-nm node through 90-, 65-, and 45-nm nodes. For 1-cm lines, five-lamina 3D was again the strongest option, with WLP-LC lines and optical approaches comparing favorably. This work shows the importance of reducing global interconnect lengths for the reduction of delay in current and future ICs, and the impact of 3D stacking with through-die interconnects.

Reduction of the interconnect line length has the strongest impact on reducing the interconnect component of the delay. Reduction in the longest lines is achieved through appropriate design of stacked thinned strata, making 3D integration one of the strongest candidates for performance, speed, bandwidth, and, to a lesser extent, power improvement in current and future ICs.

15.3 APPROACHES TO 3D

A variety of approaches to 3D ICs have been demonstrated, all having various advantages and disadvantages. Approaches that use singulated die will be discussed first, followed by wafer-level approaches.

15.3.1 Singulated Die 3D

An approach to 3D integration that has seen recent application in industry is to stack dies. The SiP panels in Fig. 15.1 illustrate a schematic and a demonstration of approaches to this type of integration [8,9]. Kada and Smith [16] reported early work in this field, and their approach has been called a stacked chip-scale-package (S-CSP). A variety of products such as cell phones, hearing aids, and flash drives use similar technology. Flip-chip stacked die combined with through-silicon vias (TSV) provides higher I/O density than overedge approaches such as wire bonding, but current technology limits TSV approaches to 10–30 μm diameter connections, an order of magnitude (two orders in area) larger than those offered by wafer-level 3D. An approach that is capable of improving TSV density through wafer thinning has been reported [9,17] and will be discussed further in Section 15.4.3.

Stacked die packaging approaches include those that use die-to-die stacking in one package, package-to-package stacking, "system-in-a-cube" shown in Fig. 15.2, and folded package methods that allow die-to-die interconnection through interconnections or solder bumps on a flexible substrate [18–22]. Approaches such as the system in a cube offer reduced form factor and weight similar to the S-CSP [16] and added flexibility in the integration of heterogeneous systems. The folded package approaches demonstrated by Pinkerton [19,20] and Solberg et al. [21,22] offer the ability to carry more processes in parallel until final assembly; however, one nonyielding device can render the entire folded package unusable. Additionally, connections between dies are limited to the edge of the die.

Die-on-wafer approaches to 3D offer the ability to do some processes at the wafer level while taking advantage of singulation by using KGDs. Tong et al. [23–25] have demonstrated work related to a 3D approach in this genre and used die pick-and-place equipment to align and bond singulated KGDs to KGDs on the wafer as illustrated in Fig. 15.3. Since only KGDs are used in final assembly, there is no cumulative yield loss where bad dies are stacked on top of

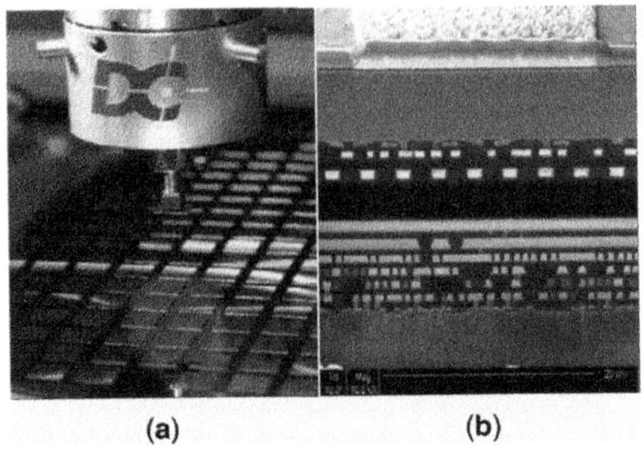

FIGURE 15.3 Die-on-wafer approach to 3D integration demonstrated by Markunas (Ref. 25) and Tong et al. (Ref. 23). (**a**) A die pick-and-place machine populates a metallized wafer. (**b**) Cross section of stacked die on wafer-level interconnect.

good dies. Fully fabricated KGDs with both silicon dioxide and copper-exposed area have been aligned and bonded to the wafer at room temperature using an approach that produces covalent bonding between the oxide fraction of each die. The fully assembled die-to-wafer stack is later annealed at 350 °C to produce copper-to-copper–die-to-wafer interconnections. This approach is being developed for a wireless communication die stack that integrates memory, processor, and programmable logic.

15.3.2 Wafer-Level 3D

Another approach to 3D integration is to use wafer bonding to stack die before singulation; this approach is referred to as wafer-level 3D. There have been a variety of approaches to wafer-level 3D that have been demonstrated, which can be categorized by the wafer-bonding approach used: oxide-to-oxide, copper-to-copper, polymer-to-polymer (or adhesive bonding), and mixtures of these approaches (such as redistribution layer bonding). Each of these four approaches will be introduced in this section, with emphasis placed on their application to 3D. The bond unit processes are described further in Section 15.4.2, and their associated CMP issues are discussed in detail in Section 15.5.

15.3.2.1 Wafer-Level 3D Using Oxide–Oxide Bonding An approach using oxide-to-oxide direct wafer bonding on silicon-on-insulator (SOI) wafers has been reported by Guarini et al. [26] and Topol et al. [27]. This approach uses interwafer via etching, metallization, and planarization after bonding and thinning, with copper used for interwafer interconnection as depicted in Fig. 15.4. For oxide bonding, early BEOL compatibility is perceived as an advantage because it allows integration of devices that are not yet fully fabricated but are past all of the front-end-of-line (FEOL) fabrication steps. This group has also demonstrated the use of UV releasable adhesive for bonding an IC wafer to a temporary glass handle wafer for transfer of circuit

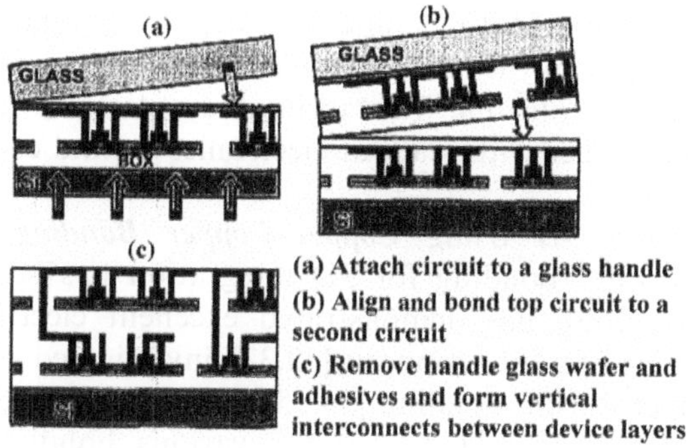

FIGURE 15.4 Representation of key steps in the layer transfer approach to 3D demonstrated by Guarini et al. and Topol et al. This approach uses a glass handle wafer and oxide-to-oxide bonding at a temperature of 300 °C (from Ref. 27, 28).

structural layers [27,28]. The use of a transparent glass handle wafer allows direct inspection and monitoring of layers during critical wafer-to-wafer alignment. Device characteristics were assessed before and after circuit transfer and found to have acceptable performance for long-channel devices and slight degradation for short-channel devices. The degradation in those devices was attributed to a change in the buried oxide (BOX) thickness (i.e., BOX in SOI wafer) after layer transfer.

Another oxide-to-oxide 3D approach features low-temperature oxide bonding and alignment of 150-mm wafers, thinning to a BOX etch stop, high-aspect-ratio (HAR) etching for the interwafer via, and chemical vapor deposited (CVD) tungsten for the HAR metallization (interwafer interconnection), as reported by Burns et al. and Warner et al. in References 29 and 30. The use of oxide fusion bonding allows tungsten to be deposited at 475 °C. Fig. 15.5 shows a scanning electron microscope (SEM) cross section of a high-density imager fabricated using this approach. The low-temperature oxide–oxide bonding approach has advantages in that it can withstand the

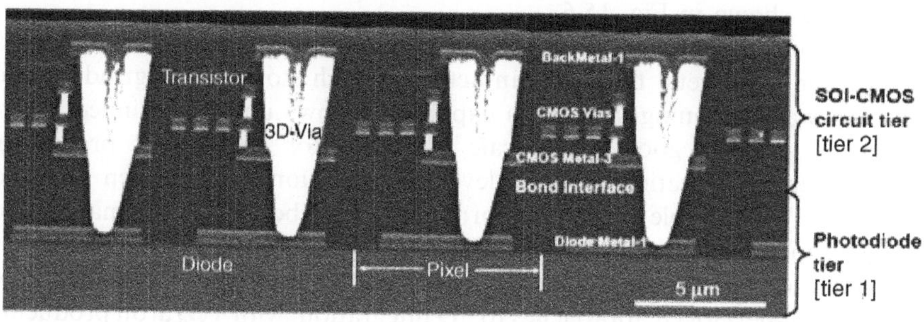

FIGURE 15.5 Cross-sectional SEM of the imager demonstrated by Suntharalingam et al. The oxide–oxide bonding approach is capable of withstanding the 400–475 °C temperatures for via plug fill and defect state hydrogenation (from Ref. 31).

temperature required for HAR metallization, and also in that thin layers can be used for the bonding, implying that a smaller critical dimension is possible for the interwafer via etch and fill. This technology platform has been used to realize a three-tier imager with readout electronics behind each pixel [31].

15.3.2.2 Wafer-Level 3D Using Copper–Copper Bonding Research in copper-to-copper wafer bonding for 3D integration has been conducted by Chen et al. This group has demonstrated excellent electrical behavior of bonded copper interwafer interconnects [32,33], investigated the morphological evolution of copper during bonding [34,35], and investigated the mechanical strength of bonded copper structures for various bonding and annealing schedules [36]. These will be discussed in detail in Section 15.4.2.

Recent work by Morrow et al. demonstrated 3D integration on 300-mm silicon wafers [37,38] using a copper-to-copper bond with an oxide recess. 4096 link via chains were shown to exhibit lower resistance variation than what would have been produced by a 5% variation in the copper film thickness, implying that bonding had little net effect on the via chains. Ring oscillators fabricated using 65-nm technology were tested and were shown to have characteristics within 2–5% of oscillators fabricated in a control sample [37]. Strained-Si and low-*k* dielectric processes were included in a further demonstration [38], which also showed no degradation in performance or functionality as a result of stacking and TSV integration processing.

A metal-to-metal bonding approach for pilot manufacturing of multiple ICs has been demonstrated by Patti [39]. Damascene-patterned copper/oxide interconnects are fabricated in each memory wafer, followed by the oxide recess. The wafers are aligned and brought into contact, followed by formation of copper-to-copper bonds (under BEOL-compatible conditions), for both structural integrity and interwafer electrical interconnection [40]. After top-wafer thinning to expose embedded contacts, further strata can be added. The option of tungsten metallization has also been demonstrated for this process, presumably as a HAR fill capability to bring signals to the thinned wafer surface. A cross-sectional image of interconnected wafers fabricated using this approach is shown in Fig. 15.6.

An important component in the process flow is a through-wafer via with a nitride liner (and metal fill) that can act as an etch stop for the grinding and polishing wafer-thinning step. This capability allows uniform thinned layers and could provide good wafer-scale planarity for subsequent processing (although characterization of wafer-level planarization has not been reported to date) [41]. Available product information describes the performance and specifications of 3D components [40,42].

Bonding of wafers using direct copper-to-copper interconnections is one of the most well-developed approaches to 3D integration, with Tezzaron producing product prototypes at this time [39–42]. Furthermore, Intel has reported a copper-bonding process compatible with 65-nm technology on 300-mm wafers [37,38].

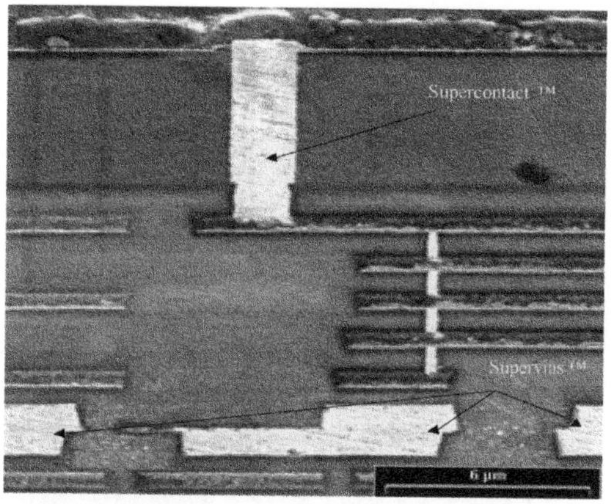

FIGURE 15.6 Cross-sectional view of interwafer interconnects fabricated by Patti et al. Modified from Ref. 40.

15.3.2.3 Wafer-Level 3D Using Adhesive Bonding Adhesive wafer bonding is another notable platform for 3D integration as several research groups have demonstrated functional integrated systems using this approach. Here we briefly summarize the work done by three groups using polymeric films as bonding layers.

Work done by Sailer et al. in 1997 and Burns et al. in 2001 included an adhesive bonding layer and an SOI etch stop for layer transfer [43,44]. The demonstration of a 3D ring oscillator and an imager with 3D integrated control electronics was a significant milestone for this approach. The thickness of the adhesive layer limited the interwafer via size, and the bonding material did not allow the use of tungsten as a metallization material for high-density applications. For these reasons, the adhesive approach was later substituted by the low-temperature oxide–oxide bonding approach discussed in the previous section, which mitigated these two problems.

Another 3D demonstration that used adhesive bonding included In–Au microbumps for interwafer interconnection and was reported by Lee et al. in 2000 [45]. A large shared cache memory was stacked above a processor to enable memory-intensive applications such as multiple processor computing, using 1.5-µm gate-length technology and doped polysilicon.

3D integration using a dielectric adhesive has also been demonstrated by Lu et al., and is schematically represented in Fig. 15.7. Benzocyclobutene (BCB) is used as one of the adhesive materials and provides a strong bond at relatively low temperature (250 °C). Via chains with specific contact resistance $\sim 5 \times 10^{-6}\,\Omega\,cm^2$ were demonstrated [46] by integrating wafer-to-wafer alignment, bonding, HAR etching in inductively coupled plasma (ICP), HAR filling using CVD copper, and thinning to an etch stop. This approach has demonstrated that active devices and passive copper/low-*k* structures can survive this bonding process and an aggressive thinning process several times

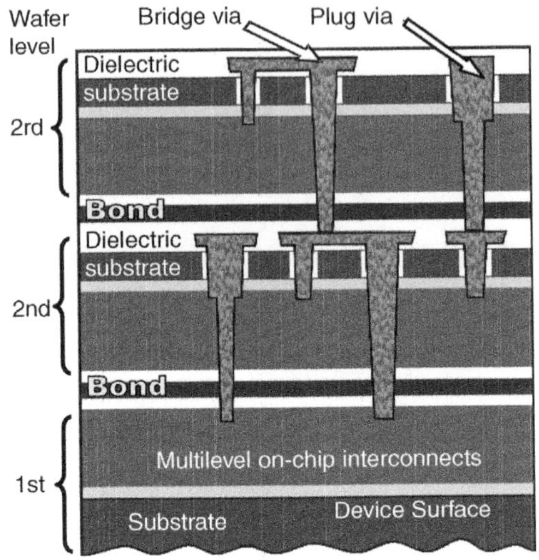

FIGURE 15.7 A schematic of the wafer-level dielectric adhesive bonding via-last approach to 3D demonstrated by Lu et al. (from Ref. 46–50).

over [47], thermal cycling to 400 °C [48], liquid-to-liquid thermal shock, and autoclave tests [49,50], and that the advantageous stress-buffering features of BCB remain after bonding [50].

15.3.2.4 3D Integration Using Redistribution Layer Bonding Finally, an approach to 3D that uses a redistribution layer for bonding has been demonstrated by McMahon et al. [51]. This approach is represented schematically in Fig. 15.8 and features a damascene-patterned bonding layer, so it involves both copper-to-copper bonding and polymer-to-polymer bonding. By using this redistribution layer as a bonding surface, interwafer interconnection and mechanically robust wafer bonding can be achieved in a single unit process step. Three challenges for realizing this approach are (1) the need for a dielectric that is sufficiently mechanically stable to allow damascene patterning but has a surface that is sufficiently active to promote strong bonding, (2) the competing factors for providing a surface activation step that promotes bonding in the dielectric while promoting clean electrical contact between the interwafer interconnect, and (3) the planarization at different length scales required during CMP (discussed in Section 15.5.2) for bonding of the dielectric adhesive and good electrical interwafer interconnect.

15.3.2.5 Summary of Wafer Level 3D Approaches Table 15.1 presents the perceived pros and cons of the 3D approaches discussed in this chapter, with respect to various characteristics offered by 3D integration. If an aspect is viewed as challenging, then it is currently limiting some application of 3D or requires significant process modification to address; a rating of good implies

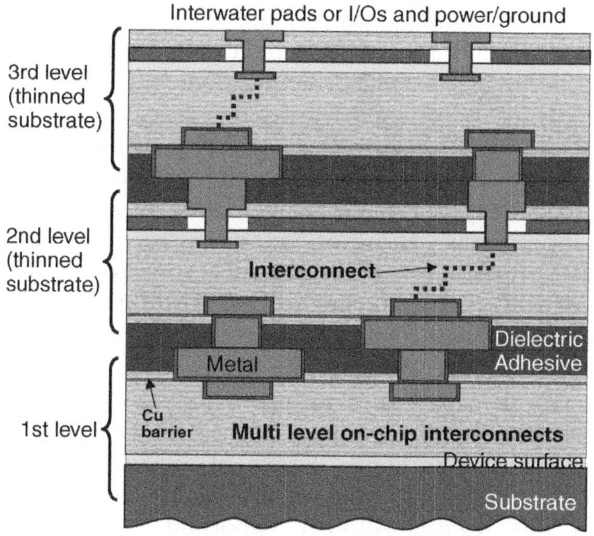

FIGURE 15.8 Schematic representation of the redistribution layer bonding approach to 3D integration demonstrated by McMahon et al. (from Ref. 51).

that the characteristic is acceptable for current prototype integration, and a description of best implies a comparative advantage over the other approaches. This comparison is intended to summarize the capabilities of each approach.

Alignment, bond strength, and bond temperature are three of the most important aspects of any wafer-level approach to 3D integration. The approach using BCB–BCB bonding as a dielectric adhesive suffers from bond-induced misalignment, which can be abated with self-alignment techniques (discussed in the next section), but these techniques add complexity to the process by increasing the number of process steps. Bond strength must be considered for each approach, and BCB-to-BCB bonding has the highest quantifiable bond strength (as measured by four-point bending). Bond temperature has aspects that are conflicting in various applications: use of

TABLE 15.1 Comparison of Features of Four Approaches to Wafer-level 3D Integration.

	Oxide–oxide	Cu–Cu	BCB–BCB	Redist. Layer
Alignment	Good	Good	Challenging	Good
Bond strength	Good	Good	Best	Good
Bond temperature	Challenging/best	Good	Best	Good
Capacitive coupling	Challenging	Best	Good	Good
Stress buffering	Good	Good	Best	Best
Step accommodation	Challenging	Challenging	Best	Good
SiRE usage	Good	Good	Challenging	Good
Thermal management	Good	Good	Good	Best

CVD tungsten requires higher temperature capability than current conventional copper/low-k BEOL requirements but can be integrated into 3D using an oxide–oxide bonding approach. For applications requiring high-performance integration (implying use of copper/low-k interconnection), a bond temperature below 400 °C is required. Usage of low-k dielectric is an important component in BEOL integration for high-performance applications, and the Cu-to-Cu approach may be capable of integrating air-gap dielectrics for small feature size applications. However, as discussed previously, the largest advantage in performance offered by 3D integration is in reducing long lines, which implies that capacitive coupling will be a less critical issue because larger critical dimensions are used where long lines are being replaced. Integration at these upper levels of metal alleviates the stringent requirement for alignment and interwafer via density, but often has needs for stress-buffering and step coverage capabilities, making the BCB-to-BCB and redistribution layer approaches strong in this arena. Usage of silicon real estate (SiRE) is important for any process, and the via-first approaches outshine the via-last approaches in this regard. Another important issue in 3D integration, discussed in detail elsewhere [52–54], is thermal management: the Cu-to-Cu and redistribution layer approaches offer the ability to tailor the fraction of high heat conductivity and low heat conductivity areas, whereas the oxide-to-oxide and BCB-to-BCB approaches need to use copper interwafer vias as thermal as well as electrical conductors.

15.4 WAFER-LEVEL 3D UNIT PROCESSES

As discussed above, a variety of approaches to wafer-level 3D integration have been demonstrated. All have a range of advantages, target applications, and challenges. Common to each of these approaches are some key unit process steps that must be undertaken to realize the end result of stacked, interconnected active devices. Among those that need to be discussed are BEOL-compatible CMP, wafer alignment and bonding, wafer thinning, and TSV processing. BEOL-compatible CMP will be discussed in greater detail in Section 15.5 of this chapter and so will not be covered here. Various approaches push the limits of each unit process step in different ways, so it is useful to consider each step individually.

15.4.1 Wafer-to-Wafer Alignment

Wafer alignment is clearly an important step in wafer-level 3D integration. Currently available technology provides consistent wafer-to-wafer alignment of 1 μm using optical approaches. Work reported by Mirza [55] discusses many optical approaches to wafer alignment including infrared, wafer through-holes, transparent substrates, backside alignment, interwafer alignment, and SmartView™ alignment. 3D interconnect approaches that aim to integrate devices in the early stages of fabrication, where 1–2 layers of metal are fabricated, have the strictest requirements in wafer alignment, as feature sizes can be in the sub-100 nm range. Another application requiring deep-submicron alignment is nanoimprint lithography (NIL). Various optical [56,57] approaches have been pursued for submicron alignment in this arena and can involve balancing images of diffraction orders or shadow moiré patterns to achieve high levels of alignment accuracy. These techniques have yet to be proven in a manufacturing environment but do offer outstanding capabilities.

Mechanical structures have also been investigated for alignment improvement. Niklaus et al. [58] have demonstrated mechanical structures for the maintenance of alignment while employing polymer bonding. This approach uses a metallic structure around the edge of the wafers that touches down just as the wafers are aligned and remains in contact during the bond process. This contact friction prevents bonding-induced misalignment that can be caused by thermal coefficient of expansion (TCE) mismatch or sliding of the wafers with respect to one another during polymer cross-linking. This approach was shown to reduce bonding-induced misalignment by about an order of magnitude (from more than 10 μm to several microns). Fig. 15.9 illustrates another mechanical approach to alignment improvement called keyed alignment. This work, which has been demonstrated by Lee et al. [59], used structures with sloped sidewalls that mate together during alignment. Soft baked BCB can be used over the keyed structures to allow sliding during the initial stages of bonding, producing wafers aligned with submicron tolerance. Although both of these approaches require extra processing steps to produce mechanical structures that are raised with

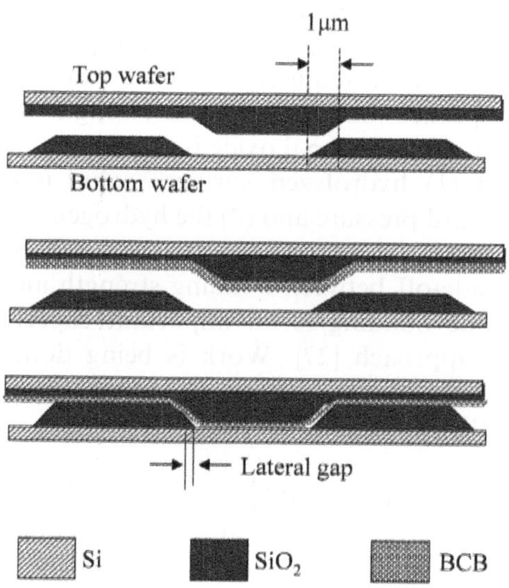

FIGURE 15.9 Schematic representation of wafer-level self-keying alignment approach using raised mechanical structures and liquid polymer to attain submicron registration (from Ref. 59).

respect to the device surface, they may provide a means to provide the alignment required for 3D applications in the FEOL or early BEOL stages of processing.

15.4.2 Wafer-to-Wafer Bonding

Although the bonding of wafers using blanket films is useful in a number of applications other than 3D integration (such as SOI and sensors or actuators requiring large mass or large cavities), discussion here will focus more on the wafer-bonding mechanisms rather than applications as discussed in Section 15.3.2.

15.4.2.1 Oxide–Oxide and Silicon–Oxide Wafer Bondings Silicon direct bonding (SDB) has been an important IC process since the inception of Smart Cut silicon-on-insulator (SOI) technology in 1997 [60]. The manufacturing technology developed for SOI wafer bonding has turned out to be very beneficial for all bonding techniques, and a large body of work exists because of this development [12,61,62]. Some restrictions are required for BEOL-compatible wafer-level 3D integration; (1) CMOS compatibility implies that the materials used must be accepted by IC foundries; (2) planarization is a critical factor for SDB as atomically smooth surfaces and/or wafer-level planarization are required and is difficult to achieve at late stages of fabrication, which will be discussed in greater detail in Section 16.5; and (3) BEOL compatibility implies that temperatures must remain below 400 °C. Furthermore, the bonding process must provide enough mechanical bond strength for large-scale silicon wafers (200–300 mm) to survive the processing and packaging subsequent to bonding.

The temperature restriction imposed by BEOL compatibility is nearly as important as the planarization requirement in the oxide bonding approach to 3D, because this approach has better bond strength when annealed at a temperature of 1100 °C when thermal oxide is used alone. This is because in the mechanism for SDB (1) hydrolyzed surfaces come into contact at room temperature and standard pressure and (2) the hydrogen diffuses away during a high temperature anneal. This produces strong Si–O–Si bonds across the interface [61]. The trade-off between bonding strength and bonding temperature for oxide-to-oxide bonding is an important aspect of the technology platforms using this approach [27]. Work is being done in plasma-assisted oxide bonding to improve mechanical bonding strength while keeping temperatures in the BEOL-compatible range [63].

15.4.2.2 Copper–Copper Wafer Bonding Copper bonding for 3D integration has been investigated by several groups. While Section 15.3.2 discussed the demonstration of use in 3D applications, this section will focus on the fundamental mechanisms of the bonding and characterization of the unit process step. When patterned structures are used in this approach, the copper pads are raised with respect to the dielectric to promote bonding under high pressure and elevated temperature. Key advantages of this approach include direct copper interwafer interconnections (minimization of high aspect ratio etching and filling in comparison to dielectric adhesive bonding) and relaxation of planarization requirements (in comparison to oxide-to-oxide bonding) through the use of a dielectric recess.

Work done by Chen et al. has included (1) electron beam characterization of bonded copper-to-copper structure for varying bonding time and temperature [34,35], (2) electrical characterization, such as contact resistance, for bonded copper [32,33], and (3) mechanical testing through dicing of bonded wafer pairs [36]. For these studies, tantalum and copper were evaporated onto thermally oxidized silicon to thicknesses of 50 nm and 300 nm, respectively, before bonding. Transmission electron microscopy (TEM) of bonded films showed that twin boundaries cross the original bonding interface with a bonding temperature of 400 °C. The mechanism for copper bonding is discussed in Reference 64. Grain reorientation at elevated temperature under mechanical downforce produces intermingling of the copper layers. In this scenario, surface roughness and geometrical placement of contact asperities contribute to the postbonding structure. When surfaces with similar roughness are bonded and the asperities line up peak to peak, a noninterface structure results because high contact pressure results from the geometry. When asperities line up in a peak-to-valley fashion, a zigzag interface results because the contact pressure is lower. If two surfaces with different roughness are bonded, a distinct interface is observed, but it is predicted that this interface would disappear for longer bonding and annealing times [64]. Electron dispersion spectroscopy (EDS) also showed that oxygen content was evenly distributed at a weight percentage of ~3 % within the copper film after bonding [34]. Furthermore, grain size was shown to increase from ~0.3 to ~0.8 μm after the as-deposited and bonded copper was annealed; the process typically included a bond time of 30 min plus an anneal time of 60 min. One finding from this study was that the dominant grain orientation changes from (111) to (220) after bonding and annealing at 400 °C [35]; unfortunately, the (220) grain structure has disadvantageous electromigration properties compared to (111) grain structure [65]. In the dicing test, bonded structures were diced into 5 mm by 5 mm pieces, and the number of dies that failed at the copper-to-copper bonding interface were counted. A 50 % failure rate was found for a 225 °C process that included bonding for 30 min and annealing for 60 min. A 1 % failure rate was found for a 300 °C process that included 30 min of bonding time and 30 min of annealing time. Excellent bonding conditions were achieved with a process that includes 30 min of bonding at 350 °C followed by 60 min of annealing at 350 °C, or 30 min of bonding at 400 °C followed by 30 min of annealing at 400 °C [36]. A low specific contact resistance of $\sim 1 \times 10^{-8}\,\Omega\,cm^2$ was measured for fabricated structures designed to minimize the effect of wafer-level misalignment [32,33].

Tadepalli and Thompson focused on the bonding strength of copper-to-copper bonded structures using four-point bending characterization [66]. Adhesion energy was characterized for three different surface preparation methods and at temperatures in the range of 250–400 °C. Highest adhesion energy of 28 J/m^2 was found for a surface preparation method that included an acetic acid cleaning for 10 min at 35 °C, and then bonded for 30 min and annealed for 30 min. By varying the temperature alone, adhesion energies of 17 J/m^2 at 300 °C, 10 J/m^2 at 275 °C, and 3 J/m^2 at 250 °C were obtained for copper-to-copper bonded films. Two values for comparison are the adhesion energy of SOI wafers, noted to be 10 J/m^2 in Reference 66, and the lower limit for BEOL processing compatibility defined at 5 J/m^2 by Scherban et al. [67]. The impact of bonding copper parts of different diameters and pitch necessary to establish design rules to achieve adhesion energy > 5 J/m^2 over patterned areas has not been reported but is assumed to be acceptable when dummy structures are included.

15.4.2.3 Polymer Adhesive Wafer Bonding Dielectric adhesive bonding is an attractive approach to 3D integration because of the multitude of polymeric materials currently being researched in the semiconductor industry and the ability to tailor the properties of these materials through synthesis of the components that make up the final structure. Many of the polymer dielectrics that have been used in the past have advantageous adhesive properties, including polyimide [68,69], epoxy [70], FLARE [71], and BCB [72–74]. A review paper on adhesive wafer bonding is also available [75].

Properties that are advantageous for the application of 3D integration include strong cohesive bond strength, strong adhesive bond strength to silicon-containing materials (often through use of an adhesion promoter), a cure process that produces no gaseous by-products, a film that can be bonded well using pressure and temperatures within BEOL limits but is not tacky at room temperature (to facilitate wafer-to-wafer alignment), and ease of integration into an interconnect structure without affecting yield or reliability of the final structure (i.e., low stress, reasonable patterning capability, and coefficient of thermal expansion similar to silicon, glass, and copper). Here we describe the research to date on the use of BCB as a wafer-bonding layer.

Work using BCB as a bonding layer has shown that a strong bond (adhesion energy >30 J/m^2 as measured by four-point bending) can be achieved at a bonding temperature of 250 °C; this bond remains stable at temperatures as high as 400 °C [48,50]. The mechanism for bonding in this case is proposed to be cross-linking across the original bonding interface [72,75]. An adhesion promoter is also required for strong bonds to be produced over Si and SiO_2 surfaces. The bonding temperature of this process is lower than that of oxide-to-oxide, copper-to-copper, or polyimide-bonding approaches, producing some improvement in bond process wafer throughput. Bond strength was evaluated for a wide variety of bonded surface combinations including silicon, oxide, and low-*k* dielectrics, under a wide variety of reliability test conditions such as autoclave, liquid-to-liquid thermal shock, thermal cycling as high as 400 °C, and bond layer thicknesses from 0.6 to 2.6 μm. Four-point bending tests have shown that this dielectric provides a strong bond that can easily

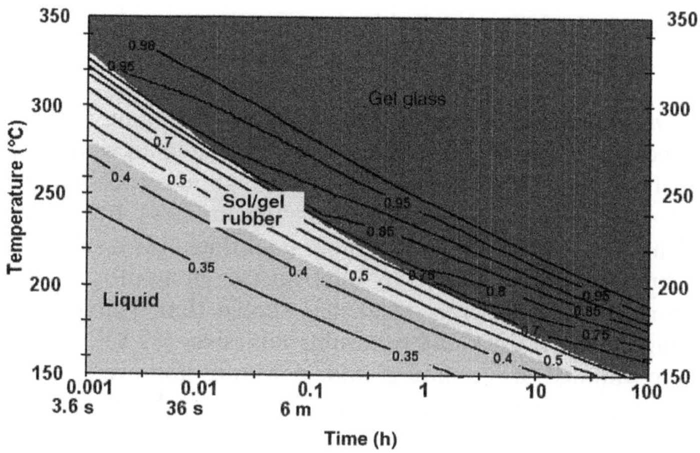

FIGURE 15.10 Time–temperature transformation chart illustrating various levels of cross-linking for given temperature and time treatment for BCB (from Ref. 79).

surpass the BEOL lower limit of 5 J/m^2 when used as a bonding material. Since BCB acts as a planarizing material during spin coating and can be coated over topography, this approach offers the most accommodating planarization requirements when compared to oxide bonding or copper bonding. Therefore, BCB is ideally suited for the integration of fully fabricated circuits using a wafer-level packaging technique such as redistribution layer bonding.

Further work using BCB as a bonding layer has focused on partially curing the BCB film prior to bonding [51,76–78]. A number of advantages have presented themselves in using partially cured BCB: reduced bonding-induced misalignment [78], mechanical rigidity requisite for copper damascene patterning [76], and enough free bonds at the surface to continue to produce strong bonding [51]. Fig. 15.10 shows a time–temperature transformation (TTT) chart for dry etch formulation BCB [79]. As shown in the chart, an as-spun film that is baked at 170 °C for 3–5 min will not be further cross-linked (the material is supplied at a 35 % degree of cross-linking); however, a film baked at 190 °C for 30 min will be partially cured to about 45% cross-linking, and a film cured at 250 °C for 1 min will be cured to about 55 % cross-linking. Fully cured BCB is typically baked at 250 °C for 30–60 min, resulting in cross-linking of about 99%.

15.4.3 Wafer Thinning for 3D

Wafer thinning is a critical component of 3D integration because one of the most important advantages gained is the shortening of interconnect wire lengths. When face-to-face bonding is employed, the length of I/O interconnect is defined by the degree of thinning, and when face-to-back bonding is used, the length of all interwafer interconnects are defined by the degree of thinning. Two important approaches to thinning are used in 3D integration: timed control for removal of silicon and thinning to a stop layer (either in etching or CMP) to improve uniformity of the final layer thickness.

15.4.3.1 Timed Removal Thinning Approaches Grinding and polishing of silicon has become a popular option for thinning wafers in wafer-level 3D and thinning wafers for die stacking [80]. High-throughput processes that are able to thin 200-mm wafers to a thickness of 100 μm have been demonstrated [47,81]. Pushing this capability to a thinner final thickness results in remaining silicon thickness being dominated by removal rate nonuniformities.

Work done by Landesberger et al. has shown that dicing by thinning is feasible using carrier frames and releasable adhesives [82,83]. Wafers in these studies were thinned by grinding and polishing followed by subsequent thinning in wet chemical spin etching. The spin-etching capability allows wafers to be thinned with a noncontacting wafer handler to minimize edge damage. As with grinding and polishing, high removal rate processes generally produce higher removal rate nonuniformities and higher surface roughness. Furthermore, because this process etches isotropically, edge loss becomes more of an issue for processes requiring greater silicon removal.

Low-pressure wafer-thinning processes are possible using atmospheric downstream plasma (ADP) etching, XeF_2 etching, and SF_6 etching. ADP has been investigated by Francis in 1999 [84], and results showed that high etch rates could be attained with relatively small sacrifice in etch rate nonuniformity; a 200-mm wafer could be thinned 200 μm with 1 μm variation at 3σ. Xenon difluoride offers high selectivity to many common films (SiO_2, SiN_x, and Al) but suffers from loading effects on larger diameter wafers. Sulfur hexafluoride can offer higher etch rates than XeF_2 but does not have a suitable etch stop to allow easy integration. Both of these processes have etch rate nonuniformities that would not allow them to be used without an etch stop but offer low damage alternatives to grinding.

15.4.3.2 Thinning to Either an Etch or Polish Stop Mentioned only briefly in Section 15.4.2, the Smart Cut process [60] has been extremely effective in producing thin single-crystal silicon layers on insulating films and substrates. Typically, this process involves hydrogen implantation, wafer bonding, layer splitting during a high temperature anneal, and subsequent polishing to remove defects. Although this process cannot be used to create 3D IC structures because of the high-temperature layer-splitting process, the buried oxide layer can be utilized as an etch stop to alleviate etch nonuniformities from other processes.

As illustrated in Fig. 15.11, wafers can be bonded face-to-face, the handle of the SOI wafer can be thinned to stop on the buried oxide layer, rebonded to another handle wafer, thinned again to stop on the bonding layer, and then tested. Lu et al. have used this approach to demonstrate process compatibility on passive structures [85], Gutmann et al. have used this method to demonstrate process compatibility using active electrical structures [49], and

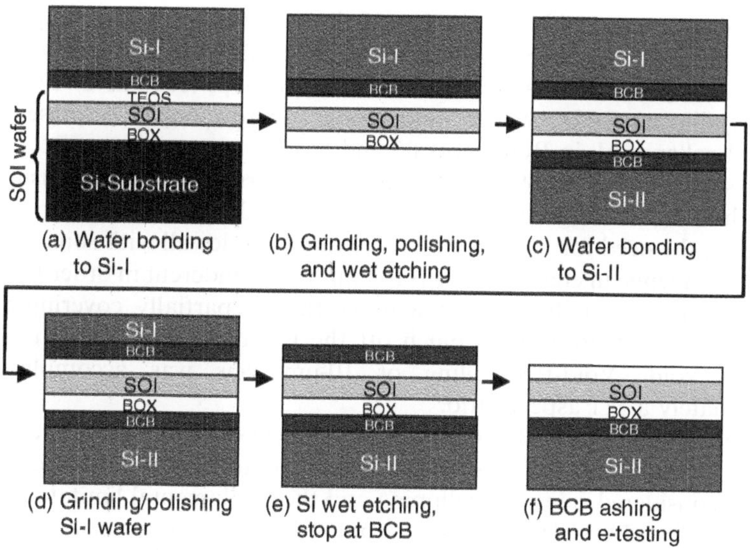

FIGURE 15.11 Process flow for thinning to a burried oxide etch stop, rebonding, and repeating to allow electrical testing of the original structures (from Ref. 49).

Pozder et al. have employed this approach to demonstrate packaging compatibility of dielectric adhesive 3D with active electrical structures [50].

Another approach to thinning that uses a stopping layer employs metal-filled vias, such as metal/TaN/nitride/oxide (the so-called Super-Via^{TM} process [41]), or Cu/TaN/CVD oxide filled deep trench isolation (DTI) regions (the so-called copper-nail process [9,17]). The copper-nail approach uses a 15 μm deep DTI etching and subsequent filling with interconnect, diffusion barrier, and CMP stop to provide a patterned etch stop for wafer thinning. These Cu-nails are placed around the active devices between FEOL fabrication and BEOL fabrication. This approach does not require an SOI starting wafer and so provides a greater degree of IC process compatibility, but the DTI regions consume SiRE. Furthermore, both the selectivity and the resultant planarity of the thinning process are pattern dependent: the more exposed nitride area, the higher the selectivity versus silicon and the more planar the surface.

15.4.4 Through-Silicon Vias

TSVs are an important component of all four wafer-level 3D technology platforms discussed above. In the via-first approaches, the I/O connections from the bond layer to the top of the stack are accomplished by TSVs, and in the via-last approach, these interconnects as well as the interwafer interconnects are defined by TSVs. In general, to accomplish high-performance high-density interconnection, TSVs should be as short and small as possible.

Two challenges are uniform thinning of the top wafer and high aspect ratio etching of the thinned layer. An approach to high-density TSV interconnection using a variety of CMOS-compatible fabrication steps (such as CVD tungsten and polysilicon, ECD copper, and solder micro-bumps) is presented by Knickerbocker et al. in References 86 and 87.

Spiesshoefer et al. [88,89] have discussed the formation of TSVs using RIE. While the need for high-density interconnection drives the work to produce HAR vias, the need for void-free filling requires a sidewall with some tapering. Another challenge in etching TSVs is to limit mask undercut in order to achieve void-free filling; metallizing with a mask that is partially covering the via promotes metal deposition to pinch off the top of the via prematurely and create a void. Void-free filling of 10-μm vias was accomplished at approximately a 1:1 aspect ratio.

ICP etching using the BOSCH process is discussed by Chambers et al. [90]. This work focused on variation of parameters to address the above challenges and also considered sidewall scalloping and the fundamental limitations of the etch rate. The BOSCH process alternates between an isotropic etch and polymer deposition process, allowing unparalleled control over sidewall angle but also producing sidewalls that are rippled. Minimization of this roughness can be accomplished by varying etch-to-deposition switching parameters.

Research in filling of TSVs with electroplated copper for 3D applications has been accomplished by a number of groups [9,17,86,87,91,92]. As mentioned above, geometry of the via is critical for obtaining void-free copper filling, but a number of issues exist in the plating process as well: seed deposition, bottom-up filling, and bonding using electrochemically deposited copper. Seed deposition in high aspect ratio filling can be an issue because the most standard approach is sputtering, which has limitations in step-coverage of HAR vias. For through-wafer vias, Nguyen et al. [91] used evaporation to create a backside source for bottom-up filling. Uniformly fabricated copper plugs as small as 20 μm by 20 μm with an aspect ratio of 7:1 have been reported using this technique on 100-mm wafers. In work reported by He et al., copper bonding using electroless-plated copper was demonstrated with subsequent annealing at 250 °C for large copper plugs [92]. This process allows bonding at temperatures lower than those previously reported for evaporated copper [32–36,64] and sputtered copper [51].

15.5 PLANARITY ISSUES IN 3D INTEGRATION

The improved planarity offered by CMP allows the integration of multiple interconnect layers in current ICs. However, the processes used in wafer-level 3D integration are sufficiently different from conventional IC fabrication that investigation of post-CMP surface topology is warranted. This section begins with a brief description of CMP planarization capabilities and then discusses how planarity affects each of the four previously discussed wafer-level 3D approaches. This section focuses on bonding of the first two wafers in a stack, but planarization issues are similar to TSVs.

15.5.1 CMP Planarity Capabilities

While CMP can have significant advantages in terms of the nanoscale roughness improvements and feature-scale planarity offered to BEOL interconnects, nonplanarites such as dishing, erosion, and dielectric loss need to be well understood for any high performance interconnect approach. Furthermore, it needs to be well understood that multiscale planarity may have a dramatic effect on the bonding capability of the surfaces used. Wafer-level 3D integration has planarity issues on length scales that span at least eight orders of magnitude (from tens of nanometers to hundreds of millimeters). A brief discussion of these nonplanarities is presented before considering their effect on wafer bonding.

15.5.1.1 Nano- and Microscale Planarization Surface roughness of IC materials can result from a wide variety of sources, and these sources can vary based on the material preparation technique (diamond cut silicon, sputtered or electroplated copper, vapor-deposited silicon oxides, etc.). Regardless of the source of the surface roughness for these materials, all show improvement in surface roughness after CMP processing [93–95]. The removal of nonplanarities in CMP is the topic of a number of models, and a great deal of literature exists on the reduction of surface roughness using CMP [94,96,97].

Feature-scale nonplanarities in damascene-patterned copper are of particular interest because of their applicability to mainstream IC processing and their projected continued use in future IC interconnect processing. Dishing, erosion, and dielectric loss are depicted in Fig. 15.12 [98]. Dishing is defined as the depression created in large-scale features, typically greater than 10 μm, which results from the pad deforming during CMP. W-shaped dishing can result from a pad that deforms over a feature and presses nonuniformly between the edge and center of that feature. Erosion is a result of a higher density of copper relative to dielectric in an area. This pattern density produces a higher removal rate for that area resulting in lower height for features in that area relative to areas with a lower areal density of copper. Dielectric loss is a result of the removal of the dielectric, most common during barrier CMP. This parameter must be closely monitored when soft or mechanically fragile dielectrics are used.

15.5.1.2 Wafer-Scale Planarity Wafer-scale nonplanarities are not often well characterized in IC processing because their effect on yield is often low compared to other processing parameters; however, work has been reported on its effects on electrical performance [99,100]. For the characterization of 3D integration and wafer-level packaging (WLP) approaches such as the use of redistribution layers, wafer-level planarization requirements have not been

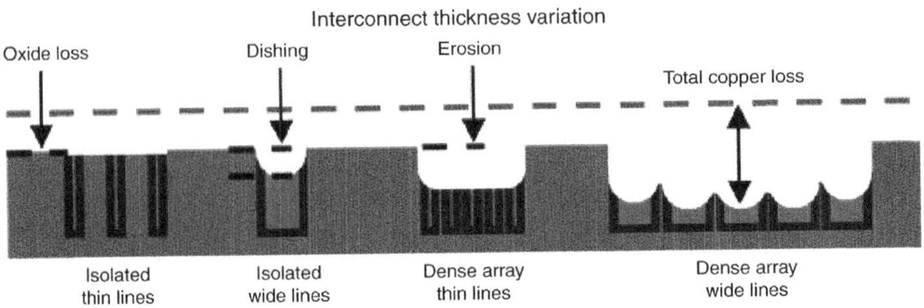

FIGURE 15.12 Schematic representation of common feature-scale nonplanarities arising from damascene patterning electroplated copper into silicon dioxide trenches (from Ref. 98).

explored to the level required for the integration of high-yield bonding processes other than direct silicon bonding [101–103]. These issues are particularly problematic for patterned wafer-level 3D technology platforms. For the purpose of discussion in this chapter, a broad definition of wafer-scale nonplanarities will be used: any nonplanarities that act over distance scales larger than 5 mm, as those nonplanarties are beyond typical planarization lengths for copper damascene CMP [104]. This can often include within-wafer nonuniformity, wafer bow, warp, total thickness variation, die-scale pattern-dependent effects (such as planarization over dicing streets), and edge effects. Although edge effects have been marginalized by assuming yield loss at the edge in the past, this trend is changing [105], and significant defects such as cracks and unsupported areas can lead to problems during wafer thinning [47].

15.5.2 Planarity Issues for Various 3D Approaches

Although various 3D integration approaches have been demonstrated, the vast array of advantages and disadvantages for each makes it difficult to predict which will be adopted for different applications. This section will outline the issues associated with CMP for each approach. In general, two types of issues present themselves in this discussion: (1) issues of planarity at the bonding interface and (2) CMP for interwafer interconnection. Both issues will be discussed for each of the wafer-level approaches introduced in Section 15.3.2.

15.5.2.1 CMP for Via-Last Approach to 3D Using Oxide-to-Oxide Bonding 3D integration using oxide-to-oxide bonding has planarity advantages at the interwafer interconnection stage, because oxide layers are easily integrated into damascene interconnection approaches for standard BEOL processing. Thus interwafer interconnection does not appear to be any more complicated than copper/oxide CMP for this approach. However, there are challenges associated with this approach (aside from the issue of achieving high bond strength at low temperature) because the planarity required at the bonding interface is extremely strict and has been documented by a wide variety of sources [27,61,62,106]. Tong and Gosele describe bondability contact mechanics for this approach, which can be represented by Equations 15.1 and 15.2 [61]:

$$P < \lambda^2 \left(\frac{2Eh^3}{3(\Delta U/\Delta A)} \right)^{-1/2} \qquad \text{for } \lambda > 2h \qquad (15.1)$$

$$P < \frac{7}{2} \sqrt{\frac{\lambda(\Delta U/\Delta A)}{E}} \qquad \text{for } \lambda < 2h \qquad (15.2)$$

where P is the gap thickness that can be closed during bonding, λ is the spatial periodicity of the gap, $\Delta U/\Delta A$ is the surface energy, E is the elastic modulus, and h is the thickness of the substrate.

Fig. 15.13 shows a bonding map for silicon wafer bonding: combinations of topography and spatial wavelength that fall below the traces will bond; those above the traces will produce voids [101]. This and further work by Turner et al. have explored parameters for direct wafer bonding under clamped conditions [107], addressed wafer bow and etch pattern considerations [108], as well as nanoscale roughness considerations [101]. This work shows that

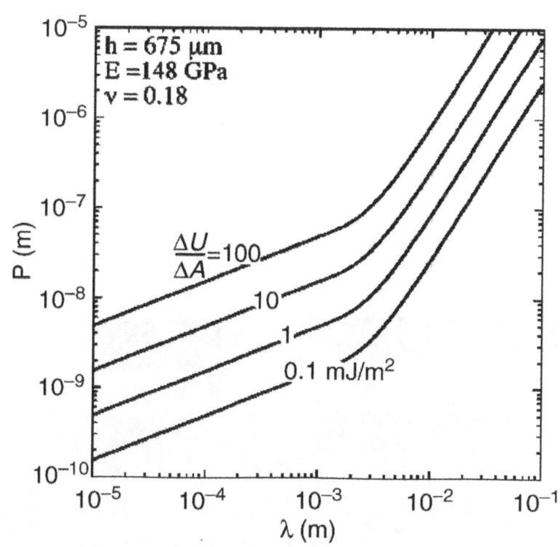

FIGURE 15.13 Direct silicon bonding map showing gap-closing amplitude as a function of wavelength and $\Delta U/\Delta A$ for standard thickness 150-mm-diameter silicon wafers (from Ref. 101). Step heights below the curve will be closed by surface forces; those above the curve will generate voids in the bonding interface.

bonding wafers with 1 mJ/m^2 surface energy will be affected by nanotopography with spatial wavelengths of 1 and 10 mm and height of 5 and 90 nm, respectively, and that polished silicon wafers exhibited height variations less than 20 nm from peak to valley across wavelengths of 10 mm. From this work, the critical role of small-scale roughness is also apparent: 0.5 nm peak-to-peak roughness control is required at 10 μm wavelengths when bonding surfaces at 1 mJ/m^2 surface energy.

15.5.2.2 CMP for Via-Last Approach to 3D Using Polymer Adhesive Bonding Using polymer adhesive bonding for 3D integration often provides smooth surfaces after polymer coating, so CMP of these materials before bonding will not provide significant improvement in planarity for blanket surfaces. Additionally, many spin-cast polymers, such as BCB, provide excellent planarity even when deposited over step heights [109], meaning that nonplanarities are accommodated during the polymer adhesive bonding step. In these respects, the planarization required at the bonding interface for via-last adhesive approaches is not a challenge for realizing 3D-integrated structures. However, the adhesive bonding approach does require that an HAR via be etched, filled, and milled (i.e., damascene patterned) for interwafer interconnection. This HAR etch, fill, and mill can be quite challenging due to (1) the nature of the etch: multiple materials need to be etched through several microns of depth, (2) the nature of the fill: atomic layer deposited (ALD) or chemical vapor deposited (CVD) copper is the best choice for HAR fill, and (3) the nature of the mill: large copper overburden is required for deep features and removal can be challenging. Figure. 15.14 shows a CVD TaN/Cu

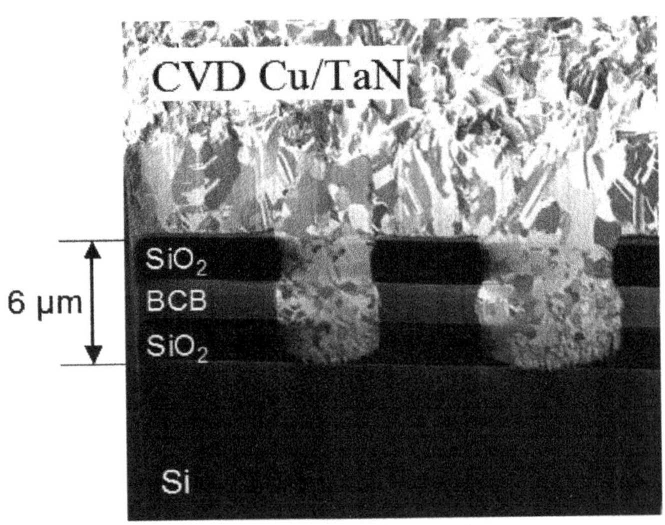

FIGURE 15.14 Void-free fill of a via using CVD copper and liner (from Ref. 110).

fill process for vias etched in a film stack common for this approach [110]. Further complicating this approach is the challenge of providing clean interconnection through a high aspect ratio via: Standard in situ contact cleaning approaches use sputter cleaning before or during deposition [111,112]. Removing material from HAR vias is challenging because the material is physically removed during sputter cleaning rather than chemically volatilized. Even with these challenges in interwafer interconnection, interwafer via chains have been fabricated using a via-last adhesive-bonded 3D platform [46].

15.5.2.3 CMP for Via-First Approach to 3D Using Copper-to-Copper Bonding Demonstrated copper-to-copper bonding approaches to 3D integration [38] use a dielectric recess to assure that copper interwafer interconnect pads touch down on one another without interaction between field surfaces. As a result, the planarity requirement at the bonding interface is different from most wafer bonding techniques because voids (sometimes referred to as air or vacuum gap dielectric) are an expected, and desired, result of the approach. More recent work suggests that the shape of the copper post above the dielectric recess can affect the ability of patterned copper to form copper-to-copper bonds [113]. The planarity requirements for the interwafer interconnect are also somewhat different from other approaches, because the dielectric recess can be done after damascene patterning copper in oxide.

15.5.2.4 CMP for Via-First 3D Using Redistribution Layer Bonding In the approach to 3D integration that involves bonding using redistribution layers, planarity considerations for the bonding interface and for the interwafer interconnect are unified into one surface preparation step. Fundamentally, the contact mechanics are similar to the oxide bonding approach above, as solid surfaces on a silicon substrate are coming into contact to bond in both cases. However, in this case the partially cured BCB is significantly softer and more flexible than oxide; as such, it accommodates more nanoscale roughness than oxide-to-oxide surfaces.

Nonplanarities arising from damascene patterning of copper into partially cured BCB in the redistribution layer bonding approach can lead to bond defectivity. As discussed in Section 15.5.1, factors like dielectric loss, dishing, and erosion will cause feature scale step heights. Work reported by Gutmann et al. [114] discusses the magnitude of these step heights and their effect on bonding. Good bonding is reported for areas where feature scale step heights are below 100 nm, implying that the BCB is flexible enough to accommodate these nonplanarities. A die layout with unpatterned area between each column was employed in further work to exaggerate these nonplanarities, and step heights were measured using a mapping profilometer as shown in Fig. 15.15. The nonplanarity caused by the nonideal die layout was determined to be 0.5–1.5 μm, and the nonplanarity over the entire wafer was determined to be 2.0–2.5 μm.

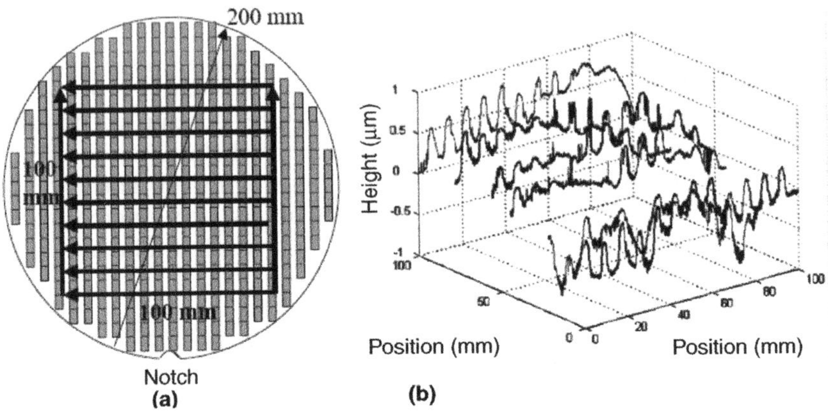

FIGURE 15.15 (**a**) Layout of dice and profilometry scans over a 200-mm wafer for wafer-level height distribution study. (**b**) Wafer-level height distribution for CMP over a nonideal die layout (four of the ten scan lines have been removed for clarity) (from Ref. 114).

Since this surface topography acts over distances greater than the thickness of the wafer, Equation 16.2 implies that the substrate will need to flex to accommodate these heights. Using Fig. 15.13 and a surface energy of 30 mJ/m^2 for BCB [115], a topography of 2 μm can be expected to be accommodated over lengths in the range of 20 mm. This implies that the contact forces arising from placing the BCB films in contact are on the same order of magnitude as the elastic forces required to overcome the topography measured in this wafer-level nonplanarity study.

15.6 CONCLUSIONS

3D integration promises to provide a new paradigm in the fabrication capabilities of ICs: from reduction of form factor and the integration of heterogeneous processes by stacking ICs vertically to increased signal-processing speed, bandwidth, and power performance through shortening the interconnect length.

The fundamental nature of wafer-level 3D requires the integration of wafer bonding into standard BEOL- processing flows, and CMP has been shown to reduce bond defectivity in many wafer-bonding approaches. The significance of CMP at various points of several 3D integration approaches has been discussed, generally focusing on the surface planarity required to (1) complete all of the steps to form a good bond and (2) fabricate high-quality damascene-patterned interwafer interconnects using CMP. While Table 15.1 in Section summarizes the comparative advantages and challenges of each wafer-level

TABLE 15.2 Summary of Planarization Requirements for Processing Steps Required to Form Both the Bond and the Interwafer Interconnect for Various Wafer-Level 3D Integration Technology Platforms.

	Oxide–oxide	Cu–Cu	BCB–BCB	Redistribution Layer
Bonding	Challenging	Good	Best	Good to challenging
Interconnection	Good	Good	Challenging	Good

3D platform, Table 15.2 delineates the salient features of each approach with respect to these aspects of CMP. 3D integration using oxide bonding has strict planarization requirements at the bonding interface because of the rigidity of the materials used and the contact mechanics involved but relaxed requirements for interwafer interconnection because the materials used are easily integrated into BEOL processing. Owing to the use of a dielectric recess, 3D using copper bonding has moderate planarization requirements, both at the bonding interface and for interwafer interconnection. 3D using dielectric adhesives (such as spin-on BCB or polyimide) have relaxed planarization requirements at the bonding interface due to the planarizing nature of the coating; however, planarization challenges for the interwafer interconnection remain because of the need for HAR processing. 3D integration using redistribution layer bonding has new planarization requirements at the bonding interface, since both Cu–Cu connection and BCB–BCB bonding are needed. However, due to the soft BCB bonding material, only moderate planarization requirements for interwafer interconnections are required.

QUESTIONS

1. Discuss the fundamental mechanism for oxide-to-oxide bonding, copper-to-copper bonding, and adhesive bonding.
2. Compare the advantages of die stacking to the advantages gained by wafer-level 3D.
3. How do alignment, thinning, and HAR processing affect the density of interconnects possible in a wafer-level 3D approach?
4. Discuss the advantages of using 3D integration at the early stages of BEOL fabrication (with 0–2 levels of metal present) for each wafer-level 3D approach discussed. Repeat for late stages of BEOL fabrication (with as many as 10 levels of metal present).
5. Define the planarization length. What are typical planarization lengths for IC fabrication? How does this relate to wafer-level 3D and wafer bonding?
6. How do planarity considerations at the feature level and the wafer level affect bond defectivity for various bonding approaches?

7. What are the issues associated with manufacturing (such as yield, reliability, and cost) when nonplanar structures are bonded for various wafer-level 3D approaches?

REFERENCES

1. Davis JA, Venkatesan R, Kaloyeros A, Beylansky M, Souri SJ, Banerjee K, Saraswat KC, Rahman A, Reif R, Meindl JD. Interconnect limits on gigascale integration for the 21st century. Proc IEEE 2001;89 (3):305–324.
2. Rahman A, Fan A, Chung J, Reif R. Wire-length distribution of three-dimensional integrated circuits. IEEE International Interconnect Technology Conference;1999.p 233–235.
3. Zhao Y. Standardization of mobile phone positioning for 3G systems. IEEE Communications Magazine;July 2002. p 108–116.
4. Barton J, Delaney K, Bellis S, O'Mathuna C, Paradiso J, Benbasat A. Development of distributed sensing systems of autonomous micro-modules. Proceedings of the IEEE Electronic Components and Technology Conference;2003. p 1112–1118.
5. Cook B, Lanzisera S, Pister K,SoC issues for RF smart dust. Proc IEEE 2006;94 (6):1177–1196.
6. Krishnaswamy D, Stevens R, Hasbun R, Revilla J, Hagan C. The Intel© PXA800F wireless internet-on-a-chip architecture and design. IEEE Custom Integrated Circuits Conference; 2003. p 39–42.
7. http://www.internationalsensor.com
8. Kuo A-Y, Mu Z. The IC package: missing link between nanometer silicon and multi-gigabit PCB systems. Advanced Packaging; Feb 2006. p 16–18.
9. Swinnen B, Beyne E. 3-D-stacked ICs with copper nails allows system size reduction. Advanced Packaging; Feb 2006. p 27–28.
10. Gann K,Neo-stacking technology. High Density Interconnect Magazine; Vol. 2, 1999.
11. Gianchandi YB, Ma KJ, Najafi K. A CMOS dissolved wafer process for integrated P++ microelectromechanical systems. In Proceesings of The 8th International Conference on Solid State Sensors and Actuators; 1995. p 79–82.
12. Parameswaran L, Hsu C, Schmidt M. A merged MEMS-CMOS process using silicon wafer bonding. Proceedings of the IEEE International Electron Devices Meeting; 1995. p 613–616.
13. Smith J, Montague S, Sniegowski J, Murry J, McWhorter P. Embedded micromechanical devices for the monolithic integration of MEMS with CMOS. Proceedings of the IEEE International Electron Devices Meeting; 1995. p 609–612.
14. Baltes H, Brand O, Hierlmann A, Lange D, Hagleitner C. CMOS MEMS—present and future. Proceedings of the IEEE 15th International Conference on MEMS; 2002. p 459–466.
15. Bamal M, List S, Stucci M, Verhulst A, M VanHove, Cartuyvels R, Beyer G, Maex K. Performance comparison of interconnect technology and architecture options

for deep submicron technology nodes. Proceedings of IEEE Interconnect Technology Conference,; 2006. p 202–204.

16. Kada M, Smith L. Advancements in stacked chip scale packaging (S-CSP), provides system-in-a-package functionality for wireless and handheld applications. Proceedings of Pan Pacific Microelectronics Symposium Conference; 2000. p 1–7.
17. Beyne E. The rise of the 3rd dimension for system integration. Proceedings of the IEEE International Interconnect Technology Conference; 2006. p 1–5.
18. Goldstein H. Packages go vertical. IEEE Spectrum 2001;38(38):46–51.
19. http://www.valtronic.ch/home.html
20. Pinkerton G. Miniaturized electronics: driving medical innovation. Medical Device & Diagnostic Industry, March 2003.
21. http://www.tessera.com
22. Solberg V, Mitchell C. Practical and cost effective solutions for 3D IC packaging. Pan Pacific Microelectronics Symposium Proceedings,2003.
23. Tong Q-Y, Enquist P, Rose A. Room temperature metal direct bonding. US patent0161795 A1,2005.
24. Singer P. New 3-D chip interconnect technology. Semiconductor Int 2005;28 (12):26.
25. Markunas B. 3D architectures for semiconductor integration and packaging. Presented at the RTI International Technology Venture Forum, Burlingame, CA,2004.
26. Guarini KW, Topol AW, Ieong M, Yu R, Shi L, Newport MR, Frank DJ, Singh DV, Cohen GM, Nitta SV, Boyd DC, O'Neil PA, Tempest SL, Pogge HB, Purushothaman S, Haensch WE. Electrical integrity of state-of-the-art 0.13 mm SOI CMOS devices and circuits transferred for three-dimensional (3D) integrated circuit (IC) fabrication. Digest of International Electron Device Meeting; 2002. p 943–945.
27. Topol AW, Furman BK, Guarini KW, Shi L, Cohen GM, Walker GF, Enabling technologies for wafer-level bonding of 3D MEMS and integrated circuit structures. Proceedings of the IEEE 55th Electronic Components and Technology Conference (ECTC); 2004. p 931–938.
28. Topol AW, Guarini KW, Yu R, Shi L, Newport MR, Tornello J, Melick D, O'Neil PA, Colburn M, Singh DV, Cohen GM, Krishnan M, Ruiz N, Pogge HB, Ieong M, Purushothaman S, Haensch WE. A demonstration of wafer-level layer transfer of high-performance devices and circuits for three-dimensional integrated circuit fabrication. Proceedings of the 4th International Conference on Microelectronics and Interfaces; 2003. p 5–7.
29. Burns J, Aull B, Chen C, Chen C-L, Keast C, Knecht J, Suntharalingam V, Warner W, Wyatt P, Yost D-R. A wafer-scale 3-D circuit integration technology. IEEE Trans Elect Device 2006;53(10):2507–2516.
30. Warner K, Chen C, D'Onofrio R, Keast C, Poesse S. An investigation of wafer-to-wafer alignment tolerances for three-dimensional integrated circuit fabrication. Proceedings of The IEEE International SOI Conference; 2004. p 71–72.
31. Suntharalingam V, Berger R, Burns J, Chen C, Keast C, Knecht J, Lambert R, Newcomb K, O'Mara D, Rathman D, Shaver D, Soares A, Stevenson C, Tyrell B, Warner K, Wheeler B, Yost D, Young D. Megapixel CMOS image sensor

fabricated in three-dimensional integrated circuit technology. Proceedings of the IEEE International Solid-State Circuits Conference; 2005. p 356–357.

32. Chen KN, Fan A, Tan CS, Reif R. Contact resistance measurement of bonded copper interconnects for three-dimensional integration technology. IEEE Elect Device Lett 2004;25(1):2004.
33. Chen KN, Tan CS, Reif R. Abnormal contact resistance reduction of bonded copper interconnects in three-dimensional integration during current stressing. Appl Phys Lett 2005;86 (011903):011903-1–011903-3.
34. Chen KN, Fan A, Rief R. Microstructure examination of copper wafer bonding. J Electr Mater 2001;30 (4):331–335.
35. Chen KN, Fan A, Tan CS, Reif R. Microstructure evolution and abnormal grain growth during copper wafer bonding. Appl Phys Lett 2002;81(20):3774–3776.
36. Chen KN, Tan CS, Fan A, Rief R. Morphology and bond strength of copper wafer bonding. Electrochem Solid State Lett 2004;7(1):G14–G16.
37. Morrow P, Kobrinsky MJ, Ramanathan S, Park C-M, Harmes M, Ramachandrarao V, Park H-M, Kloster G, List S, Kim S. Wafer level 3D interconnects via Cu bonding. Advanced Metallization Conference 2004; 2004. p 125–130.
38. Morrow P, Park C-M, Ramanathan S, Kobrinsky MJ, Harmes M. Three-dimensional wafer stacking via Cu–Cu bonding integrated with 65-nm strained-Si/low-*k* CMOS technology. IEEE Elect Device Lett 2006;27(5):335–337.
39. Patti R. Three-dimensional integrated circuits and the future of system-on-chip designs. Proceedings of the IEEE 2006;94 (No. 6):1214–1222.
40. http://www.tezzaron.com
41. Gupta S, Hilbert M, Hong S, Patti R. Techniques for producing 3D ICs with high-density interconnect. In Proceedings of 2004 VLSI Multilevel Interconnection Conference; 2004. p 93–97.
42. Advance Data Product Information:1/2/4Mb FaStack™ Synchronous Burst SRAM, Tezzaron Semiconductor Corp., Rev. 1.2, August 20, 2004.
43. Sailer PM, Singhal P, Hopwood J, Kaeli DR, Zavracky PM, Warner K, Vu DP. Creating 3D circuits using transferred films. Circuit Device 1997;13(6):27–30.
44. Burns J, McIlrath L, Keast C, Loomis A, Warner K, Wyatt P. Three-dimensional integrated circuits for low-power, high-bandwidth systems on a chip. Proc IEEE ISSCC 2001;453:268–269.
45. Lee KW, Nakamura T, One T, Yamada Y, Mizukusa T, Hasimoto H, Park KT, Kurino H, Koyanagi M. Three dimensional shared memory fabricated using wafer stacking technology. Digest of International Electron Device Meeting; 2000. p 165–168.
46. Lu J-Q, Jindal A, Kwon Y, McMahon JJ, Lee K-W, Kraft RP, Altemus B, Cheng D, Eisenbraun E, Cale TS, Gutmann RJ. 3D system-on-a-chip using dielectric glue bonding and Cu damascene inter-wafer interconnects. International Symposium on Thin Film Materials, Processes, and Reliability, at the 203rd Meeting of The Electrochemical Society, Inc., Mathad S, Cale TS, Collins D, Engelhardt M, Leverd F, Rathore HS. editorsVol. PV2003-132003. p 381–389.
47. Jindal A, Lu J-Q, Kwon Y, Rajagopalan G, McMahon JJ, Zeng AY, Flesher HK, Cale TS, Gutmann RJ. Wafer thinning for monolithic 3D interconnects. In

Materials, Technology, and Reliability of Advanced Interconnect, MRS Symposium Proceedings; 2003. Vol. 766,21–26.

48. Kwon Y, Seok J, Lu J-Q, Cale TS, Gutmann RJ. Thermal cycling effects on critical adhesion energy and residual stress in benzocyclobutene-bonded wafers. J Electrochem Soc 2005;152 (4):G286–G294.
49. Gutmann RJ, Lu J-Q, Pozder S, Kwon Y, Menke D, Jindal A, Celik M, Rasco M, McMahon JJ, Yu K, Cale TS. A wafer-level 3D IC technology platform. Proceedings of Advanced Metallization Conference; 2003. p 19–26.
50. Pozder S, Lu J-Q, Kwon Y, Zollner S, Yu J, McMahon JJ, Cale TS, Yu K, Gutmann RJ. Back-end compatibility of bonding and thinning processes for a wafer-level 3D interconnect technology platform. Proceedings of International Interconnect Technology Conference; 2004. p 102–104.
51. McMahon JJ, Lu J-Q, Gutmann RJ. Wafer bonding of damascene-patterned metal/adhesive redistribution layers for via-first 3D interconnect. In Proceedings of the IEEE Electronic Components and Technology Conference; 2005. p 331–336.
52. Kleiner M, Kuhn S, Ramm P, Weber W. Thermal analysis of vertically integrated circuits. In Proceedings of IEEE International Electron Devices Meeting; 1995. p 487–490.
53. Goplen B, Sapatnekar S. Thermal via placement in 3D ICs. In Proceedings of the American Computing Machinery International Symposium on Physical Design; 2005. p 167–174.
54. Wilkerson P, Raman A, Turowski M. Fast, automated thermal simulation of three-dimensional integrated circuits. In Proceedings of Inter Society Conference on Thermal Phenomena; 2004. p 706–713.
55. Mirza A. One micron precision, wafer-level aligned bonding for interconnect, MEMS and packaging applications. Proceedings of the Electronic Components and Technology Conference; 2000. p 676–680.
56. White D, Wood OII. Novel alignment system for imprint lithography. J Vacuum Sci Technol B 2000;18 (6):3552–3556.
57. Li W, Bing-heng L, Yu-cheng D, Zhi-hui Q, Hong-zhong L. A nano-scale alignment method for imprint lithography. Fron Mech Eng China 2006;1 (2):157–161.
58. Niklaus F, Enoksson P, Kalvesten E, Stemme G. A method to maintain wafer alignment precision during adhesive wafer bonding. Sensors Actuators A 2003;107:273–278.
59. Lee S, Niklaus F, McMahon J, Yu J, Kimar R, Li H-F, Gutmann R, Cale T, Lu J-Q. Fine keyed alignment and bonding for wafer-level 3D ICs. Proceedings of MRS Spring Meeting; 2006. Forthcoming.
60. Bruel M. Silicon-on-insulator material technology. Electr Lett 1995;31 (14): 1201–1202.
61. Tong Q-Y, Gosele U. Semiconductor Wafer Bonding. New York:Wiley;1999.
62. Gosele U, Tong Q-Y, Schumacher A, Krauter G, Reiche M, Plobl A, Kopperschmidt P, Lee T-H, Kim W. Wafer bonding for microsystems technologies. Sensors Actuators 1999;74:161–168.
63. Dragoi V, Ferrens S, Lindner P. Low temperature MEMS manufacturing processes: plasma activated wafer bonding. Proceedings of Materials Research Society Spring Meeting; 2005. p J.7.1.1–J.7.1.6.

64. Chen K, Fan A, Reif R. Interfacial morphologies and possible mechanisms of copper wafer bonding. J Mater Sci 2002;37:3441–3446.
65. Ryu C. Microstructure and reliability of copper interconnects. Ph.D. Thesis, Stanford University, 1998. p 100.
66. Tadepalli R, Thompson CV. Quantitative characterization and process optimization of low-temperature bonded copper interconnects for 3-D integrated circuits. Proceedings of The IEEE International Interconnect Technology Conference; 2003. p 36–38.
67. Scherban T, Sun B, Blaine J, Block C, Jin B, Andideh E. Interfacial adhesion of copper-low *k* interconnects. Proceedings of The IEEE International Interconnect Technology Conference; 2001. p 257–259.
68. Khun SA, Kleiner MB, Ramm P, Weber W. Interconnect capacitances, crosstalk, and signal delay in vertically, integrated circuits. Proceedings of The International Electron Devices Meeting 1995. p 10.3.1–10.3.4.
69. Kleiner MB, Kuhn SA, Ramm P, Weber W. Performance improvement of the memory of RISC-systems by application of 3-D technology. IEEE Trans Comp Pack Manuf Technol B 1996;19 (4):709–718.
70. S van der GroenRosmeulen M, Jansen P, Baert K, Deferm L. CMOS compatible wafer scale adhesive bonding for circuit transfer. Proceedings of The International Conference on Solid-State Sensors and Actuators; 1997. p 629–632.
71. Kwon Y. Wafer bonding for three dimensional integration. Ph.D. Thesis, Rensselaer Polytechnic Institute, 2004.
72. Kwon Y, Seok J, Lu J-Q, Cale T, Gutmann RJ. Critical adhesion energy of benzocyclobutene-bonded wafers. J Electrochem Soc 2006;153 (4):G347–G352.
73. Niklaus F, Andersson H, Enoksson P, Stemme G. Low temperature full wafer adhesive bonding of structured wafers. Sensors Actuators A 2001;92:235–241.
74. Niklaus F. Adhesive wafer bonding for microelectronic and microelectromechanical systems. Ph.D. Thesis, Royal Institute of Technology (KTH), Stockholm, Sweden.
75. Niklaus F, Stemme G, Lu J-Q, Gutmann RJ. Adhesive wafer bonding. J Appl Phys 2006;99:031101-1–031101-28.
76. McMahon JJ, Niklaus F, Kumar RJ, Yu J, Lu J-Q, Gutmann RJ. CMP compatibility of partially cured benzocyclobutene (BCB) for a via-first 3D IC process. Mater Res Soc Symp Proc 2005;867:W4.4.1–W4.4.6.
77. McMahon JJ, Kumar RJ, Niklaus F, Lee SH, Yu J, Lu J-Q, Gutmann RJ. Unit processes for Cu/BCB redistribution layer bonding for 3D ICs. In Proceedings of the Advanced Metallization Conference; 2005. p 179–183.
78. Niklaus F, Kumar R, McMahon J, Yu J, Lu J-Q, Cale T, Gutmann R. Adhesive wafer bonding using partially cured benzocyclobutene for three-dimensional integration. J Electrochem Soc 2006;153 (4):G291–G295.
79. Dibbs MG, Townsend PH, Stokich TM, Huber BS, Mohler CE, Heistand RH, Garrou PE, Adema GM, Berry MJ, Turlik I. Cure management of benzocyclobutene dielectric for electronic applications. In Proceedings of SAMPE Electronic Materials and Processing Conference; 1992. p 1–10.
80. Topper M, Scherpinski K, Sporle H-P, Landesberger C, Ehrmann O, Reichel H. Thin chip integration (TCI-Modules) — a novel technique for manufacturing three dimensional IC-packages. Proc Int Symp Microelectr, SPIE 2000;4339:208–211.

81. http://www.aptekindustries.com
82. Landesberger C, Scherbaum S, Schwinn G, Spohrle H. New process scheme for wafer thinning and stress-free separation of ultra thin ICs. Proceedings of Microsystems Technologies; 2001. p 431–436.
83. Landesberger C, Klink G, Schwinn G, Aschenbrenner R. New dicing and thinning concept improves mechanical reliability of ultra thin silicon. Proceedings of the IEEE International Symposium on Advanced Packaging Materials; 2001.
84. Francis D. Thinning wafers for flip chip applications. High Density Interconnect Magazine 1999;2(5):22–25.
85. Lu J-Q, Jindal A, Kwon Y, McMahon J, Cale TS, Gutmann RJ. Evaluation procedures for wafer bonding and thinning with interconnect test structures for 3D ICs. Proceedings of International Interconnect Technology Conference; 2003. p 74–76.
86. Andry P, Tsang C, Sprogis E, Patel C, Wright S, Webb B, Buchwalter L, Manzer D, Horton R, Knickerbocker J. A CMOS-compatible process for fabricating electrical through-vias in silicon. Proceedings of IEEE Electronic Components and Technology Conference; 2006. p 831–837.
87. Knickerbocker J, Patel C, Andry P, Tsang C, Buchwalter L, Sprogis E, Gan H, Horton R, Polastre R, Wright S, Cotte J. 3-D silicon integration and silicon packaging technology using silicon through-vias. IEEE J Solid-State Circuit 2006;41(8):1718–1725.
88. Spiesshoefer S, Schaper L. IC stacking technology using fine pitch, nanoscale through silicon vias,in Proceedings of the IEEE Electronic Components and Technology Conference; 2003;p 631–633.
89. Spiesshoefer S, Schaper L, Burkett S, Vangara G, Rahman Z, Arunasalam P. Z-Axis interconnects using fine pitch, nanoscale through-silicon vias: process development. Proceedings of the IEEE Electronic Component and Technology Conference; 2004. p 466–471.
90. Chambers A, Ashraf H, Hopkins J, Pink J. Through-wafer via etching. Advanced Packaging ;April 2005. p 16–20.
91. Nguyen N, Boellaard E, Pham N, Kutchoukov V, Craciun G, Sarro P. Through-wafer copper electroplating for three-dimensional interconnects. J Micromech Microeng 2002;12:395–399.
92. He A, Osborn T, Bidstrup S, Allen P, Kohl P. Low temperature bonding of copper pillars for all-copper chip-to-substrate interconnects. Electrochem Solid-State Lett 2006;9 (12):C192–C195.
93. Mendel E. Polishing of silicon. Solid State Technol 1967;10:27–39.
94. Steigerwald J, Murarka S, Gutmann R. Chemical Mechanical Planarization of Microelectronic Materials. Wiley; New York 1997.
95. Tong Q, Lee T-H, Kim W-J, Tan T, Gosele U. Feasibility study of VLSI device layer transfer by CMP PETEOS direct bonding. Proceedings of IEEE International SOI Conference; 2001. p 36–37.
96. Borst C, Gill W, Gutmann RChemical–Mechanical Polishing of Low Dielectric Constant Polymers and Organosilicate Glasses. Kluwer Academic Publishers; Boston 2002.
97. Babu S, Danyluk S, Krishnan M, Tsujimura M. Chemical mechanical polishing – fundamentals and challenges. MRS Symposium Proceedings; Vol. 566, 2000.

98. Singer P. Copper CMP: taking aim at dishing. Semicond Int 2004;27 (11):38–42.
99. Kwak B, Sun S-S, Falk C, Burke P. Global uniformity optimization and its impact on the distribution of physical and electrical properties of Cu damascene lines. Proceedings of the Advanced Metallization Conference; 2005. p 495–499.
100. Ramachandran V, Hagan J, Hong D, Jamin F, Kaltalioglu E, Kim S, Li W-K, Marokkey S, Naujok M, Park K-C, Siew Y, Cowley A. Impact of CMP-induced topography on a 65 nm technology copper damascene interconnect module. Proceedings of the Advanced Metallization Conference; 2005. p 533–537.
101. Turner K, Spearing S, Baylies W, Robinson M, Smythe R. Effect of nanotopography on direct wafer bonding: modeling and measurements. IEEE Trans Semicond Manuf 2005;18 (2):289–296.
102. Lu J-Q, Rajagopalan G, Gupta M, Cale T, Gutmann R. Planarization issues in wafer-level three-dimensional (3D) integration. MRS Symp Proc 2004;816:K7.7.1–K7.7.10.
103. Gutmann RJ, McMahon JJ, Lu J-Q. Gobal planarization requirements for 3D ICs. Proceedings of the 3rd World Tribology Congress; 2005.
104. Lee B, Gan T, Boning D, Hester P, Poduje N, Baylies W. Nanotopography effects on chemical mechanical polishing for shallow trench isolation. Proceedings of IEEE/SEMI Advanced Semiconductor Manufacturing Conference; 2000. p 425–432.
105. Braun A,The wafer's edge. Semicond Int 2006;29 (3):44–48.
106. Bengtsson S, Ljungberg K, Vedde J. The influence of wafer dimensions on the contact wave velocity in silicon wafer bonding. Appl Phys Lett 1996;69 (22):3381–3383.
107. Turner K, Thouless M, Spearing S. Mechanics of Wafer Bonding: Effect of Clamping. J Appl Phys 2004;95 (1):349–355.
108. Turner K, Spearing S. Modeling of direct wafer bonding: effect of wafer bow and etch patterns. J Appl Phys 2002;92(12):7658–7666.
109. Stokich T, Fulks C, Bernius M, Burdeaux D, Garrou P, Heistand RH. Planarization with Cyclotene^{TM} 3022 (BCB) polymer coatings. MRS Symposium Proceedings; Vol. 308,1993.
110. Lu J-Q, Lee K, Kwon Y, Rajagopalan G, McMahon J, Altemus B, Gupta M, Eisenbraun E, Xu B, Jindal A, Kraft R, McDonald J, Castracane J, Cale T, Kaloyeros A, Gutmann R. Processing of inter-wafer vertical interconnects in 3D ICs. Proceedings of the Advanced Metallization Conference; 2002.
111. Aruanchalam V, Smith G, Kailasam S, Knorr A, Hettiaratchi K, Rozbicki R, Pfeifer K, Ho P, Pyun J. Comparison of barrier-first and argon pre-clean first processes for copper metallization in ultra low-*k* (ULK) dual damascene integration. Proceedings of the Advanced Metallization Conference; 2005; p 413–419.
112. Alers G, Rozbicki R, Harm G, Kailasam S, Ray G, Danek M. Barrier-first integration for improved reliability in copper dual damascene interconnects. Proceedings of the IEEE International Reliability Physics Symposium; 2003; p 27–29.
113. Chen K, Tsang C, Topol A, Lee S, Furman B, Rath D, Lu J-Q, Young A, Purushothaman S, Haensch W. Improved manufacturability of Cu bond pads and

implementation of seal design in 3D integrated circuits and packages. Proceedings of the VLSI Multilevel Interconnection Conference; 2006; p 195–202.

114. Gutmann R, McMahon J, Rao S, Niklaus F, Lu J-Q. Wafer-level via-first 3D integration with hybrid-bonding of Cu/BCB redistribution layers. Proceedings of the International Wafer Level Packaging Congress; 2005.

115. Luo S, Wong CP. Surface property of passivation and solder mask for flip chip packaging. Proceedings of the IEEE Electronic Components and Technology Conference; 2001.

16

POST-CMP CLEANING

Jin-Goo Park, Ahmed A. Busnaina and Yi-Koan Hong

16.1 INTRODUCTION

With the decrease in device feature sizes, planarization of both front- and back-end layers by the chemical–mechanical planarization (CMP) process has become necessary for integrated circuit (IC) fabrication technologies smaller than 0.35 μm. As much as the industry has benefitted tremendously from the introduction and implementation of CMP, the very process has also brought along a set of new challenges to the fabs. For example, surface defects or imperfections such as leftover particles, organic residues, metallic contaminations, scratches, and corrosion spots can often be found on the polished wafers after the CMP process. These CMP-induced defects can arise from the slurries, the polishing pad, the pad conditioner, and to a lesser extent, the polisher. The leftover particles can physically attach to the wafer surface or, in the worst case, even partially embed into the top layer of a wafer. The CMP process can also leave metallic contamination typically in the range of 10^{11}–10^{12} atoms/cm^2. These contaminants mainly arise from the abraded metal lines, metal ions in the slurries, and polishers. In front-end applications such as a shallow trench isolation (STI) process, the control of metallic contamination levels is very critical because the following high-temperature process may lead to a complete incorporation of the metal ions into the lattice. In the case of back-end processes, if not removed, these metal contaminants can lower the breakdown voltage of devices. Also, fast diffusers such as copper can reach the active area from the backside surface or edge during the following thermal process [1]. During CMP, slurry additives and pad debris can leave organic residues on the

Microelectronic Applications of Chemical Mechanical Planarization, Edited by Yuzhuo Li

wafer, which may require extra cleaning steps such as plasma ashing or SC1 (standard cleaning 1) that are not desirable in mass production. With the ever-increasing demand for lowering surface defectivity, the requirement for post-CMP cleaning has become more and more stringent at each technology node, which in turn has accelerated the tremendous growth of the field in market size (see Chapter 2), innovative equipments, more effective cleaning solutions, and better understanding of the interfacial behaviors during the cleaning process. In this chapter, the importance of post-CMP cleaning will be first introduced. The unique aspects of post-CMP cleaning process for each important application will be discussed, such as copper, STI, oxide, and poly-Si cleaning. The fundamental rationales behind some of the postcleaning strategies will be presented. Some of the post-CMP cleaning techniques will be examined along with some case studies.

16.2 TYPES OF POST-CMP CLEANING PROCESSES

Wet cleaning process was the first to be adopted and is still the most commonly used platform for post-CMP cleaning processes for removing leftover surface particles, organic residues, and metallic contaminations. In this section, various wet cleaning methods used in semiconductor process will be reviewed.

16.2.1 Wet Bath Type Cleaning

In a wet batch cleaning process, a cassette of 25 wafers is usually processed together in wet processing stations to remove specific contaminants. One advantage of the wet bath type cleaning is the lowest cost of ownership (COO). However, in CMP, it should be pointed out that batch-type cleaning process is not always required or optimal if the total number of wafers to be processed is low in each polishing step. In order to remove all types of particles, residues, and contaminants from the wafer surface, the wet cleaning process is typically composed of several steps. As shown in Fig. 16.1, in the first step, Piranha or SPM (sulfuric–peroxide mixture, $H_2SO_4 + H_2O_2$) cleaning targets the organic contaminants including organic particles. Because of the high viscosity of SPM, QDR (quick dump rinse) is usually used after the SPM step. An SC1 (using a NH_4OH, H_2O_2, and H_2O mixture) step is commonly applied at an elevated temperature to remove particles with light organic contaminations. An SC2 (standard cleaning 2 using a mixture of HCl, H_2O_2, and H_2O) process is designed to solubilize metallic contaminants by forming complexes with its Cl^- ion. DHF (dilute HF, typically less than 0.5 vol%) has been used to remove a thin layer of oxide from wafer surfaces, which in turn could effectively remove metallic contamination adsorbed or trapped in the oxide. It should also be mentioned that H_2O_2-containing chemistry often leads to metal corrosion. Therefore, precaution must be taken if metal layers are exposed after CMP.

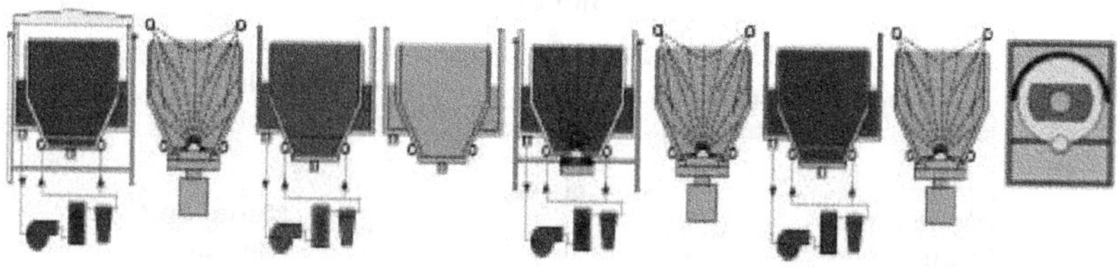

FIGURE 16.1 Typical batch-type RCA wet cleaning procedure.

16.2.2 Single-Wafer Cleanings

To meet the stringent requirements set by the design rules for sub-100-nm technology node at 300-mm wafer, a single-wafer cleaning system (a single-bath wet station or a single spin processor) is often used in multitype small lot production. Both single bath and spin processor are usually equipped with megasonic transducers to enhance the particle-removal efficiency. Compared to the concerns about pattern damages in the FEOL (front end of the line) or BEOL (back end of the line) processes, a post-CMP cleaning process can tolerate greater physical stress because the polished surface often does not have patterned features or collapsible unsupported patterned features on the top layer. One of the key features of a single-wafer cleaning process is the use of a brush scrubber. The scrubber is usually made of polyvinyl alcohol (PVA). The advantages of single-wafer cleaning systems include high cleaning efficiency, low consumption of cleaning chemicals, and small footprint. In the following sections, several popular and commercially available systems are described.

16.2.2.1 Immersion-Type Single-Wafer Post-CMP Cleaning System An immersion-type single-wafer cleaning system offers effective cleaning of both sides of the wafer with effective control of chemical concentration and temperature. In addition, the standard immersion process can also be coupled with a Marangoni-type IPA (isopropyl alcohol) drying option by moving either the wafer or the air–liquid interface. Furthermore, the inclusion of a megasonic unit has also become very common in single-bath-type cleaners. Figure 16.2 shows a schematic diagram of a commercially available single-bath cleaner [2] in which a triple transducer megasonic cleaning unit is optimized for single-wafer cleaning. The system uses three megasonic transducers with the top two focused at the liquid–air–wafer triple boundary point. The system scans the triple boundary point across the face of the wafer. It has been reported that such a cleaning procedure causes little or no damage to the cleaned wafers because of the use of megasonic force [2].

16.2.2.2 Single-Wafer Spin Cleaner Another type of single-wafer cleaning involves a spinning wafer next to a megasonic transducer as shown in Fig. 16.3.

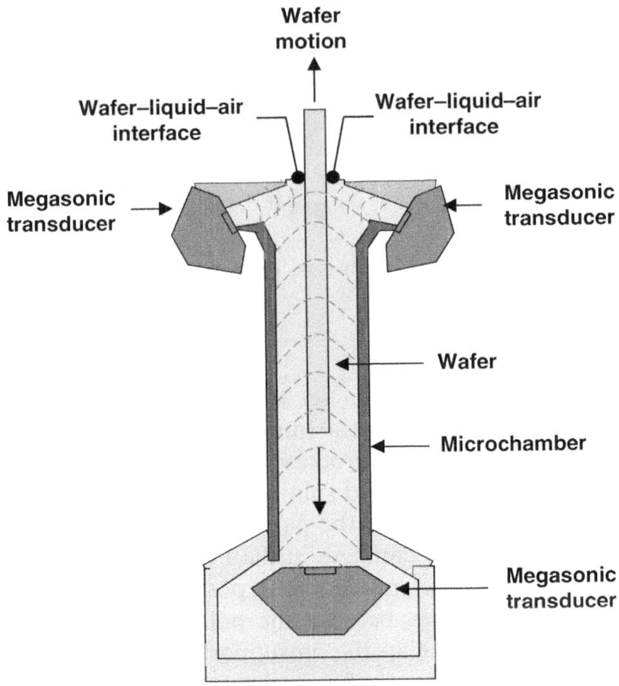

FIGURE 16.2 Triple transducer megasonic single-wafer cleaning system (from Ref. 2).

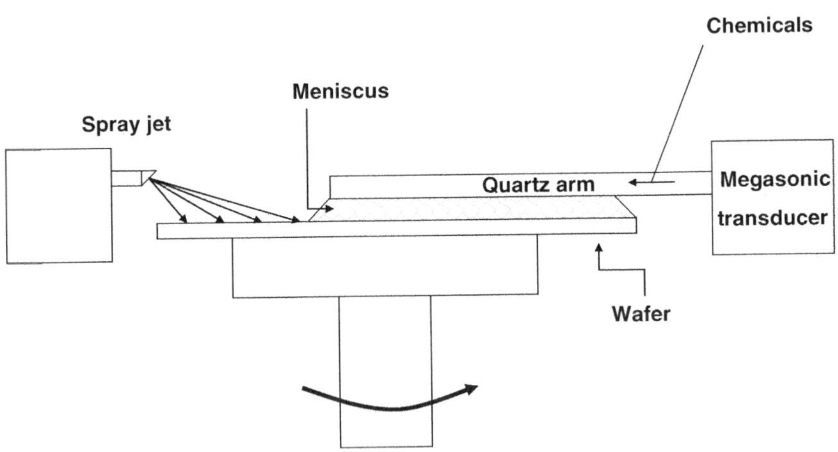

FIGURE 16.3 Schematic diagram of a single-wafer megasonic cleaner (from Ref. 3).

Unlike the bath-type cleaner, a simple spin processor usually cannot provide enough uniform cleaning power over the entire wafer surface, which results in low efficiency in the removal of fine particles from the wafer surface. The use of megasonic power over the wafer while it is spinning significantly enhances the particle-removal efficiency because the spinning is not enough to remove fine particles from wafer surfaces. The position and the design of a megasonic transducer is the key technology in a wafer spin processor for post-CMP cleaning. Figure 16.3 shows the schematic diagram of a single-wafer megasonic cleaner [3]. This cleaner uses a quartz arm to which a megasonic transducer is attached. The arm is swept across the wafer so close enough that deionized (DI) water or chemicals such as a NH_4OH solution sprayed at the wafer form a meniscus between the wafer and the quartz arm. This meniscus acts as an acoustic coupler and a waveguide between the wafer and the transducer.

Sonic cleaning is classified by the applied frequencies, that is, ultrasonic (50–150 kHz), high ultrasonic (150–800 kHz), and megasonic (above 800 kHz) [4]. In ultrasonic cleaning, the sound waves create cavities and then collapse the cavity as the sound waves go through the positive and negative phases of the wave in the cleaning solution as shown in Fig. 16.4. The ultrasonic cleaning is very effective in promoting the removal of micrometer-size particles. However, as the typical abrasive and critical particle sizes reduce to the submicrometer ranges, ultrasonic cleaning is no longer effective. The critical particle size refers to a value above which the particles will be counted as a defect. When cleaning frequencies are increased to the megasonic frequencies, two changes occur in the action of the sonic energy. At first, the cavitation strength decreases and is replaced by acoustic streaming—high-velocity pressure waves in the cleaning fluid. Secondly, the thickness of the boundary layers shrinks as shown in Fig. 16.5, which allows the effective removal of particles [5–7]. In reality, a combination of acoustic streaming, cavitation, and the level of dissolved gases and oscillatory effects all contribute to particle-removal performance [4]. It

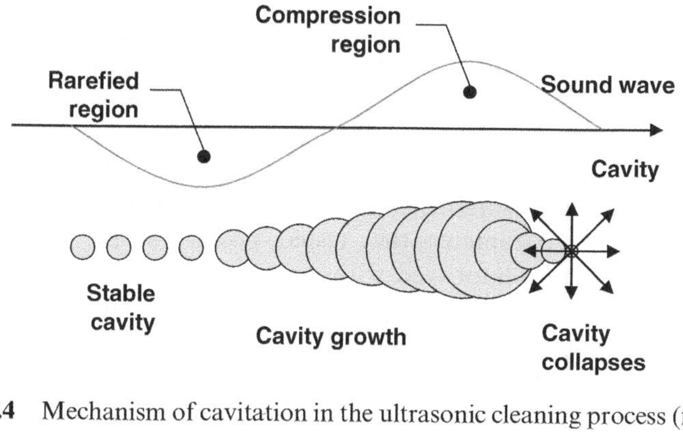

FIGURE 16.4 Mechanism of cavitation in the ultrasonic cleaning process (from Ref. 4).

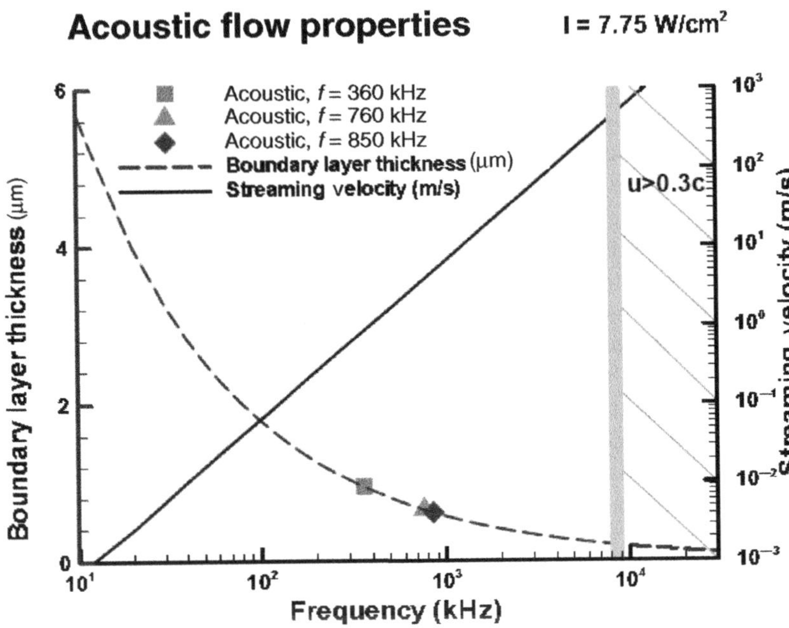

FIGURE 16.5 Boundary layer thickness as a function of acoustic frequency.

should be mentioned that a time-consuming processing is often necessary to optimize the efficiency of the megasonic generator. For sub-90-nm technology node and beyond, a new problem with megasonic cleaners has emerged. The energy required to remove a particle is now in the same realm as the energy required to damage the pattern. It thus requires careful optimization of single-wafer megasonic cleaning to lower the damages done by megasonic power [8,9].

Megasonic cleaning applications were first described in detail by RCA (Radio Corporation of America) engineers [10–12]. McQueen [13,14] later recognized the importance of acoustic streaming in decreasing the boundary layer thickness based on his studies on removing small particles from surfaces. Busnaina et al. [15–22] studied ultrasonic and megasonic particle removal, focusing on the effects of acoustic streaming. They showed that removal efficiency increased with power and frequency. Syverson et al. [23] reported a significant increase in particle-removal efficiency when the wafer was cleaned in the presence of megasonic stream and SC1 solutions. Wang and Bell [24] performed experiments using megasonic cleaning after RIE (reactive ion etch) planarization. Of the parameters they tested, power had the greatest influence on the results. Cleaning improved with increased power up to the maximum tested value of 300 W, which is consistent with what was observed by Kashkoush, Busnaina, and Gale [15–22]. A recent study by Busnaina et al. [25] shows the effects of the three key parameters (power, time, and temperature) on cleaning process using a dilute SC1. A very low defect count was obtained after the cleaning of a polished thermal oxide wafer, when the optimum power, temperature, and cleaning time are used.

16.2.2.3 Brush Cleaning Since the introduction of the CMP process, the most commonly used material for brush cleaning has been PVA. In order to receive thoroughly clean surfaces, the wafer is usually sandwiched between the double-sided brush scrubbers so that the front and back sides of the wafer can be cleaned simultaneously (Fig. 16.6). In addition, a lateral brush is used to clean the edges of the wafer.

A scrubber cleaning process can be optimized by adjusting the brush and wafer rotation speed, the DI water/chemical flow rate, and the gap between the brush and the wafer surface. Because brush cleaning is based on the mechanical pressure applied to the wafer surface, the brush must be compressed to the wafer surface to have either direct or semidirect contact with the wafer to effectively remove fine particles.

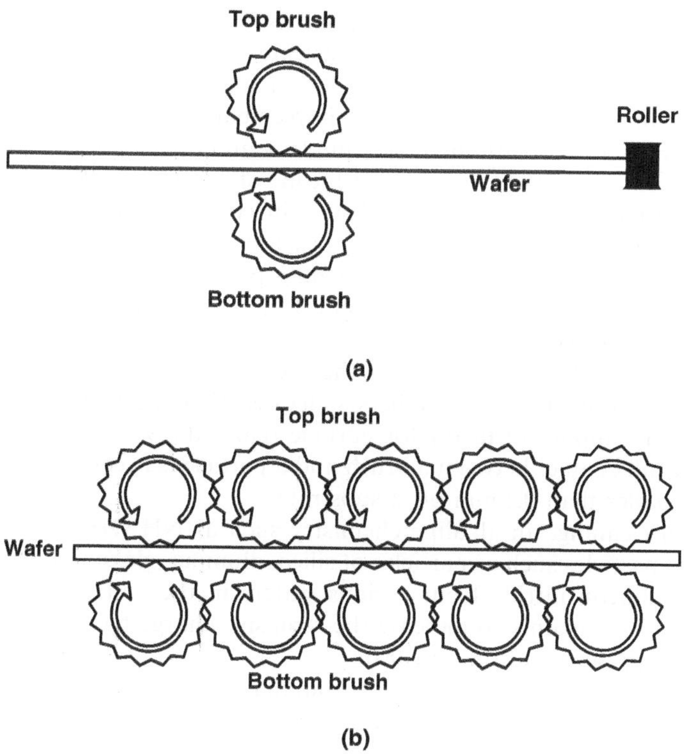

FIGURE 16.6 Cross-sectional views of post-CMP cleaning using double-sided roller brushes: (**a**) a single brush and (**b**) multiple brushes (from Ref. 26).

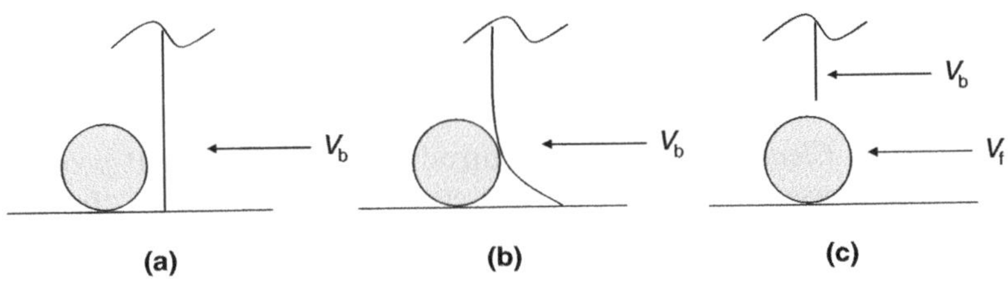

FIGURE 16.7 Mechanisms of brush cleaning: (**a**) ideal, (**b**) fiber deformed, and (**c**) noncontact mode (from Ref. 27).

Although widely used, limited theoretical work has been published in the field of brush cleaning. Figure 16.7 shows a proposed working mechanism [27] for brush cleaning. Figure 16.7a shows the ideal brush-cleaning implementation. The brush surface travels at a velocity (V_b) equal to the tip velocity (V_t) multiplied by the brush radius. The brush impinges upon the contaminant and removes it from the surface. The full force of the brush is used for the impact. Figure 16.7b shows a more realistic scenario. The brush contacts the contaminant at an angle and above the midpoint. The brush force, delivered in this manner, is divided into two components F_{bx} and F_{by}. F_{bx} is the force that works to remove the particle from the substrate. F_{by} actually works to drive the contaminant into the substrate. In order to minimize the negative vertical component of the force, a noncontact implementation of brush cleanings is shown in Fig. 16.7c. Although this implementation eliminates the risk of surface deformation, it is an inefficient means of energy transfer. The brush is used as a force to move the fluid that is in contact with the particle. The difference between the tip velocity of the brush, V_b, and the velocity of the fluid, V_f, can be quite dramatic. In brush-cleaning processes, the number of brush cycles, the brush pressure, and the brush rotating speed are the key parameters in the removal of particles.

A tribology study [28] focusing on the coefficient of friction by varying the instrumental parameters of scrubbers suggested that the hydrodynamic drag force is a very important factor for particle removal [29]. Busnaina et al. [30] found that full contact between a brush and a particle is necessary to lift or roll particles smaller than 0.1 mm off a substrate.

In brush cleaning, an alkaline chemistry such as NH_4OH is often used to remove the particles such as particles of silica, alumina, glass, polystyrene latex (PSL), and silicon nitride from various wafers in the first brush. The basic chemistry is used mainly to increase the repulsive charge by the zeta potential between the particle and the substrate.

In the case of W or Cu CMP where alumina slurries are used, the pH of the solution must be greater than 9 or lower than 2 to avoid adhesion of the slurries in the porous structure of the brush. This phenomenon, called the loading effect, increases the final particle levels on the wafer and therefore drastically reduces the brush lifetime [1].

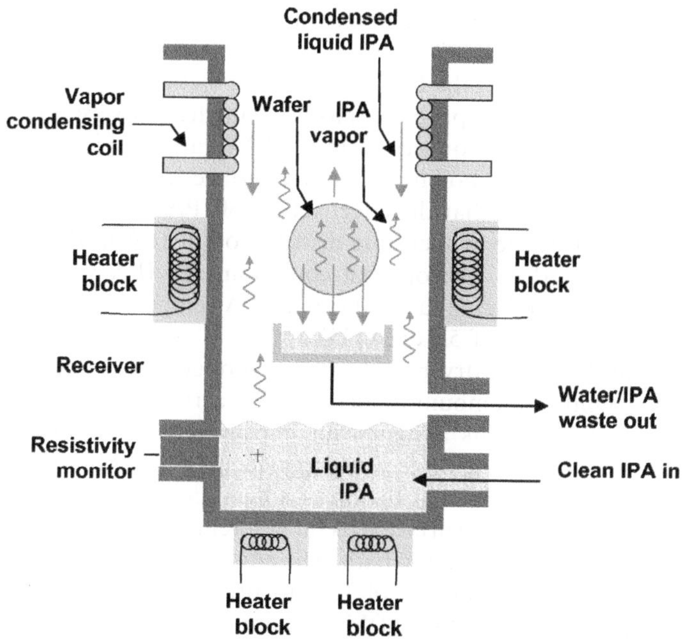

FIGURE 16.8 Schematic diagram of a hot IPA dryer system (from Ref. 31).

The diluted HF (DHF) at 0.5–1% injected in the second brush of the scrubber is known to effectively remove native oxide and particles by an underetching mechanism. Any metallic impurities in the oxide can be removed with native oxide during DHF cleaning. Dilute NH_4OH and HF chemicals are well accepted as standard chemicals in post-CMP brush cleaning. Double-sided brush scrubbing followed by a spin–rinse–dry step dominates the industry. It should be mentioned that H_2O_2 (higher than 5%) cannot be used in brush cleaning owing to the degradation of PVA by H_2O_2 chemical.

16.2.2.4 Drying Wafer drying is the last step of wet cleaning process. Generally, wafers can be dried with one of the methods such as spin drying, vapor drying, nitrogen blowing, hot air blowing, IR lamp drying, vacuum drying, slowly drawing from warm water, or a combination of the above. For example, single-wafer spin drying has routinely been used by combining IR lamp irradiation during spinning. Alternatively, the spin drying process can also include a stream of heated nitrogen gas directed toward a wafer that is spun at high speed. During spin–rinse drying, the following issues can arise: static adhesion of particles caused by a charge on the high-speed rotating wafer, particle emission from the transfer mechanism part, adhesion of contaminated mist, generation of watermarks on patterned wafers, poor efficiency in removing moisture from the trench-contact holes, and wafer chipping. In addition, a spin-rinse dryer can impose significant stress on large wafers at high spin speeds [31].

In comparison, however, the IPA vapor drying method (Fig. 16.8) that does not exhibit any of the above problems has been an alternative solution to spin–rinse drying in semiconductor wet cleaning. IPA vapor drying method takes advantages of IPA's lower boiling point and surface tension relative to water. Mishima et al. [32] found that the effectiveness of IPA drying is a function of water content in IPA, the temperature distribution of IPA in the vapor dryer, and the velocity of IPA. The optimal conditions for IPA vapor drying are water content in IPA of less than 1000 ppm, IPA temperature of 82 °C at the wafers, and IPA velocity of 5.0 cm/s [33].

Although an IPA vapor dryer works well in producing dry and clean wafers, it consumes a significant amount of alcohol and carries a risk of fire hazard. A Marangoni dryer represents a significant improvement over the classic IPA dryer. Figure 16.9 shows the basic working principles of a Marangoni dryer in which IPA is readily absorbed at the tip of a meniscus where the DIW surface tension is low. The resulting surface tension gradient pulls water away from the wafer as it is lifted out of the DIW bath. The advantage of Marangoni dryers is that they use very small amount of IPA to produce a clean and dry wafer surface. One potential limitation of Marangoni drying is the drain speed (the rate at which the boundary layer moves across the wafer). Typical drain speeds today are 1–1.5 mm/s [34], which translates to a low throughput. The Marangoni drying technique is widely accepted for Cu/low-*k* BEOL post-CMP cleaning and drying because of the hydrophobic nature of low-*k* surfaces [35].

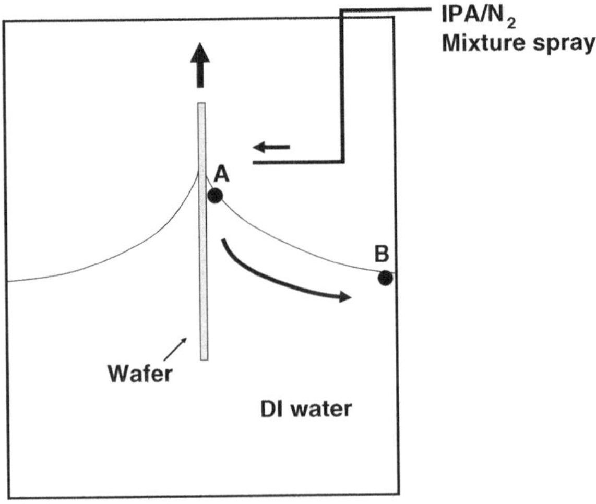

FIGURE 16.9 Marangoni drying mechanism (from Ref. 36).

16.3 POST-CMP CLEANING CHEMISTRY

16.3.1 Conventional Wet Cleanings

In the semiconductor manufacturing processes, a variety of wet cleaning chemicals have been used to remove particulate and metallic contaminants. Table 16.1 provides a list of typical wet cleaning solutions. SC1 is one of the batch-type RCA cleaning processes and has superior particle-removing effectiveness. SC1 undercuts the adsorbed contaminants by etching the Si or SiO_2 layer that supports or hosts the contaminants. The etching is a direct result of hydroxide attack on the silicon-oxygen bond in the silicon dioxide matrix. Therefore, the etching rates depend on the concentration of its key component NH_4OH, the source of OH^-. For silicon surface, the attack by the hydroxide is usually after the oxidation of silicon to silicon dioxide by H_2O_2.

In an aqueous or liquid environment, a wafer or particle can form an electrical double layer on its surface. The zeta potential of such a surface layer depends on the pH of the environment. At low pH (acidic solution), depending on the surface function groups, the zeta potential is likely to be positive because of protonation. In a basic solution, the zeta potential is more likely to be negative because of deprotonation or addition of hydroxide ions. For effective particle removal, the polarity of the wafer and particle surfaces should be the same to have a repulsive electrostatic force. Furthermore, the particles should be either subsequently dissolved or lifted up from the surface by slightly etching the surface underneath the particle. Figure 16.10 shows the typical ways to remove particles from the surface either by dissolving in oxidizing solutions or by slight etching of the substrate surface in alkaline solutions. The particle-removal mechanism of SC1 solution can be attributed to both slight etching of substrate and electric repulsion by NH_4OH and oxidation and dissolution by H_2O_2.

The oxidizing and etching effects that help in particle removal may lead to a rougher surface on the substrate such as poly-Si [37,38]. Therefore, a balance must be maintained between the need for greater etching depth to remove

TABLE 16.1 Typical Cleaning Solutions Used in the Semiconductor Manufacturing Process.

Cleaning Solution	Composition	Purpose
APM (SC1)	$NH_4OH/H_2O_2/H_2O$	Removal of organic contaminants and particles
HPM (SC2)	$HCl/H_2O_2/H_2O$	Removal of metallic ions, surface passivation by native oxide formation
SPM	H_2SO_4/H_2O_2	Removal of organic contaminants and metallic ions
DHF	HF/H_2O	Removal of native oxide film and metallic ions

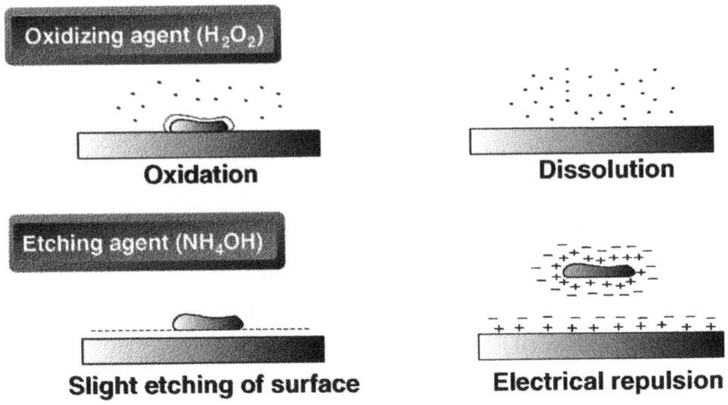

FIGURE 16.10 A schematic diagram of the particle removal mechanism in SC1 solution.

particles and the minimum concentration of ammonium hydroxide to achieve desirable surface smoothness.

In the case of SC2 cleaning, metallic contamination can be removed by the complexation of Cl^- with metal ions. An SPM solution can used to remove various organic contaminations. It has been hypothesized that there is equilibrium between $H_2SO_4 + H_2O_2$ and $H_2SO_5 + H_2O$. H_2SO_5 is a highly reactive species and reacts with hydrocarbon to form H_2SO_4. The DHF solution removes the native oxide films by etching.

In addition to these chemical reactions described above, many physical parameters need to be considered as well in terms of post-CMP cleaning. Extensive investigations have been conducted on the correlations between post-CMP cleaning efficiency and relevant physical forces such as the van der Waals force, electrostatic force, particle adhesion, chemical adsorption, surface charge modification, and wettability. It is expected that these factors strongly influence the particle-removal capability of post-CMP cleaning solution.

16.3.2 Chemicals Used in Post-CMP Cleaning and Their Roles

16.3.2.1 NH_4OH In single-wafer-type post-CMP cleaning, polished wafers are usually sprayed with dilute NH_4OH solution in the presence of megasonic radiation or followed by a PVA brush cleaning [39]. The mechanisms by which ammonia (or dilute ammonia) solution removes the particles are slight etching and creating negative zeta potential on wafer surfaces [40].

16.3.2.2 HF In the semiconductor manufacturing process, buffered HF (BHF) and DHF are widely used for the etching of oxide layers and the removal of native oxide. Most polished surfaces will have a thin layer of oxide: poly-Si, oxide, W, Cu, and so on. The removal of oxide by DHF could also help to remove metallic ions in oxide. DHF cleaning is also known to be very effective in removing particles trapped in recessed and eroded features.

TABLE 16.2 General Properties of Citric and Oxalic Acids.

	Citric acid ($C_6H_8O_7$)	Oxalic acid ($C_2H_2O_4$)
Chemical structure	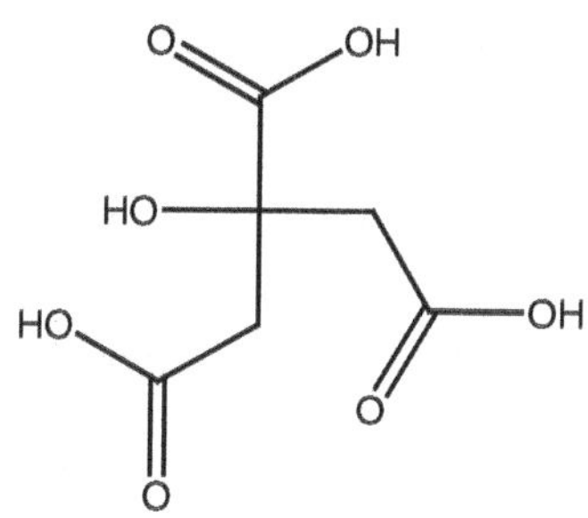	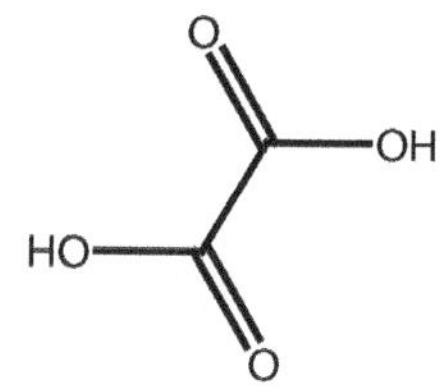
MW	192	90
Dissociation constant (pK_a) at room temperature	$H_3Cit \rightarrow H^+ + H_2Cit^-$, $pK_1 = 5.8$ $H_2Cit^- \rightarrow H^+ + HCit^{2-}$, $pK_2 = 4.3$ $HCit^{2-} \rightarrow H^+ + Cit^{3-}$, $pK_3 = 2.9$	$H_2OX \rightarrow HOX^-$, $pK_1 = 1.2$ $HOX^- \rightarrow H^+ + OX^{2-}$, $pK_2 = 4.2$

16.3.2.3 Organic Acids In addition to HF, organic acids can also be used as an etchant. Some organic acids may also carry complexing functionality and influence the zeta potentials of surfaces [41,42]. In chemistry, organic acids (also called carboxylic acids) are acids characterized by the presence of a carboxyl group (–COOH). Most organic acids are weak acids, with only about 1% of R–COOH molecules typically dissociated into ions at room temperature in an aqueous solution (where R is the organic group). Organic acids react with bases to form carboxylate salts, in which the hydrogen of the –OH group is replaced with a metal ion. The carboxyl groups can also react with amines to form amide bonds and with alcohols to form esters. In this section, citric and oxalic acids are used as examples [43].

The physical properties of citric acid are summarized in Table 16.2. Citrates make excellent buffers for controlling the pH of acidic solutions and can form complexes with many metal ions. Oxalic acid (ethanedioic acid) is a stronger acid compared to other organic acids. Oxalic acid also combines with metals ions.

16.3.2.4 Surfactants In addition to organic acids or bases, a post-CMP cleaning solution may also contain surfactants [44] that can profoundly influence the zeta potentials of the surfaces and alter their mode of interaction. For example, as shown in Fig. 16.11, the attraction between a positively charged abrasive particle and the negatively charged wafer surface may lead to particle deposition in the absence of a surfactant (Fig. 16.11a). In the presence of an anionic surfactant in the solution, particle adhesion is controlled owing to the electrostatic force by the adsorption of negative charge (Fig. 16.11b). When

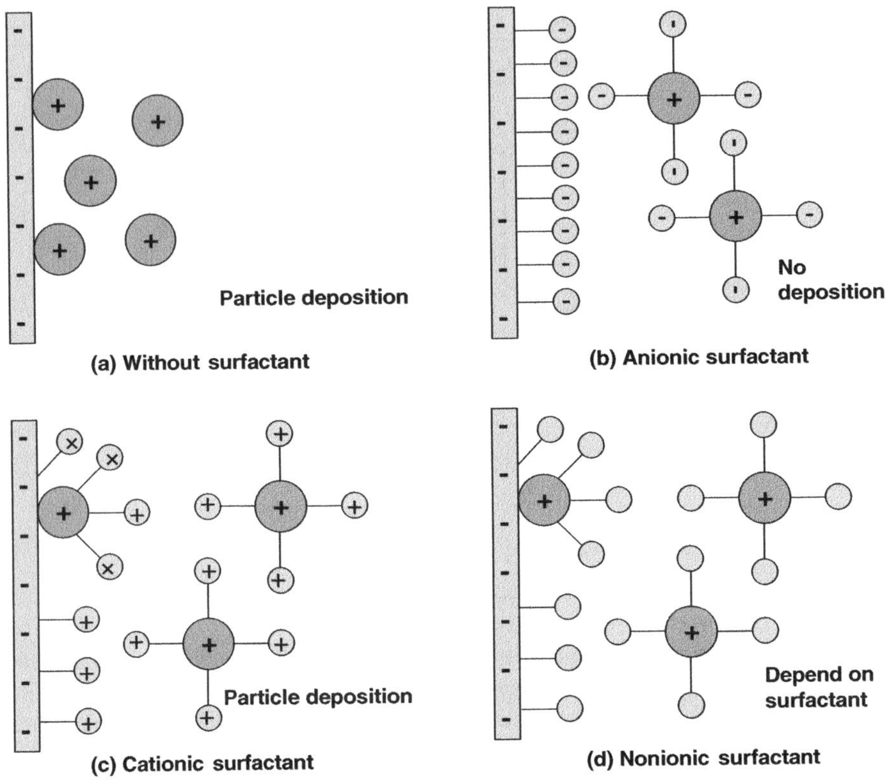

FIGURE 16.11 Model of particle adhesion control by addition of surfactants (from Ref. 45).

a cationic surfactant is added to cleaning solutions, there is some particle adhesion between abrasive particles and polished wafer surface due to the time of immersion and the surfactant orientation as shown in Fig. 16.11c. Although they are controlled as positive charges, the same surface charges of the abrasive particles and wafers can reduce the particle adhesion and contamination. With the addition of a nonionic surfactant as shown in Fig. 16.11d, the results similar to those of anionic surfactants or cationic surfactants are achieved depending on the surface charges of the particles and polished wafer surfaces [45].

16.4 POST-CMP CLEANING ACCORDING TO APPLICATIONS

16.4.1 Post-Oxide CMP Cleaning

In oxide CMP process, colloidal or fumed silica slurries with high pH values are often used. After the polishing, the wafer must be washed effectively to remove residual slurry and prevent them from redeposition. In principle, a stand-alone SC1 station can be used for this purpose. However, most CMP tools now have integrated post-CMP cleaners with a double-sided scrubber and a spin–rinse dryer. Typically, there are two brush stations, one can be used with NH_4OH and the other with DHF. It has been found that a combination of ammonium hydroxide and brush scrubbing, in general, provides efficient cleaning for most oxide surfaces. The cleaning efficiency is pH dependent, a diluted NH_4OH solution of pH 12 usually gives the highest particle-removal efficiency [46]. It is important to point out that this cleaning process cannot remove any microscratches caused by the polishing step. As a matter of fact, the etching nature of the cleaning solution may actually enlarge the scratch (see Chapter 17 for details).

16.4.2 Post-W CMP Cleaning

Even though Cu will be the dominating interconnect material, tungsten (W) is still commonly used in metal interconnection down to a 90-nm node. CMP is the only way to form a W plug in metal interconnection. In general, W CMP is the process that removes the W on the top of dielectric by controlling the selectivity between W and the dielectric, leaving the contact/via completely filled [47]. It is very important to control the W CMP in two aspects for a robust process; one is the finished topography such as erosion, dishing, plug recess, and seam, and the other is the level of defects such as slurry particles, microscratch, and metallic impurity. After W CMP, the recessed plugs can easily retain the slurry particles trapped during the CMP process. These particles are particularly difficult to remove by brush scrubbing because the brushes are not able to reach these particles easily. Those particles trapped inside tungsten plug deteriorate the electrical characteristics of the device by increasing the contact resistance on the plug and even reducing the reliability because those particles block the contact between the tungsten plug and the upper metal line. One solution to this problem is to use DHF chemistry in combination with a brush scrubber [48]. HF lifts off the particles by etching the plug during scrubbing and effectively removes them. In some cases, however, the device may not be able to tolerate the presence of HF [49]. To overcome this challenge, another buffing step with oxide slurry is usually added after the regular W CMP steps. Owing to its high selectivity over W, the oxide buffing step leads to slightly protruded tungsten plugs. The raised profiles make the brush contact much more effective and result in a particle-removal efficiency even in NH_4OH cleaning similar to that in HF brush scrubbing. However, certain devices are not allowed to expose HF chemistry on W plug due to poor electrical characteristics [49].

16.4.3 Post-STI CMP Cleaning

For STI CMP, both silica and ceria slurries have been used. Newer ceria-based slurries can stop on the nitride layer effectively and hence allow the

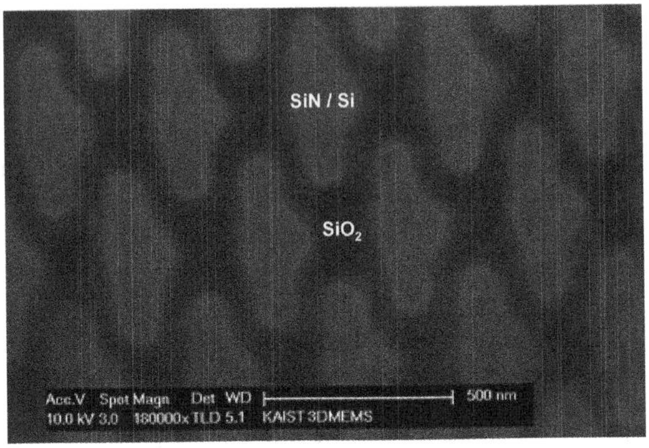

FIGURE 16.12 STI patterned wafer surfaces in the DRAM process.

integration of a direct STI process (Fig. 16.12) without the need for a reverse mask. After STI CMP, the nitride areas are often protruded above the oxide due to differences in polishing rate. Abrasive particles have been observed next to the nitride edge. These abrasive particles on a nonplanar surface will be held by the topography leaving particulate contamination after CMP. Generally, NH_4OH- or HF-assisted brush scrubbings can be applied in post-STI CMP cleaning. After the brush cleaning, a H_3PO_4-based nitride strip process is a necessary step in the formation of shallow trenches. Additional wet cleaning processes are usually performed to reduce the particulate and organic contamination from polished surfaces. In principle, the post-STI CMP cleaning process described above can eliminate a majority of the residue particles. Again, it is worthwhile to point out that the etching nature of the cleaning process may help to reveal other types of defects such as scratches and chatter marks.

16.4.4 Post-Poly-Si CMP Cleaning

Poly-Si CMP has been implemented to reduce the step height of gate poly-Si in the construction of RCAT (recess channel array transistor) and FinFET three-dimensional structures. Poly-Si CMP uses either the same or similar pads or slurries as those used for oxide CMP. Poly-Si surface is hydrophobic in nature. Therefore, after CMP, the Poly-Si surface attracts more hydrophobic organic residues than the oxide surface. These organic defects are very hard to remove by the general post-CMP cleaning method. Pan et al. used TMAH (tetramethyl ammonium hydroxide) and EDTA (ethylene diamine tetraacetic acid) as chelating agents with dilute ammonium hydroxide solution for reducing organic particles and metallic impurities [50,51], but some organic defects remained.

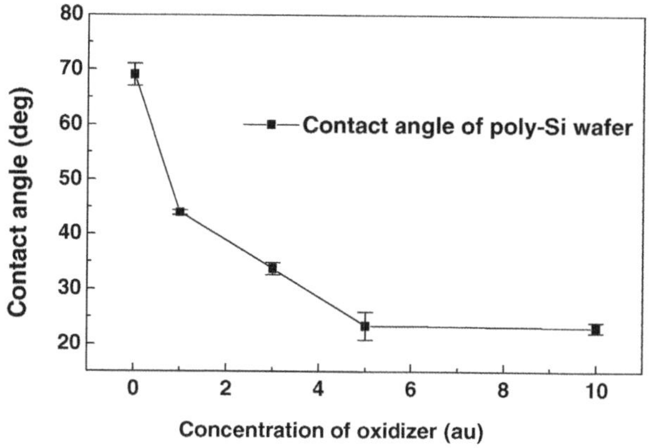

FIGURE 16.13 Contact angles of poly-Si as a function of oxidizer concentration.

Figure 16.13 shows the contact angle of poly-Si surface in DI water as a function of oxidizer concentration. Contact angles of poly-Si surface decreases from 69° to 23° as the concentration of oxidizer increases. A high contact angle indicates a poor surface wetting whereas a low angle shows a good wetting. The control of poly silicon wettability during polishing could reduce the attraction force of organic particles, thus leading to lower organic defects after CMP.

The adhesion force of pad particles on the poly-Si wafer surfaces was measured in the KOH solution (pH 11) as a function of polysilicon wettability, which was varied by treating the surface with different concentrations of the oxidizer. As shown in Fig. 16.14, the adhesion force decreases and then levels

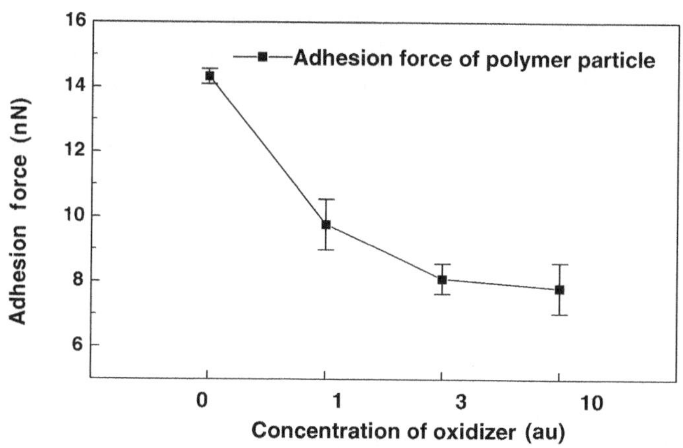

FIGURE 16.14 Adhesion force of polymeric particle on poly-Si as a function of oxidizer concentration.

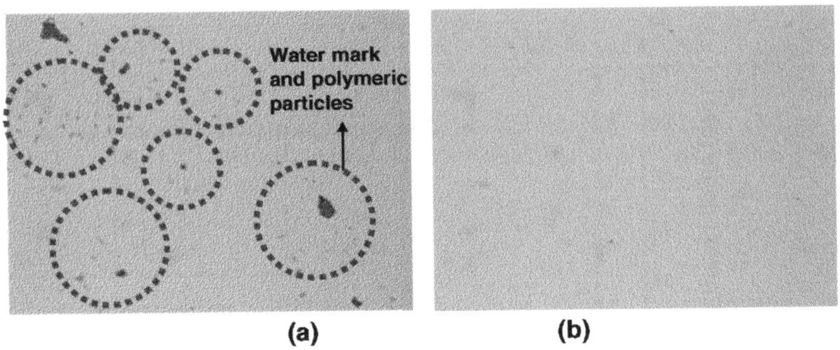

FIGURE 16.15 Optical images of polymeric particle contamination on (**a**) hydrophobic and (**b**) hydrophilic poly-Si.

as the contact angle decreases. Therefore, the change in surface wettability has a direct impact on the interaction between the organic materials and Poly-Si surface and the contamination level [52].

In order to further investigate the particle adhesion and removal on the poly-Si surface, organic pad particles (abraded pad materials) were dispersed in the KOH solutions. The suspension was then brought to a poly-Si substrate that is either untreated (hydrophobic) or treated with an oxidizer (hydrophilic). Figure 16.15 shows the optical microscope images of the Poly-Si surface with pad particle contaminations and watermarks. It is clear that the hydrophobic polysilicon surfaces attracted much more pad particles with watermarks than their hydrophilic counterpart. Figure 16.16 shows the defect maps of poly-Si surface as a function of oxidizer concentration after poly-Si CMP. When enough oxidizer was added to the commercial silica-based slurry, the particle contamination (mainly organic) was gradually reduced. Figure 16.17 shows a schematic hypothesis of how polymeric particle contamination occurs on a hydrophobic surface. A hydrophilic surface does not usually create any watermarks. On the contrary, a hydrophobic surface tends to attract water droplet that contains organic materials (dissolved or particulate). When the water droplet was removed from the hydrophobic poly-Si surface during a wafer drying process, watermarks remained and organic residues formed.

16.4.5 Post-Cu/Low-k CMP Surface Cleaning

During Cu CMP, wafer surfaces are exposed in two steps to at least two different slurries. During the first step, bulk Cu is removed at a high rate and polishing is stopped when the barrier layer is exposed or thin Cu layer is left. The remaining Cu and barrier layers are removed in the second step. The second-step slurry should have low selectivity between Cu and barrier (~1:1) and yield minimum dishing and erosion. The most commonly used abrasives in Cu CMP slurries are silica (fumed and colloidal silica) and alumina (Al_2O_3)

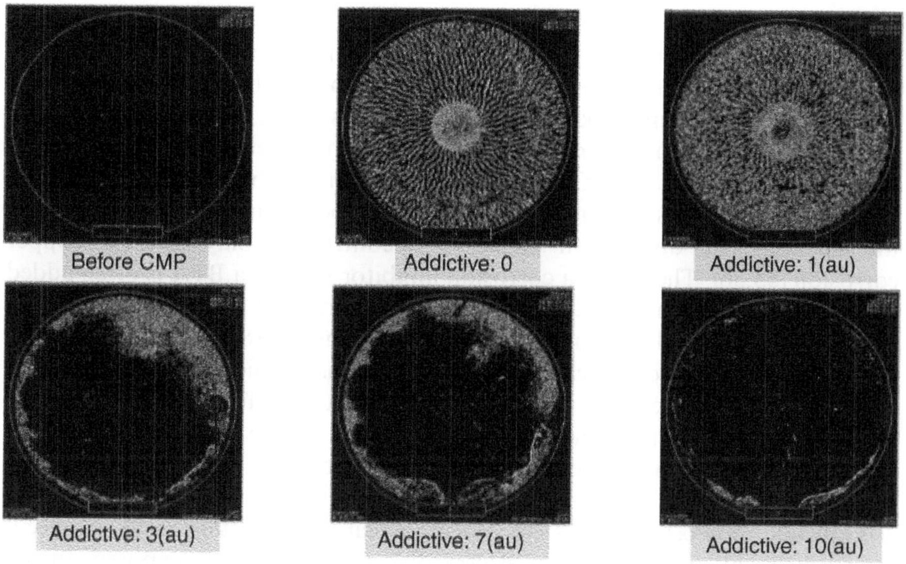

FIGURE 16.16 Change of particle contamination on poly-Si as a function of oxidizer concentration.

particles. The alumina particles are mainly used in the first step to obtain a high removal rate of Cu. In the second step colloidal or fumed silica particles are used to reduce the defects on the polished surface. Because of defect concerns, colloidal silica replaces alumina even in the first step of copper polishing. These particles were considered as a significant contaminant during post-Cu CMP cleaning process.

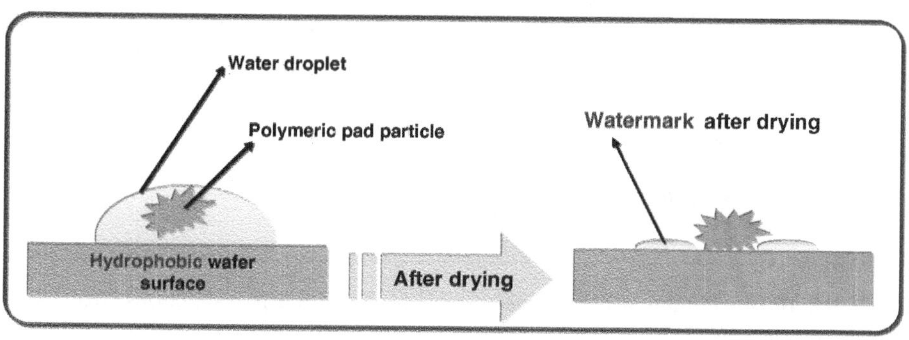

FIGURE 16.17 Schematic of polymeric particle contamination on hydrophobic surface.

16.4.5.1 Corrosion Carboxylic acids including amino acids have a tendency to form a complex with Cu^{2+} ions. Depending on their solubility in an aqueous environment and diffusivity away from the copper surface, the complex can play a key role in forming passivation films or dissolving copper oxides [53,54]. In a copper CMP slurry, carboxylic acid can be employed to perform either or both functions. In a post-CMP cleaning solution, carboxylic acid is mainly used to assist the dissolution of various copper species and remove them from the surface. The dissolution effect, if uncontrolled, may lead to copper corrosion [55,56]. Therefore, a corrosion inhibitor such as a BTA is often added to post-Cu CMP cleaning solutions as well. The corrosion inhibitor also helps in avoiding the so-called photoinduced corrosion (Figs. 16.18 and 16.19) [57]. This photoassisted corrosion mechanism results in catastrophic displacement of Cu from the lines connected to the p-doped regions to the lines connected to the n-doped regions as in a solar cell. This corrosion can be reduced by adding a corrosion inhibitor or reducing light impinging on the wafer surface during cleaning process.

Galvanic corrosion tends to occur when two metals with different electrochemical potentials are electrically connected and exposed in an electrolyte. As a result, the less noble metal will suffer from accelerated corrosion [58]. When excess copper is polished away by copper CMP, copper and barrier metal are exposed to the CMP slurry simultaneously. Copper and barrier metal have different electrochemical potentials and thus trigger galvanic corrosion at the interface between copper and barrier metal at a certain kind of slurry composition. In this galvanic corrosion, electrons are transferred from titanium anode to copper cathode. During overpolishing of the patterned wafer, titanium near the copper structure is recessed owing to dissolution ($Ti \rightarrow Ti^{2+} + 2e^-$) and Cu^{2+} ions are preferentially deposited onto

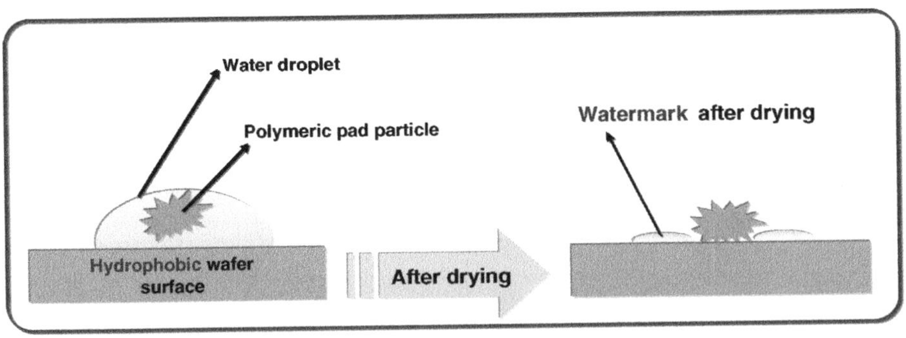

FIGURE 16.18 Photoassisted corrosion mechanism in the p–n junctions (from Ref. 57).

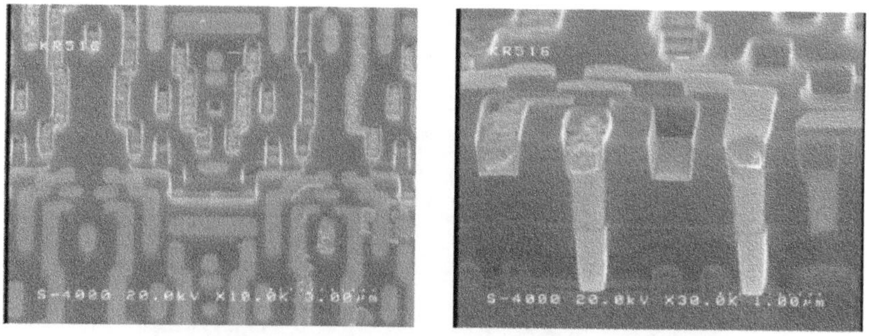

FIGURE 16.19 Images of photoassisted corrosion in the p–n junctions.

the copper line adjacent to the titanium layer and titanium continues to dissolve as shown in Fig. 16.20. The titanium metal acts as a local anode, whereas the copper metal acts as a local cathode. Figure 16.21 shows the image of galvanic corrosion on copper damascene line after Cu CMP [59].

16.4.5.2 ***Organic Residue*** Although it is well known that organic chemicals are involved in CMP and post-CMP processes, the understanding of organic contamination is very limited. This may be attributed to the fact that the detection of surface organics is not done as routinely as that of particles and metallic contaminants. In addition to surfactants and corrosion inhibitors, a buffering agent is often added to a post-Cu CMP cleaning solution to regulate and stabilize the pH of the solution. Some of these buffering agents are organic in nature.

The use of these organic additives increases the chance of organic residues after post-CMP cleaning. Furthermore, there are many sources of organic residues from the polishing step, such as hydrophobic slurry additives (BTA or alike) and polymeric residues of abraded pad material, retainer ring, and other consumables. In order to minimize organic contaminants, various cleaning solutions have been developed for use on the last platen in the CMP tool with a

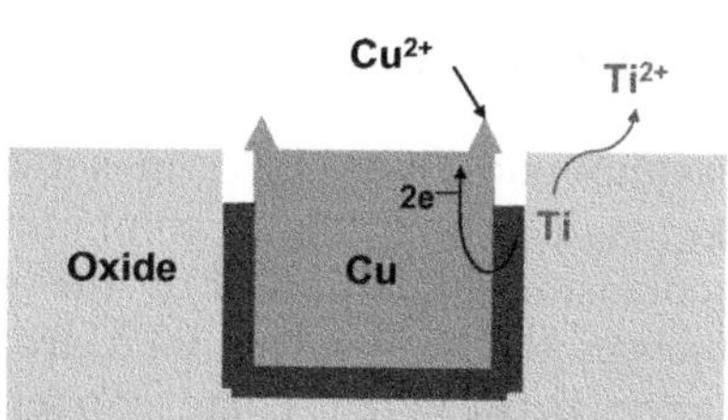

FIGURE 16.20 Galvanic corrosion of Ti with Cu.

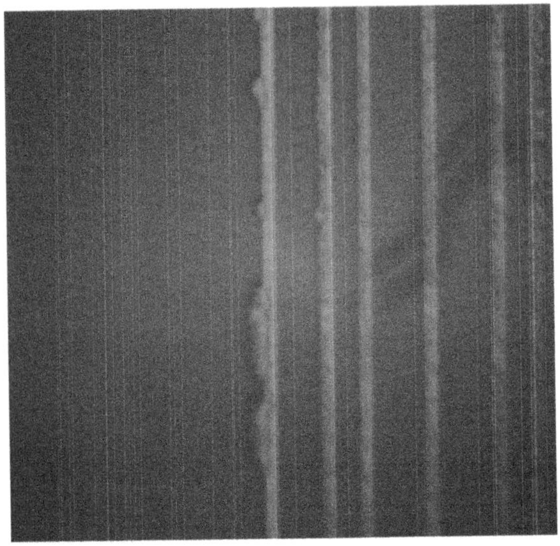

FIGURE 16.21 Image of galvanic corrosion on copper damascene line (from Ref. 59).

buffing pad and a mechanical brush scrubber. More specifically, various strategies have been explored and implemented to remove hydrophobic species such as BTA. If the neutral BTA molecule is referred to as BTAH, the protonated and dissociated species are $BTAH_2^+$ and BTA^-, respectively. Some equilibrium reactions between BTAH and copper can be written as follows [60]:

$$CuBTA + 2H^+ + e \Leftrightarrow Cu + BTAH_2^+$$

$$CuBTA + H^+ + e \Leftrightarrow Cu + BTAH$$

$$CuBTA + e \Leftrightarrow Cu + BTA$$

$$2CuBTA + H_2O \Leftrightarrow Cu_2O + 2BTA + 2H^+$$

$$Cu_2^+ + BTAH_2^+ + e \Leftrightarrow CuBTA + 2H^+$$

$$Cu_2^+ + BTAH + e \Leftrightarrow CuBTA + H^+$$

Copper does form compounds with BTA in solutions. The pH of the solution influences the stability of these compounds and thereby the effectiveness of their corrosion inhibition. For the same reason, for this chemical, pH also influences the removal of BTA from a copper film after copper CMP. The reaction between BTA and copper favors the release of BTA at low pH. In order to minimize organic contamination, the interaction between the organic chemical and the film to be polished has to be understood. Additionally, it is also important to understand the solubility of organic compounds because organic contamination can occur in the form of insoluble organic compounds. In practice, it is possible that the thin organic coat on the film is physically removed during a typical buffing step with ultrapure water.

To remove the BTA–copper complexes and other organic and inorganic defects, three different cleaning mechanisms have been used. The undercut-and-lift-off approach uses chemicals that dissolve the copper oxides in the BTA–copper complexes and in the underlying oxides, effectively lifting off defects. The dissolution approach penetrates, swells, and dissolves the organic film. The displacement approach relies on a copper complexing agent to displace the BTA on the copper oxide surface following barrier removal on the last platen. The resulting organocopper complex is more readily removable in subsequent cleaning steps than BTA.

16.4.5.3 Low-k Materials The IC industry has already implemented some low-*k* materials in the advanced devices. In some forms of low-*k* materials, the Si–O bonds of silica have been replaced with less polar Si–F or Si–C bonds. A greater *k*-value reduction can be achieved by using virtually all nonpolar bonds, such as C–C or C–H pure organic polymers.

Figure 16.22 shows that silica (SiO_2) has a tetrahedral elementary unit. To reduce the *k* value of silica, some oxygen atoms are replaced with F, C, or CH_3. The addition of CH_3 not only introduces less polar bonds but also creates additional free volume. Therefore, materials such as silicon oxycarbides (SiOCH) are constitutively porous [61].

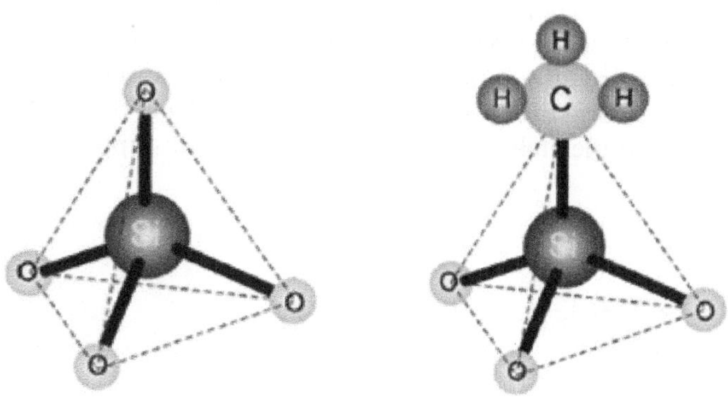

(a) (b)

FIGURE 16.22 Schematic representation of a tetrahedral (**a**) silica and (**b**) SiOCH (from Ref. 61).

All low-k materials are hydrophobic in nature owing to their nonpolar or less polar bonds. Water has extremely polar O–H bonds and a k value close to 80. Even a small amount of absorbed water significantly increases the total k value. As water is abundant in air, a low-k material should be as hydrophobic as possible to prevent the deterioration of its k value. This is especially important for porous materials, as they have a large surface area per unit volume where water could potentially be adsorbed. Hydrophobicity is usually achieved by the introduction of Si–H or Si–CH_3 bonds, and hydropobicity and low-surface energy are attributed to the stable nature of aromatic hydrocarbon bonds of most organic low-k dielectrics and stable terminating and functional groups in the structure of SiOCH- and SSQ-based materials [62].

However, hydrophobicity of a low-k material poses great challenges to an aqueous-based cleaning process. The hydrophobic surface of a material causes poor wettability and leads to defects such as watermarks and organic residues. A watermark can act as a leakage source and cause electrical short. Figure 16.23 shows the typical map of watermark defects and the image of watermarks. Most of the watermarks have a shape of circular droplet of water and dried solute can be seen in defect images [57].

In general, post-CMP cleaners used citric or oxalic acid solutions. Because of their low pH, these compounds rapidly dissolve both CuO (the outermost oxide) and Cu_2O (the underlying oxide), undercutting particles or organic defects in the oxide. If the zeta potentials of the substrates and particles have the same sign and are nearly equivalent, the particles will not be redeposited. However, for typical substrates and typical slurry abrasive particles, this equivalence is difficult to achieve at a low pH (e.g., approximately 4) [64]. Consequently, this approach required the use of a mechanical agent such as megasonic agitation or brushes to assist in particle removal. The removal of passivating oxides using citric acid solutions yielded a very reactive copper surface that sometimes required passivation with BTA to avoid corrosion while the wafers awaited the next process step. To achieve appropriate adhesion and contact resistance and also to prevent potential delamination resulting from the decomposition of buried BTA films after thermal cycling, the BTA film had to be removed before the etch stop or the barrier film was applied.

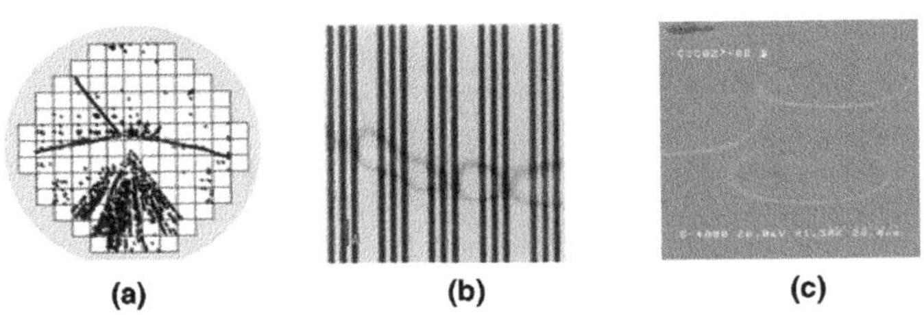

FIGURE 16.23 (**a**) Defect map; (**b**) optical and (**c**) SEM images of watermark defects (from Ref. 63).

In post-CMP cleaning, a balance must be maintained between particle or organic defect removal and surface roughness. General post-CMP cleaners use the undercut-and-lift-off cleaning mechanism that relies on surface etching to undercut residues and lift them off. To reduce surface decoration and problems associated with increased surface roughness, novel post-CMP cleaners incorporating water-soluble organic solvents have been developed [65]. Using a dissolution mechanism to remove organic residues, these solutions control and reduce the surface etch rate, thus decreasing surface roughness while undercutting and lifting off abrasive particles and some organic defects. Because of their alkaline character, they have a large negative zeta potential, resulting in excellent particle-removal efficiency.

16.4.5.4 Effect of Other Additives on Cleaning In general, a lower adhesion force between the particles and the polished wafer surface is highly desirable, which minimizes the chances of particle redeposition during and after post-CMP cleaning. As a case study, the particle adhesion forces in three different cleaning solutions were compared with that in DI water to elucidate the effect of pH adjustor (NH_4OH versus TMAH) in citric acid based cleaning solutions [43]. The lowest adhesion force was observed to be 0.0124 nN in citric acid with NH_4OH at pH 6. On the contrary, the largest adhesion force of 8.87 nN was measured in citric acid with TMAH. The adhesion force in a solution with NH_4OH is two orders of magnitude lower than that in a solution with TMAH (Fig. 16.24). These results clearly show that the pH and its adjustor are very important in the cleaning solution chemistry design.

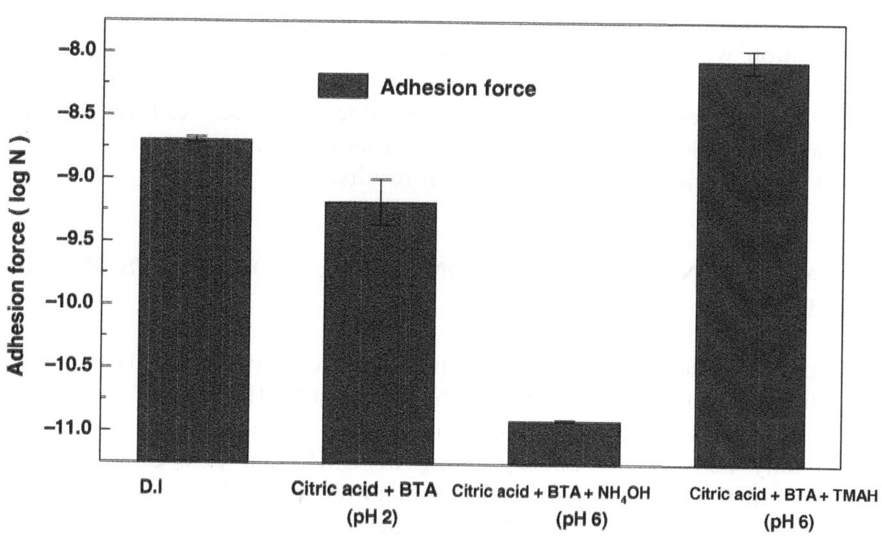

FIGURE 16.24 The adhesion forces between Cu wafers and spherical silica particles in different solutions.

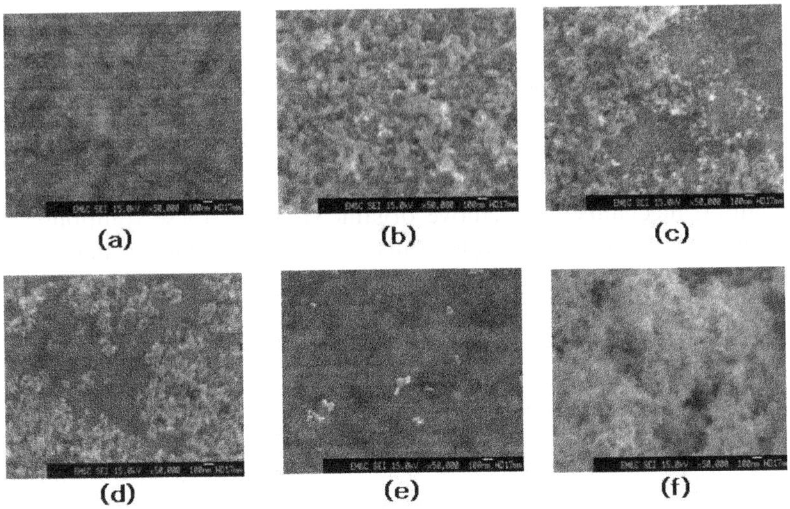

FIGURE 16.25 FESEM images of the Cu surfaces (**a**) before and (**b**) after dipping in slurry cleaned with (**c**) DI water, (**d**) citric acid solution with BTA, (**e**) citric acid and BTA solution with NH_4OH, and (**f**) citric acid and BTA solution with TMAH (from Ref. 43).

Furthermore, these cleaning solutions were used on copper wafers that are treated or contaminated with silica-based copper slurry. Figure 16.25 shows the FESEM images of Cu surfaces after cleaning in different solutions. Large numbers of residual particles were observed on Cu surfaces cleaned with DI water, citric acid solution, and citric acid solution with TMAH. However, citric acid solution with NH_4OH showed complete removal of particles from Cu surfaces. The magnitude of adhesion force measured by AFM was directly related to particle-removal results. Higher adhesion forces resulted in lower removal of particles. The particle-removal experiments also matched well with DLVO total interaction force calculation results.

16.5 ADHESION FORCE, FRICTION FORCE, AND DEFECTS DURING Cu CMP

As described in the last section, the adhesion forces between the particles and the polished surface can provide insight into post-CMP cleaning mechanism and efficiency. As a matter of fact, measurements of adhesion force and alike can directly help in the understanding of CMP and post-CMP cleaning processes. In this section, some basic principles and applications of adhesion and friction force measurements in copper CMP will be presented. The discussion is certainly applicable to post-CMP cleaning processes when similar systems are involved.

16.5.1 Adhesion Force of Silica and Alumina on Cu

The atomic force microscope (AFM) has been widely used to study phenomena such as abrasion, adhesion, cleaning, corrosion, etching, friction, lubrication, and polishing. By using AFM, one can image the surface in atomic resolution and measure the interaction force at nN scale. The force versus distance (*F* versus *D*) curves can be used to measure the vertical force that the tip applies on the surface while a contact AFM image is being taken [66–68]. In practice, *F* versus *D* curves can be quite complex and the conclusions based on these curves may be applicable only to the system under study. The force curves show the deflection of the free end of the AFM cantilever as the fixed end of the cantilever is brought vertically toward and then away from the sample surface. The cantilever positions at several points marked along the force curve are shown in Fig. 16.26.

The adhesion forces of silica and alumina particles in the DI water and slurry solution were measured by AFM and are shown in Fig. 16.27 [66]. The smallest adhesion force, 0.38 nN, was observed between the copper surface and alumina particles in a citric acid solution at pH 6. The largest adhesion force of alumina particles, 5.83 nN, was measured in DI water.

The largest adhesion force of alumina particles in DI water was attributed to a stronger electrostatic attraction between alumina particles and copper surface in DI water owing to their opposite signs of zeta potentials. The smallest adhesion force of alumina particles in the citric acid slurry was attributed to the

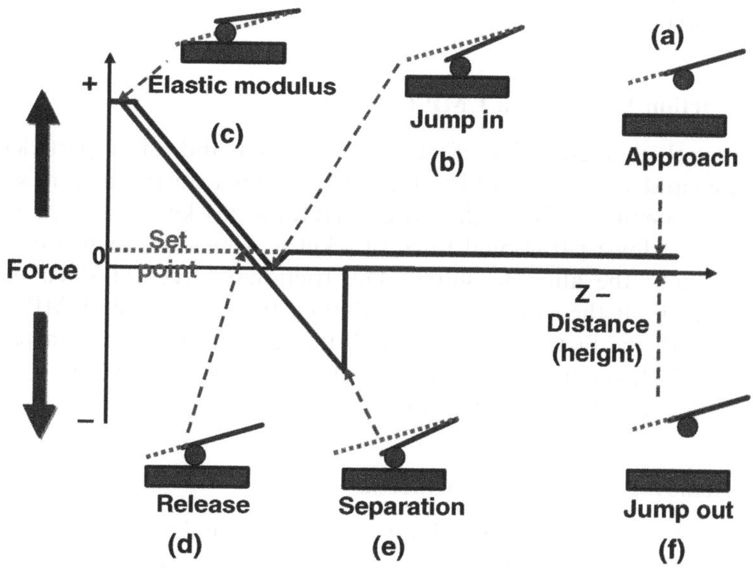

FIGURE 16.26 The change in cantilever shape as a function of applied force on a cantilever.

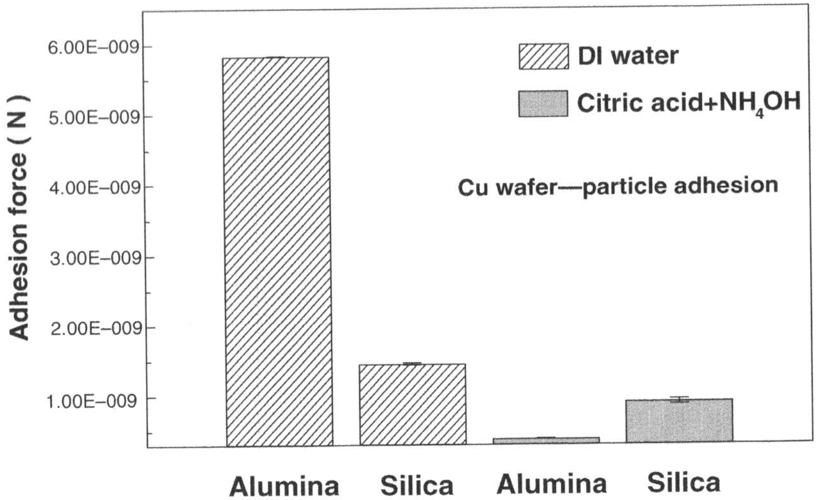

FIGURE 16.27 The adhesion forces of the particles on copper in DI water and citric acid solution.

selective adsorption of citrate onto to the alumina surface. However, the presence or absence of citric acid did not change the adhesion forces of the silica particles. This indicates a weak adsorption of citrate onto silica particles and a strong adsorption onto alumina. The extent of surface adsorption for citrate may also be related to the fact that both silica and citrate are negatively charged at pH 6.

16.5.2 Friction Force in Cu CMP Process

The frictional characteristics of abrasive alumina and silica particles were investigated and are shown in Fig. 16.28. The alumina slurry was very sensitive to the slurry chemistry. The highest frictional force of 9 kgf was observed in DI water, and the lowest frictional force of 4 kgf was measured when citric acid was added into the alumina slurry. The frictional forces of silica particles (6 kgf) were about the same with or without citric acid during CMP. This is consistent with the fact that citrate has little adsorption onto silica particles. Yoon et al. [69] reported that higher adhesion force between two surfaces caused higher friction force on them.

16.5.3 Removal Rates of Cu Surface in Cu CMP

Figure 16.29 shows the removal rate of copper under different polishing conditions. Regardless of the abrasive type, removal rates lower than 100 nm/min were observed in DI water based slurries. The low copper removal rates are attributed to the purely mechanical action of the abrasive particles.

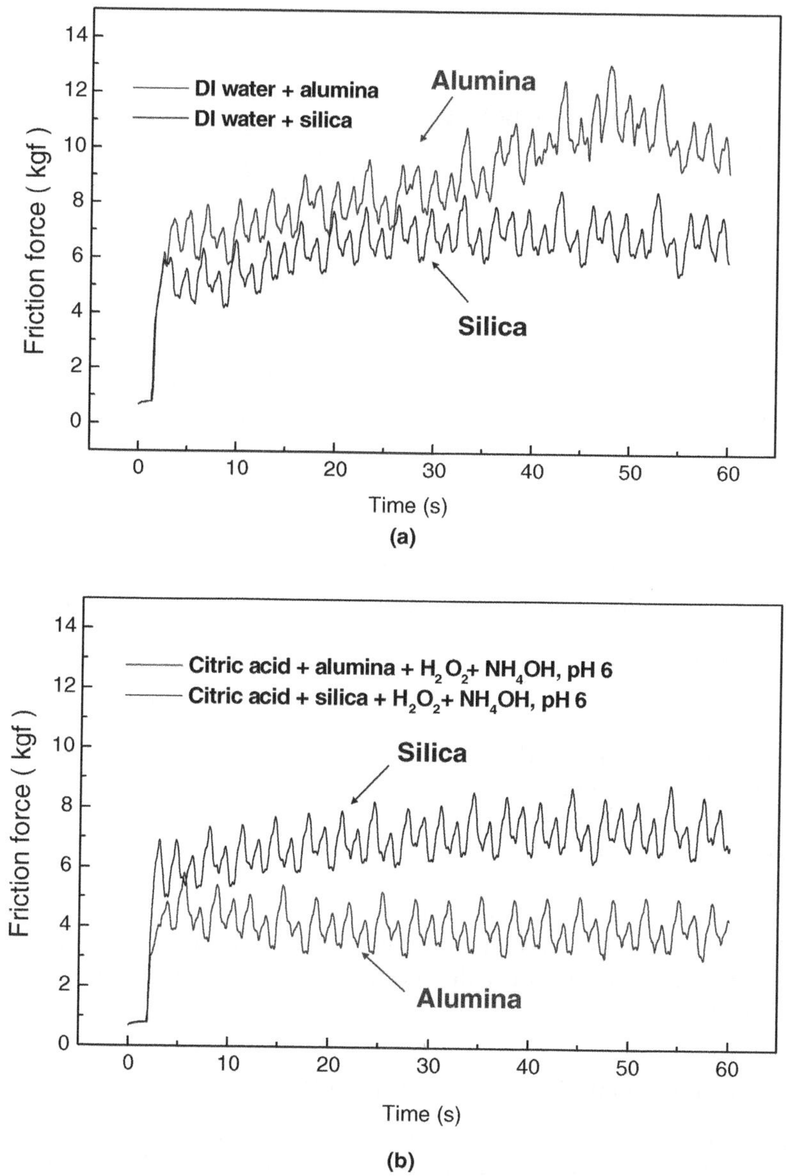

FIGURE 16.28 The friction curves of abrasive particles in (**a**) DI water and (**b**) a citric acid based solution during copper polishing.

Without significant chemical reactions (in DI water based slurries), the friction and adhesion forces are well correlated with the magnitude of removal rates. For example, four times higher removal rate as well as friction force was observed for alumina-based DI water slurry than that for silica.

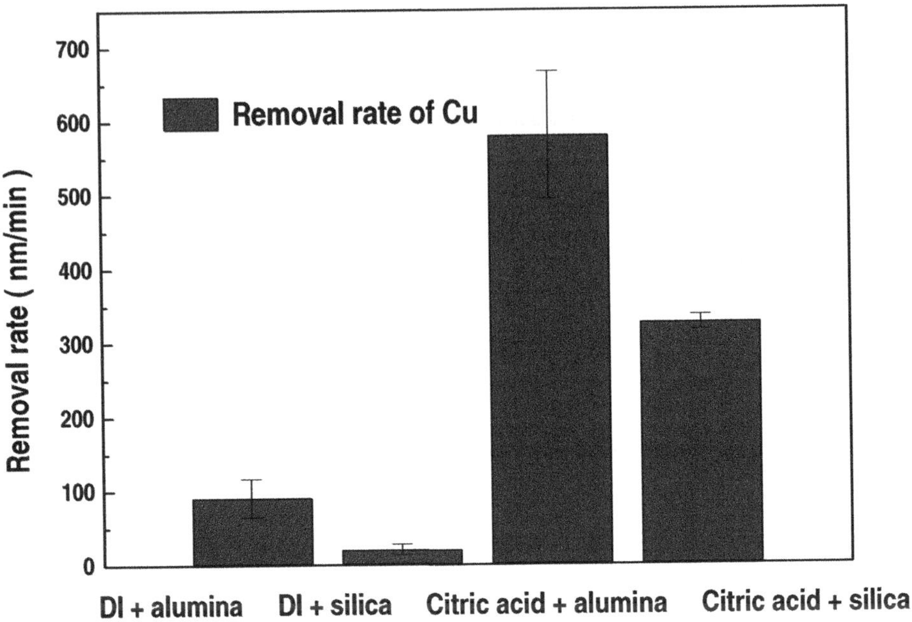

FIGURE 16.29 The removal rate of copper in various slurry solutions.

As shown in Fig. 16.27, the drastic decrease in adhesion force was measured for alumina slurry containing citric acid. A slight decrease in adhesion force was measured on silica particles. Higher adhesion and friction forces were measured in silica slurry with citric acid than in alumina because of the adsorption of citric ions onto alumina. However, the highest copper removal rate, 600 nm/min, was observed in alumina slurry containing citric acid, H_2O_2, and NH_4OH. This is consistent with the fact that polishing with a high copper removal rate requires not only the mechanical reaction but also the chemical reaction [70,71].

16.5.4 Surface Quality of Cu After Cu CMP Process

In addition to its correlation with material removal rate, the friction force can also be related to the surface quality of the polished surface. At a given removal rate, an increase in friction force often leads to higher level of scratches. A slurry with low copper removal rate and having high frictional force may result in severe scratches on the polished surface. In order to investigate the relationships among the magnitudes of particle adhesion, frictional forces, and scratching during the CMP process, AFM was used to observe the copper surfaces after the copper CMP process. Figure 16.30 shows the magnitude of particle contamination and scratches on the copper surfaces. The scan area of copper surface was $45 \times 45\ \mu m^2$, and each $R_{p\text{-}v}$ value of copper surface was

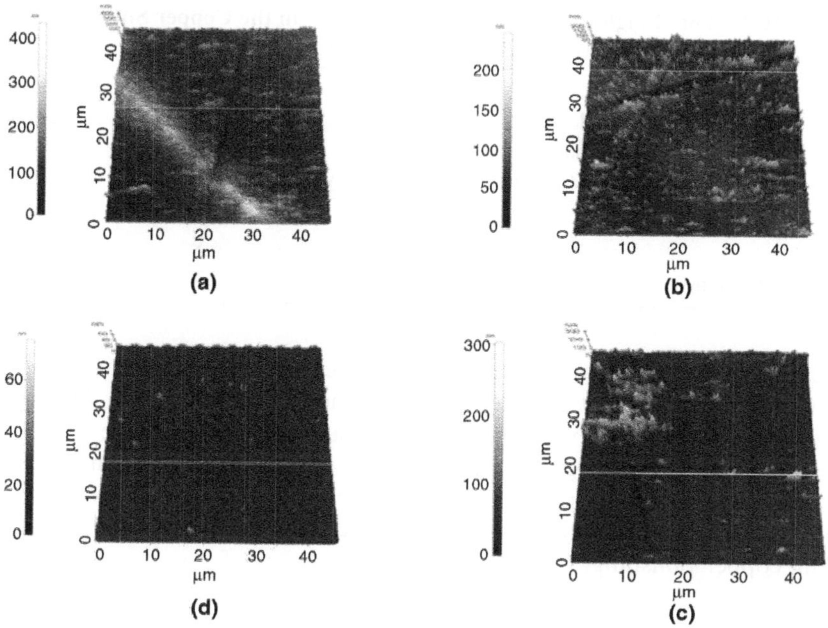

FIGURE 16.30 AFM images of the copper surface after polishing in (**a**) a DI water based alumina slurry, (**b**) a DI water based silica slurry, (**c**) a citric acid-based alumina slurry, and (**d**) a citric acid based silica slurry at pH 6.

obtained through the analysis of the cross-sectional height of polished surfaces. The depth of the scratches and magnitude of particle contamination on the copper surface are shown in Fig. 16.30 and summarized in Table 16.3. Large numbers of residual particles and scratches were observed on the polished copper surfaces in DI water with alumina particles. Silica particles also generated particle contamination and scratches on the copper surface when in either a DI water based or a citric acid based slurry. The depth of the scratches on the copper surface was dependent on the magnitude of the friction force. Higher frictional forces correlated with the observation of deeper scratches on the copper surfaces.

Table 16.3 shows R_a (average roughness), $R_{p\text{-}v}$, and depth of scratches on the copper surface after polishing. The largest values of R_a and $R_{p\text{-}v}$ were observed on the polished copper surfaces in a DI water based alumina slurry. On the contrary, the polished copper surface in the citric acid solution and H_2O_2 with alumina particles had the smallest R_a and $R_{p\text{-}v}$. Adhesion force is known to influence not only the magnitude of friction force but also the level of particle contamination on substrates, which directly relates to the surface roughness and number of scratches after CMP. Even though similar friction forces were measured in silica slurries both with and without citric acid, lower adhesion force was observed in citric acid based silica slurry. A lower adhesion force

TABLE 16.3 The Roughness and Depth of Scratches on the Copper Surface After Polishing with Aluminabased and Silicabased Slurry Solutions.

(nm)	DI water based alumina slurry	DI water based silica slurry	Citric acid based alumina slurry with H_2O_2	Citric acid based silica slurry with H_2O_2
R_a	33.8	12.0	0.8	3.9
$R_{p\text{-}v}$	271.1	106.7	6.8	74.0
Depth of scratches	113.2	72.4	3.4	54.5

indicates lower friction during polishing, which actually results in smoother surfaces in citric acid based silica slurry as shown in Table 16.3. This might suggest that the friction measurement was not as sensitive as AFM measurements. These roughness results clearly show that the amount of citrate ions adsorbed onto the particle surfaces significantly affected the frictional behavior and the adhesion forces of the particles as well as the surface quality during copper polishing.

16.5.5 Correlation Among Friction, Adhesion Force, Removal Rate, and Surface Quality in Cu CMP

The frictional and adhesion forces between the abrasive particles and wafer surfaces were experimentally measured using alumina and silica slurries with and without citric acid. Although citric acid did not affect the zeta potential of the silica particles, it resulted in a more negative zeta potential of the alumina particles due to the adsorption of the negatively charged citrate ions onto the alumina surfaces. The highest particle adhesion force was measured in an alumina slurry without the addition of citric acid. However, the alumina slurry with the addition of citric acid had the lowest particle adhesion force due to the adsorption of citrate ions onto the alumina surfaces. Although citrate ions could easily adsorb onto alumina particles, the silica particles did not appear to benefit in terms of reduced frictional force when in citric acid solutions.

The frictional curves between the abrasive particles and the copper surfaces were measured using alumina and silica slurries with and without citric acid. The highest frictional force was observed in a DI water based alumina slurry. On the contrary, the smallest friction force was measured in an alumina slurry containing citric acid. Regardless of the presence or absence of citric acid in the slurry, the frictional curves of the silica particles were not significantly changed during the CMP process. These results clearly show that the magnitude of citrate ions adsorbed onto the particle surfaces affects the frictional behavior as well as the adhesion force during the CMP process.

The low copper removal rate was observed in the DI water based slurry due to the purely mechanical action of the abrasive particles. However, the highest copper removal rate was observed in alumina slurry containing citric acid,

H_2O_2, and NH_4OH. This suggests that a high copper removal rate requires not only the mechanical reaction but also the chemical reaction. Abrasives and chemicals in the slurry solutions continuously accelerate oxidation, etching, and abrasion of copper surface by chemical and mechanical effects. Although the lowest friction force was observed in the alumina based slurry with the addition of citric acid, the highest removal rate of copper was observed in this slurry due to the chemical reaction with copper surface. The smallest adhesion force resulted in the lowest friction force in the alumina-based slurry with the addition of citric acid.

Large numbers of residual abrasive particles and scratches were observed on the polished copper surfaces in a DI water based alumina slurry. The depth of the scratches on the copper surface was dependent on the frictional force. Higher frictional forces resulted in deeper scratches on the copper surfaces. Higher particle adhesion forces generated higher frictional forces, abrasive particle contamination, and scratches on the copper surfaces during the CMP process. This indicates that the magnitude of particle adhesion on the wafer surfaces in slurries can be directly related to the frictional behavior and surface quality during the CMP process.

16.6 CASE STUDY: MEGASONIC POST-CMP CLEANING OF THERMAL OXIDE WAFERS

In this section, a case study is presented in which polished thermal oxide wafers were cleaned using megasonic cleaning and SC1 chemistry. The effects of sonic power, temperature, and oxide etching on cleaning efficiency are examined.

16.6.1 Experimental Procedure

A set of thermal oxide wafers are polished using a silica-based slurry (Cabot SC-112) on an IPEC polisher equipped with an IC1000 pad (Rohm & Hass). The polishing time is set at 30 s with a slurry flow rate of 150 ml/min and downpressure of 9 psi with 2 psi backpressure. The rotational speeds of the wafer and the pad are 25 rpm and 13 rpm, respectively. After polishing, the wafers are dried and then taken to the clean room where the STEAG bench is located.

The STEAG bench used in the study is fully automated with initial manual loading. It consists of two SC1 baths, two diluted HF baths, two DI water rinse tanks, and two Marangoni dryers. However, only the SC1 bath, rinse tanks, and dryer are used in this study. At the beginning of each cleaning experiment, one *dry* polished wafer is put in the carrier (at the center) with 49 dummy wafers (since the STEAG tool is loaded with 50 wafers). The cleaning conditions (power, temperature, and cleaning time) are set before cleaning. Once the cleaning process is accomplished, the two boats of dry wafers are unloaded and the test wafer is collected. The test wafer is then scanned using a

TABLE 16.4 Megasonic Power Conversion Table for STEAG Tool.

Power (W)	Power (level)
100	1
200	2
250	3
300	4
350	5
400	6
450	7
500	8

KLA-TENCOR 6200. The SC1 solution concentration used in the study is $H_2O/H_2O_2/NH_4OH$: 40/2/1. The megasonic power input in the STEAG bench uses the level number instead of a power value in watts. Table 16.4 gives the relationship between the input megasonic (frequency = 850 ± 10–15 kHz) power in watts and the level number for the STEAG bench. A PCT system tank, also used in this study, uses power percentages between 1% and 100% to express megasonic input power.

16.6.2 The Effect of Megasonic Input Power

The megasonic power in these experiments varies from 200 (level 2) to 500 W (level 8). Figure 16.31 shows the particle-removal efficiency as a function of input power. The figure shows that the megasonic power (in the range used) has little effect on particle removal. Experimental results on the cleaning of dried silica slurry on thermal oxide wafers (Fig. 16.32) using a PCT megasonic cleaning tank (megasonic power levels are shown in percentage for the PCT system) show that the removal efficiency increases suddenly from no removal

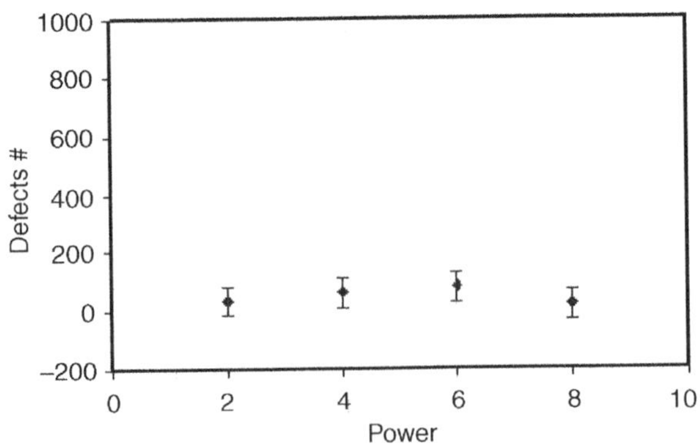

FIGURE 16.31 Number of defects as a function of megasonic power (power level) at 65 °C after 8 min using SCI chemistry using a STEAG bench.

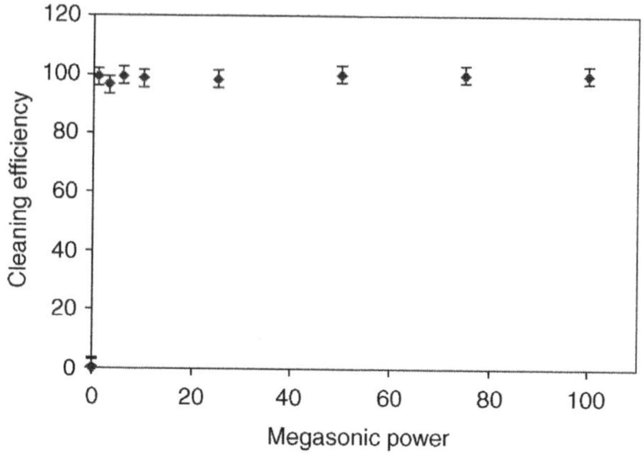

FIGURE 16.32 Cleaning efficiency as a function of megasonic power (power percentage) of dipped wafers (in silica slurry) for 8 min using SC1 at 35 °C using a PCT Systems tank.

when the power is turned off (0 W) to 99% removal efficiency with the use of only 6 W. This high removal efficiency remains constant (about 100%) while the power is increased until the maximum power is reached (640 W). This is inconsistent with the effect of power on megasonic cleaning using DI water, where power plays a very important role in the removal of particles and where at least 400–500 W is required to reach 100% cleaning efficiency.

It is important to point out that this phenomenon is not an isolated incident. As a matter of fact, it is very reproducible. It is also worthwhile to mention that most previous studies did not explore the effect of very low power on megasonic cleaning using SC1. The question now is what is the cleaning mechanism at a very low megasonic power? Compared to DI water cleaning, two important factors are uniquely associated with SC1 cleaning. The first phenomenon is the enhancement of cleaning by the etching action, and the second factor is the double-layer effect in increasing the particle repulsion at higher pH. As the temperature increase at low sonic power is minimal and the etching efficiency at low temperature is low, the phenomenon is unlikely related to the temperature variation. The effect of the double layer on cleaning in this case, however, is more apparent where the charge (zeta potential) on the particles (as well as the wafer surface) typically decreases from −40 mV at pH7 to −80 mV at pH11. This significantly increases the double-layer repulsive force between the particles and the wafer [72], which in turn lowers the required physical removal force supplied by the acoustic streaming. It has been shown that particles with a low van der Waals constant $h\varpi$ (also known as the Liftshitz–van der Waals constant) that is less than 1 eV and a repulsive double layer may separate upon immersion without applying an external force [73]. Particle detachment by simply changing the pH has also been shown using the packed column technique by Matijevic et al. [72–74].

The van der Waals constant is related to the Hamaker constant A by a factor of $3/4\pi$ ($A = 3/4\pi h\varpi$). The van der Waals constant $h\varpi$ for the current system (silica on a silicon oxide surface in water) is slightly larger than 1 (1.25 eV). This shows that only a small physical force (drag) is needed to overcome the total adhesion force. If we calculate the double layer force between a 0.1 silica particle and a silicon wafer, we get a repulsive force of 1.9×10^{-10} N using an ionic strength of 10^{-3} M. The van der Waals force for the same particle and wafer is an attractive 4.0×10^{-10} N. Although the power used (at a low power setting) is far lower than that used for effective cleaning using DI water at high power, acoustic streaming still exists with a lower streaming velocity. The resulting force at the lowest power (1% using 850 kHz) is of the same order of magnitude at 1×10^{-10} N. This shows that the double layer reduces the overall DLVO interaction by a factor of 2 and the adhesion force is larger than the removal force by a factor of 2. If we consider the drag force due to the lowest power used in the STEAG bench (level 1), the drag force will be 7×10^{-10} N, which can easily overcome the particle's total adhesion force. However, at the lowest power (1% of the total power), the particle is left with an attractive adhesion force of 1×10^{-10} N. Another force is needed to provide the final force to overcome what is left of the overall DLVO interaction.

Another important force that exists even at low sonic power is the stable cavitation, which does contribute to particle removal [20]. Cavitation at high sonic frequencies has been shown to be present even at a very low power (down to 1%) [20]. Stable cavitation bubbles do not implode violently and, therefore, do not cause substrate damage, as is the case with transient cavitation at low frequencies (25–100 kHz). Bubbles formed during cavitation also play a role in cleaning. Streaming that occurs near bubbles in the field is called micro-streaming. The bubble surface vibrates as a result of the negative and positive pressures in the sound waves. In this powerful type of streaming, the bubbles scatter sound waves and generate remarkably swift localized currents and vortices that contribute to the removal of small particles from surfaces. The currents are most pronounced near bubbles undergoing volume resonance and located along solid boundaries [20]. These currents may create a removal force higher than 1×10^{-10} N when the bubble surface is within a micrometer of the particle. In SC1, the number of bubbles is much higher than in DI water at the same power [20].

Another factor that does not depend on power and is shared by DI water and SC1 cleaning is the acoustic boundary layer thickness, which is a function of frequency and viscosity only. The boundary layer decreases from thousands of micrometers to a fraction of a micrometer when megasonics is applied at any power. The acoustic boundary layer thickness is inversely proportional to the frequency and directly proportional to the viscosity of the fluid. For example, a flow with a velocity of 4 m/s (maximum streaming velocity for the considered equipment) has a hydrodynamic boundary layer thickness of 1500 μm at the center of the wafer. By contrast, the acoustic boundary layer thickness on a wafer in a 850-kHz megasonic cleaning tank is 0.61 μm. The effect of a very

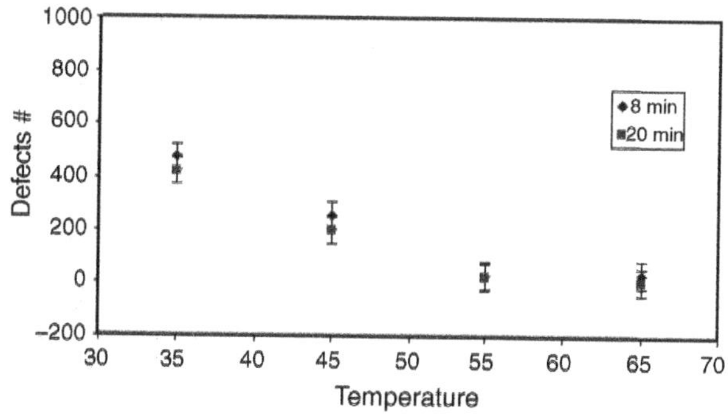

FIGURE 16.33 Number of defects as a function of temperature a at megasonic power level of 2 using SCI for 8 and 20 min using a STEAG bench.

thin boundary layer thickness in SC1 is two folds; the first is that the reduction of thickness exposes the particles to a much higher streaming velocity as compared to typical hydrodynamic boundary layer at the same velocity. The second effect has to do with the concentration gradient affected by higher convection near the surface as a result of the thin boundary layer thickness. The higher convection serves to accelerate the chemical and physical interaction due to SC1, although this is only significant above 45°C.

16.6.3 The Effect of Temperature

In this case study, the SC1 solution temperature ranges from 35 to 65 °C. Figure 16.33 shows the number of defects left after SC1 cleaning as a function of the temperature of SC1 solution at different cleaning times and megasonic powers. The number of defects decreases drastically when the SC1 temperature increases from 35 to 55 °C and then levels off. The decrease in the number of defects above 45 °C can be attributed to the SC1 etching effects [18–21]. In the lower temperature regime between 35 and 45 °C, this is similar to the cleaning experiments involving DI water, in which stable cavitation activity has been shown to increase with temperature from 23 to 65 °C (at 760 kHz). Streaming that occurs near cavitation bubbles in the field is called microstreaming, which plays a role in cleaning [75–83].

16.6.4 The Effect of Etching on Cleaning

In order to study the effect of etching of the oxide film on cleaning, the etching rate was measured as a function of temperature, cleaning time, and megasonic power (Fig. 16.34). The result shows no etching when the temperature is below 45 °C, regardless of the cleaning time. The etching increases slowly when SC1 temperature is above 55 °C. At 65 °C, the etching reaches a high value after 20 min of cleaning. This is because SC1 is a mixture of ammonium hydroxide as the etching agent and hydrogen peroxide as the oxidizing agent. At high

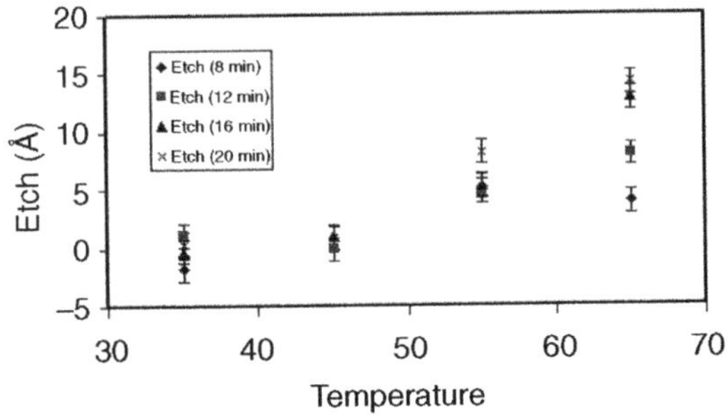

FIGURE 16.34 Oxide etching as a function of temperature at a megasonics power level of 2 using SC1 for 8, 12, 16, and 20 min using a STEAG bench.

temperatures, the etching takes over oxidation owing to the high reduction in the concentration of H_2O_2 and the low solubility of O_2 produced by the decomposition of H_2O_2. However, at low temperatures no etching is observed; the oxidation and the etching seem to have an antagonist effect or the etching rate is too low to be detected by the film thickness measurement tool. Figure 16.35 shows an increase in oxide etching with an increase in the megasonic power. When the megasonic power (intensity) increases, the streaming velocity increases very close to the surface because of the thin acoustic boundary layer. This increases the convection of the chemical etchant species near the surface. The convection of chemicals at the wafer surface and the removal of the etch product cause an increase in the etching rate.

A previous work demonstrated that megasonics has a catalytic effect on the transportation of ionic species from and to the wafer surface from the bulk solution [75]. However, etching alone is not enough to remove the particles. Megasonic convection coupled with thin acoustic boundary layer is needed to remove the particles. Etching does enhance the cleaning at high temperature, but it has to be coupled with megasonics.

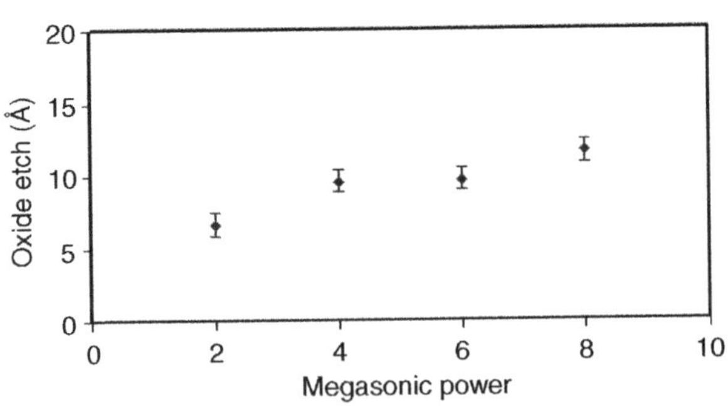

FIGURE 16.35 Oxide etching as a function of megasonic power (using power level) at 65 °C for 8 min using a STEAG bench.

The results show that noncontact megasonic cleaning with SC1 is a very effective technique for post-CMP cleaning. The cleaning results show less number of defects despite the fact that the wafers were dried before cleaning. The results show that power has no major effect on cleaning. However, the SC1 temperature has a major impact on the cleaning results. This effect is mainly due to the SC1 high etching rate at high temperatures, which was demonstrated. However, the etching is effective only when coupled with megasonic cleaning. Etching alone is not sufficient for removing any particles. The removal of particles at very low power is due to the reduction in the total adhesion force by increasing the electrostatic repulsion. Etching is negligible at temperatures below 45 °C and does not account for the decrease in the number of defects at low temperature.

16.7 SUMMARY

Post-CMP cleaning is as important as CMP itself because it directly affects the device yield. The cleaning chemistry and mechanical force must be strong enough to remove all unwanted materials from the polished surface, such as remaining particles. organic residues, and watermarks. Yet post-CMP cleaning must not introduce any new defects such as corrosion, structural damage, or metal cross-contamination.

Among all major CMP applications, post-CMP cleaning protocols involving NH_4OH and HF are adequate in addressing concerns related to oxide, STI, and W CMP processes. However, because of its hydrophobicity, polysilicon tends to attract more organic residues, hence requiring a hydrophilization process to reduce the magnitude of organic defects. In terms of Cu CMP, specially designed chemistry may be needed to address the corrosion issues.

It is important to understand the adhesion force of particles on polished surfaces, which can be correlated with the polishing and cleaning characteristics of a CMP process. Higher adhesion force often leads to higher polishing rate, higher level of scratches, and greater number of particles left on the polished surface.

When SC1 is coupled with megasonic protocol, the etching effect provided by SC1 dominates the cleaning mechanism at temperature between 45 and 55 °C, where the etch rate is the highest. At temperature below 45 °C, the microstreaming that occurs near cavitation bubbles in the field and produced by megasonic energy plays a significant role in the cleaning process.

ACKNOWLEDGMENT

HYU acknowledges the financial support from the Ministry of Education and Human Resources Development (MOE), the Ministry of Commerce, Industry and Energy (MOCIE), and Ministry of Labor (MOLAB) through the fostering project of the Lab of Excellence, post BK21 program, and Samsung Electronics Co., Ltd., Korea.

QUESTIONS

1. List the key attributes for a post-CMP cleaning solution in order to remove the silica slurry residue from silicon, silicon dioxide, silicon nitride, copper, or tungsten surfaces.
2. What are the key attributes for a post-CMP cleaning solution that is designed to remove the hydrophobic organic residues from silicon, silicon nitride, silicon dioxide, copper, or tungsten surfaces?
3. What are the key considerations for a post-CMP solution to remove silica slurry particles a surface that has both silicon dioxide and copper features?
4. What type of precautions one should take in order to avoid further defects while cleaning silicon, silicon dioxide, copper, silicon nitride, or tungsten surfaces?

REFERENCES

1. Tardif F. Semicond Semime 2000;63:183–214.
2. Rosato JJ, Yalamanchili MR. Solid State Technol. 2005;48 (10):50–55.
3. Hu A, Zhang X, Sachs E, Renteln P. IEEE Proceeding of the 15th International Electronics Manufacturing Technology Symposium; 1993. p 235–240.
4. Harman J, Hamm E. The impact of ultrasonic frequency on particle removal. Technical notes. Branson Ultrasonic Corp., Available at http://www.bransoncleaning.com/pdf/Particle.PDF.
5. Busnaina A, Park J, Bakhtari K. Particle deposition and adhesion. In: Reinhardt K, Kern W, editors. Handbook of Silicon Wafer Cleaning Technology. 2nd ed. Norwich, NY: William Andrew Publishing; 2006.
6. Busnaina A, Bakhtari K. Nanoscale defects and surface preparation in nanomanufacturing. In: Busnaina A, editor. Handbook of Nanomanufacturing. Boca Raton (FL): Taylor & Francis Group LLC; 2006.
7. Busnaina AA. Surface cleaning: particle removal. In: Kanegsberg B, editors. Critical Cleaning Handbook. CRC Press; 2001.
8. Prasad J. Front-End Wafer Cleaning Challenges. IBID (International Baccalaureate) Press; 2004.
9. Aaron Hand. Damage-Free Cleaning Beyond 65 nm. IBID Press; 2005.
10. Schwartzman S, Mayer A, KernW. RCA Rev 1985;46:81.
11. Mayer A, Schwartzman S. U.S. patent 3,893,769. 1975 July 8.
12. Mayer A, Schwartzman S. J Electron Mater 1979;8:855.
13. McQueen DH. Ultrasonic 1986;24:273.
14. McQueen DH. Ultrasonic 1990;28:422.
15. Kashkoush I, Busnaina A, Kern F, Kunesh R. In: Mittal KL, editor. Particles on Surfaces. Volume 3, Detection Adhesion, and Removal. New York: Plenum Press; 1991. p 217–237.
16. Busnaina AA, Gale GW. J Particulate Sci Technol 1997;15:361–370.
17. Kashkoush I, Busnaina A. J Particulate Sci Technol 1993;11:11.

18. Busnaina A, Kashkoush I. Chem Eng Commun 1993;125:47.
19. Gale GW. PhD thesis.Potsdam, NY: Clarkson University; 1995.
20. Busnaina AA, Kashkoush II, Gale GW. J Electrochem Soc 1995;142(8):2812–2817.
21. Busnaina AA, Gale GW. J Particulate Sci Technol 1999. Forthcoming.
22. Gale GW, Busnaina AA. J Particulate Sci Techno 1995;13:197–211.
23. Syverson W, Fleming M, Schubring P. Second International Symposium on Cleaning Technology in Semiconductor Manufacturing, Electrochemical Society Proceedings PV92-10;1992. p 10.
24. Wang P, Bell D. Third International Symposium on Cleaning Technology in Semiconductor Manufacturing, Electrochemical Society Proceedings PV94-7;1994. p 132.
25. Moumen N, Piboontum J, Busnaina AA. The Adhesion Society Proceedings, 22nd Annual Meeting; Panama City FL;1999 Feb 21–24.
26. Bibby T, Holland K. Semicond Semimet 2000;63:5–45.
27. Taylor J, Busnaina A. IEEE/SEMI Advanced Semiconductor Manufacturing Conference; 2000. 14–17.
28. Philipossian A, Mustapha L. Solid State Phenom 2003;92:275–280.
29. Philipossian A, Mustapha L. J Electrochem Soc 2004;151:G456–G460.
30. Busnaina A, Lin H, Moumen N, Feng J, Taylor J. IEEE Trans Semicond Manuf 2002;15:374–382.
31. IC Knowledge LLC. Wafer cleaning and photoresist stripping. 2006. 42–59.
32. Ohmi T, Sudoh S, Mishima H. IEEE Trans Semicond Manuf 1994;1(4):440–446.
33. Park JG, Pas MF. J Electrochem Soc 1995;142(6):2028–2031.
34. Funkhanel J, Schupke K. Wafer drying in wet processing: the challenge of the future. Semiconductor International; Oct 2004.
35. Stein N, Shirazi G, Tang J, Jackson R, Viloria G, Achkire Y, Hsu W. Post-CMP Marangoni drying eliminates defects. European Semiconductor; Apr 2004.
36. Tang J, Lu W, Xie B, Martinez E, Ko A, Endo R, Todd C, Lee JT. Eighth International Symposium on Ultra Clean Processing of Semiconductor Surfaces; 2006. p 197–198.
37. Kawahara H, Yoneda K, Murozona I, Todokoro Y. IEICE Trans. Electron 1994;E77-C (3):492–497.
38. Kawahara H. Ultraclean Surface Processing of Silicon Wafers. Berlin: Springer-Verlag; 1998. p 456.
39. Shieh MS, Chen PS, Tsai MJ, Lei TF. J Electrochem Soc 2006;153(2):G144–G148.
40. Itano M, Kern FW Jr, Kawanabe I, Miyashita M, Rosenberg RW, Ohmi T. Particle deposition and removal in wet cleaning processes for semiconductor manufacturing. IEEE Trans Semicond Manuf 1992;5:114–120.
41. Luo J, Dornfeld DA. IEEE Trans Semicond Manuf 2003;16(3):469–476.
42. Obeng YS, Forshoefel KM, Richardson KA, Burton RH. Proceedings of Chemical Mechanical Planarization, PV98-7 Electrochemical Society; 1998. p 255.
43. Hong YK, Eom DH, Lee SH, Kim TG, Park JG, Busnaina AA. J Electrochem Soc 2004;151(11):G756–G761.

44. Ishikawa A, Shishida Y, Yamanishi T, Hata N, Nakayama T, Fujii N, Tanaka H, Matsuo H, Kinoshita K, Kikkawa T. J Electrochem Soc 2006;153(7):G692–G696.
45. Itano M, Kezuka T. Ultraclean Surface Processing of Silicon Wafers. Berlin: Springer-Verlag; 1998. p 125.
46. Tardif F, Semicond Semimet 2000;63:200.
47. Kaufman FB, Thompson DB, Broadie RE, Jaso MA, Guthrie WL, Pearsons DJ, Small MB. J Electrochem Soc 1991;138(11):3460–3465.
48. de Larios JM, Zhang J. Evaluating chemical mechanical cleaning technology for post-CMP applications. Micro Magazine 1997. p 61.
49. Cho GS, Kim HS, Lee JK, Jeng JD, Kim DY. CMP-MIC 2005;9:263.
50. Pan TM, Lei TF, Chen CC, Chao TS, Liaw MC, Yang WL, Tsai MS, Lu CP, Chang WH. IEEE Electron Device Lett 2000;21:338–340.
51. Pan TM, Lei TF, Ko FH, Chao TS, Chiu TH, Lee YH, Lu CP. IEEE Trans Semicond Manuf 2001;14:365.
52. Hong YK, Kang YJ, Park JG, Han SY, Yun SK, Yoon BU, Hong CK. Effect of poly silicon wettability on organic type defects in poly CMP. 210th ECS Meeting; Cancun Mexico; Oct 302005.
53. Overbeek JThG. Electrochemistry. Volume I, MIT Video Course. MIT Center for Advanced Engineering Study, No. 61-2102; March 1961.
54. Pourbaix M. Altas of Electrochemical Equilibria in Aqueous Solutions, Houston, TX: NACE (National Association of Corrosion Engineers); 1974. p 384.
55. Tamilmani S, Huang W, Raghavan S, Small R. J Electrochem Soc 2002;149 (12):G638–G642.
56. Jones DA,Principles and Prevention of Corrosion. 2nd ed. New Jersey: Prentice-Hall; 1996. Chapter 3, p 76.
57. Homma Y, Kondo S, Sakuma N, Hinode K, Noguchi J. J Electrochem Soc 2000;147(3):1193–1198.
58. Kondo S, Sakura N, Homma Y, Ohashi N. Japan J Appl Phys 2000;39:6216–6222.
59. Miller AE, Fischer PB, Feller AD, Cadien KC. Proceedings of the IEEE 2001;2001. June 4–6 p 143.
60. Tromans D. J Electrochem Soc 1998;145:L42–L45.
61. Shamiryan D, Abell T, Lacopi F, Maex K. Low-*k* dielectric materials. Mater Today 2004;7(1):34–37.
62. Maex K, Baklanov MR, Shamiryan D, Iacopi F, Brongersma SH, Yanoviskaya ZS. J Appl Phys 2003;93:8793–8841.
63. Han JH, Koo JE, Choi KS, Park BL, Chung JH, Hah SR, Lee SY, Kang YJ, Park JG. A study on water- mark defects in copper/low-*k* chemical mechanical polishing. Mater Sci Forum 2007. Forthcoming.
64. Lee SY, Lee SH, Park JG. J Electrochem Soc 2003;150(5):G327–G332
65. (a) Buley T, Epshteyn Y, Kulus M. Rohm and Haas Electronic Materials CMP Technologies. (b) Tran C, Bartosh K, Peters D, Watts C. ATMI Surface Preparation Products "Wet Surface Technologies". Available at http://www.micromagazine.com/archive/05/10/buley.html.

66. Hong YK, Han JH, Kim TG, Park JG, Busnaina AA. J Electrochem Soc 2007. Forthcoming.
67. Butt HJ. Measuring electrostatic, van der Waals, and hydration forces in electrolyte solutions with an atomic force microscope. Biophys J 1992;60:1438–1444.
68. Butt HJ, Jaschke M, Ducker WA. Measuring surface forces in aqueous solution with the atomic force microscope. Bioelectrochem Bioenerg 1995;38:191–201.
69. Yoon ES, Yang SH, Han HG, Kong H. Wear 2003;254:974–980.
70. Preston F. J Soc Glass Technol 1927;11:214–256.
71. Krupp H. Ad Colloid Interface Sci 1967;1:111–181.
72. Kuo R, Matijevic E. J Chem Soc Faraday Trans I 1979;75:2014.
73. Kallay N, Matijevic E. J Colloid Interface Sci 1981;83:289.
74. Ryde NP, Matijevic E. J Colloid Interface Sci Forthcoming. 2000.
75. Suni II, Gale GW, Busnaina AA. J Electrochem Soc 1999;146(9):3522–3526.
76. Busnaina AA. In:Kanegsberg B, editor. Critical Cleaning Handbook. CRC Press; 2000.
77. Busnaina AA, Piboontum J, Moumen N. Proceedings of the Material Research Society Meeting; San Francisco; 1999. Apr 5–9
78. Busnaina AA, Moumen N, Piboontum J. Proceedings of the VLSI Multilevel Interconnection Conference (VMIC), Santa Clara; CA; 1999. Feb 8–12
79. Busnaina AA, Elsawy TM. J Electron Mater 1998;27(10):1095–1098.
80. Busnaina AA, Dai F. J Adhesion 1998;67:181–193.
81. Busnaina AA, Dai F. Semiconductor International; Aug1997.
82. Busnaina AA, J Acoust Soc Am 1996;100(4):2775.
83. Gale W, Busnaina AA, Dai F, Kashkoush I. Semiconductor International; Aug 1996.

17

DEFECTS OBSERVED ON THE WAFER AFTER THE CMP PROCESS

Paul Lefevre

17.1 INTRODUCTION

This chapter gives an overview of the defects that can be generated or revealed during chemical–mechanical planarization (CMP). Most defects are specific to the type of CMP operation the wafer has just experienced. For this and other obvious reasons, this chapter is organized by CMP applications. However, there are some families of defects that are common to most CMP processes. Indeed, scratches [1–5], remaining particles [6], and surface residues [7] could be found in all CMP applications. Therefore, the materials are presented in the order of complexity, from the simplest such as oxide CMP to the most sophisticated such as copper CMP. Almost all defects presented in oxide CMP could exist in the other applications such as poly-Si, W CMP, and Cu CMP.

This chapter provides general definitions of the defects present after CMP with specific examples and images whenever available. The likely causes for the defects are explained in light of formation mechanism whenever possible. In addition to the defects visible to optical and electron microscopes, such as scratches, corrosion [8,9], and particles, this chapter also deals with residues and nuisances that are not visible yet in some cases more important, such as metal contamination [10] on the wafer surface. For example, a trace of radioactive element left from a CMP [11] process would be detrimental to the transistor structures. At the same time, it is one of the least visible contaminations under conventional defect detection methods.

Microelectronic Applications of Chemical Mechanical Planarization, Edited by Yuzhuo Li

The last section of this chapter presents the techniques and tools used to observe and analyze the nature for these defects. In some cases, a discussion will also be given on the quantification of these defects.

17.2 DEFECTS AFTER OXIDE CMP

17.2.1 Introduction

The silicon dioxide chemical–mechanical planarization (oxide CMP) technique is used in at least four applications:

- STI (shallow trench isolation) [12–14]
- PMD (premetal dielectric) [15]
- ILD (inter/intrametal dielectric) [16,17]
- MEMS (microelectromechanical systems) [18]

ILD CMP represents the first application of CMP technology in semiconductor industry. PMD CMP is very similar to ILD CMP except that PMD CMP planarizes topography over device structures, whereas ILD CMP planarizes oxide over interconnect structures. It is not uncommon that PMD and ILD may use different type of oxides. One possible combination may include BPSG (boron phosphorus doped silicon glass) or PSG for PMD and TEOS (tetraethyl orthosilicate) for ILD. Unlike other competitive non-CMP technologies that could perform only local planarization and leave global step heights behind, oxide CMP enables planarization at a local (submicrometers to micrometers) and global scale (several millimeters). It prepares a planar oxide surface and allows the implementation of higher resolution lithographic operations for the formation of sub-half micrometers features. The true benefit is an ability to build more layers of interconnect metal lines and improve the IC (integrated circuit) performance.

STI CMP has been introduced after oxide CMP. Before the implementation of STI CMP, the isolating structures were limited in size because of the edge of thermally grown oxide structures [19]. A damascene technique that uses STI enables the formation of vertical oxide edge and smaller structures (< 0.25 μm). The first generation of STI CMP processes were a copy of ILD CMP processes with low oxide-to-nitride selectivity and poor planarization efficiency [20,21]. To compensate such a deficiency, a scheme can be implemented in which a reverse mask is first applied. This resist mask protects the low structures. The following etch operation will remove a portion of the oxide on the top of the structures [22]. In essence, such a reverse mask scheme preplanarizes the STI layer before CMP. The high-performance "direct STI CMP" uses a high (oxide over nitride) selectivity slurry that usually contains ceria abrasives [23,24]. The obvious advantage of direct STI is the elimination of the need for "preplanarization." The main challenges of direct STI are in the areas of topography and defect control.

The conventional STI CMP is a proven process with low and stable defect performance that translates to very high and stable yield.

Oxide CMP for MEMS fabrication represents the most recent oxide CMP application. It deals with very thick oxide films and large step heights. Typically, PMD or ILD CMP would remove 0.1–1 μm of oxide. In MEMS applications the amount of oxide to be removed is on the order of 1–10 μm. The step height to be eliminated is in the same range and the structure size ranges from a micrometers few to millimeters. Most MEMS CMPs use processes similar to ILD CMP. As these tools and consumables were originally developed for conventional CMP, it usually takes much longer time to process MEMS wafers (see Chapter for details).

Most oxide CMP applications except STI consist of polishing one material only: oxide. STI CMP involves two types of materials: silicon dioxide (SiO_2) and silicon nitride (Si_3N_4). Highly doped silicon dioxide such as BPSG is usually more reactive than the oxide film made from TEOS [25]. Depending on their deposition methods, different oxides also have varying degree of hardness. Therefore, the types and extent of surface defects are a function of the CMP process as well as the nature of the substrate. This section will review all possible defects left on the oxide surface after CMP and post-CMP cleaning.

17.2.2 Scratches

An oxide scratch is the most commonly seen defect after oxide CMP. The scratches on the oxide surface are mostly due to a very small amount of large particles in the slurry. As a typical oxide CMP slurry may contain as much as 10 % by weight of abrasives, there are ample opportunities for defect-causing oversized particles to reach a wafer surface. As shown in Fig. 17.1, some of the scratches are large enough to be seen under a regular optical microscope. Others may require an electron microscope.

The three most common abrasives for oxide CMP slurries are fumed silica, colloidal silica, and ceria. The advantage of fumed silica is that the purity of the

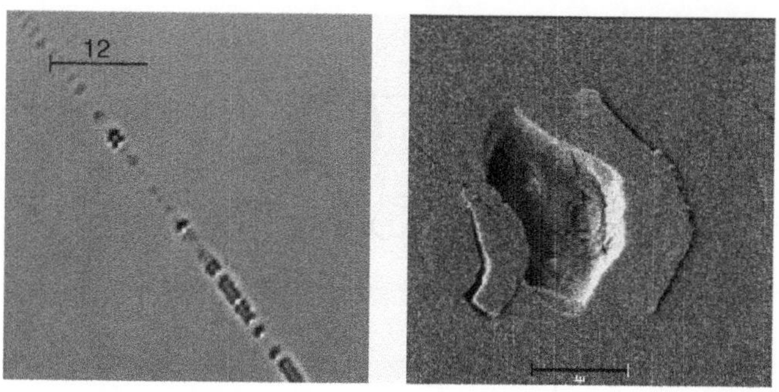

FIGURE 17.1 A very large scratch on oxide. The left picture shows the scratch under an optical microscope. On the right is an SEM image of the same scratch.

abrasive tends to be higher than that of standard colloidal silica [26]. The disadvantage of fumed silica is that the particles easily agglomerate. The agglomeration can be facilitated by the shearing force. Colloidal silica typically leaves fewer and smaller scratches. The ceria particles have three major advantages over silica particles. Because ceria particles have higher density and hardness, a more angular morphology, and greater reactivity than silica, they offer a higher removal rate at lower solid concentrations. They naturally offer higher oxide-to-nitride selectivity. Most important of all, they planarize oxide topography with much higher efficiency than silica particles [27]. This is probably due to the very effective mechanical abrasion of ceria on the oxide. The down side of the ceria is that they tend to leave scratches on the oxide surface. Therefore, it is desirable to formulate silica slurry with very high oxide-to-nitride selectivity and manufacture ceria slurry with low scratch counts by means of particle morphology control, size control, and particle surface treatments or modification.

Silicon dioxide is a nonplastic material. It goes from elastic deformation to the breaking point without having a plastic deformation regime [28]. Therefore, the scratches left in silicon dioxide do not look like smooth continuous trenching. Rather oxide scratches typically leave local fractures along the scratching line. These fractures are curved surfaces "orthogonal" to the scratching particle trajectory as illustrated in the cartoon (Fig. 17.2).

There are three levels of a scratching event:

1. The stress induced by the scratching particle is high enough to make nanodislocation on the oxide but nothing is visible unless a long enough etch (e.g., dilute HF) reveals the seams (Fig. 17.3).
2. The stress is high enough to break the oxide and remove some glass around the dislocations. The result of this event is visible under a scanning electron microscope (SEM) (Fig. 17.4). It can also be revealed by an optical microscope after a dilute HF process that removes some of the broken materials.

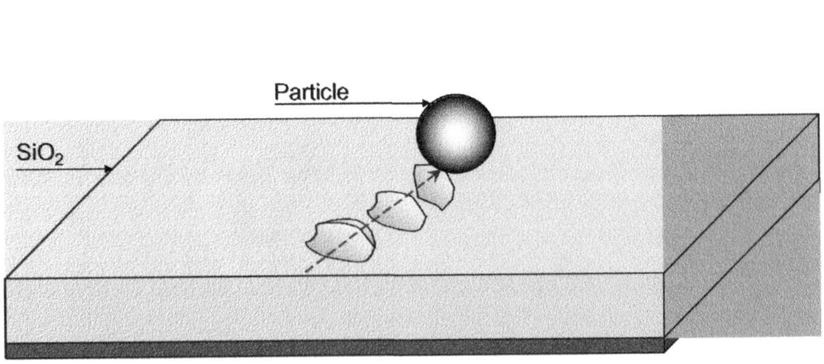

FIGURE 17.2 Schematic illustration of fractures in SiO_2 left behind by a large scratching particle.

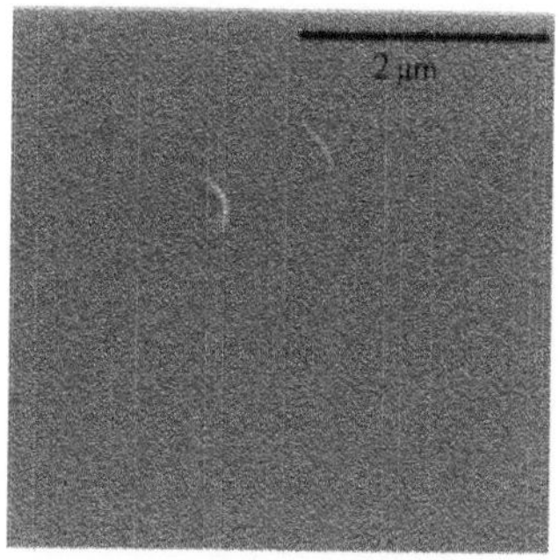

FIGURE 17.3 SEM image of a scratch on the oxide surface at level 1. The polished wafer was etched with dilute HF, which results in a 500 Å oxide loss. This etch process is often referred as 500 Å oxide etching with dilute HF.

3. The stress is so high that the oxide is broken and pieces of the glass are removed at the same time. This is visible under optical and SEM microscopes without any treatment (Fig. 17.5).

Very small scratches are not seen by a defect inspection tool. A conformal etch process such as a wet etch using dilute HF will enlarge the small scratches and make them visible under the defect inspection tool. The longer the etch process, the more visible the scratches become as shown in Fig. 17.6. The etch process was conducted step by step to reveal more and more of the level II and possibly some level I scratches.

The scratching mechanism described above is very typical for oxide CMP. In order to minimize the scratching level, slurry manufacturers have explored various abrasive types and formulations. Using ultrahigh-quality colloidal silica with the right combination of chemical additives, it is possible to produce scratch-free slurries. With such a slurry, the dilute HF has no effect on the scratch count as shown in Fig. 17.7.

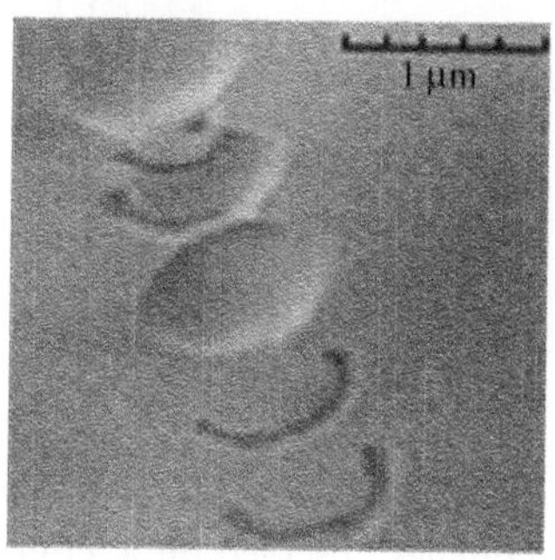

FIGURE 17.4 SEM image of a scratch on the oxide surface at level 2 after 500 Å oxide etching with dilute HF.

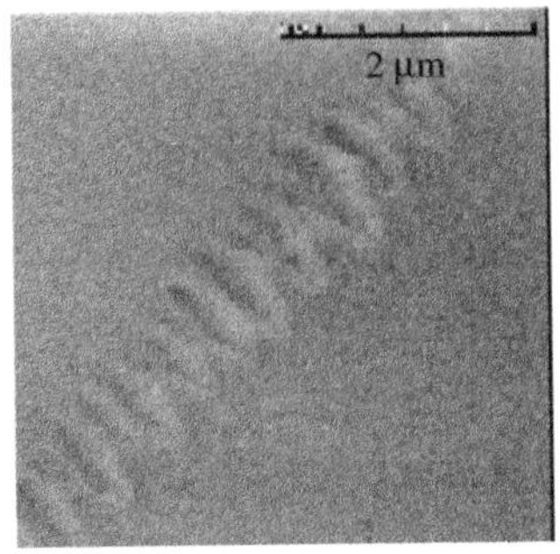

FIGURE 17.5 SEM image of a scratch on the oxide surface at level 3 (no etching is needed to reveal the scratch).

17.2.3 Color Variation—Oxide Thickness Variation

The color of silicon dioxide is a function of its thickness. Color variation across the dies and the wafer indicates oxide thickness nonuniformity. Ideally, there should be no color variation over the same type of underneath structures. Global color variation across the wafer indicates a CMP uniformity problem. Local color variation at a structure level or arrays of structures within a die reveals a lack of planarization. There is a fine difference between planarization deficiency and nonuniformity. Nonuniformity is revealed in the form of a very gradual thickness variation over 10 mm or more. It is usually not pattern dependent. A rapid thickness variation across arrays of structures less than 5 mm wide is indicative of a planarization problem. This is a pattern-dependent

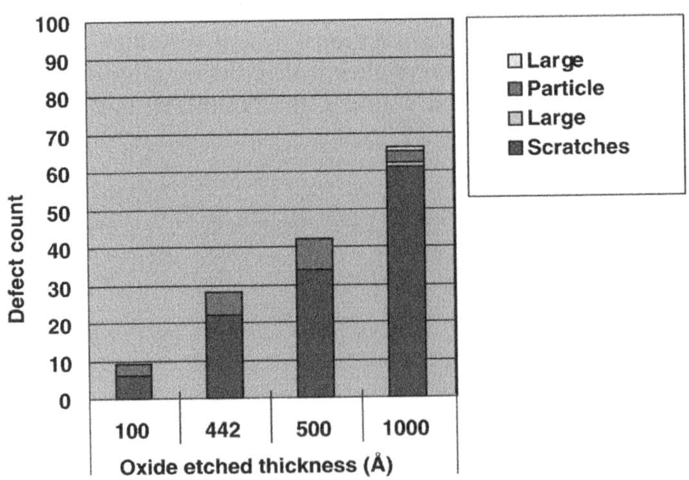

FIGURE 17.6 Number of scratches observed on a TEOS blanket wafer using KLA-Tencor 6420. The more the small scratches are etched, the more visible they are. At the same time, the other defects such as particles are slightly reduced.

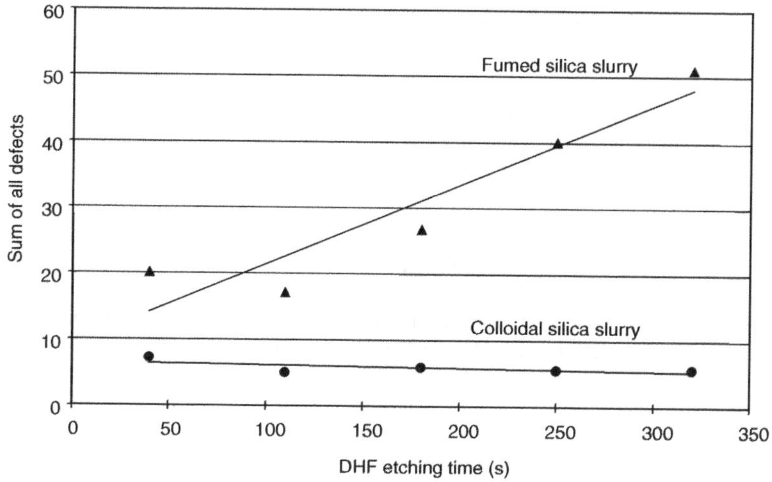

FIGURE 17.7 Slurry abrasive effect on scratching performance. Ultrahigh-purity (TEOS base) colloidal silica slurry has an excellent oxide scratch behavior. On the contrary, the fumed silica has a tendency to scratch the oxide surface. A step-by-step dilute HF etching reveals smaller and smaller scratches.

mechanism. The only case where uniformity can be confused with planarization is where they overlap. That often happens at the edge of the wafer when the typical planarization distance of 3 mm for oxide CMP is better than uniformity performance. Such confusion can be easily resolved by remembering that planarization is pattern dependent; the color variation follows pattern symmetry. Uniformity is not, the color variation follows a radial symmetry. With naked eyes it is easy to see color variation due to lack of good uniformity but more difficult to see color variation due to lack of planarization. Under an optical microscope, it is the opposite. Under a microscope, a color change in the oxide around the underneath structure will reveal a lack of planarization. For more detailed discussions on oxide CMP planarization issues, readers are referred to the publications by Boning and his team at MIT [29–31].

The planarization efficiency is defined as the unitless formula one minus the ratio of removal rate for the bottom over the removal rate for the top of the structure. It is actually difficult to achieve planarization efficiency over 99% for all structure types, all feature sizes, all density variations, and all step heights at the same time. The easiest features to planarize are small individual structures (submicrometers or a few micrometers in size). Over such a structure it is unlikely to see any color variation in the oxide around the structure. A large array of structures adjacent to another large array of features or large field area (size in millimeters) with different relative average height of oxide create an oxide step that is difficult to planarize. Before CMP, there is no color variation within each individual array. After CMP, it is likely to see color variation at the edge of each array or in the adjacent field area.

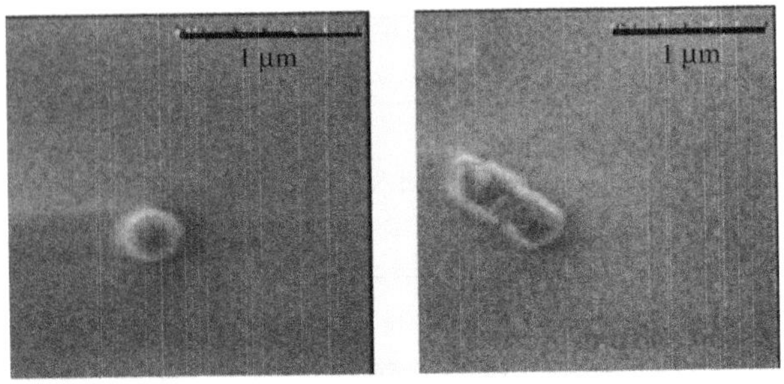

FIGURE 17.8 Slurry abrasive residues. In case the particle adhesion on the oxide wafer is stronger than the cleaning scrubbing force, abrasive particles remain on the wafer surface.

17.2.4 Slurry Residues and Organic Residues

Typically, slurry residues are due to strong adhesion of some abrasive particles to the oxide surface of the wafer. For some more complex slurry formulations, it is possible to leave organic residues [32,33] on the wafer surface as well. In some rare cases, it is also possible to leave pad residues, pad debris on the wafer surface, especially when an improper pad break-in procedure is combined with an inadequate post-CMP cleaning. Figures 17.8 and 17.9 show some representative residues on a polished oxide surface.

In an ideal case, the post-CMP cleaning that follows oxide CMP should remove all the residues. Most post-oxide CMP cleaning processes simply use a PVA brush to scrub the wafers with DI water. In some cases dilute ammonia is used, and in very rare cases dilute HF is used. Obviously, dilute HF will be very effective in removing the silica particles on the surface. But that process can also etch the scratches and make them larger. Another method commonly used to reduce the residues on the wafer is to perform water buffing on the wafer after oxide CMP using a very soft pad. The buffing step may enhance the efficiency of the brush cleaner [34,35]. On the contrary, it adds a polishing step. Not all tools are configured to do so.

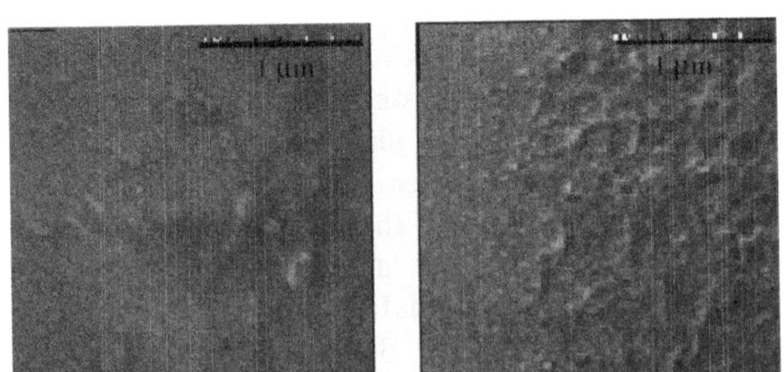

FIGURE 17.9 Example of organic contamination on the TEOS surface.

17.2.5 Other Particles

Particles can come from many different sources:

- The polishing and cleaning tools and their consumables (slurry, pad, brush, etc.).
- Metrology tools.
- The environment: people, building, and other tools in the area.

There are a large number of potential sources for particles to fall onto a wafer. Advanced tool design and proper selection of clean room technology can be very effective in reducing or eliminating this type of defect. In addition, it is desirable to eliminate manual intervention. The use of minienvironment around the tools and closed wafer-handling cassettes such as FOUP (front opening unified pod) are proven effective. These types of particles are more frequently seen in small clean rooms where wafers are handled manually.

17.2.6 Crystal Formation

When highly doped oxide such as BPSG is used as a substrate, beautiful looking crystals may develop on the surface after the CMP. It takes only a few days for these soft crystals to grow to a few micrometers in size. These crystals are water soluble. With regular TEOS, it is not likely to see crystal formation.

17.2.7 Trace Elements

The trace metal left on the surface of the wafer is mostly due to the slurry. The quality of the chemicals and abrasive used in the slurry is never perfect. An average oxide CMP slurry has trace metals in the range of 1–10 ppm per element. There are very few ultrahigh-purity slurries with trace metals below 10 ppb (actually under the metrology detection limit). What matters most is not the amount of trace metals in the slurry but the amount left on the surface after post-CMP cleaning. The most effective technique to remove the trace metals from the wafer surface is to perform a very thin etching of the oxide surface using dilute HF, for example.

17.2.8 Radioactive Contamination

This type of defect is very rare. Radioactive contamination is effective at destroying the transistors [11]. Some ceria particles may contain radioactive impurities depending on the source of raw materials. Therefore, it is important to screen for radioactive impurities for slurries that may contain radioactive raw materials. In addition, a radioactive contaminated packaging material may also cause irreversible damage if used to store CMP consumables.

17.2.9 Defects Existing Before Oxide CMP

The defects existing prior to oxide CMP are usually visible through the transparent oxide film before CMP. Unlike metal CMP process, these defects cannot hide under an opaque metal film. The defects existing before oxide CMP may include

- defects in dielectric or embedded particle, or contamination in the oxide;
- defects in the underlayers: litho error (missing patterns, bad patterns, etc.) defects in underlayer thin films, and defects created by previous polishing steps (Si wafer polishing, STI CMP, poly-Si CMP, W CMP, PMD CMP, ILD CMP, etc.).

17.2.10 Source of Defect-Causing Large Particles

The most challenging issue in oxide CMP is the instability of the defect level and the fluctuation of the uniformity across the wafer and the die. The stability of a CMP process is dependent on the stability of the tool and its consumables. The pad is usually a known variant as it suffers material loss after each wafer. For example, the pad groove height is reduced after each polishing and conditioning process. The diamond pad conditioner also experiences wear during each conditioning step. The microchanneling and active sites on the pad created during the conditioning are never exactly the same before each wafer. Even the slurry is rarely identical every day. The extremely large surface area of the nanoparticles in the slurry by nature drives the particles to aggregate in order to reduce the surface energy. It is ironic that the very attempt to redisperse the aggregated particles, such as agitation, may actually accelerate further aggregation due to the possible use of excessive shear. In addition, the particle aggregation may also be facilitated by the settling force. It is important to point out that many defects such as scratches can be traced back to the presence of these oversized particles [36–43].

17.3 DEFECTS AFTER POLYSILICON CMP

17.3.1 Introduction

Depending on its crystalline structure, silicon can exist in single-crystalline, polycrystalline, or amorphous form. Polysilicon (p-Si) is short for polycrystalline silicon, which is a form of silicon composed of many crystals, as opposed to amorphous silicon (a-Si), which is an unordered form with a random internal structure. The polysilicon CMP process can be used for several applications such as

- polysilicon contacts
- polysilicon capacitors

- polysilicon gate
- Polysilicon floating gate (underneath the control gate, nonvolatile storage application).

In addition to the applications listed above, there are other proprietary applications that deal with monocrystalline silicon epitaxy in a damascene scheme that could involve a polysilicon CMP or a silicon CMP process. The main challenge of the polysilicon CMP comes from the structure design. Some designs involving very large structures (several hundreds of micrometers in size) demand extremely stringent dishing requirements. Another challenge in most polysilicon CMP applications is the defect density and metal contamination. Because, the polysilicon CMP process is integrated at very close proximity of the active area of the transistors, ultralow trace metal contamination is critical. Some integration schemes use a sacrificial layer that would be removed later on. This allows a wider margin in metal contamination management. But not all integrations methods can use a protective film against metal contamination. Some polysilicon CMP processes require ultrahigh-purity CMP consumables with less than ppb level of trace metal in the slurries. In some cases, the polysilicon CMP process has similar defect and contamination requirements as the final silicon polishing step used in the manufacturing of the bare silicon wafers. In addition to that, not all the polysilicon materials are the same. The deposition method, the doping concentration, and subsequent thermal treatments can make subtle differences in the polysilicon. These differences can make noticeable impact on the CMP performance. For example, the removal rate of the polysilicon is very sensitive to the polysilicon crystalline structure and so is the dishing [44] severity. From a defect point of view, a small change in the doping concentration may lead to a large difference in the defect level.

17.3.2 Scratches

Most poly-Si CMP processes involve patterned oxide filled with polysilicon. The generation mechanism and physical appearance of the scratches for polysilicon CMP are the same as in the oxide CMP described earlier. Most Polysilicon CMP processes are rather selective toward silicon oxide. More specifically, the polysilicon removal rate is usually much higher than that for oxide. Such slurries usually contain low concentration of abrasive, which usually translates to low defect density. At low magnification, scratches on the poly-Si film after CMP usually look like a continuous line rather than chatter marks. At higher magnification, the scratch usually looks like irregular trench with multifaceted walls.

17.3.3 Polysilicon Residues

Thin polysilicon is transparent to visible light. Therefore, it is not always easy to see the poly-Si residues. With experience, one can detect the slight color

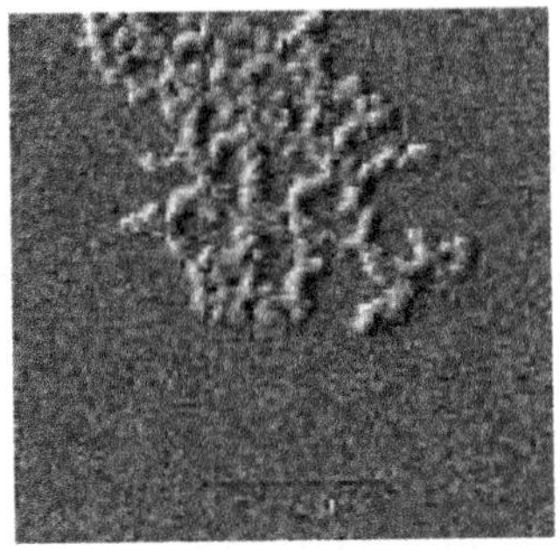

FIGURE 17.10 Slurry particle on the poly-Si surface.

variation that is created by poly-Si residues. The common cause for poly-Si residues is lack of good within-wafer uniformity associated with insufficient overpolishing time. The second cause for poly-Si residues comes from the patterned structure topography. Some designs have recessed local areas that are difficult to clear. Also, the topography inherited from steps prior to poly-Si CMP could be difficult to planarize. Large poly-Si bumps over a wide structure may not be planarized by the time the poly-Si clears. These last two causes leave poly-Si residues that are pattern dependent.

17.3.4 Particles

The polysilicon is relatively hydrophobic and its oxide is hydrophilic. The two surfaces may attract different types of particles or residues. Figure 17.10 shows slurry particles left on the poly-Si surface after CMP. Therefore, it is more difficult to clean the poly-Si surface than its oxide counterpart when using an aqueous cleaning solution. A key to resolve this problem is to chemically change the poly-Si surface hydrophobicity. This is a very effective method to clean the wafer surface. As the surface treatment usually involves the use of organic chemicals such as a surfactant, one challenge is how to completely remove the organic chemical at the end of the cleaning process.

17.3.5 Residues

The natural consequence of using organic chemical to clean the poly-Si wafer is the risk to leave organic residues on the wafer surface. Some representative organic residues left on the poly-Si surface after CMP are shown in Fig. 17.11. Ironically, sometimes these organic residues pose even greater challenges than those residues or particles being cleaned off by these organic chemicals.

17.3.6 Trace Elements

The poly-Si CMP process is very sensitive to trace metals. Copper, iron, nickel, and other similar metals are forbidden during this process as they can poison

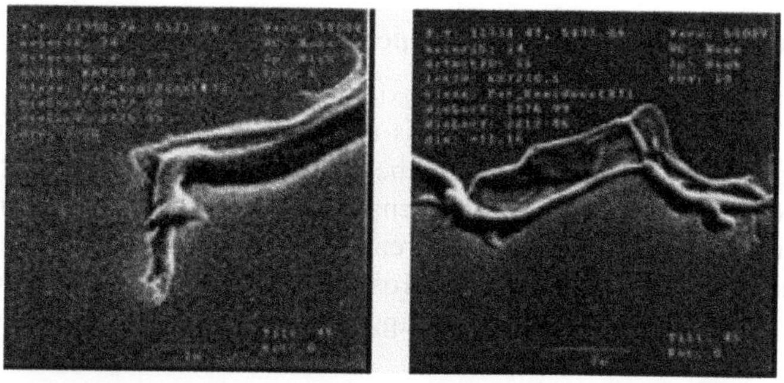

FIGURE 17.11 Organic residues on the poly-Si surface.

the transistor. CMP consumables especially the slurry must use high-purity raw materials and must be kept away from any exposure to these unwanted metal ions.

17.3.7 Polysilicon Pitting and Voids

Pitting or void in the poly-Si is a common mode of defect. The severity of such a defect depends on the type of poly-Si material involved. More specifically, the crystalline structure and the level of doping elements can have a significant impact on the tendency of the poly-Si film to form pits and voids during the CMP process.

17.3.8 Discoloration at the Edge of the Structure or Edge of the Arrays

Oxide, silicon nitride, and thin poly-Si are all transparent films. Slight gradual thickness variations in any of these films will induce corresponding color disparity. The film thickness variation can come from different origins. The within wafer nonuniformity (WIWNU) in material removal can lead to remaining film thickness variation across the wafer. Topography resulting from lack of planarization can also create local thickness variation. For example, a dishing in a large poly-Si structure will result in a color difference between the outside and the inside of the poly-Si structure. Two adjacent arrays with different pattern densities could be polished with different removal rates. This may also translate to gradual thickness variation at the border area between these arrays, which results in color variation at the edge of the arrays.

17.3.9 Defects Existing Before and Revealed After Polysilicon CMP

The types of defects existing before polysilicon CMP are similar to the types of defects that can be observed before oxide CMP. The main difference is that the poly-Si film is not as transparent as an oxide film. The thicker the polysilicon film, the less transparent it is.

17.3.10 Influence of Processing Temperature

There are many similarities between oxide CMP and poly-Si CMP. The main difference between the two processes is that poly-Si CMP slurries contain less abrasives and are, in general, more chemically active. Therefore, the poly-Si CMP process is by nature very sensitive to the polishing temperature. Temperature has a direct effect on removal rate, topography removal, and defect density (pitting and voids). Most poly-Si CMP slurries use colloidal silica that is less likely to form large aggregates than the fumed silica.

17.4 DEFECTS AFTER TUNGSTEN CMP

17.4.1 Introduction

Tungsten (W) CMP is mainly used to form two types of microstructures (Fig. 17.12):

- Vertical W contacts (low or high aspect ratio)
- W lines.

A typical W CMP process consists of several polishing materials: W, W liner (Ti and/or TiN), and Oxide [45–47]. In some unique W CMP applications, the process may involve the removal or stoping of other materials such as for SiN and poly-Si. The focus of this section will be on the popular stacks (W, Ti, TiN, and oxide) and the common defects left on the oxide surface after CMP and post-CMP cleaning [48,49].

17.4.2 Corrosion, Pitting, and Void

There are many possible causes for local W loss. The most common one is W corrosion. The root causes for W corrosion may include

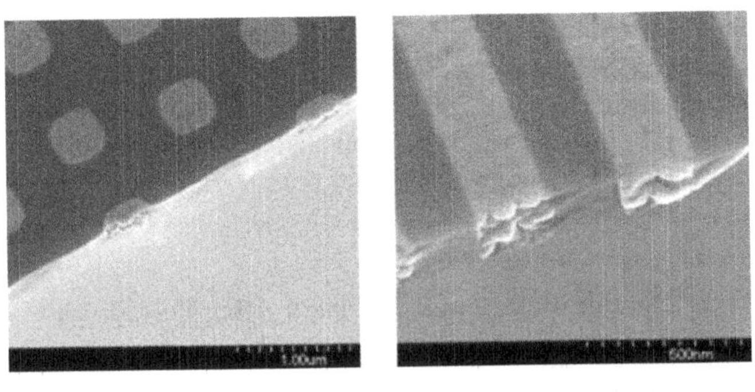

FIGURE 17.12 SEM top view with cross section of two different W structures: contact (left) and lines (right). It appears that the metal density is much higher for the line structures.

- massive electrochemical attack (galvanic corrosion),
- local electrochemical attack (maybe scratch induced),
- watermark,
- void in the W structure,
- incomplete W filling.

The massive electrochemical attack or galvanic corrosion will result in loss of tungsten in many structures. This can occur when the wafer experiences prolonged exposure to an electrolyte. In such an aqueous solution, the W surface can be slowly dissolved, which leaves corrosion marks. Unlike copper, the tungsten surface is usually not very reactive. Therefore, it is possible to develop an effective W CMP slurry without using a corrosion inhibitor. However, some tungsten structures are much more susceptible to corrosion than a simple film. Indeed, some W structures that are connected to transistor contacts could have a different electrical potential than the wafer surface. This can result in a local galvanic corrosion of the W contact. The other design-related source for corrosion is that some W contacts are very difficult to fill completely, which could leave a pit or corrosion defect. A good W CMP slurry and process should be able to reduce the sensitivity to these design variations and yield a corrosion-free surface, regardless of the tungsten structure layout.

In addition to the massive corrosion described above, there are some cases in which the W CMP process is directly responsible for the small and local loss of tungsten in the contacts. A very early stage of global galvanic corrosion could leave many pits on the W surface. This can be prevented by not leaving the wafers for too long in an electrolyte. A quick cleaning and drying process can prevent or minimize this type of corrosion. By the same token, a watermark could leave a pitting pattern on the same spot. Therefore, an effective post-CMP process is a prerequisite for a defect-free surface.

Some of the defects found during and after CMP have origins prior to the CMP process. For example, a small void in the middle of the W contact can be traced back to poor tungsten filling during the W deposition process. This deposition defect is revealed and enhanced during the W CMP process. After W CMP, this defect looks like a small hole or a pit at the center of the W contacts. The higher the aspect ratio, the more difficult is the W filling of the contact, and more likely there will be a void in the middle of the contact. There are two main reasons that W CMP can have such amplified caged defects. The slurry can be trapped in the void at the center of the W contact which result may in accelerated etching at that location. The second cause is that the thinning of the W contact results in revealing the existing void in the W contact as presented in Figs. 17.13 and 17.14.

17.4.3 Tungsten Recess and Rough Tungsten Surface

Tungsten recess in a small structure is usually visible when examined using SEM angle illumination. There is a visible shadow on one side of the W

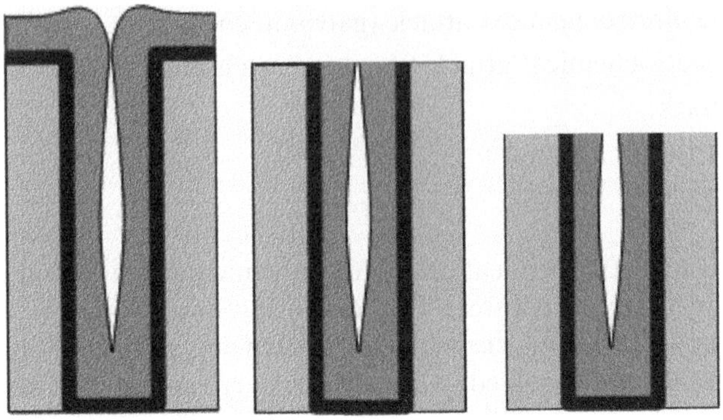

FIGURE 17.13 W deposition with no overfill (left). W deposition with overfill does not change anything to the void formation. At just the metal clearing of stage W CMP, the void is not opened (center). W overpolishing will result in oxide erosion and oxide loss. This will tend to open the void and make it visible (right).

contacts (Fig. 17.15). The W recess is common with a high-selectivity W CMP process in which the material removal rate is higher on W than on the dielectric. W continues to be removed during the overpolishing phase. The most effective action to reduce W recess is to reduce the static etch of tungsten through process optimization or slurry reformulation. As a W CMP polishing process can bring the slurry temperature in between pad and wafer surfaces to over 40°C, one easy process solution is to reduce the polishing temperature toward the end of the CMP process using a soft landing technique during the overpolishing phase. Alternatively, one can also modify the slurry formulation in order to reduce the static etch at high temperatures.

On the large structures, a W CMP process may leave a very rough W surface behind as a result of insufficient W filling before CMP. The CMP process simply did not reach down to these recessed areas. Normally, the W structures

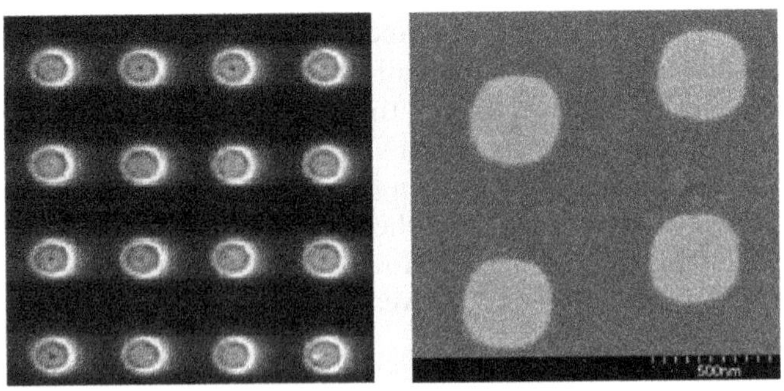

FIGURE 17.14 SEM picture of an array of high-aspect-ratio 90-nm W contacts (left). It is visible that some voids are opened after W CMP. Low-aspect-ratio 250-nm W structures have good W filling (right). The filling is much easier; no void is created.

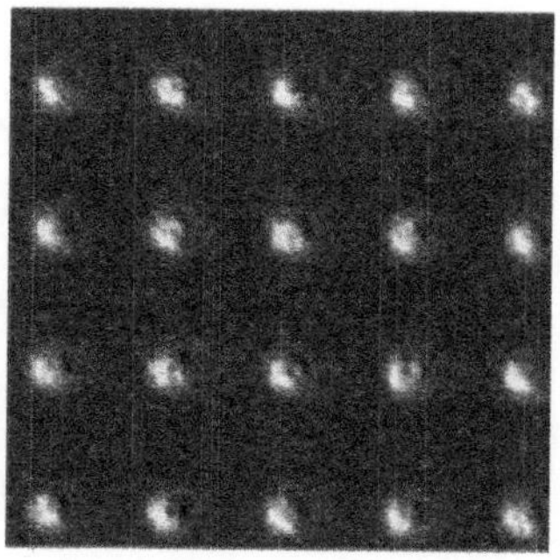

FIGURE 17.15 Shadow on the W contact characteristic of W recess.

are very small and close to the critical dimension (half pitch). However, in some cases, the structures in between the dies, such as the lithoalignment mask, are typically much wider. It is possible that such large trenches could be left incompletely filled without harming the subsequent layer of the metallization process (Figs. 17.16 and 17.17). In some cases, it is not acceptable to leave

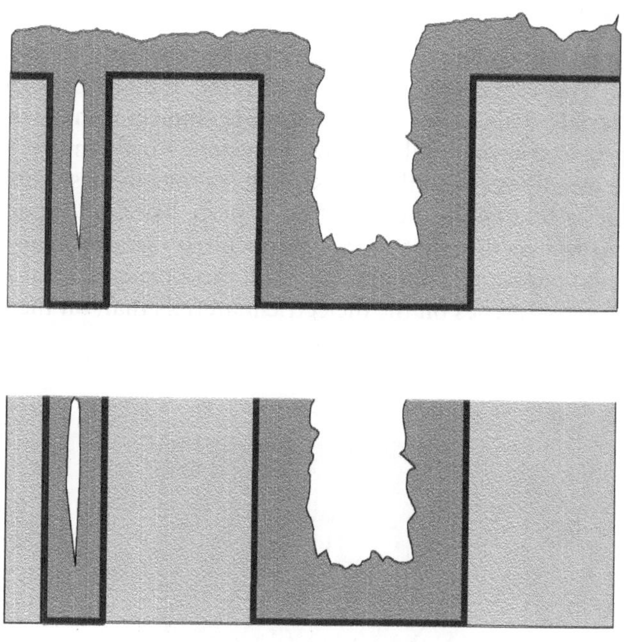

FIGURE 17.16 Two cartoons of a cross section of W structures. The cartoon on the top is a view before W CMP. The bottom cartoon shows the same structures after W CMP. In case the structure width is more than twice the W deposition thickness, the structure height must be smaller than the W deposition thickness in order to be completely filled.

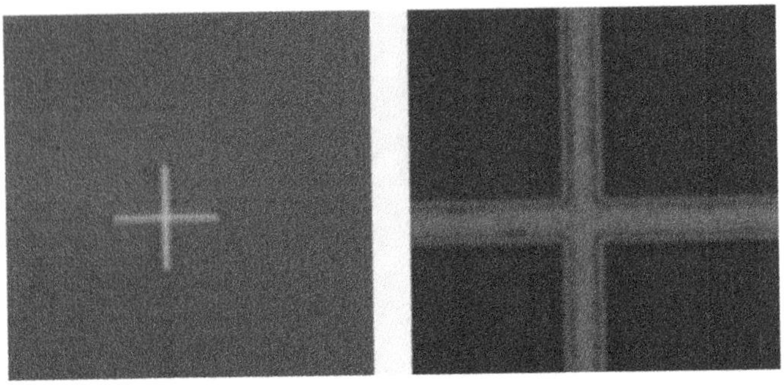

FIGURE 17.17 Top view of two tungsten crosses. Both images have been taken with the same magnification (100× objective). The small cross has good W filling, and the large cross has incomplete W filling. Note that these structures are not part of the product and are not creating any problem.

rough W or high local dishing behind. Either the mask design must be modified or the W filling on these large trenches must be complete.

17.4.4 Scratches

On a typical product wafer, the tungsten density is usually low, that is, about 20% or less. The remaining surface is covered mainly by oxide or other dielectric materials. This leaves four times more chances to have the scratch on the oxide or dielectric material than on tungsten. Furthermore, the tungsten structures are usually very small and a scratch is typically less visible. On a very small structure, a scratch on the W will look more like a pit. In addition, W is one of the hardest metals that is more difficult to scratch than other softer materials such as copper or aluminum. For all these reasons, after W CMP, it is more likely to see scratches on the dielectric materials than on the W structures (Fig. 17.18).

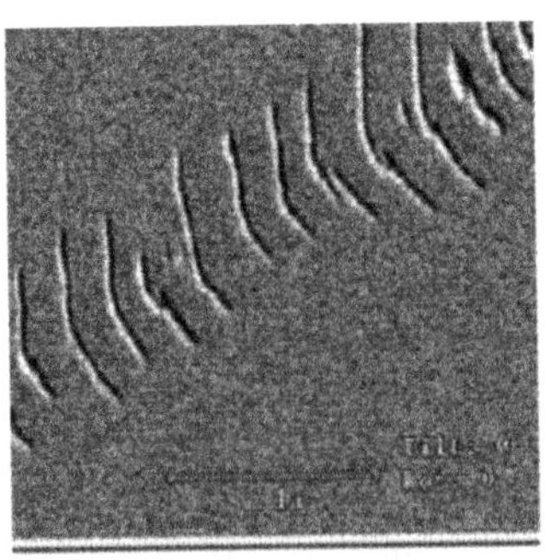

FIGURE 17.18 Scratches left on BPSG after W CMP.

17.4.5 Discoloration—Edge Overerosion (EOE)

Color variation at the edge of arrays is often a sign of thickness variation in the oxide. In most cases, this thickness variation is due to overerosion at the edge of the arrays. Just before the clearing of the tungsten, the patterned wafer surface can reach a stage that is perfectly flat. Indeed, one can easily verify this by simply taking an underpolished patterned wafer and measuring the topography in an area that is about to clear. On further polishing, for a CMP process that has 1:1:1 (metal:barrier:dielectric) removal rate selectivity, the topography usually remains completely flat. The the EOE appears when the removal rate of the metal is much higher than that of the dielectric. The removal rate ratio of the metal liner over the metal or the dielectric is generally not significant in terms of the EOE phenomenon. When the W metal is polished faster than the dielectric and the recess in the metal structure during overpolishing is significant, the overpolishing will likely create erosion of the dielectric between the metal structures and lead to edge overerosion. In the case of extremely high selectivity of metal over dielectric ($>$1000:1), there is practically no dielectric removal and therefore no dielectric erosion. In the case of common selectivity (between 10:1 and 1000:1 metal:dielectric), the higher the metal density, the lower the dielectric density and the faster the dielectric is eroded. The erosion of the array creates a step relative to the field area. This step slowly creats a trap for the abrasive particles. Locally, the concentration of abrasive particles is much higher at the edge of the array than at the center of the array [50]. This results in a locally higher removal rate on the edge of the array. When the EOE height relative to the array center is about the average size of the abrasive particle, the EOE stops growing and remains constant during the overpolishing phase (Fig. 17.19). In addition to array edge overerosion, there are other slurry properties that could be responsible for color change variation. A worst scenario may occur when the removal rate oxide (e.g., BPSG) is significantly higher than that for the metal liner (e.g., TiN). As soon as the liner is locally removed, the oxide underneath will be polished at a very high rate, which will leave uneven thickness across the die

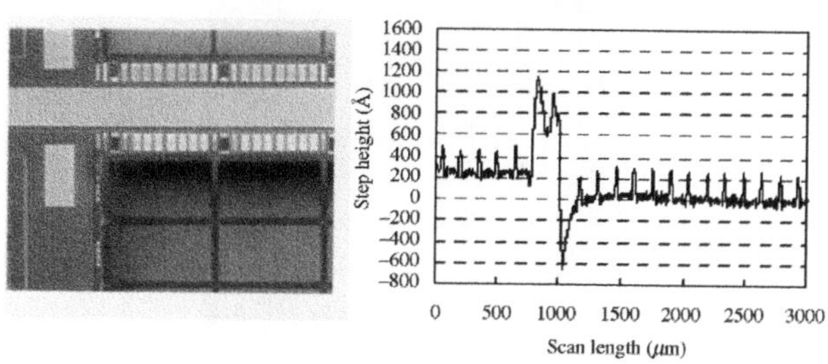

(a) (b)

FIGURE 17.19 Discoloration at the edge of a cell array: (**a**) top-down view; (**b**) step height profile from a periphery to a cell array in the *y*-axis direction. *Source*: Samsung paper on edge array erosion at the 204th meeting of the Electrochemical Society, 2003.

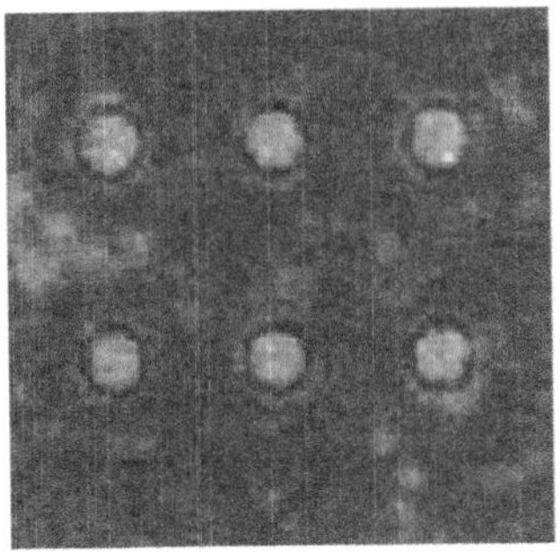

FIGURE 17.20 Metal residues around W contacts left by underpolishing.

and the wafer. A color variation across the die and wafers is usually visible by naked eye and under an optical microscope at low magnification.

17.4.6 Tungsten and Metal Liner Residues

During W CMP, if the WIWNU is poor, some areas may clear sooner than the others. This may leave tungsten and/or tungsten linear residues (Fig. 17.20).

If the oxide or the premetal dielectric (PMD) layer was not planar enough before W contact patterning, the metal deposition will fill the topography or the valleys. The tungsten deposited in these valleys will be very difficult to remove during tungsten CMP and is often left as puddles, unless the entire wafer is severely overpolished. If the topography left after the PMD process is relatively shallow, a moderate W overpolishing should be sufficient to completely clear all the metal residues. Figure 17.21 shows an example of such shallow topography left after PMD CMP. The worst case is when there is no PMD CMP before tungsten deposition. In this case, the W CMP process has to perform the PMD CMP task at the same time when tungsten and its linear are removed. This can be accomplished by using a low-selectivity

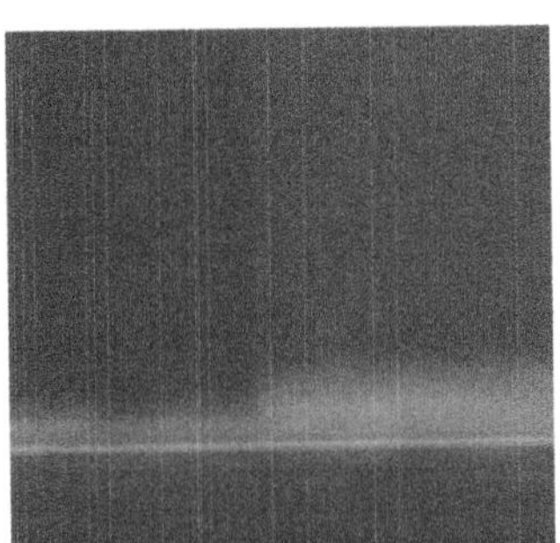

FIGURE 17.21 The W residue on this image is pattern dependent. The W structure is above the edge of some underneath arrays of structures. These arrays have different topography that still remains after PMD deposition and even after not 100% efficient PMD CMP.

(W over dielectric) CMP slurry or a multiple-step CMP process combining a high-selectivity W CMP slurry with an oxide CMP slurry.

17.4.7 Particles, Slurry Residues, and Trace Metals

Similar to oxide CMP, particles and slurry residues are of great concern for tungsten CMP processes. The factors that influence the level of particles and slurry residues on the oxide surface are essentially the same for W CMP. In addition, the topography on the W structures left after the W CMP process can act as an anchor for slurry particles and slurry residues. In the case of a high-selectivity W CMP process, the topography left on the W contacts is mainly the normal recess or dishing. These small recessed W contacts are not easy to clean because they trap particles and residues. In the case of a conventional oxide touch-up after W CMP, the W contacts are usually left slightly protruded. These protruded W contacts are also good anchors for particles and slurry residues. The most effective approach to avoid the accumulation of these particles and slurry residues is to ensure a CMP process that leaves no W topography. Typically, a low-selectivity W CMP process or a regular CMP process followed by a low-selectivity W CMP touch-up leaves a planar surface that is easy to clean.

A post-W CMP cleaning process is more challenging than oxide CMP cleaning due to the fact that there is a risk of W structure corrosion. Obviously, this risk does not exist for post-PMD or ILD CMP. Therefore, the post-W CMP cleaning chemistry must be carefully screened for any sign of corrosion potential. To be safe, efforts should be made to avoid any prolonged exposure of polished wafers to any wet environment, regardless of the characteristics of the cleaning chemistry.

In most cases, the trace metal left on the wafer surface after W CMP can be removed by a conventional post-CMP cleaning. Under some circumstances, a more aggressive post-W CMP cleaning that etches the oxide surface is needed to completely remove any trace metal left on the patterned wafer surface.

17.4.8 Delamination

Unlike copper CMP, it is rare to see delamination during W CMP. For W integration, most of the delamination can be traced to a poor adhesion between the metal and the linar or the liner and the dielectric film.

17.4.9 Preexisting Defects Revealed After Tungsten CMP

As discussed in the previous section, some preexisting defects can be revealed after oxide CMP. As a matter of fact, these preexisting defects can be detected before the CMP process due to the transparency of the oxide film. For tungsten CMP, the metal film is opaque to practically all defect inspection wavelengths. The preexisting defects, if any, are usually masked by the metal film. The only defects that can be seen before W CMP are the ones creating abnormal

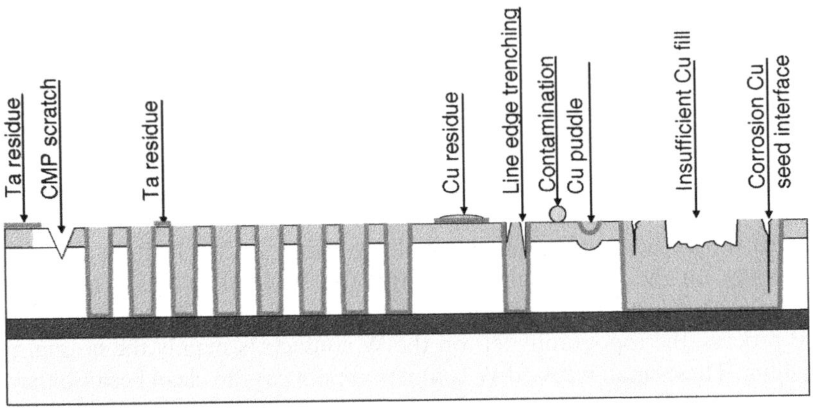

FIGURE 17.22 Schematic illustration of possible defects after copper CMP in the case of underpolishing.

topography such as litho error, holes in the wafer, or large contamination in the metal film. Most defects existing before W CMP are revealed after W CMP.

17.5 DEFECTS AFTER COPPER CMP

17.5.1 Introduction and Summary on Copper CMP Defects

From defectivity point of view, copper CMP is a challenging process because many types of undesirable features could be left on the surface after CMP. Figures 17.22 and 17.23 illustrate some of these possible defects.

In addition to the fact that copper CMP is more complex than the other CMP processes in terms of defect types, there are also many more integration

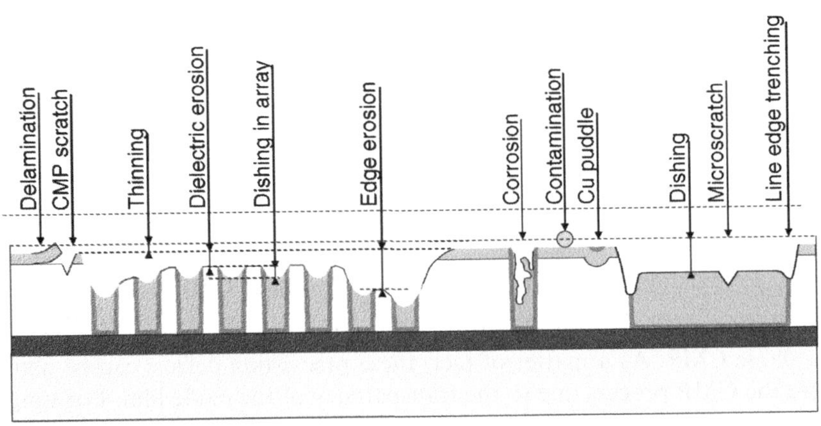

FIGURE 17.23 Schematic illustration of possible defects after copper CMP in the case of overpolishing.

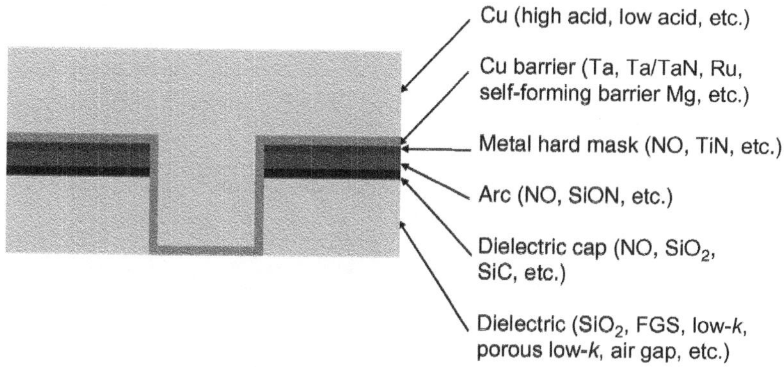

FIGURE 17.24 Schematic illustration of some integration options to make a copper damascene structure.

solutions to make a copper damascene or a dual-damascene stack. Indeed, we can consider that there are at least several of copper deposition processes [51,52] (low or high acid, low or high leveler). There are over four different types of copper barrier materials to choose from (Ta, TaN, Ru, self-forming Cu–Mn barrier) [53–55]. Some integration schemes contain several additional layers: metal hard masks, antireflective coating layers, and different low-*k* dielectric cap layers. In addition to that, there are over four different kinds of dielectric materials (TEOS, FSG, CDO, and porous low-*k*) [56,57]. Each dielectric material has several different possible deposition processes. A random combination of all these options could give literally thousands of different solutions to design a copper damascene or dual-damascene layer (Fig. 17.24). The scenario would be even more complex if one takes into account that there are infinite numbers of choices on film thickness that can be used for each layer. Obviously, the choice of the integration scheme and the process methods has a very important effect on the defect modes and density.

To simplify the discussion, from this point onward, this section assumes that nonporous low-*k* dielectric material and conventional Ta or TaN copper barrier film are employed in the integration scheme. The focus of the discussion shall be on the influence of integration scheme on the type and level of defects.

17.5.2 Copper Corrosion

Copper CMP process is a very delicate balance between selective copper removal at the protruded area and targeted copper surface protection at the recessed area. To enhance the removal, an oxidizer and a complexing agent are commonly used in the copper CMP slurry in addition to the mechanical force provided by the abrasive particles and the pad. To protect the copper in the recessed area or avoid isotropic dissolution of copper, a corrosion inhibitor such as benzotriazole (BTA) is usually added to the slurry. BTA is very effective in protecting the copper surface from the corrosive attack. On the

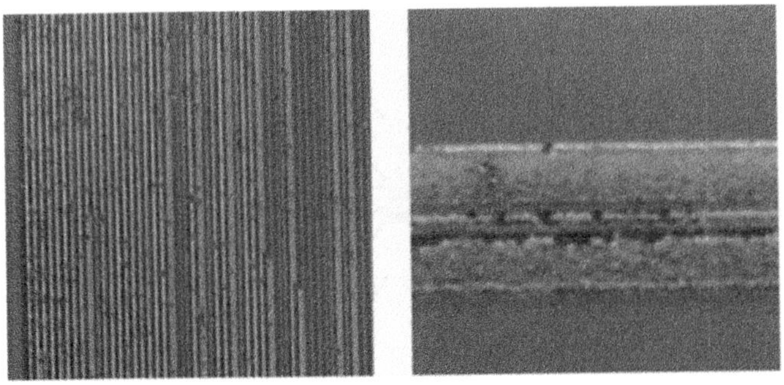

FIGURE 17.25 Massive galvanic corrosion in the case of experimental copper on Ti barrier. In the picture on the left, some lines have lost all the copper.

contrary, at high concentration, it can also severely lower the copper removal rate and leave an undesirable organic residue on the copper surface. An effective copper CMP slurry needs to balance the need for adequate removal rate and avoidance of corrosion effects [8,9].

Massive electrochemical attack known as galvanic corrosion [58,59] is the most severe form of copper corrosion. It can completely remove the copper from the structures (Figs. 17.25 and 17.26). It can occur when the wafers are exposed to a corrosive electrolyte for an extended period. It can also occur if the slurry does not contain enough or effective corrosion inhibitor. The source of such a galvanic potential on the patterned copper surface may be due to the fact that some copper structures connected to transistors have a different electrical potential than the rest of the wafer surface. Another possible cause of this type of galvanic potential is related to the barrier material induced metal–metal battery effect. Most copper CMP slurries have been developed for Cu structures with Ta or TaN as a barrier material. In some cases, other metals may also be used in addition to the barrier metal. For example, a metal hard mask could contribute to the galvanic corrosion effects. It is also possible that some types of copper are more susceptible to corrosion that others. The grain

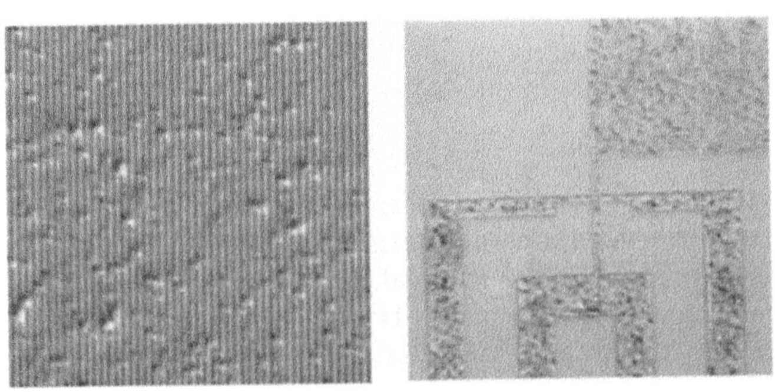

FIGURE 17.26 Corrosion at the surface in an array of small lines (left) on large copper lines (right).

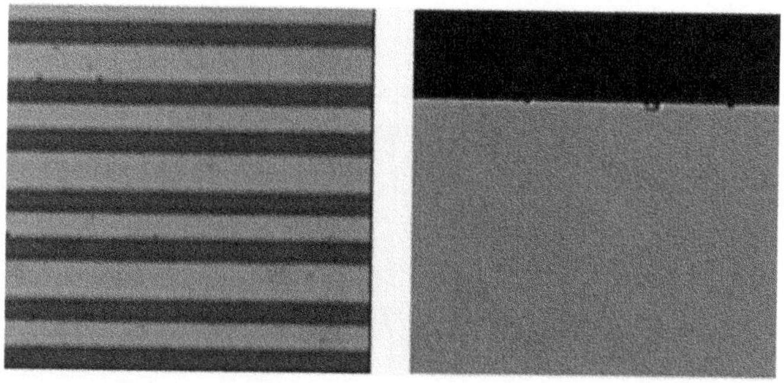

FIGURE 17.27 Copper pitting on the copper structure after Ta CMP.

size and mechanical strain in the copper line can have a positive or a negative effect on the corrosion.

17.5.3 Copper Pitting

Pitting is a mild form of copper corrosion. Copper pitting looks like dark spots under an optical microscope (Fig. 17.27). Both galvanic corrosion and pitting result in localized copper loss and lead to "voids" on the surface. The major difference is that pitting defects leave smaller voids and less number of voids than copper corrosion. It is important to point out that not all dark spots under the optical microscope can be positively identified as copper pitting. SEM is often necessary to differentiate copper pitting from other defects that leave small dark spots on the copper surface. The root cause for copper pitting, like galvanic corrosion, is a localized chemical etching of copper. The difference is that pitting needs a local seeding event that is mostly mechanical, such as localized high stress impinged upon the copper. This local stress seed for copper pitting can have multiple origins. The most common one is scratch-induced stress (Fig. 17.28). The scratch-induced corrosion has a signature. The pitting points follow a perfect line that is independent of the copper structure design and the copper grain shape. The second common cause for such a

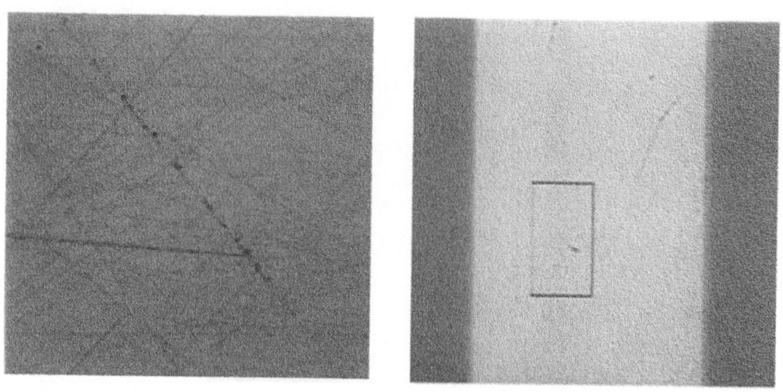

FIGURE 17.28 Scratch-induced voids after copper CMP (pitting).

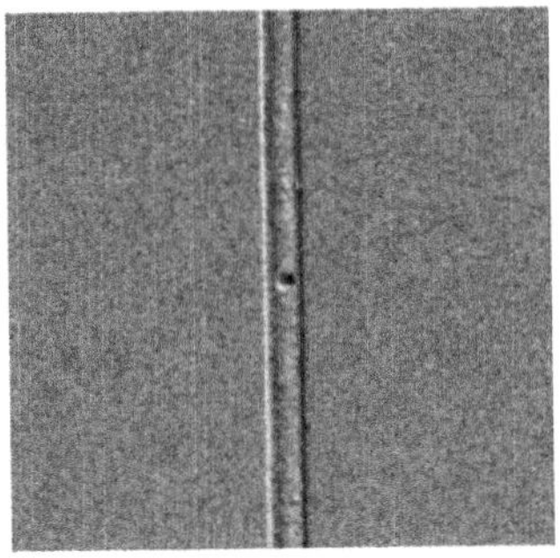

FIGURE 17.29 Stress-induced pitting. The copper grain growth makes copper to reduce its volume slightly and leave some voids. It is visible in the middle of a long copper line.

localized stress can be traced to the contraction of copper during the grain growth in the copper structure. This creates a stress-induced void that looks like pitting already. This type of defect is structure dependent. It occurs more often in the areas where large grains can be formed and therefore larger copper contraction can occur. It has been demonstrated in a study performed at SEMATECH that the copper grain in the large structure can grow larger than the copper grain in the small structures. Stress-induced void can also be visible in a long small structure where grain size growth is limited by the structure size but the aspect ratio of the very long structure induces stress during the grain growth (Fig. 17.29). It also occurs on copper blanket wafers where the grain growth is not well controlled. Stress-induced void can occur at many different places. It can be buried at the bottom of the copper structure. Obviously, the buried defect will never be revealed by CMP. The signature of a stress-induced void is its randomness compared to the scratch-induced void. Another less common source for copper pitting comes from a filling failure. This defect is pattern dependent; it usually occurs on a given linewidth and is located in the middle of the copper line (Fig. 17.30). Of these three pitting mechanisms, the last two have a root cause independent of CMP.

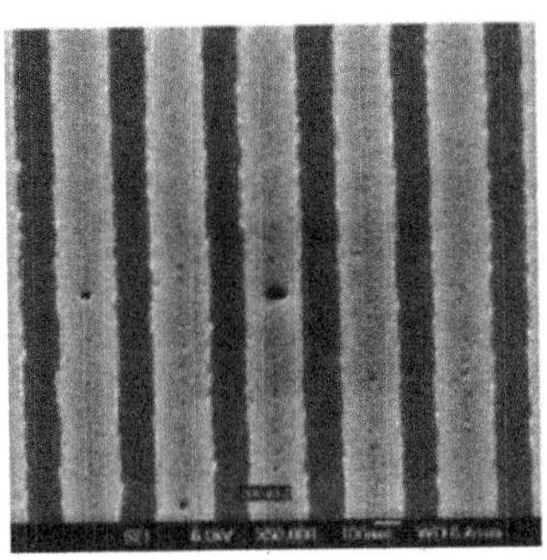

FIGURE 17.30 Copper-filled failure-induced pitting after Ta CMP.

17.5.4 Trenching at the Copper Line Edge

Trenching at the edge of the copper lines is a systemic loss of copper at the edge of the copper line. This defect is only a few nanometers wide and a few nanometers deep. It is too small to be observed with an optical microscope but it is very visible with an SEM or atomic force microscope (AFM). This defect is usually ignored as long as it has no negative effect on the copper line resistance. A severe form of this defect can have a negative effect on the defect inspection technique; it can blind the defect inspection tool. It may have a measurable effect on the electrical performance of the copper line. The trenching at the copper line edge has three possible causes. The most common form of trenching is a loss of dielectric around the copper line. This form is commonly called "fang" on an individual structure or EOE [60,61] in the case of an array of small structures. On a topography profiler, the dielectric at the edge of the large copper structure is overeroded and it looks like two fangs. This occurs on large structures such as a large array of small lines or at the edge of a large copper structure. The width of the trench in the dielectric is usually on the order of 1 μm up to several micrometers; the typical height ranges from 5 to 50 nm. This phenomenon is more often seen after Ta CMP. However, a low-selectivity copper CMP process with visible Ta erosion and, possibly, oxide erosion can lead to some form of oxide trenching. A more detailed discussion on EOE can be found in Section 17.5.6. The second most common form of trenching is a global and uniform copper corrosion only at the edge of copper line (Fig. 17.31). There is no dielectric loss present in this case. It is not visible on a normal profiler because it is too small in size, only a few nanometers in width and height. A galvanic corrosion at the interface of copper and tantalum is responsible for this type of copper trenching. Effective slurry must protect this fragile area at the copper line edge, which is more susceptible to corrosion. The selection of observation method is very important for this defect. Indeed, if the dielectric structure is made of a fragile low-*k* material, the electron bombardment from SEM on the low *k* material can break down the dielectric and result in very visible shrinkage of the structure. This phenomenon can

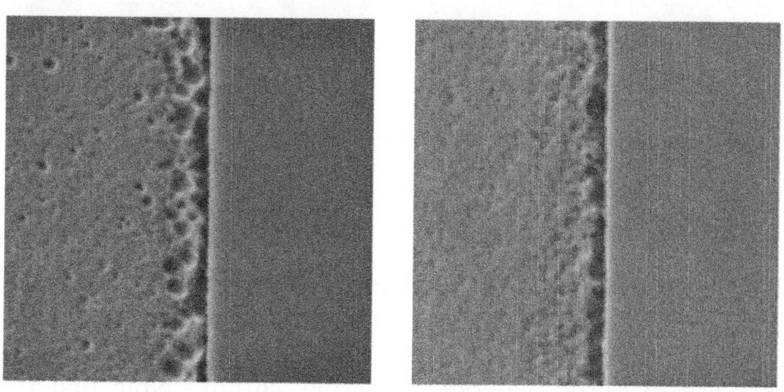

FIGURE 17.31 Edge of the line trenching after copper CMP. The left image shows that the copper line edge is an accentuated form of copper pitting.

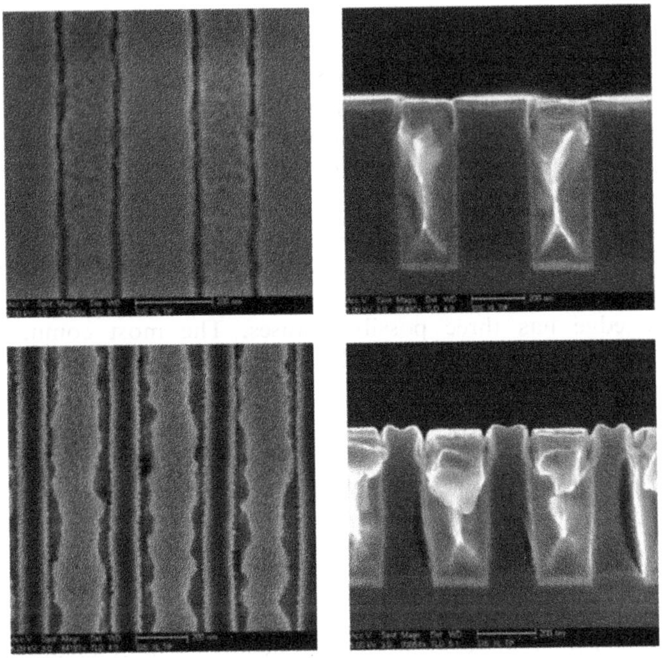

FIGURE 17.32 The top images show copper line edge trenching with TEOS dielectric. The bottom images show a very severe form of copper line edge trenching. The cross-sectional image shows that the SEM has visibly shrunk the low-*k* dielectric and exaggerated the trenching.

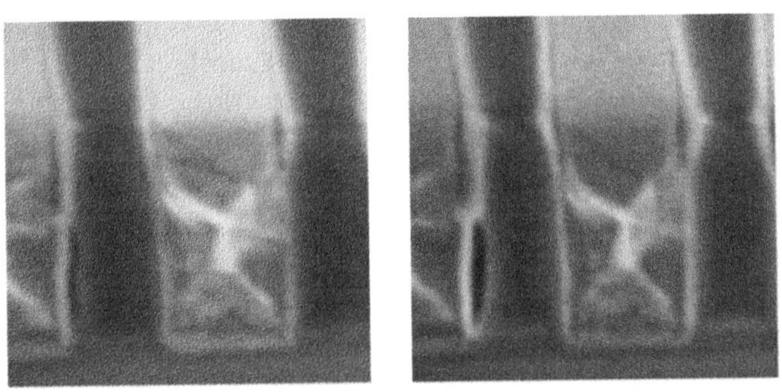

FIGURE 17.33 Effect of electron beam exposition time on the low dielectric. The left picture is taken after 2 s of electron bombardment (minimum time to make acceptable focus). The right picture presents the same structure as on the left after 8 s of electron bombardment. The dielectric material has visibly shrunk. Note that these are low-resolution SEM images because the acquisition time of a high-resolution SEM image takes longer than 8 s.

FIGURE 17.34 Copper line edge trenching due to a weak zone created during the copper deposition. The top pictures are optical microscope images of very severe form of trenching in the case of 100% PVD copper deposition. The AFM images at the bottom show a milder form of this uncommon kind of trenching. A first-generation copper electroplating technique was used with top-down filling and not optimized annealing method.

enhance or even create nanotrenching at the copper line edge (Figs. 17.32 and 17.33). In comparison, however, the AFM is the safest observation method. The last form of copper line edge trenching occurs in relatively large copper lines near the line edge. This is due to a copper deposition issue that could occur with conventional top-bottom filling PVD deposition and old ECD techniques. A conventional deposition technique leaves a weak zone at the bottom corner of the trench that grows weak following a 45° angle all the way to the top of the structure (Fig. 17.34). This normally does not happen with the new bottom-up filling techniques combined with optimized copper annealing.

17.5.5 Rough Copper and Copper Recess

The copper surface roughness can be a problem for the defect inspection tools. When the sensitivity of the defect inspection tool is increased in order to detect smaller defects, the tool may view copper roughness as a defect. The real defects may be lost in the crowd. In order not to miss the smaller defects, it is important to lower the copper surface roughness as much as possible during

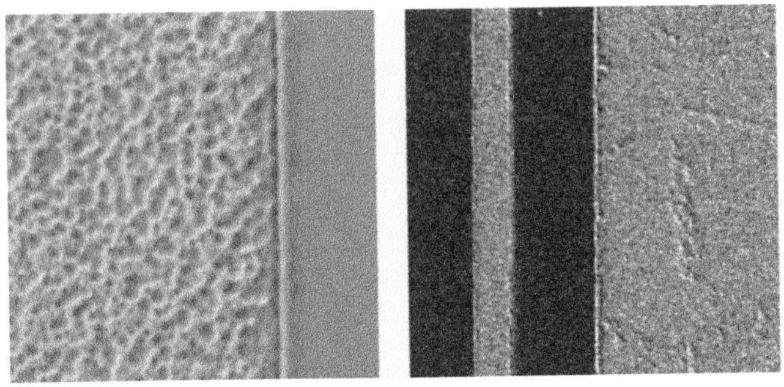

FIGURE 17.35 On the left is an example of a very rough copper surface after copper CMP. This roughness is due to the copper CMP slurry. The picture on the right shows a better roughness after Ta CMP. In this case, the fact responsible for the copper roughness was an unlucky combination of the Ta CMP slurry and post-CMP cleaning chemical. It results in a light etch of the copper grain interface and some roughness of the copper grain depending on the grain orientation. This gives unacceptable roughness.

CMP and post-CMP cleaning. The second driving force to reduce copper line surface roughness is related to the fact that electrical resistance of copper lines can be significantly influenced by surface roughness. The smaller the copper structures, the more important is the copper line conductivity. Studies show that a smoother copper surface yields better electron reflection, which translates to lower loss in electric conductivity. Obviously, copper CMP can address only one of the surfaces for a copper line: the top surface. Next-generation 32-nm devices and beyond may require copper roughness lower than 0.5 nm after copper CMP cleaning. The most important contributor to the copper surface roughness is the Ta CMP process due to the fact that the slurry chemistry is geared to a mechanically dominated removal process for Ta. In some cases, the copper CMP and the post-Ta CMP cleaning process may also have an influence on the final roughness of the copper (Fig. 17.35). As expected, the final surface roughness of the copper lines is also a function of the effectiveness of the post-CMP cleaning process (Fig. 17.36).

17.5.6 Discoloration—Metals Thickness Variations and/or Dielectric Thickness Variation

Similar to the earlier discussions on the color variation after oxide and tungsten CMP processes, discoloration of the patterned wafer surface after copper CMP is also an important concern. The edge overerosion mechanism presented in Section 17.4.5 is very much applicable to copper CMP. In addition, copper CMP may also have other forms and causes for color variation. For example, metal erosion (copper or barrier), metal loss, and metal pullback can lead to color variations. A complete loss of copper in the middle of a very large copper structure, due to massive copper dishing, can cause a color change. This may occur only in extremely large and very thin copper structures, like the design on SEMATECH mask 862 (Fig. 17.37). The latest generation of copper CMP

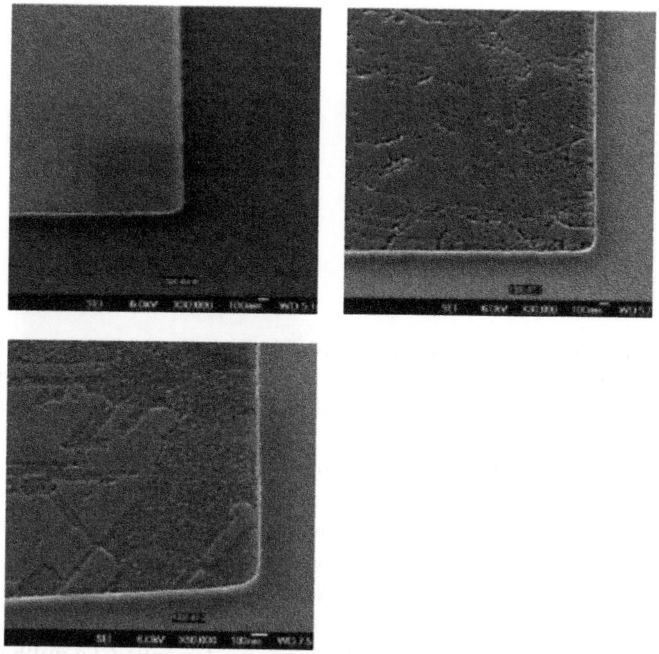

FIGURE 17.36 This series of pictures is taken at the same magnification. It shows the effect of three different post-CMP cleaning chemistries on the copper surface roughness. The same copper CMP and Ta CMP processess have been used. Only the post-CMP cleaning chemistry has been changed. The picture on the top left shows the best post-CMP cleaning chemistry for this given copper and barrier CMP process. The shadow on that picture is an effect of the SEM local charging.

FIGURE 17.37 Very high dishing on a 100 μm square pad structure (left image). The copper in the middle of the copper structure is completely removed; this copper CMP process has a very poor topography performance. On the contrary, the picture on the right shows a 200-mm wafer with SEMATECH mask 862. Only the largest 25-mm copper pad is dished down to the bottom. This copper CMP process exhibits excellent planarization performance.

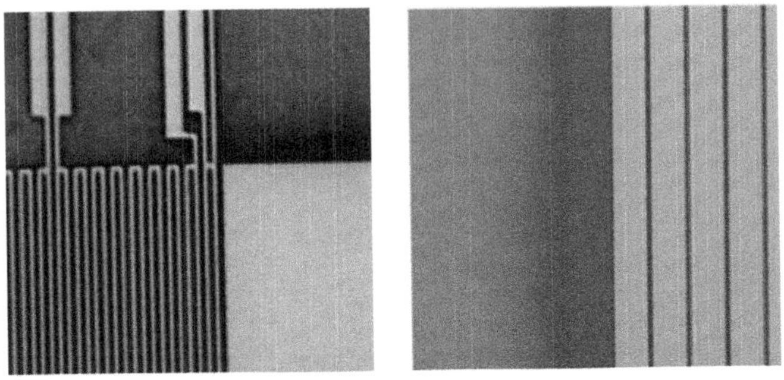

FIGURE 17.38 Ta erosion after copper CMP. The Ta in the field area at the edge of high metal density copper structures is completely removed. There is no color variation in the array of copper lines because the Ta is completely removed. This is called Ta pullback.

processes can keep the dishing over 100 μm size structure under 50 nm with a very good overpolishing margin. However, for a poor copper process that gives high dishing and has a narrow overpolishing margin, it is possible to dish the copper down to the bottom of very large structures upon overpolishing. This will lead to color variation after the CMP.

There are two common forms of Ta erosion that have a similar root cause. The first form of discoloration due to Ta erosion, also called Ta pullback, is when the Ta in the field area at the edge of the high metal density arrays is locally completely removed (Fig. 17.38). The second form of discoloration due to Ta erosion is when only the inside of the arrays of very fine copper lines presents color variation. In such a case, the Ta at the edge of the array is usually not completely eroded and does not show any color variation (Fig. 17.39). It is possible to observe both forms of Ta erosion variation simultaneously.

17.5.7 Copper Electromigration

Copper electromigration refers to a movement of copper atoms or copper ions under an electric field [62]. Obviously, copper atoms in a solid crystal move much slower than copper ions in an electrolyte. Over a period of time, even such a slow movement can cause redistribution of copper materials around the device. The copper migration occurs only when copper lines are connected to transistor structures that have sufficient doping level, which creates a strong electrical current. Copper migration typically becomes more severe after the device is powered for a long period of time. The migration of the copper can be responsible for electrical opens or electrical shorts in the circuits, which can result in the failure of the device. During copper CMP, due to the semiconductor nature of silicon, the exposure to light can induce higher activity of the electrons and holes in the transistor, which could be sufficient to create copper migration. The end result of copper migration is the overflowing and spreading of copper atoms on the top of the dielectric in some areas and the depletion of copper in other structures (Fig. 17.40). In the depleted areas,

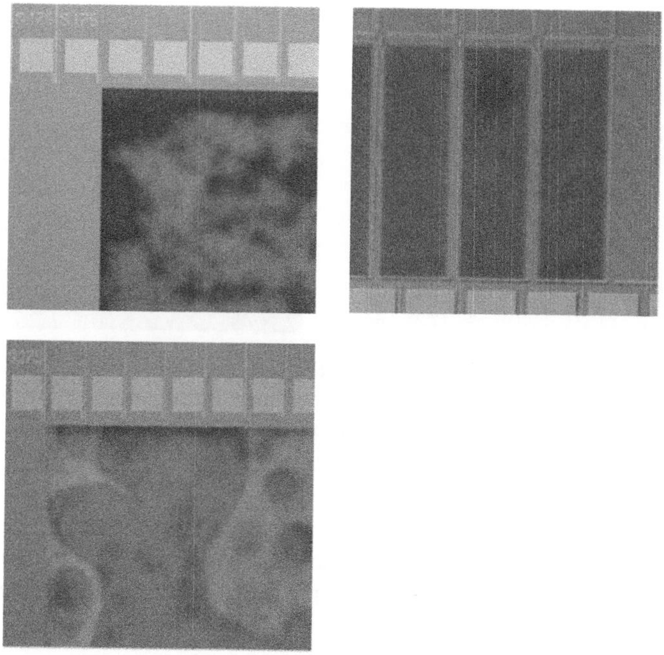

FIGURE 17.39 Ta erosion after copper CMP. The Ta in the field area is not completely eroded. Inside the area of fine copper line, the Ta is locally completely eroded. The arrays of very fine lines have an uneven coloration with stained appearance. This type of Ta erosion can be confused with copper residues. It is possible to have both Ta erosion and copper residues simultaneously as shown in the bottom picture. In arrays of very fine copper lines, SEM and AFM are needed to remove the doubt.

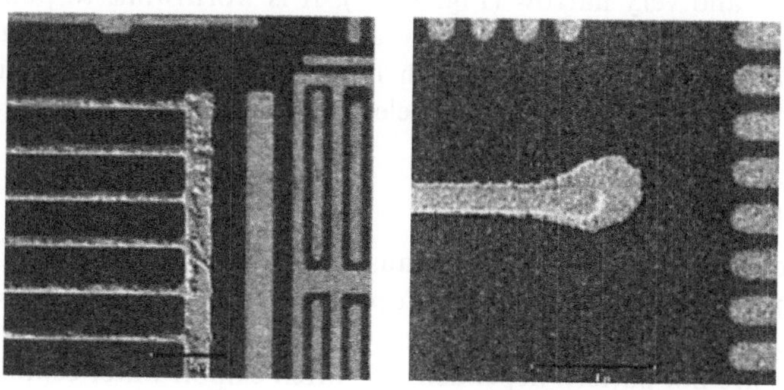

FIGURE 17.40 Electromigration of copper. In the left image, copper is depleted only in certain structures. In the right image, copper is overflowing on the top of the dielectric. *Source*: Petitdidier S. Effect of an organic inhibitor in high-pH chemical rinse on the platen for Cu-CMP. UCPSS 2006.

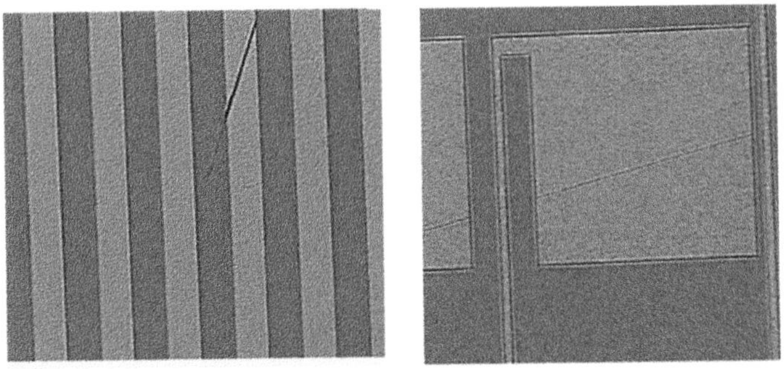

FIGURE 17.41 A large scratch in copper. The left picture shows a rather large scratch that has also marked the Ta between the copper lines. Both pictures show that a scratch marks the copper deeper than the surrounding Ta.

the defect has the same effect as copper corrosion. In some literatures, this type of electromigration-induced defect is also classified as a form of corrosion. This is why most copper CMP tools including post-CMP cleaners have shields in order to prevent the photoactivation of the transistors.

17.5.8 Scratches

On patterned copper wafers, after CMP, the surfaces are covered mainly by dielectric and copper features. The large scratches on the dielectric such as TEOS oxide will have similar shatter mark characteristics as described in Section 17.2. The scratches on the copper lines or features, however, have a very different signature. As the copper is a soft material with large plastic deformation area, it is very easy to scratch copper (Fig. 17.41). The scratches on copper usually show well-defined continuous lines. A copper scratch can be very shallow and very narrow (Fig. 17.42). It is worthwhile to point out that the extent of damage by scratch is also a function of the underlying dielectric. As a low-*k* dielectric is usually much more fragile than silicon dioxide, the damage on copper lines with low-*k* dielectric may be more severe (Fig. 17.43).

17.5.9 Metal Residues

Metal residues can have several root causes. The first one is related to WIWNU during copper and/or barrier CMP. Copper residues left on the wafer edge are typical examples of this phenomenon. Copper residues accumulated by this mechanism are not pattern dependent but often exhibit wafer center symmetry (Fig. 17.44). The simplest cure for this type of defect is to improve the WIWNU. Because WIWNU is never perfect, all high-volume manufacturing copper CMP processes have a built-in overpolishing process in order to remove the copper residues. A multizone or a scanning end-point detection system can

FIGURE 17.42 Super-shallow scratches. Some of these scratches can be below the defect detection limit. The left picture shows multiple shallow scratches using experimental slurry. It is believed that a solid form of by-product generated during the CMP process is responsible for this very high density of very shallow scratches. The right picture shows an example of an isolated scratch observed on a patterned wafer using commercial copper slurry.

detect how long it takes to clear the wafer from the first zone to the last area. In the case of a drift of the WIWNU, this time interval will increase. Some automation software could monitor and correct such process drift. The second reason for copper residues is a self-stopping copper CMP process. Most copper CMP self-stopping processes use copper ions as a catalyst for the copper removal reaction. When most of the copper is cleared, the lack of copper ion shuts the removal reaction. Such a self-stopping process is very effective in preventing or minimizing copper dishing. The disadvantage of this approach is that it is very difficult and in some cases impossible to remove the last copper residues. The third origin for copper residues is a result of inadequate

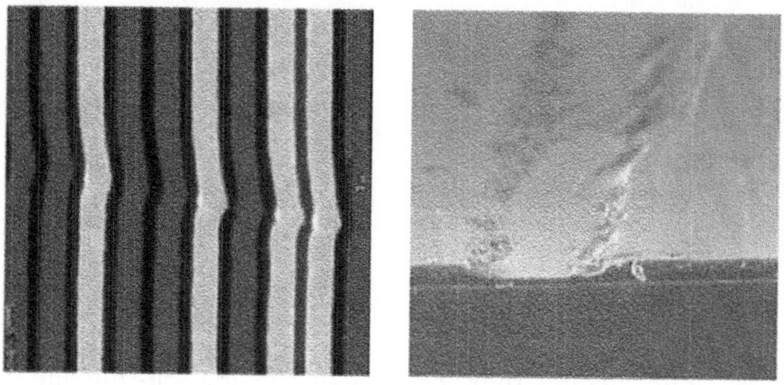

FIGURE 17.43 A scratch after copper barrier CMP on a wafer with fragile low-*k* material. The left picture shows that the low-*k* dielectric is about as weak as the copper line and deforms under the stress of the scratch. *Source*: Fisher P. CAMP 2006. The right picture shows a soft scratch in the field area where the porous low-*k* is locally completely removed. The underlayer of TEOS is not damaged.

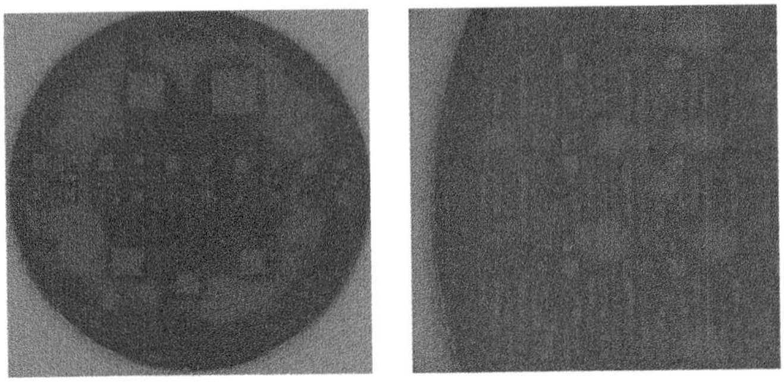

FIGURE 17.44 The picture on the left shows copper residues that are due to WIWNU. The residues are symmetrical in the middle radius. The right image shows copper residues that are pattern dependent. This wafer has been plated using a first-generation bottom-up fill copper electroplating technique without levelers.

planarization of certain areas. Some copper filling processes leave recess or protrusion over certain features. If a copper CMP process is unable to planarize all features at the same rate, it leaves pattern-dependent copper residues on the wafer (Fig. 17.45). One solution to this problem is to minimize the pattern dependency of the planarization process. In addition to changing pad or slurry, a lower polishing downforce is often very effective in equalizing the planarization efficiency over a wide range of features. If possible, one may also consider increasing the initial copper thickness, which gives more time to planarize the copper film. For conventional CMP, it is recommended that the initial copper film thickness should be at least 1.5 times the trench height. Truly, the best solution is to improve the incoming topography by better matching the copper deposition process and the design rules. The last cause for copper residues is the incoming topography. In case the topography of the

FIGURE 17.45 Both pictures show residues due to the self-stopping copper CMP process. The picture on the left shows some uniform residues that are pattern dependent. The copper is cleared around the copper structures. The picture on the right shows copper residues that are not dependent on the pattern. The slurry used in this case provides perfect topography. Very large overpolishing does not change the topography but at the same time is not able to clear these residues.

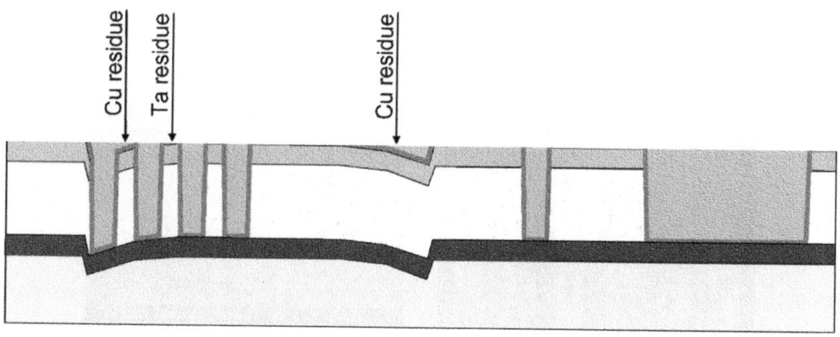

FIGURE 17.46 A cartoon of possible defects after copper CMP in the case of underlayer topography.

underlayer is nonplanar, the subsequent conformal deposition process will transfer that topography to the top of the dielectric. The metal deposition will fill all the valleys. As a normal planarization process will not be able to distinguish between a valley on the dielectric filled with copper and a real copper structure, the copper and/or Ta residues accumulate in those valleys (Fig. 17.46). Some of these copper residues can be cleared during barrier CMP, especially by those processes that remove some dielectric materials that in turn expose the metal residues in the process. Because barrier CMP process usually has much lower copper removal rate than the copper CMP step, it is difficult to remove thick copper residues. Only thin copper residues can be removed. In general, it is not easy to see residues in the large arrays of fine lines. Subwavelength structures of 65 nm, 45 nm, 32 nm, or smaller can only be reviewed with electron microscopes such as SEM. As SEM is a time-consuming method to review large areas of structures individually, a simpler test is to measure the electrical leakage performance of these critical structures.

17.5.10 Particles, Residues, and Trace Metals

An effective post-CMP cleaning process should be able to completely remove the abrasive particles from the polished surface (Fig. 17.47). It also should remove all the organic residues that are used in the copper and barrier CMP slurries (Fig. 17.48). It should also improve the copper surface roughness if possible and at the same time clean up the dielectric layer. In addition, the post-CMP cleaning process is supposed to remove all trace metals that could increase the electrical leakage and affect the performance of the dielectric. Because of the complexity presented by the diversity in various dielectric materials and integration schemes, a post-copper CMP cleaning process does not have to work universally well on all types of dielectric materials. More specifically, a relatively hydrophilic TEOS will require a different post-CMP cleaning solution and process from that for a surface based on a hydrophobic low-k material. For the same reason, a post-CMP cleaning process that works

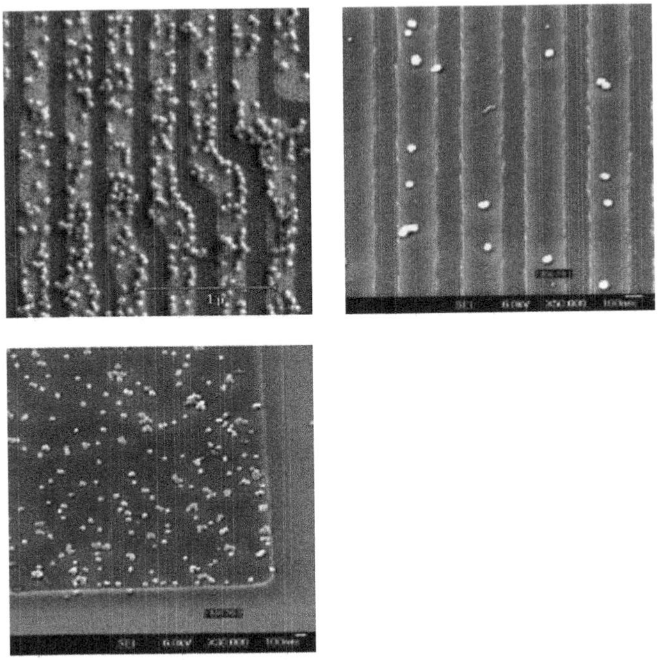

FIGURE 17.47 SEM pictures of silica abrasives strongly attached onto the copper surface. This is the result of an unoptimized post-CMP cleaning process. The post-CMP cleaning process needs to be optimized for the Ta CMP slurry in order to offer best performance in terms of particle removal, organic removal, and at the same time offering good surface roughness.

well for an acidic Ta CMP slurry may not be applicable for a process involving on alkaline barrier CMP slurry. Therefore, it is very important that the post-copper CMP cleaning process is specifically tuned for a certain type of dielectric and Ta CMP slurry.

17.5.11 Delamination

Delamination can start from a failure at the interface of two dissimilar materials or from a flaw in the bulk of a weaker material [63]. The extent of delamination is sometimes reduced when the interface of the two materials is a part of patterned structures. This is a strong indication that the delamination starts at limited spots and then propagates during the polishing. The patterned structures prevented such propagation from occurring (Figs. 17.49 and 17.50). This is consistent with the fact that delamination usually starts during the first few seconds of polishing (Fig. 17.51). The polishing debris generated during the polishing leads to a widespread delamination (Fig. 17.52). It is understandable that the initial friction and shear force are usually the highest at the beginning of a polishing event. Early studies on the integration of low-*k* materials show

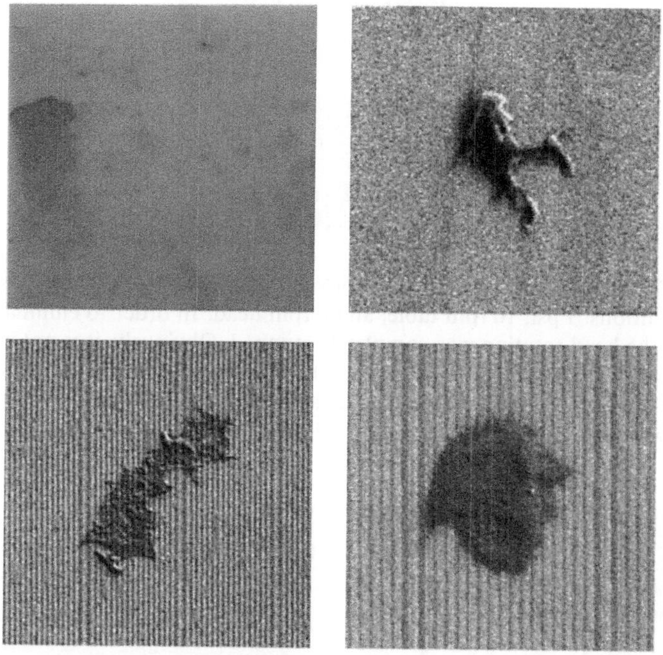

FIGURE 17.48 Pictures of watermark and organic residues. The top left picture is an optical image of watermark. The other pictures are SEM images of organic contamination of filled area and patterned area.

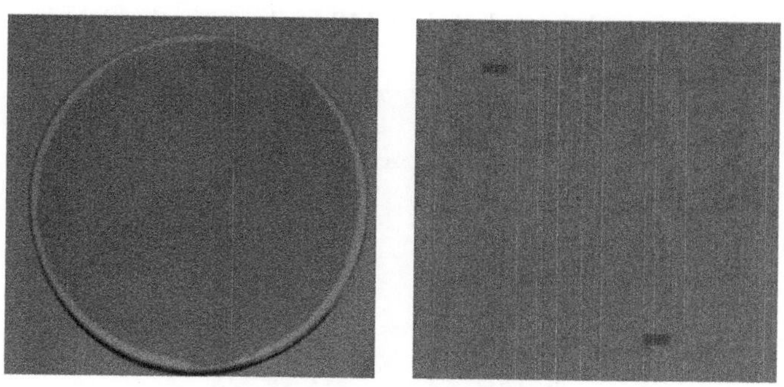

FIGURE 17.49 The left image shows a 300-mm blanket wafer with massive delamination of the porous low-*k* film during CMP when using only 1 psi and 40 rpm. The same CMP process condition has been used on the patterned wafer in the right picture. As the picture shows, the same film behaves much better when patterned with copper structures. Only small areas of the patterned wafer present signs of delamination.

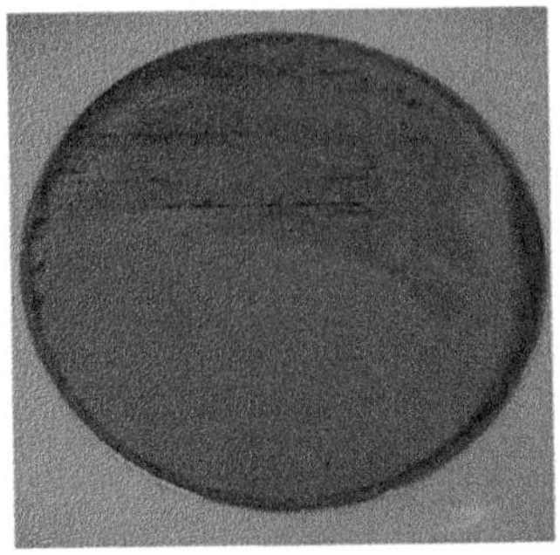

FIGURE 17.50 A wafer polished on conventional CMP tool with the most gentle process conditions: 1 psi, 10 rpm table, and 0 rpm head. In order to eliminate the wafer edge effect, the head rotation was completely stopped. Obviously, it is not practical and only experimental. The front edge on the left presents massive delamination, whereas the trailing edge on the right presents little delamination. The film at the center of this wafer was weaker than in the middle and edge areas.

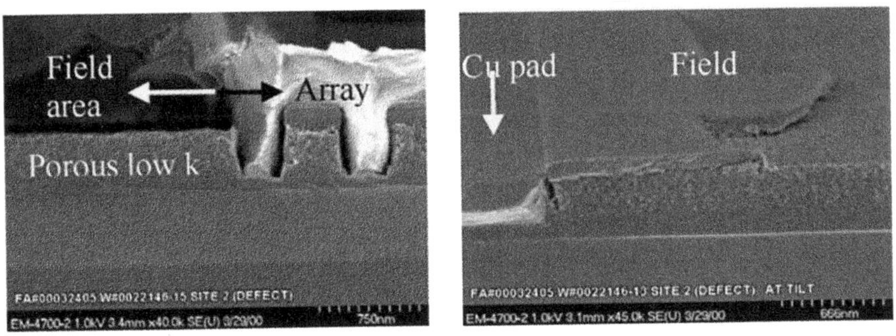

FIGURE 17.51 The delamination occurs at the early stage of copper CMP, only after a few seconds of CMP. In the present integration scheme, the delamination occurs at the low-k interface and the low-k capping material. The copper structure stops the cap material delamination until the end of the CMP as shown in the right image.

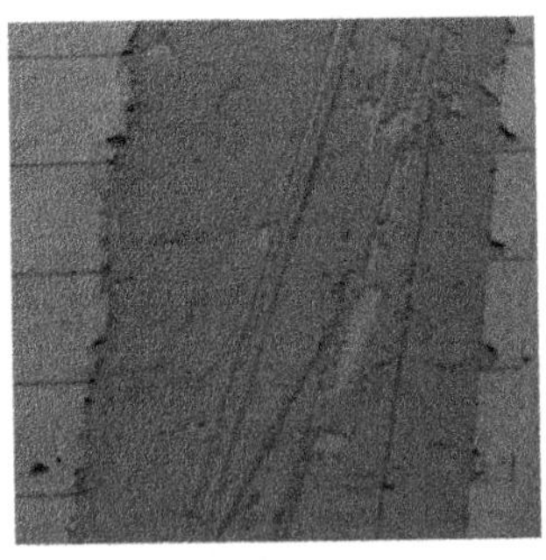

FIGURE 17.52 This picture shows that a scratch on a porous low-k with a cap results in the delamination of the cap film. The fragments of the cap material create scratches and more delamination. The porous low-k material mostly remains but it is scratched down to the bottom in some places.

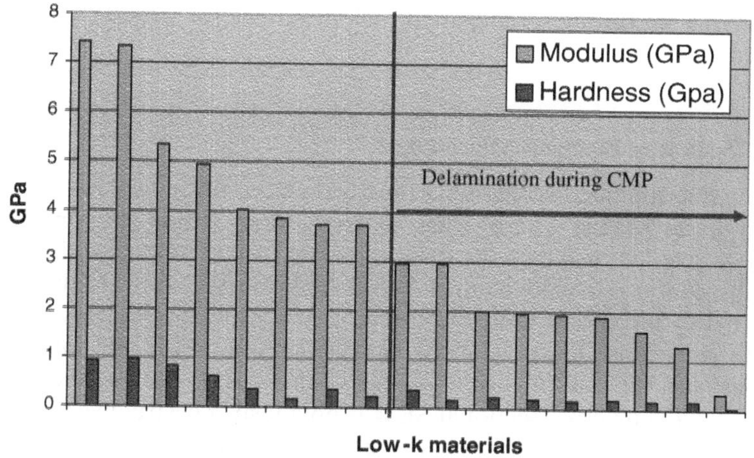

FIGURE 17.53 CMP delamination function of the low-*k* material and Young's modulus. In general, the lower the *k* value, with higher porosity, the lower the modulus is. *Source*: International SEMATECH 2001, study of 17 low-*k* materials.

that the nonporous low-*k* film did not cause prominent delaminate issues during CMP. However, the introduction of porosity in order to lower the *k* value comes with new mechanical challenges for copper and barrier CMP. This is certainly the main reason why the International Technical Roadmap for Semiconductor (ITRS) has pushed the introduction of effective *k* value of less than 2 to a later date. As a practical guideline, a low-*k* dielectric material must have sufficient strength (Young's modulus) to survive the CMP processes without delamination (Fig. 17.53).

17.6 DEFECT OBSERVATION AND CHARACTERIZATION TECHNIQUES

17.6.1 Optical Microscope

An optical microscope is a very useful tool for defect observation [64]. The resolution is limited to about 180 nm with a normal white light or down to about 80 nm with a deep UV (DUV) light source and a DUV camera. Therefore, optical microscopes are clearly inferior compared to electron microscopes in resolving the small features printed on the patterned wafers. This is a significant limitation. However, it is important to point out that, in many cases, what matters most is to see the defects not the features where the defects are located. In addition, optical microscopes can observe certain things that SEM cannot. Indeed, an optical microscope allows the observation of color variation that qualitatively indicates thickness variation. It also has the capability to see through transparent dielectric films for underlayer structures or defects.

17.6.2 Scanning Electron Microscope

Scanning electron microscopes have mostly replaced the optical microscope for detailed defect review. An SEM has an excellent resolution (1–5 nm) that can provide a detailed image of the defects [65,66]. An SEM that is dedicated to defect inspection is usually equipped with sophisticated software that links an optical defect inspection tool to a defect review SEM and quickly moves the wafer to the precise location of the defects. In addition to the top view, the SEM has been the tool of choice to observe the cross section of the structures and defects. Doing cross-sectional SEM is obviously a destructive method that may not be used if the wafer needs to go through reprocessing. An ion beam machine could do local etching and an imaging of a section with a tilted angle. This technique is destructive only locally. The wafer could continue with further processing steps and be re-inspected after additional processing steps. An SEM can have two detection modes: secondary electrons (most common) and backscattered electrons (need a special detector). Secondary electrons have low energy (<50 eV). Because of that, they have a shallow escape depth of a few nanometers. The light intensity of each spot of an SEM image is proportional to the number of secondary electrons emitted and detected by the sensor. There are much fewer backscattered electrons than secondary electrons. They have much higher energy that can go deeper in the sample. Most defects have a clear image in both modes. But certain defects are visible only in one mode. As mentioned in Section 17.5, the electron beam energy can damage the surface (e.g., the low-*k* material). Certain SEM instruments have a mechanism to slow down the electron energy when they reach the surface keeping a high-resolution and high-quality image.

17.6.3 Energy Dispersive X-Ray Spectroscopy (EDX)

During the operation of an EDX [67], an electron or a photon beam is aimed at the sample. At rest, an atom within the sample has ground-state "unexcited" electrons. The incident beam excites an electron in an inner shell, ejecting it and resulting in the formation of an electron hole. An electron from an outer higher energy shell then fills the hole. The excess energy of that electron is released in the form of an X-ray photon that is then detected by EDX. EDX is a useful technique to identify the chemical compositions of the defects. As the technique may have a penetration depth up to about 2 μm, the elements in the underlayer films will also be included in the energy spectra. For example, an EDX of a defect (less than 2 μm thick) on an oxide film will show Si and O elements as well from the film underneath. In addition, elements with low atomic number below 11 (Na) are difficult to detect by EDX.

17.6.4 Scanning Auger Microscope (SAM)

In Auger spectroscopy [68], like SEM, a highly focused and energetically well-defined electron beam is directed to the sample. The surface is irradiated with electrons having energy in the range of 2–50 keV. The incident electron beam excites an electron of the inner shell around the nucleus. This prompts its ejection resulting in a hole. An electron of higher energy from the outer shell will fill the hole. The excess energy of the second electron can do more than emitting an X-ray. Instead of X-ray, the excess energy is transferred to a third electron from a further outer shell prompting its ejection. This third electron is called an Auger electron. Auger electrons emitted from the sample are analyzed in terms of their kinetic energy and quantity. The Auger electrons are the fingerprint of the atoms from which they are emitted. Because of the low kinetic energies of the ejected Auger electrons, their escape depth is limited to a few atomic layers such that one can analyze only the top atomic layers of the sample of interest. Therefore, Auger is a surface-sensitive technique. It is believed that this technique has a very good potential for the future observation method for the smaller semiconductor structures. But this technology requires ultrahigh vacuum and still needs further developments to become as robust as SEM.

17.6.5 Atomic Force Microscopy

Although the use of AFM [69] has gained tremendous growth over the last decade in almost every technology field including its use to replace a stylus profilometer in wafer topography measurement, it is still not common to use AFM to observe defects. The main reason is probably due to the speed of the equipment. It takes minutes of scanning to get a high-resolution AFM image and only seconds to get a high-resolution SEM image. In addition, the preview on AFM uses optical microscopy. This makes the localization of a small defect very difficult. The resolution of an AFM depends on the tip radius. The conventional low-cost tip has a radius of about 10 nm. AFM has an excellent resolution on the *Z* axis. Indeed, the piezoelectric element of an AFM is capable of controlling a fraction of an angstrom in vertical motion. But the final resolution is largely affected by the environment noise. Vibration insulation techniques must be implemented to lower the noise level, otherwise less than 1 Å resolution is not easy to obtain. The last element of the vertical resolution is the tip dimension and geometry. Obviously, a large pyramid-shaped tip cannot reach the bottom of a deep high-aspect-ratio hole that has a width smaller than the tip radius. The AFM resolution on the *X* axis can be very good. It is primarily a function of the scanning speed divided by the data sampling frequency. But it is also limited by the tip shape. The common tip has a pyramid shape. When scanning deep and narrow trenches or contact holes, the top of the tip cannot follow vertical walls. Only special AFM tip with “L” shape can follow the vertical wall contour. AFM is mostly used for the observation of systemic defects or defects that have wide enough size to be resolved by an optical microscope. AFM works best when topographical information of the defect is important.

17.7 ENSEMBLE DEFECT DETECTION AND INSPECTION TECHNIQUES

17.7.1 Optical Scan of Flat Film Blanket Wafers

The most common techniques to detect and count defects on a blanket wafer is to use an optical scanning technique [70]. It scans the wafer surface using a light beam or laser. A defect on the wafer surface will scatter the light. Photosensors detect the scattered light and count it as a defect on the wafer surface. This technique has been well developed and is capable of detecting smaller and smaller defects. The limitation of a detectable defect is the roughness of the wafer. Obviously, it is difficult to detect a defect with vertical size similar to the surface roughness in the background. In addition, a defect that does not reflect much light will be even more difficult to detect. The corollary is that a small defect that reflects most of the light in the direction of the photosensor will appear larger than it really is. In order to avoid misinterpretation of the results, one needs to understand the tool principle capability and the calibration method. It is commonly accepted that a defect that has a size larger than half of the minimum feature size is capable of creating yield loss damage and therefore should be detected. This means that any defect larger than 32 nm should be detected for the 65-nm technology.

17.7.2 Optical Scan of Patterned Wafers

For pattern wafers, the physical pattern etched on the mask (the die) observed by the optical scanning technique is compared to a data file that was used to write the mask. Any difference larger than the specification limit between the real pattern on the physical mask and the data file will be flagged as a defect. This method is called die-to-data defect inspection. The technique, in principle, should be able to detect all defects larger than the specification with 100% confidence. This method is, however, slow. It can take over an hour to inspect one die on the photomask. Owing to the fact that the image on the wafer is four or five times smaller than that on the mask, an optical defect inspection tool is incapable of resolving the small features on the wafer. Only the SEM defect inspection tool could possibly perform die-to-data inspection on the wafer with 65 nm, 45 nm, or smaller features. Such SEM inspection is to make sure that the pattern on the wafer is free from systemic defects (same defects on the same structure or same location on all the dies). In most cases, photolithography process engineers rely on the mask inspection only. The patterned wafer defect inspection will focus only on random defect. Random defect could be a local or a global defect. For random defect inspection, the fastest technique is to compare one die on the wafer to another die on the same wafer. Any difference larger than the specification between the two images in the same area of the two different dies that are compared is flagged as a defect. This technique is called die-to-die defect inspection, which is commonly used to inspect defects after CMP. It is desirable to acquire images that resolve 100% of the pattern including the smallest features for die-to-die defect inspection. In case this is not possible, the first option is to exclude the arrays or area that contains features that are too small to be resolved by the defect inspection tool. This is the most common solution. The other solution is to treat the area of subresolution structure as one feature. In this case, a defect in such an area can be detected only if its size creates a local change in contrast or color.

17.7.3 Defect Classification

For semiconductor manufacturing, a simple defect count is inadequate without knowing the defect types. For example, one corroded copper line can create more potential damage than thousands of spots with color variation. Until recently, the defect classification was done manually until the implementation of the so-called automatic defect classification (ADC) [71]. The working principle of ADC is to first sort the random defect (defect of interest), the repetitive defect (usually not relevant to CMP), and the defect cluster (group of many defects in a small area). Then images are processed through a pattern recognition step. Each image is compared against a catalog of defects of different size, shape, and contrast. The first generation of ADC was not reliable, and for some applications, process engineers still prefer to perform manual defect classification. Obviously, for new consumable and process development, a preloaded ADC database will not be able to identify and classify the new defect mode properly. For high-volume manufacturing involving well-known processes, the current version of ADC is mature enough to provide effective and fast defect classification. Furthermore, the latest defectivity tool can not only use SEM images but also take advantages of its associated EDX capability for more detailed classification.

17.8 CONSIDERATION FOR THE FUTURE

As long as CMP is used, it is likely that the CMP defects presented in this chapter will still exist. It is believed that CMP technology will be used for many years to come. Furthermore, CMP will likely be extended to more applications such as the ones listed below:

- Metal CMP (Al, Au, Co, Ir, Ge, Pt, Sb, Te, combinations of these elements, and others).
- Dielectric CMP (low-*k*, nitride, resist, and others).

There will be new types of defects accompanying these new applications. But the main source for new defects will be most likely due to the fact that the structure dimension shrinks at the speed predicted by Moore's law. The reduction in the feature size and the increasing capability open us to newer defects not picked up by older technologies. In addition, existing defects will move from undesirable category to unacceptable category. For example, copper pitting is now tolerated but may be not for too long. At this moment, the specification for copper surface roughness after CMP is fairly relaxed. New and higher resolution defect inspection tools will be implemented. Also, AFM is likely to transfer from optional to necessary for CMP performance control in the future.

New technologies, like CMP, usually go through multiple phases of process development and implementation. For the CMP, these phases can be linked to how defects are handled. For example, the pads used for CMP were essentially the same as the soft pads used for silicon polishing, and it still is the case of many applications. The pads used in phase 1 usually give low defect and low planarization. Phase 2 uses harder pads that give high planarization efficiency with increased defect density. The third-generation pads should offer higher planarization with lower defects. The CMP slurries have also followed the trend. For example, copper CMP slurries are reaching the third phase. Slurries developed during the initial phase give high friction, high removal rate, and high dishing. The slurries used in phase 2 give high friction, lower removal rate, and medium dishing. In the current phase, slurries tend to give medium friction, medium removal rate, and low dishing. The targets for the next phase will be low friction, high removal rate, and low dishing. Some believe that ECMP will accelerate the development for phase 4 slurry and phase 3 pads. Others believe that new CMP pad and slurries will push ECMP away. ECMP has it own technical challenges such as clearing the metals (copper and barrier). The ECMP process is not yet defect free. As a matter of fact, ECMP is accompanied by new types of defects.

ACKNOWLEDGMENTS

I want to give special thanks to the people who helped me prepare this chapter. First to Yuzhuo Li, who has given me the honor and opportunity to write this chapter for his book. Certainly, Fujimi R&D has been the largest source of information and pictures. Special thanks to Kazusei Tamai for many discussions and all the pictures he gave me for this chapter.

QUESTIONS

1. What steps would you implement in order to minimize the possibility of scratches in a CMP process?

2. What is the issue with a visual difference of color at the edge of an oxide wafer? How could the problem be eliminated?
3. What does a visual difference in color at the edge of a large array structure under anoptical microscope mean? How can you eliminate the problem?
4. One lot of 25 wafers has been polished on two different oxide CMP tools using two different oxide CMP slurries. The 13 wafers polished on the second CMP tool show zero final test yield. You have inspected the wafers and even did defect classification and have not seen any difference at the time and still see no difference in revisiting all the data you have collected. What could have happened? Find and explain two different possible scenarios.
5. In the case of excessive tungsten contact recess, what action would you take to improve the process?
6. What AFM tip radius would you use to check the recess on a W contact at 45-nm process technology?
7. What do you think is the reason behind Cu residues being found at the edge of the wafer? How could the problem be fixed?
8. What do you think is the reason behind Cu residues being found at the field areas (around the structures)? How could the problem be possibly eliminated?
9. What do you think is the reason behind thin copper residues being found over the array of very fine copper lines? What are the possible solutions for the problem?
10. What could be the possible reason behind the occurrence of copper corrosion right after the CMP process? What could be the possible reasons in case you eliminate slurry being the cause?
11. What is the reason and how would you eliminate the problem of copper corrosion during slurry development? List two different kinds of chemical components in addition to the copper inhibitor that could affect copper corrosion.
12. What is the benefit of defect classification? Assuming you being an R&D engineer working on the next technology process, would you use ADC? Give the reasons and the consequences of your choice.
13. Is it possible to observe scratches on the wafer that is polished with abrasive-free Cu CMP slurry? Explain the reason behind your answer.
14. You just performed ADC on a sample of 100 defects on one lot that has been split on two CMP tools with two different copper CMP processes? The first split shows about 200 defects with just one evidence of copper line corrosion. The second split shows about 2000 defects with no evidence of corrosion? What process do you like better at that point? What action should you take to confirm that your initial opinion is correct?
15. What is the reason behind a possible scenario of delamination being observed at low-*k*/cap interface? How could the problem be solved?

16. What would you suspect being the reason behind your AFM images exhibiting large amounts of noise? If after controlling the noise in the AFM images, they exhibit poor quality, what can be wrong?

17. Your SEM image of low-*k* shows severe nanotrenching at the edge of the copper lines. What could be the reason? Elaborate two different possible scenarios, one of them involving local oxide loss. Now develop a third scenario, where you suspect that your CMP processes not responsible for the observation. What would you do to prove it?

REFERENCES

1. Seo Y, Kim S, Lee W. Reduction of process defects using a modified set-up for chemical–mechanical polishing. Microelectr Eng 2003;65:371–379.
2. Seo Y, Kim S, Lee W. Advantages of point of use (POU) slurry filter and high spray method for reduction of CMP process defects. Microelectr Eng 2003;70:1–6.
3. Seo Y, Lee W, Yeh P. Improvements of oxide chemical–mechanical polishing performances and aging effect of alumina and silica mixed abrasive slurries. Microelectr Eng 2004;75:361–366.
4. Wang MT. Process to remove microscratches. US patent 6,537,919. 2003 Mar 25.
5. Tiwari R, Soucek M, Strupp J. Development and implementation of 300 mm Cu manufacturing systems. Fut Fab Intl 2002;12; p 546.
6. Fournier B. Method of cleaning semiconductor wafers after CMP planarization. US patent 6,280,899. 2001 Mar 13.
7. ESC. Cu-based interconnect post-CMP cleaning technology update. CMPUG Meeting; 2003.
8. Lu CF, Ho CH, Chen ML, Huang LK. Method for preventing Cu CMP corrosion. US patent 6,555,477. 2003 Apr 29.
9. Keleher J, Tyre E, Her R, Babu SV, Li Y. Hydroxyl radical formation and copper line corrosion in Cu-CMP. Proceedings of the Fifth International CMP for ULSI Multilevel Interconnection Conference (CMP-MIC); Santa Clara, CA;2000 Mar 2–3.66–72.
10. Peterson ML, Small RJ, Shaw GA, Shen ZJ, Truong T. Investigating CMP and post-CMP cleaning issues for dual-damascene copper technology. Micro magazine 1999.
11. Chakraborty K, Mazumder P. Fault-Tolerance and Reliability Techniques for High-Density Random-Access Memories.Prentice-Hall; 2002.
12. Park J-G, Katoh T, Lee W-M, Jeon H, Paik U. Surfactant effect on oxide to nitride removal selectivity of nano-abrasive ceria slurry for chemical mechanical polishing. Japan J Appl Phys 2003;42:5420–5425.
13. Bu K-H, Moudgil BM. Colloidal silica based high selectivity shallow trench isolation (STI) chemical mechanical polishing (CMP) slurry. Proceedings of the MRS, Symposium W; Spring 2005.
14. Schlueter J. Trench warfare: CMP and shallow trench isolation, Semicond Int October1999.

15. Kim S, Hwang I, Choi K. Hard-pad-based CMP for premetal dielectric planarization. J Electrochem Soc 2003;150(8):G450–G455.
16. Jeong Y, Kim SY, Seo YJ. System facility factors for hot spot reduction of inter-level dielectric (ILD) CMP process. Proc ISEIM 2001; p 95.
17. Bath SH, Legegett R, Maury A, Monning K, Tolles R. Planarizing interlevel dielectrics by chemical mechanical polishing. Solid State Technol 1992;35; p 87.
18. Tang BD, Xie X, Boning DS. Damascene chemical–mechanical polishing characterization and modeling for polysilicon microelectromechanical system structures. J Electrochem Soc 2005;152(7)G582–G587.
19. Lin CF, Tseng W, Fengand M, Wang Y. A ULSI shallow trench isolation process through the integration of multilayered dielectric process and chemical–mechanical planarization. Thin Solid Films 1999;22:248–252.
20. Mills CR, Grover GS, Mueller BL, Steckenrider JS, Ganeshkumar S, Leach GW, Huang CK, Grillaert J. Proceedings of the Second International CMP for ULSI Multilevel Interconnection Conference (CMP-MIC); Tampa, FL; 1997. p 179.
21. Detzel T, Hosali S, Sethuraman A, Wang J-F, Cook L, Grillaert J. Proceedings of the Second International CMP for ULSI Multilevel Interconnection Conference, (CMP-MIC); Tampa, FL; 1997. p 202.
22. Hwee LL, Balakumar S, Mahadevan S, Sheng ZM, See A, Rahman M, Senthilkumar A. Dishing and nitride erosion of STI-CMP for different integration schemes. J Electronic Mater 2001;30(12).
23. Kang H, Katoh T, Kim S, Paik U, Park H, Park J. Effects of grain size and abrasive size of polycrystalline nano-particle ceria slurry on shallow trench isolation chemical mechanical polishing. Japan J Appl Phys 2004;43(3A):L365–L368.
24. Lim DS, Ahn JW, Park HS, Shin JH. The effect of CeO_2 abrasive size on dishing and step height reduction of silicon oxide film in STI-CMP. Surf Coat Technol 2005;5–6:1751–1754.
25. Tang SK, Vassiliev VY, Mridha S, Chan LH. Investigation of borophosphosilicate glass roughness and planarization with the atomic force microscope technique. Thin Solid Films 1999;1–2:77–84.
26. Barnett RJ, Mezner MB. Production of fumed silica. US patent 6,217,840. 2001 Apr 17.
27. Treichel H, Frausto R, Srivastan S, Whithers B. Process optimization of dielectrics chemical–mechanical planarization processes for ultralarge scale integration multilevel metallization. J Vac Sci Technol A 1999;17.4:1160–1167.
28. Lin M-T, El-Deiry P, Chromik RR, Barbosa N, Brown WL, Delph TJ, Vinci RP. Temperature-dependent microtensile testing of thin film materials for application to microelectromechanical system. Microsyst Technol 2006;12(10–11); p 1045–1051.
29. Divecha RR, Stine BE, Ouma DO, Boning D, Chung J, Nakagawa OS, Oh S-Y, Hetherington DL. Comparison of oxide planarization pattern dependencies between two different CMP tools using statistical metrology. VLSI Multilevel Interconnect Conference; Santa Clara, CA;1996. p 427–430.
30. Stine B, Boning D, Chung J, Camiletti L, Equi E, Prasad S, Loh W, Kapoor A. The role of dummy fill patterning practices on intra-die ILD thickness variation in CMP processes. VLSI Multilevel Interconnect Conference;Santa Clara, CA;1996. p 421–423.

31. Chang E, Stine B, Maung T, Divecha R, Boning D, Chung J, Chang K, Ray G, Bradbury D, Oh S, Bartelink D. Using a statistical metrology framework to identify random and systematic sources of intra-die ILD thickness variation for CMP processes. International Electron Devices Meeting;Washington DC; 1995. p 499–502.
32. Miyamoto M, Hirano S, Chibahara H, Watadani T, Akazawa M, Furukawa S. Enhancement of post-Cu-chemical mechanical polishing cleaning process for low-*k* substrate. Japan J Appl Phys 2006;45(10A)7637–7644.
33. Eissa MM. Post copper CMP clean. US patent 6,383,928. 2002 May 7.
34. Chen P-L, Chen J-H, Tsai M-S, Dai B-T, Yeh C-F. Post-Cu CMP cleaning for colloidal silica abrasive removal. Microelectron Eng 2004;75:352–360.
35. Shen JJ, Costas WB, Cook LM. The effect of post chemical mechanical planarization buffing on defect density of tungsten and oxide wafers. J Electrochem Soc 1998;145(12)4240–4243.
36. Johl B, Buley T. Proceedings of the Seventh International CMP Conference; San Jose, CA; 2002.
37. Johl B, Singh R. Proceedings of the VMIC Conference; Santa Clara, CA; 2001.
38. Singh R, Johl B. Proceedings of the VMIC Conference; Santa Clara, CA; 2001. p 557.
39. Singh RK, Roberts BR. Proceedings of the VMIC Conference; Santa Clara, CA; 2000. p 545.
40. Bare JP, Johl B. Proceedings of the AVS N CMPUG Annual Symposium; CA, USA; 1997. p 297.
41. Singh RK, Roberts B R. Proceedings of the CMP-MIC; 2001. p 441.
42. Bare JP, et al. Proceedings of the Semicon West Workshop; 1998.
43. Remsen EE, Anjur S, Boldridge D, Kamiti M, Li S, Johns T, Dowell C. Analysis of large particle count in fumed silica slurries and its correlation with scratch defects generated by CMP. J Electrochem Soc 2006;153(5):G453–G461.
44. Miyashita N, Uekusa S, Kodera M, Matsui Y, Katsumata H. Development of dishing-less slurry for polysilicon chemical-mechanical polishing process. Japan J Appl Phys 2003;42:5433–5437.
45. Seo Y-J, Lee W-S. Effect of different oxidizers on the W-CMP performance. Mater Sci Eng B 2005;118:281–284.
46. Lim G, Lee J-H, Kim J, Lee H-W, Hyun S-H. Effect of oxidants on the removal of tungsten in CMP process. Wear 2004;257:863–868.
47. Seo Y, Lee W. Effects of oxidant additives for exact selectivity control of W- and Ti-CMP process. Microelectr Eng 2005;77:132–138.
48. Lefevre P. Next generation W CMP slurry. CAMP 9th International Symposium on Chemical–Mechanical Planarization; Lake Placid, NY; 2004.
49. Lee W, Kim S, Seo Y, Lee J. An optimization of tungsten plug chemical mechanical polishing (CMP) using different consumables. J Mater Sc; 2004;12(1):2001.
50. Vacassy R, Chen Z. Edge-over-erosion in tungsten CMP. Proceedings of the 11th International CMP for ULSI Multilevel Interconnection Conference (CMP-MIC); Fremont, CA; 2006. p 632–639.
51. Zetrath D. Modified electroplating solution components in a low-acid electrolyte solution. US patent 20,050,077,181. 2005 Apr 14.

52. Zetrath D. Modified electroplating solution components in a high-acid electrolyte solution. US patent 20,050,077,180. 2005 Apr 14.
53. Moffat TP, Walker M, Chen PJ, Bonevich JE, Egelhoff WF, Richter L, Witt C, Aaltonen T, Ritala M, Leskelä M, Josell D. Electrodeposition of Cu on Ru on Barrier Layers for damascene processing. J Electrochem Soc 2006;153(1):C37–C50.
54. Qu X, Tan J. Comparison of Ru/Ta and Ru/TaN as barrier stack for copper metallization. Proceedings of MRS Symposium F; 2006.
55. Koike J, Wada M. Self-forming diffusion barrier layer in Cu–Mn alloy metallization. Appl Phys Lett 2005;87(4)
56. Fayolle M, Passemard G, Assous M, Louis D, Beverina A, Gobil Y, Cluzel J, Arnaud L. Integration of copper with an organic low-k dielectric in 0.12-mm node interconnect. Microelectr Eng 2002;60(1–2):119–124.
57. Schulze K, Schulz SE, Frühauf S, Körner H, Seidel U, Schneider D, Gessner T. Improvement of mechanical integrity of ultra low k dielectric Stackand CMP compatibility. Microelectr Eng 2004;76(1–4): 38–45.
58. Ernur D, Kondo S, Shamiryan D, Maex K. Investigation of barrier and slurry effects on the galvanic corrosion of copper. Microelectr Eng 2002;64:117–124.
59. Kondo S, Sakuma N, Homma Y, Ohashi N. Japan J Appl Phys 2000;39:6216–6222.
60. Cheemalapati K, Bundi D, Duvvuru V, Li Y, Hann S, Li H. Investigation of edge over erosion (EOE) in Cu CMP. Proceedings of 11th International CMP for ULSI Multilevel Interconnection Conference (CMP-MIC); Fremont, CA; 2006. p 650–658.
61. Shin S. Proceedings of the Second PacRim International Conference on Planarization CMP and its Application Technology; Seoul, Korea; 2005. p 89–94.
62. Tao J, Cheung NW, Hu C. Electromigration characteristics of copper intrerconnects. IEEE Electr Dev Lett 1993;14(5):249–251.
63. Balakumar S, Chen XT, Chen YW, Selvaraj T, Lin BF, Kumar R, Hara T, Fujimoto M, Shimura Y. Peeling and delamination in Cu/SiLK process during Cu-CMP. Thin Solid Films 2004;462–463:161–167.
64. Basics on optical microscope. Available at http://en.wikipedia.org/wiki/Microscope
65. Li Y. CMP slurry developments. CMP for ULSI Multilevel Interconnection Short Course; Fremont, CA; 2005 (With courtesy from P. LeFevre).
66. Basics on scanning electron microscope (SEM). Available at http://www.mos.org/sln/SEM/.
67. Basics on energy dispersive X-ray spectroscopy. Available at http://en.wikipedia.org/wiki/Energy-dispersive_X-ray_spectroscopy.
68. Basics on scanning auger microscopy. Available at http://www.uksaf.org/tech/sam.html.
69. Basics on atomic-force microscope. Available at http://en.wikipedia.org/wiki/Atomic_force_microscope.
70. Overview on principle and working of optical defectivity tools. Available at http://www.kla-tencor.com/j/servlet/HomePage?version = flash.
71. Automatic-defect classification. Available at http://www.ornl.gov/sci/ismv/research_industrial_adc.shtml.

18

CMP SLURRY METROLOGY, DISTRIBUTION, AND FILTRATION

Rakesh K. Singh

Chemical–mechanical planarization (CMP) is one of the most tightly controlled, key enabling processes in integrated chip (IC) manufacturing. With the industry's transition from 90 to 65 nm and then further to 45 and 32 nm, CMP processes are becoming much more complex while demanding stringent management of CMP slurry quality during its manufacturing, blending, and distribution. Slurry distribution in IC fabs is accomplished using chemical blending and delivery systems designed on different liquid dispense technologies. Handling in these systems may cause changes in abrasive and/or chemical properties of the slurry. To maintain slurry health during its usage and replenishment, it is essential to monitor and adjust its chemical properties (e.g., oxidizer and additive levels and their replenishment needs) and the abrasive characteristics (e.g., composition, large and mean particle size distributions, weight percentage of solids (wt% solids), density or specific gravity, and settling and redispersion behavior). Bench-top and handling characterizations can provide useful insights into the slurry metrology parameters and health management challenges. Efficient CMP slurry metrology tools should be able to measure the chemical and/or oxidizer concentration(s), large particle counts (LPC), and mean particle size distribution (PSD) on a real-time basis and use this information for CMP process control.

The removal of large, defect-causing particles without changing constant slurry flow and mean PSD of abrasives is critical for CMP slurry health. LPC growth in slurries is typically managed using postblending, global loop, and point-of-use (POU) filtration. Filtration can effectively remove large particles if

Microelectronic Applications of Chemical Mechanical Planarization, Edited by Yuzhuo Li

the size of particles to be removed (e.g., >0.5 or $>1.0\,\mu m$) is at least one order of magnitude larger than the mean working particles to be retained in the slurry. In slurry-handling situations involving even small increases in the mean size of the abrasives, it may be very difficult to selectively remove particles that are only slightly larger than mean size particles to restore handled slurry PSD and LPC to the fresh slurry. Tighter POU filter flow rate, pressure drop, and lifetime can be significantly affected because of mean PSD changes in extensively handled slurries. Using optimum rating filters in the global loop and POU and limiting the total number of slurry turnovers before consumption should help in slurry health maintenance, providing stable CMP performance in such cases. Current slurry manufacturing processes target 90% and higher reduction in cumulative LPC at ≥ 0.56 or $1.01\,\mu m$. Single-pass filtration solutions using graded density, multiple thin layer, pleated depth filter media, or membranes can provide required large particle retention depending on the slurry abrasive and chemical composition. Results of particle retention, flow rate, and pressure drop from filtration tests using tighter depth filters show very different behavior in silica, alumina, and ceria abrasive slurries, demonstrating that CMP slurry filter optimization still remains empirical in nature.

This chapter focuses on CMP slurry metrology and characterization approaches, slurry health monitoring and control, blending and distribution, slurry large particle management through filtration, and pumping effects on slurries. Shearing effects caused by slurry delivery technologies are studied with a consideration of the pump speed, slurry turnover rate, number of slurry turnovers before consumption, global distribution loop pressure, slurry temperature control requirements, abrasive settling and redispersion, and filter lifetime. This information must be considered for optimum slurry health management. With growing complexity of the next-generation CMP processes, more collaborative research and joint development would be required between the slurry and slurry distribution system manufacturers, filtration solution providers, and the end users earlier in the cycle of development to meet the significant low-defectivity and fine-tuning demands of newer CMP slurries. Such efforts are likely to optimize process parameters in a shorter time, reduce repetition of efforts and cost of development, improve process knowledge, and lower the cost of ownership.

18.1 INTRODUCTION

The quality of polishing, which is critical to yield, depends upon the quality and consistency of the CMP slurry and the integrity and cleanliness of the polishing tool set. Contamination of the process by large particles that may have agglomerated during transportation, storage, and handling of the slurry may cause microscratches on the polished wafer. CMP processes and consumables must continue to improve to meet much higher performance specifications demanded by the onset of 45 nm and further shrinking feature sizes, the introduction of copper, ultra-low-*k* and high-*k* dielectrics, along with noble metals and larger wafers [1–4]. Low-*k* process integration needs a CMP process, which offers very low mechanical impact (i.e., very low downforce in CMP process) to the target materials. Also, the process must be able to work on the wafers with minimal overpolish to achieve metal loss and defectivity targets [2]. Advanced metal dielectric slurries for 45 nm may use sub-1% abrasive content by weight. Copper CMP slurries are targeted to produce global planarity and adequate copper removal rate for high throughput at low downforce, and high planarization efficiency to minimize dishing and erosion.

More stringent quality management of CMP slurries is being driven by the increasing yield requirements per wafer, lowest defectivity, development of complex higher density devices, use of CMP in new devices (e.g., hard disk, SOI, 3D memory, optoelectronics, LEDs, biochips, GaAs MMIC, SiGe, and MEMS), and application of new materials to improve device electrical characteristics. New uses of CMP include FinFET transistor formation, tri-gate transistor, 3D integration, metal FUSI gates, new photodiode creation (requiring dissimilar materials CMP), novel metal gate approach, and single-grain thin-film transistor. Further challenges of CMP include applications in new materials, improved yield, excursion control and metrology, reduced consumable development, chip and wafer uniformity, in-line consumable monitoring, and reduced cost of ownership (CoO).

An efficient CMP slurry management system should take into account (i) monitoring and control of changes in the slurry properties (e.g., LPC, PSD, pH, density, wt% solids, etc.) due to handling and limited postblend useful life, (ii) abrasive settling and slurry foaming behavior, (iii) variability in the slurry and additive properties of different product lots and age, (iv) effects of environmental changes on slurry during the recommended storage period and use, (v) uncertainties of oxidizer and organic additives decay during slurry delivery and their required adjustments with time, (vi) effective filtration schemes to maintain slurry abrasive consistency through LPC control, and (vii) true online real-time LPC, PSD, oxidizer level, and process control needs. Improper mixing and extensive handling without regular usage and replenishment can cause changes in the abrasive particle characteristics and agglomeration in some slurry [5–7]. Large particles in slurries can form because of various reasons and may produce wafer scratches. Particles that have a higher degree of coalescence and more rigidity are responsible for such defects. The efforts should focus on not creating such particles during slurry distribution and selectively removing them from employing filtration if they are created due to less than optimum slurry handling.

Most CMP slurries are sensitive to extreme temperatures. For incoming slurry lots, it is important to ensure that the slurry container has not been exposed to low temperatures (e.g., $< 10\,^{\circ}C$). The minimum and maximum allowable exposure temperatures are usually specified by the slurry suppliers on the slurry container label. Even short exposure to freezing temperatures may cause irreversible damage to the slurry abrasives and result in agglomeration or gel formation [5]. Slurry supply or storage container cleanliness must be ensured before opening the container cap for slurry transfer. Long-term storage may cause drying of the slurry on the inside of the container headspace. Mixing of even small amounts of dry slurry particles can be very harmful to the slurry health, because of the potential of these large particles to cause an increase in slurry LPC and wafer scratches during polishing. Such dried slurry residue in the cap thread region of the supply containers could be carefully removed. Similar care should be taken in the operation of daytanks connected to the slurry delivery systems.

To have a stable CMP process, the slurry should be stored and handled at a near-ambient constant temperature. The slurry mixing in the supply containers should be optimal and enough to redisperse all abrasive particles, but not to shear slurry excessively using extended pump recirculation or a very high-speed propeller. Any dead legs in the piping and sudden changes in the slurry flow direction, velocity, and cross-sectional area should be minimized to reduce handling-related shear (especially in shear-sensitive silica abrasive slurries), for example, by using a vacuum-pressure-dispense system (VPDS) and maintaining the required minimum flow velocity in the global distribution loop, which is essential to keep abrasive particles dispersed. Also, the slurry supply tank and the daytanks should be blanketed with nitrogen with high relative humidity (>90%) nitrogen to maintain consistency in slurry wt% solids over time [5]. Further, CMP slurry mixing and delivery systems require more frequent maintenance and cleaning than ultrapure chemical distribution systems. The nature and extent of cleaning required depends on the slurry type, handling system design, and slurry usage characteristics. The slurry delivery system including metrology module and global distribution loop should be periodically cleaned following scheduled preventive maintenance (e.g., twice per year).

Slurry LPC control through filtration can help in improving global and local planarity as well as significantly reducing defectivity in the CMP process. More integrated slurry filtration solutions are needed to reduce critical defect size and number of defects [8–11]. Appreciable changes in slurry abrasive particle and chemical behavior during blending and distribution can pose challenges to the stability of filtration process and LPC measurements [12–16]. Reliable LPC characterization in the next-generation CMP slurries requires innovative instruments capable of measuring particle size down to 0.20 μm reliably as compared to the current level of 0.50 μm. Modular or open architecture LPC instruments with the capability to fine tune the measurement in different slurries (ceria abrasive based in particular) and possibly detecting the soft or hard (defect causing) nature of agglomerates or aggregates would be beneficial. Because of the complex nature of abrasives and chemistry, next generation CMP slurries may need multipass filtration to achieve target retention even with tighter graded-density depth filters. To improve the understanding of newer slurry filtration, more research is needed in filtration media development, alternative particle separation technologies, and filter clogging mechanisms in innovative CMP slurries, containing modified, mixed, and composite abrasives.

18.2 CMP SLURRY METROLOGY AND CHARACTERIZATION

To conduct feasibility studies with any new CMP consumables, today's IC fabs demand more thorough characterization of CMP consumables and a conclusive data set. The key objectives of slurry metrology include identification of the sensitive measurement parameter(s) to create and maintain an accurate slurry blend, measurements of abrasive PSD and LPC, and monitoring and control of oxidizer and additives concentration in the slurry blend during its useful lifetime. CMP slurry characterization studies focus on slurry chemistry and particle optimization, abrasive mixing, redispersion and dispersion stability behavior, slurry pot-life and shelf-life determination, effect of slurry delivery system on the slurry health, identification of slurry system cleaning chemistry and protocol development, and slurry waste disposal and related government regulations. Different levels of slurry characterizations may include screening, first-pass optimization, baseline characterization, fine tuning, and process sensitivity analysis. Monitoring and controlling slurry health (i.e., its stability and contamination level over a lifetime, quantified using some form of metrology tools) is essential to reduce CMP process variability due to slurry changes.

CMP slurry quality can be monitored by measuring LPC, PSD, mean particle size (mean, mode), oxidizer concentration, zeta potential, pH, conductivity, viscosity, total dissolved solids (TDS), oxidation–reduction potential (ORP), density or specific gravity at 25 °C, wt% solids, and metal impurities. Removal rate (RR) and defect performance during CMP may depend on slurry batch-to-batch and within batch, as well as variation (during supplier recommended storage lifetime) of the abrasive over time and its chemical characteristics. Slurries may also have instabilities due to their aggregation characteristics and may change in PSD and LPC behavior with time. A slurry stability ratio (SSR) has been reported to quantify slurry stability as a function of chemical environment [17]. SSR is the ratio of rate constants for rapid aggregation versus condition of interest. The former constant can be calculated or measured, whereas the latter is measured using light scattering instruments. RR can be driven by the strength of interaction between the abrasive and the film, and stronger interactions tend to result in higher RR. Large agglomerates in slurry may be formed depending on the strength of interaction, particle size, and relative number of particles. Slurry heterostability provides a measurement of the strength of interaction and propensity to form large agglomerate [17], and heterostability ratio depends on the size of silica particles and the number ratio of silica to ceria.

18.2.1 Slurry Health Monitoring and Control

Mean PSD determines bulk RR of the material being polished, whereas LPC relates to the wafer scratches. LPC measurements are more important in slurry quality management due to their significant effects on defectivity and yield in CMP processing. However, mean particle size should be well controlled in application requiring tighter filtration, which will be discussed later. Measurement of LPC can detect excursions in large particles due to various sources such as temperature fluctuations during shipping and transportation, slurry abrasive drying and settling, pH shock during blending, and slurry pumping and distribution [5,18]. In many CMP slurries, it is essential to monitor and control the characteristics of slurry abrasive particles and the oxidizer level during slurry usage [7]. Not many technologies can provide the oxidizer concentration as well as the abrasive mean size and LPC information of the slurry blends, especially on a real-time and in-line basis, without sampling and dilution. Single particle optical sensing (SPOS) technique based instruments are commonly used for obtaining LPC data in slurries. These use the principle of light obscuration to count and size particles one at a time. Slurry PSD measurements are typically made using dynamic light scattering (DLS), multiwavelength light extinction, electroacoustics, hydrodynamic fractionation, disk centrifuge, elliptically polarized light scattering, and laser diffraction and light scattering based analyzers. DLS-based techniques require sampling and dilution and are sensitive to both small and large particles. A detailed comparison of common PSD analyzers is provided elsewhere [12].

The oxidizer concentration (e.g., H_2O_2 wt%) in the slurry + oxidizer blend can be measured using autotitration or bench-top titration, ultrasonic analyzers, FT-NIR (Fourier transform near infrared), filter NIR, conductivity, pH, UV/Vis and IR absorption, and refractive index (RI). These technologies may also be employed for concentration measurements in ultrapure chemicals as well as wet etch and clean chemistry blends such as HF/HCl, SC1 or RCA-B ($NH_4OH/H_2O_2/H_2O$), SC2 or RCA-A ($HCl/H_2O_2/H_2O$), and DSP (dilute H_2SO_4/H_2O_2). In general, titration-based methods require large quantity of consumables (especially if used frequently) and periodic maintenance, and provide intermittent (not in real time) data of concentration, whereas ultrasonic- and RI-based methods can provide continuous data but may have limitations with the required accuracy in the oxidizer level measurements of multicomponent blends (e.g., slurry + H_2O_2 + water). Conductivity measurement also has limited application even when it has good sensitivity to the blend ratio in different slurries, since it is usually not controlled by slurry vendors in different lots of slurry and may also vary with time elapsed after manufacturing, during the recommended storage time of the slurry. FT-NIR is a complex technology with higher price and slower measurement speed. Chemical concentration can be continuously measured using UV/Vis or IR absorption. Measurement of the electric surface properties of abrasives such as zeta potential, which affect particle–surface and particle–particle interactions, is equally important.

An efficient slurry health monitoring tool should be able to provide both chemical as well as abrasive particle information on a continuous basis. There have been some efforts in this direction using an NIR absorption spectrum based analyzer [19]. This unit can provide oxidizer concentration and abrasive particle information in CMP slurry and operates on the principles of chemometrics, which is a two-phase process. In the first "calibration" phase, samples with known property values are measured by the system. A mathematical procedure then determines the correlation between the measured spectra and the true property values. The output of this phase is a "model" that optimally calculates the parameter values from the measured spectra of the calibration samples. In the second "measurement" phase, unknown samples are measured by the system, employing a model to produce estimates of the property values.

Results of a large particle index (LPI) obtained using the above NIR analyzer data show a linear correlation with the cumulative LPC measured employing a SPOS analyzer. This real-time NIR analyzer has an in-line, noncontact flow cell with a polyetheretherketone body and sapphire windows, and has the capability of providing real-time data. On the basis of the above oxidizer concentration measurement in the slurry blend, a spiking dispense system may be activated in the slurry delivery system, dynamically controlling the oxidizer level within target limits. Further, using the identified LPC trend, an appropriate filter change-out time may be scheduled. Another area of great interest in CMP slurry abrasive characterization is the aggregate stability of the slurry, that is, identification of aggregation and flocculation. Aggregation is detrimental to polishing and may result in significant spikes in defectivity and yield losses, whereas loose flocs are unlikely to cause much damage to the wafer-polished surface. Intelligent technologies enabling measurements of small changes in abrasive behavior (including softness and hardness of the large agglomerates) and chemical properties of the slurry in real time could prove to be a very cost-effective tool in reducing CMP consumable cost of ownership and yield loss through continuous slurry health management.

18.2.2 CMP Slurry Blend Control

CMP slurry blend accuracy and quality are of paramount importance in achieving efficient and consistent wafer planarization. Slurry blend quality can vary significantly after blending and during fab distribution, based on the slurry usage and replenishment cycle and the number of slurry turnovers before consumption. The quick settling behavior of abrasives [20], limited postblend useful life, tighter blend accuracy and control requirements, variability in the CMP slurry properties of different lots, and oxidizer decay with time may pose unique challenges for traditional slurry delivery approaches [21]. Continuous monitoring and control of slurry health is becoming more and more essential with tighter specifications of the CMP processes. Slurries can be continuously or batch blended and monitored online for consistency using various approaches as discussed in the following sections.

18.2.2.1 Two-Component Blend Control Monitoring and control of two-component blends of CMP slurries can be achieved by online or bench-top measurements. The control parameters depend on the chemistry of the slurry blend. Commonly used blend control parameters include density, wt% solids, concentration of oxidizers and other additives (hydrogen peroxide, periodic acid, benzotriazole, etc.), pH, ORP, and conductivity. For many CMP slurries, conductivity has very good sensitivity to the blend ratio. However, as discussed earlier, it often cannot be used as an independent control parameter, since its value is frequently not controlled by slurry manufacturers in different lots of the same slurry. Conductivity of a slurry may also vary with aging of the same slurry. A majority of slurries are chemically buffered and do not show significant variation in pH with the change in the blend ratio. Similarly, ORP does not have good sensitivity in most slurry blends. Therefore, density, wt% solids, and oxidizer assay are the most commonly used blend control parameters.

Examples of common slurry blends that are amenable to the above approach include: Cabot Microelectronics Semi-Sperse® 25 (SS-25) and DI water blend with density as a control parameter, Cabot Semi-Sperse® W2000 and H_2O_2 blend using titration to measure wt% H_2O_2 level in the mix as a control, Cabot EPC-5001® and H_2O_2 blend controlled by using autotitration to measure wt% H_2O_2 level and wt% solid measurements to monitor settling of abrasive particles, Hitachi Chemical HS-8005® and HS-8102GP® blend control using wt% solids, and Hitachi Chemical abrasive-free chemical polishing solution HS-C430® and H_2O_2 blend control using refractive index and pH measurements.

Figures 18.1–18.4 present the distributions of pH, conductivity (μS/cm), ORP (mV), TDS (ppm), density (g/cc), and wt% solids in SS-25, W2000, and EP-C5001 slurry blends. These bench-top blending and measurement

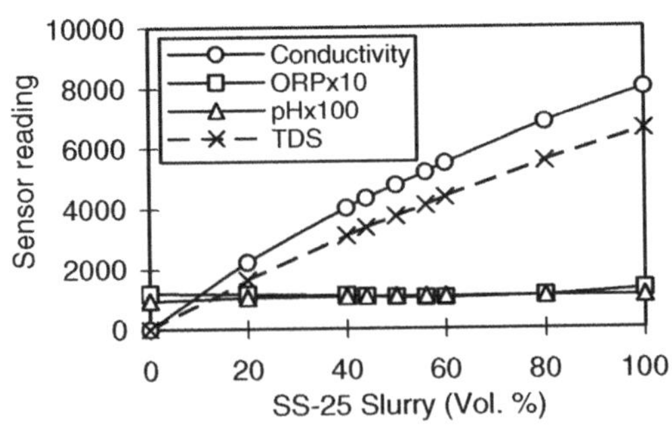

FIGURE 18.1 SS-25 and DI water blend data for pH, conductivity, ORP, and TDS.

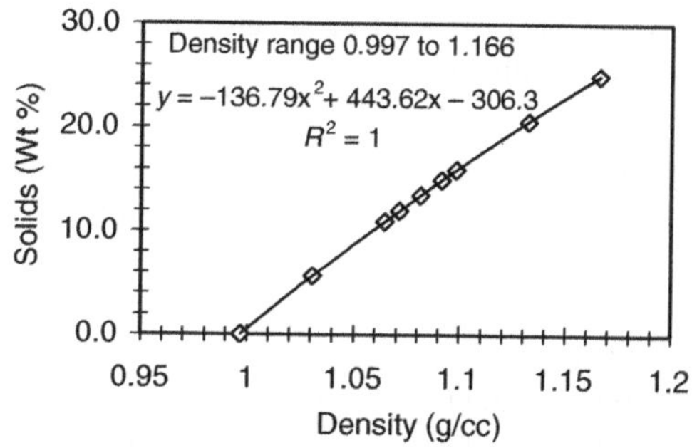

FIGURE 18.2 SS-25 and DI water blend data for density and wt% solids.

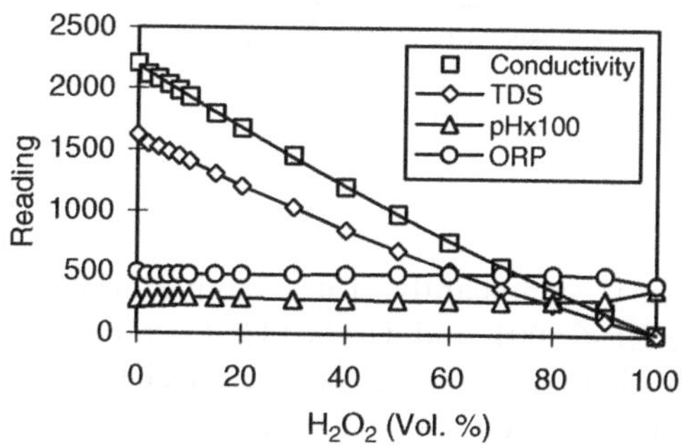

FIGURE 18.3 W2000 and H_2O_2 blend data for pH, conductivity, ORP, and TDS.

experiments were conducted using 30% concentration H_2O_2. A single parameter is usually sufficient to control two-component blend. Table 18.1 illustrates the sensitivity of various parameters for EP-C5001 and H_2O_2 blend. It is important to verify the oxidizer level (H_2O_2 wt%, using autotitration) in this blend for consistent CMP performance. The requirement of oxidizer

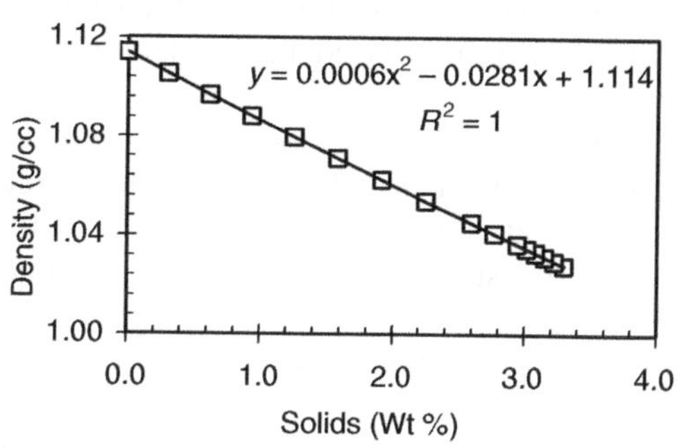

FIGURE 18.4 EP-C5001 and H_2O_2 blend data for density and wt% solids.

TABLE 18.1 Sensitivity of Measurement Parameters to H_2O_2 wt% Concentration in Cabot EP-C5001 and H_2O_2 Blend.

Target H_2O_2, wt%	Density, g/cc	Density, change	Solids, wt%	Conductivity, μS/cm	pH	Hydrogen peroxide, vol%	Change, vol%
2.4	1.03429	0.00026	3.036	14,357	7.598	7.426	0.311
2.5	1.03455	0.00026	3.025	14,313	7.591	7.736	0.311
2.6	1.03481	0.00026	3.014	14,270	7.584	8.047	0.311
2.7	1.03508	0.00026	3.003	14,227	7.577	8.359	0.311
2.8	1.03534	0.00026	2.992	14,183	7.570	8.670	0.311
2.9	1.03561	0.00026	2.981	14,140	7.563	8.982	0.312
3.0	1.03587	0.00026	2.970	14,096	7.557	9.293	0.312
3.1	1.03614	0.00026	2.959	14,052	7.550	9.605	0.312

The above calculation is based on slurry EP-C5001 supply wt% solids = 3.30%, pH = 7.81, density = 1.028 g/cc, and conductivity = 15,170 μS/cm; H_2O_2 pH = 3.54, density = 1.114 g/cc, and conductivity = 12.5 μS/cm.

monitoring may vary significantly; for example, H_2O_2 does not degrade appreciably in the EP-C5001 and H_2O_2 blend, unlike in W2000 and H_2O_2 blend. In the latter blend, continuous significant H_2O_2 degradation with time (e.g., ~0.1 to 0.2 wt% per day) makes it essential to monitor oxidizer level more frequently and adjust oxidizer concentration as needed.

18.2.2.2 Three-Component Blend Control Slurries requiring three or more component blending pose significant challenges to blend-ratio control. Polishing with diluted Cabot Semi-Sperse® W2000, DI water, and a sufficient concentration of H_2O_2 in the blend reduces the total defect counts (over 60%), with a reduced cost of ownership (~30%), small oxide erosion (thinning), and lower tungsten plug recess [22]. In three-component blends, it is usually necessary to measure two independent parameters to resolve the mix ratio. In the above application, a method to dynamically control the blend ratio of the above mix to acceptable tolerances using a combination of autotitrator and density measurements (in a pump-based batch-blending slurry delivery tool) could be developed [7].

In the above method, an autotitrator repeatable to ±0.5% relative error (or 0.01 wt%) was employed to measure H_2O_2 wt%, independent of the relative amounts of DI water and slurry. A densitometer with an accuracy of ±0.0005 g/cc was used to estimate the percentage volume of slurry in the mixture (repeatability = ±3% slurry by volume), for a specific concentration of H_2O_2. In the above blend, the measured H_2O_2 wt% concentration was used to control the speed (or strokes/batch) of the W2000 slurry pump, whereas the density, obtained using a coriolis mass flow meter, controlled the speed of the DI water pump. On the basis of the property curves obtained from bench titration and the outputs from autotitrator and densitometer, an empirical algorithm dynamically controlled the stroke ratio for the bellows pumps in the blend module of an on-demand pump (pressure) pressure-dispense continuous blending and distribution system [21]. Using this approach, the H_2O_2 and slurry concentrations in the mix could be controlled within ± 0.1 wt% and ± 3.0 vol%, respectively.

18.2.3 CMP Slurry Characterization

CMP slurry characterization can be performed at various levels of detail. Bench-top characterization aims to identify sensitive measurement parameters for blend monitoring and control. This metrology sensitivity test is conducted by measuring the slurry components, target blend, and other blend ratio. DI water dilution test is performed to identify parameters to detect DI water fab leak. Slurry characterization may include (i) extensive handling tests in slurry delivery systems and pump loops, (ii) filtration characterization, (iii) slurry stirring and redispersion studies for the supply containers, day tank and global loop, and (iv) slurry handling-system cleaning or preventive maintenance protocol development. Slurry system blending, handling, and filtration studies are performed to determine the stability of slurry blend creation and handling over time. Such characterization may provide valuable information on optimum handling conditions (e.g., recommended minimum flow velocity to keep abrasive particles fully dispersed) and filtration essential to maintain slurry quality during simulated slurry consumption. These accelerated studies, typically performed for 1–2 weeks, can generate quantitative data on slurry metrology parameters and the effects of handling system on slurry properties [6,15]. The filtration tests are typically conducted as part of such studies to develop the POU and global distribution loop slurry filtration recommendations for the end users.

It is common to generate slurry delivery system cleaning or preventive maintenance information at the end of slurry handling and filtration characterization studies. In these tests, the slurries are removed from the delivery system and the system is rinsed and subsequently filled with ambient temperature or slightly heated DI water, or more commonly with *mother liquor* based DI water solution (e.g., KOH solution in DI water at pH ~11–12, if the slurry has pH ~11 with KOH as main ingredient). Recirculation of the solution in various parts of the slurry delivery system usually removes slurry abrasive deposits. The solution recirculation time and temperature may vary depending on the preventive maintenance schedule (e.g., on a monthly, quarterly or half-yearly basis). The extent and frequency of maintenance depends on the slurry composition and usage. Slurry vendors generate such system cleaning information for the slurry end users. In some slurry, by using DI water in the slurry system with residual slurry, content may generate pH shock and precipitation in the system rather than dissolving and removing the slurry residue.

Hot water or chemical recirculation may help in the cleaning, but may cause loosening of the plastic tubing joints and damage the plastic parts and instruments due to thermal cycling. Hence, it is important to use good judgment based on the past experience in the identification of the correct chemical and protocol for the CMP slurry system cleaning experiments. These systems are commonly cleaned using KOH, NH_4OH, HCl, HNO_3, or oxalic acid solutions in DI water. Typical solution concentrations are in the range of 1–3% by weight of the above chemicals. Chemical cleaning is typically used to clean slurry mixing and dispense modules, pressure-vacuum vessels, daytanks and agitation system, metrology module and related piping, and global distribution loop tubing on a regular basis, (and more frequently) especially for the slurries (e.g. silica based) with higher susceptibility to large particle agglomeration and drying. System cleaning tests provide useful insight on the composition, concentration, and temperature of the cleaning solution as well as the recirculation time and protocol needed to achieve required cleaning.

Slurry characterization tests may demonstrate the suitability of a specific distribution system in handling slurry over an extended period of time while maintaining the quality of the slurry. These tests may also provide information for the optimization of CMP slurry metrology and quality management and help in the identification of compatible slurry systems for a specific slurry. A majority of silica abrasive slurries show sensitivity to shear stress application. These slurry blends typically produce large particle agglomerates when exposed to high level shear stress during handling. Agglomerated particles may contribute to higher defectivity and yield loss in the CMP process. In general, the chemical composition of slurries maintains a high repulsive electrostatic surface charge on abrasive particles to resist agglomeration. Aggressive handling may cause shear to the slurry abrasives due to strong velocity gradients produced by slurry flow by suddenly changing flow path and geometry.

The above conditions may cause breakdown in the abrasive repulsive forces balance by imparting significant levels of energy, bringing particles too close, and activating the van der Waals attractive intermolecular forces, resulting in particle agglomeration explained by the DLVO theory (named after Derjaguin, Landau, Verwery and Overbeek). Other factors resulting in particle agglomeration may include lack of humidification, pH shock due to improper dilution, and air entrapment during blending and delivery. Figs 18.5 and 18.6 show representative results of a silica slurry handling characterization using a VPDS and bellows pump recirculation loop. Similar results of a test for another silica slurry with lower solid content are presented in Figs. 18.7 and 18.8. Both of these studies demonstrate the advantage of using a VPDS for silica slurry handling. These are typical results found in silica slurry. Slurry characterization studies are usually performed jointly by the slurry and slurry delivery system suppliers, and the filtration solution provider to create in-depth information for the fab end users.

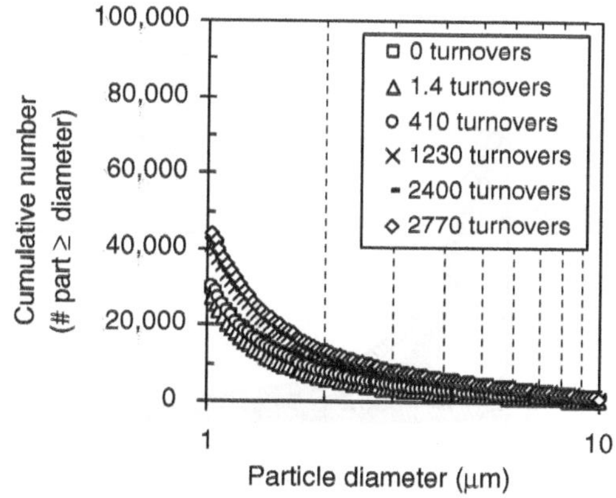

FIGURE 18.5 LPC distribution during handling of a ~ 30 wt% solids silica abrasive slurry in a VPDS for slurry delivery (BOC Edwards P2200).

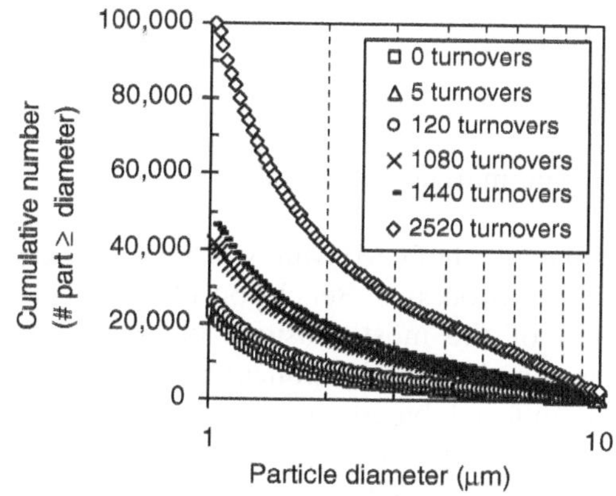

FIGURE 18.6 LPC distribution during handling of a ~ 30 wt% solids silica slurry in a bellows pump (Nippon Pillar PE-20MAN) recirculation loop.

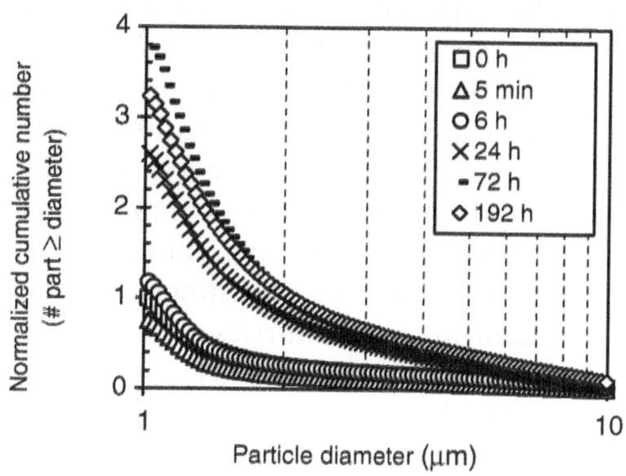

FIGURE 18.7 Normalized LPC distribution during handling of a ~ 10 wt% solids silica abrasive slurry in a VPDS for slurry delivery.

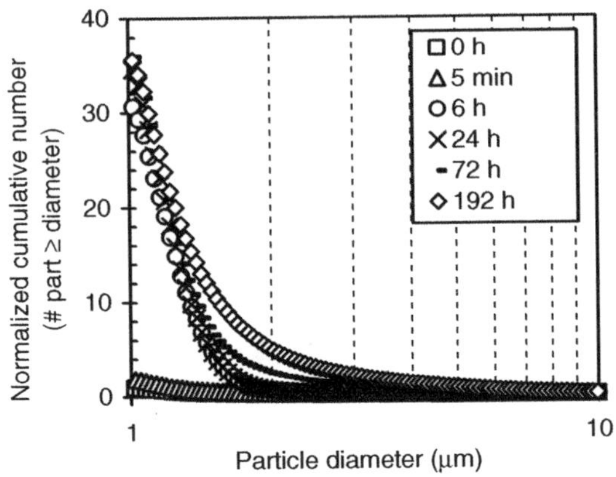

FIGURE 18.8 Normalized LPC distribution during handling of a ~ 10 wt% solids silica slurry in a bellows pump (PE-20MAN) recirculation loop.

18.2.4 Summary

Key findings may be summarized as below:

- Monitoring and control of CMP slurry properties is essential for effective and uniform CMP processes. Bench-top blend sensitivity analysis helps in identification of the most sensitive blend monitoring and control parameter. Two- and three-component blends of CMP slurries can be created and monitored based on the measurements of density, wt% solids, refractive index, pH, and oxidizer level. Typical silica oxide slurry blend ratio is controlled using density as a control parameter, whereas tungsten and copper CMP slurries usually need an autotitrator for periodic monitoring of the oxidizer level.
- Slurry handling characterization may provide useful insights on slurry health maintenance challenges, which may occur during actual slurry delivery and usage in the fab. Slurry distribution management challenges include: tighter blend accuracy control requirements of newer slurries, quick settling of alumina and ceria abrasives, limited postblending useful life of the slurry, variability in the slurry components and blend chemical properties of different lots over time, uncertainties of oxidizer and additives decay in slurry and their adjustments with time, stringent LPC and mean PSD specifications of slurries, tighter filtration requirements, and detection and removal of large particles at very small concentrations, and not clearly defined newer slurries, requiring fine tuning for the specific processes.
- Common slurry health monitoring parameters include: LPC (≥0.56 or 1.01 μm), mean particle size, PSD, zeta potential, pH, conductivity,

viscosity, refractive index, total dissolved solids, wt% solids, density or specific gravity, oxidizer and additives concentration, and ionic contamination. Silica-based oxide slurries are typically monitored for particle agglomeration, wt% solids, LPC and PSD, whereas, tungsten, copper, and STI slurries are observed for abrasive settling, oxidizer level, density, and LPC.

18.3 CMP SLURRY BLENDING AND DISTRIBUTION

Rapid growth in CMP processing has increased the demand on CMP slurry blending and delivery systems. These systems employ a pumping device to recirculate slurry in the global distribution loop. Silica-based oxide CMP slurries are known to agglomerate and generate defect-causing large particles ($>1\ \mu m$ size) when subjected to repeated shearing during pump handling. Alternative systems (e.g., VPDS) can reduce pump cycling and thus the effects of shear [5]. An example of a pump-based system is the pump (pressure)-pressure-dispense system (PPDS), which uses bellows pumps for blending and humidified N_2 for dispense [21]. In this blending configuration, the number of slurry passes through the pump are kept to an absolute minimum before the slurry is transferred to the dispense module. The distribution loop is then supplied with slurry as needed, using the dispense vessels under humidified nitrogen pressure. In a silica-based oxide CMP slurry handling study [21], PPDS did not generate significant number of large particles during its slurry handling operation over an extended period of time, similar to the results seen in VPDS.

In Figs. 18.5–18.8, the PSD of silica slurries illustrate distinct, developing characteristics (increase in LPC accompanied with mean particle size and a rightward shift of PSD) as a result of repeated shearing during pump recirculation [15]. Unlike most silica-based oxide slurries, alumina- and ceria-based slurries do not seem to generate the same level of large particles due to pump handling over extended periods of time [18]. The large soft slurry flakes in some of these slurries are believed to break into smaller particles as a result of repeated pump shearing. Metal CMP slurries in particular present challenges for maintaining oxidizer concentration, managing suspended solids, and providing reliable metrics to monitor slurry quality. On demand, continuous blending replenishes slurry blend in the delivery system as the slurry is consumed by the CMP tools, providing a stable supply of fresh slurry and maintaining a consistent concentration of oxidizer. Continuous blending eliminates the need for daytank and mix tanks, which can be areas of slurry drying, agglomeration, and aging. Data from bench scale laboratory studies, R&D scale pilot testing, and production scale field trials show that continuous blending can be extremely stable and reliable while creating slurry blends with relative accuracy of 1% or better [21].

18.3.1 Slurry Delivery Technologies

CMP slurries can be blended and distributed employing various approaches including the following.

(i) *Manual blending and distribution* has limitations of scalability and blend consistency, contamination risks, safety risks, higher traffic in the polishing area, and abrasive settling and pot-life, making it impractical for many applications.

(ii) *Integrated bulk slurry distribution* employs pumping from remote centralized location and has the features of consistent slurry quality to all point-of-use (POU) locations, slurry health monitoring and control, system integration with the fab process tools, on-demand blending and delivery, and round the clock availability of slurry through the double contained global distribution loop.

(iii) *VPDS* is based on the principle of volumetric blending and with characteristics of minimal handling shear on slurry. This method is ideal for shear-sensitive silica slurries, but may have limitations on slurries with sensitive volatile or decomposing components (e.g., NH_4OH) which may be affected by repeated vacuum application during handling over an extended period of time.

(iv) *PPDS* is based on metering pump principle, employing diaphragm, bellows, centrifugal, or vacuum-pressure-dispense pumps, and may be used for shear sensitive slurries with minimal recirculation during slurry usage and replenishment, with dynamic monitoring and control of slurry blend ratio and volatile/decomposing components employing integrated slurry metrology.

(v) *POU blending* uses a weight or pump based blender at the CMP tool. It is most suitable for limited pot-life slurry blends and can provide very consistent blend to the tool, but may have limitations in monitoring and controlling the slurry blend quality just before consumption.

18.3.2 Continuous (On-Demand) Slurry Dispense and Metrology

CMP slurry delivery systems follow design similar to bulk chemical dispensing tools. Modifications such as wet nitrogen blankets, circulating loops, and densitometers are typically added to account for slurry's specific properties. However, as CMP processes become more critical to production, slurry delivery should more fully account for special requirements in instrumentation to assure mix accuracy and control of "time" sensitive components (e.g., hydrogen peroxide, ammonium hydroxide, etc.) and system integration. CMP of metal layers presents a specific challenge to maintain a consistent oxidizer concentration and to accommodate alumina-based slurries, which tend to stratify more readily [23]. On-demand blending may be used to most effectively distribute such slurries. In this approach, slurry is mixed in small volumes as it is consumed by the CMP tools. In a full production environment, the constant use of slurry provides a stable average residence time for slurry in the system, minimizing the impact of "pot-life" issues. On-demand mixing eliminates the need for external daytanks where the slurry has a large exposed surface area that is subject to drying and the slurry can stratify. Settling can be prevented through mechanical mixing; however, mechanical mixing is likely to cause fluid shearing, leading to particle agglomerations.

A pump-based PPDS may employ the following components: precision bellows pumps, which produce high volume with repeatable stroke volumes; a multiple nozzle contactor; an in-line static mixer; and a multiple-port injection system into a mix vessel. Multiple chemical constituents are metered and delivered to the contactor by the bellows pumps. Previously mixed slurry is combined with raw constituents in this nozzle and the static mixer so that chemical and pH shock is minimized and rapid homogenization is achieved. Recirculation through the mix vessel averages blend variations, producing a stable mixture. Blending and transfer of chemical can occur simultaneously and continuously. Large mix tanks are usually not required, so held volume is minimized. Some continuous dispense technologies use a set of three pressure vessels, which function in an alternating fashion in dispense, return, and fill modes. Typically one vessel delivers the slurry to the loop at controlled pressure, one vessel accepts slurry from the dispense loop at controlled pressure, and a third vessel receives fresh slurry from the blend system. Pressure in each vessel is controlled via gas regulators, based on pressure measured at one or more locations in the dispense loop (valve boxes). Dispense and return pressure are dynamically controlled such that dispense pressures at the tools remain nearly constant and flow rates are maintained within the desired range. This control automatically compensates for the changes in slurry flow to the tools.

A typical slurry delivery tool can be a redundant five module system capable of supplying up to 19 lpm (5 gpm) of slurry to up to 20 CMP tools [21]. The installation consists of two blend modules with on-demand mixing, one control module containing the instrumentation and wet nitrogen generation system and two dispense modules using pressurized wet N_2 for distribution. The slurry blend quality is verified by the control module using a conductivity sensor, densitometer, pH sensor, or other analytical instruments to achieve better than 1% relative accuracy. In tungsten and copper CMP, H_2O_2 concentration directly impacts the removal rate. H_2O_2 may not be stable in slurry distribution loop over time, because it may degrade to water and oxygen in slurries. Bench-top blend testing data indicate that sensor response to variations in H_2O_2 concentration in DI water is nearly identical to those seen for the slurry. Therefore, density and conductivity cannot distinguish between water and H_2O_2 as the diluting liquid. Hence, the desired method for determining H_2O_2 concentration in the slurry is titration. Similarly, for silica-based oxide slurries, parameters such as density, conductivity, and pH may be able to provide an indication of the slurry quality in the system. In most tools, PSD and LPC measurements are conducted periodically with the samples collected from different blend and dispense module of the slurry delivery system.

18.3.3 Slurry Turnovers in Fab Distribution

In a typical production-scale fab, slurry circulates through a bulk slurry dispense system and fab piping approximately 100 times before being consumed. This number is based on usual fab slurry consumption, storage tank size, and flow rate through a global distribution loop. Storage tank size is 800 l; slurry consumption rate is 200 l/day (one 200-l drum/day). Therefore, slurry life is 4 days; global loop flow rate is 13 lpm (~3.4 gpm);

$$\begin{aligned}\text{Number of turnovers} &= (\text{flow(lpm)}/\text{consumption rate}) \times 1440\,\text{min/day}\\ &= (13\,\text{lpm}/200\,\text{lpd}) \times 1440\,\text{min/day}\\ &= 94\\ &= \sim 100 \text{ before consumption.}\end{aligned}$$

18.3.4 Slurry Abrasive Settling and Dispersion

CMP slurry abrasive particles should be uniformly suspended during blending and distribution for consistent polishing process. A minimum flow velocity must be maintained in the slurry global loop to keep abrasive particles well dispersed. Settling rate (SR) information for slurries may be used to specify the agitation requirements for the storage tank and daytanks as well as the required minimum flow velocity (RMFV) for the slurry in the global loop. In the present study, a number of oxide, tungsten, copper, and STI CMP slurries were analyzed using a liquid dispersion optical analyzer to measure SR in terms of changes in turbidity (transmission and back scattering signals) of various layers in the slurry sample. RMFV was obtained from the slurry handling tests in a VPDS or pump-based delivery system using a simulated global loop. The transmission and back scattering raw data obtained using the above method were analyzed employing the absolute thickness and mean value graphs to study the settling behavior of the slurries. The settling rate could be obtained from the absolute thickness plots. In this section, settling and redispersion characteristics related data of twelve commercially available CMP slurries will be evaluated. The analyzed slurries (see Table 18.2) include: five silica slurries (silica-1 to silica-5), four alumina slurries (alumina-1 to alumina-4), one ceria slurry (ceria-1), and two slurries with silica and alumina abrasives (Si&Al-l and Si&Al-2). Undiluted slurries (abrasive component or one-component slurries) and slurry blends (abrasive and additive mix) were analyzed to determine the changes in the settling characteristics of mixtures as compared only to abrasive component.

18.3.4.1 Slurry Settling Rate Quantification The settling rate measurements were conducted using a Beckman Coulter QuickSCAN™ Liquid Dispersion

TABLE 18.2 Required Minimum Flow Velocity (RMFV) of Slurry Blend in the Global Loop, Abrasive Settling Rate (SR), and Stokes Number (St) for Tested CMP Slurries. SR Values Depend on Selected Transmission and Back-Scattering Zones. (A) = Slurry Abrasive Component Only and (B) = Slurry Blend.

Slurry	Abrasive solids, wt%	SR, mm/h	RMFV, ft/s	Stokes Number, $St \times 10^4$
Silica-1	25	< 1	0.5	0.3
Silica-2	13	< 1	0.5	1.3
Silica-3	30	< 1	0.5	0.2
Silica-4	5	< 1	0.5	1.2
Silica-5	20	< 1 (A), < 1 (B)	0.5	1.1
Alumina-1	3	15 (A), 15 (B)	2.0	4400
Alumina-2	3	15	2.0	4400
Alumina-3	7	16	2.5	2.5
Alumina-4	8	12	1.5	53.7
Ceria-1	3.5	14–30 (A)	2.5	240
Si&Al-1	11	< 1 (A), 13 (B)	2.0	2.8
Si&Al-2	20	< 1 (A), 9 (B)	1.5	28.3

Optical Analyzer. This technique allowed detection of minute concentration and particle-size variations in the slurry sample earlier than observation by the naked eye [24]. The slurry samples were analyzed in a cylindrical glass measurement cell. The detection was composed of a pulsed near-infrared light source ($\Lambda = 850$ nm) and two synchronous detectors. The transmission detector received the light that passed through the sample (0°), while the back scattering detector received the light back scattered by the sample (135°). The detector head scanned the entire length of the sample (~65 mm) vertically, acquiring transmission and back scattering data every 40 μm, or 1625 acquisitions in transmission and in back scattering per scan.

The above technique could be used for the slurry dispersion samples ranging from slightly turbid to concentrated and opaque (0 to 60 wt% solids) with particle size ranging from 0.1–1000 μm and without prior dilution. If a slurry blend remains stable over time, the transmission and back scattering graphs do not change and different time plots superimpose on the zero-hour reference line. Progressive changes in the graphs indicate mixture destabilization. An increase in transmission and/or decrease in back scattering values in the top layers (for example) of the slurry sample would illustrate the abrasive particle migration to the bottom of the sample, as a result of settling. Insignificant changes in the overall transmission and back scattering values along the entire length of the graph suggest insignificant settling during the test. As discussed earlier, RMFV information was collected from slurry handling tests in a VPDS for slurry delivery equipped with a 200-foot long simulated global distribution loop.

18.3.4.2 Settling Behavior of Different Abrasive CMP Slurries Representative results of transmission and back scattering (the profile graphs), and absolute

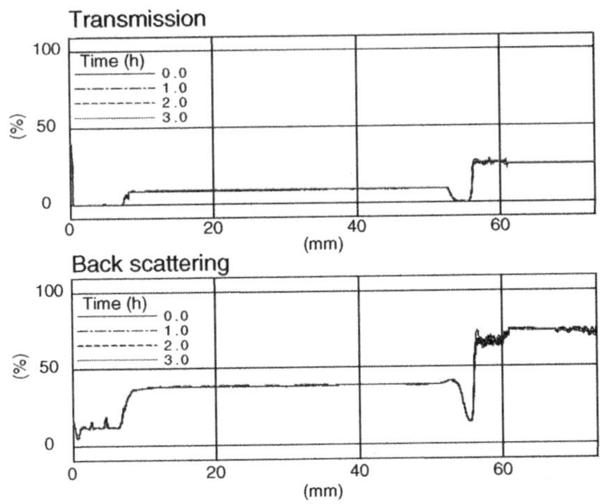

FIGURE 18.9 Transmission and back scattering plots for silica-1 oxide CMP slurry (silica particles).

thickness (the kinetic graphs) are presented in Figs. 18.9–18.11, for silica-1, alumina-1, and ceria-1 slurries, respectively. In these experiments, the data acquisitions were executed once every minute in an automatic mode. The transmission and back scattering profiles as a function of the sample height were measured once every minute for 200 min for each sample. The transmission and back scattering variations with time were plotted starting at zero time (the beginning of the experiment). In Figs. 18.9, 18.10a and 18.11, profiles are presented for 0, 1, 2, and 3 h, for clarity. The slurry samples were filled up to a height of ~55 mm in the glass tube. In these figures, the sample height from 0 to 7 mm (approximately) represents the opaque plug region of glass tube bottom. The time shown is the time elapsed from the beginning of the settling tests.

The transmission and back scattering plots for silica-1 in Figs. 18.9 show insignificant settling during a 200-min test. A similar behavior was also observed for the other silica slurries: silica-2 to silica-5. Figures 18.10a and 18.11 for alumina-1 and ceria-1, respectively, illustrate the sediment layer formation (a progressive increase in the back scattering values) at the sample bottoms and a clarification phenomenon (a decrease in the back scattering values) at the slurry sample top layer. The settling behavior of alumina-2 to alumina-4 slurries was very similar to alumina-1 results seen in Fig. 18.10a. The absolute thickness profiles in Figs. 18.10b illustrate the variations in layer thickness as a function of time for alumina-1 slurry. SR values for different slurries were quantified from the slope of the curves in similar plots. These graphs show that the settling rate of CMP slurries change over time. The data were used to calculate transmission and back scattering mean value variation in the selected zone. These plots illustrate the average clarification time at a specified height in the slurry sample.

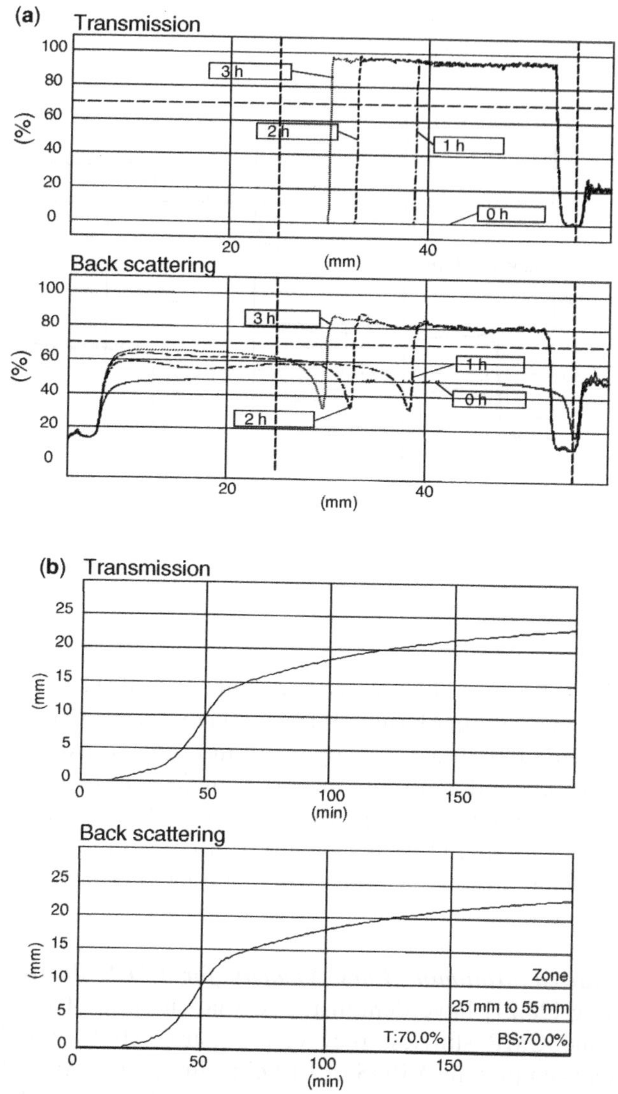

FIGURE 18.10 (**a**) Transmission and back scattering plots for alumina-1 copper CMP slurry (alumina particles); (**b**) absolute thickness plots for alumina-1 copper CMP slurry (alumina particles).

Alumina- and silica-abrasive slurries display different settling characteristics for the blend as compared only to abrasive component. In most alumina-based slurries, normal blends (created by diluting abrasive with DI water and/or chemical additive component) settled more quickly as compared only to the abrasive component, whereas in silica-based slurries the settling behavior was

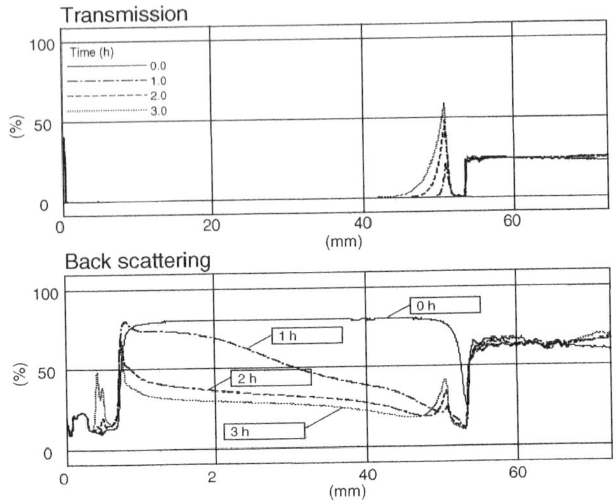

FIGURE 18.11 Transmission and back scattering plots for ceria-1 STI CMP slurry (ceria particles).

nearly the same for the abrasive component as well as the normal blend. However, in some alumina-based slurries, where chemical reactions seem to take place between the abrasive component and the additive, the slurry blend settled more quickly and at a more uniform rate as compared only to abrasive component. The estimated average SR data for different slurries from above experiments were compared with the empirically determined RMFV (to keep abrasives suspended and wt% solids within target specification) in the slurry delivery system and simulated distribution loop.

18.3.4.3 Required Minimum Flow Velocity for CMP Slurries Results of slurry handling [20,21] have demonstrated that in spite of the quick settling nature of some CMP slurries, they can be successfully handled by slurry delivery systems employing VPDS or PPDS technologies. Appropriate RMFV must be maintained in the delivery system and the global loop. SR and RMFV data for slurries in Table 18.2 suggest a possible correlation between these two parameters. This empirical relationship may be approximated as RMFV $\approx C + \phi$ SR, where ϕ is the correlation factor ≈ 0.1; constant $C \approx 0.5$ is the minimum flow velocity (in ft/s) recommended for the nonsettling slurries; and SR and RMFV are in mm/h and ft/s, respectively. The SR data, obtained by analysis of a small sample (< 100 ml) in a short time, may be used to estimate RMFV for CMP slurries without conducting time-consuming slurry handling tests in slurry delivery system and distribution loop.

A theoretical parameter to compare the abrasive settling of slurries and the response time of abrasive particles relative to response time of the fluid [25] may be calculated in terms of Stokes number (St). St is defined as $(\rho D^2 u)/(\mu L)$, where ρ is density of abrasive particles, D is mean aggregate diameter, u is relative pad-wafer linear velocity, μ is slurry viscosity, and L is hydraulic diameter of the groove. Results of St for different slurries are also included in Table 18.2. Particles with St > 0.1 have inertia and may not follow the flow field and settle down. According to this computation, only alumina-1 and alumina-2 slurry abrasives have St values above threshold. The majority of silica-based CMP slurries demonstrate slow settling behavior (settling rate < 1 mm/h), whereas alumina- and ceria-based slurries usually have quick settling characteristics (settling rate ≈ up to 30 mm/h).

Results of present study suggest that most silica abrasive slurries can be kept dispersed by maintaining a RMFV value close to 0.5 ft/s, whereas alumina and ceria slurries may require RMFV = 1.5–2.5 ft/s, depending upon the abrasive concentration and type, and the slurry chemical constituents. It is important to consider the effects of shearing on abrasives under extended handling, especially in shear-sensitive silica abrasive slurries [15,18]. These slurries should be handled optimally and in some low usage rate applications only periodically and during the slurry usage and replenishment cycle, to limit the cumulative effects of handling and related particle agglomeration behavior. These results illustrate the applicability of a quick and economical method for settling rate measurement of CMP slurry abrasives and also show an empirical correlation between SR and RMFV.

18.3.5 Summary

Key observations may be summarized as below.

- CMP slurry blending and delivery can be performed using manual mixing methods in small R&D applications, whereas integrated bulk slurry systems are used for fab global slurry distribution loops. Typical bulk slurry delivery systems employ pumping from a remote centralized location, provide consistent slurry quality to all POU, and have slurry metrology and feedback control ability. These tools can be integrated to the fab process and offer on-demand blending and delivery using a double-contained global distribution loop.
- VPDS is designed on the principle of volumetric blending. These systems cause minimal handling shear on slurry and hence are ideal for shear-sensitive silica slurries. However, special care should be exercised in handling slurries with volatile or decomposing components, which may be affected by repeated vacuum application.
- PPDS works on the principle of metering pumps and employs diaphragm, bellows, centrifugal, or vacuum-pressure-dispense pumps. These systems may be used for slurries including shear-sensitive slurries and cause minimal recirculation during slurry usage and replenishment.

Such systems may provide dynamic control of slurry blend ratio and integrated slurry metrology. POU blending is most suitable for limited pot-life slurry blends and can provide consistent blend over time, but may have limitations in metrology and advance warning on slurry health changes.

- Slurry delivery systems are similar in design to bulk chemical dispensing systems and have features to suit specific requirements of slurry metrology, blending, and handling. Modifications, such as wet nitrogen blankets in the day tank and supply containers, global circulating loops, and instrumentation like densitometers, and pH and conductivity meters are typically added to account for specific slurry needs. In on-demand blending, the slurry is mixed in small volumes as it is consumed by the CMP tools. In a production environment, the constant use of slurry provides a stable average residence time for slurry in the system, minimizing the impact of pot-life related issues.
- Present results suggest that most of the silica slurry abrasives can be kept in dispersion by maintaining a RMFV value of ~0.5 ft/s, whereas alumina and ceria slurries require 1.5 to 2.5 ft/s, depending on abrasive and chemical composition of the slurry. Next generation slurry delivery systems should be able to measure and control abrasive particle and chemical characteristics (to ensure mix accuracy and control of "time" sensitive components such as hydrogen peroxide, ammonium hydroxide, etc.) of the slurry on a real-time basis and consistently provide good slurry at the CMP tools.

18.4 CMP SLURRY FILTRATION

Filtration has gone from being nonexistent in the CMP process to now using submicron level filtration at the slurry manufacturing facilities, as well as during blending, distribution, and POU applications. CMP slurry filtration is a fine line between taking out the oversized particles and leaving the bulk working particles and other active components that are important for efficient planarization of wafer. Not many polishers have incorporated filtration modules in their basic designs and still it is up to the end users to install, characterize, and implement filtration for specific slurries. To take maximum advantage of the POU filtration, the filter needs to be the last major component the slurry comes in contact with before the slurry is dispensed. The filter's resistance increases over its lifetime, resulting in a drop in the flow rate. Also, it is difficult to identify the timing for a filter change so that the lack or significant reduction in flow does not impact the process. An arrangement employing pump, flowmeter, control-valve, pressure transducers, and feedback control mechanism from flowmeter may be used to achieve stable flow over filter lifetime.

18.4.1 Slurry Filtration Methodology

The optimal CMP slurry filter will be cost effective, compact, and have adequate lifetime and optimum particle removal, with the ability of a reasonably quick and clean change-out. Depth filtration media is the most prevalent type of filter for slurry applications with capability of handling high particle loading and solids concentration. Most slurries use 0.03–0.20 μm size abrasive particles at typical concentrations of 0.3–30 wt% solids. In CMP slurries there are enormous numbers of particles in the desired mean PSD range. If a filter starts removing even a small percentage of particles from this mean distribution, the large quantity of particles will progressively clog the filter, causing it to fail quickly. Filter clogging in CMP slurries [26] may take place by complete plugging (e.g., in typical silica-based slurries) or cake formation (e.g., in ceria and alumina slurries) or a combination of the two mechanisms depending on the slurry formulation. The challenge is to find a filter that will remove the unwanted large particles without removal of particles in the mean distribution range and provide an acceptable lifetime and pressure drop (Δp). Slurry filtration locations include slurry drum or pail intake, post-dilution or mixing, global distribution loop, and POU. The most commonly used options are POU and global distribution loop.

The concept of filters working as strainers, only by size exclusion, is simplistic. There are several widely recognized mechanisms: inertial impaction, interception, adsorption or adhesion, diffusion, and gravitational settling. Inertial impaction is when a flow stream suddenly changes direction and the particle's inertia prevents it from following the flow, causing it to impact on the filter structure and become trapped. Interception is when a flow stream passes close enough to the filter structure that a particle traveling in that flow stream hits that structure and becomes trapped. Adsorption/adhesion can be any sort of surface chemistry causing an attraction to passing particles, pulling them across flow streamlines to a structure and trapping them. All these effects work to prevent filters from providing a simple step function in retention, as will be discussed at a later point. Fractionation is the process of removing particles larger than a certain size while allowing smaller particles to pass. CMP filtration is a fractionation step, as a range of working particles are required to pass through the filter unchanged, while stopping a significant number of particles greater than a specified size. For example, in a silica slurry the mean particle concentration was $> 10^{15}$ particles/ml, whereas the cumulative counts of the large particles $\geq$1.10 μm was $\sim 10^5$ particles/ml. Similarly, the cumulative counts of the large particle $\geq$1.10 μm in a new STI CMP ceria slurry was $\sim 8 \times 10^8$ particles/ml. The filtration challenge is increased when the size difference between the particles to be passed and the particles to be retained is small (less than 1 order of magnitude). Fractionation becomes difficult due to the retention characteristics of all filtration media.

18.4.2 Filter Design Consideration

The relationship between retention and particle size for any media is not a simple step function. All media show a gradual transition between retention to passage over a range of particle sizes forming an S-shaped curve (Fig. 18.12a). The sharpness of the curve is a function of filter media structure. The range of this transition limits close fractionation. Using the CMP slurry example above, the working particle range is from 30 to 200 nm but there is a desire to control or remove all particles as small as 500 nm given the drive to thinner polished layers and more complex chip designs. This sort of fractionation becomes very difficult when the transition range of most media is over several microns. Media that will allow complete passage at 200 nm, typically does not reach retention above 80% until the particle size is well over 2000 nm. The sharpness of a media's retention curve (Fig. 18.12a), whether membrane or fiber based, can be improved by layering the media. In the silica slurry example discussed earlier, a fibrous media that allows full passage of 200 nm particles may not reach 80% retention until above 2 μm in a single layer. But by adding layers of the same media, the 200 nm particles will still substantially pass, but the larger particles will have ever increasing odds of capture. This has the effect of sharpening the retention curve. Further, layering (Fig. 18.12b) can also help in extending filter lifetime and provide more control in retention design of a filter. Filter design with layers of differing retention, more open layers at the beginning (near inlet) and tighter layers near the outlet (see Fig. 18.12b), will allow equivalent final retention with better utilization of all layers and much longer filter lifetime. Removal efficiency can be defined as the fraction of particles removed per layer [27]. Mathematically,

$$\lambda = -\frac{\delta C}{C}\frac{1}{\delta L} \tag{18.1}$$

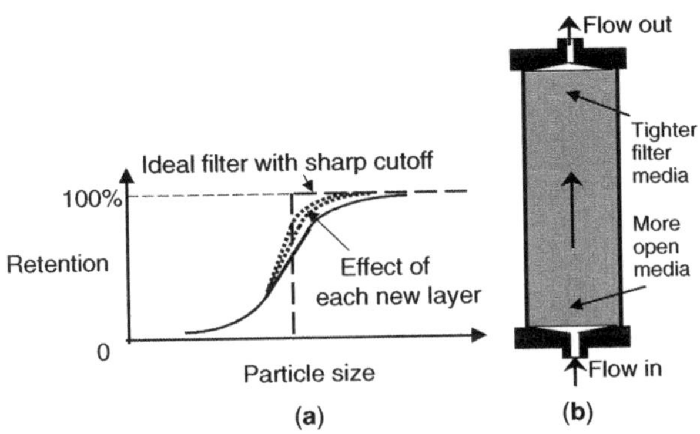

FIGURE 18.12 Effect of multiple layer depth media on filter cutoff behavior and schematic arrangement in a POU filter design.

where $-\delta C$ is the reduction in concentration of particles passing through a layer δL. This equation can be rearranged as

$$-\frac{\delta C}{\delta L} = \lambda C \qquad (18.2)$$

where λ is a retention coefficient. Integration of Equation 18.2, at the commencement of filtration ($t = 0$), when the filter is clean and therefore uniform, gives

$$\ln\frac{C}{C_0} = -\lambda L \qquad (18.3)$$

where C_0 is the inlet concentration of particles being filtered. The consequences of above Equations 18.1–18.3 show that the number of particles in a filtrate suspension declines logarithmically with the depth media thickness. However, the number of particles retained in each layer is not uniform, as the top or initial layers contain most of the retained material and the lower or latter layers contain very little. Therefore, a constant retention filter does not become uniformly clogged. This rapid clogging of the initial layers leads to a shortened filter life, with the latter layers essentially unused. The same filter design with layers of differing retention, more open layers at the beginning (near inlet) and tighter layers near the outlet (see Fig. 18.12b), will allow equivalent final retention with better utilization of all layers and much longer filter lifetime. Since each layer also contributes to the pressure drop, there is a careful balance required between filter lifetime, pressure drop, level of retention, and sharpness of the retention curve for optimal performance.

In the submicron range of particle sizes, there are at least two media types to consider: depth and microporous membrane. Typically, depth media is melt-blown fibrous nonwoven type (Fig. 18.13). Membranes are basically cast or expanded microporous films. Depth filters with a wrapped or pleated construction are used for slurry filtration (see Fig. 18.14a and b). Membrane

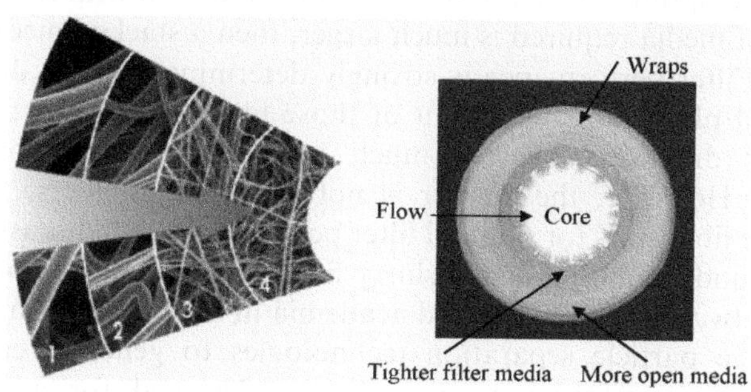

FIGURE 18.13 Typical graded-density depth media for CMP slurry filtration.

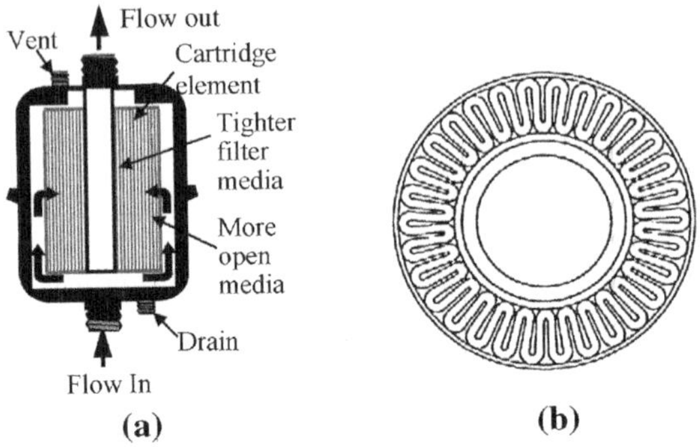

FIGURE 18.14 Typical depth filter housing arrangement and a pleated depth filter configuration.

filters are usually pleated in construction and used for particle/gel removal from ultrapure chemicals, UPW, and photoresists.

When CMP was first emerging as an important process, microporous membranes were tested in a tangential flow filtration (TFF) mode. In typical CMP slurries, TFF was found to be unacceptable due to the high solids content. Also, there was insufficient control of the retentate build-up on the membrane face, called concentration polarization. TFF works best with applications that have lower solids and a greater spread between retained and passed species. The excessive polarization we observed with high solids silica slurries prevented useful fractionation. TFF may be useful for the newer slurries with low solids content. Three critical questions to any filter design in slurry applications are (i) how sharp can the filtration curve be to enable clean fractionation, (ii) how is retained material handled to control filter life, and (iii) how does media selection affect the above two points.

One of the first steps in filter design is determining the depth of media required to stop the undesired large particles. If that depth is relatively small, then the optimal filter will maximize the surface area of this media combination by using a variation of a pleated design (Fig. 18.14b). If that depth is somewhat larger, then the optimal filter will have a wrapped design with less surface area. If the depth of media required is much larger, then a stacked media design can be effective. Filter performance is strongly determined by the distribution of fiber sizes and physical arrangement of those fibers within the media [28,29]. Smaller fiber diameters provide much improved retention/pressure drop relationships. However, the answer is not just smaller fibers, but a proper balance of the fiber sizes for optimal filter performance. As discussed earlier, to improve our understanding of the slurry filtration, more collaborative studies are needed between the industry and academia in the filter media development and alternative particle separation technologies to generate comprehensive data set for high-performance new slurries designed for advanced CMP processes.

18.4.3 Slurry Filter Characterization

Empirical CMP slurry handling and filter characterization studies are useful in identification of optimum filters for specific slurries [13–16]. CMP slurry characterization may involve measurements of conductivity, pH, density, wt% solids, oxidizer and organic concentrations, viscosity, refractive index, mean particle size, PSD, LPC, abrasive settling rate, and zeta potential. Combining this information with past experience and filtration experiments data with similar property slurries may provide insight about the potential filter type and rating that may provide targeted retention of large particles. However, due to complexities of newer slurries with very unique abrasives and multitude of background chemistries, and tighter filtration needs, it is usually essential to empirically determine optimum set of filters (e.g., in POU and global distribution loop locations) for specific slurry application using a laboratory setup (see Fig. 18.15).

CMP slurry filtration evaluations may include (i) large-particle retention behavior, pressure drop and filter lifetime tests, (ii) extensively handled slurry effects on the POU and global distribution loop filtration performance, (iii) analysis of gel content in silica-based slurries or tail-end large-particle mass concentration in other slurries, employing analytical membranes, semiquantitative filtration method for particle loading, and scanning electron microscopy (SEM) and environmental SEM (ESEM), (iv) filterability and gel or particle content quantification through slurry filtration, filtration time monitoring and weight gain, (v) used filter analysis to quantify extent of filter plugging by flow/

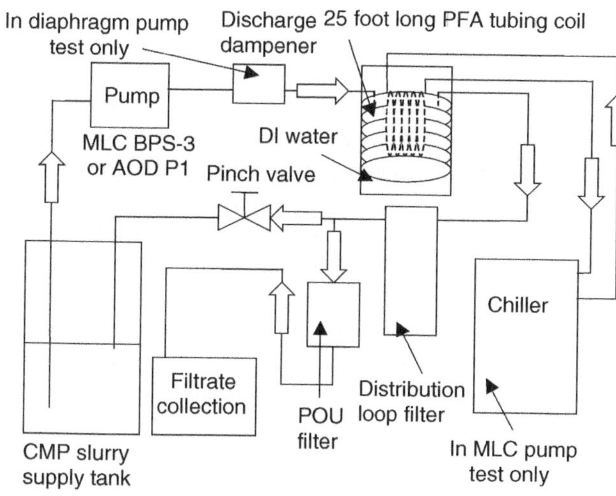

FIGURE 18.15 Test setup for CMP slurry handling and filter characterization. Experiments can be performed employing an air operated diaphragm (AOD) pump or a magnetically levitated centrifugal (MLC) pump.

Δp, weight gain and ESEM, and electron dispersive spectra (EDS) for species detection, and (vi) filtration subsystem development and optimization.

Filter benchmarking tests are conducted to quantify relative filtration performance of a series of filters, primarily to identify advantages and limitations of different designs for specific applications. In extensive collaborative filtration evaluations, LPC, PSD, zeta potential, wafer polishing rate, and defectivity data are collected to quantify the effects of filtration. Such multiparty joint studies may involve slurry manufacturers, fab end users, filtration solution providers, and slurry delivery system suppliers [15,30]. During the slurry LPC tests, it is very important to use the slurry particle analyzers in an optimum mode, especially for the extremely high particle concentration ceria-based STI slurries. Finally, since there are no commercially available slurry LPC analyzers in the market (capable of differentiating between the soft and hard aggregates), all significant LPC increases detected during slurry blending and delivery stages must be investigated with extreme care. It is always desirable to conduct wafer polishing and defectivity evaluation with the fresh, extensively handled, and the filtered slurry samples to generate conclusive information.

18.4.4 CMP Process and Consumable Trends and Challenges

With decreasing feature size, the CMP processes are becoming more challenging day by day with significantly improved planarity specifications in Cu, STI, poly-Si, and ultra-low-*k* CMP, and introduction of new devices, slurries and pad materials, revised polishing parameters (e.g., much lower pressures on pad in ultra-low-*k* CMP), and more effective and consistence post CMP cleaning (e.g., employing supercritical CO_2, molded-through-the-core PVA scrubbing brushes [31], etc.) processes. According to ITRS 2005, a total metal loss budget (at 10% of total metal) of 120 Å at 65 nm reduces to 60 Å at 60 nm at the 32-nm node [32]. When CMP processes move to lower mechanical stress methods, an improvement in CMP equipment, slurries, filters, CMP rings, pads, post-CMP clean brushes, and other CMP consumables would be required to achieve required process performance.

Photomask generation dictates how quickly new IC products can be introduced. For example, in an advanced semiconductor fab, 130 nm was qualified in 3Q02, 90 nm in 4Q04, and 65 nm in 2Q06, with 45 nm currently under development. This fab was able to change to 130-nm copper and 300-mm wafers together in July 2002. Another state-of-the-art fab worked on ramping 130 and 90 nm processes simultaneously, completing these tasks in August and December of 2004, respectively. Within the next 3 months, this fab went ahead from prototyping four new 90 nm products to their full scale production. The fab expects to ramp to 65 nm in 2007, and planning for 45 nm readiness by the third quarter of 2007. The above leading fabs have well-defined goals and unique systems and functionalities in place such as cross-functional process loop teams, sameness statistic within their facilities, zero wasted resources, zero illnesses and zero injuries in their operations, automated materials handling system and computer-integrated manufacturing that transport wafers seamlessly, advanced process control (APC), fault detection and classification, as well as recipe management.

A number of above fabs are part of common platform initiatives, which brings unprecedented freedom of choice and cross-foundry cooperation to customers, significantly improving the rate of technology evolution and product development. The first 300-mm wafer pilot line began production as a joint venture between Siemens and Motorola in Dresden, Germany. According to a recent estimate from Strategic Marketing Associates, by the end of 2006, there will be as many as seventy 300-mm fabs online, with a total capacity (when fully equipped) of ~1.6 million wafers per month. An additional twenty-one new 300-mm fabs are expected to come online in 2007, adding ~0.67 million wafers a month to the industry's capacity. Further, between 2008 and 2013, the industry should have an additional thirty-nine 300-mm fabs and the actual number may be even higher. Samsung, the industry's largest memory manufacturer, for example, has announced plans for as many as 11 new fabs that are scheduled to come online through 2013.

Controlling concentration of chemicals [33] at the process tool by real-time continuous data output (e.g., with <1 s response time) POU sensors (e.g., compact optical fluidic cell) is becoming increasingly important during chemical blending and delivery (CMP slurry, chemical baths, spiking applications in batch–wafer processes). Such data may be correlated with wafer-surface metrology data. In recent years, end users and tool suppliers have become more interested in evaluation of CMP disruptive technologies based on fixed abrasive, electrolytic CMP or ECMP [34], and electropolishing, requiring reduced CMP processing and hence slurry filtration. However, it appears that CMP will remain the enabling process for smaller feature IC device fabrication in the foreseeable future.

Slurry filtration needs are changing with rapid growth in Cu, STI, and low-*k* slurries from numerous new suppliers. Also, new manufacturing processes for MEMS industry have very different specifications for CMP and hence slurry filtration [35]. MEMS devices are much larger in size (1–100 μm or even in mm) as compared to most advanced microelectronics (<100 nm), and require much thicker material layers to be removed by CMP. To achieve higher removal rates for MEMS, mechanical or chemical aspects such as using larger, harder, and/or sharper abrasive particles or tailored polishing pads, increasing the slurry temperature and adding accelerating additives can be employed. This would result in very different CMP slurry large-particle management and filtration needs for the MEMS product fabrication. Also, process control and metrology to support new more complex CMP have gained importance in the recent years. Further, *in situ*, extended *in situ*, integrated metrology and its contribution to process control, and within wafer process control and recipe adjustments per wafer are expected to be even more critical for next-generation devices [36].

Key challenges in copper CMP [3] include chip uniformity (e.g., dishing and erosion), wafer uniformity (run-to-run and within wafer), minimal wafer edge exclusion (for higher yield), equipment and consumables evolution, material diversity, and improved metrology. Abrasive-free slurries or solutions may help in minimizing dishing and erosion as well as promoting more uniform slurry flow and pad temperature, resulting in better within wafer uniformity. Defect reduction strategies may include slurry abrasive surface charge modifications, corrosion control (e.g., using BTA), and process optimization. Process optimization variables may include slurry particle and chemistry, pad characteristics (e.g., downpressure and linear velocity), feature size and density, microscale parameters (e.g., pad contact area and pressure), particles on pad, chemical concentration, and nanoscale interactions (chemomechanical–surface layer formation, thickness uniformity, rate of formation, and chemical and mechanical–etching and mechanical removal).

There have been some efforts in slurry recycling to reduce CoO of CMP consumables. In a recent study [37], the effluent samples were characterized for pH, trace-metal levels, viscosity, specific gravity, mean aggregate particle size, and LPC ($>1.0\,\mu m$) before and after depth (melt-blown polymeric media) filtration. The study showed that the use of a recycled fumed silica slurry (recycled five times) decreased the CMP removal rate and the coefficient of friction (COF) by ~40%. A perfect relationship was observed between the removal rate and COF. It was concluded that the increase in mean aggregate particle size, which lowers the contact area between the abrasive particles and the wafer, had some impact on the removal rate data. In general, there is a stronger emphasis on slurry additives and chemical action in current CMP processing with much lower maximum defectivity performance specifications [2].

In recent years, there has also been an increasing demand for high-retention low-cost true 0.5, 0.3, and 0.2 µm filters, with stringent LPC and PSD quality control specifications. Slurry manufacturers are looking for higher flow rates, longer lifetime and higher retention, and much more repeatable filtration performance from filter-to-filter and lot-to-lot. High retention or tighter filtration is being demanded at all points, including bulk manufacturing, global loop, POU, and point of dispense (POD). End users are adopting to ceria-based STI and alumina-based Cu CMP slurries with preference for ready-to-use formulations with no blending and pot-life issues.

Slurry suppliers have introduced more complex and much smaller (nano) ceria, silica, and alumina particles. Slurry vendors and end users are continuously looking forward to more innovative abrasive-free slurries, better colloidally suspended slurries with smaller abrasives, and modified, mixed, and composite (e.g., ceria–alumina) abrasive slurries with slow settling characteristics. There is a growing need to understand the physics of filtration and filter clogging mechanisms, and to achieve better stability and longer life for the slurry filters. More collaborative studies between the vendors and end users would help in optimizing the tighter filtration needs of next-generation slurries.

Information on typical filtration products used for CMP slurry filtration follows.

- Layered high-density depth filters (Fig. 18.12) are highly efficient for broader PSD, POU slurry filtration and designed to provide much longer lifetime, lower flow, self-venting, and low hold-up volume. This design provides sharper retention curve and is most suitable for unstable slurries due to its dead-volume free design. Such filters may have nominal ratings of 0.5, 1, and 3 μm, and also have quick connectology options (e.g., 10 s changeout).
- Graded-density submicron depth filters (Fig. 18.13) are suitable for broader PSD slurry global distribution loop filtration. These filters with large surface area and low face velocity are suitable for high flow POU and POD filtration applications, are typically disposable in nature, and may have nominal ratings of 0.2, 0.3, 0.5, and 1.0 μm.
- Pleated depth media filters (Fig. 18.14b) provide effective filtration for narrow PSD slurries and have large surface area and low face velocity for high flow applications with low Δp, and typically have ratings of 0.3, 1.0, 2.0, 3.0, 5.0, 10, 20, 30, and 40 μm. Such high retention, long life, and application specific technology CMP slurry filters are used in different configurations including small capsule disposable designs. These filters can be specifically designed to meet the challenges of silica and alumina abrasives typically used in oxide and copper CMP processes. Also, these may be used with innovative housing designs enabling smaller footprint, ease of filter changeout, and lower cost of ownership.

18.4.5 Slurry Filtration-Case Studies

In this section, results of four filtration characterization tests will be presented to demonstrate the nature of filter-retention data generated in typical slurry filtration evaluations. Filter characterization studies were performed using test setup shown in Fig. 18.15. In the single-pass filtration tests, a MasterFlex® peristaltic pump with Tygon® long flex life tubing was used to feed the slurries through the filter media. Without filters in the loop, the pump passed DI water at ~500 and 535 ml/min for the Entegris Planargard CS05 and Palanargard® CMP1 test filter housings, respectively.

18.4.5.1 Silica Dispersion Single-Pass High-Retention Filtration Filtration characterization tests were performed to develop filtration recommendations for a 25 wt% solids silica dispersion for CMP slurries. This single-pass application bulk filtration target was to reduce large-particle counts 10-fold for ≥0.56-μm particles in the silica dispersion. Filtration with the Entegris Planargard CS05 filter (Figs. 18.13 and 18.14a) achieved the required reduction (90 ± 3%) in the cumulative LPC ≥ 0.56 μm, whereas filtration with the

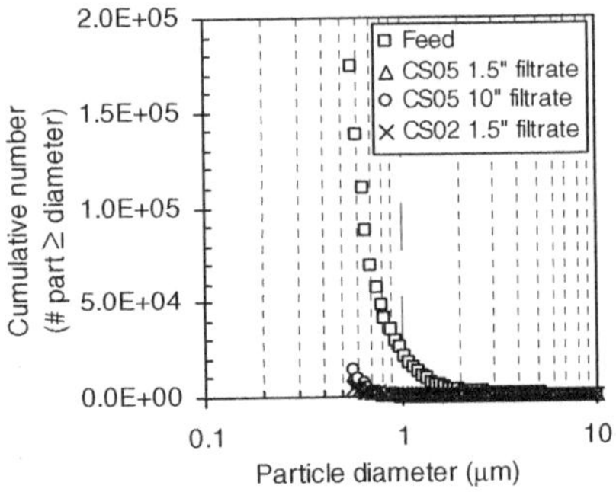

FIGURE 18.16 LPC for 0.5-μm Planargard CS05 and 0.2-μm Planargard® CS02 graded-density depth filters in 25 wt% solids silica slurry dispersion. Planargard CSO5 filter cumulative retention ~ 93% for particles ≥0.56 μm (Δp = ~28 psi, for 1-1/2″ sample, and Δp = ~ 6 psi, for 10″ sample). Planargard CSO2 filter cumulative retention ~ 97% for particles ≥0.56 μm (Δp = ~ 51 psi, for 1-1/2″ sample).

tightest CS series filter (Entegris Planargard CS02 filter) resulted in a retention of ~97% (see Fig. 18.16). Testing was performed using 1-1/2″ and 10″ sizes of Planargard® CS05 and 1-1/2″ Planargard CS02 filters.

It is a common practice to use 1-1/2″ length filters for such bench characterization of filter media large-particle retention level. As seen in Fig. 18.16, these smaller filters may have significantly higher pressure drop as compared to same filter media standard 10″ filter cartridges used in real applications. Based on limited slurry volume laboratory tests, large-scale filtration recommendations were developed. Calculations suggest that for a typical 1000 gallon dispersion single-pass large size filtration experiment, a total of 15–25 Planargard CS05 (30″ size) filters would be required, depending on the objectives of limiting maximum Δp in the system and the filter-life usage level.

A large-scale pilot run of the above silica dispersion resulted in retention levels close to those observed in small-scale filtration tests, illustrating the usefulness of in-depth laboratory filtration characterization studies. This study demonstrates that graded-density depth filters can be employed to achieve tighter retention of large particles in next-generation CMP slurry POU or bulk filtration applications. Large-particle cumulative retention, pressure drop, and flow rate data were obtained from the above filtration experiments. Measured large-particle retention levels for the 1-1/2″ and 10″ sizes of Planargard CS05 filters were similar, within experimental variability.

18.4.5.2 Silica Slurry POU and Recirculation A silica slurry (silica-A with ~12 wt% solids) was evaluated to generate filter recommendations for distribution loop and POU filtration. In the first test, the slurry was recirculated through a Planargard CMP5 5.0-μm nominal rating 10″ cartridge graded-density depth filter (Fig. 18.14a) for 5 h at 4.3 l/min. In the second test, the slurry was filtered using an Entegris Solaris®-01 (SLR1) 1.5-μm nominal rating multiple layer POU depth filter (Fig. 18.12b) at 400 ml/min, while being recirculated through the 5.0-μm loop filter. The distribution loop and POU filter's performance was evaluated by monitoring the LPC and Δp in the filter. Slurry feed and filtrate LPC were measured using a Particle Sizing Systems AccuSizer® 780 APS analyzer in the top chamber addition mode. The 5-h recirculated slurry from the above test was also filtered in single-pass tests using two graded-density depth filters: 0.50-μm Planargard CS05 and 1.0-μm Planargard CMP1.

LPC results for the POU and distribution loop filters for silica-A slurry are presented in Fig. 18.17. Figure 18.17a shows that the distribution loop LPC was stable in the 5-h run during which slurry undergoes ~170 turnovers or passes through the 5-μm Planargard® CMP5 filter (10″ length). In a real-life fab operation, the slurry typically goes through one hundred turnovers before it is consumed [15]. Figure 18.17b shows the feed and filtrate LPC for 1.5-μm Solaris -01 POU filter in removing large particles from the slurry in a single pass. The filters for these characterization tests were selected based on the silica-A slurry properties such as mean particle size and LPC, abrasive type, wt% solids, and application requirements including target retention level, flow

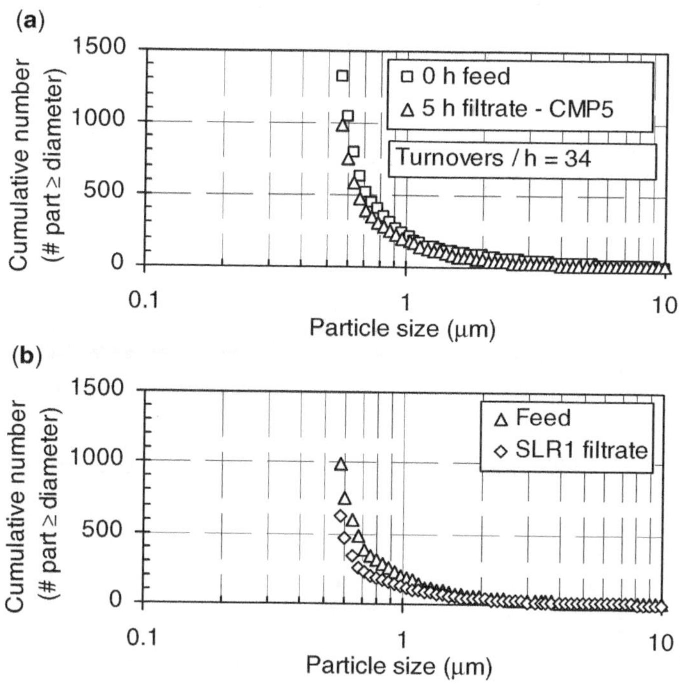

FIGURE 18.17 LPC for 5.0-μm Planargard CMP5 distribution loop depth filter (10″ length) and 1.5-μm multiple layer POU Solaris-01 (SLR1) filter in silica-A slurry. (**a**) Initial feed to distribution loop filter and filtrate from loop filter after 5-h recirculation and (**b**) 5-h recirculated slurry feed to the POU filter and filtrate from POU filter.

rate, allowable Δp, and expected filter lifetime. This study shows that the tested 1.5-μm rating POU and 5.0-μm distribution loop filters are suitable for filtration of this silica slurry. However, if the slurry is expected to see very large number of turnovers before usage, a more open 7.0-μm nominal rating Entegris Planargard® CMP7 filter may be employed for the distribution loop filtration.

The results from the filtration tests of 5-h recirculated slurry using 0.50- and 1.0-μm depth filters are presented in Fig. 18.18. The cumulative percentage retention of the particles ≥ 0.56 μm in single-pass tests using Planargard CS05, CMP1, and SLR1 were close to 55%, 38%, and 37%, respectively. Typical single-pass target retention percentage may range from 30% to 90% in CMP slurry filtration. In some slurries, more than one pass through the filter may be essential to achieve the target retention level. It is important to note that extremely high retention level usually results in limited filter lifetime and very high Δp. In most real-life slurry filtration applications, the objective is to have a reasonably tighter retention of large particles with acceptable filter lifetime. The selection of filters for specific slurry requires empirical study. Although prior experience with similar slurries may indicate the filter requirements, present results demonstrate that the performance of a filter can vary greatly across slurries due to abrasive particle morphology, LPC, PSD, wt% solids, particle settling characteristics, chemical composition, and the nature of additives and oxidizers.

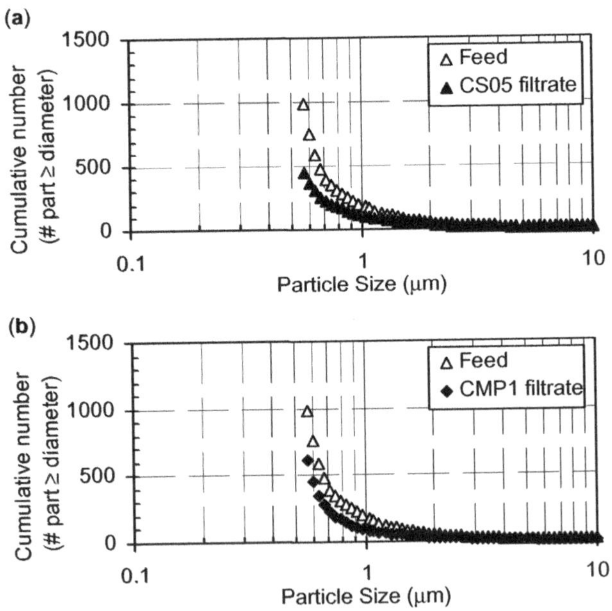

FIGURE 18.18 LPC for 0.5-μm (Planargard CS05) and 1.0-μm (Planargard CMP1) nominal rating (1-1/2″ length) depth media filters in single-pass filtration test.

18.4.5.3 Silica, Ceria, and Alumina Slurry Tighter Filtration Experiments were conducted to obtain filter retention, flow rate, and pressure drop data for Planargard CS05 and Planargard CMP1 filters (1-1/2″ length) in different abrasive slurries. Each filter was used to process a ~25 wt% solids silica-based slurry for oxide CMP (silica-1), a < 1 wt% solids ceria-based slurry for STI CMP (ceria-1) and a < 1 wt% solids alumina-based slurry for copper CMP (alumina-1). In addition, tests were conducted with the tighter Planargard CS05 media (1-1/2″ length) to filter a different ~25 wt% solids silica-based dispersion (silica-2) and another < 1 wt% solids alumina-based slurry for copper CMP (alumina-2).

Figures 18.19 and 18.20 present the filtration test results for Planargard CS05 and CMP1 filter media, respectively, with silica-1, ceria-1, and alumina-1

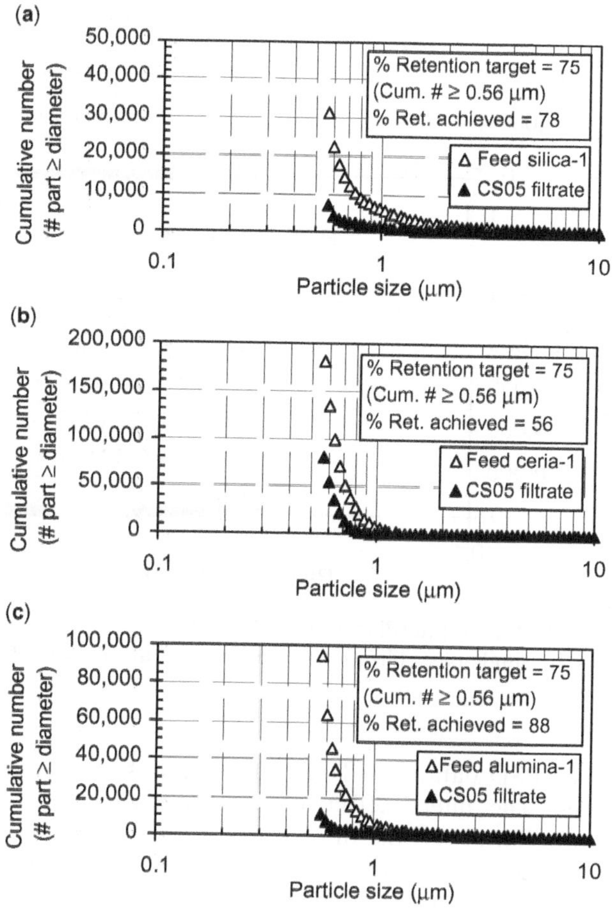

FIGURE 18.19 LPC for 0.5-μm (Planargard CS05) nominal rating depth media (1-1/2″ length) filters in single-pass filtration experiments. (**a**) silica-1, (**b**) ceria-1, and (**c**) alumina-1 slurry.

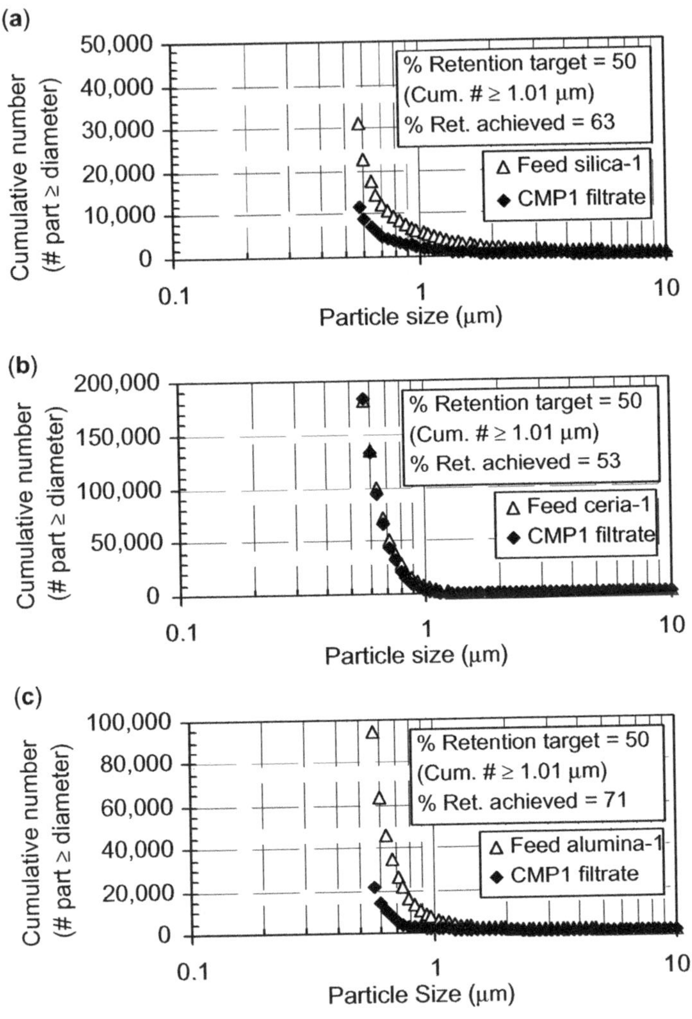

FIGURE 18.20 LPC for 1.0-μm (Planargard CMP1) nominal rating depth media (1-1/2″ length) filters in single-pass filtration experiments. (**a**) silica-1, (**b**) ceria-1, and (**c**) alumina-1 slurry.

slurries. The feed PSD data show large differences as may be expected from different slurries. The filters' effectiveness in reducing LPC demonstrates considerable variation across the three slurries. For example, the CS05 filter seems to be very effective in removing large particles from the alumina-1 slurry, but less so for the ceria-1 slurry, despite both slurries having comparable low wt% solids. The silica-1, ceria-1, and alumina-1 slurries source samples were measured to have $\sim 8 \times 10^5$, 2400×10^5, and 130×10^5 particles/ml (for size $\geq$0.56 μm), respectively, suggesting that these ceria and alumina slurries have

TABLE 18.3 Filters Show Large Slurry Dependent Variations in LPC Retention Level, Δp, and Flow Rate Data.

Slurry/Challenge Solution	Planargard CS05 (cumulative% LPC Reduction for Particles ≥0.56 μm)	Pressure Drop Δp (psi) 1-1/2″ Length Filter Media	Flow Rate (ml/min)
Silica-1	78	40	127
Ceria-1	56	12.7	469
Alumina-1	88	19	450
Silica-2	90	28	275
Alumina-2	83	14	458
PSL bead solution	62	11.8	500
Slurry/Challenge Solution	**Planargard CMP1 (cumulative% LPC Reduction for Particles ≥1.01 μm)**	**Pressure Drop Δp (psi) 1-1/2″ Length Filter Media**	**Flow Rate (ml/min)**
Silica-1	63	7.8	423
Ceria-1	53	5.2	519
Alumina-1	71	3.5	535
Alumina-2	69	5.2	531
PSL-bead solution	36	4.4	535

close to 300 and 17 times, respectively, higher number of particles as compared to the silica slurry. Also, the silica-1, ceria-1, and alumina-1 slurries source samples were found to have $\sim 1.4 \times 10^5$, 73×10^5, and 8.2×10^5 particles/ml (for size ≥1.10 μm), respectively.

There are considerable differences in filter performance across the different slurries for Planargard CS05 as well as Planargard CMP1 filters (1-1/2″ length). As seen in Table 18.3, alumina-1 and alumina-2 slurries with similar low wt% solids of abrasives showed similar retention and flow rate for CS05 media. A similar result of alumina-1 and alumina-2 slurries was seen for CMP1 filter. For CS05 media, silica-2 with similar wt% solids as silica-1 showed lower retention, much higher Δp, and lower flow rate. Also for CS05, alumina-1 with similar wt% solids as ceria-1 resulted in much higher retention level as well as higher Δp and a slightly lower flow rate.

The LPC retention seen in ceria-1 is much lower than alumina-1 for CMP1 filter. As expected, for Planargard CS05 as well as Planargard CMP1 filters, the PSL-bead challenge solution with negligible wt% solids as compared to slurries showed lower retention, lower Δp, and lower flow rates as compared to most of the slurries. The settling rate of various slurries [20,23] can be significantly different depending on the colloidal stability of the particles and their densities (e.g., silica, alumina, and ceria abrasive densities are ~2, 4, and 8 g/cc, respectively). The mean particle size of the tested slurries ranged from 120 to 160 nm. These results demonstrate that filter media, large-particle retention, pressure drop, and flow rate are strongly influenced by the chemical additives and abrasive characteristics in the slurry, and that empirical filter characterization and optimization is essential for current and next-generation CMP slurries.

18.4.5.4 Polystyrene Latex (PSL) Bead Solution Filtration Experiments were conducted to obtain filter retention, flow rate, and Δp data for a DI water based PSL bead mix solution prepared using particles ranging from bead diameters of 0.772 to 20 μm. It is a common practice to use PSL bead challenge solutions (created by mixing different size PSL bead standards in specific volumetric ratio to simulate slurry-like particle size distribution for the bead mix solution) to obtain relative quantitative retention data for various filters. These solutions are expected to retain stable PSD and provide more consistent information compared to real CMP slurries, which may change particle characteristics over time.

Similar to slurry samples, LPC for PSL bead samples were also measured using AccuSizer 780 APS in the top chamber addition mode. The resulting LPC reduction Δp and flow rate for each filter media are summarized in Table 18.3. The Δp and flow rates were obtained for the 1-1/2″ length filters at 10 min after start of the filtration tests. The experimental uncertainty in the measurement of LPC, flow rate, and pressure drop measurements is estimated to be ±5%, ±10 ml/min, and ±0.5 psi, respectively. The LPC reduction is presented in terms of percentage cumulative reduction of ≥0.56-μm particles for Planargard CS05 (0.5 μm rating) and percentage cumulative reduction of ≥1.01-μm particles for Planargard CMP1 (1.0 μm rating) filters.

18.4.6 Summary

Key findings are summarized below.

- Current and next-generation CMP slurries with much smaller mean particles size, lower wt% solids, and complex abrasive morphology and chemical composition pose unique challenges to the filtration process. New slurry filtration technology targets tighter retention of large particles at much smaller large-particle cutoff (e.g., 0.5 or 0.3 μm), more consistent flow rate and pressure drop behavior during usage, extended filter lifetime, and minimal effect on the mean working particles and PSD, essential to achieve highly repeatable local and global planarity in CMP processes.
- Graded-density depth filters can be effectively used to manage large-particle behavior in new CMP slurries. Optimum slurry filter design should consider slurry abrasive morphology and composition, chemical additives, LPC, PSD, wt% solids, viscosity, abrasive settling, target retention level, maximum allowable Δp, flow rate, expected filter lifetime, and most importantly the cost of ownership.
- Filtration evaluations may involve analytical membranes, particle-loading tests, measurement of gel content or tail-end large-particle mass concentration, and the effects of filtration on oxidizer and organic

components in the slurry. Postuse filter analysis can provide useful insights on the filter clogging mechanisms by quantification of filter plugging or remaining lifetime, Δp and weight gain, SEM/ESEM of retained mass, filterability, gel/particle analysis, and EDS for species detection.

- Large-particle retention, flow rate, and Δp from filtration tests using tighter graded-density depth media samples show dramatically different behavior in the silica, alumina, and ceria abrasive slurries, indicating that new CMP slurries filter optimization still remains empirical in nature.
- With the increasing complexity of CMP processes and slurries, more collaborative studies are needed between the slurry and slurry delivery system manufacturers, filtration solution providers, and the fab end users to achieve the common goal of lowest defectivity and cost of ownership in CMP processing.

18.5 PUMP HANDLING EFFECTS ON CMP SLURRY FILTRATION—CASE STUDIES

Fluid transfer pumps can be divided into positive-displacement, e.g., air-operated double-diaphragm (AOD) pump, and non-positive-displacement, e.g., magnetically levitated centrifugal (MLC) pump. Positive-displacement (PD) pumps such as diaphragm, bellows, gear, vane, or piston pumps trap and move a fixed volume of fluid employing diaphragms, bellows, or other devices. Pump stroke rate and displacement (or stroke volume) can be adjusted to control the discharge flow rate. Very shear-sensitive products are usually handled using a PD pump [38]. A study comparing the relative performance of different pumps in transferring shear-sensitive highly abrasive ammonium bromide slurry [39] demonstrated cost of lost product per day to be \$80,000 for recessed impeller centrifugal pump ($\sim$ 40% product destruction level), \$40,000 per day for a progressive cavity pump ($\sim$ 20% destruction level), and \$2,000 per day for a disc pump ($\sim$ 1% destruction level).

Effects of various pumps on shear-sensitive liquids such as blood, paints, and coating solutions have been widely investigated in biomedical, chemical, and food industries [38–41]. The effective shear stress acting on the fluid particles passing through a pump is three dimensional in nature. Many studies have been conducted on the flow field and shear stress distribution in centrifugal pump (CP) impellers to reveal the effects of blade geometry on shear stress distribution, and hemolysis and thrombus formation in biological flows [40–43]. Comparative shear stress can be an indicator of hemolysis and thrombosis. The onset of hemolysis is a function of the factors including frequency of loading, mean shear rate, and shear rate amplitude. One of the major concerns in cardiac assist devices is the possible blood cell damage due to high shear forces of the rotating vanes and thrombosis due to flow stagnation. These damaging effects are known to be dependent on shear forces and exposure times. Investigation of comparative shear stress distribution in a MLC pump revealed its maximum magnitude to be lower than 500 N/m^2 at pump rotational speed of 2000 rpm, which doubled at 3000 rpm [40]. There have been several studies on extensive pump handling effects on CMP slurries [6,15,18,44]. However, there have been only few evaluations using CP in such applications [13, 45–47].

The pumps that produce smaller increases in LPC for similar number of slurry turnovers and maintain the abrasive mean PSD, usually provide extended filter throughput and lifetime, more consistent flow, and ease in LPC management. In general, graded-density depth filters are able to provide good retention of large particles in slurry from pump handling tests. However, tighter filters are very sensitive to even small increases in the mean particle size and PSD. Since most pumps cause distinct shear to the handled slurry, it is essential to consider pump type and operational speed with the slurry composition and usage cycle for optimum slurry health management. Other useful parameters may include cumulative slurry turnover before usage, global loop backpressure, slurry temperature rise monitoring and control needs, and target filter lifetime and pressure-drop characteristics. In typical fab applications, the slurry goes through only ~100 turnovers before getting consumed [15]. However, in R&D or pilot scale slurry distribution, the slurry may reside in the system much longer, and older slurry in the system can get mixed with newly replenished slurry and go through a much larger number of turnovers before consumption. This makes it important to develop an insight into the slurry behavior under extensive handling.

18.5.1 Pump Technologies and Applications

Centrifugal, AOD, and bellows pumps are among the most commonly used pumps in the process industries. AOD pumps can handle fluidized dry powder in transfer applications and shear-sensitive liquids such as pigment slurries, latexes, abrasive TiO_2, paint, ink, and clay slip, solvents, and food products such as milk, beer, wine, and oils, and semisolids such as tomatoes, strawberries, olives, and cherries. These pumps are economical, noisy during operation, have deadhead capacity, can run dry indefinitely and self-prime, have no dynamic seal, and are less affected by air entrapment in the suction line (have mix-media pumping capability). AOD pumps have no electrical or heat generating components and pump discharge pressure is limited to pump-driving air pressure. These pumps typically operate at slower speeds and have less sensitivity to changes in the discharge and suction conditions as compared to CP. AOD pumps are typically used with a discharge pulsation dampener and suction stabilizer for hydraulic shock absorption and to smooth the pulsatile nature of flow in the pump delivery and suction lines. Bellows pumps have characteristics very similar to those of the AOD pumps.

CP transfers energy to a fluid through a spinning impeller, imparting kinetic energy to the fluid handled, which is converted into pressure through diffusive action in the pump casing. These pumps are pressure and fluid dependent and typically not used in metering applications with changing pump inlet and discharge conditions. Generally, radial-flow CP is designed to deliver relatively low discharge flow at high heads, and axial-flow pumps produce relatively high flow rates at low discharge pressures. MLC pumps typically have a permanent magnet embedded on one side of the impeller. A motor with magnets of different polarity drives the impeller with the electromagnets on the other side of the impeller keeping the impeller suspended. MLC pumps have significant advantage of seal-less design due to their air floating magnetically levitated impeller design and can be operated dry for limited time at lower speeds.

To maintain impeller balance in MLC pumps, it is essential to operate these pumps with suction line and pump casing filled with the liquid handled and without any air entrapment. It is usually not recommended to run these pumps dry at higher speeds. MLC pumps usually require cold water jacketing or a chiller and heat transfer system to maintain the handled liquid temperature in limited volume, high-turnover rate extended recirculation applications. CP does not posses self-priming feature and should be started with filled suction line and casing. Multistaging is commonly used to boost the delivery pressure in these pumps. AOD pumps operate well in parallel, but not in series, whereas centrifugal pumps work well in series to boost delivery pressure and can be used in parallel operation. New diaphragm metering pump technologies with advanced microprocessor controlled stepper-motor can provide near pulseless flow. These electronically controlled pumps (using precisely controlled motor speed throughout a single stroke cycle) are basically oscillating PD pumps.

18.5.2 Pump Shearing Effects on Slurry Abrasives

A shear-sensitive liquid is one that is altered as it passes through the shearing motion of the pump. Pumps employing mechanisms that repeatedly cut or cause rapid relative motion (e.g., in a high-velocity CP impeller vanes and stationary casing) between different layers of the handling liquid being transferred cause shear and can generate energy and heat built-up. AOD pumps are widely known in the paint, chemical, and coating industries for low shear abrasive chemical handling without any heat built-up. A CP introduces turbulence at the tip of the impeller vane and it creates shear between the vane and the casing. Even a large-diameter, slow moving CP introduces a relatively high amount of shear [38]. One way to use a CP for shear-sensitive abrasive slurry is to oversize the pump and slow it down. However, the larger pump usually costs much more and generally operates with lower efficiency.

CMP slurry quality must be maintained during pump handling [5] and filtration is critical in achieving the lowest defectivity in CMP processing through slurry large particle management [16]. A good CMP slurry pump should maintain the mean and large-particle characteristics of the abrasives during slurry delivery [48]. In slurry handling where minor gradual increases take place in the LPC, it is usually possible to employ an optimized filter to selectively remove large particles with least effects on the mean working particles. In addition to the diaphragm, bellows, and centrifugal pumps, CMP slurries can also be handled using a VPDS [15,30,47], peristaltic pump, and gravity-based system. Details of these approaches are not included here for brevity. LPC and PSD of silica-based oxide CMP slurries illustrate distinct, evolving characteristics as a result of repeated shearing during most pump handling, whereas alumina and ceria slurries do not show mean particle size or LPC increase [18]. Different pumps cause varying degrees of shearing force during slurry handling and high-shear pumps can alter the mean particle size and effective viscosity of the shear-sensitive abrasive slurries under extensive handling.

18.5.3 Pump Handling and Filtration Data

The slurry was recirculated in a closed loop configuration employing an AOD (Wilden P1) or MLC (Levitronix BPS-3) pump over extended periods of time (Fig. 18.15). Slurry samples from the loop were collected and analyzed at different time points. During this testing the slurry passed through pump, valves, bends and fittings, and 25-ft long length of PFA tubing to provide back pressure similar to those expected in the fab global distribution loop. Experiments were conducted using silica-M (~12 wt% solids), silica-H (~25 wt% solids), alumina (<1 wt% solids), and ceria (<1 wt% solids) CMP slurries. Results of handling alumina and ceria abrasive slurries are presented in References 18, 47, and 49 and will not be included here for brevity. All mean PSD measurements used a laser scattering particle-size and distribution analyzer (Horiba LA-930). This analyzer provided a wide particle size measuring range of 0.02–2000 μm. The slurry samples were measured for LPC using a SPOS analyzer (PSS AccuSizer 780 APS), employing a top injection dilution chamber with single stage autodilution [12]. Cumulative number of particle sizes ≥0.56 μm were measured to monitor the increase in LPC in the slurry samples. During LPC measurements, the slurry sample was added to the 30-ml DI water in the dilution chamber to achieve <9000 particles/ml concentration at 60 ml/min flow rate, to ensure that the measurements started immediately with the start of autodilution process. This approach is likely to provide more reliable LPC data with the minimum chances of losing any large particles in the usual online autodilution process (to get <9000 particles/ml concentration from much higher levels) before start of measurement, especially in extremely high particle concentration samples. The temperature of the slurry was maintained within 3 °C of the ambient using a chiller for the MLC pump tests. There were no metallic wetted parts in the flow path of recirculated slurry.

The cumulative LPC is a more sensitive measure of the defect-causing large-particle growth than the PSD measurement. Such large particles could cause scratches on the wafers and issues with the tighter retention of the slurry, whereas any changes in the mean PSD may change the CMP removal rate and other operating parameters. Present results demonstrate that MLC or AOD pumps can be used for handling shear-sensitive silica CMP slurry, in low turnover applications. However, none of these pumps would be very useful if slurry is recirculated over extended periods of time without regular usage and replenishment. To maintain shear-sensitive slurry's quality or health it would be desirable to recirculate the slurry in the delivery system and global loop only during the periods when slurry is needed for polishing at the CMP tools. In general, graded density depth filters with relatively open and tighter ratings provide appropriate large-particle retention in such applications, in the global distribution loop and POU locations, respectively.

18.5.4 Test Cases

To quantify the effects of handling on slurries, a series of experiments were performed. Tests 1–6 included handling and filtration experiments performed using silica-M slurry, whereas Tests 7–13 included experiments with silica-H slurry. Tests 14–15 and 16–17 involved handling and filtration tests conducted using low wt% abrasive alumina and ceria slurries, respectively. LPC and PSD from Tests 1–11 will be presented and discussed, and key findings of Tests 12–17 will be summarized as follows.

Test 1: Silica-M slurry was recirculated in the MLC pump at 8000 rpm, 28 psi back pressure, 31.7 turnovers/h, 8 lpm, for 330 h (10,461 turnovers). This represents extremely intensive working of the slurry, greatly in excess of any handling expected in practical blending and distribution systems in the laboratory testing or fab production environment. Figure 18.21a presents the LPC data for the first 45 h of the test. There is a significant increase in the cumulative LPC for the particle sizes $\geq$0.56 µm for slurry turnovers higher than 143. It is important to note that in a typical fab operation, slurry may go through $\sim$ 100 turnovers before consumption. Figure 18.21a shows negligible increase in LPC over 1 µm, but a large growth in LPC of particles between $\sim$ 0.56 and 0.60 µm for higher turnovers.

It is important to note that present AccuSizer measurements were obtained with a particle-size threshold of 0.56 µm (the lowest possible size threshold available for this analyzer) and any LPC growth behavior for particles smaller than 0.56 µm could not be seen in these data. MLC pump extensive handling showed a slight increase in mean particle size and PSD of silica-H slurry, as will be discussed later. The above behavior of LPC may be attributed to the cumulative effects of low-intensity continuous shear application to the bulk volume of slurry handled at very high pump speeds in MLC pump. In Fig. 18.21b, the LPC trend for the complete 330 h of Test 1 is very similar to the first 45 h.

Test 2: Silica-M was recirculated in MLC pump at 8000 rpm, 28 psi back pressure, 63.4 turnovers/h, 8 lpm, for 20 h (1270 turnovers). Figure 18.22 presents the LPC data with very similar characteristics as seen for Test 1.

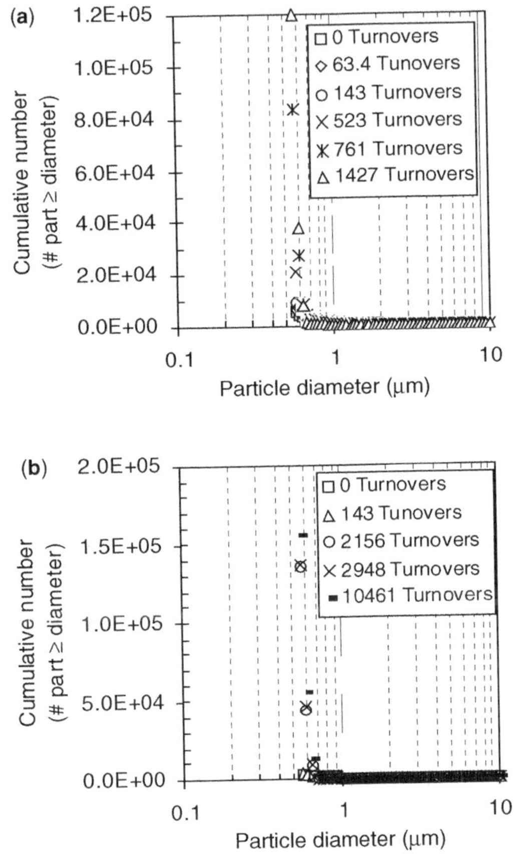

FIGURE 18.21 LPC data for silica-M slurry recirculation in MLC pump at 8000 rpm and 31.7 turnovers/h (Test 1).

Results of comparable turnovers shown in Figs. 18.23a and 5.3b show a very similar and small growth in LPC for the first ~ 140 turnovers of the slurry, demonstrating that the shearing effect on the slurry pumped by a MLC pump is a function of the number of turnovers rather than the recirculation time.

Test 3: Silica-M was recirculated in an MLC pump at 5000 rpm, 10 psi back pressure, 39.6 turnovers/h, 5 lpm, for 24 h (950 turnovers). In contrast to Figs. 18.21–18.23 for the MLC pump tests at 8000 rpm, Fig. 18.24 for test at 5000 rpm does not show any increase in LPC, suggesting that the rate of shear or pump speed may be a factor in the LPC increase (i.e., in the particle-size range ~ 0.56–0.60) in the tests at 8000 rpm.

Silica-M handling in an MLC pump at 5000 rpm does not show the peculiar increase in LPC seen for 8000 rpm tests, suggesting that pump speed and wt %

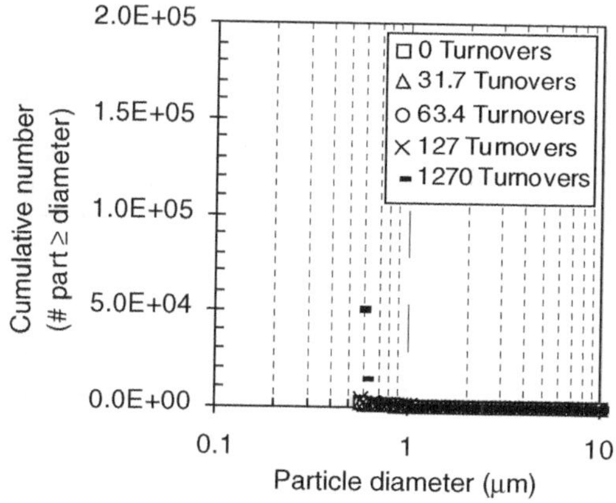

FIGURE 18.22 LPC data for silica-M slurry recirculation in MLC pump at 8000 rpm and 63.4 turnovers/h (Test 2).

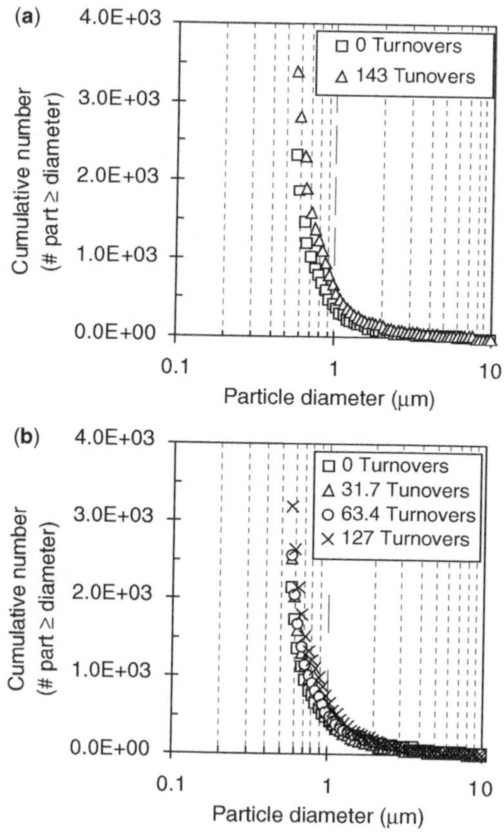

FIGURE 18.23 (**a**) LPC data for silica-M slurry recirculation in MLC pump at 8000 rpm and 31.7 turnovers/h (Test 1); (**b**) LPC data for silica-M slurry recirculation in MLC pump at 8000 rpm and 63.4 turnovers/h (Test 2).

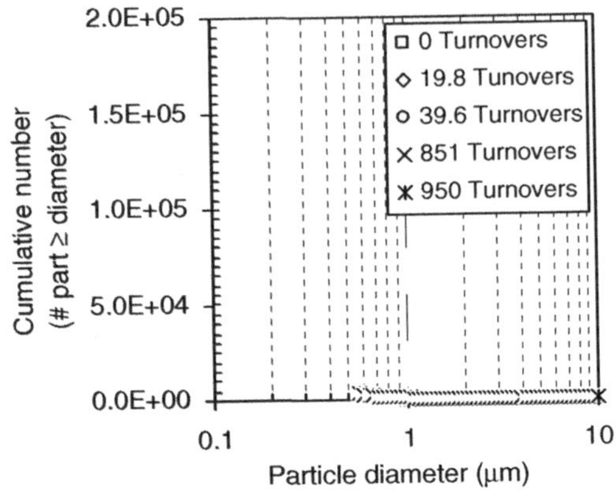

FIGURE 18.24 LPC data for silica-M slurry recirculation in MLC pump at 5,000 rpm and 39.6 turnovers/h (Test 3).

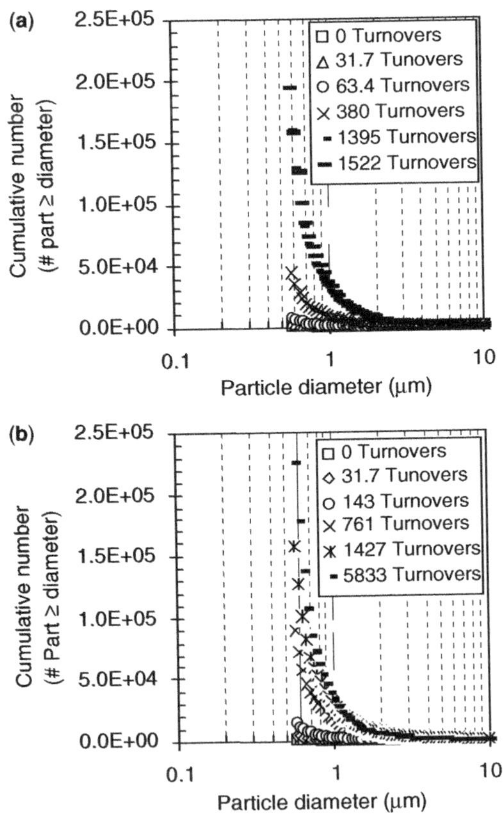

FIGURE 18.25 **(a)** LPC data for silica-M slurry recirculation in AOD pump at 28 psi backpressure, 63.4 turnovers/h (Test 4); **(b)** LPC data for silica-M slurry recirculation in AOD pump at 28 psi backpressure, 31.7 turnovers/h (Test 5).

solids of the slurry may be important factors in large-particle agglomeration under extended handling (Tests 1 and 2). In silica-H tests, an increase in the LPC was seen at the pump speed of 5000 rpm, as will be discussed later. Silica-M seems to be less affected by pump handling at 5000 rpm pump speed as compared to ~ 2× higher wt% solids silica-H.

Test 4: Silica-M was recirculated in an AOD pump at 28 psi back pressure, 63.4 turnovers/h, 8 lpm, for 24 h (1522 turnovers). Figure 18.25a shows the LPC distribution, which is significantly different as compared to MLC pump tests. AOD handling for similar turnovers resulted in a large increase in LPC throughout from ~ 0.56 to 10 μm, but a smaller increase in LPC between ~ 0.56 and 0.60 μm as compared to MLC pump Test 2, especially at high total turnovers, demonstrating the dramatic contrast in the type of shear resulting on slurry when it is handled by different pumps. AOD pump does not appear to cause shear to the bulk volume of the CMP slurry being pumped, but it does seem to cause extremely high level shear to the limited volume of slurry passing through the valve and seat interface during instantaneous valve opening and closing cycles, producing very large size slurry agglomerates. This phenomenon can result in a generation of a relatively smaller number of widely distributed large size abrasive particles, without significantly changing the mean PSD of slurry during shear-sensitive slurry extended recirculation.

Test 5: Silica-M was recirculated in AOD pump at 28 psi back pressure, 31.7 turnovers/h, 8 lpm, for 184 h (5833 turnovers). Fig. 18.25b presents the LPC distribution, which is very similar to Fig. 18.25a for 2× higher turnover rate. Again, the results of comparable turnovers (1395 and 1427) shown in Fig. 18.25a and b for Tests 4 and 5, respectively, show a very similar LPC growth behavior, illustrating that the shearing effect on the slurry pumped by an AOD pump is also a function of the number of turnovers rather than the recirculation time.

Test 6: Experiments were conducted to obtain large-particle retention, pressure drop, and flow rate data for Entegris Planargard CMP3 and Planargard CMP5 graded-density depth filters (1-1/2″ length) with nominal ratings of 3 and 5 μm, respectively. Filtration tests were conducted with the fresh silica-M (Fig. 18.26a) as well as extensively handled slurries (final samples) from Test 2 (Fig. 18.26b), Test 3 (Fig. 18.26c), and Test 4 (Fig. 18.26d). A peristaltic tube pump was used to feed the slurries through the filter media. Cumulative large particle retention curves, Δp, and flow rate data are presented in Figs. 18.26a–d. Relatively open filters were used to obtain comparative filtration data, avoiding immediate plugging of the stronger 0.5 and 1 μm filters with highly agglomerated slurry samples. Depending on the filter rating and extent of large-particle agglomerates in slurries, the filters were able to reduce LPC in all cases.

Filtration tests for recirculated silica-M slurry from Test 1, using 1 and 3 μm rating depth filters resulted in immediate plugging, whereas tests with 5 and 9 μm nominal filters (Planargard CMP5 and CMP9, respectively) resulted in Δp

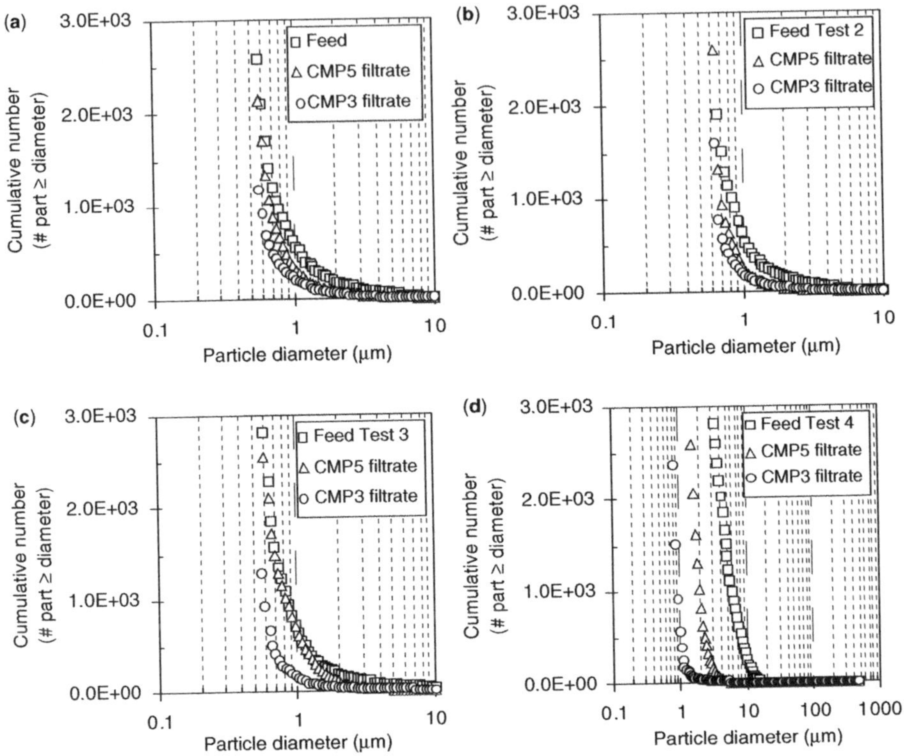

FIGURE 18.26 (**a**) LPC data for fresh silica-M slurry single-pass filtration tests using 3 and 5 μm depth filters. Δp at ~557 ml/min flow rate: CMP5 ~1.8 psi, CMP3 ~4.4 psi (Test 6); (**b**) LPC data for recirculated silica-M slurry from Test 2 filtration tests using 3 and 5 μm depth filters. Δp at ~529 ml/min flow rate: CMP5 ~1.7 psi, CMP3 ~3.9 psi (Test 6); (**c**) LPC data for recirculated silica-M slurry from Test 3 filtration tests using 3 and 5 μm depth filters. Δp at ~548 ml/min flow rate: CMP5 ~1.9 psi, CMP3 ~4.1 psi (Test 6); and (**d**) LPC data for recirculated Silica-M slurry from Test 4 filtration tests using 3 and 5 μm depth filters. Δp at ~546 ml/min flow rate: CMP5 ~2.3 psi, CMP3 ~4.3 psi (Test 6).

of ~1.9 psi at ~529 ml/min (for CMP 5) and Δp of ~1 psi at ~524 ml/min (for CMP9).

Test 7: Silica-H slurry was recirculated in the MLC pump at 7600 rpm, 28 psi back pressure, 63.4 turnovers/h, 8 lpm, for 6 h (380 turnovers). This represents intensive working of the undiluted slurry in an R&D small-scale handling system. Figure 18.27a presents the LPC data at different turnovers. Results have very similar behavior as seen for silica-M in Test 1. Again, this behavior of LPC may be attributed to the cumulative effects of low-intensity continuous shear application to large percentage of slurry handled at high speeds in the MLC pump.

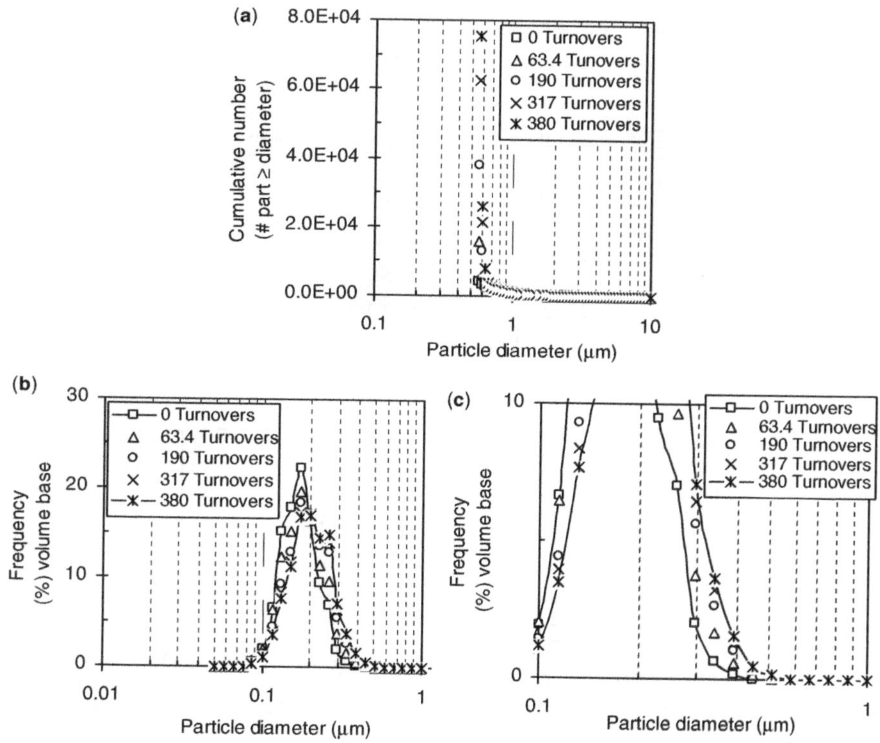

FIGURE 18.27 (**a**) LPC data for silica-H slurry recirculation in MLC pump at 7600 rpm and 63.4 turnovers/h (Test 7); (**b**) PSD data for silica-H slurry recirculation in MLC pump at 7600 rpm and 63.4 turnovers/h (Test 7); and (**c**) PSD data for silica-H slurry recirculation in MLC pump at 7600 rpm and 63.4 turnovers/h (Test 7), magnified view.

There is a significant increase in the cumulative LPC for the particle sizes ≥0.56 μm for slurry turnovers higher than 63.4, especially in the size range of ~0.56–0.60 μm, whereas there is negligible increase in LPC over 1 μm. The LPC data were consistent with the PSD data behavior presented in Fig. 18.27 and magnified large particle tail region in Fig. 18.27c. It can be seen that there is a slight increase in mean particle size (Fig. 18.27b–c) for the high turnovers in this test. A small reduction in the pump flow rate and a very small increase in slurry viscosity were also noticed in the latter part of this test.

Test 8: Silica-H was recirculated in an AOD pump at 28 psi back pressure, 63.4 turnovers/h, 8 lpm, for 6 h (380 turnovers). Figure 18.28a shows the LPC distribution, which is significantly different than that of MLC pump Test 7. These LPC results are very similar to those results seen for AOD pump silica-M slurry test earlier (Test 4). There is a larger increase in LPC throughout from ~ 0.56 to 10 μm, but a smaller increase in LPC between ~ 0.56 and 0.60 μm as

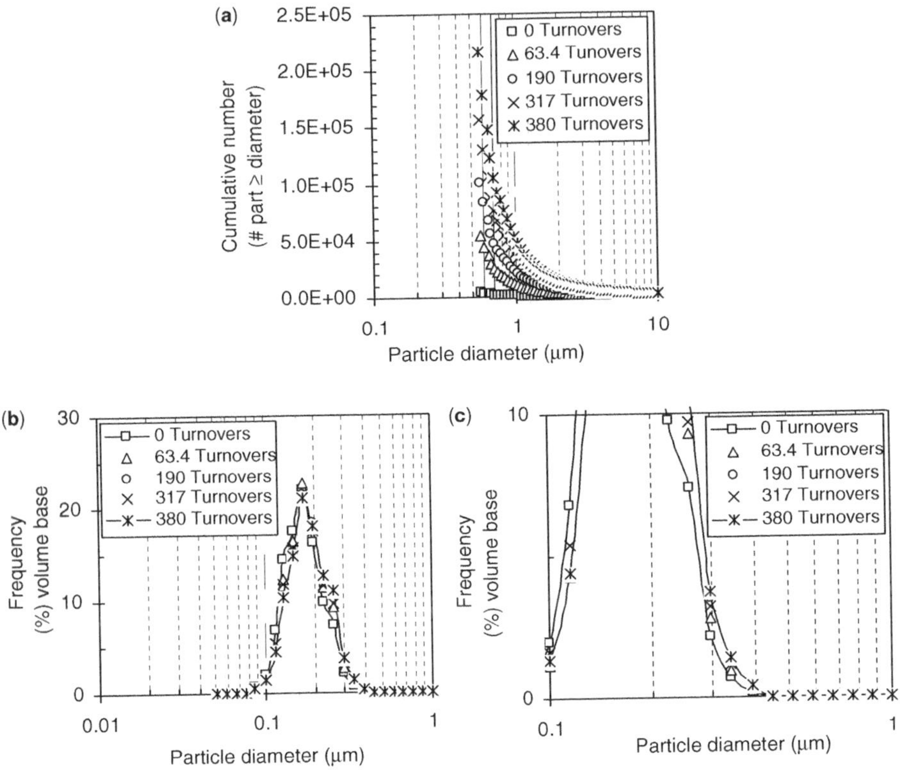

FIGURE 18.28 (**a**) LPC data for silica-H slurry recirculation in AOD pump at 28 psi backpressure, 63.4 turnovers/h (Test 8); (**b**) PSD data for silica-H slurry recirculation in AOD pump at 28 psi backpressure, 63.4 turnovers/h (Test 8); and (**c**) PSD data for silica-H slurry in AOD pump at 28 psi backpressure, 63.4 turnovers/h (Test 8), magnified large particle tail.

compared to MLC pump Test 7, especially at high total turnovers. As discussed earlier, AOD pump does not seem to cause shear to the bulk volume of the CMP slurry being pumped, but does seem to cause extremely high level shear to the limited volume of slurry during opening and closing of valves and producing very large agglomerates.

These results of LPC are supported by mean PSD measurements. PSD data for the AOD pump test samples presented in Fig. 18.28b and c demonstrate relatively smaller changes in PSD during this 380 turnover run. A very small growth of particles in the larger size tail region can be observed, but the mean PSD peak value remains nearly the same (Fig. 18.28b and c). In this test, the slurry viscosity and pump discharge flow rate remained constant within the measurement uncertainties.

Test 9: Final samples of extensively handled silica-H slurry from Tests 7 and 8 were filtered to obtain large-particle retention, Δp, and flow rate data for 0.5-μm (Planargard CS05), 1.0-μm (Planargard CMP1), and 3-μm (Planargard CMP3) rating depth filters (1-1/2″ length). The results of cumulative LPC, Δp, and flow rate are presented for the MLC and AOD pump slurry samples in Figs. 18.29a–c and 18.30a–c, respectively.

The silica-H filtration data have a strong similarity with the related tests for the silica-M slurry (Test 6). Filtration of recirculated silica-H from MLC pump Test 7 using 0.5-μm CS05 filter resulted in immediate plugging (0 flow at 83 PSI), whereas tests with 3- and 1-μm rating filters resulted in Δp of ~5.1 psi at ~490 ml/min (for CMP 3), and Δp of ~7.2 psi at ~490 ml/min (for CMP1).

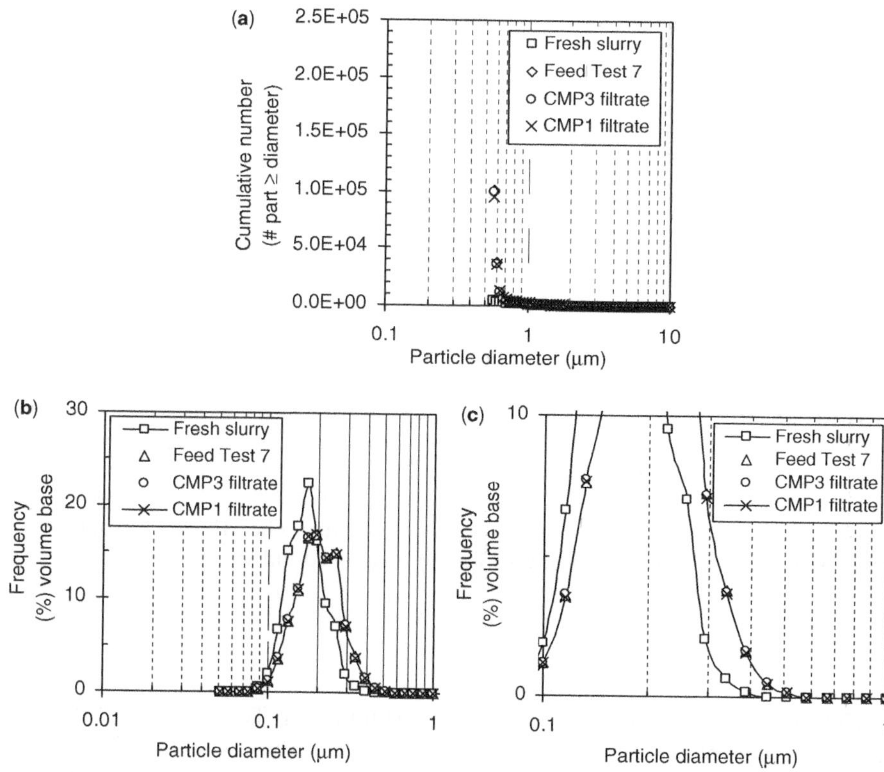

FIGURE 18.29 (**a**) LPC data for recirculated silica-H slurry from MLC pump Test 7 filtration using 1 and 3-μm filters. Δp at ~ 490 ml/min flow: CMP3 ~ 5.1 psi, CMP1 ~ 7.2 psi, CS05 filter clogged immediately (0 flow), 83 psi (Test 9); (**b**) PSD data for recirculated silica-H slurry from MLC pump Test 7. Results of single-pass filtration tests using 1 and 3-μm nominal rating graded-density depth filters (Test 9); and (**c**) PSD data for recirculated Silica-H slurry from MLC pump Test 7. Results of single-pass filtration tests using 1 and 3-μm nominal rating graded-density depth filters (Test 9), magnified large particle tail.

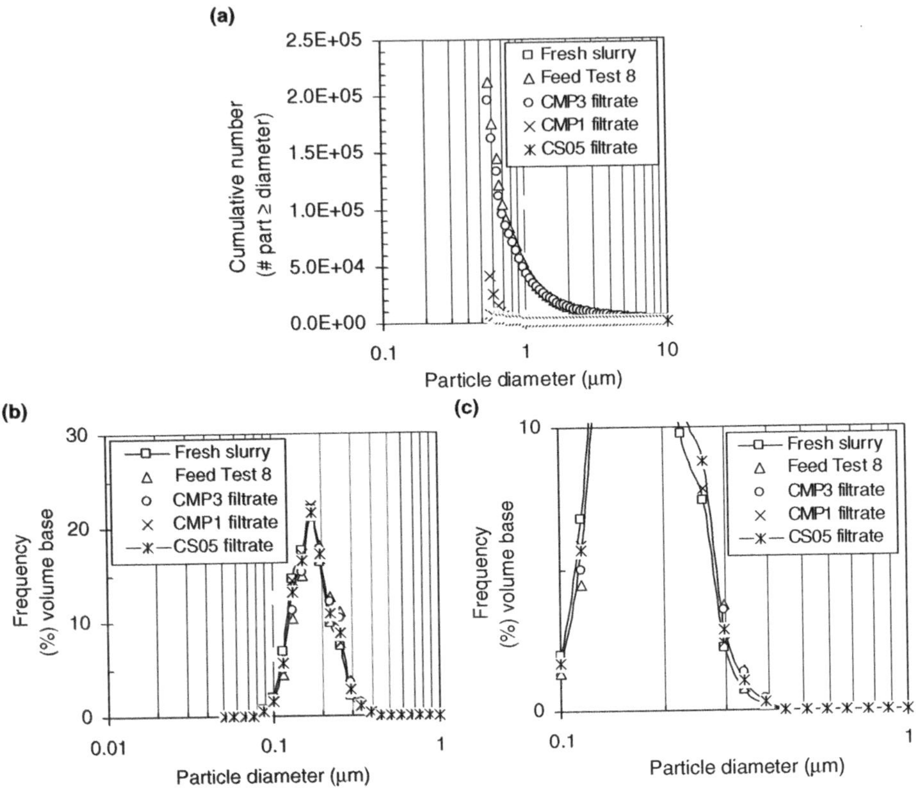

FIGURE 18.30 (**a**) LPC data for recirculated silica-H slurry from Test 8 filtration using 0.5, 1, and 3 μm filters. Δp at ~ 490 ml/min flow: CMP3 ~ 7.5 psi, CMP1 ~ 18 psi, CS05 ~ 190 ml/min (filter clogged after 2 min), 83 psi (Test 9); (**b**) LPC data for recirculated silica-H slurry from Test 8 filtration using 0.5, 1, and 3 μm graded-density depth filters (Test 9); and (**c**) LPC data for recirculated silica-H slurry from Test 8 filtration using 0.5, 1, and 3 μm filters (Test 9), magnified tail view.

Filtration of recirculated silica-H from AOD pump Test 8 using CS05 filter resulted in complete plugging after 2 min (~ 190 ml/min), whereas tests with 3- and 1-μm rating filters resulted in Δp of ~7.5 psi at ~490 ml/min (for CMP 3), and Δp of ~ 18 psi at ~490 ml/min (for CMP1). CMP1 filtration test reached Δp of ~33 psi after 7 min from start of filtration for the Test 8 recirculated slurry sample.

Test 10: Silica-H was recirculated in MLC pump at 5000 rpm, 11 psi back pressure, 31.7 turnovers/h, 4 lpm, for 24 h (761 turnovers). LPC and PSD measurements (Figs. 18.31a–c) show results very similar to Test 8, with slight increase in mean particle size and an increase in LPC for size ~0.56–0.60 μm.

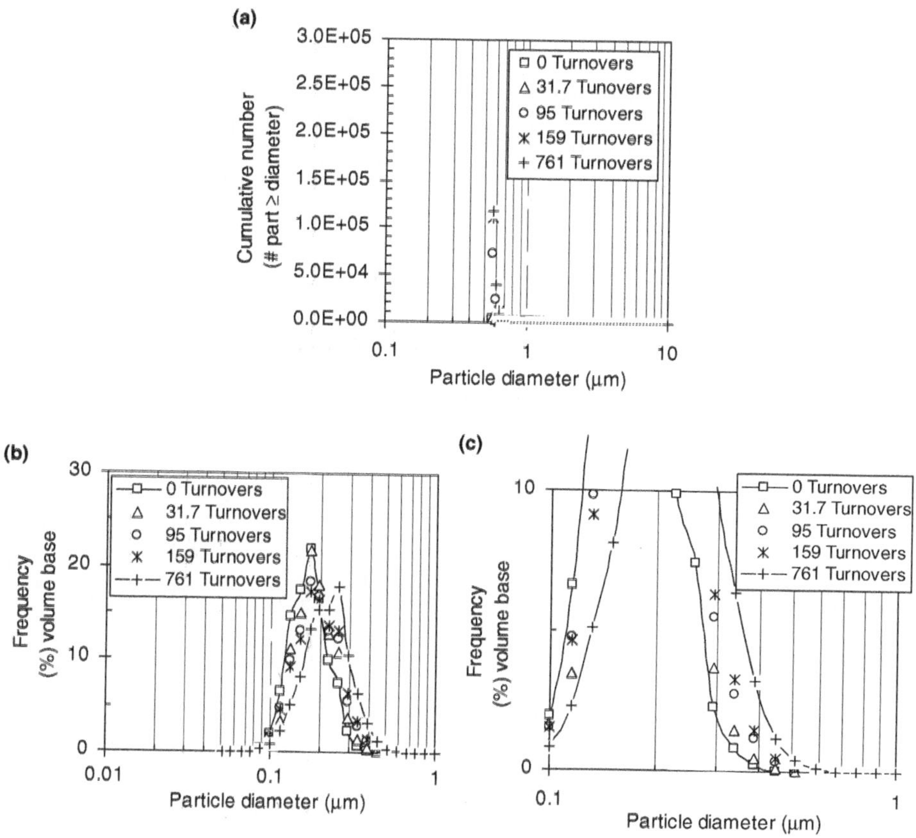

FIGURE 18.31 (**a**) LPC data for silica-H slurry recirculation in MLC pump at 5000 rpm, 11 psi backpressure, and 31.7 turnovers/h (Test 10); (**b**) PSD data for silica-H slurry recirculation in MLC pump at 5000 rpm, 11 psi backpressure, and 31.7 turnovers/h (Test 10); and (**c**) PSD data for silica-H slurry in MLC pump at 5000 rpm and 31.7 turnovers/h (Test 10), magnified tail view.

This is in contrast to the results of Test 3 for silica-M slurry, which did not show an increase in LPC. Results presented in Fig. 18.31a–c suggest that pump speed in combination with wt% solids of the silica slurry may be a factor in LPC and PSD evolution behavior even at slower pump speed of 5000 rpm.

Test 11: Silica-H recirculation in AOD pump at 11 psi back pressure, 31.7 turnovers/h, 4 lpm, for 24 h (Figs. 18.32a–c) showed behavior very similar to Test 8.

Tests 12 and 13: These were replica tests similar to Tests 7 and 8, respectively, using silica-H and confirmed the findings of the latter tests.

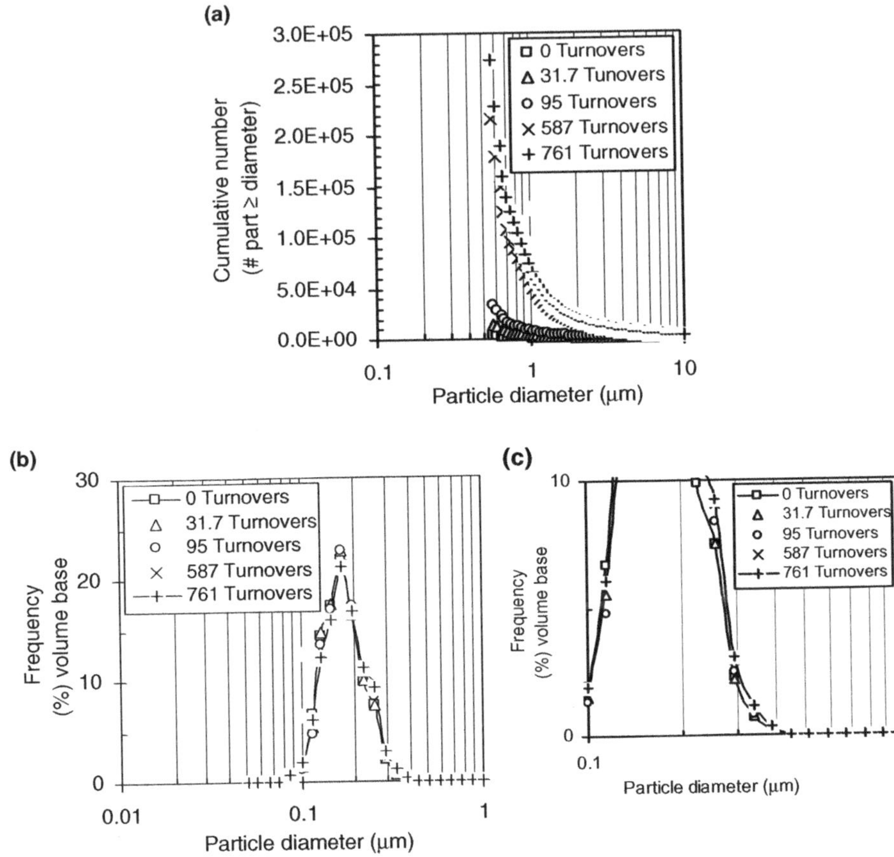

FIGURE 18.32 (**a**) LPC data for silica-H slurry recirculation in AOD pump at 11 psi backpressure, 31.7 turnovers/h (Test 11); (**b**) PSD data for silica-H slurry recirculation in AOD pump at 11 psi backpressure and 31.7 turnovers/h (Test 11); and (**c**) PSD data for silica-H slurry in AOD pump at 11 psi backpressure and 31.7 turnovers/h (Test 11), magnified tail view.

Tests 14 and 15: These were tests similar to Tests 7 and 8, respectively, using an alumina-based CMP slurry. Insignificant changes in LPC and PSD were observed in both MLC and AOD pump handling tests [49]. These results are not included here for brevity.

Tests 16 and 17: These tests were similar to Tests 7 and 8, respectively, using ceria slurry and resulted in only minor changes in the LPC and PSD. Above tests demonstrated the insignificant effects of pump handling on the alumina and ceria slurries, supporting the findings of earlier studies [18,30,47,49].

A Novel Method for Detecting Small Changes in Slurry PSD: Results of LPC were further analyzed to understand the disparities in LPC growth behavior seen for the MLC and AOD pumps in Tests 7 and 8, respectively, with Silica-H slurry. Number of particle variation in different size bins for slurry recirculation in MLC pump (Test 7) is presented in Fig. 18.33a. As discussed earlier, particle counts in bins of 0.56–0.60 μm as well as of 0.60–0.69 μm for the MLC pump test show linear increases with number of slurry turnovers. The rate of increase in particle counts in the bins of 0.56–0.60 μm is ~3.75 times as compared with the bins of 0.60–0.69 μm. It is interesting to compare a similar plot for the AOD pump Test 8 in Fig. 18.33b. Again, particle counts in bins of 0.56–0.60 μm and of 0.60–0.69 μm for the AOD pump test show linear increase, however in this case the particle increase rate in bins of 0.56–0.60 μm is only ~0.70 times as compared with the bins of 0.60–0.69-μm, i.e., more particles are being created in the slightly larger size 0.60–0.69 μm bins. This behavior is opposite to the behavior observed in the MLC pump case (Fig. 18.33a). Relatively larger number generation of 0.56–0.60-μm particles

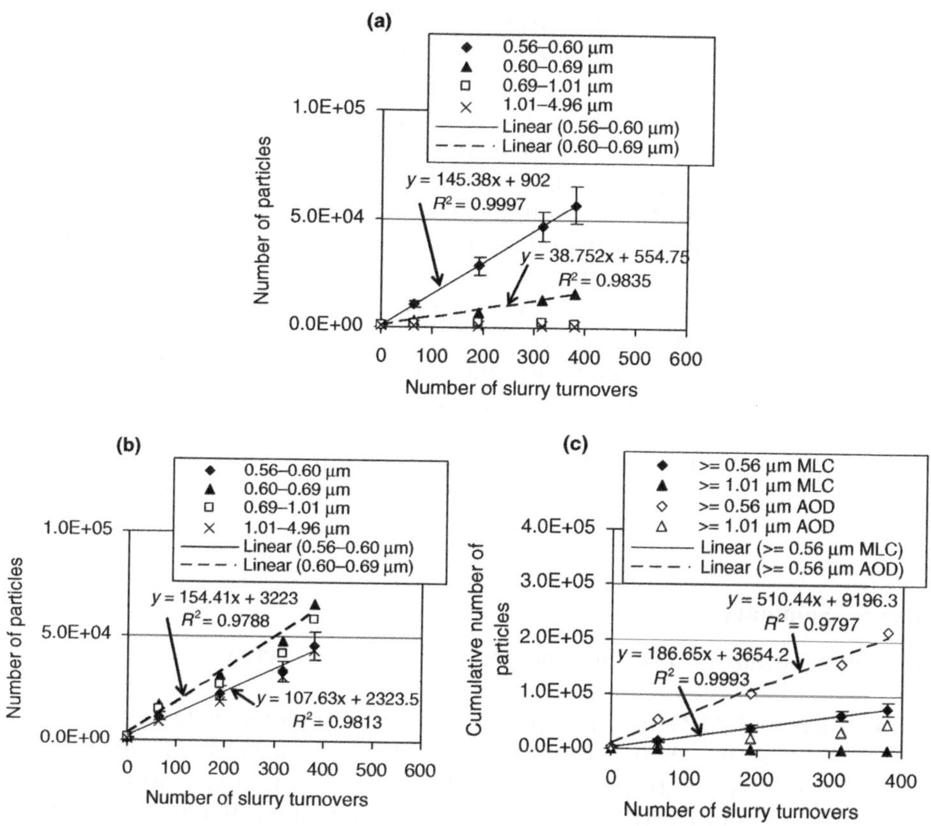

FIGURE 18.33 (**a**) Large particle counts variation in different size bins for silica-H slurry recirculation in MLC pump at 7600 rpm and 63.4 turnovers/h (Test 7); (**b**) Particle counts variation in different size bins for silica-H slurry recirculation in AOD pump at 28 psi backpressure, 63.4 turnovers/h (Test 8); and (**c**) Cumulative particle counts variation in different bins for silica-H slurry recirculation in MLC pump Test 7 and AOD pump Test 8.

seem to be the cause of rapid plugging of 0.5-μm depth filter with recirculated slurry from Test 7, as discussed earlier for Test 9.

Above described approach of monitoring and comparing large-particle counts growth in the 0.56–0.60, 0.60–0.69, 0.69–1.01, 1.01–4.96, and 4.96–9.99 μm size or similar bins can be a very useful tool for detecting progressive changes in the mean working particles and large particle during slurry recirculation, based on the LPC data. During CMP slurry handling, measurable LPC changes usually take place much earlier than detectable and repeatable mean PSD changes. The above LPC-based method can be used for comparing shearing characteristics of different pumps as well as slurry delivery systems in handling CMP slurries. This method allows the use of SPOS tool such as AccuSizer in detecting progressive small changes in the right side tail of PSD distributions. Data in Figs. 18.33a and b suggest that particle counts in the size range of 0.56–0.60 μm increase at a rate of ~145 and 108 particles per turnover during silica-H handling in MLC pump (Test 7) and AOD pump (Test 8), respectively.

Cumulative particle counts variation in different bins for silica-H slurry recirculation in MLC pump Test 7 and AOD pump Test 8 are presented in Fig. 18.33c. Rate of increase of cumulative large-particle counts in term of the particles ≥0.56 μm as well as ≥1.01 μm is much higher for the AOD pump. There is almost negligible variation in the number of particle ≥1.01 μm for the MLC pump, and some growth in number of particle ≥0.56 μm. However the rate of latter particle growth in MLC pump (Test 7) is only about ~0.37 times of that seen in the AOD pump Test 8. This shows the significant advantage of MLC pump in generating fewer large particles as compared to an AOD pump at comparable handling conditions during shear-sensitive silica slurry handling. It is important to note that the above observed behavior of abrasive particles growth (for silica-H slurry) can be significantly different for other shear-sensitive slurries with lower wt% solids and/or different abrasive particle morphology.

18.5.5 Summary

Key observations of this study may be summarized as below.

- Selecting CMP slurry pumps requires a thorough understanding of the unique characteristics of the abrasives being transferred and the pump shear behavior. In the present study, extensive handling tests of shear-sensitive silica CMP slurries using an air-operated diaphragm (AOD) pump generated significant number of large-particle agglomerates. In a similar test, a magnetically levitated centrifugal (MLC) pump was found to generate far fewer large particles (>1 μm) than that of the AOD pump for comparable turnovers. Insignificant large-particle changes were observed for the MLC pump recirculation test at moderate speed (e.g., 5000 rpm) with silica-M (~12 wt% silica abrasive) slurry.

- In extreme handling situations with limited amount of slurry in the system and at high pump speed (e.g., 7600 rpm), significant growth in number of large particles of size 0.56–0.60 μm was observed in the MLC pump silica-H (~25 wt% silica abrasive) slurry and silica-M slurry tests at large total number of turnovers. The above behavior of LPC in the MLC pump tests may be attributed to the cumulative effects of low-intensity continuous shear application to large percentage of slurry handled at high pump speeds.
- This study shows the significant benefits of an MLP pump in handling shear-sensitive CMP slurries in single-pass applications and under normal turnovers (~100) expected in a typical fab operation. Since, the MLC pump generated far fewer >1 μm particles in shear-sensitive slurries, the filter lifetime for MLC pump-based slurry delivery systems should be longer than the other AOD or bellows pump-based systems, especially when relatively open (>1 μm nominal rating) filters are used in the global loop and point-of-use (POU) locations.
- When shear-sensitive CMP slurries go through extended handling in a distribution system, and even when small increases take place in the mean size of the abrasives, it may be very difficult to selectively remove only slightly larger than mean size particles to restore handled-slurry mean PSD to fresh-slurry PSD. POU filters flow rate, pressure-drop, and lifetime can change significantly due to small mean abrasive size changes. This study illustrated that due to large concentration of 0.56–0.60 μm particles, a 0.5-μm rating graded-density depth filter was plugged immediately with 380 turnovers recirculated silica slurry in the MLC pump.
- This study presents a novel method of detecting small changes in slurry PSD based on LPC measurements. This approach could be used to understand the disparities in LPC growth behavior for the MLC and AOD pumps. Employing this method, particle counts in bins of 0.56–0.60 μm as well as of 0.60–0.70 μm for the MLC and AOD pumps show linear increases with number of slurry turnovers. Monitoring and comparing particle counts continuously in these bins can be a useful tool for detecting progressive changes in the slurry mean working particles, based on just LPC data. This approach can also be used for comparing shearing characteristics of different pumps and slurry delivery systems.
- An MLC pump can provide stable and low pressure-pulsation slurry delivery with limited LPC growth in optimally designed limited turnover applications. This pump type may require cold-water jacketing employing a chiller to maintain slurry temperature in small volume recirculation applications. This study also shows that AOD pumps with pulsation dampener can be used for handling silica slurries in low turnover and single-pass applications. Tighter graded-density depth filters could provide appropriate large-particle retention in the slurry handled using this pump.
- Present results demonstrate the importance of monitoring changes in mean PSD during slurry delivery because these changes can cause

variation in CMP polishing rate as well as significant challenges for tighter POU filtration. Mean PSD changes occur in shear-sensitive slurries only after extended number of handling turnovers, depending on abrasive and chemical composition. In R&D pilot scale slurry systems, such changes may take place during useful lifetime of the slurry, due to limited slurry volume and relatively higher turnover rates. Using appropriate rating filters in the global loop and POU locations and limiting the total number of turnovers slurry goes through the system before consumption should provide optimum large-particle management in such applications.

- CMP slurry delivery system employing filtration for LPC control should consider slurry characteristics including—abrasive type(s) and composition, LPC, PSD, wt% solids, viscosity, chemical composition and the distribution system characteristics—specific pump type and the pumping effects on slurry abrasive, pump size and speed, global distribution loop backpressure, slurry usage and replenishment cycles, slurry turnover rate and typical turnovers before consumption, filter ratings for various locations, allowable pressure drop for filters, and the slurry flow and temperature consistency needs.

ACKNOWLEDGMENTS

The author would like to thank Dr. Deepak Mahulikar, Dr. Yuzhuo Li, Dr. Ashwani Rawat, Christopher Wargo, Scott Moroney, Budge Johl and Dr. Ben Roberts for insightful discussions related to this work. The contributions of CMP slurry manufacturers by providing slurry samples and Levitronix GmbH by offering a magnetically levitated centrifugal pump for the slurry characterization setup are gratefully acknowledged.

QUESTIONS

1. What are the main objectives of CMP slurry laboratory characterization? Describe various methods and levels of slurry characterization with examples of information generated from these studies. How can such data be beneficial for CMP end user as well as CMP consumable developer and slurry delivery system manufacturers?
2. Identify CMP slurry health management challenges and discuss the pros and cones of various metrology approaches employed for monitoring and control of slurry quality during usage. Make a list of the requirements, challenges, and opportunities in online, real-time slurry health measurements.
3. Research various particle-size distribution and chemical concentration measurement technologies and related instrumentation. Create a list of their capabilities and limitations, measurement ranges, and accuracy and repeatability. Evaluate the applicability of these in the CMP slurry measurements. Discuss the possibilities of zeta-potential changes as a result

of slurry blending and handling and how such changes may influence CMP and post-CMP cleaning processes.

4. What specific slurry health concerns are seen in oxide, tungsten, and copper/low-*k* CMP processes, and what metrology approaches are appropriate in these applications to actively monitor and maintain slurry quality? Identify potential areas of improvement in the next-generation CMP slurry management. Review various approaches employed for slurry reuse and CMP waste management. What potential may they have in reducing CMP cost of ownership?
5. What are the main considerations in selecting a suitable slurry delivery system for specific slurry? How do slurry chemical and abrasive properties, blending and metrology needs, and slurry consumption and replenishment behavior can influence this selection? What operating conditions and flow geometries may cause significant shear to slurries, and what types of slurries are more likely to be affected by extended handling? Discuss various approaches employed for slurry delivery system maintenance.
6. What are the benefits and limitations of using filtration during CMP slurry distribution. Identify key slurry filter design considerations. Describe various methods used for slurry filter characterization. What slurry properties should be considered in selecting suitable filters for usage during slurry manufacturing, and blending and distribution?
7. Discuss various filtration approaches used in CMP slurry abrasive particle management. How are the requirements of new slurries filtration likely to change for future low-defectivity CMP processes? What considerations should be made to get stable and extended lifetime from CMP slurry filters? How may next-generation slurries low abrasive content and/or complex abrasives change filtration optimization?
8. Review benefits and limitations of common pumping technologies employed in chemical process industries. Identify key objectives in selecting a pump for CMP slurry blending and delivery. Describe how the slurry content and usage cycle can influence pump selection and its mode of operation and preventive maintenance schedule. Summarize typical pump handling effects on different slurries and identify the suitable pumps for shear-sensitive slurry. What precautions should be taken in such slurry distribution to limit large-particle agglomeration and related increased defectivity in CMP processing?

REFERENCES

1. Derbyshire K. Copper at a crossroads. Semiconduct Manuf 2005;6 (10):17–20.
2. Carpio R, Pham J, Tolic F, Hymes S, Bajaj R. CMP pad design for ultra-low-*k* compatible Cu CMP process. Proceedings of the 23rd International VMIC;2006.p 438–443.
3. Singh RK, Bajaj R. Advances in chemical–mechanical planarization. MRS Bulletin 2002;27(10):743–747.

4. Moinpour M, Tregub A, Oehler A, Cadien K. Advances in characterization of CMP consumables. MRS Bulletin 2002;27(10):766–771.
5. Johl B, Singh RK. Optimum process performance through better CMP slurry management. Solid State Technol August 2003;63–66.
6. Singh RK, Roberts BR. Behavior of CMP slurry properties in continuous blending and distribution systems. Proceedings of the 17th International VMIC; 2000. p 545–547.
7. Singh RK, Roberts BR. CMP slurry metrology: various approaches. Proceeding of the 2nd International AVS ICMI Conference;2001 Feb 5–9; Santa Clara, CA.
8. Seo Y-J, Kim SY, Lee W-S. Advantages of point of use (POU) slurry filter and high spray method for reduction of CMP process defects. Microelectron Eng 2003;70: 1–6.
9. Singh RK, Patel C, Conner G, Yang HJ, Towle T. Advances in CMP filtration technology. Proceedings of the Semicon China 2004 SEMI Technology Symposia (CMP); 2004 Mar 17–19; Shanghai.
10. Vasilopoulos G, Lin Z, Johl B, Joshi S, Chatterjee B. Copper CMP defect reduction using POU filtration. Proceedings of CMP Technology for ULSI Intercon. Semicon West; 2000 Jul; San Francisco, CA.
11. Derivendt K *et al.* (IMEC), Lin Z CMP defectivity and slurry filtration. Proceedings of MRS Meeting;1999 Apr; San Francisco, CA.
12. Nicholes K, Singh RK, Grant DC, Litchy MR. Measuring particles in CMP slurries. Semiconduct Int July 2001;24(8):201–206.
13. Singh RK. Pumping effects on CMP slurry abrasive particle characteristics. Semiconduct Manuf 2006;7(1):40–43.
14. Singh RK. High-retention filtration of CMP slurries. Semiconduct Int 2005;28(9):47–53.
15. Singh RK, Conner G, Roberts BR. Handling and filtration evaluation of a colloidal silica CMP slurry. Solid State Technol November 2004;47(11):61–66.
16. Singh RK, Patel C, Conner G, Towle T, Viscomi R, Federau M. Efficient filtration of new-generation CMP slurries: challenges and solutions. Semiconduct Manuf 2004;5(6):70–84.
17. Sampurno Y, Philipossian A, Choi HK, Moinpour M, Rawat A, Kim A. Stability of CMP slurries. Proceedings of the CAMP's 11th International Symposia on CMP; 2006 Aug 13–16; Lake Placid, NY.
18. Singh RK, Roberts BR. On extensive pump handling of chemical–mechanical polishing slurries. Proceedings of the 12th Annual IEEE/SEMI ASMC; 2001 Apr 23–24; Munich, Germany.
19. Juta T, Bigman J, Singh RK. Monitoring and control of CMP slurry oxidizer concentration and particle characteristics using NIR absorption spectroscopy. Proceedings of the CAMP's 10th International Symposia on CMP; 2005 Aug 14–17; Lake Placid, NY.
20. Singh RK, Roberts BR. On sedimentation and redispersion of abrasive particles in CMP slurries. Proceedings of the 6th International CMP-MIC. 2001. p 441–448.
21. Roberts BR, Singh RK, Pierce A, Blachier O. Slurry quality and consistency in a continuous slurry blending and distribution system. Proceedings of the 16th International VMIC. 1999.p 495–500.

22. Lin BT, Lee SN, Tseng SM, Ho PH. Novel tungsten CMP process with diluted slurry. Proceedings of the 16th International VMIC. 1999. p 636–641.
23. Singh RK, Hagerman S, Wargo C, Roberts BR. Effective dispersion of CMP slurry abrasive particles. Proceedings of the 23rd International VMIC. 2006. p 143–146.
24. QuickSCAN™ Beckman Coulter Product Reference Manual 8309550 Rev. A; 2000.
25. Philipossian A, Racz L, Lu J, Rogers C. Selected process consumable technology requirements for advanced CMP processes. Technical Program Presentation on CMP Technology for ULSI Interconnection, SEMICON West. 2000.
26. Lin Z, Vasilopoulos G, Adrian K, Blum R, Friedman P. CMP filter plugging mechanisms and filter lifetime optimization. Proceedings of the 4th International CMP-MIC. 1999.
27. Ives KJ. Mathematical models of deep bed filtration. In Ives KJ, editor. The Scientific Basis of Filtration. Noordhoff, Leyden; 1975.
28. Lee KW, Liu B Y H. Theoretical study of aerosol filtration by fibrous filters. Aerosol Sci Technol 1982;1:147–161.
29. Spielman L, Goren SL. Model for predicting pressure drop and filtration efficiency in fibrous media. Environment Sci Technol 1969;2:279–287.
30. Singh RK, Roberts BR, Viscomi R, Maxim M, Diaz M, Conner G. Behavior of EP-C600Y-75B copper CMP slurry under extensive handling. Proceedings of the 8th International CMP-MIC. 2003.
31. Singh RK, Wargo CR, Rosales-Yeomans D, Philipossian A. Shear force and deformation studies of poly vinyl alcohol (PVA) brush rollers during post-CMP scrubbing for copper applications. Proceedings of the 3rd International Surface Cleaning Workshop; 2004. Oct 4–5; Boston.
32. International Technology Roadmap for Semiconductors. 2005.
33. Chiarello R, Boyd CE, Wacinski C, Thomas K, Elkind J, Presley B, McDermott R, Harakas G. Using a real-time, point-of-use sensor to control liquid-chemical concentration. Micro 2005;5:33–40.
34. Singer P. Low-downforce planarization combines electropolish and CMP. Semiconduct Int 2004;27(9):24.
35. Von Trotha L, Mörsch G, Zwicker G. Advanced MEMS fabrication using CMP. Semiconduct Int 2004;27(9):54–56.
36. Karuppiah L, Swedek B, Thothadri M, Hsu WY, Brezoczky T, Ravid A. Overview of CMP process control strategies. Proceedings of the 11th International CMP-MIC. 2006. p 45–54.
37. Philipossian A, Shadman F, Levy P, Tousi S, Gotlinsky B, Rader WS, Lefevre P, Koshiyama I. Characterizing recycled fumed silica slurries in ILD CMP applications. Micro 2005;5:71–82.
38. Kuntz LA. Dealing with shear. Food Product Design, Nov 1995.
39. Glazebrook S. Pumping delicate and shear-sensitive products. Plant Services; Aug1997.
40. Chan WK, Akamatsu T, Li HD. Analytical investigation of leakage flow in disk clearance of a magnetically suspended centrifugal impeller. Artif Organs 2000;24(9):734–742.

41. Paul R, Apel J, Klaus S, Schügner F, Schwindke P, Reul H. Shear stress related blood damage in laminar couette flow. Artif Organs 2003;27(6):517–529.

42. Yadav R, Singh RK. Prediction of flow field in impellers of centrifugal pumps. Proceedings of the 5th International Conference on Numerical Methods in Laminar and Turbulent Flow. Swansea, U.K: Pineridge Press; 1987 p1977–1988.

43. Park CH, Nishimura K, Akamatsu T, Tsukiya T, Matsuda K, Ban T. A new magnetically suspended centrifugal blood pump: in vitro and preliminary in vivo assessment. Artif Organs 1996;20(2):128–131.

44. Bare J, Johl B, Lemke T. Comparison of vacuum-pressure vs. pump dispense engines for CMP slurry distributionProceedings of SEMICON/West Contamination in Liquid Chemical Distribution Systems Workshop; 1998 July.

45. Litchy M, Schoeb R. Effects of shear stress and pump methods on CMP slurry. Semiconduct Int November 2004;27(12):87–90.

46. Singh RK. CMP pump effects on filter life. Proceedings of the Levitronix CMP Users Conference;2005 Feb 17; Santa Clara.

47. Singh RK, Wargo CR, Roberts BR, Viscomi R, Federau M. Handling and filtration characterization of SiLECT 6000™ STI CMP slurry. Proceedings of the 22nd International VMIC. 2005.

48. Pohl MC, Griffiths DA. The Importance of particle size to the performance of abrasive particles in the CMP process. Electron Mater 1996;25:1612–1616.

49. Singh RK, Pumping technologies for CMP slurries. Proceedings of the 11th International CMP-MIC. 2006. p 429–432.

19

THE FACILITIES SIDE OF CMP

John H. Rydzewski

19.1 INTRODUCTION

Until now, the focus of the discussion has dealt with numerous topics pertaining to the wafer and the process tools within the cleanroom. To truly appreciate CMP, one must also have an understanding of the overall life cycle of the slurries used during processing. Process development and manufacturing engineers—when considering new or revised CMP chemistries—must also take into consideration the facility systems that treat CMP waste. A failure or a shutdown of these systems could eventually shut down the wafer manufacturing process. Therefore, this chapter will focus on key topics that must be understood and addressed when developing new or optimizing the existing CMP chemistries.

As semiconductor processing delves deeper into the more exotic regions of the periodic table, CMP waste treatment becomes more complex. In early generations of semiconductor manufacturing, the layers that were planarized (before the evolution of "CMP") were relatively benign, and the waste stream did not require special treatment prior to being discharged to the publicly owned treatment works (POTW), or city sewer. As wafer manufacturing chemistries evolved in complexity, the need for specialized waste treatment systems has become greater than ever. In addition, more consideration must be given to factory environmental discharge permits and the effect the different chemistries could have on the operation of the POTW. As a result, most contemporary semiconductor-manufacturing facilities now have multiple specialized waste treatment systems to segregate specific contaminants, such as copper and/or suspended silica, prior to discharging the waste.

Microelectronic Applications of Chemical Mechanical Planarization, Edited by Yuzhuo Li

In some cases, the dwindling global supply of fresh water is forcing some semiconductor manufacturers—whose factories each consume up to 1 million gallons of fresh water per day—to think proactively about water conservation through recycling and reuse. To provide a consistent supply of high-quality water back to the clean room through the ultrapure water production system, and/or to reclaim the CMP wastewater (which can consist of almost 40% of the water consumption in the wafer manufacturing process), the CMP waste treatment system must be effective and robust to prevent negative impacts on wafer processing or on facility systems (boilers, chillers, scrubbers) that would otherwise use fresh city water.

Each CMP waste system will have two primary components:

- Collection piping
- Treatment system

A collection system is an array of piping under the manufacturing tools that accumulates the waste at a single point, such as a buffer or storage tank. The contents of the tank are then pumped to the unit operations of treatment system. The treatment system is a series of unit operations that modify the waste to produce a desired result, such as the removal of trace metals from the liquid stream.

This chapter will describe the various aspects that must not only be appreciated by the facilities engineer designing the CMP waste treatment system, but also be taken into consideration by the process engineers developing new CMP chemistries.

19.2 CHARACTERIZATION OF THE CMP WASTE STREAM

Specialized collection and treatment systems are designed and operated based on the specific properties of the CMP waste drained from the clean room. Therefore, it is paramount to understand the details of the CMP waste, such as, but not limited to, the following:

- Chemical composition of suspended solids
- Size and shape of suspended solids
- Concentration of suspended solids in waste stream
- Chemical composition and concentration of dissolved solids
- Chemical composition and concentrations of oxidizers
- Composition and concentrations of any organics including corrosion inhibitors, complexing agents, surfactants, etchants, and buffers
- Waste stream flow rates
- Temperature of waste stream

To perform due diligence to characterize the waste stream, the facilities engineer may be faced with some challenges such as the following:

- Research engineers developing proprietary CMP chemistries do not wish to reveal the constituents of the chemistry.
- CMP waste quality during the manufacturing process development phase is highly variable owing to continued experimentation and R&D activities.
- The volume of CMP waste generated during process development stages may not be sufficient to perform adequate material testing for waste treatment options.

It is important that the manufacturing process engineer attempts to minimize the variability in the characteristics of the waste to be treated. Although the facilities engineer can minimize the impact of variations to the treatment system by damping the variations through the use of a collection tank, it is still important for the manufacturing engineer to provide some level of consistency. Otherwise, wide swings in waste quality could result in process upsets that could cause the treatment system to shut down or not effectively treat the waste stream.

Another key aspect to treating CMP waste streams is to ensure that only the wastes assigned to the waste treatment system are sent down the drain. In short, a specific waste drain is not to be used as the general-purpose drain. A drain must be assigned for the collection of a specific CMP waste since the treatment system is designed around the composition of the waste.

The main drivers for understanding the properties of the waste stream are the definition of the following:

- Materials of compatibility
- Collection system methodologies
- Treatment system methodologies

19.3 MATERIALS OF COMPATIBILITY

Once the waste stream has been characterized, it is possible to select the proper materials of construction for waste treatment facility. Proper selection will prevent the waste collection and treatment systems from corroding, eroding, and/or falling apart when exposed to the waste stream. When selecting materials for the waste treatment system, one must consider an array of factors such as cost, ease of maintenance, replacement, and chemical compatibility for future potential chemical usage. It is in the best interest of the waste system design engineers, to the best of their ability, to select the proper materials of construction.

Plastic piping rather than metal piping is commonly used in CMP waste collection systems because plastic piping tends to be less expensive, easier to obtain, and more compatible to a broader range of chemicals. Commercially available plastic piping systems include polypropylene (PP), polyethylene (PE), polyvinyl chloride (PVC), chlorinated PVC (CPVC), perfluoroalkoxy (PFA), and polyvinylidene fluoride (PVDF). All the manufacturers of these piping materials have material of compatibility tables providing an overview of the compatibility of their materials with the commonly used chemistries. Unfortunately, since CMP slurry waste is not a single component, but a mixture of various chemistries, it is important to work with the piping suppliers to better understand the compatibility of the piping materials with a particular CMP chemistry.

As the industry moves toward more exotic chemistries and oxidizers, there may be a need for higher performance (and significantly more expensive!) fluoropolymers such as PFA and PVDF. To define the appropriate materials for a specific waste treatment system, it is best to design a series of experiments involving soak tests with samples of the CMP waste with the proposed materials of construction. To ensure proper material selection, data collection during the experimentation should include the following:

- Surface porosity and corrosion cracking
- Swelling and material weight gain
- Tensile stress testing
- Discoloration

In addition to selecting the proper piping materials based on soak test of the base material, it is important to note that the jointing method for these piping systems could influence material selection. For example, solvent cements are most commonly used for the installation of CPVC and PVC, whereas PVDF, PP, and PE are jointed using thermal fusion welding techniques. Although many PVC and CPVC suppliers indicate that the solvent cement used to joint PVC and CPVC contain the same materials as the PVC and CPVC piping, it must be noted that the PVC and CPVC solvent cement also contain solvents and additives (to make it a solvent cement) that may or may not be compatible with the CMP waste. Yet another complexity with PVC and CPVC piping systems is that all PVC and CPVC pipe manufacturers make their own proprietary blends of PVC or CPVC resin with various additives and compounds. Therefore, it is important to test not only each family of materials but also the PVC and CPVC pipe and solvent joints from various manufacturers.

PP, PVDF, and PE are not jointed using solvent cements, but the weld regions contain relatively high levels of intrinsic stress caused by the thermal fusion process. Chemical stress cracking could occur at these welds if exposed to certain chemicals. When performing material compatibility tests for any plastic materials, it is important to test not only the base material but also several sample welds.

The term "materials of construction" applies not only to the piping in the waste collection system and the major equipment in the treatment system but also to elastomers used as gaskets at pipe flanges and as seals in valves and major unit operations. EPDM rubber generally performs better in alkaline solutions whereas Viton™ generally performs better in acidic solutions. Expanded PTFE is compatible with the full range of pH, but it has its challenges since some chemicals (such as hydrogen chloride) permeate through the PTFE. There are also more exotic and more expensive fluoropolymers, but a technical and economic evaluation should be completed prior to use.

The facilities engineer must consider the materials used to construct the wetted parts of the unit operations used to treat the waste. The goal is to select a cost-effective material that will not degrade in the presence of the waste stream to be treated. In general, the most commonly used materials of construction for the unit operations are 316 or 316 L stainless steel. However, if chlorides, fluorides, and other halogens are present in the CMP chemistry, it may be necessary to consider nickel, titanium, Hastelloy, or other high-performance metals.

There are a wide variety of materials available for use with CMP waste treatment systems. The challenge is to match the proper materials with the CMP chemistry with a proper financial and technical analysis. Due diligence performed in the short term will prevent increased operating expenditures in the long term.

19.4 COLLECTION SYSTEM METHODOLOGIES

When waste leaves the clean room, it is discharged to a collection system under the clean room floor. This collection system must be adequately designed to handle the flow, temperature, solids content, and chemistry of the waste stream. Most collection systems that drain waste from the clean room are gravity drain systems at atmospheric conditions. It is imperative to give proper attention to the piping design of the collection system to prevent solids accumulation and blockages in the drain system. Solids will accumulate in zero-slope pipe, crevices, gaps, and zero-flow regions. A general rule of thumb is to maintain a minimum slope of 1% (1/8-in. drop per 1 ft of horizontal length). Solids content of the CMP waste must also be considered when designing the collection system. For waste streams that are more viscous or contain a high percentage of solids, a 1% slope may not be sufficient to maintain turbulent flow. Pipe jointing method must also be considered, since each jointing method will result in a gap or a slight protrusion of piping material into the flow stream. Depending on the CMP waste stream chemistry, these gaps and protrusions could play a role in solids accumulation if the turbulence in the liquid is reduced by a reduction either in flow or in the design slopes of the piping system.

Additional considerations must be given to the potential for creating precipitation by-products if the CMP waste stream is mixed with other chemistries used in the manufacturing process that could accidentally be placed in the segregated drain. For example, if the CMP waste stream contains fluoride and city water is used to flush the piping system, then the calcium in the city water could react with the fluoride to produce a scale on the pipe, thus reducing the flow capacity of the pipe. Once precipitation starts, it will always be present since all seed crystals cannot be 100% eliminated from hundreds of feet of piping.

19.5 TREATMENT SYSTEM COMPONENTS

The basic premise of CMP waste treatment is to remove the target trace metals (Cu, W, Ta, etc.). The primary challenges that add complexity to the CMP waste treatment systems is the presence of the suspended solids, organics added to stabilize metal ions that may be counteractive to some of the treatment chemistries, and oxidizers in the slurry waste.

The basic components of a CMP waste treatment system are as follows:

- Collection and pH adjust tank
- Oxidizer removal
- Organics removal
- Treatment of total suspended solids
- Removal of trace metals

Depending on the characteristics of the CMP waste and the environmental discharge permits negotiated with the local authorities, some steps (such as control of suspended solids removal) may not be required, but every CMP waste treatment system requires a method for removing the trace metals. The rest of this chapter will discuss each aspect of the waste treatment system and the pros and cons of combining the unit operations to create a functional treatment system.

19.5.1 Collection Tank and pH Adjustment

The waste flow and composition will vary for a number of reasons. Wide variation in the waste stream characteristics may cause process upsets that could take the system out of control and cause equipment failure. The primary method for reducing this variability is to install a tank to collect the waste from the clean room. The tank serves several purposes:

- A reservoir from which a pump will push the waste through the treatment system at a constant flow

- Time average and homogenize the waste composition to minimize perturbations to the treatment system
- Storage capacity in case of downstream equipment failure, which will allow the factory to continue manufacturing

The tank is the first line of defense to ensure proper treatment system operation. In some instances, a collection tank can be used as a pH adjust tank. However, to optimize the pH adjustment step, it may be value added to install another smaller tank downstream of the collection tank. A smaller pH adjust tank would result in a smaller mixer and better aspect ratios in the tank to allow for homogeneity of the waste sent to the treatment system. The pH set point will depend on the type of technology used to remove the trace metals. For example, cation ion-exchange resin will capture trace metals and will regenerate when exposed to a low pH solution. It is important that the pH adjust tank is equipped with two-sided (acid and alkali injections) control in the event the control system overshoots the set point. To ensure stable pH control in a single tank, it is important to provide adequate retention time (tank size) and mixing of the tank contents. Since a storage tank may provide numerous days of waste storage, the mixing for pH control will also prevent the solids in the slurry from accumulating on the bottom of the tank.

19.5.2 Oxidizer Removal

As noted in Chapter 7, oxidizers play a key role in the CMP process. By the same token, oxidizers, to the treatment system designers, are a necessary evil. One oxidizer that is ubiquitous in CMP processing is hydrogen peroxide (H_2O_2). H_2O_2 is used in copper CMP processing because it interacts immediately with copper to form a copper oxide (metal oxides are generally softer than the base metal; the wafer becomes easier to polish) and the non-contaminating by-products hydrogen and oxygen. It is for this reason that H_2O_2 is generally transported and stored in plastic containers. With respect to CMP waste treatment, since H_2O_2 does not react with the plastic piping systems to decompose into hydrogen and oxygen, it becomes imperative to remove the molecular H_2O_2 from the CMP waste stream prior to metals removal because H_2O_2 damage to ion-exchange resin is generally instantaneous [1]. Some strategies for H_2O_2 removal are as follows:

- Granulated activated carbon
- Injection of reducing agent
- Ultraviolet radiation

Granulated activated carbon (GAC) is commonly used in the water purification industry to remove oxidizers—such as chlorine—and some organics from municipal water treatment systems. GAC has also found a useful role in the removal of H_2O_2 from CMP waste streams due to the operational simplicity of this methodology: fill a pressure vessel with activated carbon and allow the waste to flow through the vessel. The carbon will reduce the H_2O_2 to form hydrogen and oxygen [2].

With all its explicit simplicity, GAC has several technical challenges, namely, the formation of oxygen and hydrogen gas, the possibility of clogging the bed with slurry, and additional particulate generation caused by the oxidation of the GAC. It is very important to remove the hydrogen and oxygen gas from the carbon bed, since a combination of the three, in the proper conditions, could create an explosive device. The gas can be removed using a variety of techniques, including installing commercially available gas vents on the top of the pressure vessels, or flowing the waste stream upward through the carbon bed and pushing it downstream to an exhausted atmospheric tank. Another technical challenge that may arise is when the slurry is fed directly into the GAC bed, resulting in excessive pressure drops or clogging of the bed caused by the suspended solids in the slurry. Clogging could result in higher-than-normal pressure drops, reduced flows, and eventual channeling through the bed. Channeling will minimize the contact of the GAC with the waste stream, allowing the oxidizer-laden waste to permanently damage downstream equipment. Excessive clogging may require frequent backwashing to clear the GAC bed of slurry particles. Clogging may be resolved by selecting a sufficiently coarse GAC that allows the slurry particles to flow between the carbon particles; dilution of the slurry; or even operating the GAC bed as an upward-flow fluidized bed, preventing compression of the GAC and slurry particles. A third challenge—particulate generation caused by the oxidation of the GAC by the H_2O_2—could result in carbon fines clogging the GAC bed or downstream equipment and cause channeling. The proper technical solution will depend on the specific character of the CMP waste discharged from the clean room.

Another commonly used method for removing the H_2O_2 from the waste stream also originates from the water purification industry's need to remove chlorine from the city water [3]: the injection of sodium bisulfite ($Na_2S_2O_5$). Since sodium bisulfite injection requires instrumentation, control loops, and metering pumps, it is more complex than using GAC, but it is a relatively simple and mature technology to implement. However, because of the need for a continuous supply of sodium bisulfite, operating costs may be higher since treatment systems that implement sodium bisulfite injection tend to overdose sodium bisulfite to ensure all the peroxide has been reduced. Also, prior to using sodium bisulfite, it is critical to determine if sodium bisulfite is compatible with the waste.

A third method for oxidizer reduction, namely, ultraviolet radiation, is based on recent developments made in the water purification industry to remove residual chlorine from city water used by the food, beverage, and pharmaceutical industries. In those applications, ultraviolet light at wavelengths between 254 and 310 nm is used to reduce the chlorine and chloramines into nonoxidizing chlorides [4]. With respect to H_2O_2, UV light with a wavelength between 200 and 250 nm will convert H_2O_2 into hydroxyl radicals, which will recombine into water downstream of the UV reactor. The primary challenge with CMP wastewater is the presence of suspended solids. To gain the maximum efficiency from a UV source, the UV light must be able to transmit through the bulk fluid. In slurry, the suspended solids hinder the transmittance of the light, limiting the exposure of the bulk fluid to the UV light. This is not a widely used technique for the reduction of H_2O_2 in CMP wastewater, and specific applications are still under development [5].

Oxidizer removal will continue to be one of the biggest challenges facing the development of future CMP waste treatment systems, especially if more exotic oxidizers are required by more exotic metal layers on the microchip. However, if dissolved molecular oxygen, as noted in Chapter 7, becomes a future standard practice, then there may not be a need for oxidizer removal, since dissolved oxygen poses no compatibility issues with the current toolbox of waste treatment technologies. However, if the semiconductor industry opts to move toward more exotic oxidizers such as KIO_3, ammonium persulfate, zinc peroxide, and others also noted in Chapter 7, then existing CMP waste treatment systems will need to be analyzed at various levels (piping materials of compatibility, oxidizer removal methodology, etc.) and possibly redesigned to meet the challenges posed by the new waste streams.

19.5.3 Organics Removal

Although there is much focus on the metal ion components of a CMP wastewater, one must not neglect the presence of organics. These organics may be in the form of metal chelators, dispersants, surfactants, and other proprietary chemicals the manufacturing engineers may add to the slurry recipe. These organics, depending on the reasons for which they were added to the CMP slurry, may have an impact on the CMP wastewater treatment process, and it may be necessary to remove them prior to removing the target metal ions. Although this may seem as though another unit operation is required in the CMP wastewater treatment system, smart selection of unit operations will allow for organics removal with possibly little additional capital cost. For example, not only has activated carbon demonstrated its ability to reduce hydrogen peroxide, but it is also well known in the water purification industry as a means for removing organics [6]. The same is true for ultraviolet light [7]. Also, it may be possible to use new CMP waste treatment technologies—such as electrocoagulation and photochemistry, which will be discussed later—that also oxidize the organic constituents of the CMP slurry.

19.5.4 Treatment of Suspended Solids

Suspended solids are another special characteristic of CMP wastewater. Removal or concentration of the solids in the waste stream may or may not be required, depending on the waste treatment strategy that is implemented. Owing to the sheer number and types of filters and filter applications that exist today, an extensive literature database on filtration already exists [8–11]. This section will provide a brief overview of the filtration techniques generally used for CMP wastewater treatment and discuss strategies to remove or concentrate the solid phase of the waste stream using one or more of the following:

- Microfiltration
- Ultrafiltration
- Clarifiers

Microfiltration (MF) refers to the removal of particles that range from 0.05 μm to less than 2.0 μm in diameter. Microfilters are available in various forms including cartridge and bag filters, and the materials of construction can vary from positively charged nylon to Teflon™ and ceramic membranes. The filter selection will depend on the characteristics of the slurry and the intended role of the filter. Microfiltration is a process where 100% of the liquid waste stream must pass through the filter, as opposed to a cross-flow system where the fluid being filtered runs along the filter membrane, resulting in a reject stream [9]. The filter acts as a physical barrier to prevent solids from passing downstream. Because of the challenges of filtering slurry, noted in Chapter 7, the design engineer must take care in selecting the proper filter to prevent premature blockages, excessive pressure differentials, and flow reductions that could result in very high operating costs in terms of frequent filter replacement or the failure and shutdown of the treatment system.

Ultrafiltration (UF) refers to the removal of high molecular weight colloids (10,000 MW) up to particles less than 0.05 μm in diameter [10]. Like MF, UF places a mechanical barrier into the flow stream to separate the solid and liquid phases. The most common UF is a cross-flow hollow fiber type whereby a UF module contains hundreds of hollow microfibers. Whether or not the medium to be filtered flows inside or outside, the microfibers depends on the characteristics of the waste stream. UF is different from MF not only because UF can filter very small particles and some colloids, but also because of the cross-flow dynamics inside the UF module that keeps the surface of the hollow fibers clean. Because of the cross flow, UF modules require a reject stream as well as a permeate stream. In other words, 100% of the liquid that enters the UF module does not exit as permeate. Figure 19.1 shows the differences between the UF and microfilters with respect to flow path.

In slurry service, UF is used primarily as a concentration technology. The UF reject stream can then be recycled back to a tank containing the concentrated slurry that can then be dewatered with a filter press. The permeate will contain little, if any, suspended solids and can be sent for metals removal treatment or to recycle/reclaim if metals have already been removed.

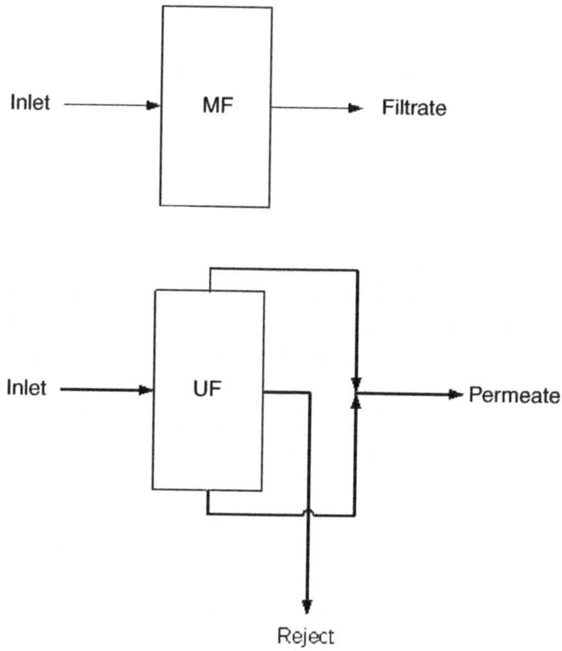

FIGURE 19.1 Schematic showing flow path difference between MF and UF technologies.

Inclined plate clarification is a traditional separation technology that has been used for decades to remove suspended solids from a liquid stream in various types of systems including traditional precipitation [12]. In the semiconductor industry, the clarifier is commonly used in fluoride waste treatment systems where calcium fluoride precipitate is concentrated prior to dewatering in a press, or in assembly/test operations to separate silicon fines from backgrind operations. The clarifier will concentrate the solid phase of slurry like a UF, but unlike the UF or MF, the clarifier may require the addition of a chemical polymer to facilitate the agglomeration of the suspended solids so that they settle and concentrate. Polymer addition adds another level of complexity to the waste treatment system. The clarifier does not provide a physical barrier to prevent the transport of solids to downstream equipment, so it may be necessary to install a UF or MF downstream of the clarifier to capture extraneous particles or to protect the downstream equipment from clarifier upsets.

The key challenges with solids removal will be determining the most cost-effective way to remove the solids. Each of the technologies discussed here—microfiltration, ultrafiltration, and clarification—have their benefits and limitations. The big challenge with solids removal, as will be discussed later in this chapter, will be to determine when in the treatment system solids removal should occur.

19.5.5 Removal of Trace Metals

If there was no metallic contamination in the CMP wastewater, then, generally speaking, the waste could be discharged directly to the POTW without any environmental issues or concern. This was demonstrated in the early days of semiconductor processing before copper interconnects became the industry standard. Since the introduction of copper processing, a whole new industry that provides waste treatment systems has blossomed.

Removal of trace metals from liquid streams is not new and has been accomplished using as ion-exchange resin, seen in Fig. 19.2, which has been used in the water purification industry since the 1940s [13–16]. Most of the challenges associated with CMP waste treatment center around pretreating the waste stream prior to the trace metals removal step. Once the waste stream has been properly conditioned, it becomes a matter of removing the trace metal contaminants. For metals removal in CMP, cation resins are used.

Ion exchange is a reversible, stoichiometric chemical reaction where a target ion with a given charge is exchanged with a similar ion with the same charge that is attached to a solid inert (polystyrene resin) substrate. Cation resin is used to remove trace metals from liquid streams and achieves this with a sulfonic functional group (SO_3H) that bridges the H^+ ion that is eventually exchanged for the metal ion, as seen in Fig. 19.3.

The ability of the resin to remove the target ion is dependent on the number of available ion-exchange sites (capacity), and selectivity, or relative propensity of the resin to capture specific contaminants. When considering common cations, their charge and relative selectivity are

$$Pb^{+2} > Ca^{+2} > Ni^{+2} > Zn^{+2} > K^{+} > Na^{+} > H^{+} > Li^{+}$$

FIGURE 19.2 A photograph of ion-exchange resin beads at approximately 15× magnification.

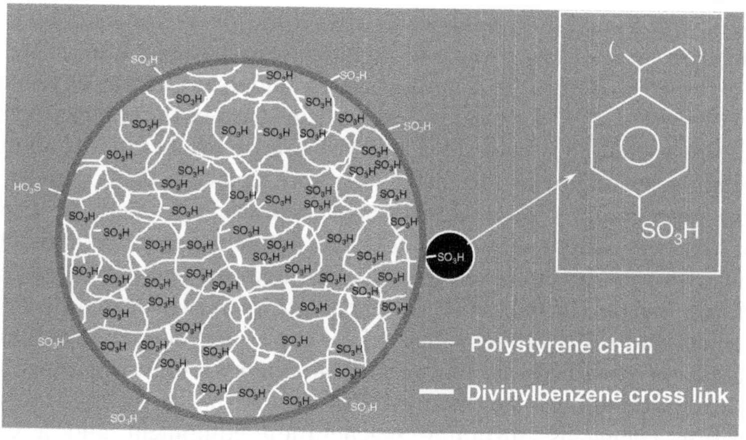

FIGURE 19.3 Schematic of a cation resin bead showing the polystyrene chains with sulfonic groups, the divinylbenzene cross-links, and the active sulfonic groups where the ion exchange occurs. Because the void spaces between the polymer chains and cross-links contain water, the effective ion-exchange surface area of the bead is not limited to its outside diameter.

In terms of capacity, since Pb^{+2}, Ca^{+2}, Ni^{+2}, Cu^{+2}, and Zn^{+2} are divalent ions, and ion exchange is a stoichiometric reaction, two H^+ sites are required for ion exchange to occur. Therefore, cation resin bed with a finite number of active sites will be able to capture only half the number of Pb^{+2}, Ca^{+2}, Ni^{+2}, Cu^{+2}, and Zn^{+2} ions when compared to the number of Na^+ and K^+ ions.

Cation resins will capture various cations in varying degrees, depending on their selectivity. A cation resin will capture both copper and nickel, but since nickel has a higher selectivity than copper, the nickel will more readily exchange with the available H^+ sites, as well as the available Cu^{+2}, Zn^{+2}, K^+, and Na^+ sites that were once H^+ sites. This would then cause the Cu^{+2}, Zn^{+2}, K^+, and Na^+ to enter the bulk fluid to be captured by another resin bead deeper in the resin bed. This phenomenon is known as the wave front. The wave front is the layer of partially exhausted ion-exchange resin in the bed that separates the fully exhausted resin from the fully regenerated resin. As the wave front reaches the bottom of the bed, some less selective target ions may be displaced by those with a greater selectivity and then flow out of the bed. This is known as breakthrough, and it can occur before the bed is completely exhausted. To prevent inadvertent or premature breakthrough, it is very important to fully understand the composition of the waste stream to be treated so that the cation beds can be properly sized, and to determine if any chemical reagents—such as pH adjust chemicals—contain high levels of metal contaminants that would interfere with the capture of the target trace metal ions, or result in premature exhaustion of the bed. In this case, it may be possible to work with the ion-exchange resin manufacturers to identify resins that may be specifically designed to be selective to one or a few elements. The challenges with using ion-specific resins arise if the CMP waste stream contains more than one type of target ion and the other ions are not captured by an ion-selective resin. Therefore, it is important to understand local environmental discharge permits as well as the types of chemistries in the waste stream.

Since ion exchange is reversible, the bed can be regenerated to close to its original state by removing the metal ions and replacing them with H^+. One common regenerant is hydrochloric acid. Sulfuric acid may be used with caution since the cations may react with the sulfate group to permanently foul the resin with solid precipitates. Once regenerated, the bed can be placed back into service. Because of the ability to regenerate and reuse the cation resin, it is a convenient and relatively easy method for capturing the metal contaminants, especially if the resin beds are regenerated off-site by a vendor. However, if the beds are regenerated on-site, the cost and complexity of storing/transporting regeneration chemicals, installing regeneration equipment and a chemical distribution system, operating the regeneration system, and disposing of a waste stream now containing concentrated acid and concentrated metal ions must be considered. In the event, the facility is having a concentrated acid collection tank whose contents are removed off-site by tanker truck for resale or reuse, it may be possible to do on-site regenerations. In this case, it becomes very important to understand the chemical composition of the concentrated chemical to ensure that it is properly dispositioned.

Because low pH solutions will regenerate cation resins, the waste stream contacting the resin must be at a relatively neutral pH. If the pH falls, then regeneration could occur, and the trace metals captured on the resin will be released into the bulk fluid (at a higher concentration than the inlet stream due to the accumulation of metal ions in the bed). If the pH is too high, metal hydroxides could precipitate out of solution and permanently foul the resin. It is for this reason that the waste stream must be properly conditioned prior to contacting the resin.

It is important to remove oxidizers prior to removing metals. Ion-exchange resin is made of divinylbenzene and styrene and will easily oxidize in the presence of strong oxidizers such as hydrogen peroxide. Hydrogen peroxide will cause permanent structural damage to the resin beads, resulting in a loss of ion-exchange capacity. Any captured metal ion may be released back into the flow stream in concentrations that would possibly violate the environmental discharge permit. Also, gaseous hydrogen and oxygen will evolve during the oxidation of the resin, which could be hazardous if in large enough concentrations. The waste stream must be properly conditioned prior to contacting the resin.

When ion exchange is implemented, there are generally space restrictions that limit the number and size of the beds. For this reason, it is important to maximize the contact time between the waste stream and ion-exchange resin while minimizing the overall footprint space of the equipment. This can be accomplished with a lead-lag rather than a single-pass bed configuration. These configurations are shown in Figs. 19.4 and 19.5, respectively.

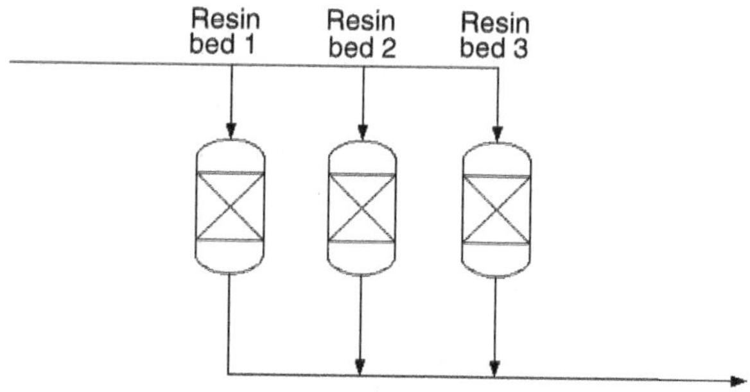

FIGURE 19.4 Schematic of cation resin beds in parallel configuration. This is a single pass configuration where the resin contact time with the waste stream is minimized. All three beds will breakthrough simultaneously, resulting in shutting the system down to replace or regenerate all of the ion-exchange resin.

When two resin beds are in a lead-lag configuration, the lead bed performs the primary ion exchange to remove the bulk of the trace metals, while the second bed acts as a polishing bed to remove trace metals that break through the first bed. The lead bed will exhaust and break through first, and the

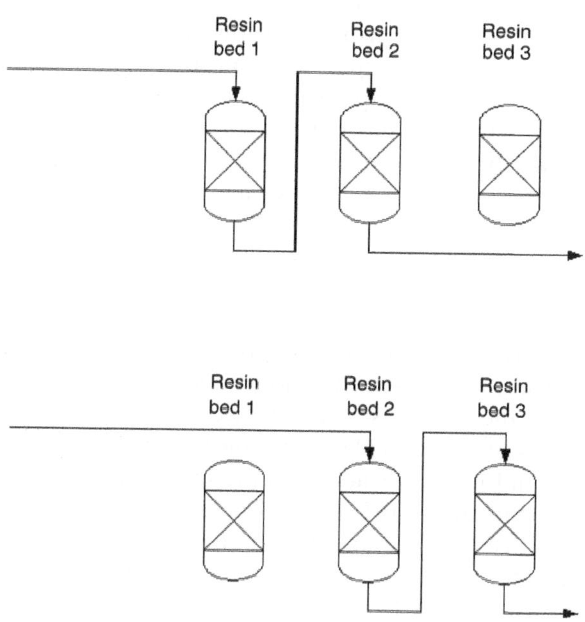

FIGURE 19.5 Schematic of cation resin beds in lead-lag configuration. This is a double pass configuration where the resin contact time with the waste stream is maximized. bed 1 sees the most concentrated trace metals in the waste stream and will break through prior to bed 2. When bed 1 breaks through, bed 2 becomes the lead bed and bed 3 becomes the lag bed. After bed 2 breaks through, bed 3 becomes the lead and bed 1 the lag bed.

minimally exhausted polish bed will capture the metal ions that break through. At this point, the lead bed is taken out of service for regeneration; bed 2 becomes the lead bed, and bed 3 becomes the lag bed. When bed 2 breaks through, bed 3 becomes the lead bed and bed 1 becomes the lag bed.

There are two other types of technologies—reverse osmosis (RO) and electrodeionization (EDI)—that remove dissolved solids from a liquid stream. Both are widely used in the water purification industry and have potential for use in the treatment of CMP wastewater. RO is a method by which water is forced through a semipermeable membrane that does not allow ions to cross. EDI removes ions from a liquid stream by means of an applied voltage. Both the EDI or RO are very effective in removing anions and cations, but the trade-off is that the feed to the EDI or RO must be preconditioned to prevent damage to the equipment. In particular, the feedwater to an RO should not have a silt density index (SDI) greater than 3.0, which may require additional filtration to ensure all the solids are removed from the liquid stream. Some EDI manufacturers recommend that the feedwater to the EDI be RO permeate or of better quality.

The most important aspect that must be considered when reviewing EDI and RO technologies is that these unit operations never remove the trace metals from the liquid stream [17,18]. Unlike ion-exchange resin, both EDI and RO employ reject streams to transport the now-concentrated contaminants out of the equipment and to another location for treatment or disposition, requiring yet more treatment equipments. This is much different from ion-exchange technology that physically removes the trace metals from the liquid stream. Even with their downside, RO and EDI may be able to play an important role in the waste treatment system if there are water reclamation targets that must be met.

Another developing technology for copper CMP waste treatment is electrodialysis (ED) [19], where ions are removed with the application of an electric potential across the liquid stream being treated, and concentrated in a reject stream. Experimentation with a copper solution created from deionized water and copper nitrate showed that electrodialysis can remove up to 90% of the copper from the waste stream. However, when hydroxylamine—an oxidizer sometimes used to replace hydrogen peroxide, which decomposes into nitrogen gas and water—is added to the solution, the removal efficiencies decreased (20% removal for pH of 7, 40% removal for pH of 3). Although electrodialysis has been studied in the laboratory, it has not yet gained enough traction to become a commercially viable technology with widespread acceptance. Since copper CMP wastewater will always contain other chemicals beyond copper and water, more research will be required to better understand the role those chemicals (such as hydroxylamine, hydrogen peroxide, and reactive silica) have on the removal efficiency of copper via electrodialysis. If electrodialysis does end up in the commercial mainstream for waste treatment, then the waste system engineers will be faced with the same type of concentrated reject stream created by the EDI and RO technologies described above.

Another method for CMP waste treatment that has been studied is electrocoagulation (EC) that also uses the Electro–Fenton method for the destruction of organics in the CMP waste stream [20]. Since first patented in 1906 to treat bilge water from ships, EC has been used where micro- and ultrafiltrations and other mechanical filtration methods have not been effective for the removal of dissolved or suspended solids such as heavy metals, colloids, and dye color from wastewater. EC consists of applying a direct current between a cathode and sacrificial anode that are submerged in a waste stream. The current is then used to enhance contact among the particles in the waste stream, fostering coagulation of the solids. For a CMP waste stream, this would allow for the concentration of silica slurry and heavy metals for sediment removal. As seen above with the electrodialysis technology, there are other chemical constituents in CMP wastewater that could negatively impact the removal performance. For example, besides oxidizers, there are organics in CMP wastewater that are generally used to stabilize the slurry particles and to prevent them from coagulating and falling out of suspension. The expectation is that these organics would reduce the effectiveness of the EC technology, but because of the electrochemistry occurring at the cathode, the organics are oxidized *in situ*. In other words, the hydroxyl radicals that oxidize the organics are formed from the Fenton-type reactions where hydrogen peroxide reacts with metal ions at the cathode in the EC device. What makes the EC technology advantageous is its ability to exploit the hydrogen peroxide already present in the waste stream to remove the organics, as well as its ability to remove the suspended silica and dissolved copper constituents. However, like the RO, EDI, and ED technologies, the EC technology does not capture the target materials but only concentrates the target ions into sludge for subsequent dewatering. Again, the facilities engineer must achieve a balance between the various waste treatment technologies and their site-specific boundary conditions.

One challenge to be considered with this means for removing trace metals is if the metal layers in wafer processing change from copper to another conductive metal. One key assumption of many CMP waste treatment systems in existence today is that the system designed is specific to a waste being treated. In other words, a copper CMP waste treatment system may not work for, say, gold CMP waste. In the event the target metal (and its oxidizers) does change from copper, systems based on ion-exchange technology have an advantage since it may be a matter of identifying only a new ion-exchange *resin to use*. For RO, EDI, EC, and ED systems, these may also be successful in concentrating the target ions, but subsequent treatment/removal of those ions will most likely be required.

In an effort to find a broader solution for the removal of numerous target ions, researchers [21] have been experimenting with photochemistry whereby a CMP waste stream (pretreated for solids removal) flows over a titanium dioxide substrate in the presence of ultraviolet light. As a result, the target ions in the CMP wastewater are then deposited on the surface of the titanium dioxide. In addition to copper removal, the use of UV light—commonly used in the ultrapure water industry to oxidize organics—allows for the *in situ* removal of the organic constituents in the CMP waste stream. In laboratory experiments, the effluent from the reaction chamber contained approximately 50 ppb copper and less than 20 ppm of organics [22]. Unlike other more selective ion removal technologies, it appears that this approach using titanium dioxide may also allow for the photoreduction of Ag, Au, Cu, Bi, Pd, and Pt [21]. However, since this approach to treating CMP wastewater is still in its infancy, additional testing will be required to determine its commercial viability.

Metals removal is the primary function of the entire CMP waste treatment system. All of the unit operations installed upstream of the metals removal step exist only to prepare the waste stream for metals removal. Therefore, it is critical to properly select the metals removal technology since each type of technology may require different pretreatment steps that could increase the overall cost of the system or provide varying degrees of effectiveness. It is imperative to fully understand the composition of the CMP wastewater to be treated and to provide the treatment system with a waste stream with little variability.

19.6 INTEGRATION OF COMPONENTS—PUTTING IT ALL TOGETHER

There are numerous equipment and system suppliers in the marketplace that use various patented methods for treating CMP waste through the integration of the unit operations described above [23–27]. Each of these configurations have pros and cons depending on the boundary conditions set forth by the end user; therefore, it becomes a function of the end user to determine which configuration works best for the lowest cost of ownership. These configurations can be broken down into three main categories (assuming pH control is given):

- Solids removal before metals removal
- Solids removal after metals removal
- No solids removal

19.6.1 Solids Treatment Before Metals Removal

Depending on the characteristics of the CMP waste stream, the oxidizer, and metals treatment technologies that are selected, removal of the solids from the waste stream prior to metals removal may be required. For example, if a reverse osmosis array or electrodeionization is used to remove the trace metals, then the suspended solids must be removed from the feed, otherwise the equipment will clog and fail catastrophically. To remove the suspended solids,

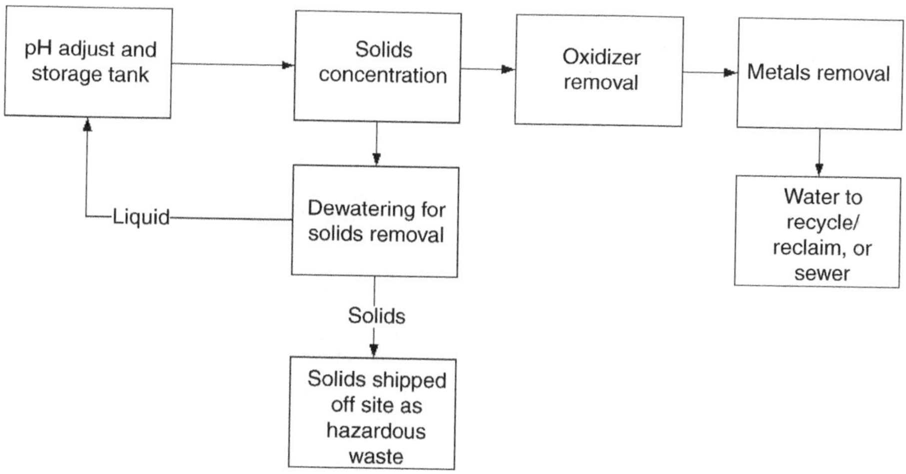

FIGURE 19.6 A schematic of a CMP waste treatment system with solids concentration and removal as the primary unit operation. Although this is a feasible method for treating CMP wastewater, it is the most expensive option because of the cost of transporting hazardous waste.

it is possible to do so as shown in Fig. 19.6 by using a preconcentration step, microfiltration, ultrafiltration, or a clarifier. The bulk liquid stream is then treated for oxidizers and trace metals prior to being recycled, reclaimed, or discharged to the sewer. The final dewatering step of the thickened slurry may be accomplished using a filter press. Since the liquid in the material forwarded to the filter press will contain trace metals, this liquid must be reclaimed and returned to the pH adjust tank. It is important to note that although there may be application-specific advantages to removing the solids before the trace metals, the resultant solid waste stream may be considered a hazardous material because of the presence of trace metals. Transporting and disposing of hazardous wastes will significantly increase the operating cost of the system.

19.6.2 Solids Treatment After Metals Removal

To avoid the expenses associated with transporting and disposing hazardous wastes, the trace metals must be removed from the CMP waste stream prior to removing the solids. This process is shown in Fig. 19.7. However, the selection of the oxidizer and trace metals removal technologies becomes critical since these unit operations must be able to process a waste stream with high-suspended solids. For example, ultraviolet radiation, which relies on the transmittance of the UV light though a fluid, may not be an effective oxidizer removal technology in the slurry. Therefore, it is important to identify technologies that will be effective as well as minimize the overall total cost of ownership of the waste treatment system.

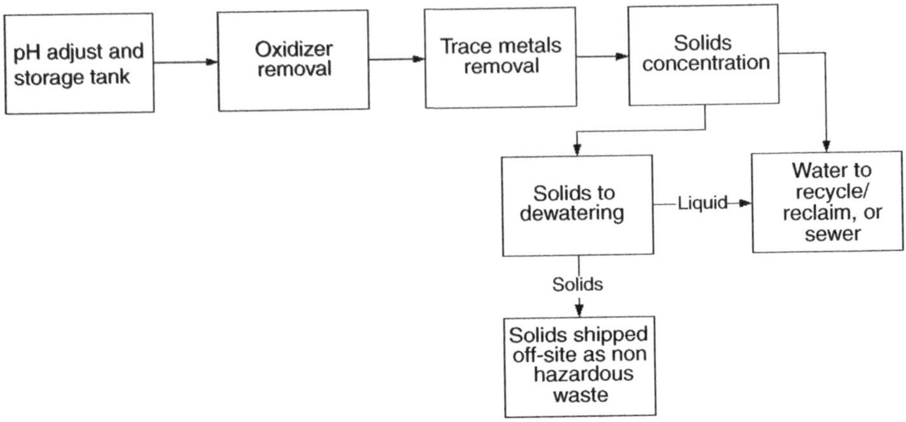

FIGURE 19.7 A schematic of a CMP waste treatment system with solids concentration and removal after metals removal. The solid waste stream generated may not be considered a hazardous waste.

Another possible way to further reduce the cost of ownership of a waste treatment system with this configuration is to forward the concentrated slurry stream to a common filter press if there are various dewatering operations (such as those used for a fluoride waste treatment system or other CMP waste streams) that exist within the facility. Since the liquid in the concentrated slurry does not contain trace metals, this stream can then be reclaimed, recycled, or discharged to sewer rather than re-processed from the start of the system.

19.6.3 No Solids Removal

The least expensive waste treatment system, if environmental regulations permit, is the discharge of the metals-free slurry waste stream to the sewer. This system is shown in Fig. 19.8. In this case, there is no need to perform the slurry concentration and dewatering steps, avoiding significant capital and operational costs. However, the costs may be realized as sewer discharge fees, or through the additional use of fresh city water since this option does not allow for the recycling or reclaiming the water stream for use elsewhere in the facility.

As more pressure is placed on the global industrial sector to become more efficient users of the fresh water supply, facilities will begin to implement

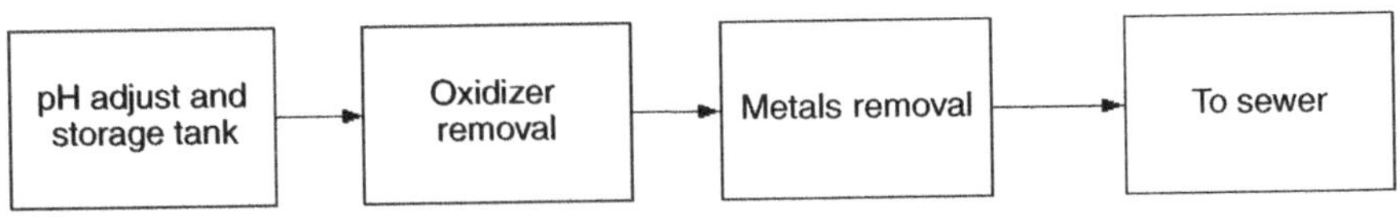

FIGURE 19.8 A schematic of a CMP waste treatment system with oxidizer and metals removal, but no solids removal. This low-cost system is feasible if it is not necessary to reclaim or recycle the CMP wastewater, and if environmental regulations permit.

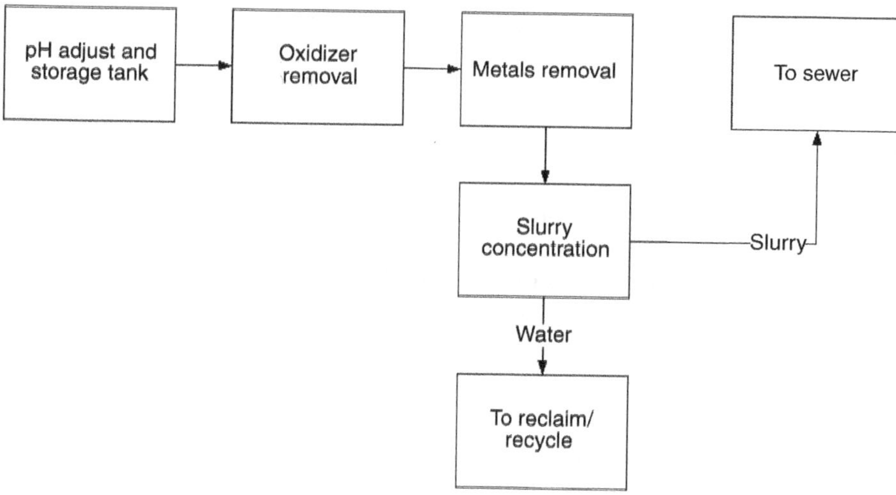

FIGURE 19.9 A schematic of a CMP waste treatment system with oxidizer and metals removal and solids concentration. This system is feasible to recycle and reclaim some water from the CMP waste stream, if environmental regulations permit.

comprehensive water recycling and reclaim initiatives that will no longer make this CMP waste treatment methodology an option. To reduce water and sewer costs, water reclaim using an ultrafilter or microfilter—after metals removal—to concentrate the slurry prior to discharging it to the sewer may be possible, as seen in Fig. 19.9.

19.7 CONCLUSIONS

CMP wastes created at the manufacturing level do not simply go to a mystical place under the clean room floor. They are treated. These facilities systems play just a critical role as the wafer manufacturing process since a failure or a shutdown of the facilities systems could eventually shutdown the entire wafer manufacturing process.

For environmental and financial reasons, CMP wastes must be treated prior to recycling, reclaiming, or discharging these waste streams to the POTW. The intent of this chapter was to provide an overview and basic understanding of the available treatment technologies and the topics that must be considered when designing not only a new waste treatment system, but also when changing or developing a new CMP chemistry. The reader is encouraged to pursue in-depth study of the topics presented here.

There are numerous commercially available CMP wastewater treatment system technologies on the market today. All of the system suppliers strive for cost-effective trace metals removal and maximum water recovery, but may differ on how they address solids control and oxidizer removal. It is a key responsibility of the facilities engineer to best understand the pros and cons of each of these systems as it applies to the specific waste stream to be treated.

QUESTIONS

1. What are the main drivers for treating CMP wastewater?
2. When designing a CMP waste treatment system, what must be considered as part of the design and operation of the system?
3. What are the five key parts to a CMP wastewater treatment system? Why is each of those parts critical to the operation of the system?
4. Where in the CMP wastewater treatment system will it be advantageous to use the new technologies being developed for CMP waste treatment?
5. What are the pros and cons of the up and coming technologies being developed for CMP wastewater treatment?

REFERENCES

1. Meltzer TH. High Purity Water Preparation for the Semiconductor, Pharmaceutical, and Power Industries. Littleton, Colorado: Tall Oaks Publishing; 1993. p 404–406.
2. Huang HH, Lu MC, Chen JN, Lee CT. Catalytic decomposition of hydrogen peroxide and 4-chlorophenol in the presence of modified activated carbons. Chemosphere 2003;51:935–943.
3. Meltzer TH. High Purity Water Preparation for the Semiconductor, Pharmaceutical, and Power Industries. Littleton, Colorado: Tall Oaks Publishing; 1993. p 344–347.
4. Free chlorine removal—UV Light oxidation of free chlorine in water. Available in the Electronic Technical Library of Hanovia Ltd., London, England, www.hanovia.co.uk, 2006.
5. Howarth C, Marsden D, McClean J. Training materials provided by Aquionics, Incorporated (www.aquionics.com), Erlanger, Kentucky, pertaining to the use of ultraviolet light in ultrapure water and wastewater treatment systems, January 2006.
6. Meltzer TH. High Purity Water Preparation for the Semiconductor, Pharmaceutical, and Power Industries. Littleton, Colorado: Tall Oaks Publishing; 1993. p 332–343.
7. Meltzer TH. High Purity Water Preparation for the Semiconductor, Pharmaceutical, and Power Industries. Littleton, Colorado: Tall Oaks Publishing; 1993. p 141–164.
8. Browne S, Krygier V, O'Sullivan J, Sandstrom E. Treating wastewater from CMP using ultrafiltration. Micro Magazine; Mar 1999.
9. Johnston P. Fundamentals of Fluid Filtration—A Technical Primer.2nd ed. Littleton, Colorado: Tall Oaks Publishing; 1998.

10. MacDougall J, Krygier V, Sandstrom E. A crossflow filtration system for heavy-metal wastewater treatment. Solid State Technology; Mar 2006.
11. Reker M, Lenart M, Harnsberger S. Treatment and water recycling of copper CMP slurry waste streams to achieve environmental compliance for copper and suspended solids. 8th Ed. Semiconductor Fabtech (www.fabtech.org); 2005. p 141–150.
12. Inclined plate clarifiers. KWM Wastewater Technologies, Evanston, Illinois, www.koch-water.com/clar.html.
13. Owens D. Practical Principles of Ion Exchange. 2nd ed. Littleton, Colorado: Tall Oaks Publishing; 1995.
14. Bornak W. Ion Exchange Deionization for Industrial Users.1st ed. Littleton, Colorado: Tall Oaks Publishing; 2003.
15. Kunin R. Amber-Hi-Lites, Fifty Years of Ion Exchange Technology.Littleton, Colorado: Tall Oaks Publishing; 1996.
16. Fatula P, Hees B. A progress report on the direct application of imidodiacetio-functional IX resins in Cu CMP waste treatment. Ultrapure Water Magazine; July 2005.
17. Byrne W. Reverse Osmosis—A Practical Guide for Industrial Users.2nd ed. Littleton, Colorado: Tall Oaks Publishing; 2002.
18. Jha AD, Gifford JD. Ultrapure CEDI for Microelectronics Applications: A Cost Effective Alternative to Mixed Bed Polishers. Presented at Ultrapure Water Asia, Singapore, March 2004.
19. Fuentevilla D. Examination of electrodialysis as a method for copper removal in CMP wastewater treatment. University of Arizona NSF/SRC Engineering Research Center for Environmentally Benign Semiconductor Manufacturing; Aug 5, 2001.
20. Kin KT, Tang HS, Chan SF, Raghavan S, Martinez S. Treatment of chemical-mechanical planarization wastes by electrocoagulation/electro-fenton method. IEEE Transactions on Semiconductor Manufacturing; vol. 18, No. 2, May 2006.
21. Li Y, Keleher JJ, Gao N. Inventors; Amia Corporation, Persee Chemical, Inc, assignee. Photo-chemical remediation of Cu-CMP waste. US Pat. 6,916,428. July 12, 2005.
22. Wu T, Wu P, Keleher JJ, Li Y, Oleng N, Guffey W, Gao N, Chen G. Photochemical degradation of organic components in a model Cu-CMP slurry. Presented at the Symposium on Chemical-Mechanical Polishing, Lake Placid, New York, Aug 2002.
23. Salamor M, Schramm T. State-of-the-art copper CMP wastewater treatment. Ultrapure Water Magazine, Nov 2004.
24. Krulik G, Kramasz K, Golden J, Small R, Shang C, Pagan L. Copper CMP wastewater chemistry and treatment. Ultrapure Water Magazine, Sept 2001.
25. Browne S, Maze J, Heid B. Assessing a system for CMP waste minimization and water recycling. Micro Magazine, Mar 2000.
26. Wismer M Woodling R. Copper CMP Treatments Using the Copper Select TM Process. 25th ed. Semiconductor Fabtech (www.fabtech.org); Jan 2005.
27. Woodling R. Treatment of Copper CMP Wastewater. 14th ed. Semiconductor Fabtech (www.fabtech.org); Jan 1999.

20

CMP—THE NEXT FIFTEEN YEARS

JOSEPH M. STEIGERWALD

20.1 THE PAST 15 YEARS

In the early 1990s, reports by some IC manufacturers on the use of silicon wafer polishers to reduce the topography in interlevel dielectric (ILD) films created by multiple levels of metallization began to surface. Dismissed at first as irrational, such reports eventually proved to be true as the CMP industry began gaining ground. By the end of the decade, feature densities approaching sub-100 nm demanded not only an ILD CMP but CMP steps in the front end of the line (FEOL) as well. Process designers referred to CMP reverently as an enabling technology because the planarity it afforded was required for submicron lithography and multilevel metallization. But CMP engineers and integrators regarded the new process as more of a necessary evil because of the many challenges associated with controlling the new technology in high volume manufacturing. These conflicting perceptions of CMP as an enabling technology that is difficult to manage have persisted throughout CMP's more than 15 years' history.

Table 20.1 shows the technology node and the year in which the manufacturing of key CMP processes was introduced at Intel Corporation. These CMP processes, while enabling different aspects of IC technology, shared several key features. First and foremost, they were all revolutionary technologies in that they (1) significantly advanced in the state of the art, (2) required significant capital investment, and (3) required significant development efforts to make them work. An additional feature that the CMP processes shared was that they were unique solutions as, for most of them, there was no viable alternative to their use—these CMP steps were required.

Microelectronic Applications of Chemical Mechanical Planarization, Edited by Yuzhuo Li

TABLE 20.1 CMP-Enabling Intel Technologies.

Node	Year	CMP	Enabling
0.8 μm	1990	ILD	Multilevel metallization
0.35 μm	1995	STI	Compact isolation
		PSP	poly-Si patterning
		W	yield/defect red.
0.18 μm	1999	SiOF ILD	RC scaling
0.13 μm	2001	Cu	RC scaling Electromigration
90 nm	2003	SiOC ILD	RC scaling
65 nm	2005		
45 nm and beyond	2007	unknown	

After the successful introduction of ILD CMP, a flood of new CMP steps was introduced in 1995 with the 0.35-μm node technology. Shallow trench isolation (STI) CMP significantly improved FEOL planarity and also compacted the area required for isolation by eliminating the bird's beak characteristic of the LOCOS isolation step that was displaced (Fig. 20.1). Poly-Si polishing (PSP) enabled sub-0.5-μm polygate patterning by further reducing FEOL topography (Fig. 20.2). Tungsten (W) CMP improved back-end-of-the-line (BEOL) defect densities by replacing W etch-back processes, a notoriously defect riddled process.

In 2001, the long anticipated conversion from aluminum metallization to copper metallization occurred, yielding lower interconnect resistance and higher immunity to electromigration. Because of the lack of suitable copper-etching processes, Cu CMP was introduced as a necessary replacement to subtractive etching processes. Borrowing from STI CMP the concept of patterning using inlaid structures, Cu metallization patterns were delineated by CMP using the so-called damascene patterning method described in Chapter 7.

The 1990s saw not only the birth of the CMP technology but also the greatest innovation in the CMP industry. As a result of the fast pace of innovation and introduction of new CMP processes, the CMP industry grew rapidly outpacing even the growth rate of the semiconductor industry as a whole (Fig. 20.3) [1]. However, since the introduction of Cu CMP, there has been a significant lull in the introduction of new CMP technologies. The 90 and 65 nm nodes brought new materials and new challenges into the CMP field, but these were extensions of previous CMP technologies and not revolutionary changes as seen in the 1990s. Hence, while the first 10 years of the CMP industry saw radical and exciting innovation, the past 5 years have seen a lull in innovation. This observation bears several questions. What will the next 15 years bring for the CMP industry? Will innovation in the CMP industry be relegated to continuous improvement of current processes or are new CMP technologies needed and if so, what barriers exist in the introduction of new CMP technologies? And the most pertinent question to the CMP technologist is that how does the CMP industry face these challenges so that the CMP industry can maintain its historical growth rate?

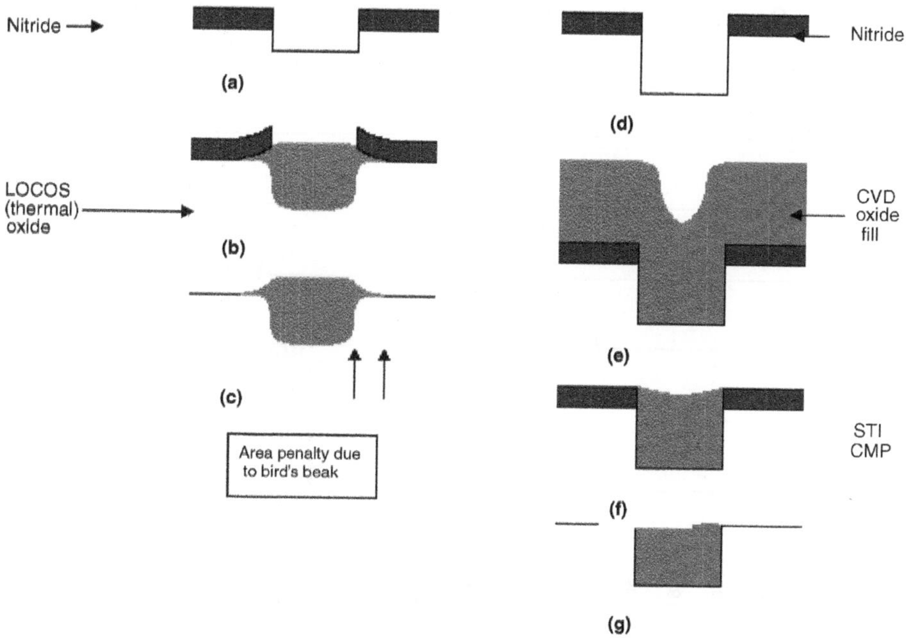

FIGURE 20.1 STI CMP. Prior to STI CMP, LOCOS isolation was the most common method of electrically isolating the MOS transistors. With LOCOS, (**a**) a nitride hardmask is patterned and (**b**) then the LOCOS oxide is grown in areas where the nitride is removed. (**c**) A large area penalty is paid because the oxide growth extended significantly below the nitride mask forming the so-called "birds beak". (**d**) With STI isolation, the nitride HM etching is extended to include a shallow trench cut into the Si substrate. (**e**) The trench is then filled with oxide and (**f**) then CMP is used to remove the oxide overburden. As compared with the original isolation pattern, the STI CMP method exerts no area penalty.

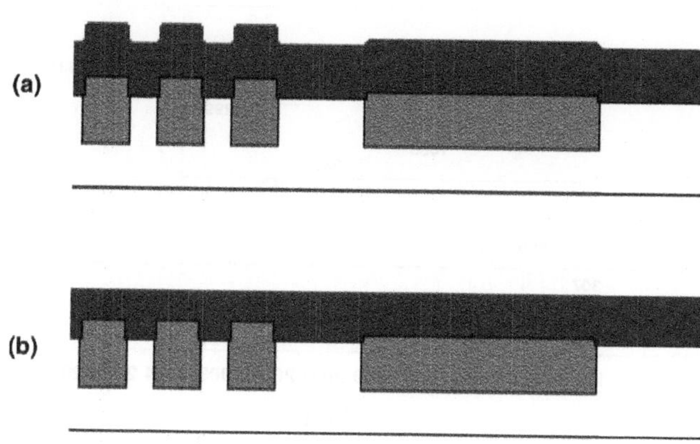

FIGURE 20.2 PSP CMP. (**a**) Topography from STI CMP and other pre-poly-Si patterning steps results in an uneven surface for lithography. (**b**) Poly-Si CMP may be used to level the surface and remove the topography.

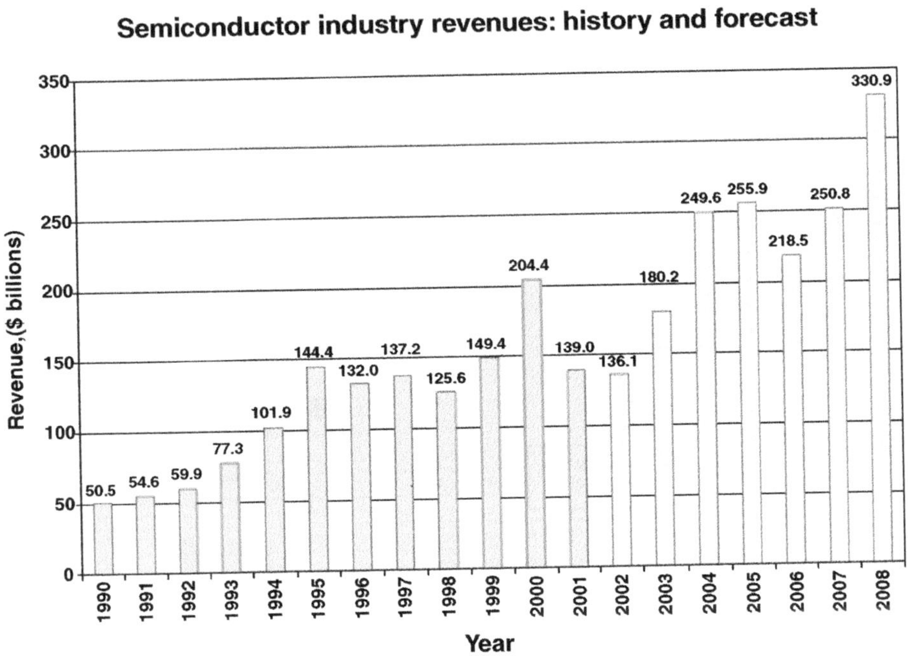

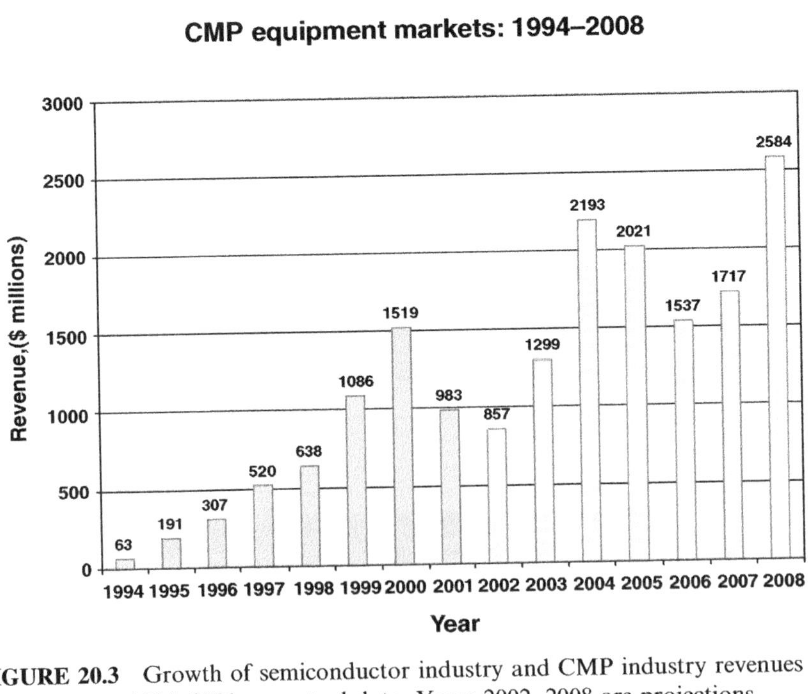

FIGURE 20.3 Growth of semiconductor industry and CMP industry revenues (from Ref. 1). Years 1994–2001 are actual data. Years 2002–2008 are projections.

20.2 CHALLENGES TO SILICON IC MANUFACTURING

The CMP processes described in the preceding section were all born of significant challenges faced by the silicon IC industry. To gain insight into how the CMP industry might grow in the remainder of the current decade and in the next decade, one should look into the challenges that the silicon industry would face in the next 15 years. Note that while CMP is also used for nonsilicon semiconductor manufacturing, the silicon side is the largest sector of the IC industry; and because silicon ICs involve a much larger scale of integration (number of transistors per circuit), CMP has been used more prolifically in silicon manufacturing than in non-Si processes. If the CMP market is to continue to grow, it will do so because new uses of CMP are found that meet silicon IC industry challenges.

Also note that this and subsequent sections use logic ICs as a basis for discussion. However, memory ICs use CMP broadly for many of the same applications as logic ICs do, as well as some applications unique to memory. In general, the challenges faced by CMP in logic IC production apply to memory IC production.

In 1965, Gordon Moore summarized the basic challenge of the IC industry. Moore predicted that by doubling the level of integration (number of transistors per circuit) every year, the IC industry would provide an exponential growth in the performance of its product offerings with exponential decay in the cost per unit function (i.e., cost per transistor) [2]. While his statement was based upon the data collected over only a few years, the rule accurately predicted the growth of the IC industry for the subsequent 40 years. Figure 20.4a [2] shows the original data Moore used to make his prediction. Figure 20.4b [3] shows a 2003 reconciliation of Moore's prediction nearly 40 years later. Figure 20.4c [3] shows the end result of Moore's Law—an exponential growth in semiconductor industry revenues over the past 40 years. While the slope of the curves show a doubling of the integration at 1.5-year pace rather than 1-year pace as Moore predicted, his prediction remains remarkably accurate and insightful. Figure 20.5 [3] shows one impact of integration scaling—the exponential improvement in circuit performance with time as measured in MIPS (million instructions per second) of Intel's microprocessors. To the end user of the IC, Moore's law means a doubling of value every 18 months. For technologists working in IC R&D or manufacturing, it means a very fast pace of innovation and learning. It is this fast pace of innovation that drives most of the challenges of the IC industry and that gives the CMP technologist opportunities to expand his trade.

Throughout the past 40 years, the industry has maintained Moore's law by shrinking the dimensions of the ICs' components. At smaller dimensions, transistors switch faster and the density of transistors increases. Hence, shrinking allows the circuit to run faster and the circuit designer to add more functionality. A basic recipe for maintaining Moore's law is to release a new technology every 2 years that scales circuit dimensions by 30% (linear) and increases transistor

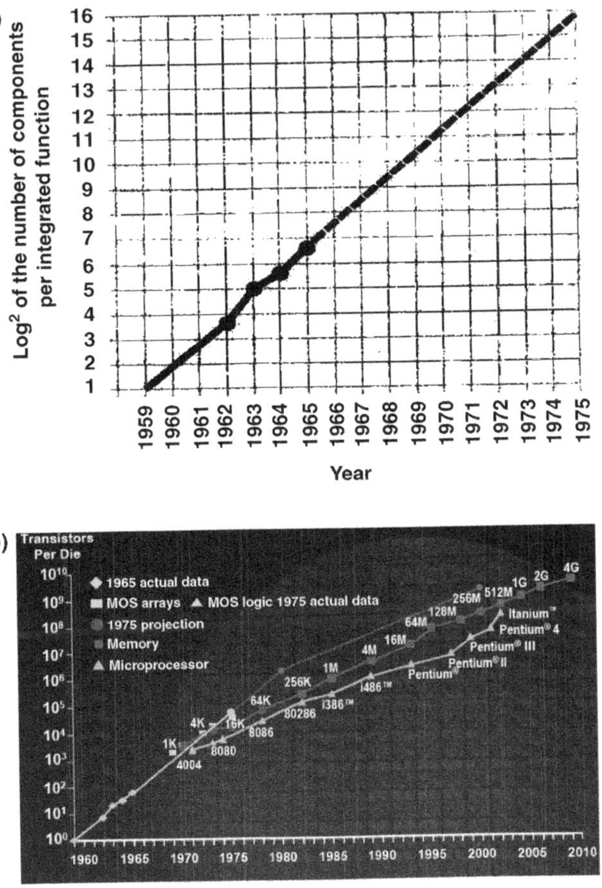

FIGURE 20.4 (**a**) Gordon Moore's prediction in 1965 that the level of integration would roughly double 18 months was an extrapolation of only five data points representing the first 6 years of the IC industry (from Ref. 2); (**b**) Gordon Moore revised his projection only slightly in 1975 to predict with great accuracy the subsequent 20 years of IC industry's growth (from Ref. 3); and (**c**) Gordon Moore's projection of exponential growth in semiconductor industry realized (from Ref. 3).

switching speed by 30% [4]. The 30% scaling of linear dimension gives a 50% scaling of aerial dimension (2× transistor density). Figure 20.5 shows the impact of the 2× density scaling on transistor integration over the past 40 years. Figure 20.6 shows that Intel's release of recent technologies is on pace with the 2-year schedule of density scaling [5].

The need for the CMP processes described in Table 20.1 arose as a direct result of shrinking circuit dimensions. In each case, a challenge arose and a CMP process was used to meet that challenge. The question for the CMP technologist is what are the future challenges to the IC industry and will CMP be used to meet those challenges? The remainder of Section 20.2 will

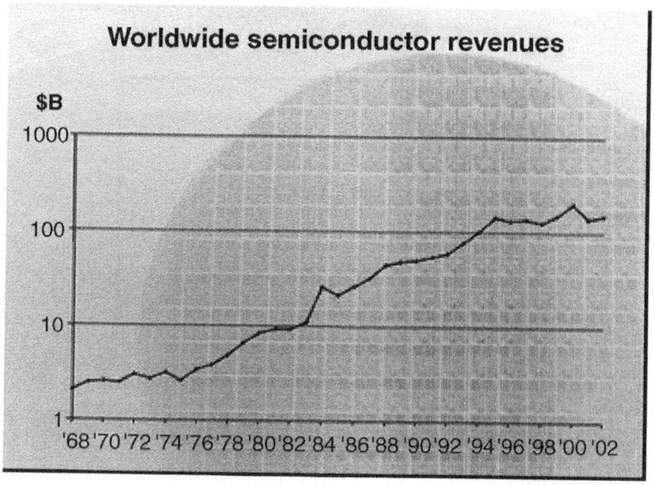

FIGURE 20.4 (*Continued*).

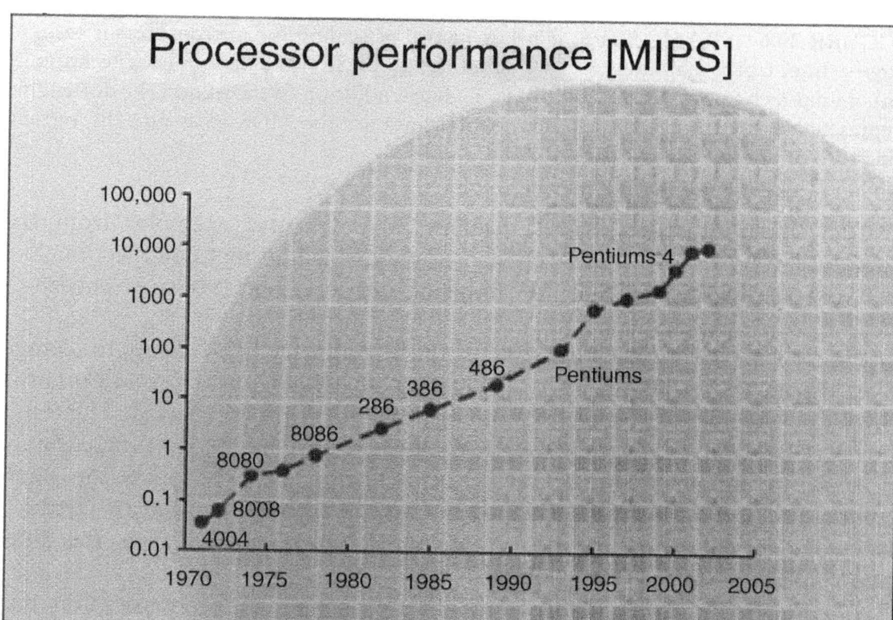

FIGURE 20.5 Exponential increase in the performance of Intel Corp. microprocessors as measured in MIPS (million instructions per second) (from Ref. 3).

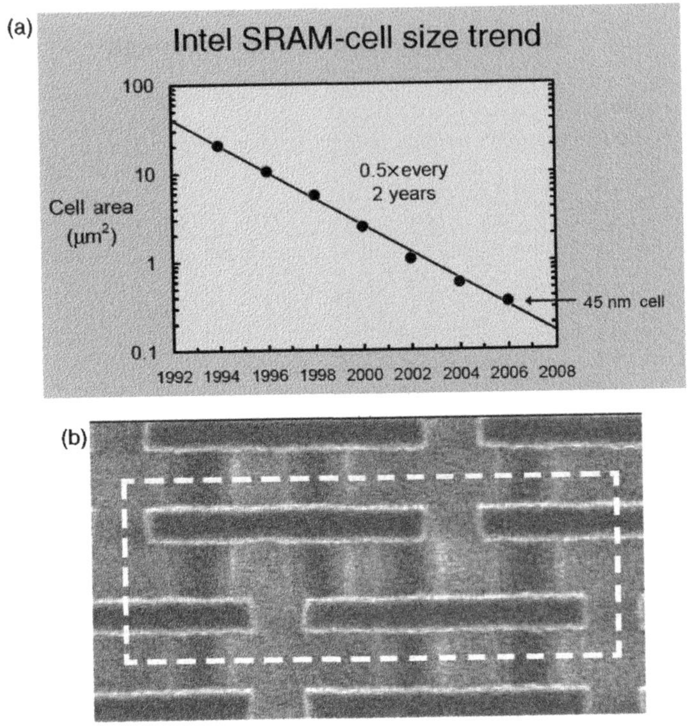

FIGURE 20.6 SRAM-cell size is a key metric of scaling for microprocessor technologies. Intel Corp. maintains a 50 % aerial shrink of SRAM cell with each technology (**a**). 45-nm technology SRAM cell (**b**) is the latest addition to the trend. The dotted line represents the six transistor cell—horizontal lines are the MOS gates and the vertical lines are diffusion regions (from Ref. 5).

discuss these future challenges; Section 20.3 will give examples from the literature of how some of these challenges may be met using CMP; Section 20.4 will discuss what actions the CMP industry must take for CMP to be chosen to meet these challenges.

Gordon Moore's basic challenge to the IC industry is not likely to change over the next 15 years. That is, a doubling of performance every 18 months driven by the release of a new technology every 2 years is expected. The trend however has been that the 30% performance increase has become increasingly difficult to obtain. More development calories must be spent on each technology and more radical changes must be made in order to create a functional circuit that operates at the new dimension and gives the 30% performance increase.

In addition to performance and density scaling, power dissipation has become a third concern that the microprocessor industry must face. Figure 20.7 [3] shows the increase in power consumption by the microprocessor also following Moore's Law. Power dissipation can be broken into two

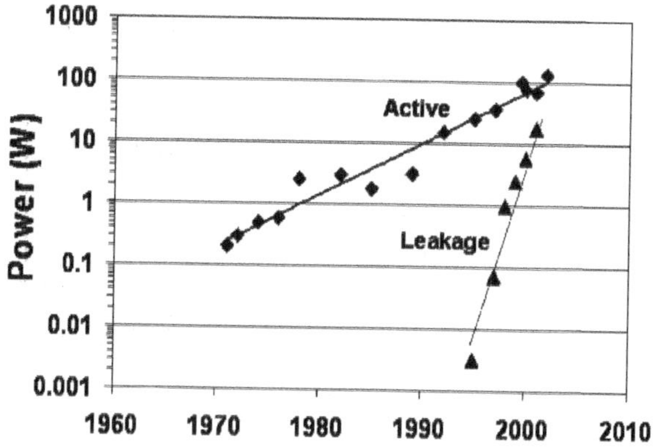

FIGURE 20.7 Power dissipation has also grown exponentially with time. Recent technology development has to contend with the rapid growth of leakage power as well as active power dissipation (from Ref. 3).

components: active power and leakage. Active power is dissipated when a CMOS gate switches; it is the energy required to turn on/off the transistors. As transistors switch faster (and performance improves), an increase in the active power is a natural result. Left alone, active power dissipation would soon prohibit future scaling of performance. Leakage power dissipation is due to currents flowing in transistors when they are not actively switching. Ideally, CMOS circuits consume no power when they are not switching. However, the narrow dimensions of modern transistors result in leakage currents through thin gate oxides and high off-state currents due to short-channel effects. As features scale, leakage currents are quickly outpacing active power dissipation as the primary source of power consumption.

To meet the three challenges, 30% faster, 2× density, and lower power, IC technologists are contemplating radical changes to the art of IC processing. The industry is poised to overhaul tried and true technologies that have served as cornerstones of the industry. Table 20.2 shows several potential changes that are debated in the literature and discussed at conferences. Each of these changes will mark a significant turning point in the industry. While the industry has had a history of changes (indeed growth of the industry is fueled by change), the changes in Table 20.2 are monumental in that they displace a cornerstone technology, that is, a technology that has served as a basic building block of the industry. Indeed, the introduction of each of these cornerstone technologies brought about a radical change in the industry itself: a change that drove Moore's law. That the industry is considering undoing these changes is a clear sign that the task of keeping Moore's law alive has become monumental.

While IC manufacturing becomes even more complex, market competition drives the need to reduce the cost. The greatest challenge for future technologies may be obtaining the performance gains demanded by Moore's law while continuing to drive down the cost of manufacturing. The cost impact of any new

TABLE 20.2 EOL of Conventional Technologies.

Technology	Year of Introduction	Replacing Technology	Year of Replacement	Reason for Replacement
Aluminum metallization	1961	Copper metallization	2001	Lower interconnect resistance electromigration
Polysilicon (refractory) gate electrode	1975	Metal gate	~2009–2011	Enable high-*k* gate oxides
SiO2-based gate oxide	1963	High-*k* oxides	~2009–2011	Increased transistor gain decreased leakage
Planar transistor	1961	FINFET	~2009–2013	Scaling performance power
Silicon-based ICs	1961	III-V semi, Si-nanowire, carbon nanotube	~2011–2015	Scaling performance power
MOS transistor	1963	unknown	unknown	

step is likely to be as great a consideration as the performance impact. Cost should be considered not only in the capital cost and cost of materials but also in the impact of the step on die yield (number of functional die per wafer at EOL test) and line yield (number of wafers making it to end of line). Die cost is influenced both by the cost of manufacturing each wafer as well the number of functional die at EOL per wafer. In many cases, it is the latter that has the stronger influence.

Figure 20.8 [6] shows the rate of yield improvement or yield learning of new technologies during its development and ramp cycles at Intel Corp. as a function of time and technology. Rapid yield learning gives a significant competitive advantage to the IC manufacturer because it decreases the time to market the advanced technology as well as reduces the unit cost of good die once the technology goes into production. Note that the slope of the curve for each new technology is becoming ever steeper, suggesting a greater pace of yield learning for each new technology. Because rapid yield learning generates significant advantage in the marketplace, the expectation is that the learning rate (i.e., the slopes of the lines) continues to increase with each new technology. Hence, for the new CMP technology to become adopted, it must not hinder yield learning (by creating yield loss modes). Better still, the new CMP process should enable higher yield learning by reducing the number of defective die. Past CMP technologies have been notorious for the high number of defects they produce and as a consequence actually slowing yield learning rather than enhancing it. This trend must be reversed if large numbers of new CMP technologies are to be adopted.

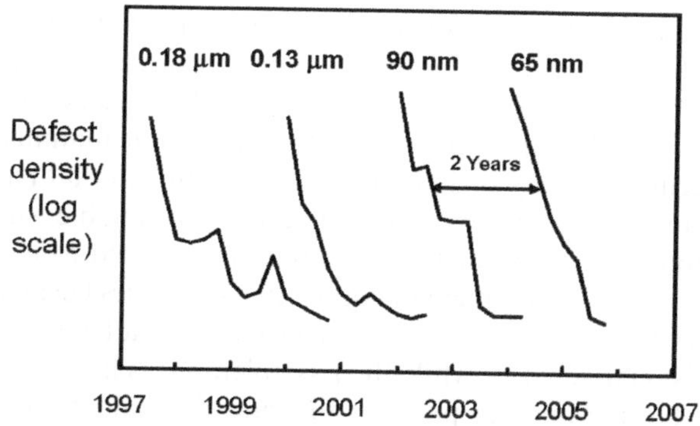

FIGURE 20.8 The number of defective die per wafer decreases exponentially during the 2-year development cycle. Such rapid yield learning is critical to maintaining competitive advantage in today's IC industry. Note that the slopes of the curves for 45 and 32 nm technologies are expected to continue the trend of increasing steepness with each new technology (from Ref. 6).

20.3 NEW CMP PROCESSES

For the CMP technologist who is worried that the recent lull in the introduction of new CMP processes is an indication that CMP growth has stagnated, there is good news. Integration and device engineers are finding creative new ways to use CMP to solve the challenges mentioned in the preceding section. Evidence to this claim can be found in the proceedings of recent device conferences. Researchers are finding novel uses for CMP. This section discusses examples of the recently published CMP uses and suggests some unique challenges that each example will bring to the CMP Engineer who is charged with the responsibility of bringing such a process from conception to manufacturing.

20.3.1 The Two-Year Development Cycle

To understand how new CMP processes are likely to be introduced, it is instructive to first look closely at the IC technology development cycle. Figure 20.9 [7] shows the technology development cycle at Intel Corp. Technology development is divided into four phases: research, pathfinding, development, and manufacturing ramp. Each phase lasts for approximately 2 years, and because Intel releases a new technology to production approximately every 2 years, there is always a technology in each one of the phases. As discussed in Section 20.1, the goal of this development cycle is to produce a new technology every 2 years, that is, a 50% density shrink of the previous technology with a 30% enhancement in the transistor performance and a reduction in power dissipation. To achieve this goal, the development cycle must produce innovations that deliver.

During the research phase, many options are evaluated for possible inclusion into the new technology. Feasibility studies are performed to determine the merits of a given option in terms of benefit, cost, likelihood of success, and readiness of the infrastructure needed to support the option in volume manufacturing. These studies may occur internally or in collaboration with external partners such as universities, equipment vendors, consortia, or consumable suppliers. Some module development may occur for items that are deemed to have a long lead time; for example, the development of copper polish slurries began in the research phase. Figure 20.9 shows that the investment made on a technology during the research stage is relatively low while the risk tolerance is high.

The role of the pathfinding phase is to critically evaluate options explored during the research phase and choose the few critical options that will be incorporated into the new technology. Data collected during the research and pathfinding phases are used to narrow the number of options. The options will be tried using the previous technology as a baseline to ensure that the option will yield the expected benefit with the anticipated cost, complexity, and risk levels. Many options are dropped or pushed out to later technologies during the pathfinding stage. For the options that remain, the pathfinding teams generate plans to integrate the options into the process flow. Spending on a technology increases slightly during the pathfinding stage in order to thoroughly evaluate each option. Risk is reduced as high risk options are dropped from consideration.

During the development stage, the enabling options are integrated into the process flow. Each option generates new modules and new process equipment that must be successfully fitted with the existing process. Much of the innovation that makes these changes successful occurs during the development

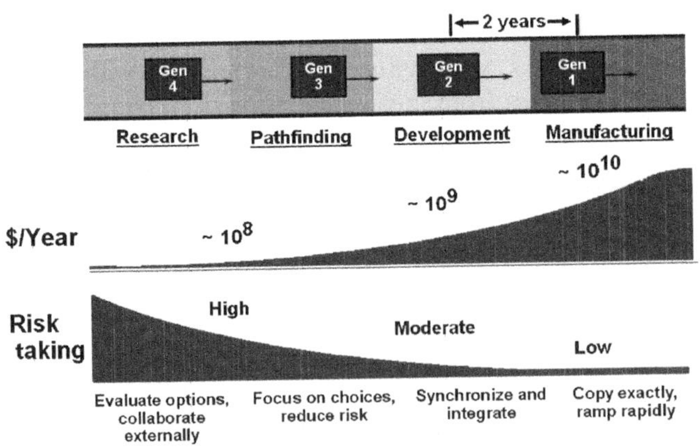

FIGURE 20.9 Technology development cycle. Each phase of the cycle lasts approximately 2 years with the amount of investment increasing exponentially with time and the amount of risk taking decreasing with time. The bulk of CMP consumables and new process development occurs between the second year of pathfinding and the first year of development (from Ref. 7).

phase. Because, while an idea may be deemed worthy during evaluation phases, its true test occurs when it is integrated into the process flow and the real impact to circuit performance and yield can be measured. Prior to the full integration of the option, these impacts can only be estimated at best. In addition, many previously unforeseen problems associated with integrating the new change become apparent during the development phase. Development teams must work to quickly solve these problems or the issues associated with integration may threaten to delay the start of production ramp. A delay in the start of ramp may result in significant revenue loss and/or competitive advantage erosion. Options that threaten the start of ramp may be abandoned during development despite their potential performance impact.

Spending on a technology increases significantly during the development stage as the number of experiments and resources committed to the technology increase. Risk is further moderated as the focus of development teams becomes to make the options work at high yield and high performance levels.

As the final stage of development ends, capital equipment is purchased and/or equipment from older technologies is converted to support the manufacturing ramp of the new technology. Development and manufacturing teams work together to bring on line the new equipment and match the performance to the established development equipment set. The focus of these teams is to minimize the variation and rapidly solve problems that arise as wafer production starts increasing. Changes in the process flow or process steps are restricted in order to minimize risk. At this stage, change occurs only in response to unanticipated problems that cannot be otherwise solved by reducing the variation in equipment performance.

Note that the stages are not independent. The development cycle can be viewed as a pipeline, each stage feeding the next. Careful evaluation and planning during research and pathfinding is critical to successful integration during development. Proper integration and characterization of the process during development leads to a successful manufacturing ramp.

A good understanding of the technology development cycle is valuable in foreseeing how the use of CMP might grow in the next 15 years. Ideas for the use of CMP may arise during either of the research or pathfinding stages. But whether or not the CMP step actually makes it to the manufacturing stage depends on how well the new process is integrated into the flow during the development stage. If the process has poor control of defects or uniformity, or if the process is costly or poorly developed, it may be abandoned in favor of a more practical solution. If, however, the new CMP process integrates easily and cost-effectively in the flow, it is likely to be accepted and survive the development cycle in manufacturing.

As we shall see in the following examples, the new uses of CMP are likely to be discretionary in future technologies. Integrators will have non-CMP options available to complete their task. In order for the CMP-based options to be considered the best options, CMP has become a valued tool in the integrator's toolbox.

20.3.2 Finfet Transistors

Table 20.2 indicates the possible end of life of the planar MOS transistor circa 2011 with a new type of MOS transistor, the Finfet. Because they are built off the plane of the silicon wafer into the third dimension, Finfets allow the continued scaling of transistor dimensions. Use of the third dimension allows for an increased area used for conduction, and hence greater transistor drive currents. Larger drive currents result in faster switching speeds. Researchers at Intel Corp. have reported a form of Finfet, termed tri-gate transistor, shown conceptually and as fabricated in Fig. 20.10 [8]. Here, the channel region of the transistor is above the plane of the wafers, and the carrier conduction along the channel occurs along three gated sides rather than the traditional single top side. In addition to the increase in the conduction of the device, many of the body effects that produce leakage currents are minimized or eliminated with this design.

That the industry is looking keenly at Finfet transistors as a potential replacement for the planar version of the MOSFET is evidenced by the large number of papers at device conferences and journals. Many of these papers discuss the use of CMP in the fabrication of Finfets. Figure 20.11 [9] shows how Toshiba researchers use CMP to planarize the surface between fin formation and gate formation. Planarization here is required in order to provide a uniform depth of focus for polygate lithography. This CMP step does not appear particularly difficult. It is an inlaid material approach and the SiN on top of the fin can serve as a good polish stoping material allowing for a high degree of process control. One disadvantage of this method to the CMP engineer is that a significant amount of topography is evolved that must later be reduced by subsequent CMP steps; otherwise higher layer patterning and CMP steps will be negatively impacted by the topography.

Samsung researchers reported on a second method of fabricating Finfets using multiple CMP steps shown in Figure 20.12 [10]. After conventional STI isolation (including STI CMP), holes are cut into the isolation oxide and a dep/CMP sequence is used to inlay nitride into the holes. The inlaid nitride is used to block the oxide etching where the sidewall gates are not desired. Next, lines

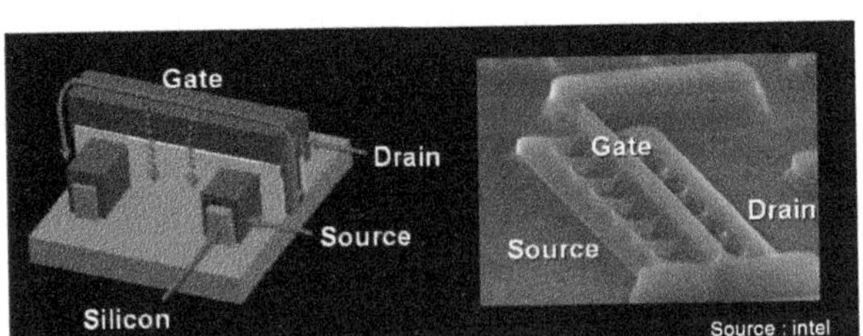

FIGURE 20.10 Intel Corp.'s schematic representation of a tri-gate transistor and an SEM image of a seven-leg tri-gate transistor. The use of three dimensions allows aerial scaling while increasing the width of transistor channel used for conduction (from Ref. 8).

- SOI wafer preparation
- Channel doping
- Hard mask SiN deposition
- **Fin patterning by SWT process**
- Plasma nitrided gate oxide formation
- Poly-Si deposition & planarization
- **Gate patterning by SWT process**
- **Selective gate sidewall spacer (SiN) formation**
- Elevated S/D extension formation with selective Si epitaxy
- S/D extension ion implantation tilted at an angle of 30°
- TEOS/SiN gate sidewall spacer formation
- S/D implantation & activation
- Ni silicidation

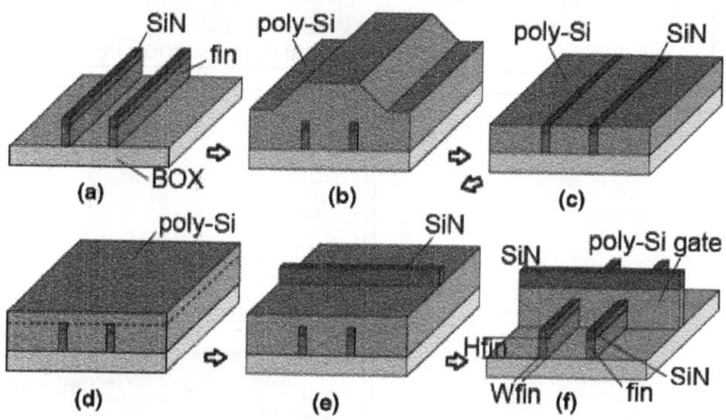

FIGURE 20.11 Example of Finfet fabrication using CMP to planarize poly-Si gate film (c) prior to patterning (from Ref. 9).

are patterned and an oxide etching is performed to define the sidewall gates (*i,j*); poly-Si is then deposited and patterned for the top gate (*n*). The use of multiple CMP steps in this method challenges the CMP engineer because of the potential for negative interaction between CMP steps. Dishing or recess of the oxide during the first CMP step could cause trapped nitride during the second CMP step. The result would be an incomplete etching of the oxide where the buried gate material is to be inlaid. The CMP engineer would need to carefully coordinate the two CMP steps to eliminate such interactions and prevent yield or performance issues associated with incomplete gate formation. The advantage of this second method of Finfet formation is that the surface is left planarized at the end of the flow.

20.3.3 High-*k* Gate Oxides

Table 20.2 shows another revolutionary change to the industry for which CMP can play a major role. The IC industry is expected to begin conversion from traditional SiO_2-based oxides to high-*k* oxides within the next 5 years. High capacitance (*k*) gate oxides yield both higher switching speeds as well as lower leakage currents, thus giving higher performance at lower power consumption.

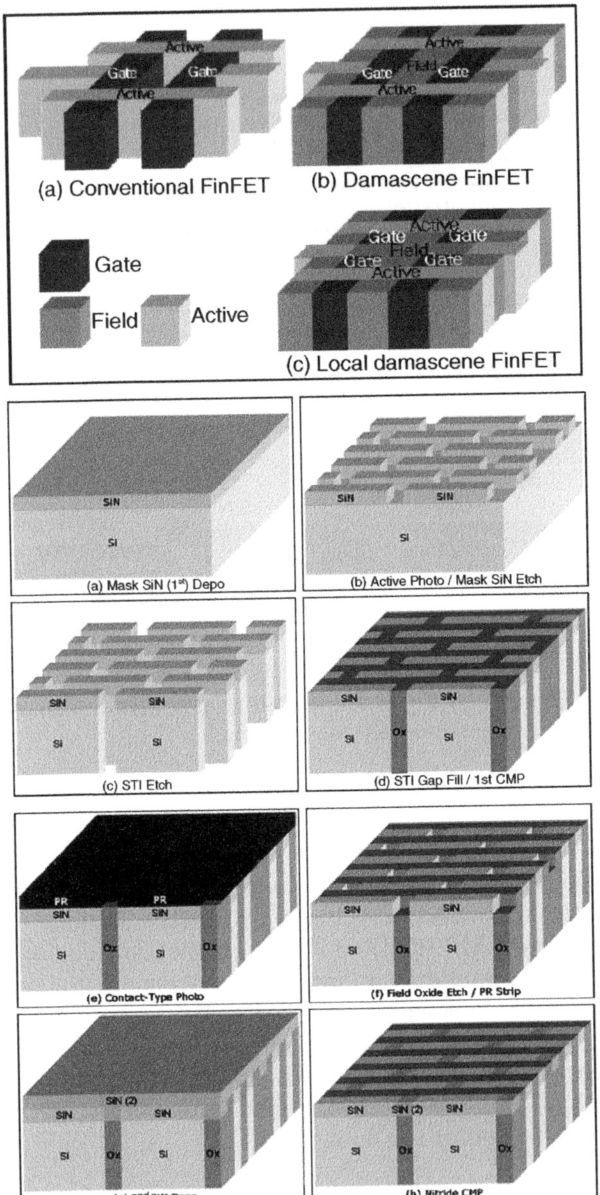

FIGURE 20.12 Example of Finfet fabrication using two CMP steps (from Ref. 10).

To see how high-k oxides impact the transistor performance, one can look at the equation for the drain current of an MOS transistor [11]:

$$I_D = \beta[(V_G - V_t)V_D - 1/2V_D^2] \tag{20.1}$$

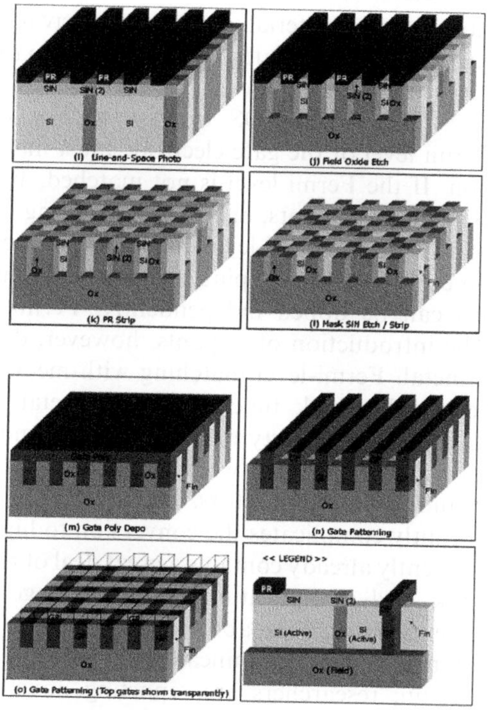

FIGURE 20.12 (*Continued*).

or

$$I_{D,sat} = 1/2 \times \beta(V_G - V_t)^2 \quad \text{where} \quad V_{D,sat} = V_G - V_t \tag{20.2}$$

where $\beta = (\varepsilon_{ox}\varepsilon_o/t_{ox})\mu(w/l)$ is the transistor gain; $\varepsilon_{ox}\varepsilon_o$ is the dielectric constant (k) of the gate oxide; t_{ox} is the gate-oxide thickness; w is the gate width; l is the gate length; V_G is the gate voltage (wrt source); V_t is the threshold voltage; V_D is the drain voltage (wrt source); and $V_{D,sat}$ is the drain voltage at which the maximum drain current is reached.

Note that the speed at which a transistor switches states is limited by the drain current of the transistor in the previous logic stage supplying the charge to switch the transistor—thus higher drain currents result in faster switching. High-k gate oxides increase drain current by increasing the gain (β) of the transistor. However, the advantage of higher k is lost if the integration of the high-k *oxide negatively affects other terms in the equation.*

The change to high-k gate oxides brings with it the need for the replacement of poly-Si gates with metal gates. Poly-Si gate materials are incompatible with high-k gate oxides because of the large number of surface states that form at the interface between the oxide and poly-Si. These states result in threshold voltage pinning (resulting in higher V_t) and phonon scattering of carriers (resulting in lower μ)—ultimately negating the effect of the higher gate capacitance. The metal gate material serves to passivate the oxide–metal interface, eliminating such surface states and hence eliminating these effects.

A significant complication to the use of metal gates is that different metals must be used for PMOS and NMOS transistors. For optimal transistor performance, the Fermi level of the gate electrode must match the Fermi level of the channel region. If the Fermi level is not matched, V_t will be either too high, resulting in lower drive currents, or too low, resulting in excess leakage in the off state. For poly-Si gates, Fermi level matching is achieved by doping the poly-gate to the same polarity and doping level as the channel. Since PMOS and NMOS regions can be doped independently, Fermi level matching is straight forward. The introduction of dopants, however, does not change the Fermi surface of a metal. Fermi-level matching with metal gates can only be performed by tailoring the work function of the metal surface. Since the work function is an inherent property of a given metal and since PMOS and NMOS transistors require different work functions, different metals are required for PMOS and NMOS transistors. The need for different metals for *p* and *n* regions significantly complicates the conversion to high-*k* gate materials, a conversion that inherently already contains a great deal of risk for the industry.

But with the risk and complication in the change to high-*k* comes opportunity. The company that is able to first master the challenges of high-*k* integration stands to gain a significant performance advantage over its competitors. Here again, researchers are reporting novel methods of using CMP to enable the integration of metal gates. Figure 20.13 [12] shows a method for realizing a bimetal gate process using fully silicided (FUSI) metallization enabled by CMP. After transistor formation using conventional poly-Si gates, a dielectric film is deposited and then polished to level the wafer surface. The surface is then masked to cover the NMOS gate regions and the dielectric is etched to expose the poly of the PMOS gates. The poly is then silicided by depositing a metal film (platinum in this case) and annealing the poly-Si and metal at high temperature to form the silicide. The dielectric deposition, CMP, photomask, etching, and FUSI steps are then repeated for the NMOS transistor. As shown, this process requires two CMP steps. However, yet a third CMP step is likely required to replanarize the surface at the end of the pictured flow.

While once again CMP can be used to solve a complex problem, that of forming differential gate metals for different regions of the wafer, the CMP engineer must be careful to ensure a low defect process as well as uniform planarization both locally and across the surface of the wafer. Small gate dimensions, both vertically and laterally, leave little room for defects or nonuniform removal rates. The consequence of high defects and/or high nonuniform removal will be a degradation in yield and transistor performances.

Figure 20.14 [13] shows a second method of converting the polygate to silicide for high-*k* applications. Here again, the CMOS transistors are formed with conventional poly-silicon gates. After ILD0 deposition, the surface is polished back to expose the poly-silicon. Metal films are then deposited and

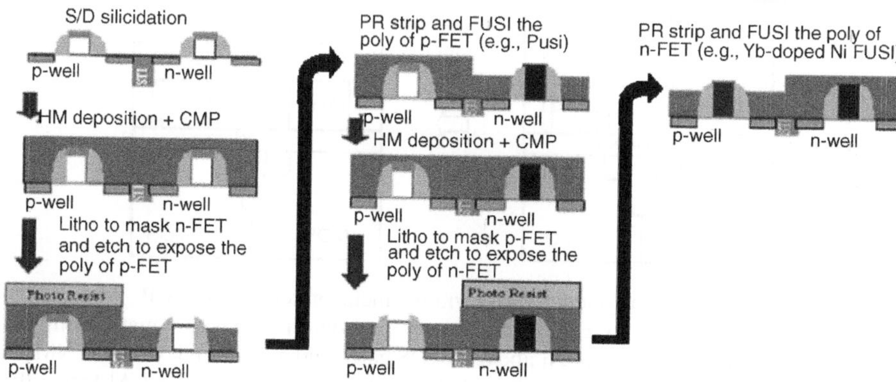

FIGURE 20.13 Metal gate transistors fabricated using CMP to enable exposure of the top of NMOS and PMOS gates (independently) followed by FUSI reaction of gates to form silicided metal gates (from Ref. 12).

patterned such that Ni covers the PMOS regions and an Al/Ti bilayer covers the NMOS regions. During thermal treatment, the PMOS poly reacts with the Ni to form NiSi while the NMOS poly diffuses through the Al layer to react with Ti at the surface causing the Al to migrate into the trench to replace the PMOS poly. After thermal treatment, the excess material above the ILD0 is removed by CMP leaving metal gates of complementary work functions for NMOS and PMOS.

While the FUSI method shown in Fig. 20.14 offers new opportunities for the CMP engineer by creating two new CMP steps, it also offers unique challenges. The first CMP step where the poly-Si gates are exposed likely requires a very high degree of process control. Removal rates and process end point must be well controlled else the thin polygates will be either overpolished, leaving not enough material for metal gate formation and hence high gate resistance, or underpolished, leaving the gate unexposed and preventing reaction with the metal. Indeed, if across-wafer nonuniformity is large, then a single wafer may contain regions of both missing gates and unreacted gates. Also, as with the previous example, defects must be well controlled in this defect sensitive portion of the line.

The second CMP step in the example of Fig. 20.14 offers yet another unique challenge. Here, three dissimilar materials, Ni, Ti, and TiSi, must all be polished at the same time with a high degree of process control. Poor rate control will bring similar overpolishing/underpolishing concerns as the first step (overpolishing leads to missing gates and underpolishing leads to residual metal and therefore shorted gates). Thus, slurry and process conditions must be found that yield high and well matched rates for the excess overburden materials (Ni, Ti, TiSi), with good selectivity to the materials that are to remain on the wafer (ILD0, Al, NiSi). Again, the consequence in this application of introducing CMP steps that have poor process control is a significant degradation in yield and/or transistor performance.

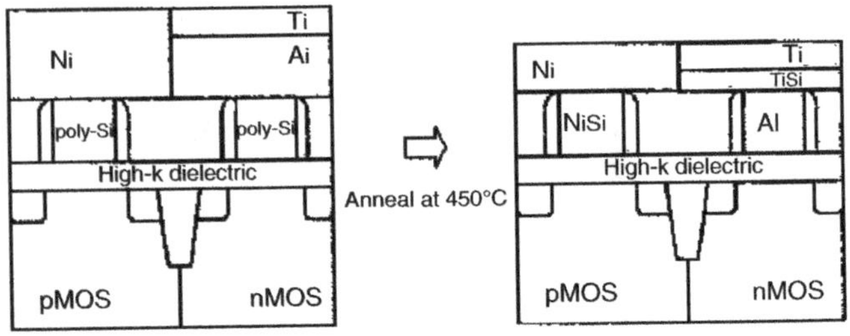

FIGURE 20.14 A second method of forming metal gates using CMP. After ILD0 deposition, CMP is used to expose the top of the poly-Si gate. Next metal films are deposited and patterned leaving Ni above PMOS gates and a Ti–Al bilayer over NMOS gates. After high temperature annealing, NiSi forms the PMOS gate and Al forms the NMOS gate. Post reaction, CMP is used to remove the excess metal (from Ref. 13).

One final point to the example of Fig. 20.14; the authors do not specify how the metal films are to be arranged on the surface, but the clever CMP engineer may recognize another potential use of CMP. After depositing and patterning one of the metal films, the second film may be deposited and a CMP step used to polish the surface, removing the second metal film in regions where it sits above the first film but leaving it in regions where the first film was earlier removed.

20.3.4 Other Examples

Many more examples of how researchers are putting CMP to use to enable new technologies can be found in the literature. Figure 20.15 [14] shows the use of CMP to planarize a substrate surface between layers of stacked chips enabling one method of 3D integration. Figure 20.16 [15] shows yet another method of 3D integration, quite different from Fig. 20.15, wherein device-quality silicon is grown using lateral epitaxial growth. This method allows for a second or even multiple planes of devices on the same wafer substrate without the need for chip bonding. CMP is used to smooth nodules that form an outcrop of the silicon recrystallization process.

Given the large number of examples of new CMP processes that researchers are investigating for use in future technologies, it is clear that the CMP industry has not reached maturity. Indeed, there appears to be more discussion of new CMP processes than in previous years. Researchers are beginning to use CMP for applications where they previously may have chosen another route, and by doing so, they have used CMP to simplify their task. There is a notable difference in the new CMP processes discussed here compared to the processes in Table 20.2. For the processes discussed in Table 20.2, there was little alternative besides using CMP. ILD, STI, and Cu CMP steps came about because there was no other option. For the examples discussed in this section, there are other methods not involving CMP. Device conference proceedings report many methods of fabricating Finfets and metal gate transistors that do

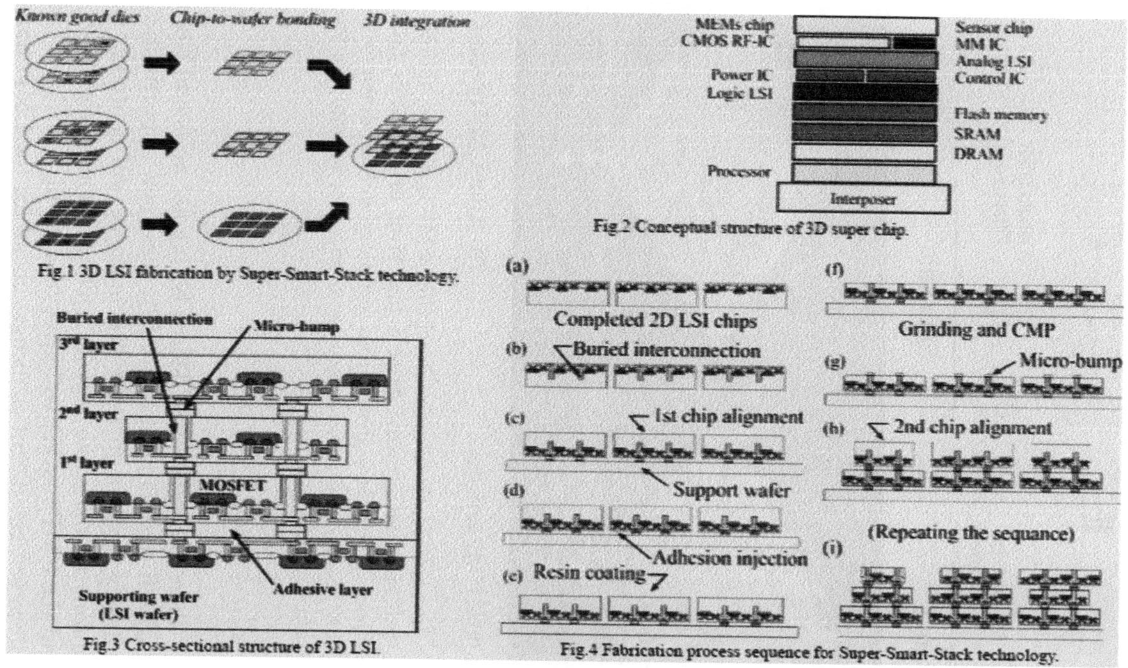

CMP used to expose chips between bonding layers in a 3D fabrication scheme (from Ref.).

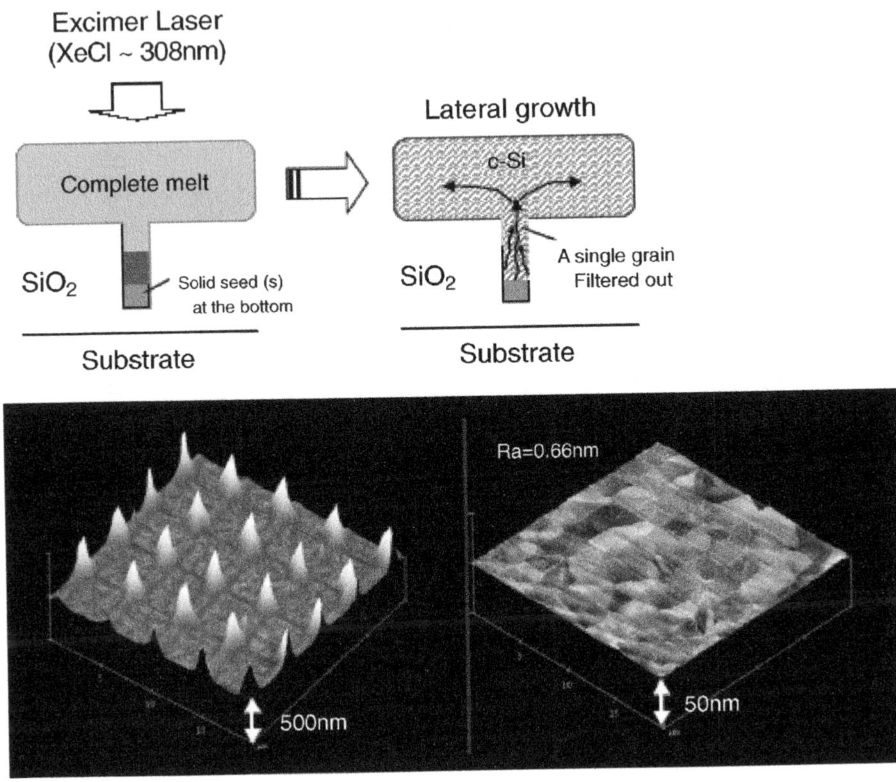

(a) Before CMP process (b) After CMP process

FIGURE 20.16 In a second 3D fabrication scheme, CMP is used to smooth the wafer surface after lateral epitaxial growth. After a first layer of transistor fabrication, a second layer of device quality Si is grown from seed holes in SiO_2 down to the original wafer surface. This process allows for transistor fabrication on multiple levels of the wafer surface (from Ref. 15).

not use CMP—there is competition. This is both good news and bad news for the CMP industry. If the challenges poised by these new opportunities are well met by the industry, growth in the CMP industry will continue. CMP will become ubiquitous in the same way as etching technologies have become ubiquitous, and the CMP market will grow as a result.

There are, however, significant challenges ahead for the CMP industry. The following section describes these challenges. If these challenges are not met, integrators working in the pathfinding and development phases will be forced to find methods to complete their task that do not involve CMP. If that happens, the CMP engineer will be relegated to continuous improvement of conventional CMP processes and growth in the CMP industry will stagnate.

20.4 CMP CHALLENGES

Many of the challenges are not new; they have been with CMP since the beginning. But if the CMP market is to continue to grow, the use of CMP must become ubiquitous. It is that potential for CMP use to become ubiquitous in combination with the sensitivity of new technologies to the faults of CMP that have increased the urgency of meeting these CMP challenges.

CMP has gained the reputation as a problematic module. As stated in the preceding section, CMP was required to enable the technologies of Table 20.2. CMP was used in those instances because there was no other choice. However, in the examples of potential new CMP uses listed in Section 20.3, there will be alternative methods for integrators to choose and they are not likely to choose a process that promises to be problematic for the job. This section describes what the CMP technologist and CMP industry must do to change the image of CMP as a difficult module that is used begrudgingly to a module that performs well and is accepted openly.

For any new technology, there are always challenges to be faced. Beyond these normal challenges, there are four areas in which CMP must make significant progress. For the CMP technologist, these are areas that should receive critical attention in the development of new processes and improvement of established processes. For the CMP industry, these are areas in which infrastructure must be developed and capital investment made if CMP is to continue to grow. These areas are (1) development time of new CMP materials, (2) reduction in CMP defects, (3) improvement in CMP process control, and (4) reduction in CMP cost.

20.4.1 Development Time of New CMP Materials

The 1990s saw rapid change and rapid growth of CMP technologies. The addition of new CMP processes was near feverish at times. However, despite the rapid introduction of new CMP processes, the development of CMP materials was slow. The same slurry and pads that were used for the first ILD CMP processes are still in use today for most of the part with only minor changes to their original formulation. Cycle time for the development of new consumables has been long. For example, the first high selectivity slurries tailored for STI have only recently found wide acceptance after nearly 10 years since the first STI CMP processes began using ILD CMP slurries. Commercially available W slurries were not ready for 4–6 years after W CMP first went into production. IC manufacturers sustained the first W CMP processes with in-house developed slurries. Even Cu CMP slurries, which were commercially available when Cu CMP went into production, had a long development lead time. Cu CMP was long anticipated prior to its widespread use, giving the CMP slurry industry 4–6 years time to develop Cu slurries before they were actually needed.

There are, however, two significant environmental changes that will require the CMP consumable industry to be more nimble in its development of new products, particularly slurries. First, IC manufacturers are becoming ever more secretive in their technology roadmaps. As stated in Section 20.1, conventional scaling methods are no longer sufficient for IC manufacturers to obtain the requisite 30% transistor speed improvements for each new technology. More innovation and more radical changes are needed, and while this has increased the risk for IC manufacturers, it has also increased the opportunity. The IC manufacturer who successfully navigates these changes stands to break away from the pack in terms of performance and cost, but only as long as it can keep its innovations secret.

The second factor driving the need for reduced cycle time of CMP consumable development is that as CMP use becomes more pervasive, many of the new CMP uses will be conceived in the pathfinding and development stages of the technology cycle rather than the research phase. While many changes are initiated in the research phase, the details of how the changes are integrated are determined in the pathfinding and development stages. Development engineers working to integrate a larger change will use CMP to simplify the change; or they may find a more elegant process flow involving CMP. In past years, such integrators were hesitant to use CMP because of the relative immaturity of the technology. As CMP becomes a more established technology, integrators are becoming more comfortable using it to realize their changes. In the example of the Finfet transistor, the core technology is explored for feasibility in the research phase, but determining exactly how the transistor may be efficiently and effectively fabricated is left to the pathfinding and development stages. As discussed in the preceding section, there are many possible methods for fabricating Finfets; some involving CMP, some not. Whether CMP will be used for Finfet fabrication may not be decided until halfway through pathfinding. Once that decision is made, the CMP engineer must develop consumables and a process to meet the schedule and requirements of the technology. Because risk is generally averted during the last year of development and during the manufacturing ramp, the CMP engineer will target the end of the first year of development to have his CMP consumables frozen. The final year will be used for small tweaks to the process and in preparing for the manufacturing ramp. It should be noted that major changes are sometimes made in the final year of development and even during the manufacturing ramp. But such late changes carry with them a great deal of risk to the technology ramp and should be undertaken only as a last resort. (Such last minute changes have contributed broadly to CMP's reputation as a problematic module.) Hence, consumable development must occur in the last year of pathfinding and the first year of development.

The CMP consumable industry must develop the infrastructure to respond to the challenge of the 2 year consumables development cycle. Because each new CMP application is likely to bring with it new and unique challenges, slurry, pad, and cleaning chemistries will need to be tailored to best suit the need of the new application. The vendor who can reliably have consumables ready in time will win the business. But to be nimble enough to develop the consumable on the 2 year cycle, the industry must increase its fundamental understanding of how the

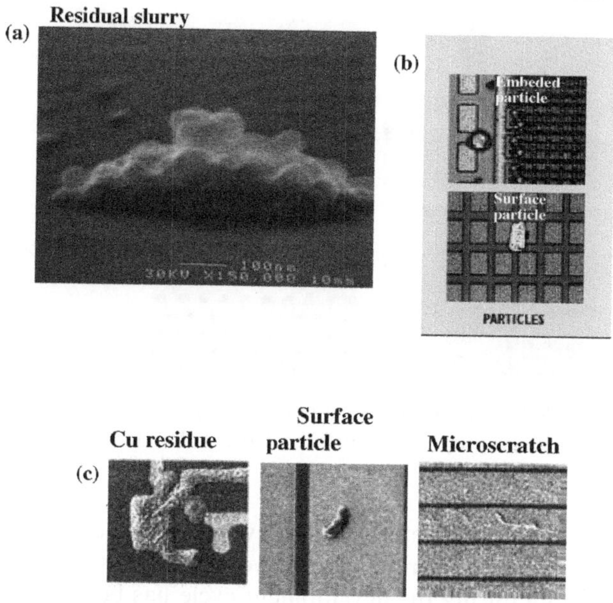

FIGURE 20.17 Examples of particle/defects commonly seen post-CMP (from Refs. 16–18).

consumable works. At the same time, the consumable vendors must develop a toolbox approach to the development of consumables. They must be able to quickly modify their product to meet the needs of each new application.

20.4.2 CMP Defect Reduction

CMP processes have historically been plagued with defects. It is not too surprising that after immersing the wafer in a slurry containing $\sim 10^{14}$–10^{17} particles/l, the particle contamination after CMP is a problem. However, residual slurry particles (Fig. 20.17a [16]) are not the only defect of concern and, in many cases, are not even the greatest concern. Organic residue from slurry or cleaning chemistry constituents, bacterial infestation, polishing pad, or scrubber brush debris are all sources of particulate contamination (Fig. 20.17b,c [17,18]). In addition, trace metal contamination has become a large concern for FEOL CMP processing. Metal ions from chemical and abrasive manufacturing can significantly degrade device performance and proper operation if left on the surface post-CMP. Damage defects are another significant source of concern. With feature sizes pushed deep into the sub-100-nm regime, small CMP chatter marks (Fig. 20.18 [17]) on the wafer surface that were once benign are now killer defects. In the BEOL, soft, low-*k* dielectric films are easily damaged by the presence of any large foreign particles in the CMP slurry or pad. Great arcing scratches across the wafer (Fig. 20.19 [16]) are not uncommon with these soft films.

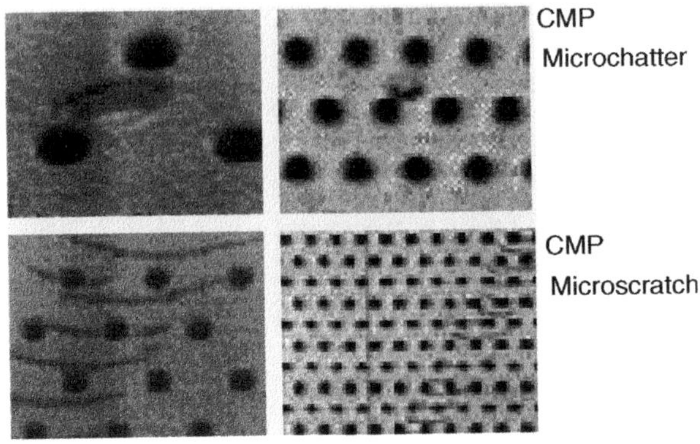

FIGURE 20.18 CMP chatter marks caused by large abrasive agglomerates and/or foreign particles impinging upon the surface of the wafer (from Ref. 17).

While concern for CMP defects is not new, their impact and the concern for them are likely to be greater in the next 15 years. As mentioned in Section 20.1, rapid yield learning during the development cycle has become a critical factor to getting a technology to market quickly, and high product yields during the manufacturing cycle are the key to keeping product cost low. Processes that limit yield learning or manufacturing yield levels due to high defects are not

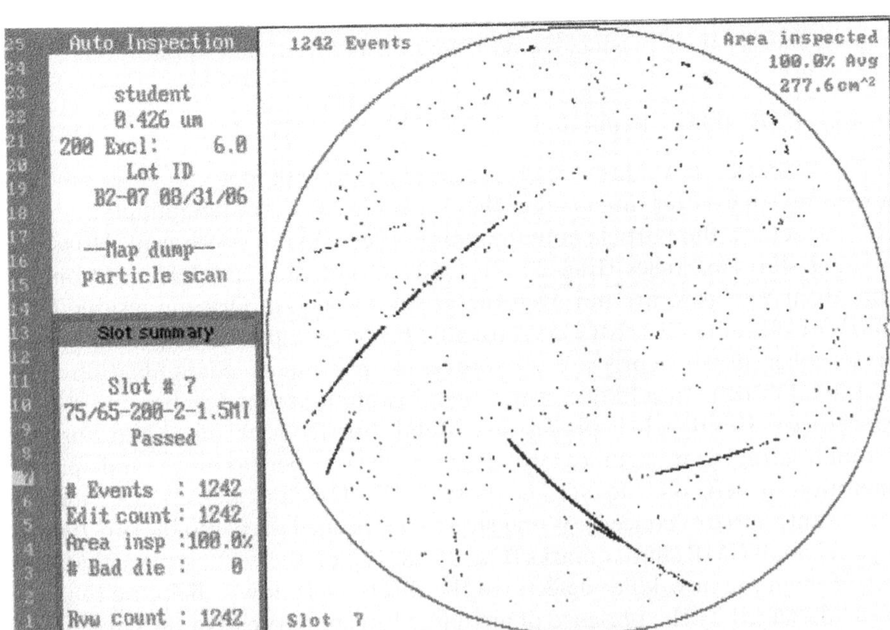

FIGURE 20.19 Arc scratches (from Ref. 16).

likely to be selected for inclusion into new technologies. Thus, for new CMP processes to make it from research to manufacturing, CMP defects must become a top consideration.

It is insufficient for the CMP development engineer alone to work on defect reduction. The polish tool manufacturer must make improvements in the cleaning technology while the consumable vendor must consider defects from the very start of their product development cycle. Because the process must be made ready within the 2-year development cycle of CMP consumables, there is no time to reengineer the consumables after they are selected. The product and process must produce low defects from the start. Processes that are seen to generate high defects during the pathfinding stage may be deselected from consideration.

Thus, the CMP industry must begin taking a more critical look at the defects generated by their products. Defect engineering is one of the most difficult aspects of the CMP process and as such, one of the most important considerations in the selection of new CMP materials. However, defect engineering is not an easy task for equipment manufacturers and consumable vendors to undertake. Defect analysis tools are expensive and require considerable expertise to fully utilize the tool's capabilities. IC manufacturers invest heavily in defect analysis tools and hire teams of full time defect engineers. It is not likely that consumable vendors will be able to make the required investment to do a proper job of engineering their product for low defects. Consumable vendors must be creative in finding partners, either in the universities, national labs, or IC manufacturers, to offset the need for capital investment. Another possibility is a consortia approach or an independent third party that specializes in defect engineering and analysis.

20.4.3 CMP Process Control

By today's IC manufacturing standards, CMP is not considered a well-controlled process. Three areas of particular concern are film thickness control, consumable materials control, and excursion prevention systems. Film thickness control afforded by CMP, whether across wafer or wafer to wafer, is poor in comparison to other IC manufacturing processes (i.e., film deposition or etching processes). If an integrator is concerned with film thickness control, they are unlikely to choose CMP. The control of consumable materials is also fragile. Excursions made in pad and slurry properties by the supplier lead to excursions in defects or film uniformity on the wafer. Such excursions cause material loss in the development cycle, leading to a delay in the introduction of the new technology, and in the manufacturing cycle, leading to material scrap and a loss in revenue. Lastly, tool excursion prevention systems for CMP tools do not provide the same real-time feedback as other processing equipments. Monitoring capability for global system parameters such as pressure and RPM exists, but methods to monitor the true processing environment are lacking due to a lack of fundamental understanding of the true CMP process environment as well as a lack of tool monitoring systems. Excursions in wafer processing are invariably caught downstream of the CMP process at subsequent metrology steps. This section discusses the key aspects of CMP process control and the infrastructure required to ensure that lack of process control systems does not limit the growth of the CMP industry.

20.4.3.1 CMP Film Thickness Control Early CMP processes had very poor thickness control. The first technologies to adopt ILD CMP compensated for poor thickness control by depositing an extra thick ILD layer and then using an excessive amount of ILD CMP to ensure that in the areas where the CMP rate was lowest, enough material was removed to ensure planarization, while in the area where CMP rate was greatest, enough material remained to ensure that there was no interlayer shorting (Fig. 20.20). Subsequent processing steps such as via etching and photolithography had to contend with large variation in film thickness and surface topography. Advancements in polishing head design, pad optimization, pad conditioner operation, slurry distribution,

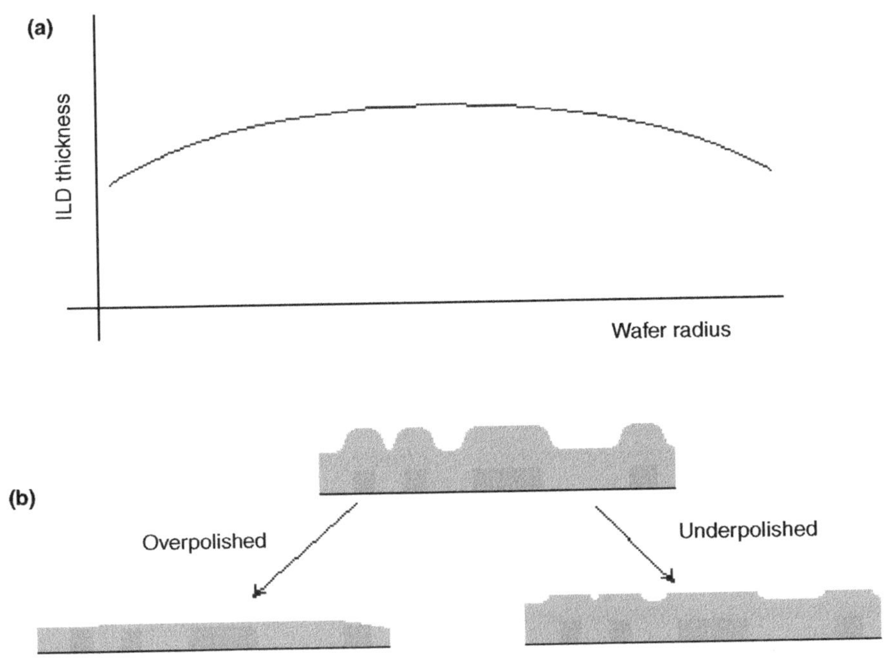

FIGURE 20.20 Poor cross-wafer thickness control in ILD CMP (**a**) leads to overpolishing in some regions of the wafer resulting in interlayer shorting and underpolishing in other regions of the wafer. The underpolished surface contains residual topography that can trap W in subsequent CMP steps.

end-point detection, advanced process control (APC), and other aspects of the CMP process have resulted in remarkable improvements in polishing rate control. However as feature sizes become even smaller and as CMP is considered for the formation of transistor structures with their ever-diminishing tolerance for dimensional variation, the requirements are outpacing the advances in CMP thickness control.

To properly understand CMP film thickness control, the CMP engineer should understand the sources of thickness variation and how they impact the total film thickness uniformity. Nonuniformity can be grouped in two categories—random variation and systematic variation. Examples of random variation include wafer-to-wafer (WTW), run-to-run (RTR), and some elements of within-wafer (WIW) variations. Elements of random variation add to the total thickness variation by their root mean square [19]

$$\sigma_{\text{total}}^2 = \sum (\sigma_{\text{random,i}})^2 \tag{20.3}$$

or

$$\sigma_{\text{total}} = [\sigma_{\text{WIW}}^2 + \sigma_{\text{WTW}}^2 + \sigma_{\text{RTR}}^2 + \cdots]^{1/2} \tag{20.4}$$

Systematic variation is a variation that maintains its structure from sample to sample. Examples of systematic variation in CMP include within-die (WID) variation and some elements of WIW variation such as edge thickness roll-off. WID and edge roll-off are considered systematic because the thickest and thinnest points on the die/wafer are predictable. Systematic variation adds directly to the total variation. Hence, systematic effects will have a greater impact on thickness variation. In general, for CMP processes, WID and edge variation values are greater than random components. For these two reasons, systematic effects tend to be the focus of CMP thickness control efforts.

For the CMP engineer, optimizing WID and edge thickness has largely been a matter of screening commercially available pads and slurries and then running experiments to optimize the process. Use of structured experimental design methodology, such as factorial experiments, and good intuition can minimize the number of experiments required, but in general, a large number of experiments are needed to optimize the CMP process. An excellent reference for experimental design of experiments is the text *Statistics for Experiments* by Box, Hunter, and Hunter [20].

New head designs (Chapter 4) have given the CMP engineer an additional knob to control across-wafer nonuniformities by varying the distribution of backside wafer pressure. However, the different pressure zones within these heads have a tendency to interact, and independent control of the wafer edge has been illusive.

WTW and RTR control of thickness are improved by the use of end-point detection systems and advanced process control. End-point detection, whether mechanical or optical, monitor the state of the wafer surface (film thickness, reflectivity, etc.) or of the entire polishing system (friction, slurry by products, etc.) in an attempt to predict when the desired amount of material has been removed (i.e., the end of process). End-point detection is most successful in processes where a change in the films on the wafer surface leads to an abrupt change in the optical or mechanical properties of the wafer surface. For example, copper CMP end point is easy to detect by optical means due to the large difference in reflectivity of the copper film compared to the barrier films. In contrast, end-point detection for small amounts of ILD removal is difficult due to the lack of change in the wafer surface or the wafer-pad interface.

APC methods use information of incoming thickness, polishing rate, postpolishing thickness, and/or historical thickness to predict polishing times. APC methods range from simple time adjustments based on targeting wafers (first, wafers polished in a lot) to sophisticated targeting algorithms that take into consideration many input variables. Such algorithms may be automated such that the polishing tool uploads input data and automatically determines the polishing time without the intervention of the operator.

20.4.3.2 Process Control Systems, Consumables Material Control, and Excursion Prevention The goal of process control systems (PCS), consumables material control systems, and tool excursion prevention systems are the same—to capture the generation of discrepant material production as close to the source as possible. When a tool or a process fails, discrepant material is produced. The discrepant material suffers from either higher defects, leading to some level of scrap, or poor performance, leading to some devaluation in the product. In either case, loss of revenue results. If only a small amount of material is affected and the offending tool or process is caught quickly, the revenue impact of the failure is minimal. But if the problem is not caught quickly, a large amount of material may be impacted leading to significant revenue loss—such an event is deemed an excursion.

Note that the product that originally became excursionary may not contain the revenue loss. For example, suppose an excursion in the raw material used to make polishing pads results in a batch of defective pads. If the defective pads are not caught by the pad manufacturer, they will be shipped to the IC manufacturer who then uses them to polish wafers. Defect metrology steps downstream of the polishing operation may then detect an unacceptable level of defects on wafers polished with the excursionary pads. If the metrology operation is many steps downstream of the polishing operation, a great many wafers may be impacted before the problem is discovered. At that point, the IC production line is shut down until the source of the problem is established. Once the source of the problem is deemed to be with the pads, unused affected pads are scrapped as are some or all of the affected wafers. Revenue loss in terms of lost material and lost productivity is large.

To prevent material loss due to excursions such as the previous example, robust process control systems are required throughout the supply chain from the raw materials manufacturer to the pad manufacturer and the CMP module. Invariably, incident reviews of such excursions reveal that the excursion could have been prevented or limited to only a small amount of material lost if the proper statistical process control systems had been in place. Invariably, the excursion could have been detected by careful scrutiny of an in-process parameter that was either monitored or should have been monitored by the subsupplier, pad manufacturer, and/or the CMP operation.

For the reader not familiar with statistical process control methodology, an excellent detailed treatment is given in Reference [21]. The essence of statistical process control is that by monitoring important variables of a process and detecting and correcting for changes in the behavior of these variables, variation in the output of the process can be minimized or eliminated and excursions to baseline performance can be rapidly detected. By minimizing variation and quickly detecting process deviations, the process owner can prevent loss due to the type of excursion described above and can also minimize the variation in the performance of the product leading to a higher quality output. The process owner must determine what variables are important to ensure the successful execution of his process and then carefully monitor those variables for any shifts or drifts in performance.

Robust process control relies on knowledge of what process variables are important and how they may be monitored. Often times, such key variables are revealed during the postmortem session of an excursion, but clearly, the postmortem is the least desirable time to discover the existence of a key process control parameter. Other means of understanding key process parameters include (1) brainstorming of failure modes, (2) process characterization studies, and (3) fundamental understanding of CMP processes. Brainstorming of failure modes involves process and equipment owners listing out all potential failure modes of the process, the tooling, and the procedures used to maintain and operate the tooling. The goal is to list every potential failure type, no mater how obscure. The fail modes are then rated in terms of their potential for material loss, the likely frequency of failure, and the ability to detect the failure. A priority is created based on these three elements and starting with the highest priority items, measures are put in place to either absolutely prevent each fail mode and/or monitor for each. In this way, effective monitoring of a process comes from a thorough consideration of how the process, equipment, or procedures can fail.

While it is important to measure the output of a production process, be it the CMP process, the production of CMP consumables, or the production of raw materials used to make CMP consumables, it is even more important to measure and control the input parameters of the process. Consider that if all of the input parameters of a process were known and all could be measured and controlled strongly, then there would be little or no variation to the process and the output need not even be measured. The advantage of measuring input variables is that deviations can be detected and corrected as soon as they occur, thereby limiting the impact of the failure to only the material that is in process at the time the deviation occurs. The key is to first identify important control parameters and then use different systems to monitor those parameters.

Given that the goal is to minimize the impact of a broken tool or process, then the key is to identify the failure as soon as possible. Production test-firings after maintenance can identify poorly performed maintenance, but if a tool breaks between maintenance cycles, a large quantity of material may be produced before the tool and process are test-fired again. Hence, there is a need for real-time monitoring of the system input parameters. Provided that the critical process input parameters can be measured, data acquisition software is available that can assimilate the input of a large volume of data from real-time measurements of key input parameters (such as pressure, velocity, fluid flows, among others). The software can process the data stream using the same process control systems methodology used to control the output variables. If a process parameter has drifted, then the software will flag the operator real time and the tool can be taken down for investigation. Such data control systems are invaluable not only in preventing excursion but also in responding to excursions because the data is archived and can be retrieved later to compare the results measured on the wafer.

Real-time monitoring fails, however, when the process tooling fails to monitor itself. If the process engineer determines that monitoring of an input parameter is critical but the tooling lacks a sensor to measure that parameter, then regardless of the capability of data acquisition software, the parameter cannot be monitored. To catch excursions at the earliest opportunity, it is critical for the tool manufacturers to build a self-monitoring process in the tool.

The process engineer must characterize his process to understand where the weaknesses are and where the process cliffs lie. Again, statistical DOE methodology is an invaluable tool. DOE methods may be used to clearly characterize the impact of variability in one or more input variables on the output of the process. The process engineer can map out the response in output variables to changes in input variables. Then if a change in an output variable is detected, the process engineer can rely on their previous characterization data to understand which input variables may be responsible for the observed change in the output. The engineer can also structure the monitoring methodology of the process around the DOE characterization to maximize the efficiency of the monitors.

Of course, in order to know what variables are important to monitor, the process engineer must have a fundamental understanding of how the process operates. With respect to fundamental understanding of the process, the CMP industry is still relatively immature. Understanding of CMP has come a long way in the past 15 years, as this book is a testament to, but more understanding is certainly needed. The industry must continue to fund fundamental studies of the CMP process.

20.4.4 Cost of CMP

If you ask the finance managers at IC manufacturing companies about their impression of CMP, they will not use glowing phrases like the enabling or key technology; they will simply say it costs too much. CMP is one of the most expensive unit processes in IC manufacturing today. Alternatively, if you ask a consumable or tool vendor about their view on CMP cost, they will tell you that the margins are too low and that the business is barely profitable. So how is it that the process costs too much yet the suppliers are not making larger profits? The reason is that the run rates and availability of CMP tools are low compared to other process tools and the consumption of consumables is high. Simply put, the CMP process is not efficient.

CMP cost is split between capital cost (cost of CMP tools and slurry delivery tools) and cost of consumables (pads, slurries, and consumable parts). By increasing the run rates (number of wafers processed per hour) of tools, both capital cost and cost of consumables can be decreased. Higher run rates require fewer tools, fewer tools require fewer consumables. Consumable cost can be further cut by extending the lifetime of consumables (if a pad can be made to last twice as long, then the pad cost will be cut in half) or the consumption rate of consumables (slurry flow rate, for example).

Just as it is imperative for the CMP industry to address the technical challenges spelled out in Sections 20.4.1–20.4.3, it is also imperative to address the cost of CMP. It may seem counterintuitive that the consumable and equipment vendors want to work to lower the CMP costs since they are the recipients of the CMP spending. However, if the CMP market is to grow, it will grow because of the increased usage of CMP in the process flow. As mentioned in the introduction to this chapter, most of the potential new usages of CMP will be discretionary because techniques using CMP must compete with techniques not using CMP. As the IC industry focuses on cost, the techniques that will survive the research and pathfinding phases will be the techniques that deliver the technical solution at an acceptable cost.

20.5 SUMMARY

CMP has come far in the 15 years since ILD CMP processes first became news. CMP technologists have gained a considerable understanding of their trade. CMP has become a topic of much interest as evidenced by the many conference sessions and entire conferences dedicated specifically to CMP, as well as abundance of university research programs. CMP growth has been fueled by the creation of enabling CMP technologies that were, for the most part, required by IC manufacturers intent on the continuous scaling of IC dimensions.

But future CMP growth cannot rely on the IC industry requiring new CMP processes. The CMP industry must prepare for the real possibility that future IC development efforts will not require CMP processes. CMP will be an option in the future; but one option of many rather than the only option.

The job of CMP technologists is to ensure that CMP will be the most attractive option.

QUESTIONS

1. What are the most significant achievements in the field of CMP for the past 15 years?
2. What would be the most challenging issues that the CMP community would face in the next 15 years?
3. What would be the breakthrough applications or processes in the field of CMP for the next 15 years that could benefit the semiconductor industry on the whole the most?
4. What could be the breakthrough technology or processes outside of the CMP filed for the next 5 years that could undermine the importance of CMP in semiconductor manufacturing process?
5. What could be the breakthrough technology or processes outside of the CMP filed for the next 5 years that could benefit the growth of CMP in semiconductor manufacturing process?

REFERENCES

1. Tucker T. CMP Market and Technology Status – 2002. Presented at CMP Users Group Meeting; 2002 Oct. 10. Proceedings published by the Northern California Chapter of the American Vacuum Society and can be found online at: http://www.avsusergroups.org/cmpug/cmpug_11proceedings.htm.
2. Moore GE. Cramming more components onto integrated circuits. Electronics. Apr 19, 1965;38(8), p 114–117.
3. Moore GE. No Exponential is Forever. ISSCC; 2003 Feb 10.
4. The International Technology Roadmap for Semiconductors. 2005. p 63–64. Prepared and published by the International Roadmap Committee, Semiconductor Association, can be found online at http://www.itrs.net/Links/2005ITRS/Home2005.htm.
5. Bohr MT. Intel First to Demonstrate Working 45 nm Chips. Intel Press Release. Jan 2006.
6. Bohr MT. Intel's Silicon R&D Pipeline. Intel Developer Forum. 2006 Apr 26.
7. Masao H, Shoji S, Koichi N. Trends of semiconductor technology for total system solutions. Hitachi Rev 1999;48(2):48–53.
8. Chau R, Doyle B, Kavalieros J, Barlage D, Murthy A, Doczy M, Arghavani R, Datta S. Advanced depleted-substrate transistors: single gate, double-gate, and tri-gate, Ext. Abstr. Int. Conf. on Solid State Devices and Materials. 2002 Sep 17. p 68.
9. Kaneko A, Yagishita A, Yahashi K, Kubota T, Omura M, Matsuo K, Mizushima I, Okano K, Kawasaki H, Inaba S, Izumida T, Kanemura T, Aoki N, Ishimaru K,

Ishiuchi H, Suguro K, Eguchi K, TsunashimaY. Sidewall transfer process and selective gate sidewall spacer formation technology for sub-15 nm finfet with elevated source/drain extension. IEDM tech digest 2005. p 844–847.

10. Kim YS, Lee SH, Shin SH, Han SH, Lee JY, Lee JW, Han J, Yang SC, Sung JH, Lee EC, Song BY, Lee DJ, Bae D-I1, Yang WS, Park YK, Lee KH, Roh BH, Chung TY, Kim K, Lee W. Local-damascene-FinFET DRAM integration with p+ doped polysilicon gate technology for sub-60 nm device generations. IEDM tech digest 2005. p 315–318.
11. Steetman BG. Solid State Electronic Devices. Englewood Cliffs, NJ: Prentice-Hall, Inc.; 1980. p 294–323.
12. Yu HY, Chen JD, Li MF, Lee SJ, Kwong DL, van Dal M, Kittl JA, Lauwers A, Augendre E, Kubicek S, Zhao C, Bender H, Brijs B, Geenen L, Absil P, Jurczak M, Biesemans S. Modulation of the Ni FUSI workfunction by Yb doping: from midgap to *n*-type band edge. IEDM tech digest 2005. p 630–633.
13. Park C, Cho BJ, Tang LJ, Kwong DL. Substituted aluminum metal gate on high-k dielectric for low workfunction and fermi-level pinning free. IEDM tech digest 2004. p 299–302.
14. Fukushima T, Yamada Y, Kikuchi H, Koyanagi M. New Three-Dimensional Integration Technology Using Self-Assembly Technique. IEDM tech digest 2005. p 348–351.
15. Shimada H, Hiroshima Y, Shimoda T. Low temperature single grain thin film transistor (LTSG-TFT) with SOI performance using CMP-flattened μ-Czochralski process. IEDM tech digest 2005. p 923–926.
16. de Larios JM, Zhang J, Zhao E, Gockel T, Ravkin M. Evaluating chemical mechanical cleaning technology for post-CMP applications. Micro Magazine; May 1997; p 61.
17. Dennison C. Developing effective inspection systems and strategies for monitoring CMP processes. Micro Magazine; Feb 1998; p 31.
18. Tiwari R, Soucek M, Strupp J. Development and Implementation of 300 mm Cu CMP Manufacturing Systems Future Fab International, Vol. 12; Feb 2002. p 546.
19. Montgomery DC. Design and Analysis of Experiments 6th ed. New York: John Wiley & Sons, Inc.;2005.
20. Box GEP, Hunter WG, Hunter JS. Statistics for Experiments. New York: John Wiley & Sons, Inc.; 1973.
21. Montgomery DC, Introduction to Statistical Quality Control. 5th ed. New York: John Wiley & Sons, Inc.; 2004.

21

UTILITARIAN INFORMATION FOR CMP SCIENTISTS AND ENGINEERS

Yongqing Lan and Yuzhuo Li

In this chapter, a collection of practical information that is useful for CMP scientists and engineers is presented. In addition to some physical and chemical properties of the key materials commonly used in CMP, some commercial sources for these materials are also included. Because of the highly dynamic nature of the industry, it is certain that the list is incomplete and the readers are encouraged to check out the corresponding Web sites for any updates or changes.

21.1 PHYSICAL AND CHEMICAL PROPERTIES OF ABRASIVE PARTICLES

Abrasive particles are a key component in CMP slurry. The most commonly used abrasive particles include silica, alumina, ceria, zirconia, titania, and diamond. Table 21.1 listed a set of information on each type of abrasive particles such as density, microhardness, and isoelectric points (IEP). It is important to point out that the specific values for these properties depend highly on the preparation techniques and the specific states of the samples. The values listed in the table represent an average of the most commonly reported data. For example, the isoelectric point for silica is a function of the number of hydroxyl groups, type and level of adsorbed species, metal impurity in the solid matrix, and the treatment history of the materials [1]. There are three major types of silica according to their preparation methods: fumed, colloidal, and precipitated. The common sources for obtaining these abrasive particles are listed in Table 21.2. As examples, some of the more specific information on

Microelectronic Applications of Chemical Mechanical Planarization, Edited by Yuzhuo Li

TABLE 21.1 Physical Properties of Abrasive Particles.

Abrasive	Density g/cc, 21 °C	Refractive Index	Microhardness[a]	IEP[b]
Silica: SiO_2	2.19	1.85	6–7	2.2
α-Alumina: Al_2O_3 (hexagonal)	3.97 (orthorhombic)	1.768	9	9
γ-Alumina: Al_2O_3 (cubic)				8
Ceria: CeO_2 (polycrystalline–cubic)	7.13	1.94	6	7
Zirconia: ZrO_2	2.9	2.2	8–8.8	4–11 6.7 (hydrous)
Titania: TiO_2	3.9–4.3	2.78	6–6.5	4.7 6.2 (hydrous)
Diamond, C	3.52	2.42	10	5.2

[a]Obtained by Rockwell hardness tests.
[b]IEP (Isoelctric point): The pH at which the particle is kinetically uncharged [2].

fumed and colloidal silica particles commercially available as Cab-o-Sol and Ultra-Sol are also listed in Tables 21.3 and 21.4.

When selecting an abrasive particle for a CMP application, many factors may influence the decision, such as the bulk and surface properties of the particles.

TABLE 21.2 Commercial Silica Products.

Type	Commercial Products
Colloidal silica	NexSil™ aqueous colloidal silica (NYACOL, www.nyacol.com)
	LUDOX® colloidal silica (SIGMA-ALDRICH, Inc., http://www.sigmaaldrich.com;
	GRACE Davison, http://www.gracedavison.com/products/ludox/overview.htm)
	TIZOX® (Ferro, http://www.ferro.com/Our+Products/Electronic/)
	AEROSIL® EG 50 (Degussa, http://www.degussa.com)
Fumed silica	Cabot fumed silica (Cabot, www.cabot-corp.com)
	WACKER fumed silica (WACKER, www.wacker.com)
	AERODISP® (Degussa, http://www.degussa.com)
Precipitated silica	SM-series (Grace Davison, http://www.gracedavison.com/eusilica/Precipitated/start_ppt.htm)
	RubberSil RS-series (H. M. Royal, Inc., http://www.hmroyal.com/silica.asp)
	ACEMATT® -series (Degussa, http://www.degussa.com)
	PERFORM-O-SIL (Nottingham Company, http://www.ppiatlanta.com/precipitated.html)

TABLE 21.3 Portion of Fumed Silica from Cabot Corporation [4]

PRODUCT GUIDE		
Product Name	Surface Area (m^2/g)	Bulk DensityTamped Density (g/l)
LM-150	160	50
LM-150D	160	125
M-5	200	50
M-5P	200	50
M-5DP	200	125
M-7D	200	125
PTG	200	50
HP-60	200	50
MS-75D	255	115
H-5	300	50
HS-5	325	50
EH-5	380	50

The most relevant bulk property of these particles is microhardness. Readers are recommended to review Chapter 7 on this topic for detailed discussion on the relative importance of bulk and surface properties. In general, for a chemically limited or dominated CMP process, the bulk properties of a particle such as microhardness become less important. Similarly, when particle sizes approach sub-50 nm, the bulk properties of a particle become less prominent than its surface properties. One of the most important surface properties is the isoelectric point. Isoelectric point is the pH at which a molecule, protein, or particle carries no net electrical charge [2]. The exact value of IEP for a particular sample may vary depending on the techniques used. It is important to realize that some commonly used techniques require significant dilution of the sample [3]. The results may represent the true property of the particles in the sample after the dilution but may hardly resemble that in the original concentrated state. CMP slurries do contain high concentrations of abrasive in most cases. The newly developed methods such as acoustic-based technology can measure concentrated samples directly and noninvasively. The drawback of these techniques, at this moment, is that they still requires a large amount of sample (> 50 ml).

TABLE 21.4 Colloidal Silica Products of Eminess Technologies, Inc. [5]

Description	Particle Size	% Solids (Mean)	pH Meas.
Ultra-Sol™ 3A	.012μ	30	10.9
Ultra-Sol™ 7A	.05μ	30	10
Ultra-Sol™ 8A	.05μ	30	10.9
Ultra-Sol™ 7H	.07μ	30	2.5
Ultra-Sol™ 2EX	.05μ	30	10.9
Ultra-Sol™ 3EX	.05μ	30	neutral
Ultra-Sol™ 555	.06μ	50	10.0

21.2 PHYSICAL AND CHEMICAL PROPERTIES OF OXIDIZERS

Oxidizers are present in almost all metal CMP slurries. Sometimes, barrier slurry, post-CMP cleaning solution, and other CMP consumables also require the use of an oxidizer. The physical and chemical properties of the commonly used oxidizers are listed in Table 21.5. Among them, hydrogen peroxide is the most commonly used oxidizer. It is important to point out that hydrogen peroxide is available in various standard and specialty grades, differentiated by the stabilizer packages appropriate for the specific end use. Most grades are available in concentrations ranging from 30% to 70%. Technical grade hydrogen peroxide is the most commonly used grade. It is usually stabilized with compounds such as tin that protect the product from decomposition during transport and storage. Some of the chemical grade hydrogen peroxides are formulated with a tin-free organic stabilizer. It is most commonly used by the chemical processing industry in applications that cannot tolerate the presence of tin. Similar to cosmetic and food industries, semiconductor grade contains even lower concentration of metal contaminants and stabilizers. These products typically are lightly stabilized or are stabilizer-free with specifications down to <100 ppt per individual cation [6]. As discussed in Chapter 7, the presence of metal ions can have a significant impact of on the performance of metal CMP slurry.

In addition to metal ions, the stability of hydrogen peroxide is also a function of environmental temperature, in general, and pH. An increase in the temperature and pH (especially at pH > 8) tends to exacerbate the impact of contaminants (especially transition metals such as copper, manganese, or iron) and, to a lesser degree, exposure to ultraviolet light. The types of stabilizers used in hydrogen peroxide vary between producers and product grades. Colloidal stannate and sodium pyrophosphate are the traditional mainstays, although organophosphonates (e.g., Monsanto's Dequest products) are increasingly common. Other additives may include nitrate (for pH adjustment and corrosion inhibition) and phosphoric acid (for pH adjustment). Colloidal silicate can also be used as a sequester for metals and, thereby, minimize H_2O_2 decomposition.

Hydrogen peroxide is weakly acidic with a p*K*a of 11.65. Therefore, if hydrogen peroxide is added into a solution that is preadjusted to a certain pH value, the pH of the solution will likely be lowered [7].

21.3 PHYSICAL AND CHEMICAL PROPERTIES OF RELEVANT SURFACTANTS

21.3.1 Classification of Surfactants

As discussed in Chapter 7, surfactant molecules are commonly used in CMP applications for colloidal dispersion stabilization to modify the wafer and particle surface properties, to control the removal rate and removal selectivity,

TABLE 21.5 Physical and Chemical Properties of Some Oxidizers.

Surfactant	Boiling Point, °C	Melting Point, °C	Density, g/cc	Water Solubility	pK
Hydrogen peroxide	150.2 ± 9	−0.43	1.4425 (25°C)	Freely soluble	$pK_a = 11.65 \pm 0.20$ (most acidic)
Ferric nitrate	< 100°C (< 212F) Decomposes	47.2	1.7	Freely Soluble	—
KIO_3	Not applicable	560°C (1040F) (partial decomposition)	3.98	32 g in 100 ml water (100°C)	—
Hydroxyl amine (50% solution)	> 100°C	Not applicable	1.078 (25°C)	Freely soluble	$pK_b = 4.7$ (basic)
Potassium persulfate	Not applicable	100°C (212F) Decomposes	2.48	4.7 g in 100 ml water	—
Ammonium persulfate	Not applicable	120°C (248F) Decomposes	1.98	80 g/100 ml water (25°C)	—

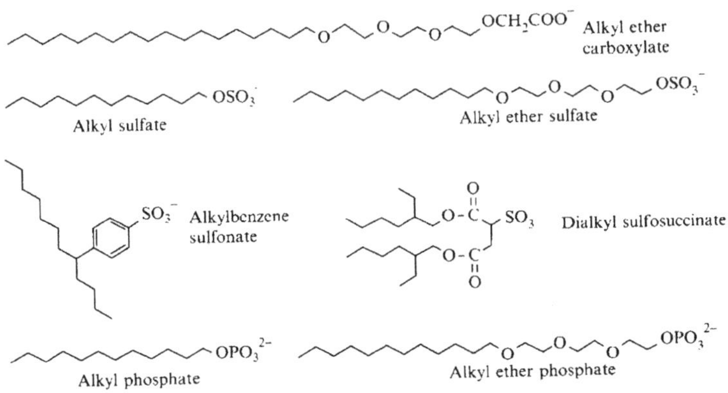

FIGURE 21.1 Structures of some representative anionic surfactants. Carboxylate, sulfate, sulfonat, and phosphoate are the polar groups found in anionic surfactants (from Ref. 8).

to passivate the metal surface to reduce corrosion, and to aid the post-CMP cleaning process. There are four types of surfactant molecules according to their ionization behavior and the charges they carry after the dissociation. The general features of these surfactants are illustrated in Figs. 21.1–21.4.

21.3.2 Critical Micellar Concentration

The most important property of surfactants is their tendency to aggregate. Such behavior is usually characterized by its unique critical micelle concentration (CMC). CMC is defined as the concentration of surfactants in free solution in equilibrium with surfactants in aggregated form. Below the critical micelle concentration, the surfactant molecules form a single layer on the liquid surface and are dispersed in the solution. At the first critical micelle concentration (CMC-I), the surfactant molecules organize into spherical micelles. Depending on the molecular shape of the surfactants and their environment, there could be secondary and tertiary critical micelle concentrations. At the second critical micelle concentration (CMC-II), the surfactant may aggregate into elongated pipes and at the lamellar point (LM or CMC-III) into stacked lamellae of pipes. The CMC depends on the chemical composition, mainly on the ratio of the head area and the tail length. When selecting a surfactant for a CMP related application, it is important to know the CMC of a surfactant, as micelles can encapsulate hydrophobic molecules and make their effective concentration much lower than the total concentration. For example, many corrosion inhibitors such as benzotriazole are hydrophobic. The presence of micelles may have an inadvertent effect on their availability or effective concentration (Chapter 2). The CMC values for some of the common surfactants are listed in Tables 21.6 and 21.7.

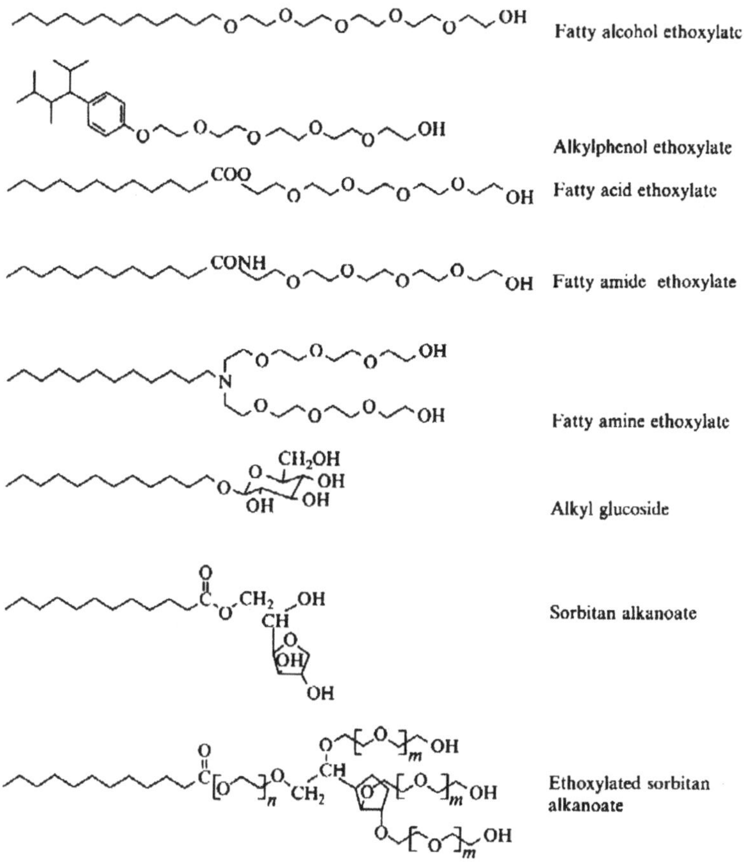

FIGURE 21.2 Structures of some representative nonionic surfactants. Nonionic surfactants have either a polyether or a polyhydroxyl unit as the polar group. In the vast majority of nonionics, the polar group is a polyether consisting of oxyethylene units, made by polymerization of ethylene oxide (from Ref. 9).

21.3.3 Ternary Phase Diagrams Involving Surfactants

As discussed in the last section, surfactant molecules can aggregate into various structures from micelles to liquid crystals. In physical or colloidal chemistry, these phenomena are described as phase behaviors of the surfactants. In order to describe the phase behavior of the surfactant, a phase diagram is often used. There have been tremendous research progress and application successes in the field of phase diagrams. The phase diagrams of the most commonly used surfactant are known [15]. As most surfactant systems can be simplified into a three-component system initially, a ternary phase diagram is commonly used. In an aqueous surfactant system, all water-soluble species can be grouped together as the aqueous phase, all hydrophobic species can be categorized as

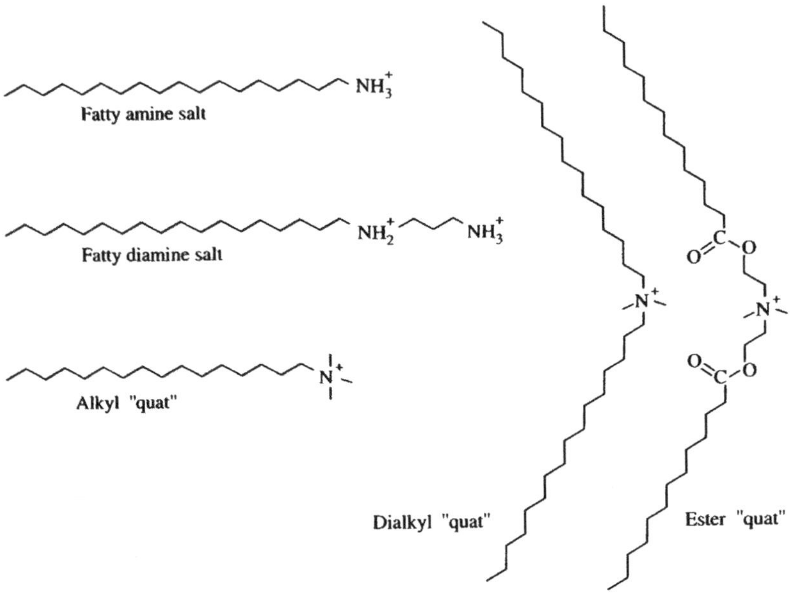

FIGURE 21.3 Structures of some representative cationic surfactants. The vast majority of cationic surfactants are based on the nitrogen atom carrying the cationic charge. Both amine and quaternary ammonium-based products are common (from Ref. 10).

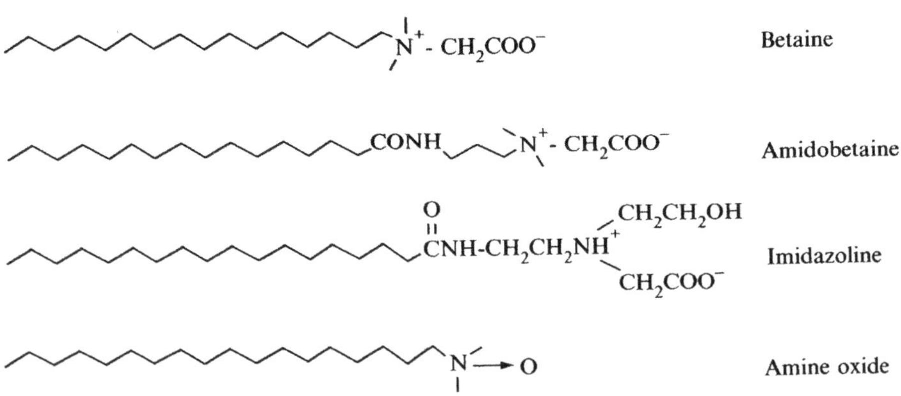

FIGURE 21.4 Structures of some representative zwitterionic surfactants. Zwitterionic surfactants contain two charged groups of different sign. The positive charge is almost invariably ammonium; the source of negative charge may vary, although carboxylate is by far the most common (from Ref. 11).

TABLE 21.6 List of CMC[a] Values for Some Common Surfactants [12].

Surfactant	CMC
Dodecylammonium chloride	1.47×10^{-2} M
Dodecyltrimethylammonium chloride	2.03×10^{-2} M
Decyltrimethylammonium bromide	6.5×10^{-2} M
Dodecyltrimethylammonium bromide	1.56×10^{-2} M
Hexadecyltrimethylammonium bromide	9.2×10^{-4} M
Dodecylpyridinium chloride	1.47×10^{-2} M
Sodium tetradecyl sulfate	2.1×10^{-3} M
Sodium dodecyl sulfate	8.3×10^{-3} M
Sodium decyl sulfate	3.3×10^{-2} M
Sodium octyl sulfate	1.33×10^{-1} M
Sodium octanoate	4×10^{-1} M
Sodium nonanoate	2.1×10^{-1} M
Sodium decanoate	1.09×10^{-1} M
Sodium undecanaote	5.6×10^{-2} M
Sodium dodecanoate	2.78×10^{-2} M
Sodium *p*-octylbenzene sulfonate	1.47×10^{-2} M
Sodium *p*-dodecylbenzene sulfonate	1.20×10^{-3} M
Dimethyldodecylamineoxide	2.1×10^{-3} M
$CH_3(CH_2)_9(OCH_2CH)_6OH$	9×10^{-4} M
$CH_3(CH_2)_9(OCH_2CH)_9OH$	1.3×10^{-3} M
$CH_3(CH_2)_{11}(OCH_2CH)_6OH$	8.7×10^{-5} M
$CH_3(CH_2)_7C_6H_4(CH_2CH_2O)_6$	2.05×10^{-4} M
Potassium perfluorooctanoate	2.88×10^{-2} M

[a]CMC (critical micelle concentration): The threshold concentration of surfactants at which micellization begins [13].

the oil phase, and surfactant molecules become the third component. For example, a clear transparent BTA solution above its usual solubility can be obtained with the aid of the surfactant (>CMC). This solution can be classified as a microemulsion. The microemulsions are clear, isotropic liquid mixtures of oil, water, and surfactant. The water phase may contain salt(s) and/or other ingredients. It is possible to prepare microemulsions from a large amount of components. In contrast to ordinary emulsions, microemulsions are formed upon simple mixing of the components and do not require high shear conditions. In ternary systems such as microemulsions, where two immiscible phases (water and "oil") are present next to the surfactant phase, the surfactant molecules form a monolayer at the interface between oil and water, with the hydrophobic tails of the surfactant molecules dissolved in the oil phase and the hydrophilic head groups in the aqueous phase.

Comparable to the binary systems (water–surfactant or oil–surfactant), self-assembled structures of different morphologies can be obtained ranging from (inverted) spherical and cylindrical micelles to lamellar phases and bicontinuous structures. To map out these regions, a phase diagram is most useful.

TABLE 21.7 List of CMC Values for Some Nonionic Surfactants [14].

Surfactant	CMC μM
C_6E_3[a]	10×10^4
C_8E_4	8.5×10^3
C_8E_5	9.2×10^3
C_8E_6	9.9×10^3
$C_{10}E_5$	9.0×10^2
$C_{10}E_6$	9.5×10^2
$C_{10}E_8$	10×10^2
$C_{12}E_5$	6.5×10
$C_{12}E_6$	6.8×10
$C_{12}E_7$	6.9×10
$C_{12}E_8$	7.1×10
$C_{14}E_8$	9.0×10
$C_{16}E_9$	2.1×10
$C_{16}E_{12}$	2.3×10
$C_{16}E_{21}$	3.9×10
$C_8\Phi E_9$	3.4×10^2
$C_8\Phi E_{10}$	3.4×10^2
$C_{12}NO$	2.2×10^3
B-D-C_8 glucoside	2.5×10^4
B-D-C_{10} glucoside	2.2×10^3
B-D-C_{12} glucoside	1.9×10^2

[a] C_6E_3 equals to $C_6H_{13}(OCH_2CH_2)_3OH$.

A basic phase diagram is easy to read. As shown in Fig. 21.5, the label at each corner represents a component, for example, A = surfactant, B = water, and C = oil. The chemical composition of the component at each corner is 100% (pure). The composition of this component for the points on the line that is opposite from the corner is always 0%. In other words, the line between the two corners is a binary phase diagram for the two components. What are the compositions for the points (1–6) indicated on the model phase ternary phase diagram? The answers for some of the points are all ready given in the table. The readers are encouraged to fill out the rest. Figure 21.6–21.8 show some typical ternary phase diagrams of different systems.

21.4 RELEVANT POURBAIX DIAGRAM

In chemistry, a Pourbaix diagram, also known as a potential/pH diagram, maps out the possible stable (equilibrium) phases of an aqueous electrochemical system. Predominant ion boundaries are represented by lines. As such, a Pourbaix diagram can be read much like a standard phase diagram with

%	A	B	C
1	0	20	80
2			
3	35	35	30
4			
5	25	25	50
6	40	0	60

(a)

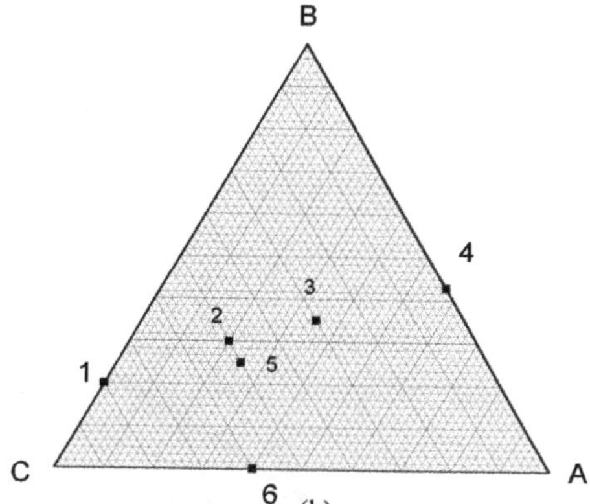

(b)

FIGURE 21.5 A basic phase diagram; the label at each corner represents a component (from Ref. 16).

a different set of axes. The diagrams are named after Marcel Pourbaix (1904–1998), the Russian-born chemist who invented them. Through the use of thermodynamic theory (the Nernst equation), the so-called Pourbaix diagrams can be constructed. These diagrams show the thermodynamic stability of a species as a function of potential and pH. Although many basic assumptions must be considered in their derivation, such diagrams can provide valuable information in the study of corrosion phenomena. Figure 21.9 represents a version of the Pourbaix diagram for the iron–water system at ambient

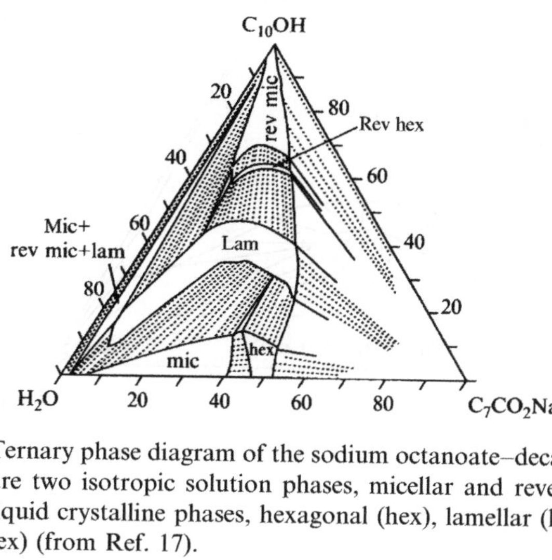

FIGURE 21.6 Ternary phase diagram of the sodium octanoate–decanol–water system at 25°C. There are two isotropic solution phases, micellar and reversed micellar (rev mic), and three liquid crystalline phases, hexagonal (hex), lamellar (lam), and reversed hexagonal (rev hex) (from Ref. 17).

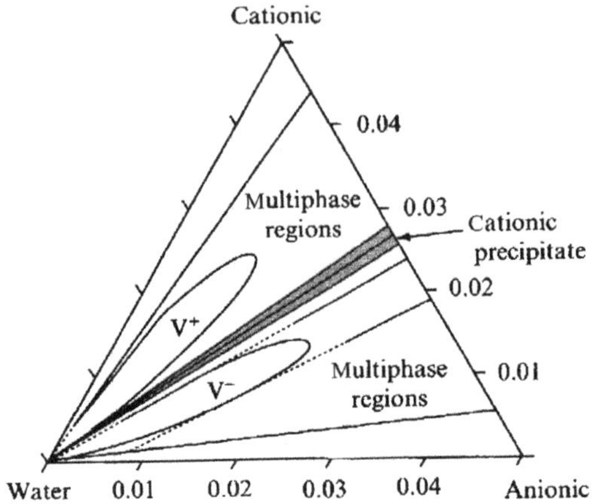

FIGURE 21.7 In a mixture of a cationic and an anionic surfactant, there are two regions of thermodynamically stable vesicles, V^+ and V^-, respectively (from Ref. 18).

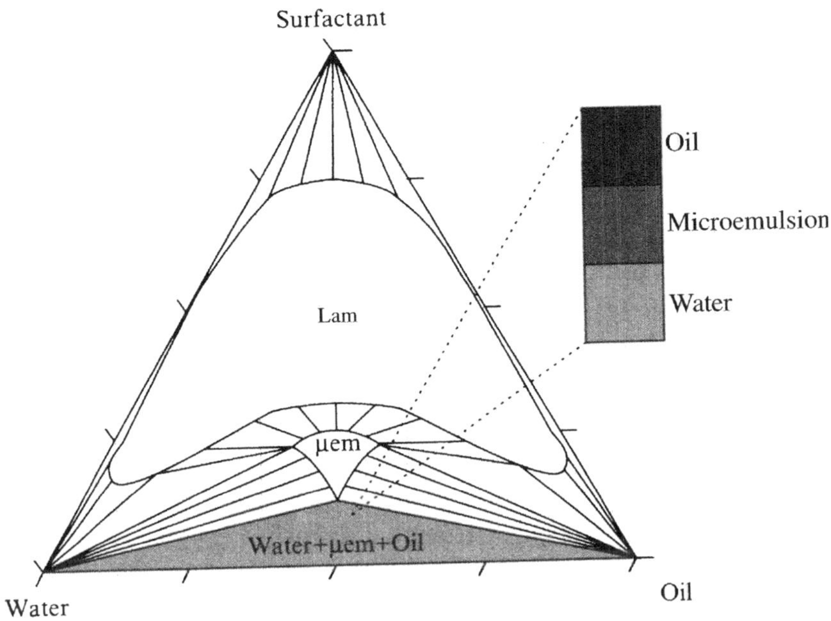

FIGURE 21.8 Phase behavior under balanced conditions of a surfactant–water–oil

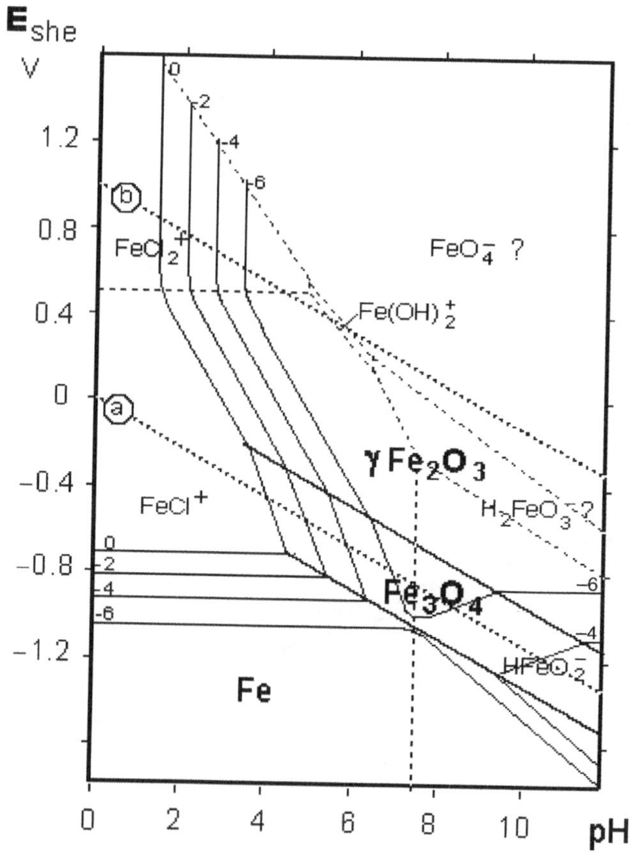

FIGURE 21.9 The Pourbaix diagram for the iron–water system at an ambient temperature (from Ref. 20).

temperature. This Pourbaix diagram enables to determine by means of potential and pH measurements whether a metal surface is in a region of immunity where the tendency for corrosion is nil, in a region where the tendency for corrosion is high, or in a region where the tendency for corrosion may still exist but where there is also a tendency for a protective or passive film to exist. Such a film can drastically affect the rate of corrosion, and, in some cases, practically stop it. Pourbaix diagrams of elements involved in semiconductor industry are listed from Figure 21.10–21.30.

Here is a brief tutorial on how to read a Pourbaix diagram:

Low E (or pE) values represent a reducing environment. High E values represent an oxidizing environment. The pE scale is intended to represent the concentration of the standard reducing agent (the e^-) analogous to the pH scale representing the concentration of standard acid (H^+). PE values are obtained from reduction potentials by dividing $E°$ by 0.059.

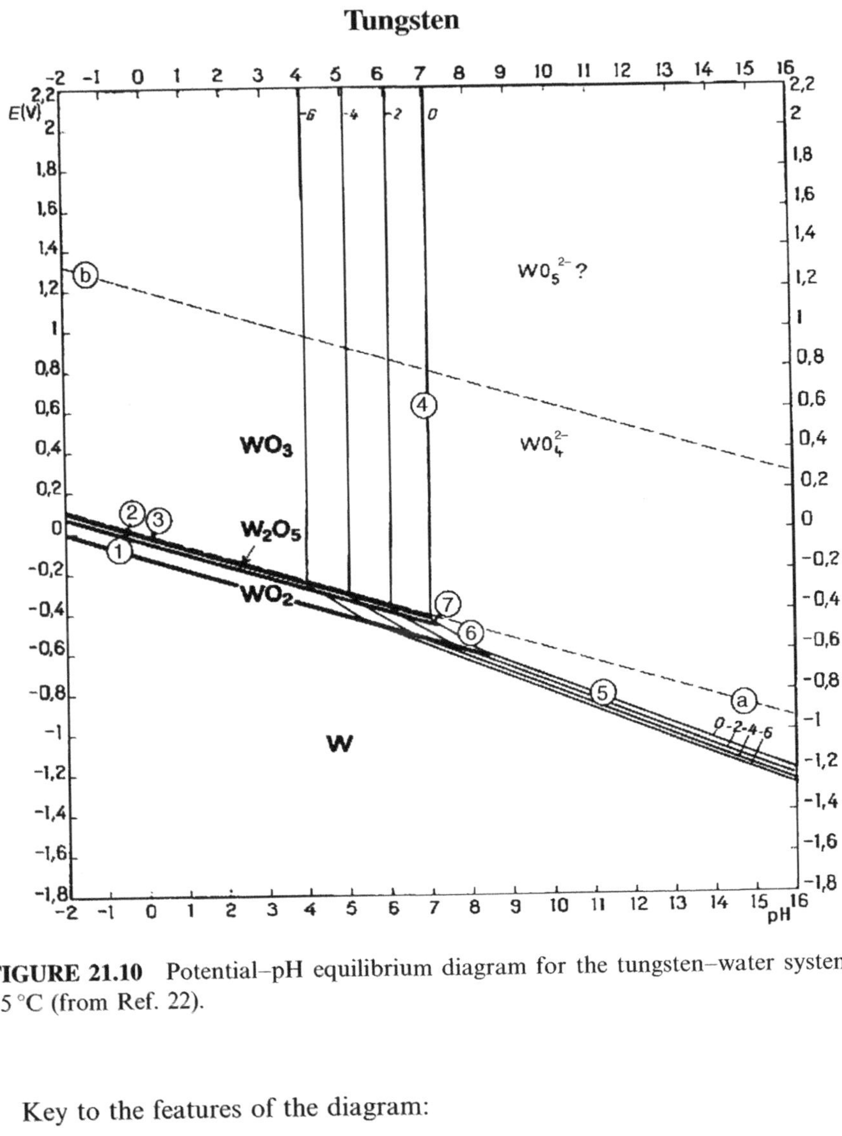

FIGURE 21.10 Potential–pH equilibrium diagram for the tungsten–water system at 25 °C (from Ref. 22).

Key to the features of the diagram:

- Solid lines separate species related by acid–base equilibria (line a):
  - Line a shows the pH at which half of the 1 M iron is Fe^{3+} and half is precipitated as $Fe(OH)_2$.
  - Pourbaix diagrams incorporate Z^1/r calculations and acid–base equilibria.
  - The position of an acid–base equilibrium is dependent on the total concentration of iron.
    - Reducing the total concentration of Fe^{3+} will reduce the driving force of the precipitation.

Copper

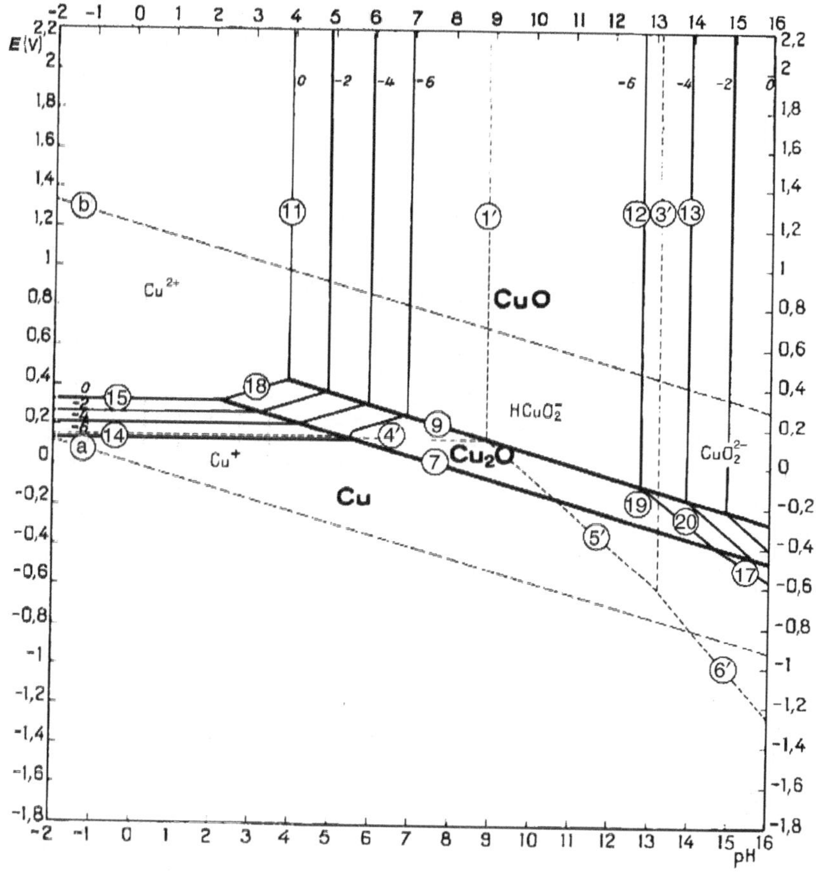

FIGURE 21.11 Potential–pH equilibrium for the copper–water system at 25 °C. [Considering the solid substances Cu, Cu_2O, and CuO; $Cu(OH)_2$ is not considered.] (from Ref. 23).

- Reducing the total iron concentration from 1 to 10^{-6} M (more realistic concentrations for geochemists and corrosion engineers) shifts the boundary from pH 1.7 to 4.2.
- In general, in more dilute solutions, the soluble species have larger predominance areas.

- Longer dashed lines enclose the theoretical region of stability of water to oxidation or reduction (lines d and f), while shorter dashed lines enclose the practical region of stability of water (e and g):
  - Dashed line d represents the potential of water saturated with dissolved O_2 at 1 atm (very well-aerated water).

Copper

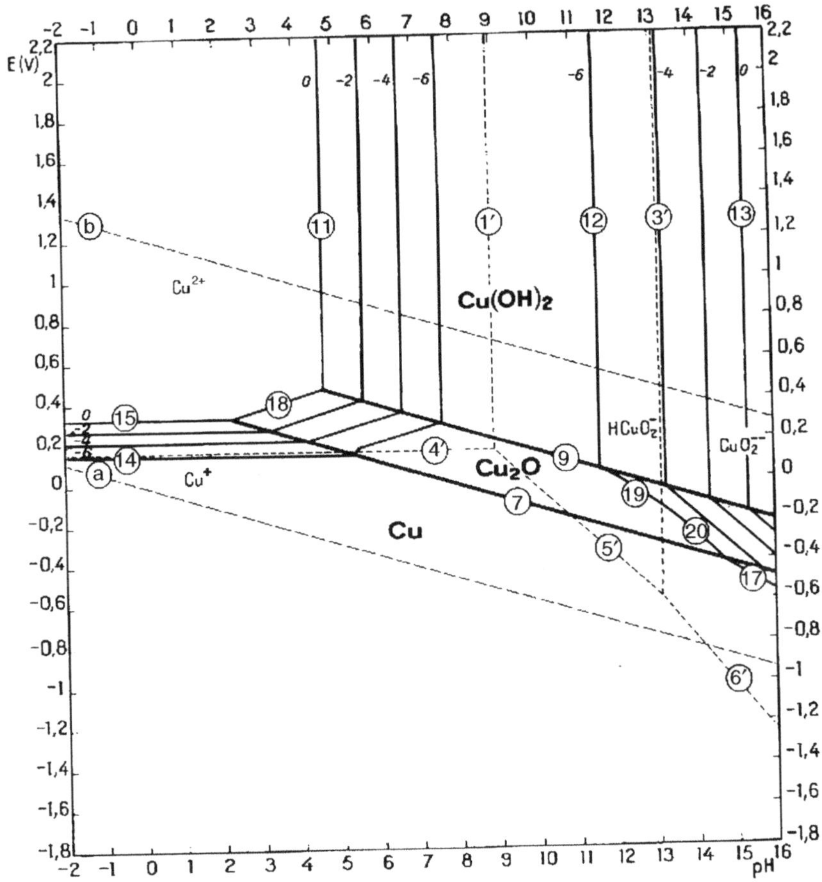

FIGURE 21.12 Potential–pH equilibrium diagram for the copper–water system at 25 °C. [Considering the solid substances Cu, Cu_2O, and $Cu(OH)_2$; CuO is not considered.] (from Ref. 24).

- Above this potential, water is oxidized to oxygen:

$$2H_2O + 4H^+(aq)O_2 + 4_e^- E^o = +1.229 \quad V$$

  - Theoretically, water should be oxidized by any dissolved oxidizing agent $E^o > 1.229$.
  - In practice, about 0.5 V of additional potential is required to overcome the overvoltage of oxygen formation (dashed line e).

Iron

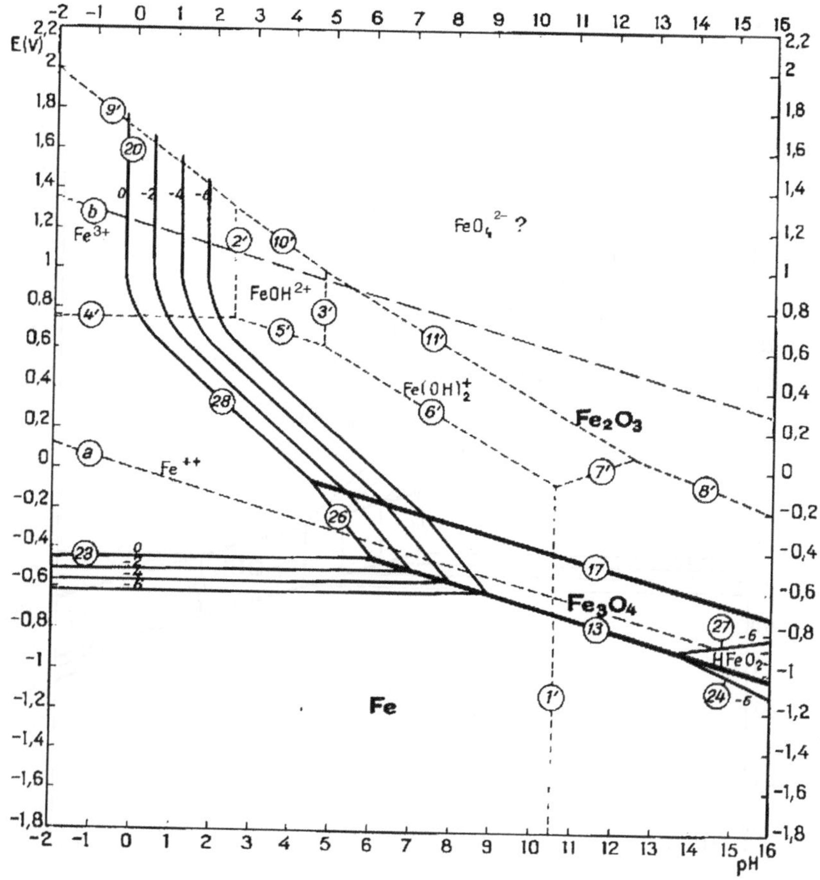

FIGURE 21.13 Potential–pH equilibrium diagram for the iron-water system at 25 °C. [Considering the solid substances Fe, Fe_3O_4, and Fe_2O_3 only.] (from Ref. 25).

- Dashed line f represents the potential of water saturated with dissolved H_2 at 1 atm (high level of reducing agents in solution).
- Below this potential water is reduced to hydrogen:

$$2H^+ + 2e^- E^o = +1.229 \quad V$$

21.5 COMMONLY USED BUFFERING SYSTEMS

For CMP related application, it is sometimes important to use a pH buffered system. Tables 21.8 and 21.9 show some of the commonly used buffering systems and their useful ranges.

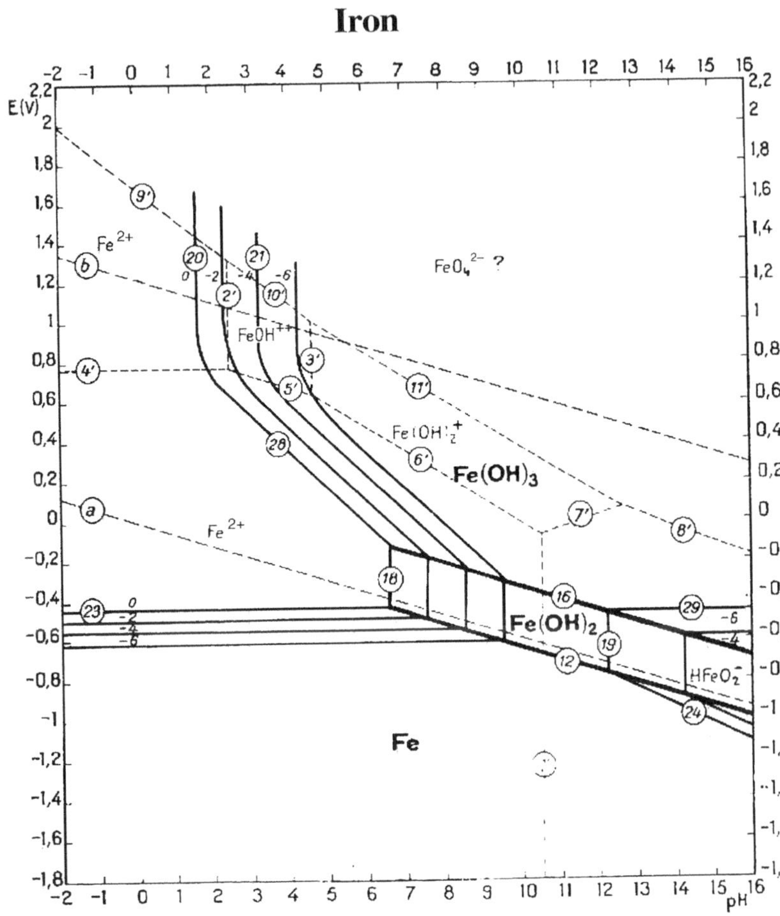

FIGURE 21.14 Potential–pH equilibrium diagram for the iron–water system at 25 °C. [Considering the solid substances Fe, $Fe(OH)_2$, and $Fe(OH)_3$ only.] (from Ref. 26).

21.6 USEFUL WEB SITES

(a) Slurry companies

- CabotCMP: http://www.cabotcmp.com/
- Fujimi: http://www.fujimico.com/
- DuPont Air Products NanoMaterials L.L.C. (DA NanoMaterials): http://www.nanoslurry.com/index_flash.htm
- Hitachi: http://www.hitachi.com/
- Toshiba: http://www.toshiba.com/tai-new/

TABLE 21.8 Useful Buffering Range for Different Buffering Systems at 25 °C [21].

Buffering System	Useful Buffering Range @ 25 °C
Hydrochloric acid–potassium chloride	1.0–2.2
Glycine–hydrochloric acid	2.2–3.6
Potassium hydrogen phthalate–hydrochloric acid	2.2–4.0
Citrate buffer	3.0–6.2
Acetate buffer	3.8–5.8
Potassium hydrogen phthalate–sodium hydroxide	4.1–5.9
Disodium hydrogen phthalate–sodium dihydrogen orthophosphate	5.8–8.0
Dipotassium hydrogen phthalate–potassium dihydrogen orthophosphate	5.8–8.0
Phosphate buffer	5.8–8.0
Potassium dihydrogen orthophosphate–sodium hydroxide	5.8–8.0

TABLE 21.9 Buffering Solutions Available for Systems with Different pH Ranges.

pH	Available Buffer Solution
1–2	Hydrochloric acid–potassium chloride
2–3	Glycine–hydrochloric acid,
	Potassium hydrogen phthalate–hydrochloric acid
3–4	Citrate buffer, acetate buffer
4–5	Citrate buffer, acetate buffer
	Potassium hydrogen phthalate–sodium hydroxide
5–6	Citrate buffer, acetate buffer
	Potassium hydrogen phthalate–sodium hydroxide
	Disodium hydrogen phthalate–sodium dihydrogen orthophosphate
	Dipotassium hydrogen phthalate–potassium dihydrogen orthophospate
6–7	Phosphate buffer,
	Disodium hydrogen phthalate–sodium dihydrogen orthophosphate
	Dipotassium hydrogen phthalate–potassium dihydrogen orthophosphate
	Potassium dihydrogen orthophosphate–sodium hydroxide
7–8	Borate buffer, phosphate buffer
	Disodium hydrogen phthalate–sodium dihydrogen orthophosphate
	Dipotassium hydrogen phthalate–potassium dihydrogen orthophosphate
	Potassium dihydrogen orthophosphate–sodium hydroxide
	Barbitone sodium–hydrochloric acid
	Tris (hydroxylmethyl) aminomethane–hydrochloric acid

- Rohm & Haas Electronics: http://electronicmaterials.rohmhaas.com/
- JSR: http://www.jsrmicro.com/pro_CMP.html
- Ferro: http://www.ferro.com/
- Praxair Surface Technologies: http://www.praxair.com/

(b) Pad companies
- Rohm & Haas: http://electronicmaterials.rohmhaas.com/
- JSR: http://www.jsrmicro.com/pro_CMP.html
- Thomas West: http://www.thomaswest.com/Presstour.html
- Fujibo: http://www.marubeni-sunnyvale.com/polishing_pads.html
- Toray: http://www.toray.com/news/elec/nr051207.html
- psiloQuest: http://www.psiloquest.com/Company.htm
- PPG CMP: http://www.ppg.com/ppgcmp/default.htm

(c) Tool companies/equipment vendors
- Applied Materials: http://www.appliedmaterials.com/
- Lam Research Corp.: http://www.lamrc.com/
- Novellus: http://www.novellus.com/
- TEL: http://www.tel.com/eng/index.htm
- Ebara: http://www.ebaratech.com/
- Strasbaugh: http://www.strasbaugh.com/

(d) Pad conditioner
- Abrasive Technology: http://www.abrasive-tech.com/Pages/CMP2.html
- Diamonex: http://www.diamonex.com/products.htm

(e) Post-CMP cleaning product companies
- ATMI: http://www.atmi.com/Products/ProductTree.a Tree.sp

(f) Metrology
- KLA-Tencor: http://www.kla-tencor.com/j/servlet/Product Category?name = browse
- Veeco: http://www.veeco.com/

(g) Specialized CMP conferences or conferences that have CMP sections
- February
  - International Conference on Chemical–Mechanical Polish for ULSC Multilevel Interconnection (CMP-MIC): http://www.imic. org/
- March
  - SEMICON China: www.semi.org
  - AVS International Confe-rence on Microelectronics and Interfaces (ICMI): www.avs.org
- April
  - MRS Spring Meetings: http://www.mrs.org/s_mrs/sec_mtgmain.asp?CID = 5&DID = 10
- May
  - National Conference (ESC): www.ecs.org
- June
  - IEEE International Interconnect Technology Conference (IITC): www.ieee.org

Aluminum

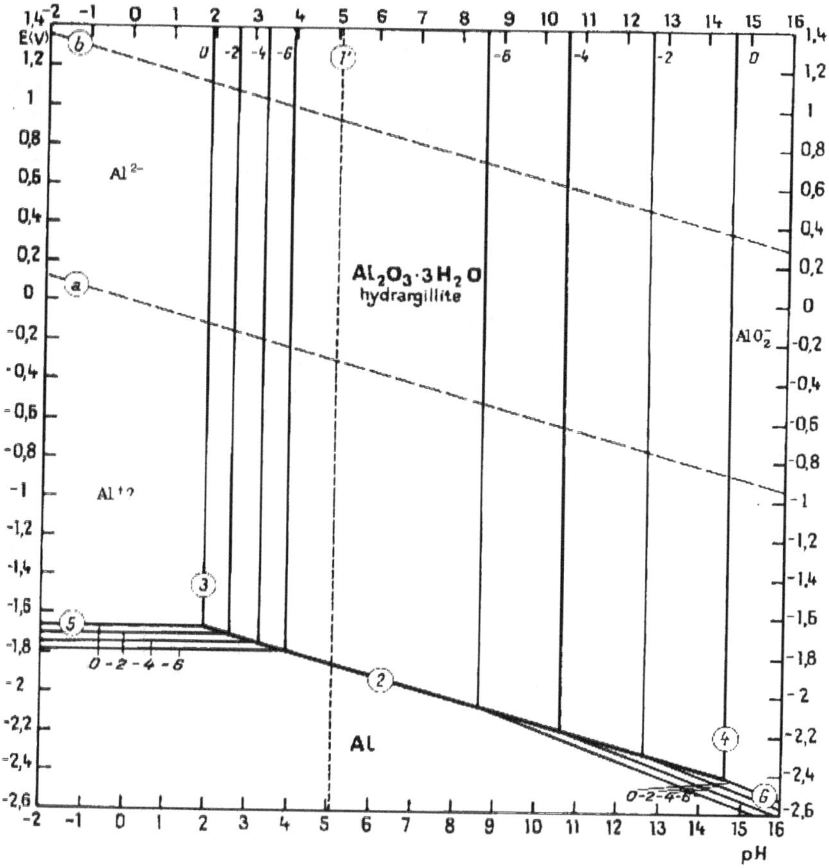

FIGURE 21.15 Potential–pH equilibrium diagram for the aluminum–water system at 25 °C (from Ref. 27).

- July
  - SEMICON West: www.semi.org
- August
  - CAMP-Clarkson Meetings: http://www.clarkson.edu/camp/
- September
  - VLSI Multilevel Interconnection (VMIC) Conference: http://www.imic.org
  - Electrochemical Society: http://www.electrochem. org/
- October
  - International Conference on Planarization/CMP Technology (ICPT): http://www.avsusergroups.org/icpt.cfm

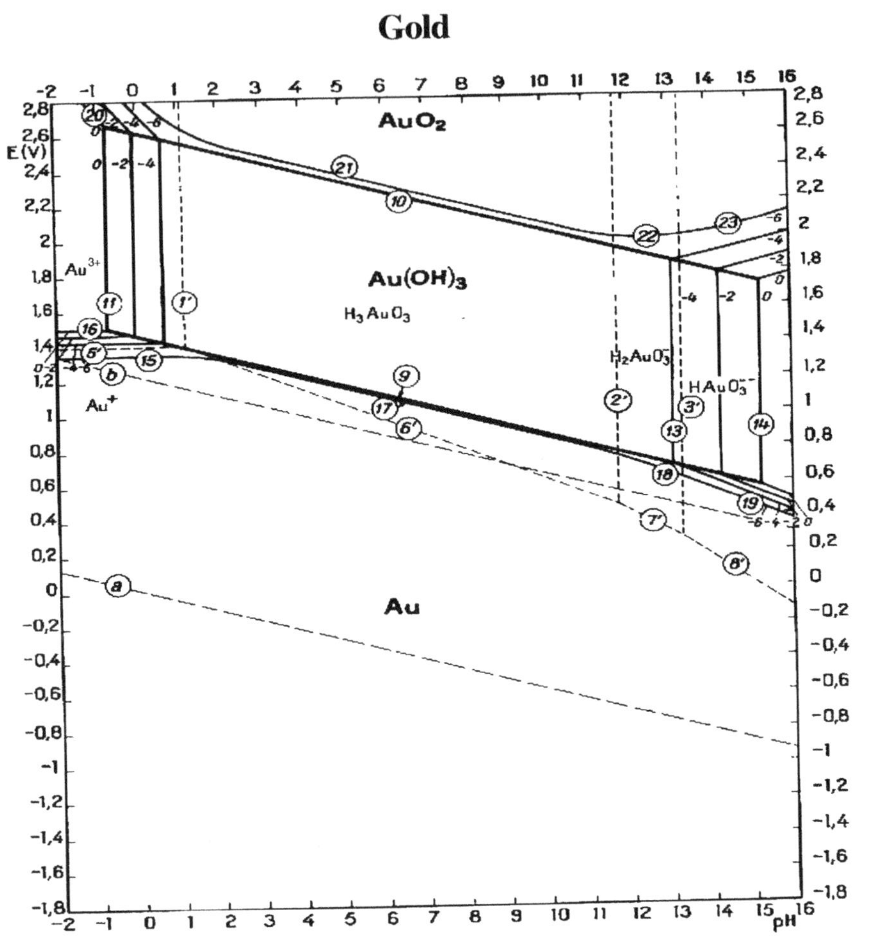

FIGURE 21.16 Potential–pH equilibrium diagram for the gold–water system at 25 °C (from Ref. 28).

(h) Journals and magazines
- IEEE Transaction on Semiconductor Manufacturing (TSM): http://ieeexplore.ieee.org/xpl/RecentIssue.jsp?punumber = 66
- Journal of Electrochemical Society: http://ecsdl.org/JES/
- Thin Solid Film: www.sciencedirect.com
- MRS Proceedings
- Micro
- Solid State Technology

(i) International technology roadmap for semiconductor.
- http://www.itrs.net/Links/2006Update/2006UpdateFinal.htm

(j) Updates and corrections related to this book
- http://www.clarkson.edu/cmpbook2007/ (maintained by the editor)

Silver

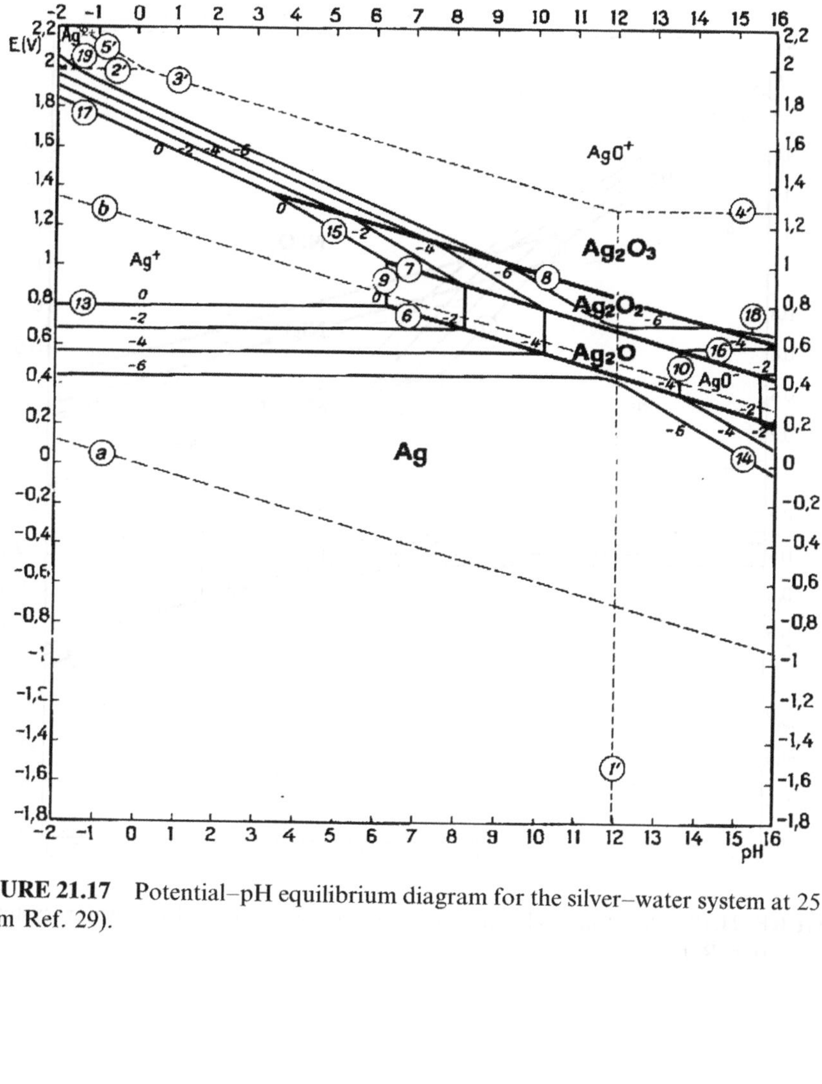

FIGURE 21.17 Potential–pH equilibrium diagram for the silver–water system at 25 °C (from Ref. 29).

Nickel

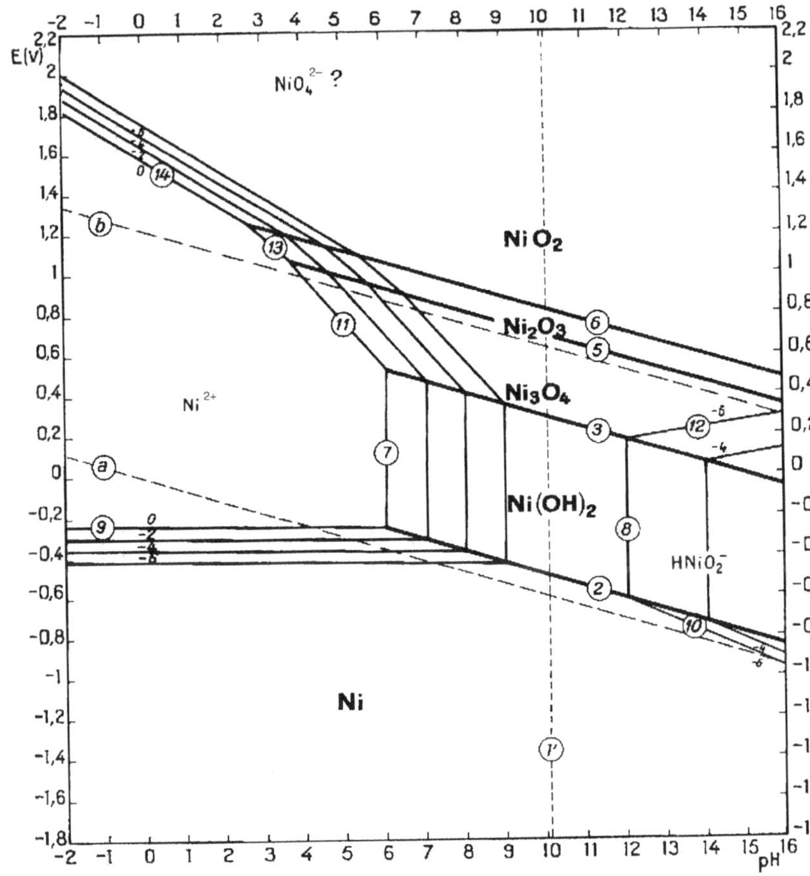

FIGURE 21.18 Potential–pH equilibrium diagram for the nickel–water system at 25 °C (from Ref. 30).

Platinum

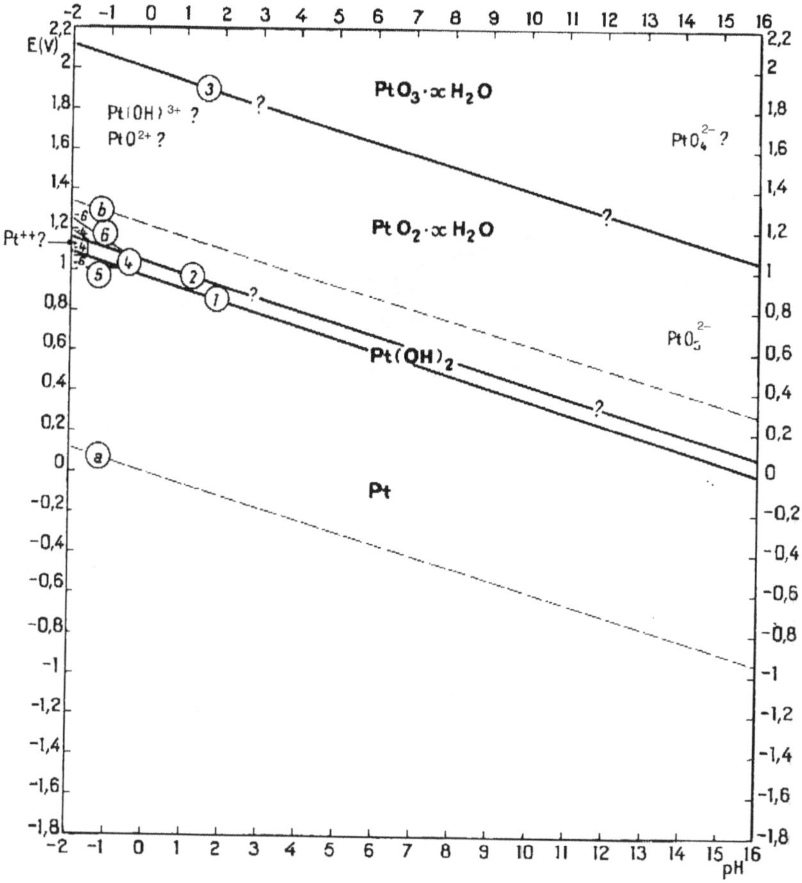

FIGURE 21.19 Potential–pH equilibrium diagram for the platinum–water system at 25 °C (from Ref. 31).

Ruthenium

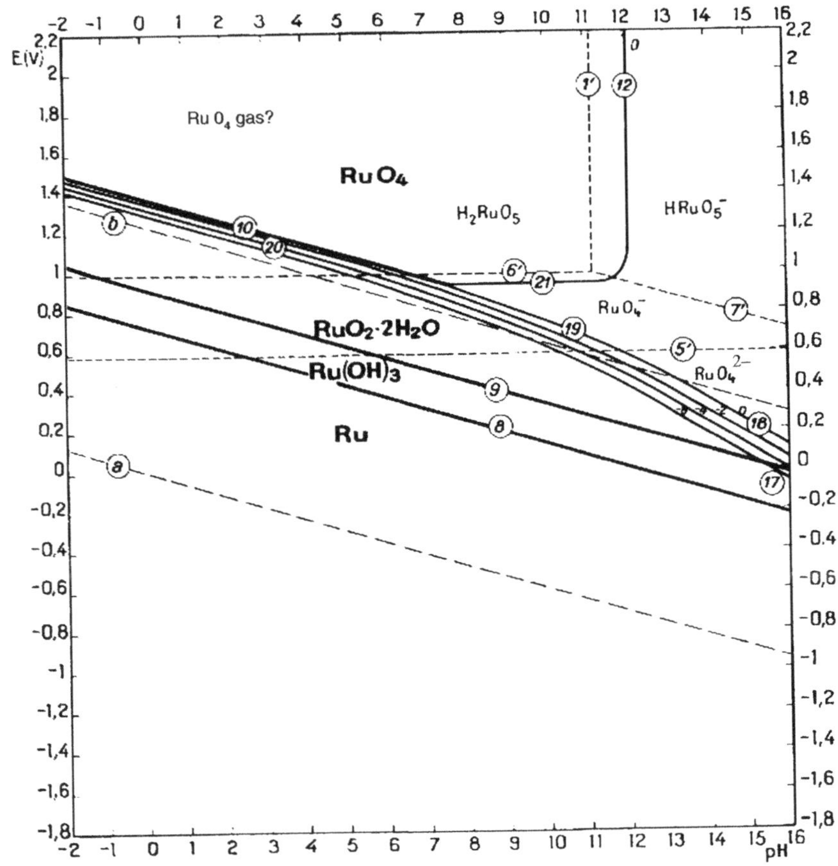

FIGURE 21.20 Potential–pH equilibrium diagram for the ruthenium–water system at 25 °C. [In the presence of solutions free from complexing substances.] (from Ref. 32).

Chromium

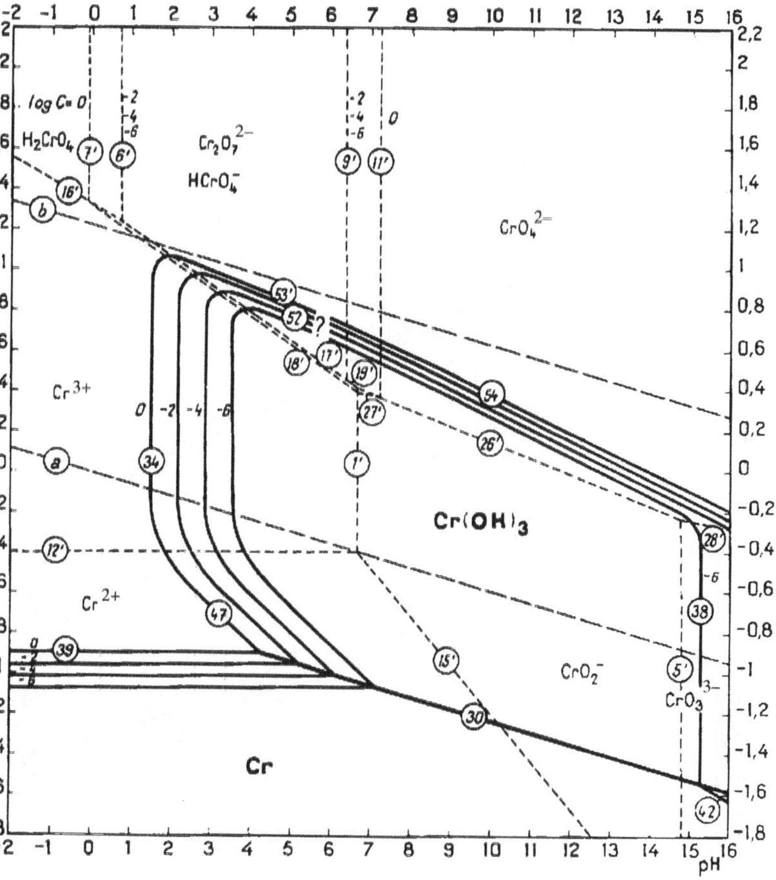

FIGURE 21.21 Potential–pH equilibrium diagram for the chromium–water system at 25 °C in solutions not containing chloride. [Figure established considering $Cr(OH)_3$.] (from Ref. 33).

Chromium

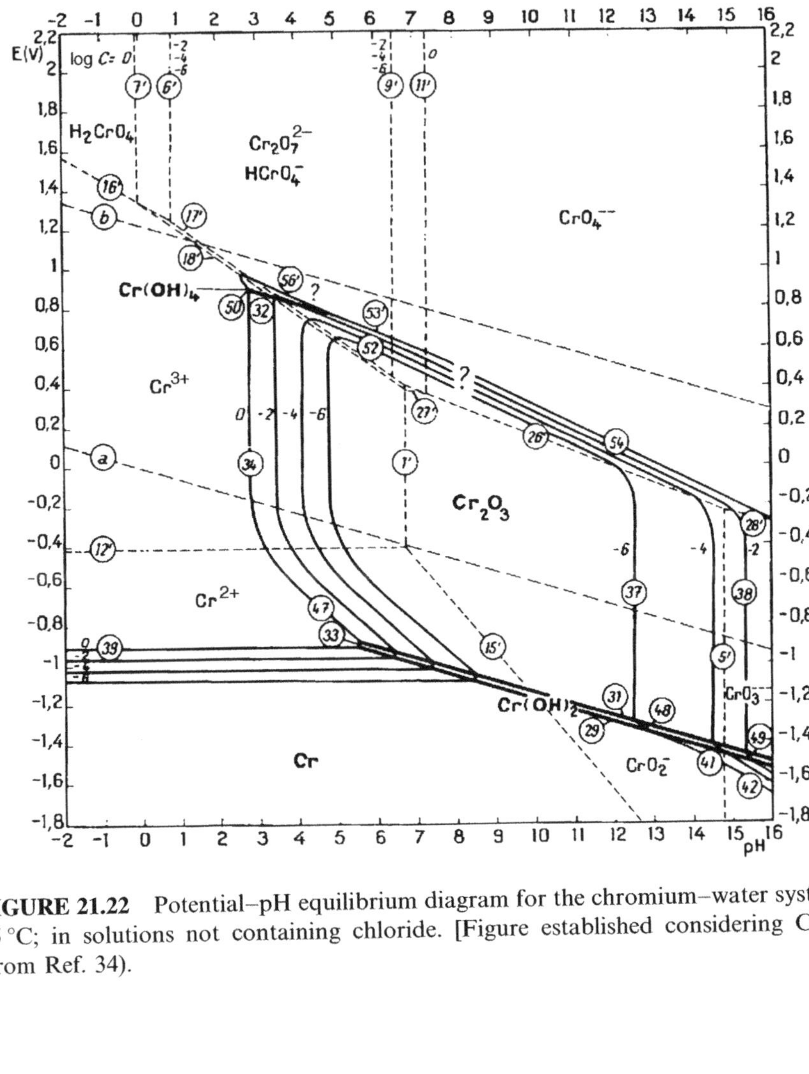

FIGURE 21.22 Potential–pH equilibrium diagram for the chromium–water system at 25 °C; in solutions not containing chloride. [Figure established considering Cr_2O_3.] (from Ref. 34).

Titanium

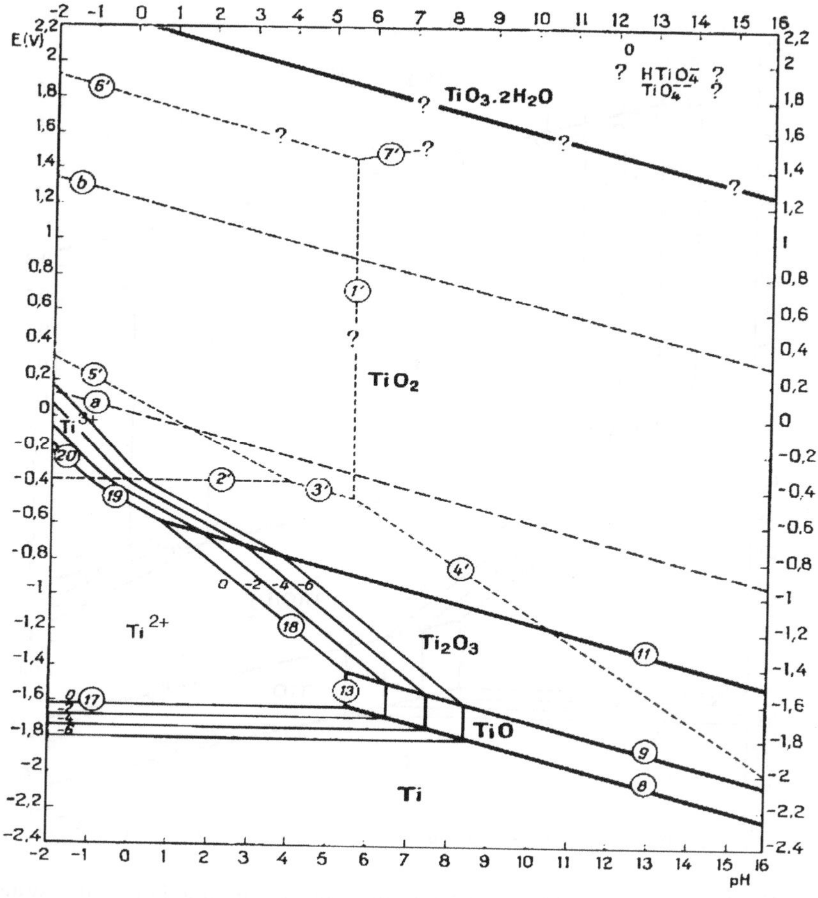

FIGURE 21.23 Potential–pH equilibrium diagram for the titanium–water system at 25 °C. [Figure established by considering, as derivatives of tri and tetravalent titanium, the anhydrous oxides Ti_2O_3 and TiO_2 (rutile).] (from Ref. 35).

Titanium

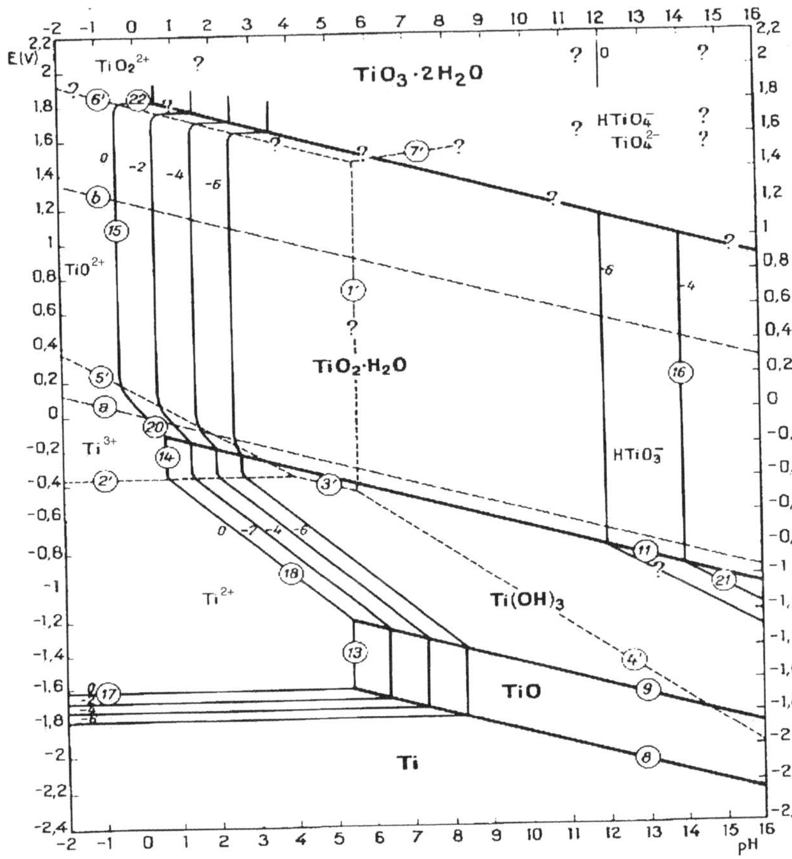

FIGURE 21.24 Potential–pH equilibrium diagram for the titanium–water system at 25 °C. [Figure established by considering, as derivatives of tri and tetravalent titanium, the anhydrous oxides Ti_2O_3 and TiO_2 (rutile).] (from Ref. 36).

Manganese

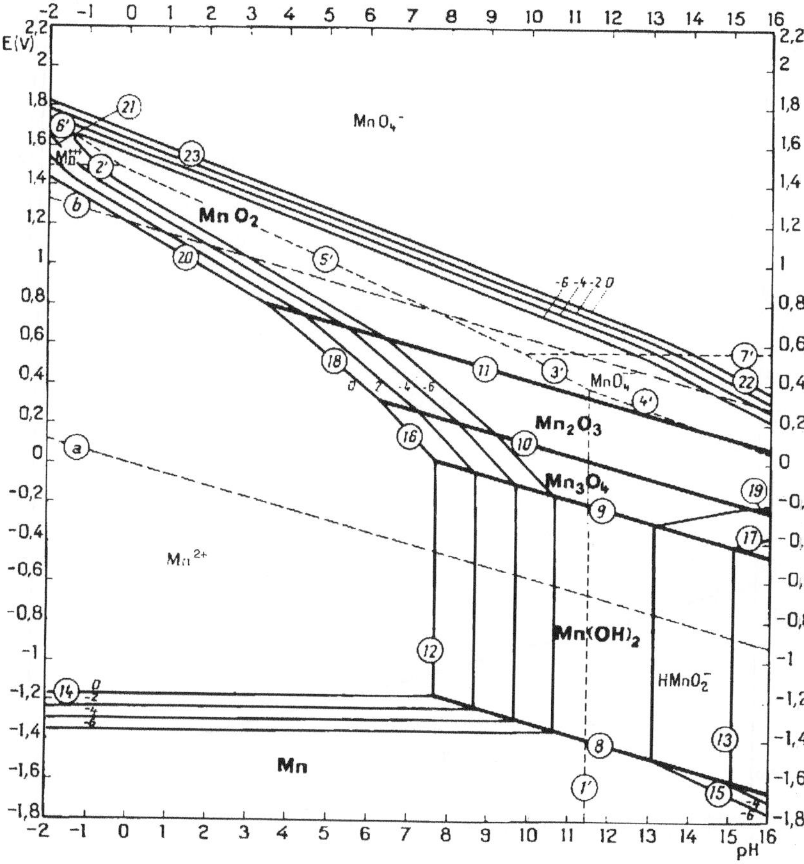

FIGURE 21.25 Potential–pH equilibrium diagram for the manganese–water system at 25 °C. [Considering MnO_2 (pyrolusite).] (from Ref. 37).

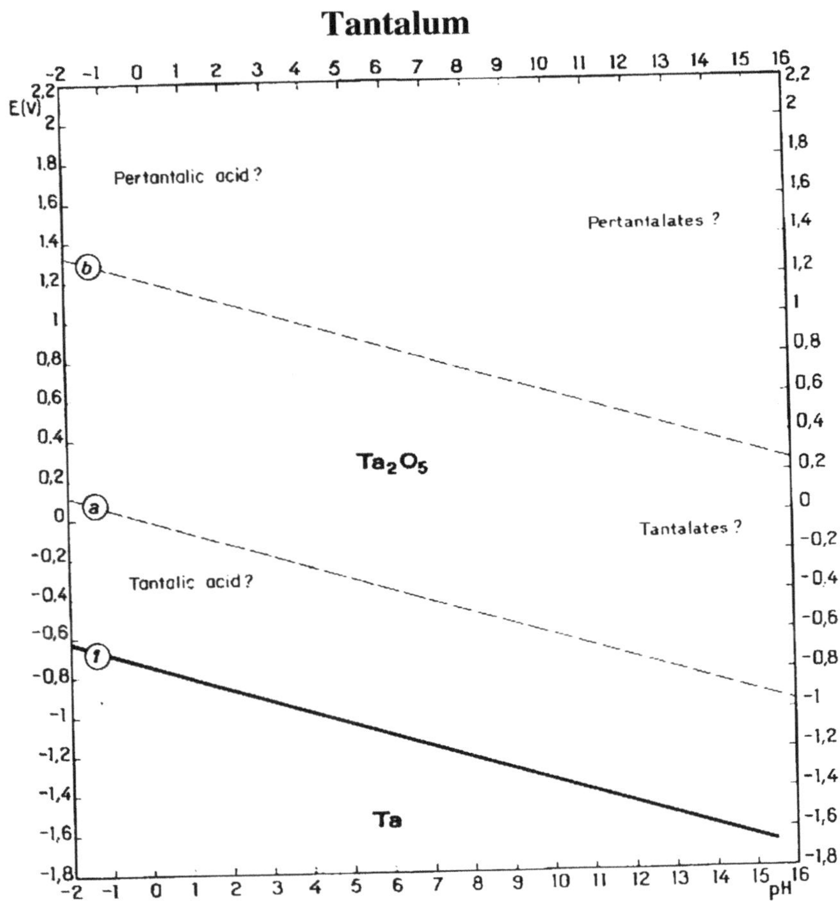

FIGURE 21.26 Potential–pH equilibrium diagram for the tantalum–water system at 25 °C (from Ref. 38).

Silicon

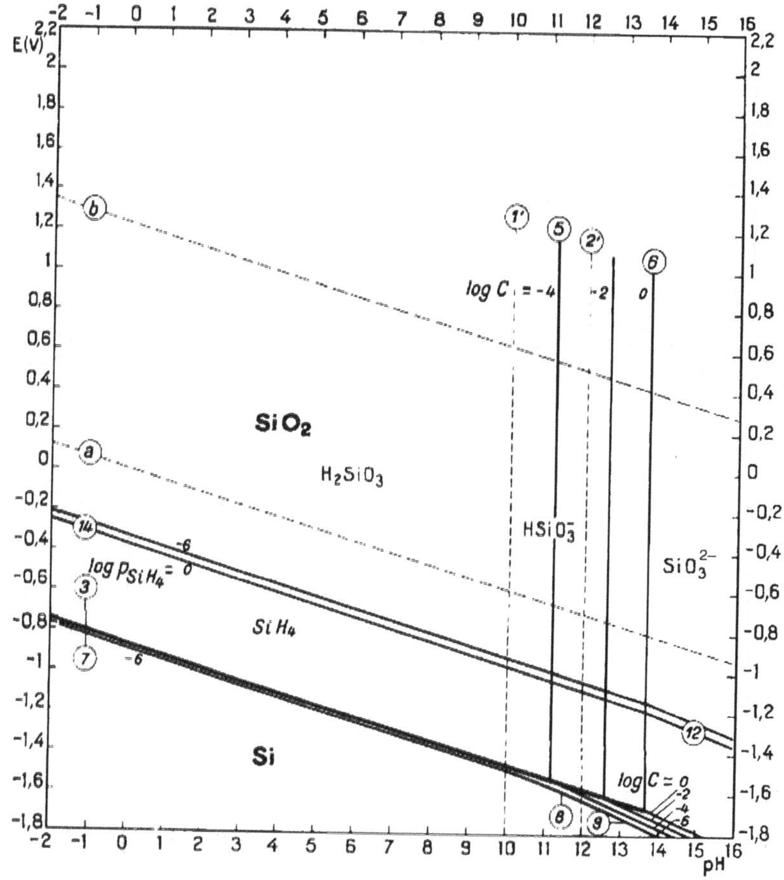

FIGURE 21.27 Potential–pH equilibrium diagram for the silicon–water system at 25 °C. [Considering SiO_2 in the form of quartz.] (Approximate diagram, from Ref. 39).

Hydrogen peroxide

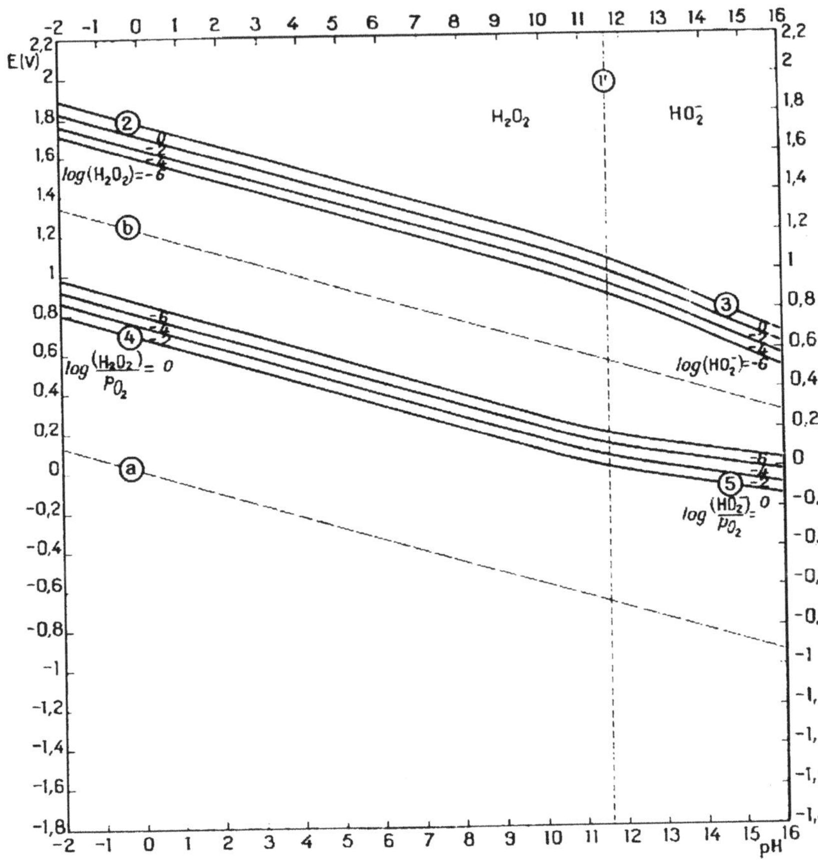

FIGURE 21.28 Potential–pH equilibrium diagram of the hydrogen peroxide–water system at 25 °C (from Ref. 40).

Iodine

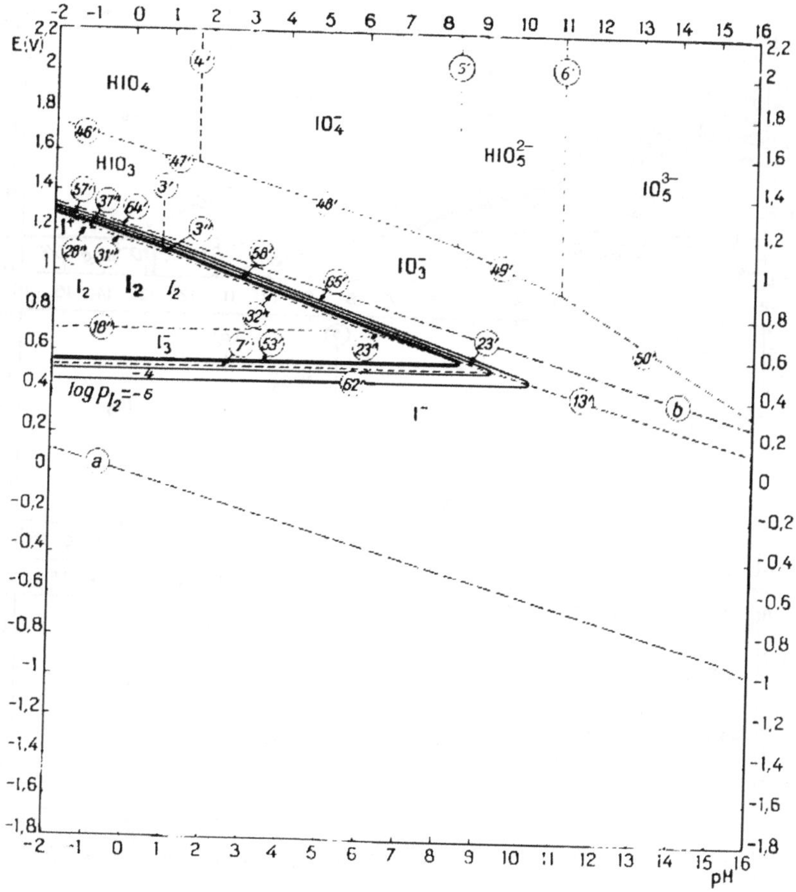

FIGURE 21.29 Potential–pH equilibrium diagram for the iodine–water system at 25 °C, for solutions containing 1g-at I/l (from Ref. 41).

Iodine

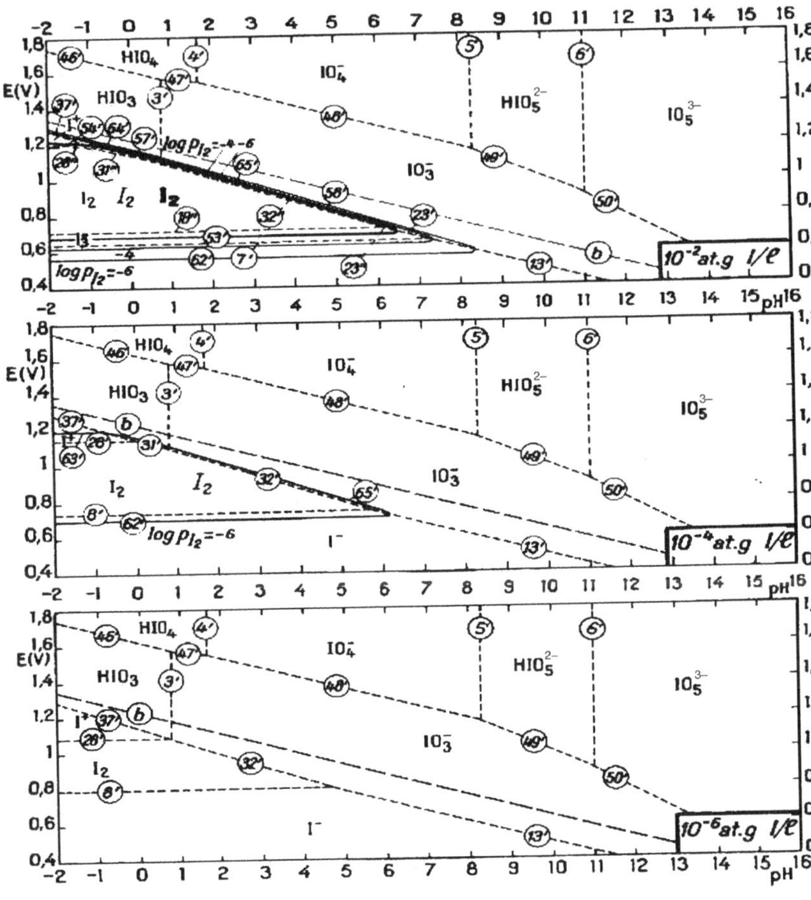

FIGURE 21.30 Potential–pH equilibrium diagrams for the iodine–water system at 25 °C, for solutions containing IO^{-2}, IO^{-4} and IO^{-6} g-at I/l (from Ref. 42).

(k) Additional CMP-related topics that are not fully covered by this book
- http://www.clarkson.edu/cmpbook2007/topics (maintained by the editor)

REFERENCES

1. Brinker CJ, Scherer GW. The physical and Chemistry of Sol-Gel Processing. Academic Press, Inc; 1990. p 667.
2. Kosmulski M, Maczka E, Rosenholm JB. Isoelectric points of metal oxides at high ionic strengths. J Phys Chem B 2002; 106: 2918–2921.
3. Kosmulski M, Maczka E, Rosenholm JB. Isoelectric points of metal oxides at high ionic strengths. J Phys Chem B 2002; 106 (11): 2918–2921.
4. http://w1.cabot-corp.com/products.jsp?N=23
5. http://www.eminess.com/ultrasolSilica.html
6. http://www.solvaychemicals.us/products/productshidden/0,40553-2-0,00.htm
7. http://www.h2o2.com/intro/faq.html#2
8. Holmberg K, Jönsson B, Kronberg B, Lindman B. Surfactants and Polymers in Aqueous Solution. Hoboken, NJ: John Wiley & Sons Inc.; 2003. p 10.
9. Holmberg K, Jönsson B, Kronberg B, Lindman B. Surfactants and Polymers in Aqueous Solution. Hoboken, NJ: John Wiley & Sons Inc.; 2003. p 16.
10. Holmberg K, Jönsson B, Kronberg B, Lindman B. Surfactants and Polymers in Aqueous Solution. Hoboken, NJ: John Wiley & Sons Inc.; 2003. p 19.
11. Holmberg K, Jönsson B, Kronberg B, Lindman B. Surfactants and Polymers in Aqueous Solution. Hoboken, NJ: John Wiley & Sons Inc.; 2003. p 23.
12. Holmberg K, Jönsson B, Kronberg B, Lindman B. Surfactants and Polymers in Aqueous Solution. Hoboken, NJ: John Wiley & Sons Inc.; 2003. p 43.
13. Hiemenz PC, Rajagopalan R, editors. Principles of Colloid and Surface Chemistry. 3rd ed. New York: Marcel Dekker, Inc.; 1997. p 359.
14. Holmberg K, Jönsson B, Kronberg B, Lindman B. Surfactants and Polymers in Aqueous Solution. Hoboken, NJ: John Wiley & Sons Inc.; 2003. p 44.
15. Langhlin RG. The Aqueous Phase Behavior of Surfactants. Academic press Inc. ; 1996. p 67.
16. Li Y, Friberg S. Course Manual for the American Chemical Society Short Course on Surfactant Micelles, Liposomes, and Liquid Crystals in Emulsions and Microemulsions. 2002.
17. Laughlin RG. The Aqueous Phase Behaviour of Surfactants. London: Academic Press; 1994. p 397.
18. Kahn A. Phase science of surfactants Curr Opin Colloid Interface Sci 1996; 1: 614–623.
19. Kabbalnov A, Lindman B, Olsson U, Piculell L, Thuresson K, Wenner-strom H. Colloid Polym Sci 1996; 274: 297.
20. Jones AD. Principles and Prevention of Corrosion. 2nd ed. Upper Saddle River, NJ: Prentice Hall; 0-13-359993-0; 1996. p 50–52.
21. http://delloyd.50megs.com/moreinfo/buffers2.html#buffer

22. Pourbaix M. Atlas of Electrochemical Equilibria in Aqueous Solutions. Long Island City, New York: Pergamon Press Inc.; 1966. p 282.
23. Pourbaix M. Atlas of Electrochemical Equilibria in Aqueous Solutions. Long Island City, New York: Pergamon Press Inc.; 1966. p 387.
24. Pourbaix M. Atlas of Electrochemical Equilibria in Aqueous Solutions. Long Island City, New York: Pergamon Press Inc.; 1966. p 388.
25. Pourbaix M. Atlas of Electrochemical Equilibria in Aqueous Solutions. Long Island City, New York: Pergamon Press Inc.; 1966. p 312.
26. Pourbaix M. Atlas of Electrochemical Equilibria in Aqueous Solutions. Long Island City, New York: Pergamon Press Inc.; 1966. p 313.
27. Pourbaix M. Atlas of Electrochemical Equilibria in Aqueous Solutions. Long Island City, New York: Pergamon Press Inc.; 1966. p 171.
28. Pourbaix M. Atlas of Electrochemical Equilibria in Aqueous Solutions. Long Island City, New York: Pergamon Press Inc.; 1966. p 402.
29. Pourbaix M. Atlas of Electrochemical Equilibria in Aqueous Solutions. Long Island City, New York: Pergamon Press Inc.; 1966. p 396.
30. Pourbaix M. Atlas of Electrochemical Equilibria in Aqueous Solutions. Long Island City, New York: Pergamon Press Inc.; 1966. p 333.
31. Pourbaix M. Atlas of Electrochemical Equilibria in Aqueous Solutions. Long Island City, New York: Pergamon Press Inc.; 1966. p 380.
32. Pourbaix M. Atlas of Electrochemical Equilibria in Aqueous Solutions. Long Island City, New York: Pergamon Press Inc.; 1966. p 346.
33. Pourbaix M. Atlas of Electrochemical Equilibria in Aqueous Solutions. Long Island City, New York: Pergamon Press Inc.; 1966. p 262.
34. Pourbaix M. Atlas of Electrochemical Equilibria in Aqueous Solutions. Long Island City, New York: Pergamon Press Inc.; 1966. p 263.
35. Pourbaix M. Atlas of Electrochemical Equilibria in Aqueous Solutions. Long Island City, New York: Pergamon Press Inc.; 1966. p 217.
36. Pourbaix M. Atlas of Electrochemical Equilibria in Aqueous Solutions. Long Island City, New York: Pergamon Press Inc.; 1966. p 290.
37. Pourbaix M. Atlas of Electrochemical Equilibria in Aqueous Solutions. Long Island City, New York: Pergamon Press Inc.; 1966. p 290.
38. Pourbaix M. Atlas of Electrochemical Equilibria in Aqueous Solutions. Long Island City, New York: Pergamon Press Inc.; 1966. p 461.
39. Pourbaix M. Atlas of Electrochemical Equilibria in Aqueous Solutions. Long Island City, New York: Pergamon Press Inc.; 1966. p 461.
40. Pourbaix M. Atlas of Electrochemical Equilibria in Aqueous Solutions. Long Island City, New York: Pergamon Press Inc.; 1966. p 108.
41. Pourbaix M. Atlas of Electrochemical Equilibria in Aqueous Solutions. Long Island City, New York: Pergamon Press Inc.; 1966. p 621.
42. Pourbaix M. Atlas of Electrochemical Equilibria in Aqueous Solutions. Long Island City, New York: Pergamon Press Inc.; 1966. p 622.

INDEX

Microelectronic Applications of Chemical Mechanical Planarization, Edited by Yuzhuo Li